AF739242

Environmental Geocomputation with R

Environmental Geocomputation with R presents theoretical and practical principles for geospatial data analysis, visualization, and modeling, covering both the individual object approach and the continuous field approach. The book addresses spatial data uncertainties and error management while integrating spatial context into analytical results for better understanding of the environment. It progressively builds complexity enabling readers to develop geocomputational projects in the real world using algorithms and big data with R and its packages. The applications are focused on environmental and agricultural analyses, helping to incorporate spatio-temporal dimensions into decision-making processes.

Features

- Explains the theory and applications of geocomputational methods for solving environmental problems involving spatial data and analysis.
- Covers big data, machine learning, methods for pattern recognition in spatial data, and tools and data analytics in the free and open-source (FOSS) R platform, making them accessible to everyone.
- Includes case studies from different environments and explains the diversity of big data and its applications.
- Provides hands-on exercises at the end of each chapter and encourages readers to understand both the potential and limitations of environmental systems.
- Supplemented by a comprehensive laboratory manual with solved exercises, homework assignments, and research suggestions in geocomputation within an environmental science context.

This textbook is intended for students in upper-level undergraduate or graduate programs using GIS, geocomputation, geoprocessing in environmental sciences and engineering, surveying, geosciences, and agriculture engineering. It is also an excellent resource for professionals in different areas who use geocomputation and image processing and wish to develop their skillset in these areas.

Environmental Geocomputation with R

Marcelo de Carvalho Alves [0000-0003-3957-328x]

CRC Press
Taylor & Francis Group
Boca Raton London New York

CRC Press is an imprint of the
Taylor & Francis Group, an **informa** business

Designed cover image: © Shutterstock

First edition published 2027

by CRC Press

2385 NW Executive Center Drive, Suite 320, Boca Raton FL 33431 and by CRC Press

4 Park Square, Milton Park, Abingdon, Oxon, OX14 4RN

CRC Press is an imprint of Taylor & Francis Group, LLC

© 2027 Marcelo de Carvalho Alves

Reasonable efforts have been made to publish reliable data and information, but the author and publisher cannot assume responsibility for the validity of all materials or the consequences of their use. The authors and publishers have attempted to trace the copyright holders of all material reproduced in this publication and apologize to copyright holders if permission to publish in this form has not been obtained. If any copyright material has not been acknowledged please write and let us know so we may rectify in any future reprint.

Except as permitted under U.S. Copyright Law, no part of this book may be reprinted, reproduced, transmitted, or utilized in any form by any electronic, mechanical, or other means, now known or hereafter invented, including photocopying, microfilming, and recording, or in any information storage or retrieval system, without written permission from the publishers.

For permission to photocopy or use material electronically from this work, access www.copyright.com or contact the Copyright Clearance Center, Inc. (CCC), 222 Rosewood Drive, Danvers, MA 01923, 978-750-8400. For works that are not available on CCC please contact mpkbookspermissions@tandf.co.uk

For Product Safety Concerns and Information please contact our EU representative GPSR@taylorandfrancis.com. Taylor & Francis Verlag GmbH, Kaufingerstraße 24, 80331 München, Germany.

Trademark notice: Product or corporate names may be trademarks or registered trademarks and are used only for identification and explanation without intent to infringe.

ISBN: 978-1-041-20818-1 (hbk)

ISBN: 978-1-041-20819-8 (pbk)

ISBN: 978-1-003-72437-7 (ebk)

DOI: 10.1201/9781003724377

Publisher's note: This book has been prepared from camera-ready copy provided by the authors.

Access the Instructor and Student Resources: www.routledge.com/9781041208181

Contents

About the Author

Dr. **Marcelo de Carvalho Alves** (Lattes curriculum vitae: 1691831453683402[1]; Orcid: 0000-0003-3957-328X[2]; Scopus Author ID: 15047587400[3]; ResearcherID: G-5223-2015[4]; SciProfiles: 1893763[5]) is full professor in the area of remote sensing and geocomputation at the Federal University de Lavras, Brazil. He graduated with a master's degree (2003), doctorate in agronomy, thesis in crop science (2006), and a post-doctorate degree in agricultural engineering at the Federal University of Lavras, Brazil (2008). His original academic thesis in the area of remote sensing and geocomputation was defended in 2024 as an application for progression to full professor in higher education. He has varied research interests and has published on surveying, remote sensing, geocomputation, agriculture and environmental applications. He has over 20 years of experience in data science, digital image processing and modeling using multiscale, multidisciplinary, multispectral and multitemporal concepts applied to different environments. Experimental field-sites included a tropical forest, savanna, wetland, and agricultural fields in Brazil. His research has been predominantly funded by CNPq, CAPES, FAPEMIG and FAPEMAT. Over the years, he has built up a large portfolio of research grants mostly relating to the agroecosystem management, broadly in the context of vegetation, soil, water, atmosphere, climate, pest-organisms, and disaster risk analysis.

[1]http://lattes.cnpq.br/1691831453683402

[2]https://orcid.org/0000-0003-3957-328X

[3]http://www.scopus.com/inward/authorDetails.url?authorID=15047587400&partnerID=MN8TOARS

[4]http://www.researcherid.com/rid/G-5223-2015

[5]https://sciprofiles.com/profile/1893763

Preface

Geocomputation is considered a revolution following the introduction of geographic information systems. Geocomputation enables data generated by geographic information systems, remote sensing and information technology to be exploited in a better and more complete way. In addition, it is possible to exploit data information beyond the original objectives of data collection. Thus, the notion of geographic information science can be expanded toward the analysis, modeling and solution of everyday problems. Geographical science is more than the art of modern surveying, also called geoinformatics. With the use of geocomputation, it is possible to work with various types of data and develop relevant geographic tools within the general context of a scientific approach to solving theoretical and practical academic problems with computer science and artificial intelligence techniques. Geocomputation is related to making useful things, containing geographic or environmental data components, modern computing technologies with a statistical, mathematical, numerical basis, as well as high-performance computing hardware.

In this book, theoretical and practical principles of data manipulation and analysis with geocomputation in geographic information systems are described and exemplified in order to effectively and critically understand environmental geocomputation with R. The text presents different ways of representing and modeling geospatial data. With the approach used by the author we can understand standard entity and continuous field approaches used in geocomputation. Ways of dealing with uncertainties in spatial data analysis enabled dealing with errors and their possible effects on the results of a project. The main focus of the book is to present concepts and theoretical literature available in the field with the aim of solving or improving data analysis by including the spatial context in the results obtained. We expected to better understand the phenomena of the analyzed environment and the most important variables that should be considered when managing and changing processes in the real world according to the nature of each existing object and dynamic process. The book format is prepared for learning and research in geocomputation. In addition, theoretical and practical principles are exemplified in order to prepare students for academia with emphasis on scientific applications for project and article writing in geocomputation using big data and geographic data science. The scientific applications of geocomputation are performed with environmental attributes and agricultural variables generally lacking spatial-temporal analysis for decision-making at different hierarchical levels of public and private institutions.

The idea behind this book is to use classic works of geocomputation and geographic information systems literature, updating them as necessary for today's world, using the R programming language but without losing sight of the basic concepts of the discipline. This enables one to further expand the use of this bibliography with applications. We have also produced a book that serves as a laboratory manual, with the aim of supporting students in developing a project in the field of geocomputation (halfway through the course) and then writing a scientific article at the end of the course. The part on integration with clouds (Google Earth Engine) is being done with the `rgee` package, which is already covered in some chapters. In addition to the computational exercises included at the end of each chapter, homework assignments have also been created that can be used to solve geographical problems related to the environmental field. The solutions to the homework assignments are included in the laboratory manual. This was done so that dedicated readers can develop a project and write a scientific article throughout the course. Some scripts and

RMarkdown files are included in a GitHub repository for use by teachers and students who are using the **main book**[6] and the **laboratory manual**[7].

The combination of big data applications, case studies, and open source software for geocomputation makes this book fill a gap on the market. While there are competitors out there, the proposed book seems to provide complementary and in some aspects additional contents about fundamentals of geocomputation. Part of the book can be used in an undergraduate course with two hours per week, and in the case of graduate studies, it can be used with four hours per week (two hours on one day and two hours on another), also covering the chapters on project development and articles that can be used for students to learn the content by conducting scientific research in the field.

It can be challenging to get an overview of alternative approaches to perform common tasks of geocomputation in R. Alternative approaches are implemented by different R libraries. This goes along with moving to new formats and paradigms. There are indeed problems with the versions of packages used and modifications to functions by some packages (e.g., discontinuation of `rgdal`, `sf` vs. `sp`, `terra` vs. `raster` vs. `stars`, `ggplot2` vs. base R plots vs. `tmap` and `RStoolbox` change versions. For example, `rgdal` was retired along with its original author, but as far as I can tell, it is being continued by the `sf` and `raster` packages. In some situations, we tried to perform the same tasks with different packages.

A book on this topic is highly needed and can be used for courses in which R is used for environmental geocomputation and geoprocessing. The breadth and uniqueness of the proposed book's niche is limited by the distribution of R in the community (Python is also emerging as an important language in this field). This book appears to provide all the necessary fundamentals in a very rigorous way. This seems to be a very valuable teaching resource and reference, perhaps even more detailed than other similar titles on the market. Our idea was to focus on well-known titles in the field, including practical geocomputation applications where these materials have certain shortcomings or are based on commercial software.

Our approach is not widely used yet, given the preference of many students for GUI interface software. However, using R, we have made significant progress in international publications of articles, which is limited when using software such as QGIS and ArcGis. Python language is already well established in the field, and many do not want to stop using Python to develop existing applications in R. I see this a lot in the field of meteorology. Our intention is to continue using R even with these barriers and limitations, as it is a language developed in universities, with scientific recognition and a language that is generally close to human language, facilitating students' understanding of the functions used. We present several practical applications in the book covering the environmental and agricultural fields, and using the laboratory manual the students can develop their work in the discipline and learn by doing.

Python has been more widely adopted than R, so we believe it is better to invest in R because of its academic development approach, scientific publications, citations, and greater detail in the literature used in package development. International journals have been very receptive to articles developed using R due to its robust statistical basis. We do not intend to move forward with Python for the time being, as there is much to study, improve, and research with R. I have noticed that the marketing industry wants you to use a language

---

[6] https://github.com/marcelocarvalhoalves/Environmental-Geocomputation-with-R-Data-Repository

[7] https://github.com/marcelocarvalhoalves/Laboratory-Manual-for-Environmental-Geocomputation-with-R-Data-Repository

because it is the most widely used in the world, but I would not choose a language solely for that reason.

Considering the way R language is used can be very easy for students to assimilate it. However, we can automate the language so that it becomes more complex. This approach to automating commands will be covered in future books that are more applied to specific content. Our approach is designed for use in classrooms with computers running R and RStudio. In class, we have been using Windows computers, but it is also possible to use R with Linux and Mac. The possibility of publishing the book online using RStudio options might be a good idea to facilitate the reproduction of commands by students, and also creating a repository of the book on the Internet can facilitate theoretical and practical classes on the material throughout an academic semester.

The items covered in the chapters are defined in a form of constructive complexity that allows the student to develop a project using geocomputation and geospatial data and, at the end, conclude the studies with a report in the form of a scientific article or scientific review on environmental geocomputation. The chapters covered are "Introduction to Environmental Geocomputation with R" (Chapter 1), "Geographic Representation with Environmental Geocomputation and R" (Chapter 2), "Geographic Database Management Systems with Environmental Geocomputation and R" (Chapter 3), "Space-time Georeferencing with Environmental Geocomputation and R" (Chapter 4), "Spatial Data Models with Environmental Geocomputation and R" (Chapter 5), "Space-time Geovisualization with Environmental Geocomputation and R" (Chapter 6), "Project Elaboration with Environmental Geocomputation and R" (Chapter 7), "Vector Spatial Analysis with Environmental Geocomputation and R" (Chapter 8), "Raster Spatial Analysis with Environmental Geocomputation and R" (Chapter 9), "Vector Geometric Analysis with Environmental Geocomputation and R" (Chapter 10), "Raster Geometric Analysis with Environmental Geocomputation and R" (Chapter 11), "Creating Continuous Surfaces with Environmental Geocomputation and R" (Chapter 12), "Geostatistical Analysis with Environmental Geocomputation and R" (Chapter 13), and "Article Elaboration with Environmental Geocomputation and R" (Chapter 14).

1 Introduction to Environmental Geocomputation with R

1.1 LEARNING QUESTIONS

The learning questions answered by reading the chapter are:

- Why is geographic information used in the earth environment?
- How is information related to data, knowledge, evidence, wisdom, and learning?
- How is geographic information used in decision-making processes?
- What are the definitions of geocomputation, geoprocessing and geographic information systems?
- What is spatial data science and its relationship to geocomputation for scientific applications?
- What are the applications and uses of geocomputation?
- In what professions has geocomputation been used?
- What was the economic investment associated with geographic information science?
- What were examples of geocomputation application in different industries?

1.2 LEARNING OUTCOMES

The learning outcomes answered by reading the chapter are:

- Know a brief history of geocomputation;
- Recognize how the science of spatial data and geographic information has been used in everyday life;
- Understand the scientific importance of geocomputation, geographic information and geographic information systems;
- Understand the need to study society, economy and nature with geocomputation.

1.3 INTRODUCTION

Based on historical records, it has been verified that the deterioration of renewable resources has affected environmental sustainability, and that population growth, pollution, and climate variation over specific periods of time have determined the collapse of civilizations without sufficient technology to reverse the natural impacts on the progress of these populations (Burrough and McDonnell, 1998; Alves et al., 2020). Thanks to the development of Geographic Information Systems (GIS), it was possible to use numerical and logical modeling, statistical and geostatistical methods to data distributed over space to contribute to management and decision-making according to the location of objects and processes that occur in the modern world (Burrough and McDonnell, 1998). Despite the ability to gain knowledge about geography just by living on planet Earth for some time, with the use of GIS, geographic information can be used by individuals in scientific research or to support decision-making (Jensen and Jensen, 2012).

DOI: 10.1201/9781003724377-1

Almost everything happens somewhere, so geographic location is an important attribute of activities, policies, strategies, and planning. Geographic information systems are a class of information systems for mapping and managing the location and occurrence of events, activities, and other events where the location aspect has been used in problem-solving, called in this case geographic problems, as in the following situations (Burrough and McDonnell, 1998; Longley et al., 2001):

- Agricultural and environmental management;
- Urban planning;
- Health management;
- Delivery companies;
- Navigation, definition of transportation routes and trips;
- Geodemographic consultation;
- Tourism;
- Military organizations and border defense;
- Archaeology;
- Epidemiology;
- Marketing;
- Registration and evaluation of urban and rural properties;
- Social studies;
- Environment.

1.3.1 GEOGRAPHIC INFORMATION SYSTEMS

The term GIS refers to systems in which geographic data have been processed (Câmara and Medeiros, 1998), and geoprocessing is the technology involved in the process of obtaining results and products using GIS. GIS has been defined as a powerful set of tools to obtain, store, retrieve, transform and display spatial data of a given region for a given purpose (Burrough and McDonnell, 1998); however, other definitions have also been made under different contexts of use (Table 1.1) (Burrough and McDonnell, 1998; Longley et al., 2001):

TABLE 1.1 Definitions of geographic information systems and use by different social groups.

Definition	Context of Use
A container of maps in digital form	General public
A computational tool for solving geographic problems	Designers and decision-makers
A spatial decision support system	Management scientists, operations researchers
Mechanized Inventory of Geographical Features	Utility managers, transportation officials, resource managers
Tool to reveal invisible issues in geographic information	Researchers and scientists
More accurate and cheaper tool to perform geographic operations on data than manually	Resource and planning managers

In this context, the main characteristics of a GIS are identified (Câmara and Medeiros, 1998):

- Integrate geographic information from cartographic data, census, urban and rural registration, satellite images, networks and digital terrain models in a single database;
- Combine information through algorithms for manipulation and analysis, consultation, retrieval, visualization of tables, graphs and maps from operational mechanisms on the geographic database;
- Analyze spatial relationships between geographic objects;
- Establish topology and structure of spatial relationships of neighborhood, proximity and relevance established between geographic objects;
- Store the topology of a map.

Therefore, GIS technology was considered different from computer-aided design (CAD) technology, i.e., in most CAD models, drawings do not have descriptive attributes, but only graphical properties such as color and thickness, without storing the topology of an object on a map (Câmara and Medeiros, 1998).

1.3.2 GEOCOMPUTATION

Stan Openshaw and his colleagues at the University of Leeds, UK, coined the term "geocomputation" to describe the application of computationally intensive approaches to problems in physical and human geography in particular and the geosciences in general. Others say geocomputation is the process of applying computing technology to geographical problems. Geocomputation just means the universe of computational techniques applicable to spatial problems. Geocomputation certainly involves the use of computers in geography, but its subject domain is not limited to geography nor is it merely another term for quantitative geography or geocomputer applications. Geocomputation refers to the science of geography in a computationally sophisticated environment. In essence, geocomputation derives from the field of artificial intelligence related to the application of a computer science paradigm to study all types of geographic phenomena, including physical and human systems, using relatively massive computing power to tackle major challenges of immense complexity. Therefore, geocomputation can be defined as the science, art and technology of solving complex spatial problems with computers (Openshaw and Abrahart, 2001).

The objectives of geocomputation can be described as (Openshaw and Abrahart, 2001):

- Accelerate existing computer-related activities so that more comprehensive experiments can be performed;
- Improve the quality of results by using computationally intensive methods to reduce the number of assumptions and eliminate shortcuts and simplifications imposed by irrelevant computational constraints;
- Enable the analysis of larger databases and obtain better results considering the ability to process data with higher resolution;
- Develop new approaches and methods based on computational technologies from disciplines such as artificial intelligence and computer vision, new ways of solving optimization problems, being opportunistic and entrepreneurial, in order to solve old problems using new technologies and innovations relevant to geographic issues, but which are currently limited by processor speed and memory size.

The environment for geocomputation is provided by geographic information systems, but what distinguishes geocomputation is the creative and experimental use of GIS that it entails. Geocomputation is the way in which new computational tools and methods are used, fundamentally also for the representation of spatial processes, and it is much more than GIS. In some important respects, the term geocomputation is synonymous with geographic information science and spatial data science, although it has often placed greater emphasis on the use of high-performance computers (Longley et al., 2001; Openshaw and Abrahart, 2001). "Spatial data science" is another recent term to deal with the amount of data used in geoprocessing systems, associated with statistical and artificial intelligence techniques for data analysis to extract meaning and insights from the data (Pebesma and Bivand, 2021a).

1.3.3 GEOCOMPUTATION AND GEOGRAPHIC INFORMATION SCIENCE

The term geographic information science was first used in a scientific paper (Goodchild, 1992). In geoinformation science, questions about geography of information in general are studied. Other terms used as synonyms, but with particularities, are (Longley et al., 2001):

- Geomatics and geoinformatics;
- Spatial information science;
- Geoinformation engineering;
- Geocomputation.

In all these terms, there are suggestions of scientific approaches to fundamental questions raised by the use of geographic information systems, with different emphasis on different geospatial problems. In addition to journal publications, geographic information science was used in university research on the following short- and long-term themes (Table 1.2)(Longley et al., 2001):

TABLE 1.2 Short- and long-term themes adopted by agenda on geographic information science by consortium of US universities.

Deadline	Research Theme
Long	Geospatial ontologies
Long	Geographic representation
Long	Acquisition and integration of geospatial data
Long	Scale
Long	Geospatial cognition
Long	Geospatial and space-time analysis
Long	Uncertainty in geographic information
Long	Geovisualization
Long	Geoprocessing and society
Long	Geographical information engineering
Short	Geoprocessing and decision-making
Short	Location-based services
Short	Identification of spatial clusters
Short	Geospatial semantic Web
Short	Incorporation of remote sensing data and information in geoprocessing
Short	Geographic information resources for management
Short	Emergency data acquisition and analysis
Short	Security of geographic information
Short	Geospatial data fusion
Short	Geographic information partnership
Short	Geocomputation
Short	Global representation and modeling
Short	Spatial prediction
Short	Geographic data mining and knowledge discovery

Deadline	Research Theme
Short	Dynamic modeling

1.3.4 SPATIAL AND GEOSPATIAL CONTEXT

The nature of geographic information can be considered dual, consisting of geographic data with geographic location with coordinates in a geographic space and descriptive attributes recorded in a conventional database. In geospatial space, geographic phenomena occur at different locations on the earth's surface. Spatial information has been related to the existence of objects with characteristics of spatial location and metric and directional topological relationship with other objects (Câmara and Medeiros, 1998).

When the geographic adjective used is the term spatial, the object or process occurred on surfaces of other planets, in the space of the cosmos, or in the space of the human body captured by medical images. With respect to the term geospatial, this denomination occurred to specify objects and processes that occurred near the Earth's surface or that were relative to planet Earth (Longley et al., 2001).

Geospatial geographic information and analysis is commonly used by society not only to determine the fastest route to a location through mapping websites, but also, for example, to assess environmental impact in natural areas and perform zoning in regions to assess the spatial distribution of existing wetlands, land use and cover, slope, terrain elevation, hydrological and transportation network (Figure 1.1) (Jensen and Jensen, 2012).

FIGURE 1.1 Cell phone with GPS technology and mapping capabilities to search geospatial databases through traditional cartography or satellite imagery.

1.4 DECISION LEVELS IN GEOCOMPUTATION

The occupation of the Earth's surface and implicit and explicit mechanisms of interrelationship between geographic phenomena have determined the occurrence of different spatio-temporal patterns. The pattern of occurrence of geographic data can be evaluated with respect to position or a known coordinate system, the attributes of the object, such as, color, cost, pH, disease incidence, and the spatial interrelationships between objects (Burrough and McDonnell, 1998). Geographic problems are classified according to the decision context in (Câmara and Medeiros, 1998; Longley et al., 2001):

- Scale or level of detail required that determines spatial correlation between geographic phenomena;
- Type of demand or purpose that determines thematic correlation between geographic phenomena;
- Occurrence over time that determines temporal correlation among geographic phenomena;
- Topological correlation such as adjacency, pertinence, and intersection between geographic phenomena.

In spatially distributed data, geographic coordinates and data attributes can be used to identify and locate attributes in geographic space. Geoinformation is disseminated in the form of maps, aerial photographs, satellite images on Internet mapping services such as (Jensen and Jensen, 2012):

- Google Earth[1];
- Google Maps[2];
- MapQuest[3];
- Yahoo! Maps[4];
- Bing Maps[5].

1.4.1 DECISION-MAKING PROCESS IN GEOCOMPUTATION

Using information systems, it is possible to manage, organize, store, access, retrieve, manipulate, synthesize and apply knowledge to solve problems. Knowing the meaning of the terms data, information, evidence, knowledge, and wisdom is useful to understand the infrastructure of the decision-making process, the ease of sharing, and examples in geocomputation. Data is described by numbers, text, or symbols without context. Data can be shared as a stream of bits, and with a set of data one can build a database (Longley et al., 2001).

Some types of geographic data used in geocomputation are (Burrough and McDonnell, 1998):

- Maps, charts, and topographic plans;
- Images obtained from satellites, aircraft, and the surface;
- Administrative boundaries and census data;
- Statistical data;

[1] www.earth.google.com

[2] www.maps.google.com

[3] www.mapquest.com

[4] www.maps.yahoo.com

[5] www.microsoft.com/maps/default.aspx

- Marketing survey data;
- Data on gas, water, electricity utilities and locations;
- Sampling data on soil, water, atmosphere, biological activity, disasters, in different spatial and temporal resolutions.

A database can be formed by the union of several datasets and the volume of data varies according to specific applications (Tabela 1.3) (Longley et al., 2001).

TABLE 1.3 Estimated database storage volumes for typical geocomputation applications, where a set of 8 bits in a single unit formed a byte.

Size	Bytes	Application
1 megabyte	1 000 024	Simple database of a small project.
1 gigabyte	1 000 000 024	Network of roads in a small town or country.
1 terabyte	1 000 000 000 024	Earth's wavelength spectral reflectance at 30m spatial resolution.
1 petabyte	1 000 000 000 000 024	Satellite image of the entire planet Earth at 1m spatial resolution.
1 exabyte	1 000 000 000 000 000 024	3D representation of the Earth at 10m spatial resolution.

The term "information" can be used as a synonym for data when there is an absence of context of the information. It is different from data when there is some degree of selection, organization and preparation of data to generate information for specific purposes, for some purpose or with some degree of interpretation. Therefore, information is usually more expensive to produce than data, but cheap to copy, reproduce and distribute. In possession of information, it can be easy to add value to and merge with other information through geocomputation. Geographic information is used especially in geocomputation for the following characteristics (Longley et al., 2001):

- Multidimensional definition of the location;
- Possibility of storing in large databases;
- Representation in different spatial, spectral, radiometric and temporal resolutions;
- Representation in different computational forms;
- Projection on a flat surface;
- Analysis by different methods and analysis times;
- Need for updating;
- Possibility of mapping large amounts of data.

Knowledge can be synonymous with information if interpretation is added for a certain context, experience, and purpose that determines some learning. Information can exist without knowledge, but for there to be knowledge there must be a learning process. Sharing knowledge between people, or quantifying it, is more difficult than obtaining information about an object or process. Knowledge about how a certain process works can be more important than its appearance, since it can be used for process prediction or new object positions. Also, knowledge is more related to assimilation than to the storage of knowledge. Evidence is considered a middle ground between information and knowledge, i.e., a multiplicity of information from different sources, related to solve several specific problems and with validated consistency. Wisdom is used in the context of decisions made based on all available evidence and knowledge, but weighted according to an understanding of likely consequences. Wisdom is defined at the top level of the decision-making infrastructure hierarchy (Figure 1.2) (Longley et al., 2001).

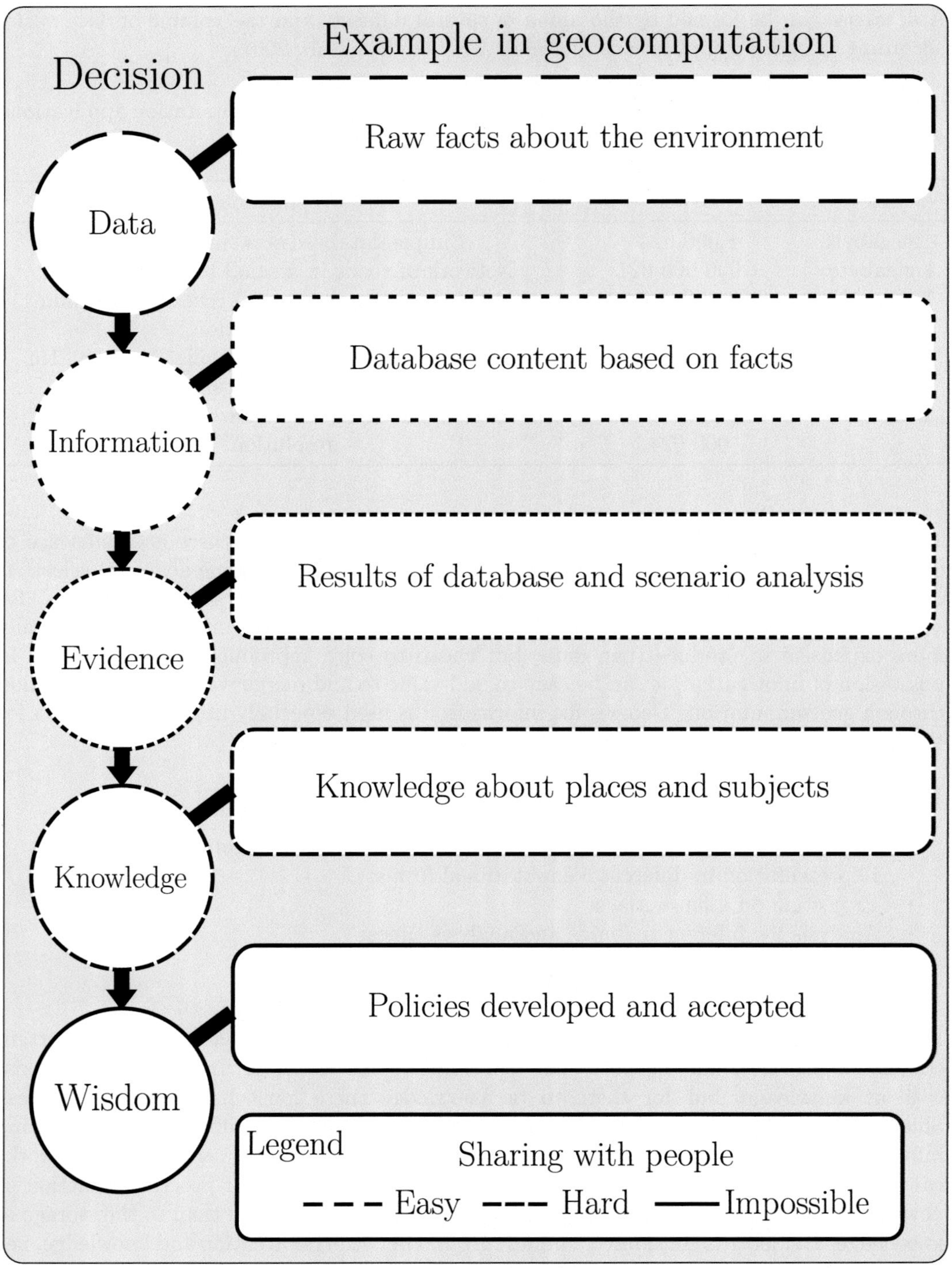

FIGURE 1.2 Decision-making infrastructure classification used in geocomputation.

1.5 HISTORY OF GEOCOMPUTATION

In Ancient Egypt and during the Roman Empire, surveyors made maps within the government organization. With the fall of the Roman Empire, geographic knowledge was expanded by scientists from Islam. With the beginning of oceanic navigations in the 14th and 15th centuries, systematic mapping was performed in territories for topographic mapping of European countries and colonies. With the development of European countries came the emergence of specific mappings of soil types, land use and vegetation. With this, the mappings started to be called thematic maps because they contain specific information about a certain subject. In the 20th century, topographic and thematic mapping was intensified, and with the advent of photogrammetry and remote sensing it became possible to map large areas with high accuracy. At that time, mathematical and statistical methods and digital image processing were also used to analyze the geographic information on maps (Câmara and Medeiros, 1998).

Geocomputation is linked by name to what is broadly termed computational science. Computational science is a relative new multidisciplinary paradigm for doing science in the late-20th century. The availability of high-performance computers, graphic workstations, speed networks, coupled with major advances in algorithms and softwares, has brought a revolution in the way scientific and engineering investigations are performed. Computational science involves using computers to study scientific problems and to complement the use of theory and experimentation in scientific investigation (Openshaw and Abrahart, 2001). The major events of the last decades determined the modern history of geocomputation were divided into eras of innovation (1957-1977), commercialization (1981-1999), and exploitation (1999-present). The intensive use of geocomputation was defined from the time there was a reduction in the price of computers below a critical threshold (Table 1.4) (Longley et al., 2001; Jensen and Jensen, 2012; Lovelace et al., 2019; Openshaw and Abrahart, 2001).

TABLE 1.4 Summary of some historical events associated with the type of use in geocomputation.

Date	Type	Event	Description
1957	Application	Automatic Mapping	Swiss Meteorologists and British Biologists
1963	Tecnologia	Technology	Geoprocessing System Development
1964	Academy	Laboratory at Harvard	Computer Graphics and Spatial Analysis Development at Harvard University
1967	Technology	Data Structure and File Development	1970 US Bureau 1970 Street Address Database
1969	Academia	First Published Papers	Geoprocessing Applications, Algorithms and Programs
1977	Academia	Conference on Topological Data Structures	Odyssey GIS Development at Harvard
1981	Commercial	Launch of ArcInfo	Development of the first commercial geoprocessing program for minicomputers

Date	Type	Event	Description
1985	Technology	Operational GPS	Start of the GPS system as a data source for navigation, surveying and mapping
1990	Science	Computational science	Computational science phenomenon gathered speed
1994	General	Birth of the OpenGIS Consortium	Establishment of consortium between vendors, government agencies and users to improve interoperability
1995	General	Physical geography	Openshaw describes various attempts at devising a computational human geography research agenda
1996	Technology	Start of geoprocessing products on the Internet	Launch of new generation of Internet products by various companies
1996	Science	Geocomputation	The word "geocomputation" first appeared during the First International Conference in Geocomputation held in Leeds
1998	Science	Geocomputation	Accelerated Strategic Computing Initiative (ASCI) and the first teraflop computers about 30 times faster than previous machines, e.g., the Cray T3E 1200 at Manchester, UK
2000	Commercial	Hardware and Software Industry	Industry with growth above 10% per year and 1 million geoprocessing users
2001	Science	Geocomputation	Increasing availability of terraflop supercomputers, hardware useful in many other areas including geocomputation
2011	Commercial	Multibillion dollar industry	Software development, data acquisition, and geoprocessing services
2020	General	Development and use of open source software in geocomputation	Development of open source software to store and access vector geometries; use of R, C++, Java and Python languages for geocomputation; geocomputation in clouds on the Internet

According to defined innovation criteria (Rogers, 1995), geocomputation was considered at the innovation stage in the 1970s, with initial use around the 1980s and, the beginning of maturity occurred in the 1990s. Currently, geocomputation was observed to be in the maturity stage greater than climax (Figure 1.3) (Longley et al., 2001).

1.6 COMPONENTS OF GEOGRAPHIC INFORMATION SYSTEMS

The components of a geographic information system are classified into (Burrough and McDonnell, 1998; Longley et al., 2001; Jensen and Jensen, 2012) (Figure 1.4):

- Digital information sharing network;
- People;
- Data;

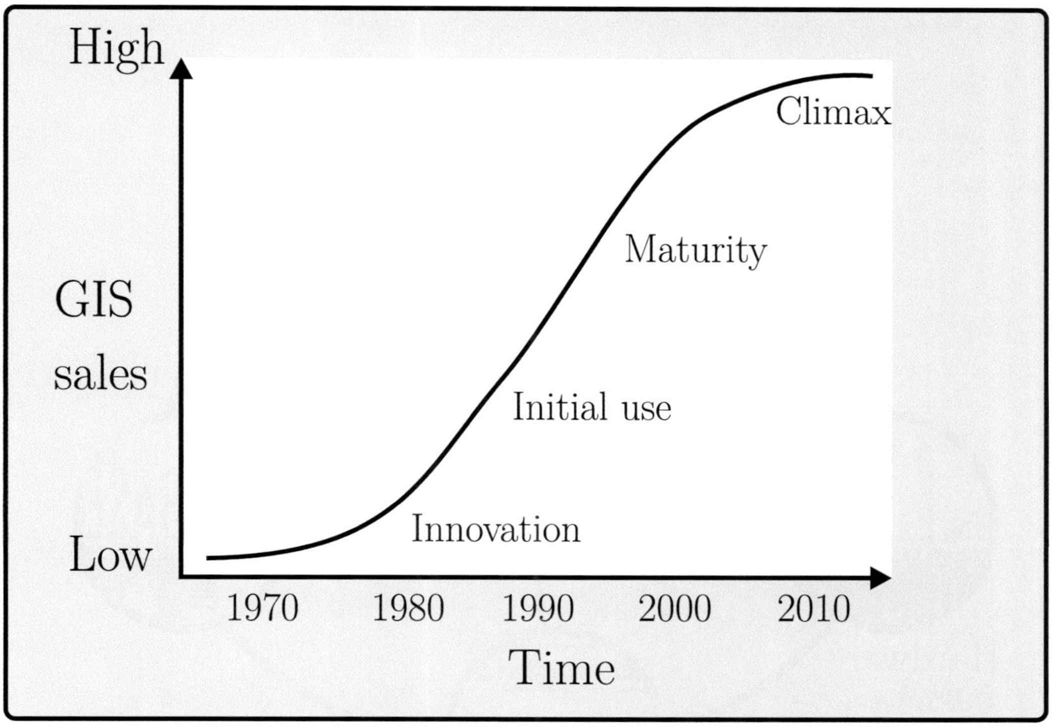

FIGURE 1.3 Geocomputation development according to an innovation model.

- Software;
- Hardware;
- Procedures and routines.

Data can be stored on the computer's hard disk, on the Internet, or on other digital data storage devices. The user controls the computer and other accessory equipment with a keyboard and mouse. Software in a geographic information system is classified according to functionality into (Burrough and McDonnell, 1998; Câmara and Medeiros, 1998) (Figure 1.5):

- Data import and verification;
- Database storage and management;
- Data export, visualization and plotting;
- Query, spatial analysis, image processing, data transformation;
- User interface.

The basic requirements for a geographic information system according to functionality are specified in (Burrough and McDonnell, 1998):

- Show the location of an A object;
- Show the location of an A object relative to B;
- Count the number of objects of type A at a distance D from object B;
- Evaluate a function f at a position X;
- Determine the area and perimeter of B;
- Determine the result of intersection between different types of geospatial data;
- Determine the least cost path along a network or continuous surface;

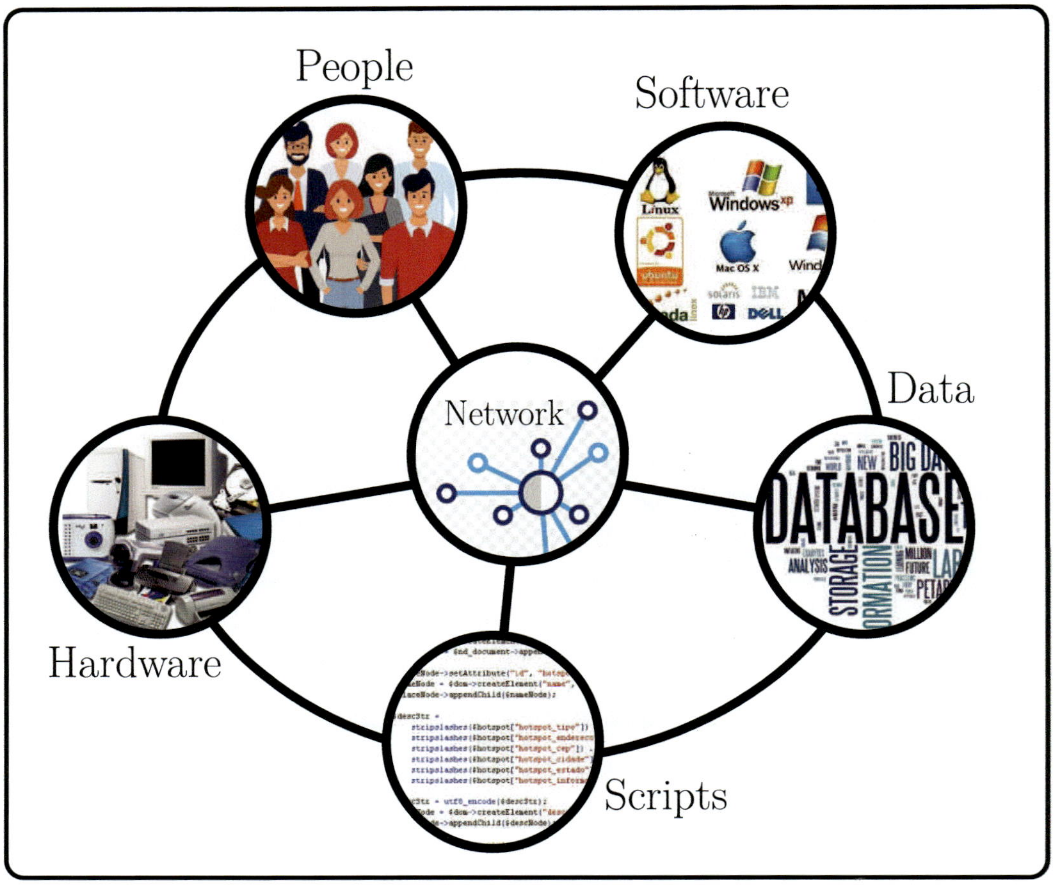

FIGURE 1.4 Components of a geographic information system.

- List the attributes of objects at locations X_1 and X_2;
- Determine objects near others according to a combination of specific attributes;
- Reclassify objects with a particular combination of attributes;
- Use geostatistical methods to model the spatial dependence of an attribute Z in a neighborhood of points at locations $X_1, X_2, ..., X_n$ and predict the value Z at locations $Y_1, Y_2, ..., Y_n$;
- Use statistical and computational intelligence methods to derive new variables from existing variables and vice versa;
- Use a digital model of the Earth and simulate the effect of a process P, at a time T, on a scenario C.

1.7 GEOCOMPUTATION AND OTHER FIELDS OF KNOWLEDGE

Geocomputation works in harmony with other geographic information sciences, including remote sensing, cartography, topography, geostatistics, digital image processing, statistics, computer science, artificial intelligence and computer vision, as well as other sub-disciplines such as engineering, mathematics, logic, physics, chemistry, biology, and sociology (Figure 1.6).

Geocomputation is concerned with computational techniques, algorithms, and paradigms

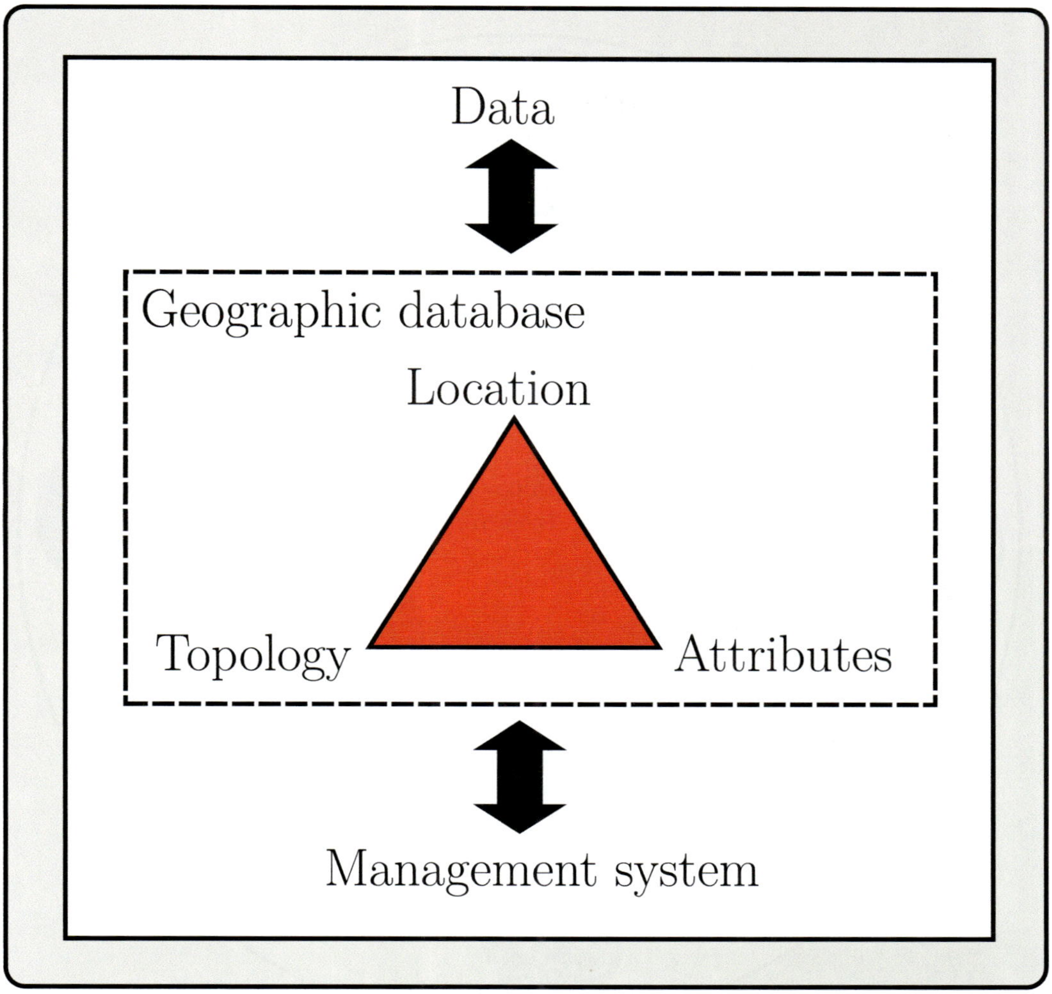

FIGURE 1.5 Components of a geographic database.

that involves four leading edge technologies (Openshaw and Abrahart, 2001):

- GIS, for data creation;
- Artificial intelligence and soft computing, to provide smart tools;
- High performance computing, to provide the power;
- Science, to provide the philosophy.

1.8 GEOCOMPUTATION SOCIETIES

In geoprocessing societies, stimulus to development and study and research in the field of geographic information systems and related areas are promoted to bring together people and entities with interest in geographic problems, development and applications of geocomputation and interaction with other technical-scientific societies, involved directly or indirectly with this area. Some related geoprocessing societies are:

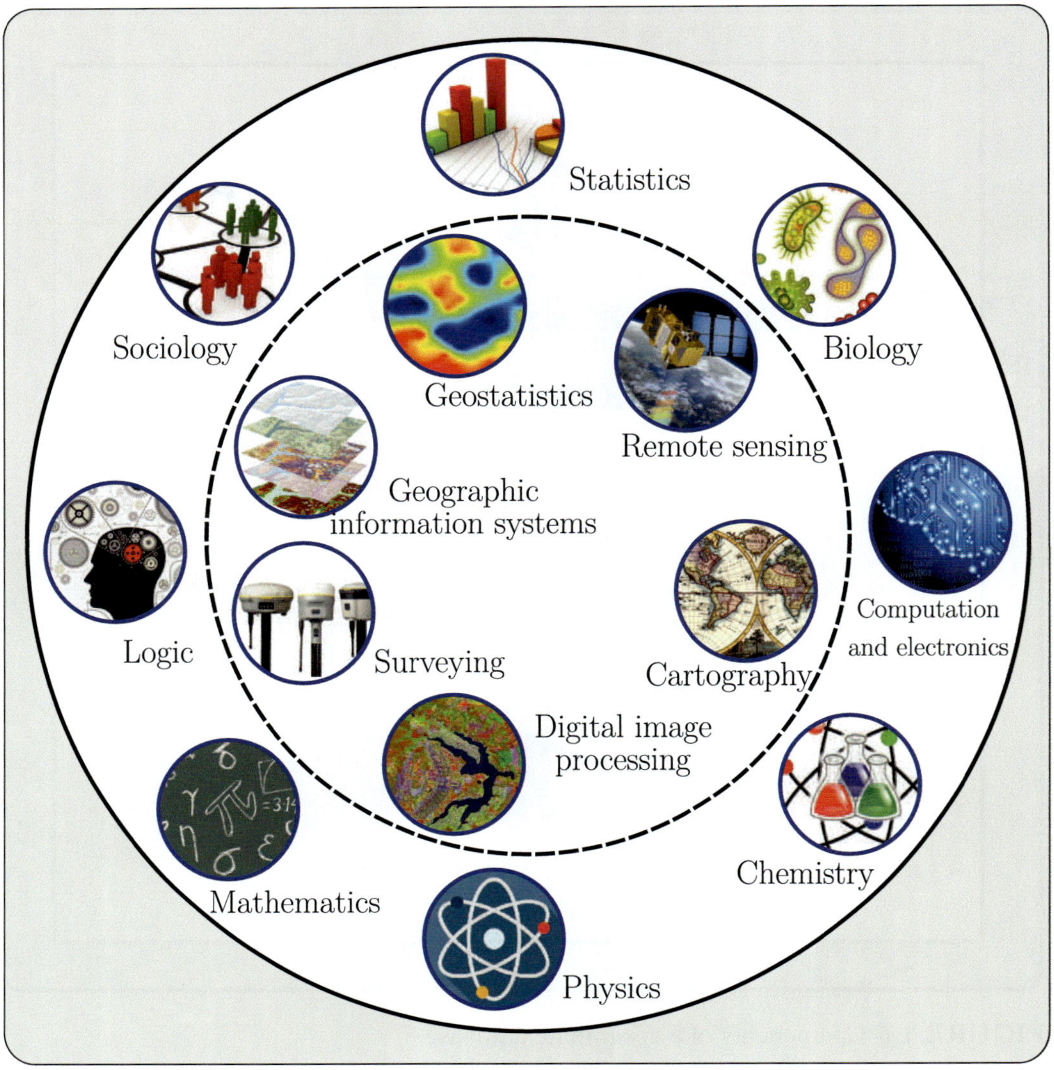

FIGURE 1.6 Geocomputation and other sciences.

- National Geographic Society[6]
- Geographical Information Systems International Group (GISIG)[7]
- Geospatial Information and Technology Association (GITA)[8]
- The Association for Geographic Information (AGI)[9]
- American Geographical Society (AGS)[10]
- Cartography and Geographic Information Society (CaGIS)[11]

---

[6] https://www.nationalgeographic.org/about-us/

[7] http://www.gisig.eu/

[8] https://www.gita.org/

[9] https://www.agi.org.uk/

[10] https://www.americangeo.org/

[11] http://www.cartogis.org/

- Canadian Association of Geographers (CAG)[12]
- The Imaging and Geospatial Information Society[13]
- University Consortium for Geographic Information Science (UCGIS)[14]
- Geospatial Professional Network (GPN)[15]

1.9 DATA SOURCES IN GEOCOMPUTATION

Government agencies and companies perform work in the area of geocomputation, where geospatial data infrastructure can be obtained for carrying out projects and research work through images and thematic maps of the territory and products derived from basic data obtained by scientific research (Figure 1.7)(Jensen and Jensen, 2012). Some geocomputation data and products can be obtained:

- Agência Espacial Brasileira (AEB)[16]
- Instituto Nacional de Pesquisas Espaciais (INPE)[17]
- Canada Centre for Remote Sensing (CCRS)[18]
- Centre National d'Etudes Spatiales (CNES)[19]
- Scientific and Industrial Research (CSIRO)[20]
- The European Space Agency (ESA)[21]
- National Aeronautics and Space Administration (NASA)[22]
- National Remote Sensing Centre (NRSC)[23]
- Japan Aerospace Exploration Agency (JAXA)[24]
- Google Earth Engine (GEE)[25]
- Instituto Brasileiro de Geografia e Estatística (IBGE)[26]
- spData[27]
- rnaturalearth[28]
- raster getData[29]
- BDMEP[30]

[12] http://www.cag-acg.ca/en/index.html

[13] http://www.asprs.org/

[14] http://www.ucgis.org/

[15] https://thegpn.org/?

[16] http://www.aeb.gov.br

[17] http://www.inpe.br

[18] https://natural-resources.canada.ca/science-data/science-research/research-centres/canada-centre-remote-sensing

[19] http://www.cnes.fr/

[20] http://www.csiro.au/

[21] http://www.esa.int/

[22] http://www.nasa.gov/

[23] https://www.isro.gov.in/NRSC.html

[24] https://global.jaxa.jp/

[25] https://earthengine.google.com/platform/

[26] https://downloads.ibge.gov.br/downloads_geociencias.htm

[27] https://cran.r-project.org/web/packages/spData/index.html

[28] https://cran.r-project.org/web/packages/rnaturalearth/

[29] https://www.rdocumentation.org/packages/raster/versions/3.1-5/topics/getData

[30] http://www.inmet.gov.br/portal/index.php?r=bdmep/bdmep

- Agência Nacional de Águas (ANA)[31]
- National Oceanic and Atmospheric Organization (NOAA)[32]

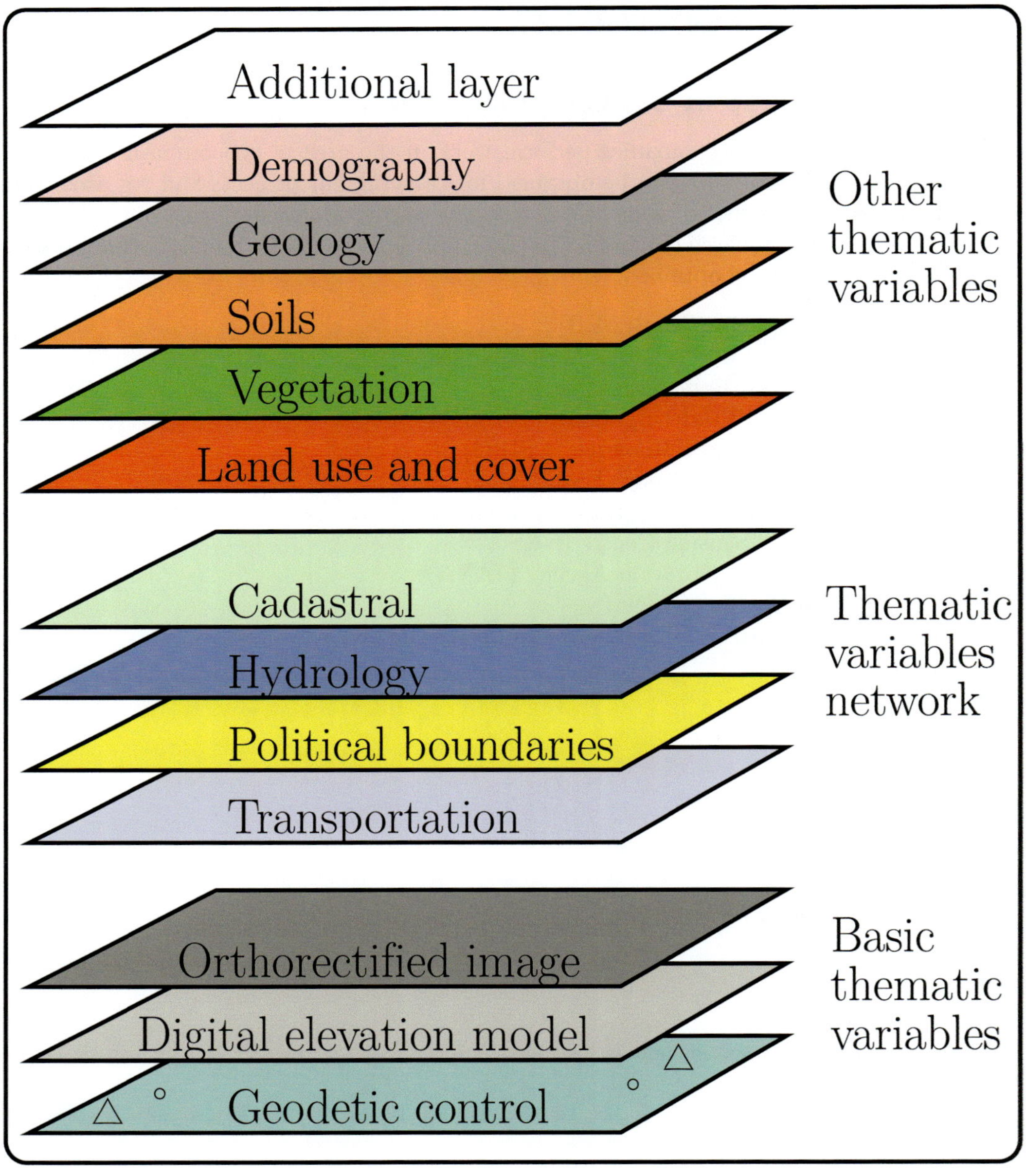

FIGURE 1.7 Geospatial data infrastructure.

1.10 GEOCOMPUTATION JOURNALS

Journals in the area of geocomputation are related to the progress of knowledge in the area, in order to facilitate communication among professionals and to update the development

[31] https://www.snirh.gov.br/hidroweb/apresentacao

[32] https://www.ncdc.noaa.gov/cdo-web/

of science and technology using geographic information systems. Some journals with a theoretical and practical approach to geocomputation can be accessed via the Internet, such as:

- International Journal of Geographical Information Science;
- Computers and Geosciences;
- Computers;
- Computers, Environment and Urban Systems;
- International Journal of Applied Earth Observation and Geoinformation;
- GeoInformatics;
- Geospatial Solutions;
- GIScience and Remote Sensing;
- IEEE Transactions on Geoscience and Remote Sensing;
- Pattern Recognition;
- Theoretical and Applied Engineering;
- Transactions in GIS;
- Annals of the Association of American Geographers;
- Cartography and Geographic Information Science;
- The Cartography Journal;
- Geographical Analysis;
- GeoInformatica;
- Journal of Geographical Systems;
- Transactions in GIS;
- URISA Journal.

1.11 GEOCOMPUTATION, SOCIETY AND INVESTMENTS

Geographic information science is linked to the following context in society (Figure 1.8) (Longley et al., 2001):

- The individual;
- The computer;
- Society.

Research on the individual is related to cognitive science to understand geospatial concepts, learning and reasoning about geographic data, and computer interaction. Research on the computer is characterized by issues of representation, adaptation to new technologies, geoprocessing and visualization. Research on society is distinguished by the theoretical and practical scope for studying relevance in the social context.

The use of geoprocessing is organized in different economic sectors of modern society. The centrality of geospatial infrastructure, is composed of sectors of (Lo and Yeung, 2007):

- Technology;
- Research;
- Services;
- Data;
- Standard Setting;
- Social policy.

The geospatial information sector, around geospatial infrastructure, is composed of (Lo and Yeung, 2007):

- Hardware and software vendors;

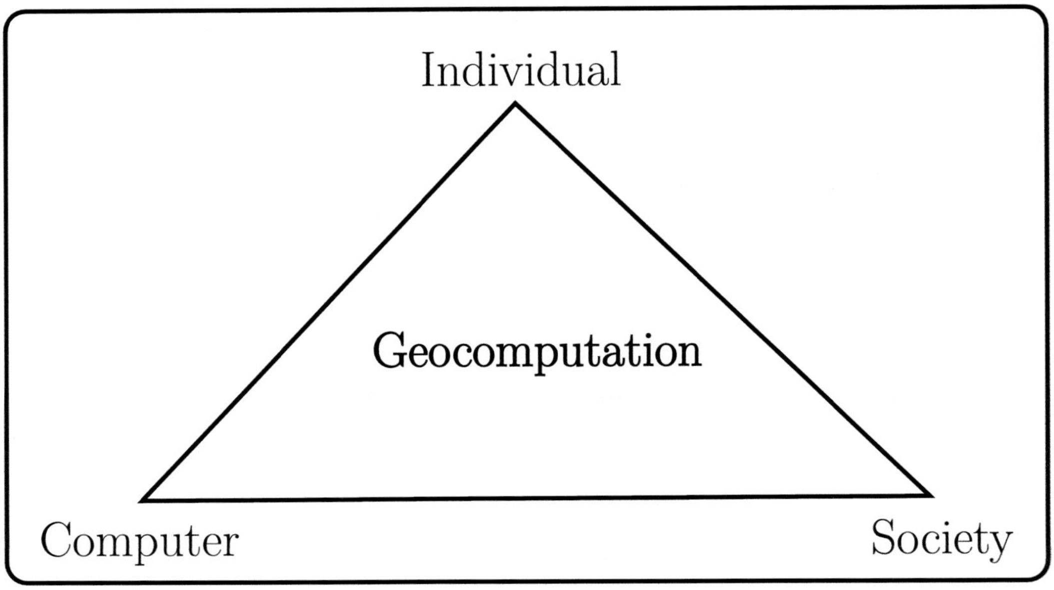

FIGURE 1.8 Contextual relationship of geoprocessing in society.

- Map production;
- Surveying;
- Remote sensing;
- Geographic information system;
- Satellite positioning system.

In the sector of consumers of products, services and spatial information, around the geospatial information and infrastructure sectors there is a subdivision of application sectors into the industrial, business, academic and public sector. In the industrial sector, applications of geocomputation are observed in (Lo and Yeung, 2007):

- Agriculture, forestry and livestock;
- Environment;
- Mining;
- Transportation and logistics;
- Construction.

In the business sector, applications of geocomputation are observed in (Lo and Yeung, 2007):

- Bank financing and insurance;
- Consumption of geographic information services.

In the academic sector, geocomputation is used in (Lo and Yeung, 2007):

- Education;
- Research and publications;
- Training.

In the public sector, applications of geocomputation are observed in (Lo and Yeung, 2007):

- Defense;
- Public health;
- Emergency response;

- Municipal management infrastructure;
- Land registration and taxation;
- Urban planning
- Democratic participation of communities.

Designing and implementing a functional geocomputation system can take more than a year, depending on the size of the organization and the amount and type of location data used (Longley et al., 2001; Jensen and Jensen, 2012). In the service industry, for example, geospatial data is used daily to answer geographic questions for customers, as the example of surveys with farmers to maximize crop yields with precision agriculture (Jensen and Jensen, 2012).

Real-time location-based services (LBS) are considered a new branch of geospatial application. In this type of service, the location of a mobile device is used to deliver services and conduct business in real time by combining geocomputation, positioning system, wireless mobile device, and Internet (Figure 1.9).

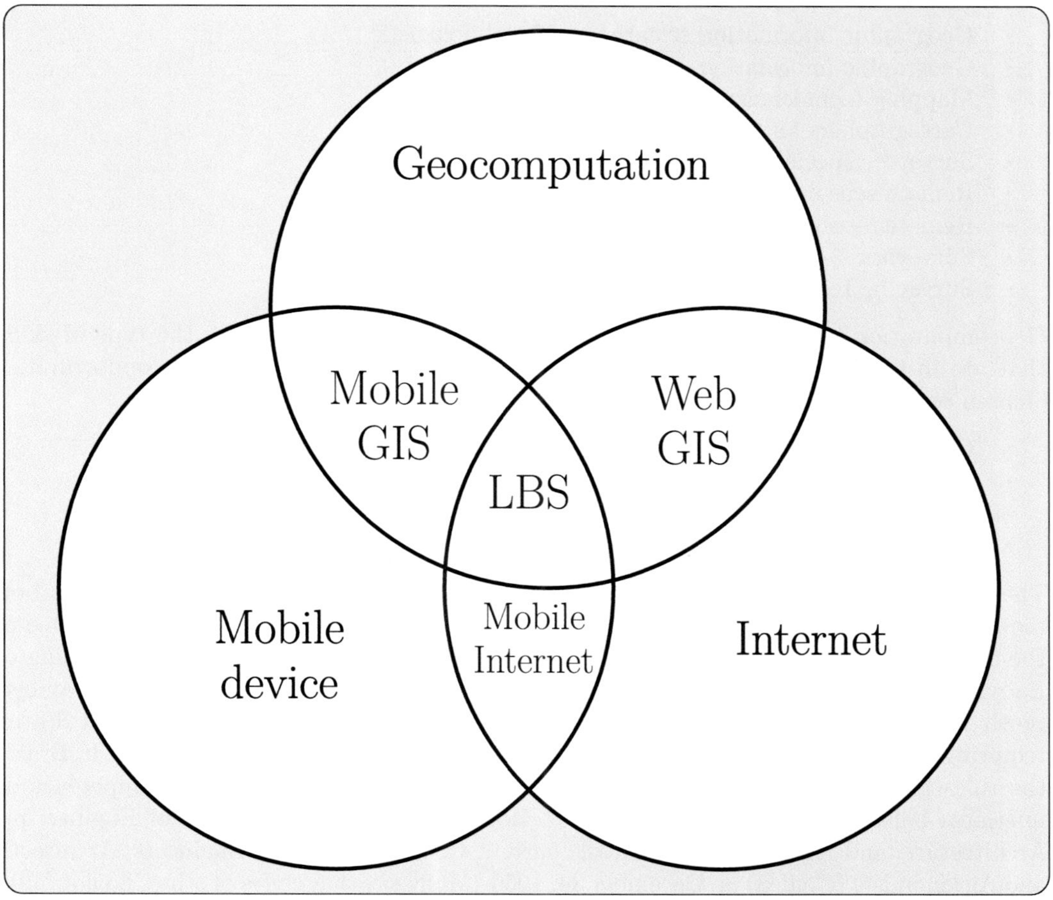

FIGURE 1.9 Technologies used in real-time location-based services with geocomputation.

1.12 GEOCOMPUTATION AS A PROFESSION

Once a geographic information system is developed, the system can be used to make geospatial decisions, perform mapping and monitoring of the environment. However, the system must be used intelligently to make good decisions (Jensen and Jensen, 2012).

Jobs in geocomputation may be available in public agencies (national, state and local governments; colleges and universities) and private commercial companies. The demand for geocomputation professionals increases as geospatial data and analysis are more commonplace (Jensen and Jensen, 2012). With the advent of satellite positioning technologies and availability of very high spatial resolution satellite imagery of much of the Earth via the Internet (Google Earth, Google Maps, MapQuest, Bing Maps), there has been increased interest from the general public in geospatial issues. In universities and colleges, geocomputation principles are covered in disciplines such as agronomy, geography, geology, engineering, natural resource management, forestry, agriculture, marine sciences, and others. Despite this, the educational system cannot keep up with the demand for qualified professionals in geoinformation science (Mondello et al., 2008; Jensen and Jensen, 2012).

In work demands related to geoprocessing, the following job positions are defined (Jensen and Jensen, 2012):

- Geographic information scientists and technologists;
- Geographic information systems technicians;
- Mapping technicians;
- Cartographers and photogrammetry specialists;
- Surveyors specializing in geodesy;
- Remote sensing scientists and technologists;
- Remote sensing technicians;
- Surveyors;
- Surveying technicians.

Geocomputation jobs can be searched on the Internet at sites to assess the type of skills that are in demand in the market, typical starting salaries, and other job opportunities (Jensen and Jensen, 2012):

- GIS Jobs[33];
- GIS Jobs Clearinghouse[34];
- GeoJobs[35];
- LinkedIn[36].

The purpose of councils, such as those for engineering and professional registration requirements, is to protect the public from unqualified or unscrupulous individuals. One of the first regulations of this type was contained in the code of laws of Hammurabi, king of Babylon in the 18th century BC. This code of ethics became famous for its "eye for an eye, tooth for a tooth" section. In 1883, registration requirements reached the United States, requiring a license for professionals in different fields to practice their professions. In Brazil, the rules for excluding geoscience professionals from registration due to incompetence or unethical behavior are not often enforced. However, the Federal Council of Engineering, Architecture, and Agronomy adopted the current Code of Ethics for Engineers, Architects, and Agronomists (Confea) on December 24, 1966. Many say that codes of ethics do not fulfill their purpose, but insofar as they postulate a goal to be achieved, they make professionals

---

[33] www.gisjobs.com

[34] www.gjc.org

[35] www.geojobs.org

[36] https://www.linkedin.com/?trk=nav_logo

look to a higher moral plane. In addition, the most important asset a professional has is their impeccable reputation. No amount of money, fame, or knowledge is an adequate substitute. Therefore, ethics can be defined as the obligations that a professional has toward the public and colleagues in order to help them understand what standards should be achieved and what they can expect from their colleagues. Therefore, environmental geocomputation professionals must be loyal, impartial, and faithful to their clients, employers, and the public. In addition, professionals should be interested in public progress, applying their knowledge for the benefit of humanity (McCormac et al., 2012; de Carvalho Alves and Sanches, 2022). In terms of acquired professional knowledge, an analogy can be made between professional performance and the development of a corn plant. Just as ears of corn without grain raise their heads disdainfully to the sky, so do the arrogant, while those full of knowledge (or fruit) lower them to the ground. In other words, empty ears of corn without grain symbolize arrogance and emptiness. Ears full of grain, on the other hand, bow to the ground in a gesture of humility, reflecting the wisdom and abundance that lead to modesty, the opposite of pride. Pride represents the haughtiness of an ear of corn, which, due to a lack of content (knowledge or fruit), stands up without reason, not bowing (de Montaigne, 1987).

1.13 THE FUTURE OF GEOCOMPUTATION

High performance computing systems have become commonplace, and cloud computing is now available as a commodity. Petabyte-scale remote sensing data is made available by Brazilian, North American and European government agencies for free. Using all the available resources requires expertise and a lot of dedication. Considering the difficulties of using information technology in relation to data acquisition and storage, data format, database management, allocation of machines, CPUs, networks, multiple geocomputation routines, virtual libraries and remote sensing data were created on the Internet. Through the virtual libraries, data catalogs and information readily available via high performance parallel computing service, there is rapid access and visualization of search results in geographic database with spatial data infrastructure via interactive interface on the Internet (Gorelick et al., 2017) (Figure 1.10).

1.14 GRAPHICAL USER INTERFACE AND COMMAND LINE INTERFACE

The development of geocomputation is closely associated with advances in computer and automatic development. The hardware used in geoprocessing has become faster, more efficient, affordable and smaller in size. There are geographic information systems with graphical user interfaces (GUI), such as QGIS and SAGA GIS software, rather than a command-line interface (CLI) typed by keyboard, as in R (Table 1.5). Reproducibility is a major advantage of CLIs, as the same results can be generated by others via publicly accessible code, leading to profound implications for teaching and science (Lovelace et al., 2019; Pebesma et al., 2012).

TABLE 1.5 Differences in characteristics between a geographic information system with a graphical user interface (GUI) and a command line interface (CLI).

Feature	GUI	CLI
Basic subjects	Geography	Computer science, statistics
Software engine	Graphical User Interface	Command Line
Reproducibility	Minimum	Maximum

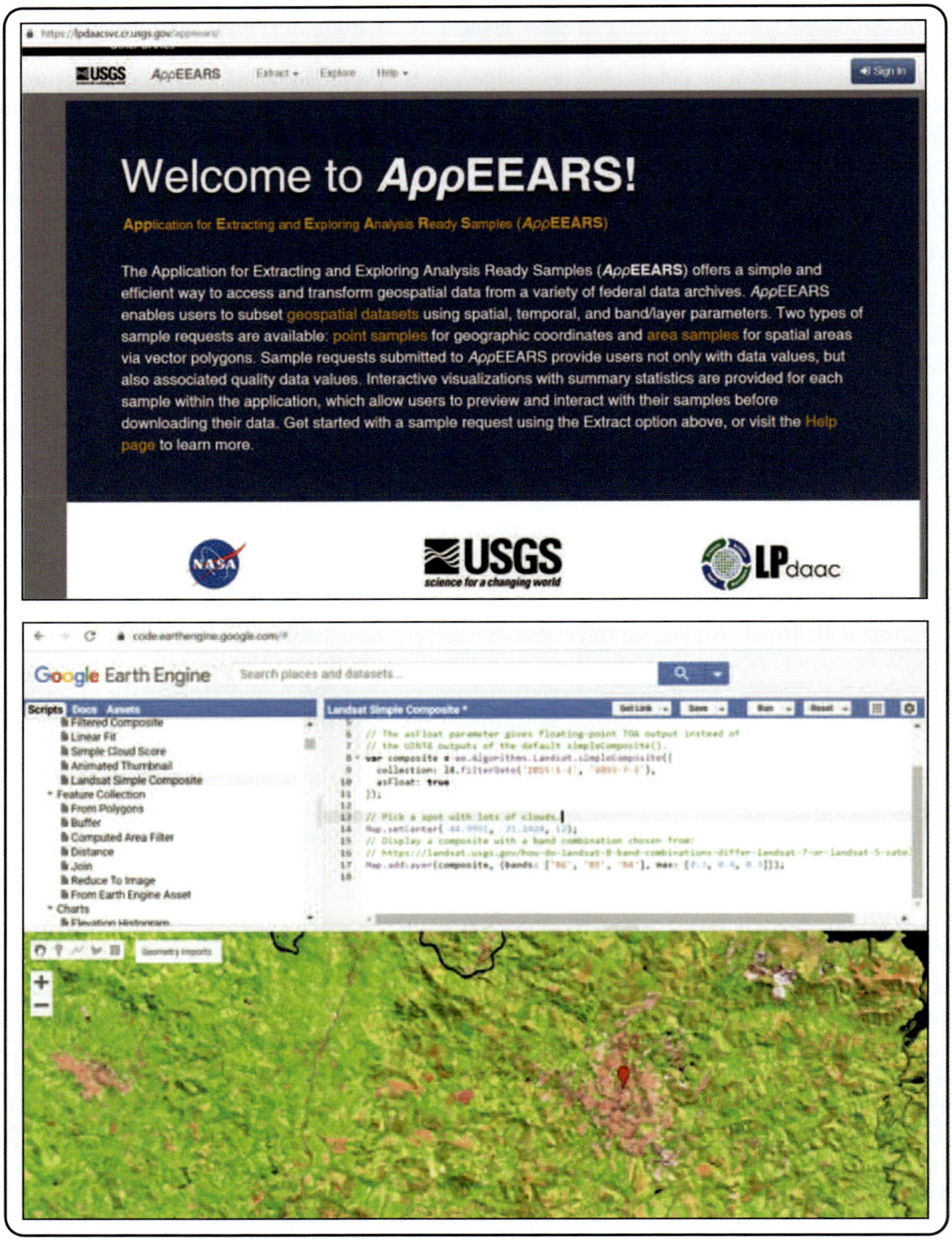

FIGURE 1.10 AppEEARS virtual libraries (top) and GoogleEarth Engine code (bottom) used for digital image processing and obtaining spatial data infrastructure.

With the sharing of different geographic information systems, geospatial analysis is broadened in terms of analysis possibilities. The Open Geospatial Consortium (OGC) started in 1994 with a group of more than 100 universities, corporations and government agencies with the goal of making geocomputation interoperable and popular. Various analysis and geocomputation data packages can be obtained on the Internet (Jensen and Jensen, 2012).

With the use of geocomputation software, the gap between research, science and technology is bridged in order to obtain algorithms and processing routines in different software with applications for spatial analysis of environmental variables (Goslee, 2011; Kwok, 2018). R consists of an object-oriented functional programming language and open source cross-platform environment for statistical computing and graphics, advanced geospatial analysis, modeling, and geovisualization, with a wide variety of packages. R is ideal for quickly creating geo-tools without the need to master lower-level languages such as C, FORTRAN, or Java. With the use of R, there is easy access to other languages, such as access to C++ and Python code through the Rcpp and reticulate packages, respectively. C++ and Python are used as the basis for the QGIS, GRASS and SAGA GIS (Lovelace et al., 2019) software (Figure 1.11).

Some free softwares used in geocomputation are:

- R and R packages[37];
- Quantum GIS (QGIS)[38];
- SPRING/INPE[39];
- GRASS[40];
- SAGA GIS[41];
- Python[42]

Some commercial softwares used in geocomputation are:

- MATLAB ®[43];
- ArcGIS ®[44];
- PCIgeomatics ®[45].

Regarding commercial software, because of the dynamics of the software business, many programs can change names according to the marketing involved in joining companies and modifications that can occur with time, interest, and market trends.

According to Lovelace et al. (2019), with the open source of R, there has been an evolution in package developments in the geospatial area. The package `sf` (Pebesma et al., 2021a), for example, is based on its predecessor `sp` (Pebesma and Bivand, 2005; Bivand et al., 2013). With `sf` we can define simple open source feature pattern and model to store and access vector geometries (Lovelace et al., 2019).

---

[37] https://www.r-project.org/

[38] https://qgis.org/pt_BR/site/

[39] http://www.dpi.inpe.br/spring/

[40] https://grass.osgeo.org/

[41] http://www.saga-gis.org/

[42] https://www.python.org/

[43] https://www.mathworks.com/products/matlab.html

[44] https://www.esri.com/en-us/arcgis/products/index

[45] https://www.pcigeomatics.com/

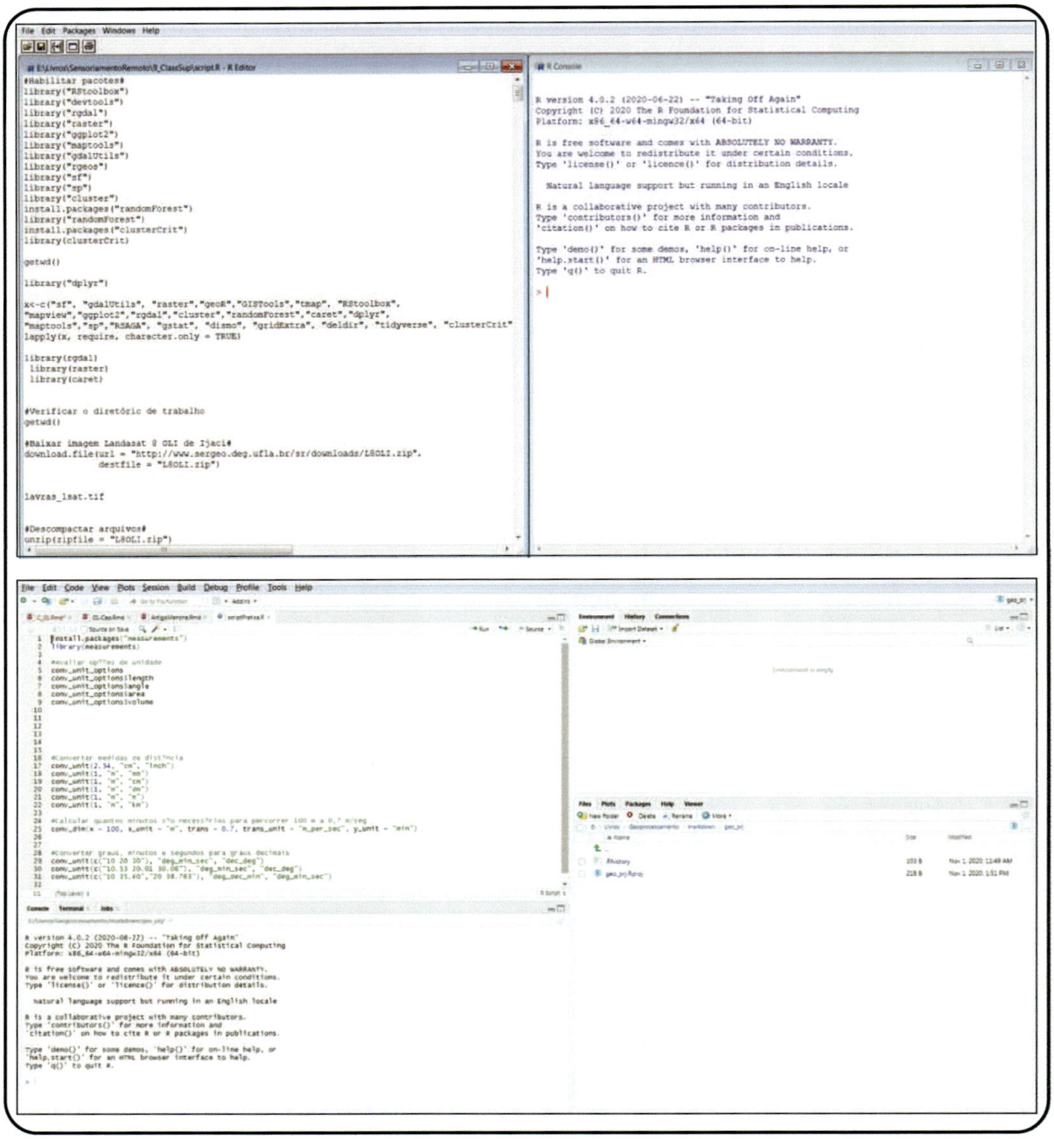

FIGURE 1.11 R (Superior) and RStudio (Inferior) softwares used in geocomputation.

`Maptools` (Bivand et al., 2020a) is used to import ESRI Shapefiles into nested lists of geometry. `rgdal` (Bivand et al., 2021), released in 2003, provides `GDAL` bindings for R, and there was support for coordinate reference systems via the `PROJ` library, reprojections, and importing vector file formats. Many of these additional features were developed by Barry Rowlingson and released in the `rgdal` codebase in 2006 (Lovelace et al., 2019). With the `sp` package (Bivand et al., 2013), released in 2005, it is possible to distinguish spatial and non-spatial objects (Pebesma and Bivand, 2005). Before 2005, geographic coordinates were treated as an arbitrary number, and with `sp`, there was the creation of generic classes and methods to analyze points, lines, polygons and meshes and attribute data (Lovelace et al., 2019). In `sp`, information is stored with coordinate reference system and attributes in geo-object slots with `S4` class system and geographic data operations. In `sp` generic summary methods are also available in the `summary()` and `plot()` functions for geographic data. In 2013 and 2018, `sp` was used in over 100 and 500 packages, respectively (Bivand et al., 2013; Lovelace et al., 2019).

In the `raster` package (Hijmans et al., 2020) a system of classes and functions is used for operations with raster type data. A key feature of raster is the ability to work with very large data-sets. In `raster` there is also the development of functions for map algebra. Parallel to these developments of systems and class methods, functions for interfaces of R with other geographic information systems were developed, as in the case of links between R and SAGA GIS by the `RSAGA` package (Brenning et al., 2018a), links of R with Python ArcPy functions via the reticulate package to access ArcGIS geoprocessing functions via the `RPyGeo` package (Brenning et al., 2018b) and links of R with QGIS, via the `RQGIS` package (Lovelace et al., 2019).

Despite methods for making maps using the base plotting and network system of the `sp`package (Pebesma et al., 2021b), the demand for advanced mapping capabilities increased after the release of `ggplot2` (Wickham et al., 2020) in 2007, with access to base map blocks from online services such as Google Maps. Raster visualization received a boost with the release of the `rasterVis` package (Lamigueiro and Hijmans, 2020) to visualize geospatial and temporal data (Lamigueiro, 2018). Starting in 2018, new packages for map production in R emerged, via `tmap` (Tennekes, 2018; Tennekes et al., 2020), `leaflet` (Cheng et al., 2021) and `mapview` (Appelhans et al., 2020) that used the `sf` class system (Lovelace et al., 2019).

1.14.1 COMPUTATIONAL PRACTICE

1.14.2 INSTALLING R

R is an open source programming, visualization and data analysis environment originating from the S (Lovelace et al., 2019) programming language. The latest version of R for Windows can be obtained for download[46]. Older versions may be more useful if a package used in the data analysis has not yet been updated to the latest version of R. Older versions can be obtained for download[47].

R can also be accessed through RStudio, an integrated development environment (IDE) for R. In RStudio there is a user-friendly interface to R with features for editing source code with colored syntax, data table viewer, Git and GitHub integration, Markdown and Bookdown output. RStudio has a free software version and also a private version for companies or individuals who need customer support. The latest version of RStudio for Windows can be obtained for download[48]. Older versions of RStudio can be obtained for download[49].

1.14.3 USING R WITH COMMAND LINE

R can be run from an R console or from an RStudio command line environment. We can assign values to objects, such as five numbers to objects x, y with an arrow (<-) or the equal symbol (=) and then perform the sum of the objects at z, with the following code in R:

[46] https://cran.r-project.org/bin/windows/base/

[47] https://cran.r-project.org/bin/windows/base/old/

[48] https://rstudio.com/products/rstudio/download/

[49] https://www.rstudio.com/products/rstudio/older-versions/

```
y <- c(5,4,3,2,1) # Create y database
y # Evaluate y results
# [1] 5 4 3 2 1
x = c(1,2,3,4,5) # Create x database
x # Evaluate x results
# [1] 1 2 3 4 5
z <- x+y # Create z database with x,y sum
z # Evaluate z results
# [1] 6 6 6 6 6
```

In the case of typing more than a few lines of code in a command prompt environment or when one wants to save a series of commands as part of the analysis of a project, one should create an R script file and save it with an R extension to make it easier to retrieve the codes to continue the project at another time.

Several packages for geospatial measurement, mapping and analysis are available in R. The packages are used in solving geographic problems in geocomputation. Searching for information about an R package can be performed based on keywords about the subject of interest by means of the package `sos` (Graves et al., 2020).

The search for relevant scientific publications can be performed automatically using the R software, using the public API of the Crossref database. To achieve this, the `cr_works()` function from the `rcrossref` (Chamberlain et al., 2022) package can be used. You can also use this package with the `dplyr` (Wickham et al., 2021a) and `readr` (Wickham et al., 2021c) packages for data processing and export.

Next, as an example, it is shown how to set up directory, install and enable packages, evaluate the version of the operating system and R on the operating system, as well as a keyword-based package search method for application in different subject areas. Finally, the `tmap` (Tennekes, 2018) package is used to map regions of the globe.

1.14.4 SET UP DIRECTORY IN R

To evaluate the working directory in R, we use the `getwd()` function, and to set up a directory in R, we use the `setwd` function with the address of the new working directory.

```
getwd()
setwd("C:/geo/c1")
```

1.14.5 INSTALLING R PACKAGES

The `install.packages` function is used in the R console to install the `sf`, `sos`, `tmap`, `devtools` (Wickham et al., 2021b), `brazilmaps` (Siqueira, 2017), and `GADMTools` (Decorps, 2021) packages. If a package is not available for installation via CRAN, it may be necessary to download the zip file of older versions of the package and its dependencies for installation.

```
install.packages("sf")
install.packages("sos")
install.packages("tmap")
install.packages("devtools")
```

```
devtools::install_github("rpradosiqueira/brazilmaps")
install.packages("GADMTools")
install.packages("rcrossref")
```

1.14.6 LOADING R PACKAGES

The `sf`, `sos`, `tmap`, `devtools`, `brazilmaps`, `GADMTools` and `rcrossref` packages are loaded with the `library` function in the R console.

```
library(sf)
library(sos)
library(tmap)
library(brazilmaps)
library(GADMTools)
library(rcrossref)
```

1.14.7 EVALUATING INFORMATION ABOUT COMPUTER PROGRAMS AND R PACKAGES USED

The `sf` package is installed and enabled to evaluate information about computer programs and R packages used in writing the book.

Using the `print(version)` function, we can evaluate information about the operating system and version of R used.

```
print(version)
```

The version of windows used, the base R packages, as well as other R packages installed on the computer are obtained with the `print(sessionInfo())` function.

```
print(sessionInfo())
```

The installation directory of R packages on the computer is evaluated using the `print(.libPaths())` function.

```
print(.libPaths())
```

The `packageVersion()` function is used to evaluate the version of a specific R package installed on the computer; in this case the version of the sf package is taken.

```
packageVersion("sf")
```

1.14.8 EVALUATE THE EXISTENCE OF AN R PACKAGE WITH KEYWORDS

Using the `findFn()` function, it is possible to perform searches on existing packages on a subject topic of interest; in this case, with the keywords "gis" and "spatial", 80 items were found in 2021.

```
findFn("gis spatial")
```

1.14.9 ACCESS CROSSREF USING KEYWORDS

Using the `cr_works` function, it is possible to perform searches on existing publications on a subject topic of interest, in this case, with the keyword "geocomputation", 398 items were found.

```
geo <- cr_works(query = "geocomputation")
geo
# $meta
#  total_results   search_terms start_index items_per_page
# 1           398 geocomputation           0             20
geo[["data"]][["url"]] # Only the first 4 items with doi url
# [1] "https://doi.org/10.1201/b17091-8"  "https://doi.org/10.1201/b17091-13"
# [3] "https://doi.org/10.1201/b17091-26" "https://doi.org/10.1201/b17091-20"
#...
```

1.14.10 CREATE BIBTEX CITATIONS

A BibTeX citation can be created for R packages using the `write_bib` function from the `knitr` package (Xie, 2018, 2023, 2015). As an example, we created the citation for the `rcrossref` package. Another option is to use the `utils` package (R Core Team, 2023), `toBibtex`, and `citation` functions to obtain the citation of an R package in BibTeX or LaTeX.

```
knitr::write_bib("rcrossref")
# @Manual{R-rcrossref,
# title = {rcrossref: Client for Various CrossRef 'APIs'},
# author = {Scott Chamberlain and Hao Zhu and Najko Jahn and Carl Boettiger and
# Karthik Ram},
# year = {2022},
# note = {R package version 1.2.0,
# https://github.com/ropensci/rcrossref},
# url = {https://docs.ropensci.org/rcrossref/},}
toBibtex(citation("dismo"))
# @Manual{R-dismo,
# title = {dismo: Species Distribution Modeling},
# author = {Robert J. Hijmans and Steven Phillips and John Leathwick and Jane
# Elith},
# year = {2024},
# note = {R package version 1.3-16},
# url = {https://CRAN.R-project.org/package=dismo},}
```

1.14.11 MAPPING AND ACCESSING ATTRIBUTE DATA IN THE GLOBE

After installing the tmap package, a World dataset is accessed with the `data` function and then mapped with the `qtm` function to generate a boundary division map (Figure 1.12).

```r
data(World) # Access data
qtm(World) # Map
```

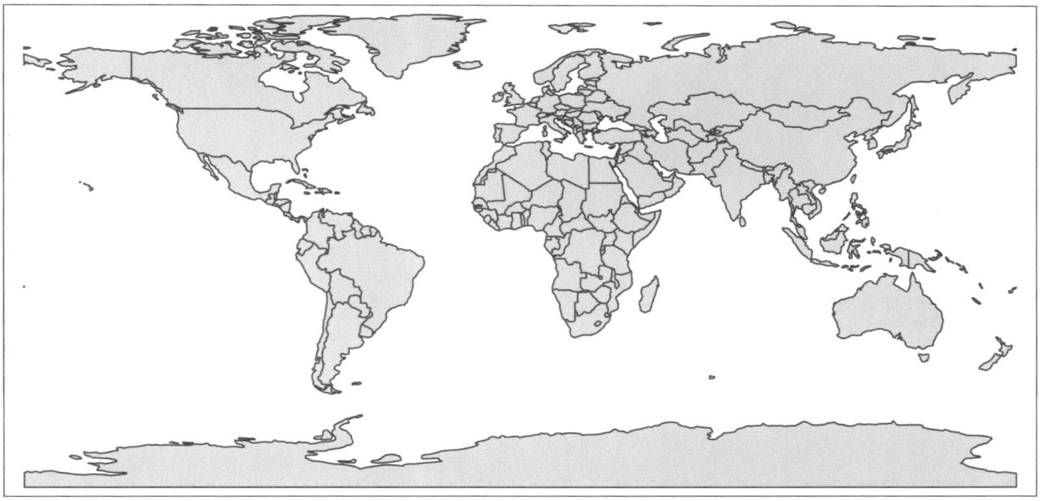

FIGURE 1.12 World map with country borders obtained with the `qtm` function.

The attribute table and some characteristics of the geometry type and coordinate system (CRS) for polygons are accessed in the first 10 lines of the file:

```r
World
# Simple feature collection with 177 features and 15 fields
# Geometry type: MULTIPOLYGON
# Dimension:     XY
# Bounding box:  xmin: -180 ymin: -89.9 xmax: 180 ymax: 83.64513
# Geodetic CRS:  WGS 84
# ...
```

The table attribute names are accessed with the `names` function.

```r
names(World)
# [1] "iso_a3"        "name"          "sovereignt"    "continent"     "area"
# "pop_est"       "pop_est_dens" "economy"
# [9] "income_grp"    "gdp_cap_est"   "life_exp"      "well_being"    "footprint"
# "inequality"    "HPI"           "geometry"
```

A choropleth map of the continents is made with the `qtm` function and filling in the data column of interest, in this case the variable continent (Figure 1.13).

```r
qtm(World, fill="continent")
```

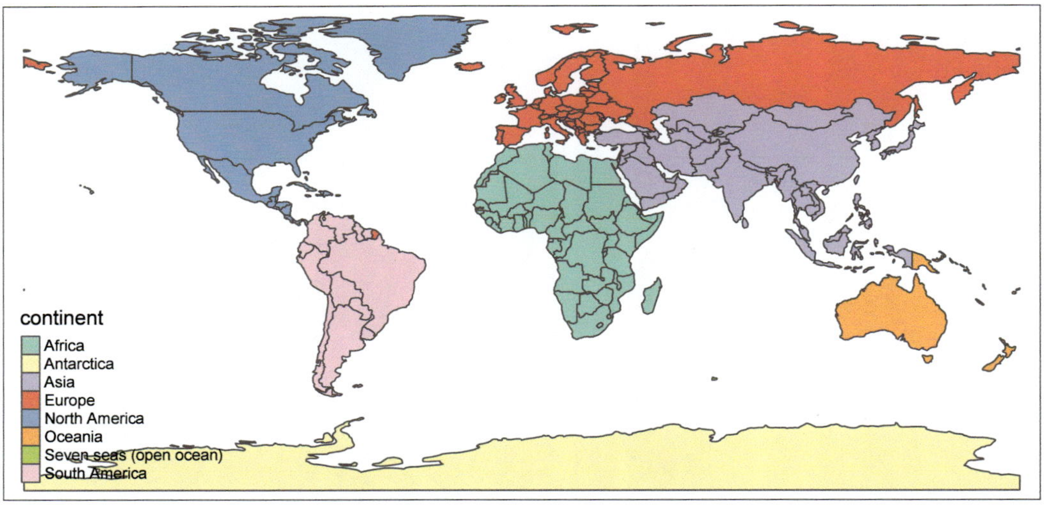

FIGURE 1.13 Choropleth world map of countries by continent.

1.14.11.1 Defining a geographical region with a county level polygon

To locate the county of origin of the rural property, the `get_brmap` function is used with the state 51 configuration at the Mato Grosso county level.

```
mt <- get_brmap(geo = "City", geo.filter = list(State = 51), class = "sf")
```

The municipalities are mapped for visual analysis using the 'plot' function (Figure 1.14).

```
plot(mt[1], main = "Mato Grosso")
```

Then, the file characteristics and the coordinate reference system were evaluated.

```
mt
# Simple feature collection with 141 features and 6 fields
# Geometry type: MULTIPOLYGON
# Dimension:     XY
# Bounding box:  xmin: -61.62788 ymin: -18.0416 xmax: -50.22481 ymax: -7.351771
# CRS:           +proj=longlat +ellps=GRS80 +no_defs
# ...
```

A subset is performed in the municipality of Cuiabá, Mato Grosso, to obtain the polygon geometry of just one municipality in the state.

```
cba <- mt[mt$nome == "CUIABÁ",]
```

The municipality of lavras is mapped for geovisualization of the municipal boundaries (Figure 1.15).

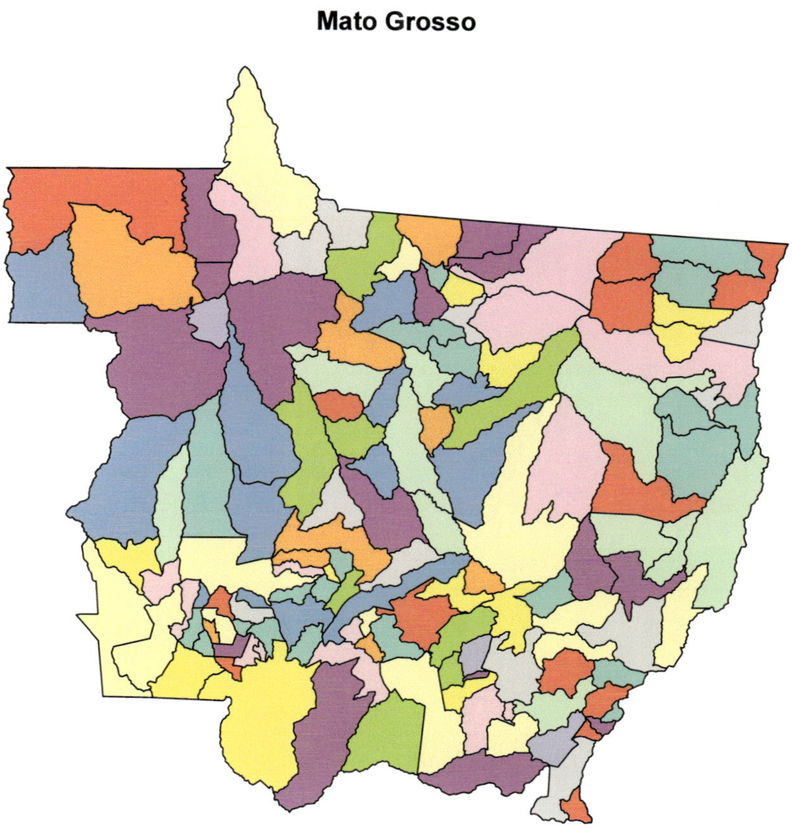

FIGURE 1.14 Instituto Brasileiro de Geografia e Estatística (IBGE) digital municipal mesh mapping for the state of Mato Grosso.

```
plot(st_geometry(cba), graticule = TRUE, col = "grey90", border = 'black',
     axes = TRUE)
```

The `lavras` polygon is exported as an ESRI Shapefile vector file for later use.

```
st_write(cba, dsn = "C:/geo/c1/cba.shp", driver = "ESRI Shapefile", "cba.shp")
```

One option for mapping the United States can be done with the `tigris` (Walker, 2025) or `USAboundaries` (Mullen and Bratt, 2025) packages from R. As an example, the `tigris`, `sf` and `dplyr` (Wickham et al., 2021a) packages are loaded into R with the `library` function.

```
library(tigris)
library(sf)
library(dplyr)
```

The borders of the United States are obtained using the `states` function, and the transformation of coordinates to WGS-84 is performed using the `st_transform` function,

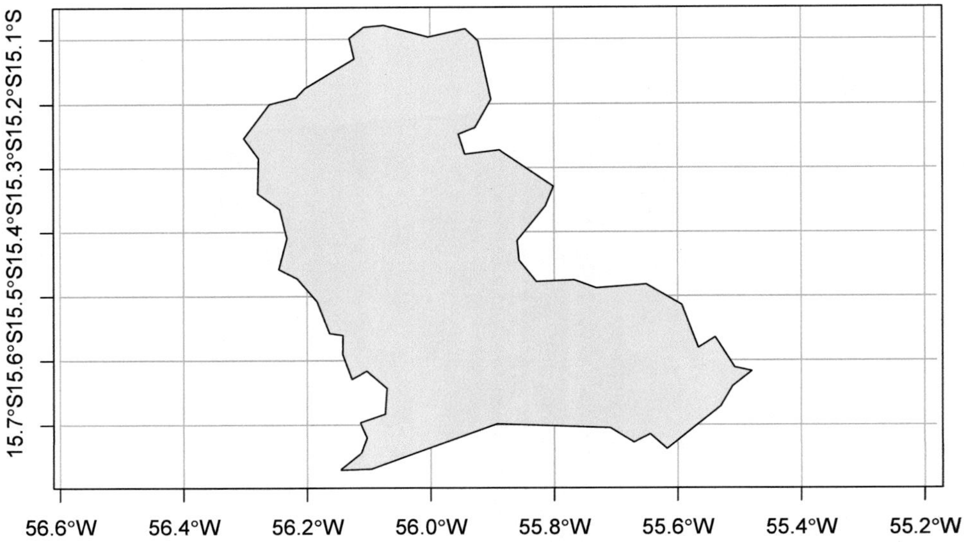

FIGURE 1.15 Mapping of the municipality of Cuiabá in Mato Grosso, Brazil.

EPSG code 4326.

```
us <- states() |> st_transform(4326)
# Retrieving data for the year 2024
us
# Simple feature collection with 56 features and 15 fields
# Geometry type: MULTIPOLYGON
# Dimension:     XY
# Bounding box:  xmin: -179.2311 ymin: -14.60181 xmax: 179.8597 ymax: 71.43979
# Geodetic CRS:  WGS 84
# ...
```

The `filter` function associated with the `!` operator is performed to remove territories with abbreviations AK, HI, MP, AS, VI, GU, and PR which represent the U.S. states of Alaska (AK) and Hawaii (HI), and the U.S. territories of Puerto Rico (PR), U.S. Virgin Islands (VI), American Samoa (AS), Guam (GU), and the Commonwealth of the Northern Mariana Islands (MP) that are not of interest, solely for mapping purposes. Note that the database now has only 49 features.

```
us <- us |> filter(!STUSPS %in% c("PR", "AK", "HI", "MP", "AS","VI", "GU"))
us
# Simple feature collection with 49 features and 15 fields
# Geometry type: MULTIPOLYGON
```

```
# Dimension:      XY
# Bounding box:  xmin: -124.849 ymin: 24.39631 xmax: -66.88544 ymax: 49.38448
# Geodetic CRS:  WGS 84
# ...
```

The filter function can also be used, for example, to obtain a subset of the state of Washington (WA). With the st_make_grid function, it is possible to define a rectangle surrounding the state. With the counties function, the counties are obtained and transformed to WGS-84 with the st_transform function.

```
wa <- us |> filter(STUSPS == "WA") # Filter state
wa_box <- st_make_grid(wa, n = 1) # Make rectangle
wa_co <- counties(state = "WA") |> st_transform(4326) # Get counties
# Retrieving data for the year 2024
```

The geometry results are mapped using functions from the sf package (Figure 1.16).

```
plot(us$geometry, axes = TRUE, col = "antiquewhite", bgc = "gray92")
plot(wa$geometry, add = TRUE, col = "gray92", border = "black")
plot(wa_co$geometry, add = TRUE, border = "blue", col = NA)
plot(wa_box, add = TRUE, border = "red3", col = NA, lwd = 2)
```

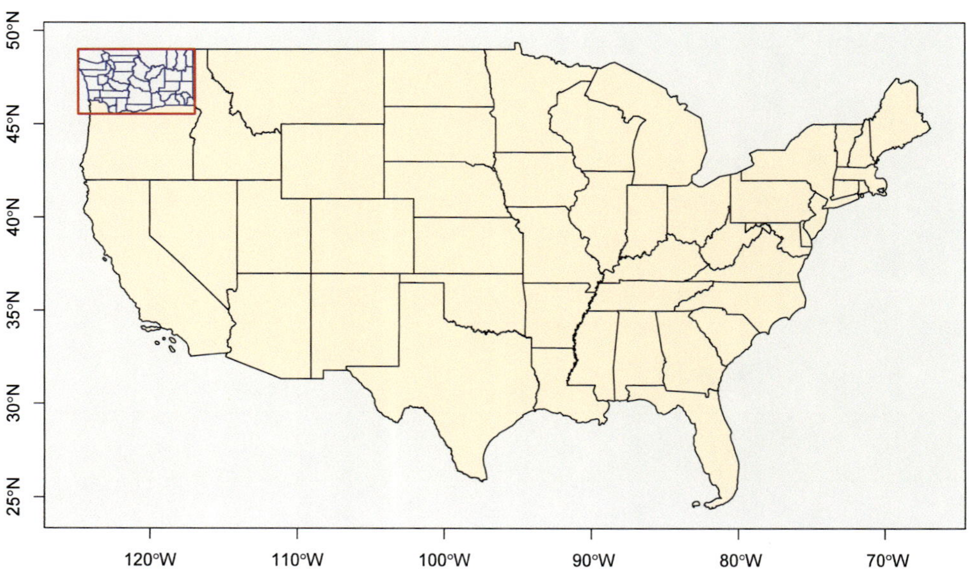

FIGURE 1.16 Mapping the United States with a subset of Washington state and counties.

1.14.11.2 Defining a geographical region using global databases

GADM[50] is a spatial database of world administrative boundaries for use in geocomputation. You can search for data on the website by country and get data at different hierarchical levels according to the subdivisions of countries into states and municipalities, according to the specifications of each region. The data can be obtained in `Geopackage`, `Shapefile`, `GeoJSON`, and `KMZ` formats.

[50] https://gadm.org/data.html

2 Geographic Representation with Environmental Geocomputation and R

2.1 LEARNING QUESTIONS

The learning questions answered by reading the chapter are:

- How do you develop a digital model to represent a feature of the Earth's surface?
- How useful are digital representations in learning, thinking, and reasoning about digits and times beyond immediate personal experience?
- How can geographic representations be used in basic planning and obtaining scientific research results?
- How does the complexity of geographic representation require strategies for representing specific details and times?
- How can an attribute be represented considering the relationship between scale and level of detail?

2.2 LEARNING OUTCOMES

The learning outcomes answered by reading the chapter are:

- Understand the importance of geographic representation in geocomputation;
- Know the concepts of fields and objects;
- Know the vector and raster geographic representations;
- Verify principles, techniques and applications of vector and raster geographic representations;
- Understand how the science, technology and art of geocomputation is used to represent objects and processes on Earth;
- Verify the relationship between scale and the level of detail of representation of a geospatial attribute.

2.3 INTRODUCTION

Features observed on Earth are basically defined as phenomena subdivided into objects and processes (Lo and Yeung, 2007). Therefore, when observing the environment, the complexity inherent in the situation is simplified with abstraction of key features to create a model representation of the area (Burrough and McDonnell, 1998). The terms representation and model imply a simplified outcome relative to the observed content. Representations or models can occur (Longley et al., 2001):

- In the human mind, when capturing information from surroundings, such as images, sounds, and memory;
- In photographs, when representing light emitted or reflected by objects on Earth through a camera lens by two-dimensional models;

DOI: 10.1201/9781003724377-2

- In spoken descriptions and written text, when describing aspects of objects and processes by some form of language;
- In numbers, by making direct and indirect measurements of characteristics of the Earth's environment.

By constructing representations, it is possible to acquire knowledge about the planet Earth for the purposes of planning, management, conservation, travel, operations, and the performance of services. By using representations, it is possible to define rules and laws applied to the unobserved in order to predict the future occurrence of attributes and processes of the environment. For example, in geocomputation spatial interpolation methods, unobserved geospatial attributes are estimated through measured neighboring attributes, based on the first law of geography attributed to Waldo Tobler, in which all digits are similar, but nearby digits are more similar than distant digits. That is, in general, there is a greater relationship in the environment between closer objects than more distant objects (Longley et al., 2001).

2.4 NATURE OF GEOGRAPHIC DATA

Human behavior in geographic space is often a reflection of previous patterns of behavior (Alves et al., 2020). Based on the relationship between consecutive events over time, the concept of temporal autocorrelation is defined. In the analysis of time series data, direct direction of causality is observed, i.e., past events are sequentially related to the present and future. Spatial autocorrelation is similar to temporal autocorrelation; however, time moves in only one direction, being one-dimensional, while in geospatial processes, events occur anywhere in two-dimensional or even three-dimensional space (Longley et al., 2001).

Spatial autocorrelation can be determined as a function of the degree of heterogeneity of the geographic region. Heterogeneity occurs both in the shape of the landscape and in the processes that acted on the landscape. The controlled spatial variation of some processes simply oscillates around an average, while in other processes there is greater, uncontrolled variation over time. Deforestation of natural areas and global warming, for example, can show uncontrolled variation. As a general rule, greater heterogeneity of spatial variation is observed in spatial data with increasing distance (Longley et al., 2001).

Similarly, gently undulating terrain landscapes are considered smooth and continuous, while others, on terrain faults or mountain ranges, are irregular and discretely bounded. Smoothness and irregularity are important distinguishing characteristics of geographic data. Some geographical phenomena vary smoothly over space, while others are extremely irregular, violating the first law of geography. Spatial autocorrelation patterns are used to construct inferential statistical models of the covariation of geographic phenomena to predict future changes or to explore the variation of a single more detailed variable in space as a function of another less detailed one, according to the cross-geospatial variation between two or more variables (Longley et al., 2001; Alves et al., 2020).

In general, measures of spatial and temporal autocorrelation are dependent on the scale and self-similarity of the observed attribute. The scale is related to the level of spatial resolution and the degree of detail of the attribute stored in an application, such as the pixel size used in a remote sensing spectral band. When there is accuracy in replicating a spatial autocorrelation pattern at a coarser scale, it can be said that there is characteristic self-similarity of the attribute (Longley et al., 2001) (Figure 2.1).

FIGURE 2.1 Individual coffee plants can be characterized in different levels of detail according to the scale size.

2.5 DIGITAL REPRESENTATION

A computerized version of a map is considered a digital image of a paper document stored and interactively displayed on the computer. With computer mapping, it is possible to obtain more options to record, store, access, analyze and visualize geospatial data than a conventional map on paper. With the use of geocomputation, tasks can be automated, digital data can be archived, and operations can be performed to explore interactions between processes and spatio-temporal patterns of different physical, natural, and social phenomena (Burrough and McDonnell, 1998).

Information technology is based on digital representation and can be seen in cell phones, music, television, newspapers, magazines, e-mail, websites, and computers in digital form. The term digital derives from digits, or fingers, and the counting system is based on ten digits of the human hand. Although in the counting system it consists of ten symbols, 0 to 9, in the representation system in digital computers only two are used, 0 and 1. Therefore, considering that in a digital system all information is represented with the combination of the two symbols 0 and 1, the term binary system is more accurate and more appropriate when referring to electronic technology based on binary representations (Burrough and McDonnell, 1998; Longley et al., 2001).

In the binary counting system only two symbols, 0 and 1, are used to represent numerical information. A group of eight binary digits is known as a byte, and storage volume is usually measured in bytes rather than bits. In the binary system, there are only two options for a single digit, but there are four possible combinations for two digits (00, 01, 10, and 11), eight possible combinations for three digits (000, 001, 010, 011, 100, 101, 110, 111), and 256 combinations for a full byte. Integers are stored in geocomputation as short integer (2 bytes or 16 bits) or long integer (4 bytes or 32 bits) (Longley et al., 2001). Using an 8 byte, it is possible to represent 256 numbers, from 0 to 255 ($2^8 - 1$). If 2 bytes are combined into 16 bits, it is possible to encode numbers from 0 to 65535. However, it may be useful to store positive and negative numbers in the geocomputation system, in order to use half the range of 65535 to encode numbers from −32767 to +32767 values, called 16-bit integers. In a 32-bit integer, there can be values from −2147483648 to +2147483648. The 32-bit or even 64-bit precision, also called double precision, is required for real numbers in which positive and negative numbers with decimal digits are stored. Numbers with decimal digits can be stored using 128 bits on most computers, but it may vary according to the number of decimal digits, i.e., 7 digits for 32 bits, 15 digits for 64 bits, and 33 digits for 128 bits (Lo and Yeung, 2007). Thus, a wrong encoding of numbers used in geocomputation can affect the accuracy of the stored number in terms of decimal digits due to rounding of decimal numbers (Burrough and McDonnell, 1998).

Therefore, in a geocomputation system, short integers can range from −65535 to +65535, and long integers from −4294967295 to +4294967295. Numbers with decimal digits are encoded by real or floating point representations. A number with decimal digits is first transformed to the power of ten and the fractional part is stored in separate sections of a 4-byte (32 bits, single precision) or 8-byte (64 bits, double precision) block. With this, it is possible to store approximately 7 significant digits in single precision files, or 14 digits under double precision. The option BLOB, which stands for binary large object, is included to associate images or sounds with digits, instead of text or numbers. In BLOB a sufficient number of bits is allocated to store image or sound, without specifying the meaning of the bits (Longley et al., 2001).

Digital data is easy to copy and can be transmitted near the speed of light, can be stored in high density in very small spaces, and is less subject to physical deterioration than paper and other physical media. With data in digital form, it became easy to transform, process and analyze data. Geographic information systems enable to measure accurately and quickly, overlay and combine, scale, zoom, and pan without limits like those using a map drawn on paper. In addition, digital representation is adopted for its simplicity and low-cost characteristics.

We are talking here about the advantages of digital representation from a software perspective. From a hardware perspective, it is necessary to assess which technology would be most useful for geographical representation in terms of deterioration, given that computer technology requires frequent updates at short intervals. This comparative perspective is quite controversial, considering that fragments of manuscripts dating back thousands of years have been found and can still be read (Wise, 1986).

A digital mapping of coffee fields in municipalities in southern Minas Gerais, Brazil, can be performed with the `sf` and `tmap` packages as described below. The library function is used to load the packages.

```
library(sf)
# Linking to GEOS 3.11.2, GDAL 3.7.2, PROJ 9.3.0; sf_use_s2() is TRUE
library(tmap)
# package ‘tmap’ was built under R version 4.3.3
```

The `read_sf` function is used to import geographic data on municipalities and coffee fields in shapefile format. It should be noted that this data was initially obtained in kml format and then converted to shapefile format, as shown in the computational practice example in the previous chapter.

```
munMG2007 <- read_sf("C:/Aulas/2021_1/Geoprocessamento/shape/municipiosMG2007.shp",
                   options = "ENCODING=WINDOWS-1252") # MG municipality data
coffee <- read_sf('C:/geo/c8/coffeeMG.shp', options = "ENCODING=WINDOWS-1252")
```

Functions from the tmap package are used to perform a vector geographic representation of the results obtained (Figure 2.2).

```
mapCoffeeMG <- tm_shape(munMG2007[4:4]) +
 tm_polygons("NOME_MUNIC", title = "Municipalities") +
 tm_fill("darkolivegreen3") +  tm_grid() +
 tm_legend(position = c("right", "top"), frame = TRUE,
 bg.color="gray92")+tm_layout(legend.outside = F, legend.height = 0.29) +
 tm_shape(coffee[1:1]) +  tm_borders("brown") + tm_polygons("brown") +
 tm_compass(type = "8star", position = c("left", "top")) +
 tm_scale_bar(breaks = c(0, 10, 20), text.size = 1, position = c("left", "bottom"))
mapCoffeeMG # Map
```

2.5.1 BINARY NUMBER ARITHMETIC

The binary system is defined with base 2 in which counting of numbers is performed using 1 and 0. Data is represented in counting order in which the first column of data is 2^0, the second is 2^1, the third is 2^2, and so on (Table 2.1) (Burrough and McDonnell, 1998).

TABLE 2.1 Arithmetic of binary numbers using a base 2 system related to decimal base system.

Binary Sequence	Equivalent Count in base 10
0010_2	$2^3 * 0 + 2^2 * 0 + 2^1 * 1 + 2^0 * 0 = 2_{10}$
1100_2	$2^3 * 1 + 2^2 * 1 + 2^1 * 0 + 2^0 * 0 = 12_{10}$
$1110_2 - 0011_2 = 1011_2$	$14_{10} - 3_{10} = 11_{10}$
$10111_2 + 10111_2 + 10111_2 + 10111_2 = 1011100_2$	$23_{10} * 4_{10} = 23_{10} + 23_{10} + 23_{10} + 23_{10} = 92_{10}$
$1011100_2/10111_2 = 100_2$	4_{10}
$2 * 0101_2 = 1010_2$	$2^3 * 0 + 2^2 * 1 + 2^1 * 0 + 2^0 * 1 = 5_{10}$
$110_2/2 = 0111_2$	$2^3 * 1 + 2^2 * 1 + 2^1 * 1 + 2^0 * 0 = 14_{10}$

2.5.2 DATA ENCODING IN THE HEXADECIMAL SYSTEM

The hexadecimal system is used in encoding text characters, diagrams, images, sounds, and other types of information. This system is also known as the American Standard Computer Information Index (ASCII). With this system, a basis is provided for encoding alphanumeric data and letters of the alphabet with numbers from 0 to 9 by means of an 8-bit byte for encoding symbols, numbers and letters written on the computer (Table 2.2) (Burrough and McDonnell, 1998).

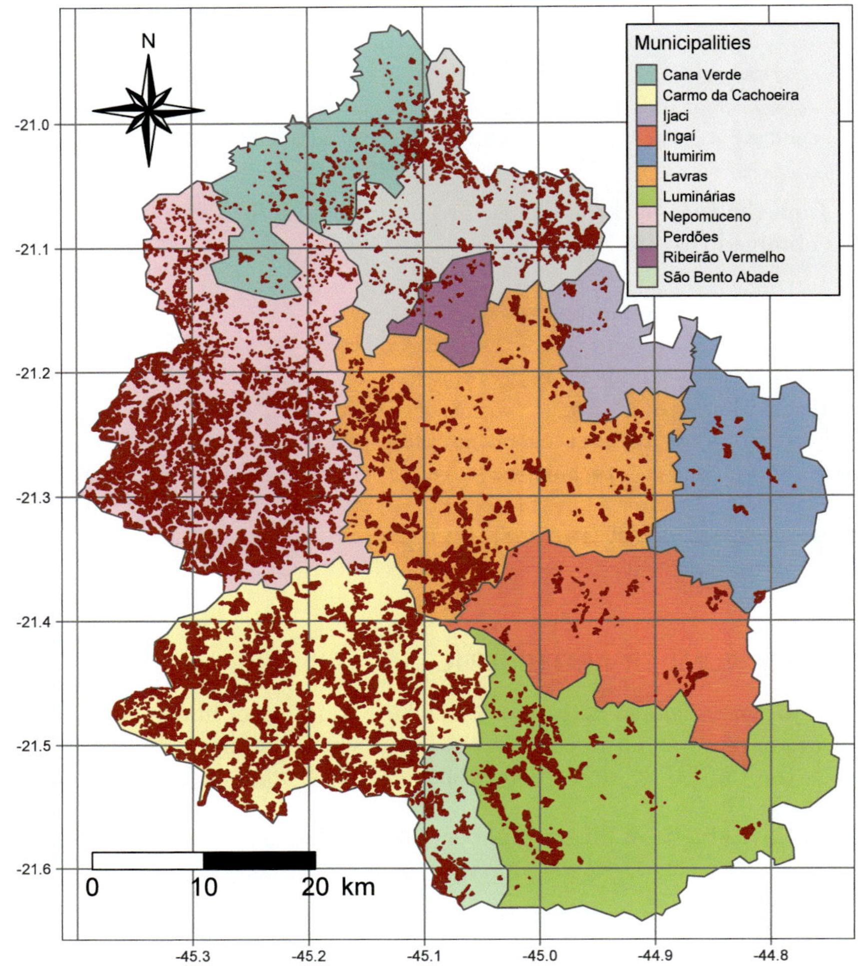

FIGURE 2.2 Representation of coffee producing areas at the scale of municipalities, in Southern Minas Gerais, Brazil, for the 2018 harvest.

TABLE 2.2 Computer encoding symbols, numbers, and letters of the alphabet in the hexadecimal, decimal, and binary systems.

Symbol, Number or Letter	Hexadecimal	Decimal	Binary
A	41	065	01000001
5	35	053	00110101
μ	82	130	10000010

2.5.3 DATA VOLUME

The volume of bits needed to encode information is expressed in thousands (kilo), millions (mega), or billions (giga or tera) bytes. For example, on an A4 page encoded in ASCII with 64 lines of text with 80 characters per line, 40,960 bits or 5,120 bytes or 5k (1k=1024 bytes) is required. Larger amounts of storage space are required for maps and other types of geospatial data, but it may be possible to save space and computational processing time depending on the type of data and number of bits required for encoding (Burrough and McDonnell, 1998).

2.6 GEOGRAPHIC REPRESENTATION

In the tradition of free and open access, much of humanity's accumulated store of knowledge about the geographic world is available in the form of paper representation, through the institution of libraries and the doctrine of copyright on the use of personal material. The Internet has become the delivery mechanism for access to geographic information (Longley et al., 2001).

One of the most important criteria for the usefulness of a representation is accuracy. Since geographic complexity can be seemingly infinite, choices must always be made about what should be represented or not in relation to what will be represented. Many features on Earth are slowly changed or considered static. Height above sea level changes slowly due to erosion and movements of the Earth's crust, as a function of processes occurring on scales of hundreds or thousands of years. On the other hand, atmospheric temperature changes daily, although climate variables such as annual mean temperature are considered static in periods of climatological normal. Therefore, the amount of attributes in geographic information is large and can vary slowly or rapidly. Some attributes are physical and environmental in nature, while others are social and economic. Some attributes are used simply to identify a place or an object, distinguishing it from others. For example, addresses, social security numbers, or the registration number of a landholding are social and economic, while others, such as air temperature and elevation are physical and environmental, and can be classified into categories, such as land use class, associated with agriculture, industry, or residential (Longley et al., 2001).

Geographic attribute types are classified in (Burrough and McDonnell, 1998; Longley et al., 2001):

- Boolean, to perform logical operations with values 0 or 1;
- Nominal, to identify or distinguish one object from another by names;
- Ordinal, for values with natural order for comparisons and magnitude, from 0 to ∞;
- Integer, for whole arithmetic numbers, from $-\infty$ to $+\infty$;
- Real, for numbers with decimals, from $-\infty$ to $+\infty$;
- Interval, when there is significance in presenting the information in intervals;
- Proportional, as in the case of variation in percentages;
- Cyclical, for data with cyclical or recurrent variation;
- Topological, for topological links between entities with integer numbers.

2.7 GEOGRAPHIC REPRESENTATION ON PAPER

The map on paper is an effective means of communicating geographic information, defined as a source of geographic data in an analog geocomputation product. One characteristic of a map on paper is the scale, defined as the relationship between the distance on the map and the equivalent distance on the ground. For example, in a map with a scale of 1:10000, each centimeter on the map corresponds to a distance of 100 meters on the ground. However, there can be misrepresentation in the case of a curved surface, since the map is laid out on the flat surface of the paper. Therefore, the scale cannot be exactly constant in a cartographic representation (Longley et al., 2001).

Printed maps were most important before the development of digital technology, although many of the ideas used in geocomputation are inherited from knowledge gained from making maps drawn on paper. Scale only makes sense on printed maps, since distance on the ground is related to a distance on paper as a function of a certain scale. With digital representations, it is possible to include information that would be very difficult to show on printed maps, such as the curvature of the Earth's surface in three-dimensional mapping, without the need for distortions associated with flattening. Digital maps also make it easier to perform updates without the need to alter the map after printing (Longley et al., 2001). Vector and raster data can be mapped together in order to exploit the large database of sensor data available. For example, it is possible to map a false-color Landsat-8 composition over variable relief represented by a three-dimensional watershed elevation model in Mato Grosso using SAGA GIS software (Conrad et al., 2015). Areas are two-dimensional (2-D) objects, and volumes are three-dimensional (3-D). However, there is also in geocomputation a dimension between 2-D and 3-D called 2.5-D. In the 2.5-D dimension, a variable with a Z value, such as the reflectance pixel brightness in a false-color composition of a Landsat-8 image, is represented in all parts of the geographic region in two dimensions, and a complete 3-D representation of that same image is obtained after overlaying the image on the digital elevation model of the area (Figure 2.3).

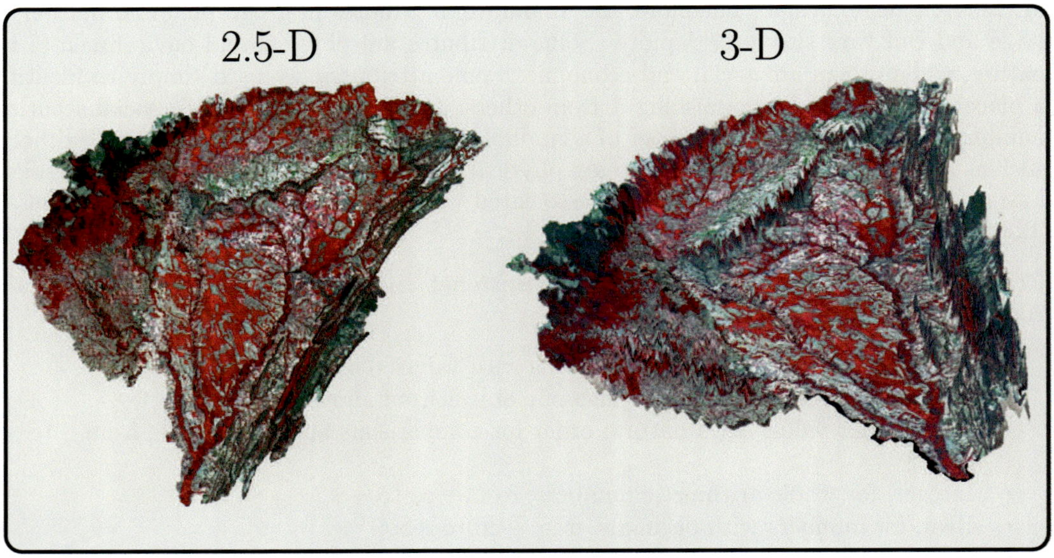

FIGURE 2.3 Landsat-8 infrared false-color composition over variable relief of watershed in southwestern Mato Grosso state, Brazil.

2.8 COMPLEXITY OF DIGITAL REPRESENTATION

Interpreting the characteristics of a landscape may require experience and cultural training in order to represent geographical phenomena on the computer. The first step in the computational representation of a phenomenon is the analyst's experience in observing the world, either directly in the field or through images to perceive fixed objects and process variations over space and time. This initial perception can influence all subsequent analysis steps performed in a geocomputation system. Therefore, the success of a geocomputation project depends not only on the technology used, but also on conceptual models of space and the geospatial interactions represented (Burrough and McDonnell, 1998).

The Earth is infinitely complex, but computer and geographic information systems are finite. Therefore, in representations, there must be some way to limit the amount of detail captured. Furthermore, the idea of recording atoms of geographic information is meaningless, because as you look closer at the world, more details will be revealed, in an *ad infinitum* process. Therefore, in practice, any representation must be biased in terms of limiting the level of detail of attributes, ignoring the change of a process over time, or ignoring certain attributes in order to simplify the representation in some way (Longley et al., 2001).

A very common way of limiting detail is accomplished by ignoring information that applies only to small areas, in other words not looking too closely (Longley et al., 2001). Soybean crops in the state of Mato Grosso, for example, were mapped in the SOJASAT[1] project with MODIS sensor data, considering the probability of occurrence of crops in areas with a minimum size of 100 ha (Chaves et al., 2018) (Figure 2.4).

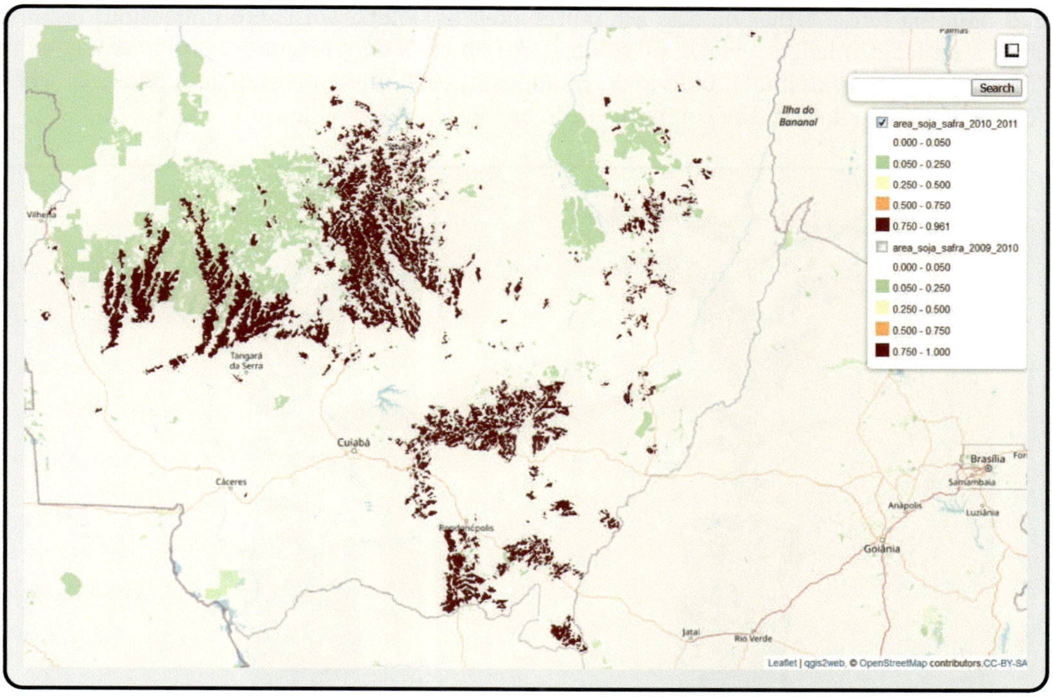

FIGURE 2.4 Map of probability of areas cultivated with soybean crop in Mato Grosso for the 2010/2011 harvest.

Geographic phenomena representation requires two basic descriptors: what occurs and where it occurs. These phenomena are described as objects (entities) or processes. The phenomena are grouped into levels of resolution (scale), hierarchy and geographic coordinate system or a map projection. With conceptual ideas of geographic phenomena it is possible to establish forms and characteristics of geospatial representation (Burrough and McDonnell, 1998).

[1]http://www.sergeo.deg.ufla.br/sergeo/Projetos/SojaSat/SojaSat.html

2.9 DISCRETE OBJECTS AND CONTINUOUS FIELDS

Discrete objects and continuous fields are the two fundamental forms of geographic representation for describing entities and the variation of an attribute over space (Burrough and McDonnell, 1998; Longley et al., 2001).

2.9.1 DISCRETE OBJECTS

Earth can be simplified by naming individual objects as instances of broader categories. This preferred view of viewing the geographic world is known as the discrete object view. In this view, the empty world is occupied by objects with well-defined boundaries that are instances of recognized categories that can be counted. Geographic objects are identified by their dimensionality. Objects that occupy area are called two-dimensional, and generally called areas. The term polygon or polygonal is also widely used to describe areas. Some objects are represented as one-dimensional lines, as in the case of roads, railroads, rivers and planting lines. Other objects are represented as points, with zero dimension, because there's no length, height, width, or volume. An object's only property is its location, as in the case of plants, animals, individual buildings, posts, and topographic landmarks when represented by points in the computer (Longley et al., 2001) (Figure 2.5).

FIGURE 2.5 Coffee seedlings can be considered discrete objects.

All three-dimensional objects perceptible to the human eye can be represented as volumes. In some geocomputation software, it is possible to insert a third (vertical) coordinate for each point location (Longley et al., 2001). The information about each discrete object can be detailed in a table, as in a situation where it is of interest to perform a cadastral survey of coffee plants in a field, where each row of the table corresponds to a different plant and each column corresponds to an attribute measured on the plant (Table 2.3).

TABLE 2.3 Example of a plane geographic information representation of coffee plants from a plantation in Minas Gerais, Brazil.

Plant ID	Coord X	Coord Y	Coord Z	Planting date	Plant height	Leaf area index
001	502391.342	7659044.650	930.30	5/7/2009	1.6	0.7
002	502395.813	7658747.478	928.40	5/7/2009	1.5	0.5
003	502407.103	7658872.969	931.50	5/7/2009	1.4	0.6
004	502369.001	7658888.791	932.70	5/7/2009	1.7	0.8

Depending on the situation, a coffee plot cultivated with the same cultivar, planted at the same time and with similar soil and relief characteristics can be represented as a discrete object, in the case of a cadastral survey of texture and occurrence of soil erosion (Table 2.4), for example, or as a continuous field, in a situation where we wish to evaluate the geospatial variability of altitude, clay and soil organic matter in the 0 to 20 cm depth layer, by means of a georeferenced irregular sampling grid, with several sampling points within a single plot cultivated with Arabica coffee.

TABLE 2.4 Soil sampling in different parts of a field used to evaluate soil physical characteristics.

Sample ID	Coord X	Coord Y	Coord Z	Sampling date	Texture	Erosion
001	502391.342	7659044.650	930.30	5/7/2009	Clayey	True
002	502395.813	7658747.478	928.40	5/7/2009	Very clayey	False
003	502407.103	7658872.969	931.50	5/7/2009	Medium	False
004	502369.001	7658888.791	932.70	5/7/2009	Sandy	True

In another similar situation, watershed flow can be represented by polygons with the geographical boundaries of watersheds or by means of points established at the centroid of the watersheds where a reference flow measurement is established for each watershed. This can be useful for mathematical modeling of the watershed flow considering neighborhood information in the model application (Figure 2.6).

2.9.2 CONTINUOUS FIELD

Continuous surfaces form the basis of the other common view of geographical phenomena, known as the continuous field of view. In this view, the geographical world is described by a series of variables, each measurable at any point on the Earth's surface, and with variation in value across the surface. In the continuous field view, the real world is represented as a finite number of variables, each at a defined position in the field (Figure 2.7) (Longley et al., 2001).

A continuous field can vary smoothly or abruptly, depending on the characteristic of the landscape (Figure 2.8). Population density can be a kind of continuous field, defined everywhere as the number of people per unit area. Continuous fields can also be created from land classifications, in categories of land use or soil type. In this case, there was abrupt variation in the boundaries between the different classes. The field can also vary along lines,

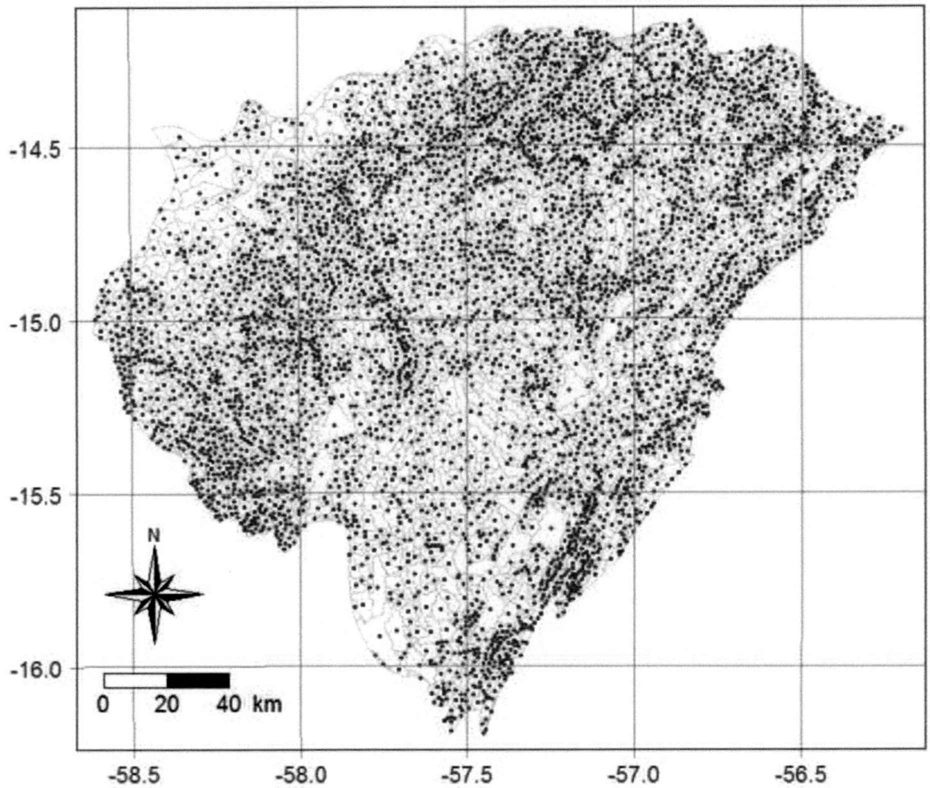

FIGURE 2.6 Polygons with geographical limits of watersheds and points established at the centroid of the polygons where a reference flow rate measurement was obtained in a watershed in southwestern Mato Grosso, Brazil.

rather than across space, as in the case of vehicle traffic density on a highway (Longley et al., 2001).

Continuous fields can be distinguished as a function of an attribute measured at each point along a sampling grid, for example. A field with two variables, magnitude and direction, at each point in space can be used to represent flow phenomena, such as winds or currents. This field is called vectorial. Fields with only one variable are called scalar. In addition, the scale used in mapping can be important in defining the best representation of the attribute in question, whether in discrete or continuous form (Figure 2.9) (Longley et al., 2001). Contour lines can be used to represent the variation of an attribute in an area, such as contour lines for terracing projects for soil and water conservation in a region, but the lines may be inefficient in situations where it is necessary to model spatial variation and interactions in an area (Burrough and McDonnell, 1998).

FIGURE 2.7 The soil with organic matter used in no-till farming can be considered a field with continuous variation.

In general, the choice of an object or field representation approach will depend on the scientific and technical disciplines pursued by the analyst. In disciplines focusing on the understanding of geospatial processes of the environment, a continuous field approach may be preferred. On the other hand, in the social sciences and economics, the administrative representation context may occur through a series of distinct object units (Burrough and McDonnell, 1998).

2.10 DIGITAL GEOGRAPHIC REPRESENTATION BY VECTOR AND RASTER

Continuous fields and discrete objects are used to define distinct concepts of views of geographic phenomena. Two methods used to represent geographic objects and processes on the computer are called raster and vector objects. In principle, both can be used to encode discrete fields and objects, but in practice, there is strong association between raster objects with the view of continuous fields, as well as between vector objects with the view of discrete fields. Raster and vector are two methods of representing geographic data in geocomputation (Longley et al., 2001).

2.10.1 RASTER DATA

In a raster representation, space is divided into an array of rectangular, usually square, cells, in which all geographical variation is expressed through attribute features in these cells. The cells are also called pixels or picture elements. One of the most common forms

FIGURE 2.8 Abrupt variation of a continuous field in the Véu da Noiva waterfall, Chapada dos Guimarães, Mato Grosso, Brazil.

FIGURE 2.9 Mountains can be seen as a continuous field, since it is difficult to define exactly where a mountain begins and ends, as in the case of the Serra da Bocaina in Lavras, Minas Gerais.

of raster data can be obtained through remotely sensed image data in different spectral regions or spectral bands. The images can vary according to spatial, spectral, radiometric and temporal resolutions. In many sensors, images are captured in several regions of the electromagnetic spectrum (Figure 2.10), in the same area, simultaneously, and a raster with several spectral bands can be obtained.

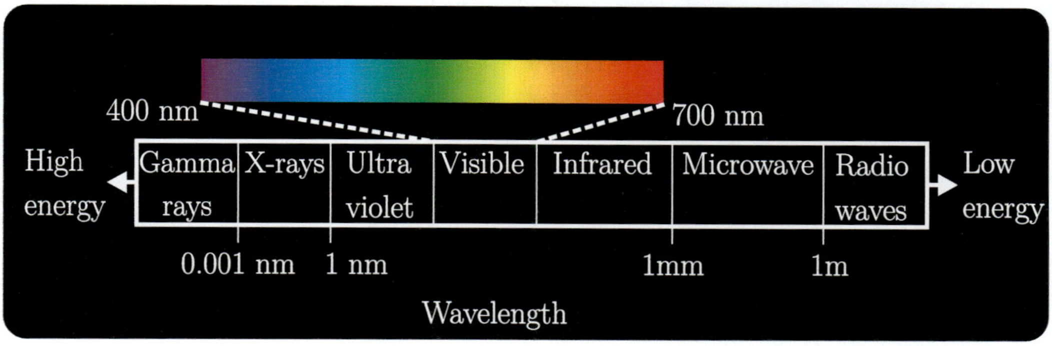

FIGURE 2.10 Electromagnetic spectrum used to determine spectral variation in remote sensing.

The square cells fit perfectly on a flat table or on a sheet of paper, but they did not perfectly represent the curved surface of the Earth. In the paper representations the Earth is flattened, in a projected manner, to contemplate representations with rasters as well. Many of the terms used in describing rasters refer to examples such as laying a ceramic tile floor, on a flat surface or tiling an area. Therefore, the raster representation is considered a type of tessellation, derived from the word for mosaic. With a mirrored ball hanging above a dance floor you can see that it is impossible to cover a spherical object like the Earth perfectly with flat, square pieces. When information is represented in raster format, all details about the variation within the cells are lost, and the pixel or cell is given a single value. A raster representation and subset of a raster image are illustrated by quadrants where average reflectance values of sunlight stored in pixels are obtained by Landsat-8 OLI imaging at Chapada dos Guimarães (Figure 2.11).

In a raster representation, a single value is stored in each cell to identify the surface with values of a given attribute. In this case, there should be a rule to establish a specific pixel value depending on the spatial resolution available or considered by sampling *in situ* the attribute in the field. In this case, there must be a rule to be applied to decide the pixel value when a cell falls on more than one heterogeneous feature defined on the terrain. Often, the rule established is to consider that the major part of the pixel with a specific feature on the terrain defines the value stored in the pixel. Another rule may be based on the situation where the value of the pixel is defined with respect to the center point of the cell, and the value at that point is assigned to the cell as a whole. The rule that a predominant value in the area is represented by the pixel value can be preferred by imagining that the predominant signal in the region is considered as the signal to be stored to represent a specific region of the terrain. A raster object is defined in the computer as a data structure with a number of columns, rows, pixels, extent, spatial resolution and reference coordinate system. In the case of imaging in different spectral regions, three bands in distinct regions of the electromagnetic spectrum are stacked to make a false-color composition from the RGB color channels of the digital image processing software (Figure 2.12).

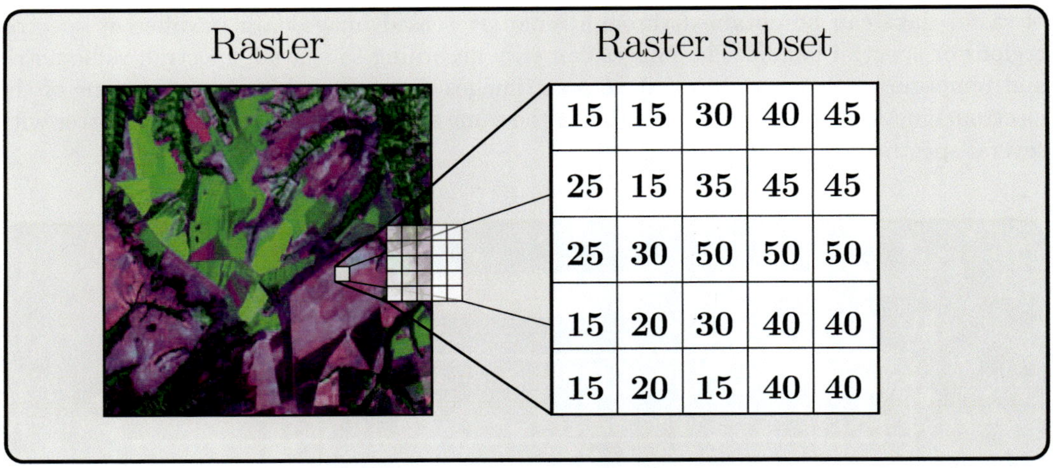

FIGURE 2.11 Raster representation and raster subset with quadrant illustration of average reflectance values of sunlight stored in pixels by Landsat-8 OLI imaging, in Chapada dos Guimarães, Mato Grosso, Brazil.

2.10.2 VECTOR TYPE DATA

Geospatial data can be discrete objects with defined boundaries and used to represent rivers, roads, countries, cities, or a rural property. In such data there must be a description of the geometry or shape of the objects, plus additional variables. For example, with a vector dataset you can describe the borders of countries (geometry) and also store names of countries and the coffee yield in 2017, or the geometry of rivers in an area, with the name and type of river. These additional variables are also called attributes. The main vector data types are characterized by points, lines and polygons. The geometry of these data structures consisted of sets of coordinate pairs (x, y) with a possibility to associate n variables. A point can represent a location where a soil sample is taken to assess clay content, as well as information about the sampling location.

In line geometry, a set of one or more connected polylines defines line segments. Lines are represented as ordered sets of coordinates defined by nodes for specific directions. The representation of a line is very similar to that of a multipoint structure; however, the order of the points is required to define the order of connection between the points that defines the geometry of the line. A network is a special type of line geometry in which additional information about flow, connectivity, direction and distance are considered. A polygon is defined by a set of closed polylines. The geometry is very similar to that of lines, but to close a polygon, the last pair of coordinates coincides with the first pair. Several polygons can be considered as a single geometry forming together a single layer of multipolygons. Coffee plantations in municipalities in Minas Gerais are described by multipolygons.

In a vector representation, all lines are captured as points connected by precisely straight lines, but in some geographic information systems, the points can be connected by curves, determining a higher density of points. A geographic region can be obtained as a series of points or vertices connected by straight lines. Since there are straight edges between the vertices, the vector representation is also called a polygon. To obtain an object referring to an area, in vector format, it is necessary to specify the locations of the points that define the vertices of the polygon. However, many geographical phenomena cannot be located with high precision, leading to problems with vector representation if this is the

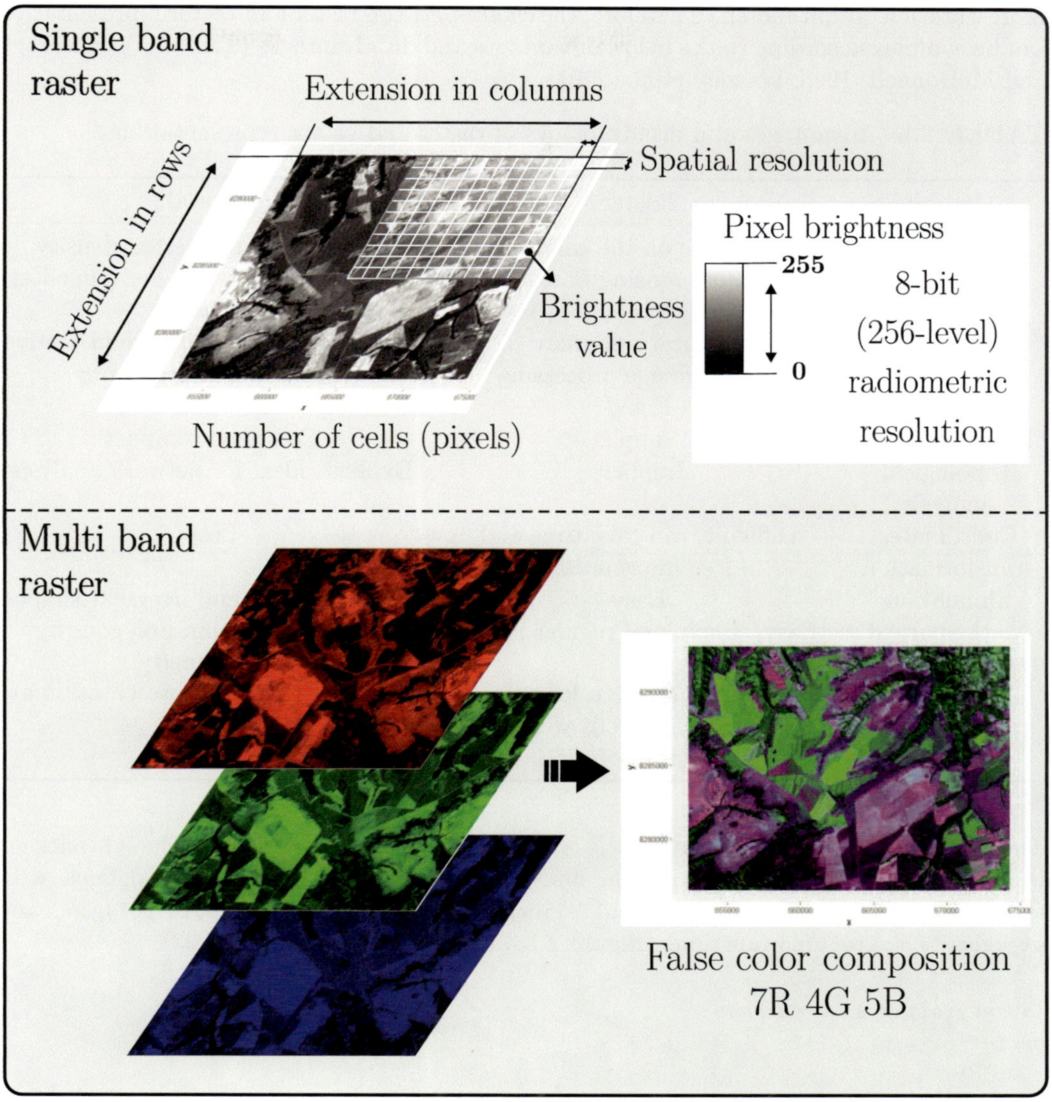

FIGURE 2.12 Example of a single-band raster with 8-bit (upper) spatial and radiometric resolution and a raster of three electromagnetic spectrum bands stacked to realize a false-color composition.

representation option chosen. Therefore, the choice between raster and vector representation can be complex according to the below advantages and disadvantages (Table 2.5) (Burrough and McDonnell, 1998; Longley et al., 2001).

TABLE 2.5 Advantages and disadvantages of raster and vector representations.

Variable	Raster	Vetorial
Data volume	Depends on the cell size	Depends on the vertex density
Data types	Remote sensing imaging	Socio-economic and environmental data
Applications	Natural resources	Social, economic and administrative
Software	Digital image processing	Automated cartography
Resolution	Fixed	Variable
Data structure	Simple	Complex, compact
Topological analysis	Implicit	Explicit, ideal for network analysis
Coordinates transformation	Difficult, can take time and lose information	Easy
Simulation	Easy	Hard, different and irregular shapes
Mathematical modeling	Easy, simple and regular shape	Impossible within polygons by homogeneity
Graphic representation	There may be a loss of information in cells with low spatial resolution	Accurate graphic representation at all scales

The seven types of geometries in the `sf` package are shown below. The `sf` package is loaded with the `library` function, and the vector geometry types are defined with the `st_point`, `st_multipoint`, `st_multilinestring`, `st_polygon`, `st_multipolygon`, and `st_geometrycollection` functions (Figure 2.13) (Pebesma and Bivand, 2023).

```
library(sf) # Load package
par(mfrow = c(3, 3))
# Point
p <- st_point(0:1)
plot(p, , pch = 21, col = "red", cex = 3, bg = "blue", axes = TRUE)
title("Point")
box(col = 'black', lwd = 2)
# Multipoint
mp <- st_multipoint(rbind(c(4, 1.5), c(2, 2.5), c(3, 1), c(3, 3), c(3, 2)))
plot(mp, pch = 21, col = "blue", cex = 3, bg = "red", axes = TRUE)
title("Multipoint")
box(col = 'black', lwd = 2)
# Linestring
ls <- st_linestring(rbind(c(2, 2), c(5,4), c(5, 4), c(4, 6), c(3, 5), c(2, 6),
                          c(1, 3)))
plot(ls, lwd = 2, col = "red", axes = TRUE)
title("Linestring")
box(col = 'black', lwd = 2)
# Multilinestring
```

```
mls <- st_multilinestring(list(
  rbind(c(2, 2), c(5, 4), c(5, 4), c(4, 6), c(3, 5), c(2, 3), c(1, 3)),
  rbind(c(3, 0), c(4, 1), c(2, 1), c(4, 2))))
plot(mls, lwd = 2, col = "blue", axes = TRUE)
title("Multilinestring")
box(col = 'black', lwd = 2)
# Polygon
pol <- st_polygon(list(rbind(c(2, 1), c(3, 1), c(5, 1), c(6, 2), c(5, 3), c(4, 4),
  c(3, 4), c(1, 3), c(2, 1)), rbind(c(2, 2), c(3, 3), c(4, 3), c(3, 2), c(2,2))))
plot(pol, border = 'black', col = 'skyblue', lwd = 2, axes = TRUE)
title("Polygon")
box(col = 'black', lwd = 2)
# Multipolygon
mpol <- st_multipolygon(list(
    list(rbind(c(2, 1), c(3, 1), c(5, 1), c(6, 2), c(5, 3), c(4, 4), c(3, 4),
      c(1, 3), c(2, 1)), rbind(c(2, 2), c(3, 3), c(4, 3), c(4, 2), c(2, 2))),
    list(rbind(c(3, 5), c(4, 6), c(5, 8), c(3, 9), c(2, 6), c(3, 5)))))
plot(mpol, border = 'black', col = 'skyblue', lwd = 2, axes = TRUE)
title("Multipolygon")
box(col = 'black', lwd = 2)
# Geometrycollection
gc <- st_geometrycollection(list(pol, ls + c(0, 5), st_point(c(2, 5)),
                                 st_point(c(5, 4))))
plot(gc, border = 'black', col = 'red', pch = 21, lwd = 2, axes = TRUE)
title("Geometrycollection")
box(col = 'black', lwd = 2)
```

2.10.3 VECTOR-RASTER INTEGRATION

The raster and vector structures can be integrated and represented together in geocomputation software in different layers (Burrough and McDonnell, 1998). Observing the spatio-temporal variation of the annual average air temperature, we observed that the temperature increases progressively in the state of Minas Gerais when comparing the climate change scenario with the current period. It is worth mentioning that there is improvement in the resolution by interpolation considering a digital elevation model, in the spatial resolution of 2.5 minutes ($\sim 85km^2$), according to WorldClim v2.1 data (Fick and Hijmans, 2017; Harris et al., 2014). In the future scenarios, the CNRM-ESM2-1 model (Séférian et al., 2019) was used, under socio-economic scenario SSP 126, for the future climate periods of 2021-2040, 2041-2060, 2061-2080, 2081-2100.

As an example of how to perform this analysis, the R packages `sf` (Pebesma et al., 2021a), `sp` (Pebesma et al., 2021b), and `raster` (Hijmans et al., 2020) are loaded. It should be noted that the `sp` package is loaded together with the `raster` package due to `raster's` dependency on the `sp` package, and it is not necessary to load the `sp` package with the `library` function.

```
library(sf)
library(raster)
# Loading required packages: sp
```

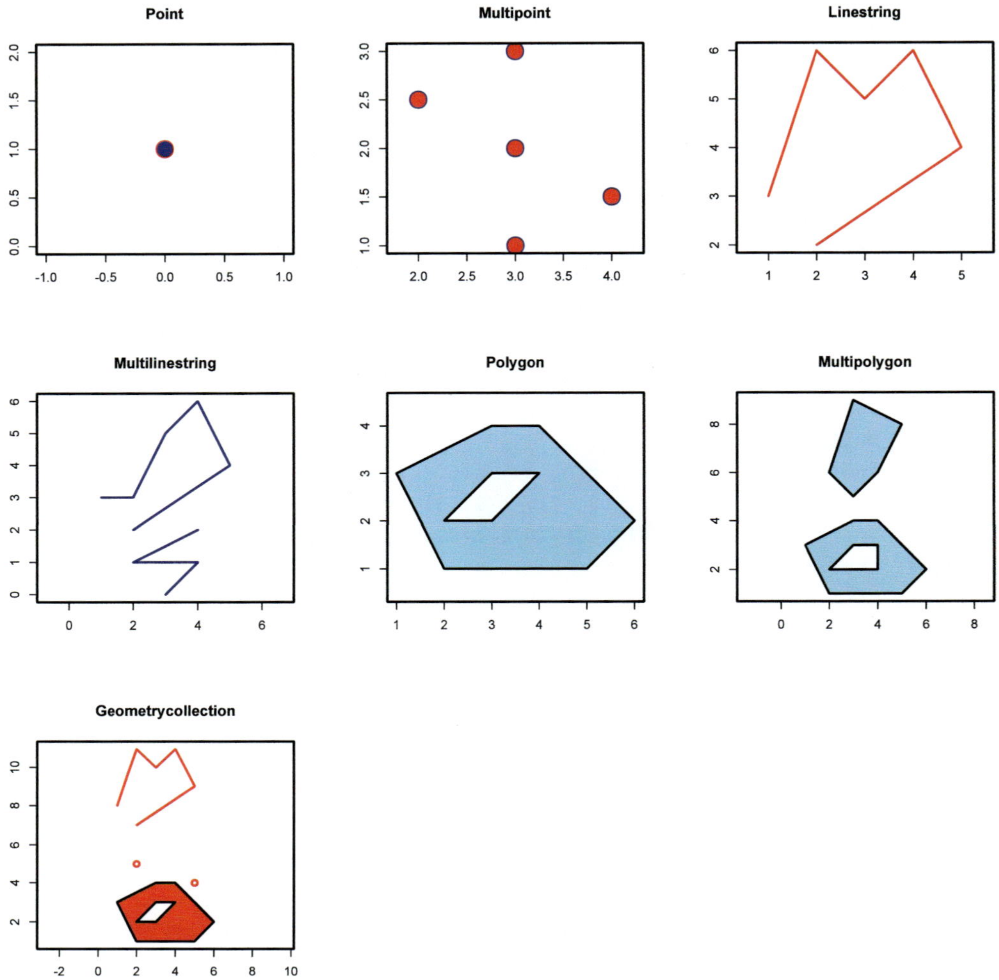

FIGURE 2.13 Representation of seven types of vector geometries used in the `sf` package.

A subset of data from municipalities around Lavras, Minas Gerais, Brazil obtained from the Internet is imported into R with the `read_sf` function. The encoding `WINDOWS-1252` is used as an argument option so that Latin characters from the database are displayed in the correct style. Part of the header of the imported file is observed when executing the name of the object created (`munMG2007`) in the R console.

```
munMG2007 <- read_sf("C:/Aulas/2021_1/Geoprocessamento/shape/municipiosMG2007.shp",
   options = "ENCODING=WINDOWS-1252") # Import data
munMG2007 # See header
# Simple feature collection with 11 features and 9 fields
# Geometry type: POLYGON
# Dimension:     XY
# Bounding box:  xmin: -45.39973 ymin: -21.64273 xmax: -44.74138 ymax: -20.92158
# Geodetic CRS:  WGS 84
```

```
# A tibble: 11 × 10
#...
```

In this case, we chose to perform a comparative mapping of the annual temperature in current periods and future scenarios with a single legend using functions from the `sp` package. This requires converting the polygons from the `sf` `data.frame` class to `sp` using the `as_Spatial` function, creating a new `mg` object of `SpatialPolygonsDataFrame` class.

```
mg <- as_Spatial(munMG2007) # Converting sf classes to sp
mg # See header
# class       : SpatialPolygonsDataFrame
# features    : 11
# extent     :-45.39973, -44.74138, -21.64273, -20.92158 (xmin, xmax, ymin, ymax)
# crs         : +proj=longlat +datum=WGS84 +no_defs
# variables   : 9
#...
```

Climate surface data with spatial variation in mean annual temperature is imported into R with the `stack` function. The `names` function is used to rename each climate surface according to the scenario considered in the time periods between "1970-2000", "2021-2040", "2041-2060", "2061-2080", and "2081-2100".

```
ty <- stack("C:/manuscript/mclimaticas/dados/GCM/CNRM-ESM2-1/ty00_100.tif") #Import
ty # See header
# class      : RasterStack
# dimensions : 208, 268, 55744, 5  (nrow, ncol, ncell, nlayers)
# resolution : 0.04166667, 0.04166667  (x, y)
# extent     : -51.04167, -39.875, -22.91667, -14.25  (xmin, xmax, ymin, ymax)
# crs        : +proj=longlat +datum=WGS84 +no_defs
# names      : ty00_100_1, ty00_100_2, ty00_100_3, ty00_100_4, ty00_100_5
# min values :   10.95500,   12.00417,   12.55417,   12.77917,   12.84583
# max values :   24.71667,   26.02500,   26.56250,   26.82917,   26.89167
names(ty) <- c("T annual 1970-2000", "T annual 2021-2040", "T annual 2041-2060",
             "T annual 2061-2080", "T annual 2081-2100") # Renaming layers
names(ty) # View modified layer names
# [1] "T.annual.1970.2000" "T.annual.2021.2040" "T.annual.2041.2060"
# "T.annual.2061.2080" "T.annual.2081.2100"
```

The `crop` and `mask` functions are used to subset data and mask temperature values around municipalities in Minas Gerais defined by a subset of data around Lavras.

```
tycrop <- crop(ty, mg) # Subset
tymask <- mask(tycrop, mg) # Mask
```

The masked raster data subset is exported to the directory of interest in the computer with the `writeRaster` function.

```
writeRaster(tymask, filename = "C:/geo/c2/tymask.tif",
            options = "INTERLEAVE=BAND", overwrite=TRUE)
```

Raster subset results are mapped using the `plot` function. This shows an increase in temperature in the region as the years considered in the analysis increase (Figure 2.14).

```
plot(tymask)
```

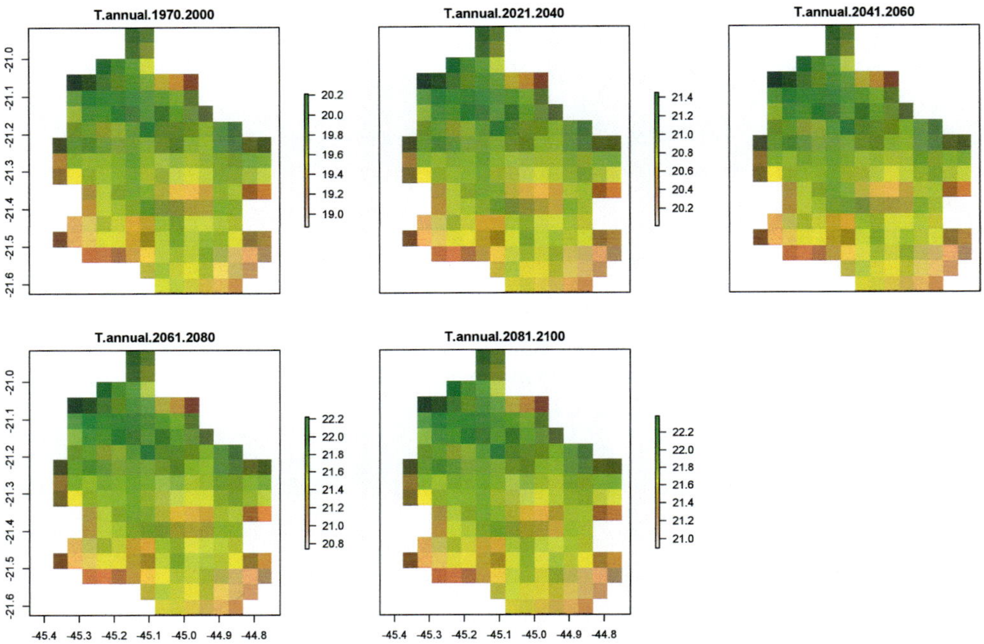

FIGURE 2.14 Raster subset mapping of the mean annual air temperature (degrees Celsius) in Minas Gerais, Brazil, from 1970 to 2100.

A file with polygon edges is created for use in mapping functions by the `spplot` function of the `sp` package.

```
borders <- list("sp.polygons", mg, fill = "transparent", col = 'black', first = FALSE)
```

A sequence of objects is created in a code chunk to carry out the spatio-temporal mapping of the annual temperature in municipalities in Minas Gerais with the `spplot` function (Figure 2.15).

```
t <- tymask
t[["a"]] = tymask$T.annual.1970.2000
t[["b"]] = tymask$T.annual.2021.2040
t[["c"]] = tymask$T.annual.2041.2060
t[["d"]] = tymask$T.annual.2061.2080
t[["e"]] = tymask$T.annual.2081.2100
```

```
spplot(t, c("a", "b", "c", "d", "e"),
names.attr = c("T annual 1970-2000", "T annual 2021-2040",
"T annual 2041-2060", "T annual 2061-2080", "T annual 2081-2100"),
as.table = TRUE, sp.layout = list(borders), scales = list(draw = TRUE),
xlim = c(-45.39973, -44.74138), ylim = c(-21.64273, -20.92158))
```

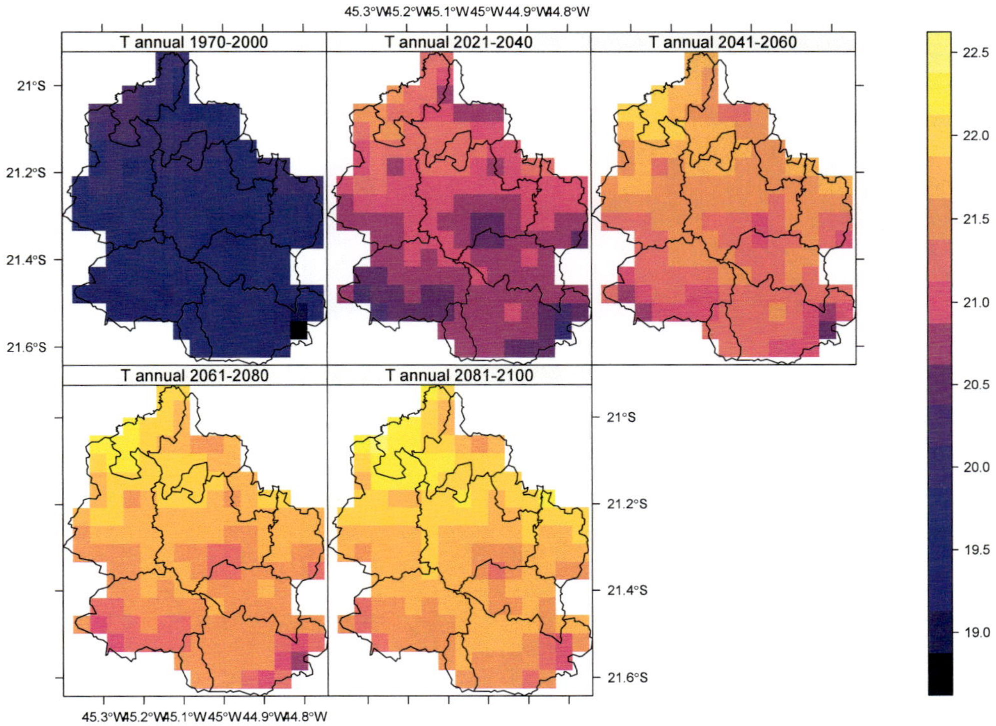

FIGURE 2.15 Mapping the mean annual air temperature (degrees Celsius) in Minas Gerais, Brazil, from 1970 to 2100.

The `disaggregate` function is used to increase the spatial resolution of mean annual air temperature in order to map the spatio-temporal variation of temperature within polygons of coffee fields. Otherwise, there would have been a loss of information in the sinuous edges of polygons to the detriment of large pixels in the temperature product originally used. It was therefore necessary to disaggregate the original information by 10 to represent the climate data within the coffee fields, obtaining a spatial resolution of 0.004166667 minutes.

```
tyd <- disaggregate(ty, fact=c(10, 10))
tyd
# class      : RasterBrick
# dimensions : 2080, 2680, 5574400, 5  (nrow, ncol, ncell, nlayers)
# resolution : 0.004166667, 0.004166667  (x, y)
# extent     : -51.04167, -39.875, -22.91667, -14.25  (xmin, xmax, ymin, ymax)
# crs        : +proj=longlat +datum=WGS84 +no_defs
```

```
# source      : memory
# names       : T.annual.1970.2000, T.annual.2021.2040, T.annual.2041.2060, ...
# min values  :          10.95500,          12.00417,          12.55417, ...
# max values  :          24.71667,          26.02500,          26.56250, ...
```

The polygons created around coffee plantations are imported into R with the `read_sf` function.

```
coffee <- read_sf("C:/geo/c8/coffeeMG.shp", options = "ENCODING=WINDOWS-1252")
coffee
# Simple feature collection with 8005 features and 14 fields
# Geometry type: POLYGON
# Dimension:     XY
# Bounding box:  xmin: -45.39288 ymin: -21.63094 xmax: -44.77784 ymax: -20.94797
# Geodetic CRS:  WGS 84
# A tibble: 8,005 × 15
# ...
```

The polygons of the object of class `sf data.frame` are converted into class `sp`, with the function `as_Spatial`, creating a new object `cafesp` of class `SpatialPolygonsDataFrame`.

```
coffeesp <- as_Spatial(coffee)
coffeesp # See header
# class       : SpatialPolygonsDataFrame
# features    : 8005
# extent     : -45.39288, -44.77784, -21.63094, -20.94797(xmin, xmax, ymin, ymax)
# crs         : +proj=longlat +datum=WGS84 +no_defs
# variables   : 14
#...
```

The `crop` and `mask` functions are used to subset data and mask temperature values around coffee plantations in municipalities in Minas Gerais.

```
tydc <- crop(tyd, coffeesp) # Subset
tydm <- mask(tydc, coffeesp) # Mask
```

The results of the raster subset inside coffee plantation polygons are mapped with the `plot` function (Figure 2.16).

```
plot(tydm)
```

The raster data subset resulting from the geometric analysis to reduce the pixel size is exported to the directory of interest in the computer with the `writeRaster` function.

```
writeRaster(tydm, filename = "C:/geo/c2/tydm.tif", options = "INTERLEAVE=BAND",
            overwrite = TRUE)
```

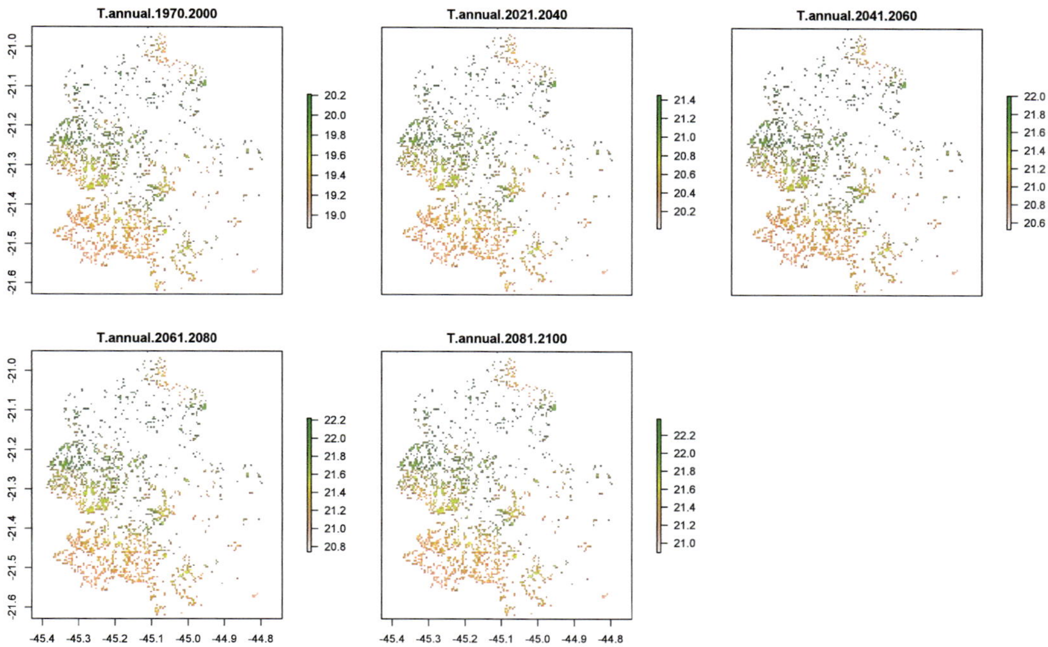

FIGURE 2.16 Raster subset mapping of the mean annual air temperature (degrees Celsius) inside coffee plantations in municipalities in Minas Gerais, Brazil, from 1970 to 2100.

A sequence of objects is created in a code chunk to perform spatio-temporal mapping of the annual temperature inside polygons of coffee plantations in municipalities in Minas Gerais with the `spplot` function.

```
t <- tydm
t[["a"]] = tydm$T.annual.1970.2000
t[["b"]] = tydm$T.annual.2021.2040
t[["c"]] = tydm$T.annual.2041.2060
t[["d"]] = tydm$T.annual.2061.2080
t[["e"]] = tydm$T.annual.2081.2100
spplot(t, c("a", "b", "c", "d", "e"),
names.attr = c("T annual 1970-2000", "T annual 2021-2040",
"T annual 2041-2060", "T annual 2061-2080", "T annual 2081-2100"),
as.table = TRUE, sp.layout = list(borders), scales = list(draw = TRUE),
xlim = c(-45.39973, -44.74138), ylim = c(-21.64273, -20.92158))
```

With the geovisualization analysis presented, the increase in temperature is also evident in the Arabica coffee-producing regions defined in municipalities close to Lavras in simulations of future mean air temperature scenarios when compared to the current temperature scenario considered in the analysis (Figure 2.17).

For many algorithms developed for vector data structure of polygon type, there is an alternative to use raster data, sometimes more efficiently. The calculation of area and perimeter of polygons and other operations can be performed by simple counting of raster data from a given radius range of a point, for example. Some questions must be answered when choosing data models and database types according to the required application

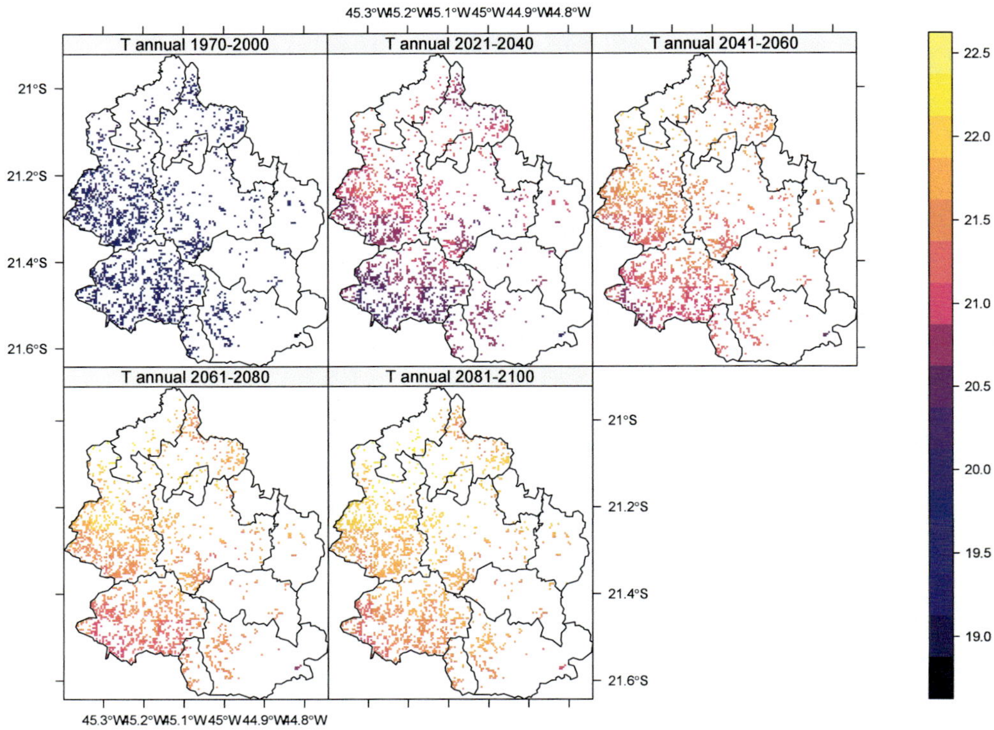

FIGURE 2.17 Mapping the annual spatio-temporal variation of mean air temperature (degrees Celsius) in Minas Gerais, Brazil, from 1970 to 2100, in coffee fields mapped in 2018.

(Burrough and McDonnell, 1998):

- Is the phenomenon under study considered simple or complex?
- Are the entity types used to describe the phenomenon described in detail or generalized?
- What are the data types used to characterize the variable, Boolean, nominal, ordinal, integer, real, or topological?
- Are the representation properties exact or vague? Deterministic or stochastic?
- Can the entities be represented by discrete physical characteristics or by continuous fields?
- Are the attributes obtained by counting or sampling?
- Is the data used for purposes of description, management or analysis?
- Is logical, empirical, or process modeling-based information used to generate a new database or to make inferences about the phenomenon studied?
- Is the process being evaluated dynamic or static?

2.11 COMPUTATIONAL PRACTICE

In the computational practice of geographic representation, the objective is to map raster and vector data in a geographic region using the packages `sf` (Pebesma et al., 2021a), `sp` (Pebesma et al., 2021b), `rgdal` (Bivand et al., 2021) and `raster` (Hijmans et al., 2020).

In the simple feature data model, a wide variety of geometry types can be represented in hierarchical form. In `sf`, all common vector geometry types such as points, lines, polygons, and their respective multiversions that group features of the same type into a single feature can be represented. In `sf` there is also support for collections of geometry in a single object. The `sp` is an R package used to represent vector geometries of points, lines, polygons, and collections of geometries. However, with the use of `sf`, compared to `sp`, there can be improved performance for reading, fast data writing and plotting. In `sp` there may also be simpler functionality to solve a problem when compared to `sf` in certain situations. So it is important to learn about both the `sf` and `sp` packages. The `raster` package is used a lot in the `raster` data representation. The `rgdal` package is used to perform coordinate transformations and map projections and has `sp` as a required package for operation, so it is not necessary to install `sp` separately with the use of `rgdal`.

The contour of a coffee crop is created using as reference the very high spatial resolution imagery available on Google Earth of a coffee crop on the campus of the Federal University of Lavras. The integrated mapping of vector and raster data is applied in a climatic study. Using the `sf`, `sp` and `raster` packages, mapping of mean annual air temperature in municipalities of Minas Gerais is performed by integrating raster and vector data in the mapping.

2.11.1 INSTALLING R PACKAGES

The `install.packages` function is used to install the `sf`, `raster` and `rgdal` packages in the R console.

```
install.packages("sf")
install.packages("raster")
install.packages("rgdal")
```

2.11.2 LOADING R PACKAGES

The `library` function is used to load the `sf`, `raster`, and `rgdal` packages in the R console.

```
library(sf)
library(raster)
library(rgdal)
```

2.11.3 DEFINE GEOGRAPHIC REGION FROM A POLYGON DRAWN IN GOOGLE EARTH

A polygon created in Google Earth with the `KML` extension can be imported into R as a vector that is referenced to an object or remotely sensed process defined relative to some location on Earth by creating a polygon drawn on top of the target of interest, such as in a color composition of satellite imagery of very high spatial resolution. Google Earth can be accessed on the Internet[2]. However, it is recommended to perform the practice on a desktop version of the program. In this case, a Windows version of Google Earth Pro can be obtained here[3].

[2] https://earth.google.com/web/

[3] https://www.google.com/earth/versions/#download-pro

To create a polygon in Google Earth do the following: (1) open Google Earth. (2) Go to a location on the map. (3) Above the map, click Add Path. (4) To add a shape, click Add Polygon. (5) A `New Polygon` dialog box will appear. You may have to move it out of the way before proceeding to the next step. (6) To draw the shape you want by connected lines, click on a starting point on the map and drag it. (7) Click an end point to close the polygon. In possession of a closed polygon geometry, you can calculate the area of the polygon by going to `Tools <- Ruler`. However the calculation of area is not our objective at this point. (8) Enter a description and properties of the polygon. The polygon will be created as a layer in Google Earth's left menu. (9) Right click on the created polygon layer and export the file to the `filename.kml` extension with `save place as` option. (10) Click on `save` (Figure 2.18).

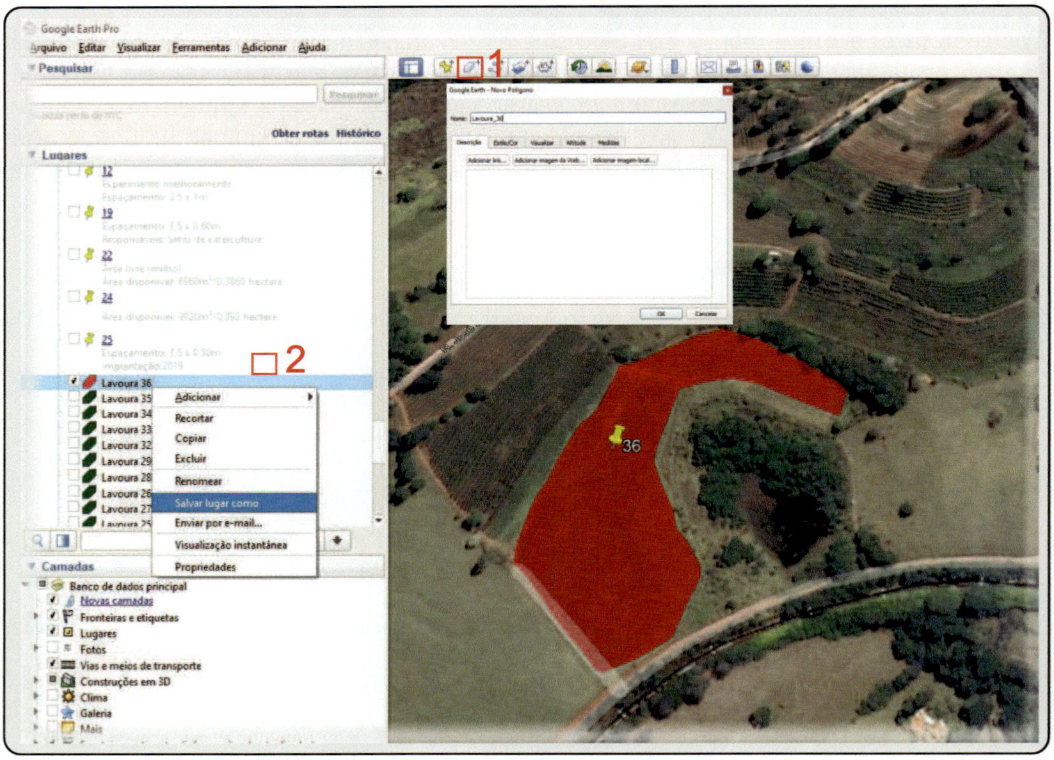

FIGURE 2.18 Creating (1) and exporting (2) a Google Earth polygon.

2.11.3.1 Get the .kml file from the Internet

The file created in the practical activity by the teacher on Google Earth is obtained from the Internet and unzipped to perform the practical activity. The file used in the practice was made available for download[4].

2.11.3.2 Import the KML file and map in R

After placing the file in a known directory and unzipping, the file `Lavoura_36.kml` is imported into R with the `st_read` function.

[4]http://www.sergeo.deg.ufla.br/sr/downloads/Lavoura_36.zip

```
p <- st_read("C:/geo/c2/Lavoura_36.kml")
# Reading layer `Lavoura_36.kml' from data source `C:\geo\c2\Lavoura_36.kml'
# using driver `KML'
# Simple feature collection with 1 feature and 2 fields
# Geometry type: POLYGON
# Dimension:     XYZ
# Bounding box:  xmin: -44.96357 ymin: -21.23146 xmax: -44.96134 ymax: -21.22897
# z_range:       zmin: 0 zmax: 0
# Geodetic CRS:  WGS 84
```

2.11.3.3 Convert vector file to ESRI shapefile

The `st_zm` function is used to extract the geometry from the KML file and add Z and M dimensions of the simple feature polygon geometry.

```
pol <- st_zm(p[1], drop = T, what = 'ZM')
```

2.11.3.4 Export the polygon as an ESRI shapefile

The polygon `p` can be exported as a vector file of type `ESRI Shapefile` for further use, as follows.

```
st_write(pol, dsn = "C:/geo/c2/pol.shp", driver = "ESRI Shapefile", "pol.shp")
```

2.11.3.5 Import and map the polygon shapefile in R

The polygon shapefile p is imported into R with the `st_read` function and then mapped with the `plot` function to observe the features of the region studied. We can observe that brackets `[]` were used in the attribute mapping operation of the class file `sf` named as `p` (Figure 2.19).

```
p <- st_read("C:/geo/c2/pol.shp") # Import shapefile
# Reading layer `pol' from data source `C:\geo\c2\pol.shp'
# using driver `ESRI Shapefile'
# Simple feature collection with 1 feature and 1 field
# Geometry type: POLYGON
# Dimension:     XY
# Bounding box:  xmin: -44.96357 ymin: -21.23146 xmax: -44.96134 ymax: -21.22897
# Geodetic CRS:  WGS 84
plot(p[1,], axes = T, col = "white", bgc = "grey92", main = "coffee field") # Map
```

2.11.3.6 Evaluate file class

The file class is evaluated with the `class` function.

```
class(p)
# [1] "sf"         "data.frame"
```

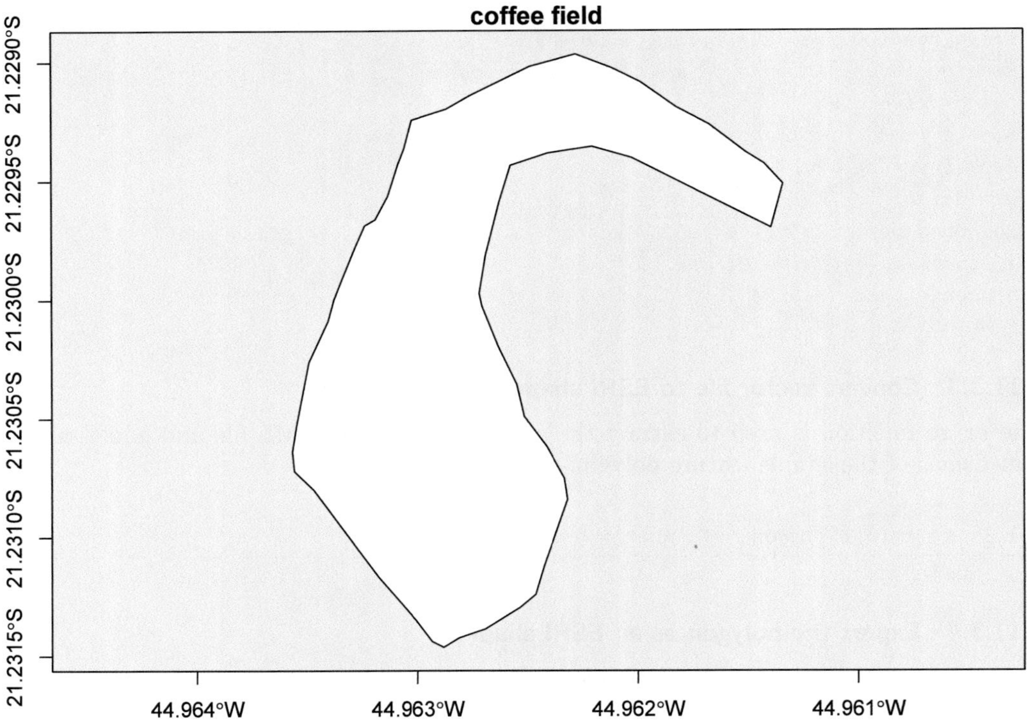

FIGURE 2.19 Mapping coffee crop on the campus of the Federal University of Lavras, Minas Gerais, Brazil, using the `sf` package.

2.11.3.7 Convert file from class sf to sp

The `as_Spatial` function is used to convert vector files of class `sf` into class `sp`. The `class` function is used to check the new class of the file.

```
psp <- as_Spatial(p) # Convert in sp
class(psp) # Evaluate the class
# [1] "SpatialPolygonsDataFrame"
# attr(,"package")
# [1] "sp"
```

2.11.3.8 Map sp class file

The `plot` function is used to map the coffee crop drawn from Google Earth (Figure 2.20).

```
plot(psp, axes = T, col="grey92", main = "Coffee field")
```

2.11.4 PERFORM REPRESENTATION INTEGRATING VECTOR AND RASTER

The polygon with the political boundaries of municipalities surrounding Lavras is obtained from a subset of the 'rnaturalearth' data package (Massicotte and South, 2023) and made

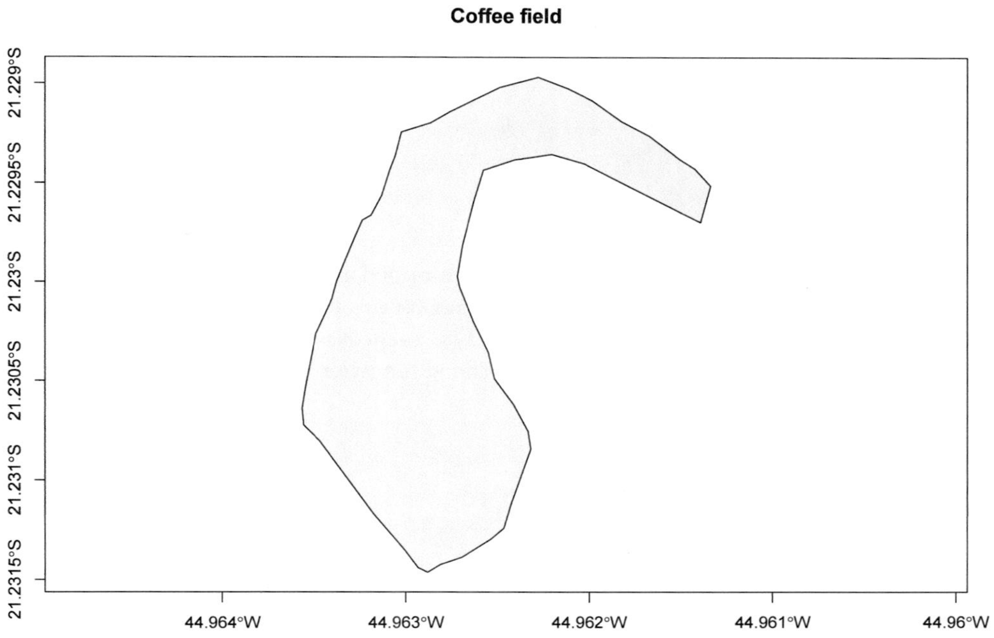

FIGURE 2.20 Mapping coffee crop on the campus of the Federal University of Lavras, Minas Gerais, Brazil, using the `sp` package.

available for download[5].

The moderate resolution climate surface of mean annual air temperature for the period from 1970 to 2000 obtained from the WorldClim project (Fick and Hijmans, 2017; Harris et al., 2014), was clipped and masked into the polygons of municipalities in the surroundings of Lavras and made available to download[6].

The annual mean air temperature data is imported into R using the `raster` function and the climate variable is renamed with the `names` function.

```
t <- raster("C:/geo/c2/ty1970_2000.tif") # Import
names(t) <- c("ty1970_2000") # Rename variable
```

Spatial polygons with attributes of administrative boundaries of municipalities are imported into R using the `readOGR` function.

```
mg <- readOGR(dsn ="C:/Aulas/Geoprocessamento/shape/municipiosMG2007.shp",
              layer = "municipiosMG2007", use_iconv = TRUE, encoding = "UTF-8")
mg
# class       : SpatialPolygonsDataFrame
```

[5]http://www.sergeo.deg.ufla.br/geo/downloads/municipiosMG2007.zip

[6]http://www.sergeo.deg.ufla.br/geo/downloads/ty1970_2000.zip

```
# features    : 11
# extent      : -45.39973, -44.74138, -21.64273, -20.92158  (xmin, xmax, ymin, ymax)
# crs         : +proj=longlat +datum=WGS84 +no_defs
# variables   : 9
# names       : GEOCODIG_M, UF, SIGLA, NOME_MUNIC,     NA., NA..1,  ...
# min values  :    3111903, 31,    MG, Cana Verde, Sudeste,  3109,  ...
# max values  :    3160801, 31,    MG, Nepomuceno, Sudeste,  3111, ...
```

The mean annual air temperature variation is mapped in layers by adding the boundary polygons of municipalities. The `extent` and `setExtent` functions are used to adjust the extent of the raster relative to the polygons. The `keepress=TRUE` parameter is used so that the spatial resolution of the raster remains the same even when adjusting the size of the polygons (Figure 2.21).

```
e <- extent(mg) # Set extension in mg
extent(t) <- e # Set extent of mg at t
t <- setExtent(t, e, keepres = TRUE) # Set the raster resolution
plot(t) # Map the temperature
plot(mg, add = T) # Map municipalities
```

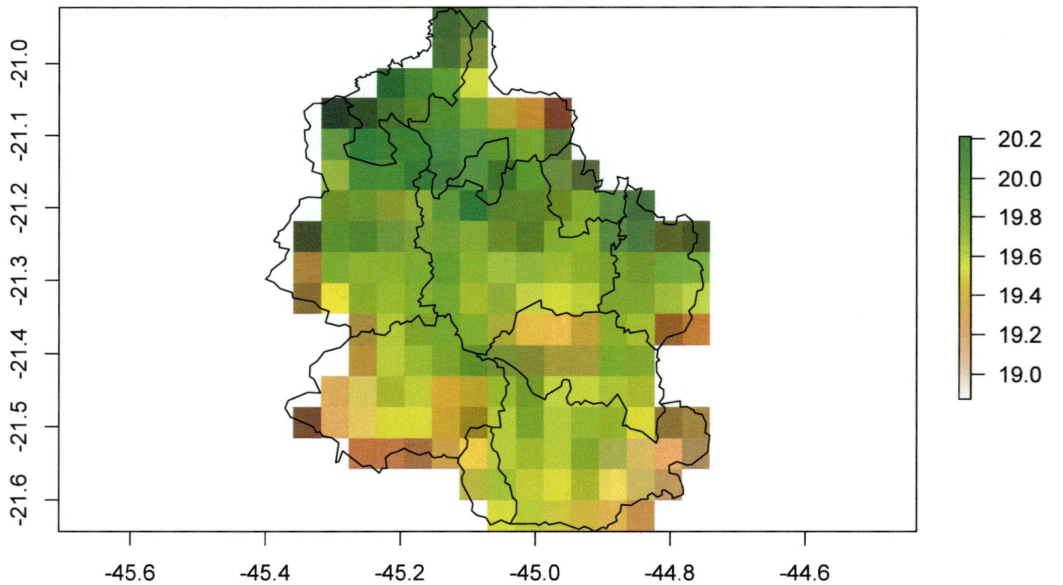

FIGURE 2.21 Mapping the mean annual air temperature in municipalities of Minas Gerais, from 1970 to 2000 using a `sp − class` vector file.

The same mapping of temperature and counties can also be done with functions from the `sf` package. In this case, it is necessary to transform spatial polygons with attributes into spatial features with attributes by the `st_as_sf` function.

```
mg_sf <- st_as_sf(mg)
```

Then, the temperature variation in the area is mapped by means of the layered sf package adding the vector feature of municipalities boundaries (Figure 2.22).

```
plot(t, axes = T, col = topo.colors(20)) # temperature
plot(st_geometry(mg_sf), col = NA, add = TRUE, bgc = "grey92") # municipalities
```

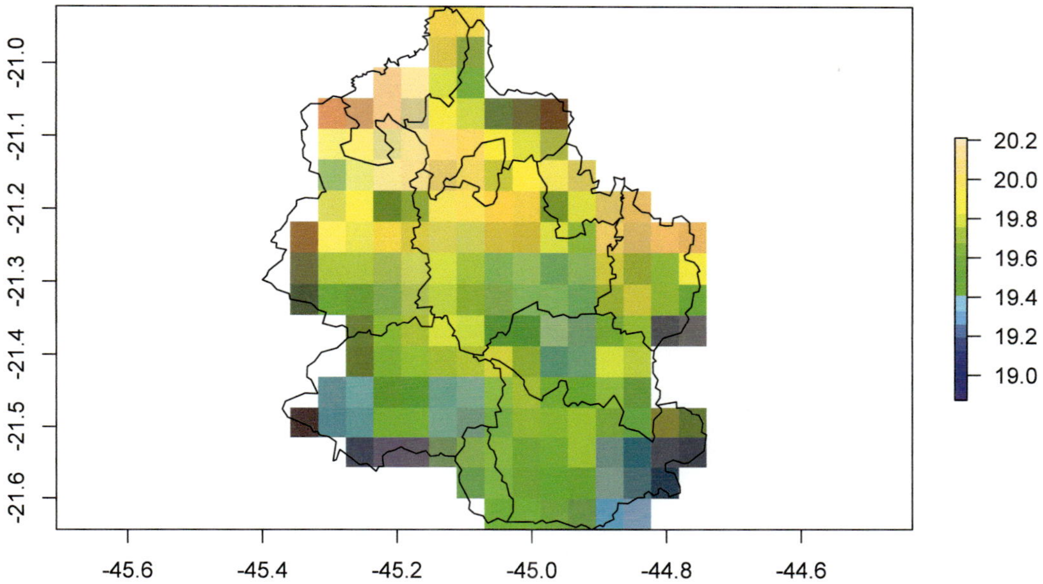

FIGURE 2.22 Mapping the mean annual air temperature in municipalities of Minas Gerais, in the period from 1970 to 2000 using sf − class vector file.

3 Geographic Database Management Systems with Environmental Geocomputation and R

3.1 LEARNING QUESTIONS

The learning questions answered by reading the chapter are:

- How does a database management system work in geocomputation?
- How can geographic data be manipulated in geocomputation?
- What are the main types and functions of geodatabases?
- What are the steps required to design a geodatabase?
- How to structure geographic information through topology and indexing.
- How can geocomputation be performed by multiple users?
- What is the relationship between geographic database management system and geocomputation software versions?

3.2 LEARNING OUTCOMES

The expected learning outcomes answered by reading the chapter are:

- Understand the function of database management system in geocomputation;
- Recognize languages for geographic data manipulation;
- Understand the main types and functions of geographic databases;
- Familiarize with the stages of database design geographic;
- Understand the main techniques for structuring geographic information through topology and indexing;
- Understand multiuser use and version control in geocomputation.

3.3 INTRODUCTION

Geographic information systems are used for different purposes, such as land use planning, environmental monitoring, agricultural crop monitoring, decision-making in mineral prospecting, electric grid management, Internet network, among other utilities. As the geospatial data infrastructure is extensive and of large volume, it is necessary to use complex data structures and interrelationship mechanisms to contemplate databases with several thematic maps, aerial photographs and satellite images with spatial and temporal interrelations in the same geographic location (Ramirez, 2007).

A database consists of a set of files structured for easy access to sets of information used to describe entities in the Earth's environment. In geographic databases, the data stored in the system are related to the location of entities, in addition to alphanumeric data, so that it is possible to perform geospatial queries and analysis (Medeiros and Pires, 1998). A database can be a set of integrated data about a particular subject, while geographic

DOI: 10.1201/9781003724377-3

databases or spatial databases are considered simple sets of databases with geographic data for a particular area in a particular context. A geographic database is a crucial part of operational geocomputation considering the impact of the geographic database on modeling analysis and decision-making. The advantages of using geographic databases are (Longley et al., 2001; Lo and Yeung, 2007):

- Reduction of redundancy and duplication of data with the same location;
- Cost savings from reduced data duplication;
- Different applications can share the same geographic database at different times;
- Data sharing is facilitated and different forms of data visualization can be achieved;
- Data security, standards and access are developed and used;
- Maintenance of data integrity with respect to accuracy, compatibility, consistency, update, backup and recovery.

Geographic data can be converted into discrete records on the computer to represent the location, presence or absence, type of object or process in the environment. When importing data into the computer a structural organization is required for data access, retrieval and manipulation. Systems for organizing data on the computer range from simple lists and indexed files to databases structured in hierarchical, network, relational, and object-oriented schemas. The type of structure determines how records are organized on the computer and how the data is manipulated. By using object-oriented database structures, it is possible to obtain the same basic units as in raster and vector systems, but in hierarchically connected entity structures. This enables a more realistic entity structure (Burrough and McDonnell, 1998).

3.4 ORGANIZING DATA ON THE COMPUTER

The essential feature of any data storage system is to be able to access and cross-reference data quickly. There are several ways, methods and situations in which different methods of managing data on the computer can be chosen in order to be able to import, export, store and edit a digital database (Burrough and McDonnell, 1998).

3.4.1 SIMPLE LISTS

The simplest form of database is a simple list of all items. Each new item in the database is placed at the end of the file, increasing the size of the file. Adding data in this system is easy; however, there may be inefficiency in retrieving the data, considering the difficulty in locating data in an unstructured list. In a list with n items, the average time required to locate an item (t_s) of interest is (Burrough and McDonnell, 1998):

$$t_s = t[\frac{(n+1)}{2}] \tag{3.1}$$

where t is the average time to locate item n in a database.

3.4.2 SEQUENTIAL ORDERING OF FILES

Words in a dictionary or names in a phone book are structured in file sequence order. Adding a new item requires the creation of an extra field, but locating an item can be faster. Files in sequence order can be located by binary search methods. In a binary search method, the average time required to locate an item (t_b) of interest is (Burrough and McDonnell, 1998):

$$t_b = t[log2(n+1)] \tag{3.2}$$

where log2 is the binary logarithm of base 2.

3.4.3 FILE INDEXING

The basic organization to build a data block can be accomplished by data items. A data item is the occurrence or instance of a particular characteristic of an entity that can be, for example, a person, thing, event, or phenomenon. In database terminology, a data item is called an attribute or data storage field. The value of an attribute can occur in a data field in the form of a numeric, character, date or logical value. A record is formed after grouping all attributes of interest into a data record line or tuple (Lo and Yeung, 2007).

In the case of applications using geographic information, in a single item, pixel, points, lines or polygons, there will be not only an identifying code or name, but also information associated with attributes, such as an ordered list of soil samples with information on soil class, pH, depth, drainage, texture and erodibility (Table 3.1) (Burrough and McDonnell, 1998).

TABLE 3.1 File indexing by soil sample.

ID Sample	Class	pH	Depth	Drainage	Texture	Erosion
1	Ultisols	5.5	Deep	Good	Clayey	No
2	Inceptisols	5.2	Shallow	Good	Silty	No
3	Ultisols	5.4	Deep	Good	Medium	No
4	Entisols	5.5	Deep	Poor	Sandy	Yes
5	Nitisols	5.9	Deep	Good	Very clayey	No
6	Inceptisols	4.2	Shallow	Good	Medium	Yes

In a sequential indexing search method for a block of data, the average time required to locate an item (t_i) of interest is (Burrough and McDonnell, 1998):

$$t_i = t_1[\frac{(n_1+1)}{2}] + t_2[\frac{(n_2+1)}{2}] \tag{3.3}$$

where n_1 and n_2 are the number of indexing classes and the number of items in a data block located by index, respectively. t_1 and t_2 are the average time to locate items n_1 and n_2, respectively.

In database indexing, we can get fast access to the data of interest; however, there can be problems of updating records due to continuous data logging. Another disadvantage is accessing the data by indexing and then performing another method of sequentially searching for records within the index itself (Burrough and McDonnell, 1998).

From a group of records, a data file can be formed. Each data file can be identified on the computer by a unique name. The use of databases instead of data files represents a change in the perception of data, in the way data is processed, and in the purpose of data use.

That is, database use is an approach to managing and processing data without a physical change in the way the data is stored (Table 3.2) (Lo and Yeung, 2007).

TABLE 3.2 Comparison between data file and database.

Data file characteristic	Database characteristic
Data collection with specific purpose	Generic data of organizations and businesses
Collection of records of the same data type and format	Collection of interrelated records, organized in one or more data files with different data types and formats
Computer processing to solve specific problems	Processing associated with an organization's database management system
Information support for specific applications	Support for decision-making and daily processing operations of an organization

3.5 DATABASE MANAGEMENT SYSTEM

Database management systems are computer programs used to organize and manage databases and can be built with a combination of hierarchical, network, relational and object-oriented structures. The purpose of a database management system is to make data available to multiple users in addition to maintaining integrity, including, removing, and updating data when necessary, among other features described below (Figure 3.1) (Burrough and McDonnell, 1998; Longley et al., 2001; Lo and Yeung, 2007):

- Enable to store, manipulate, and select data based on one or more attribute relationships;
- Manage access to and change of database content by separating storage and manipulating data;
- Provide interface between database and software based on a specific logic;
- Perform access to functions in applications independent of the physical storage structure, without applications being affected by changes in storage;
- Allow data access by users simultaneously;
- Protect the database from impermissible modifications;
- Availability of rules to remove errors and inconsistencies from the database;
- Use of metadata storage system about the class and structure of data;
- Support for multiple layered data views;
- Spatial indexing of large databases;
- Integrate graphics (maps) and description of data characteristics;
- Perform geospatial operations and spatial relationships;
- Provide geographically supported data model;
- Present query language;
- Control updates;
- Present tools for database administration for different applications;
- Present mechanisms for creating custom applications with programming language and interface;
- Integration between different software and languages used in geocomputation.

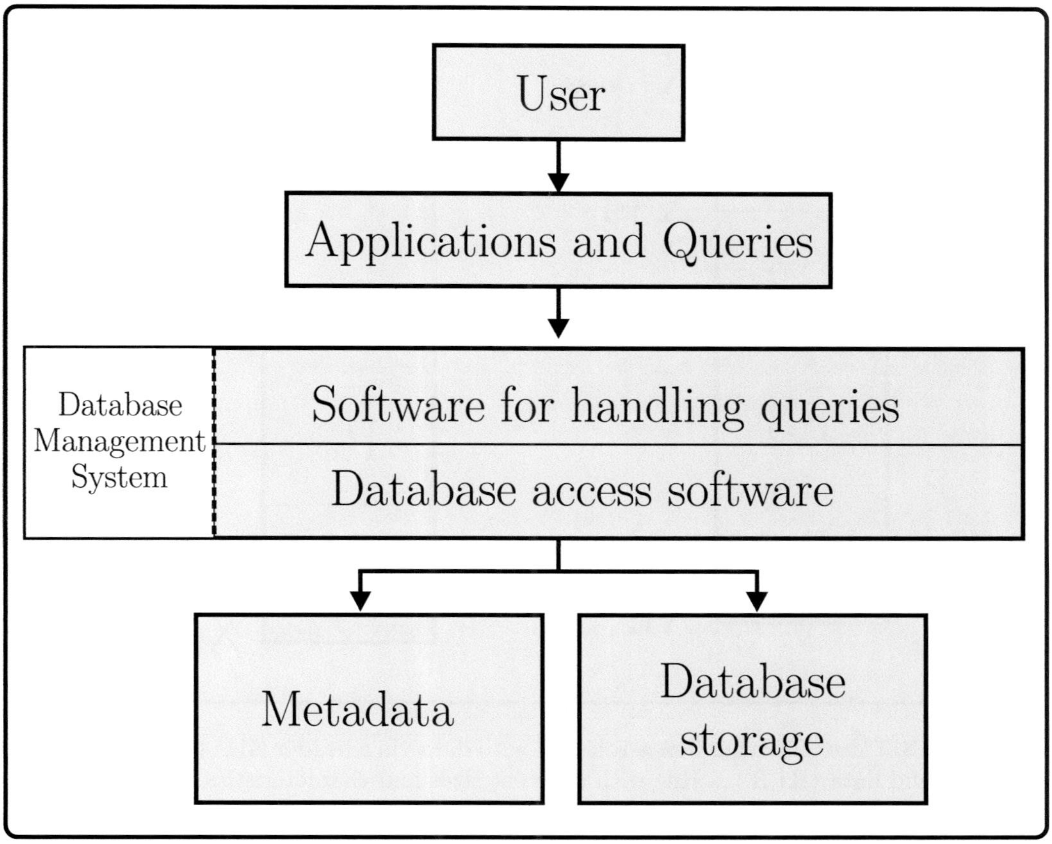

FIGURE 3.1 Components of a database management system.

3.6 DATABASE MANAGEMENT STRUCTURES

Geographic database management systems are classified according to how they store and manipulate data (Longley et al., 2001). Data from environment variables can be stored on the computer by means of simple or organized data lists. In the basic unit of a data record or tuple, there must be all the basic information of an entity. Data records can be of the same or varying size and adding information about the data in a header can improve data storage efficiency (Figure 3.2) (Burrough and McDonnell, 1998).

Database modeling refers to the task of designing an efficient database with correct information, logical structure, and the possibility of maintaining and modifying the amount of data. Modeling methods used to organize information to represent real-world structures on the computer are accomplished by means of logical model schemas (Burrough and McDonnell, 1998):

- Hierarchical;
- Network;
- Relational;
- Object-oriented.

In each logical model there is a fundamental principle of organizational structure of spatial information that can be registered as a continuous or discrete data model entity (Burrough and McDonnell, 1998).

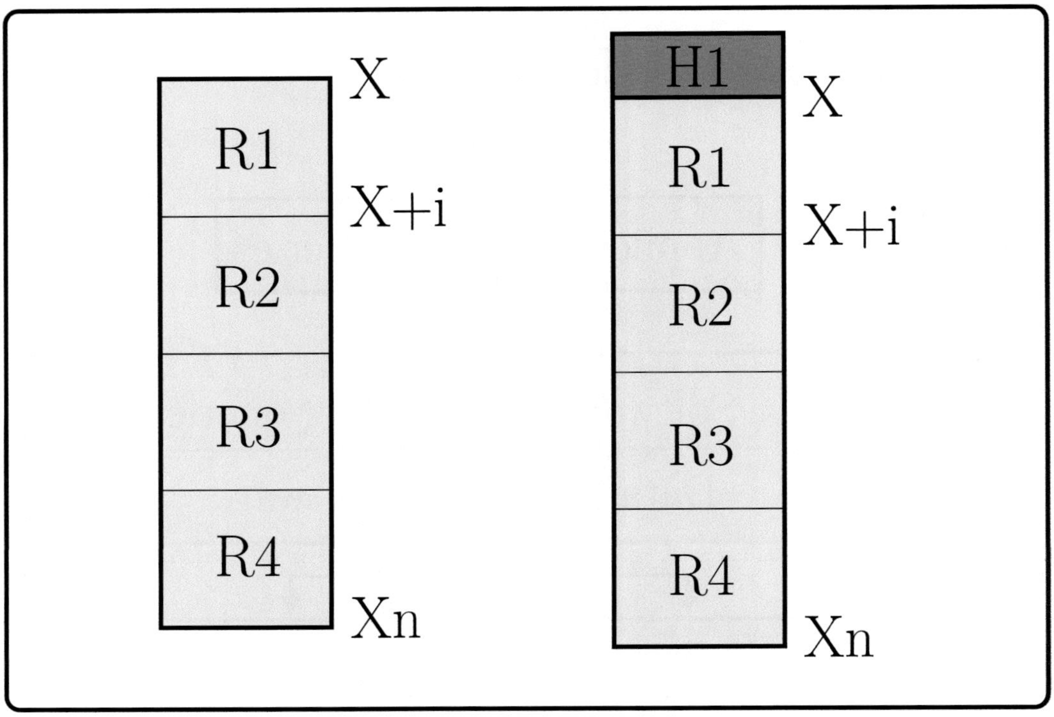

FIGURE 3.2 Use of information associated with data via a header (H1) to improve storage efficiency and data (R1,R2,...,Rn) with different sizes and characteristics.

3.6.1 HIERARCHICAL DATABASE STRUCTURE

The use of hierarchical structures facilitates quick access to soil class data or pixel data within a polygon through one-to-many type relationships. The hierarchical structure is also used in environmental science, animal, plant and organism taxonomy, and soil classification. In this type of structure, in each part of the hierarchy there is a key criterion used to describe the structure of the data as a whole with the assumption that there is correlation between a key attribute and the other attributes along the hierarchical sequence (Burrough and McDonnell, 1998).

A polygon map can be encoded according to a hierarchical model (Figure 3.3). The disadvantages of the hierarchical method are (Burrough and McDonnell, 1998):

- Difficult association with attributes;
- Inflexibility of the hierarchy to answer some types of questions about the environment;
- When maintaining and updating large files, it may be necessary to repeat some attributes several times, generating redundancy and increasing storage costs.

The advantages of the hierarchical method are (Burrough and McDonnell, 1998):

- Simplicity and easy access to the key criteria of the hierarchy;
- Fast performance;
- Ease of understanding;
- Ease of updating, expanding, and organizing data in storage systems;
- Question-based data retrieval based on knowledge of the structure.

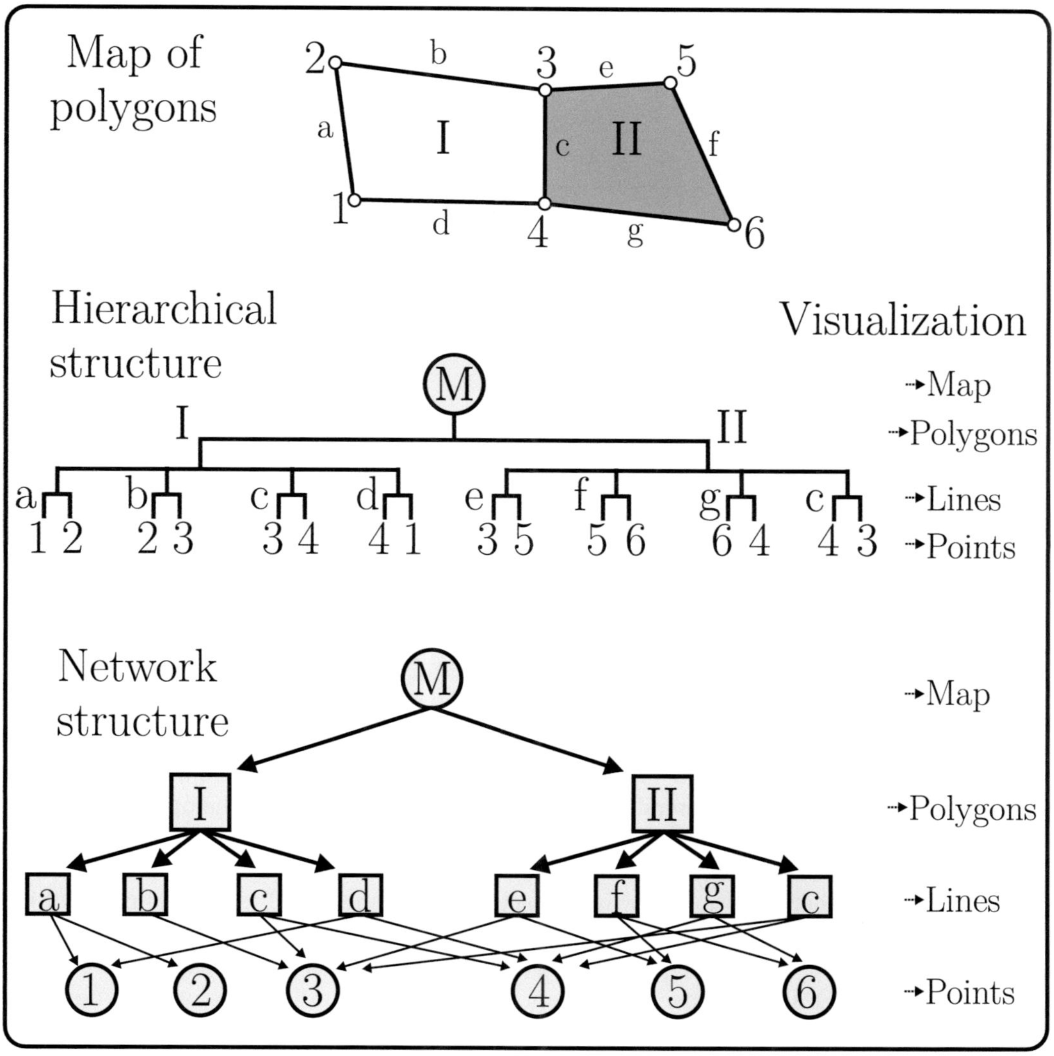

FIGURE 3.3 Hierarchical and network organization of vector type data.

3.6.2 NETWORK DATABASE STRUCTURE

The network database structure can be represented by arrow structures in sequential links used to navigate the topological structure of the network (Figure 3.3). The network database is useful for specifying the relationship between the items in the structure and avoiding redundancy of the available data. However, the database increases in size as a function of the link arrows. Maintaining the link arrows could overload the database system (Burrough and McDonnell, 1998).

3.6.3 RELATIONAL DATABASE STRUCTURE

A relational database comprises a set of tables with a two-dimensional list of object attribute records (Burrough and McDonnell, 1998) (Figure 3.4). This simple structure is flexible and used in different areas and occurs in 95% of database management systems (Longley et al., 2001).

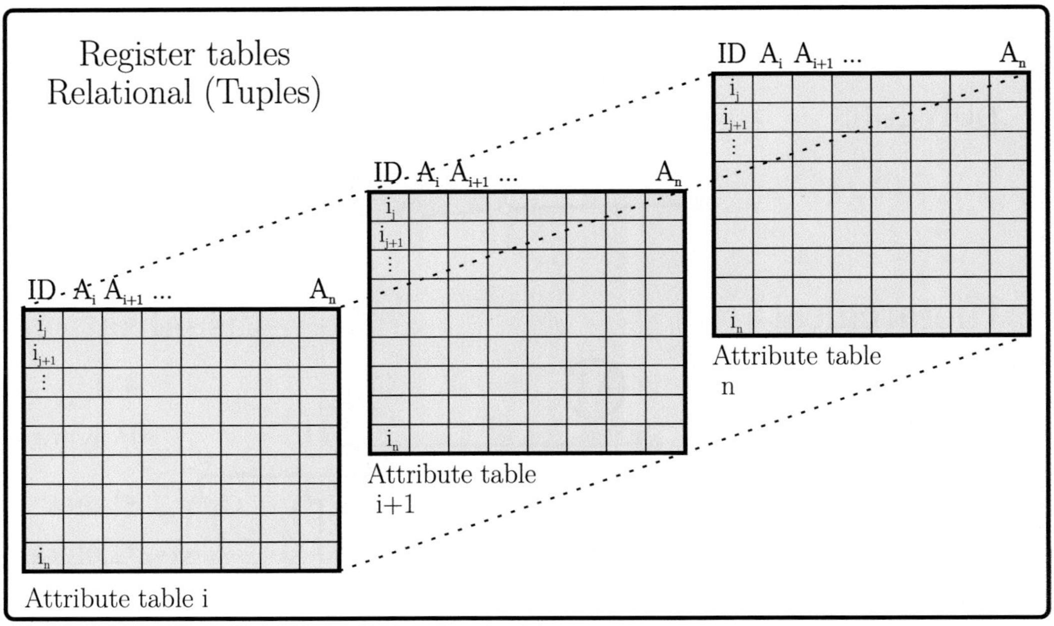

FIGURE 3.4 Data tables in a relational database management system.

The lowest level of user interaction with the database of a geographic area is realized through an organized collection of data on specific topics, such as: pipes in water and sewage networks, soil polygons in a watershed, and terrain elevation values. The object classes are stored in standard database tables as a two-dimensional matrix of rows and columns. Each object class is stored as a single database table in a database management system. In the rows of the table there are objects with object class instances and in the columns, object properties or attributes. The data is stored in individual rows with different values along column intersections. In the geographic database tables, there is a geometry column for the data. To save space and improve performance, the actual coordinate values can be stored in a highly compressed binary format. Different tables can be combined using standard database tools, based on the common value in an identifying column (the key) to create a third table. After joining, the geocomputation operations of query, display and analysis can be performed on a single table (Longley et al., 2001).

A series of rules are used for efficient and effective database design in simple and stable table structures with the following principles (Longley et al., 2001):

- In each table cell, there is a value at the intersection of row and column;
- All values in a table column are about the same subject;
- Each row in the table is unique with no duplication of records;
- There is no meaning to the sequence of columns and rows.

A standard database query language (SQL), ISO Standard ISO/IEC 9075, is adopted by some databases. The main types of SQL statements are: data definition language (DDL), data manipulation language (DML), and data control language (DCL). In the third major revision of SQL (SQL 3) in 2004, spatial types and functions are defined as part of a multimedia extension called SQL/MM. Other features of the SQL language are (Longley et al., 2001):

- It is used directly via an interactive command line interface;

- It can be compiled into a programming language of diverse use (e.g., C/C++/C#, Java, or Visual Basic);
- It can be an embedded graphical user interface (GUI).

In the relational database structure, there is no hierarchy and no link arrows. In this case, data is stored in simple records known as tuples or attribute data fields. The tuples are grouped together in two-dimensional data relationship tables like in spreadsheet software. Each table relationship is a separate record in the file. Identification codes are used in place of link arrows and key criteria (Figure 3.5) (Burrough and McDonnell, 1998).

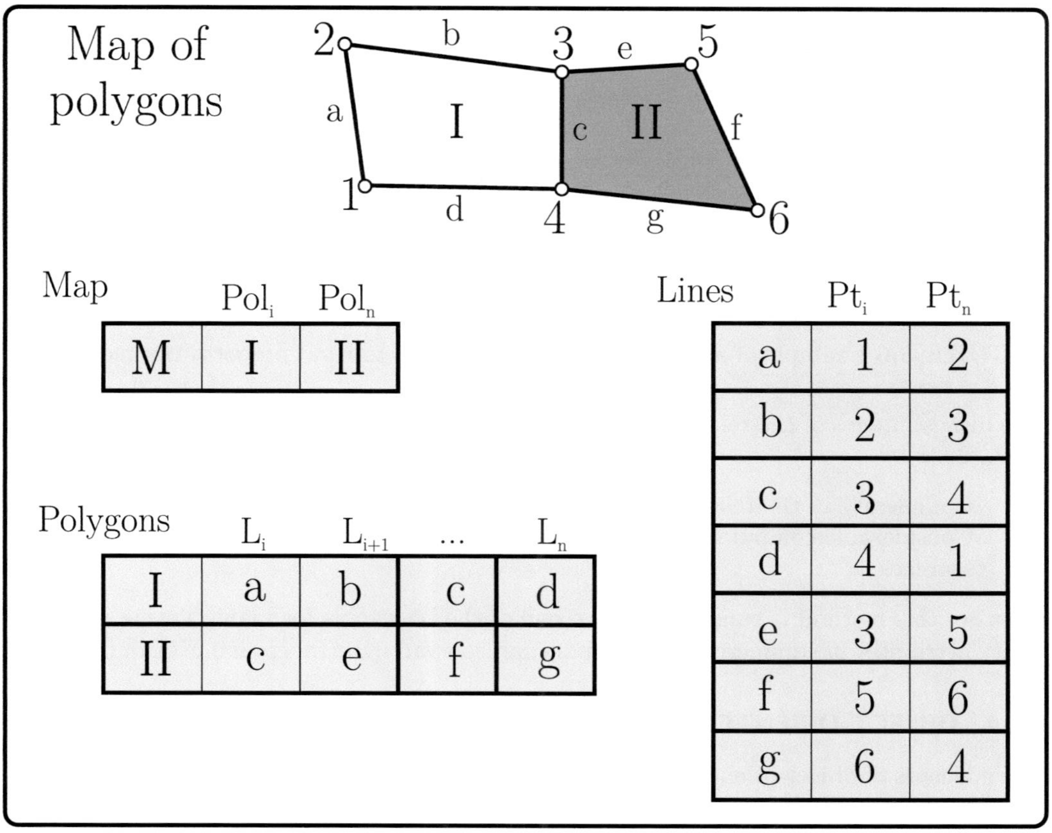

FIGURE 3.5 Relational organization of vector-type data.

Data can be extracted from the relational database according to the relationship of interest to the user. Relational algebra methods can be used to generate new tables to explore different relationships between the data by means of a coded language (Figure 3.6). Normalization methods can be used to eliminate redundancy and create more efficient code by simplifying the structure without loss of information. The advantages of using relational structure are (Burrough and McDonnell, 1998):

- Structural flexibility;
- Can meet demands of queries with Boolean logic and mathematical operations;
- Different types of data can be searched, combined and compared;
- Data input and output is made easy by editing on the tuple, row, column or entire table;
- The joining of data can be accomplished through common data fields;

- Ideal in situations where all records have the same number of attributes and no natural hierarchy.

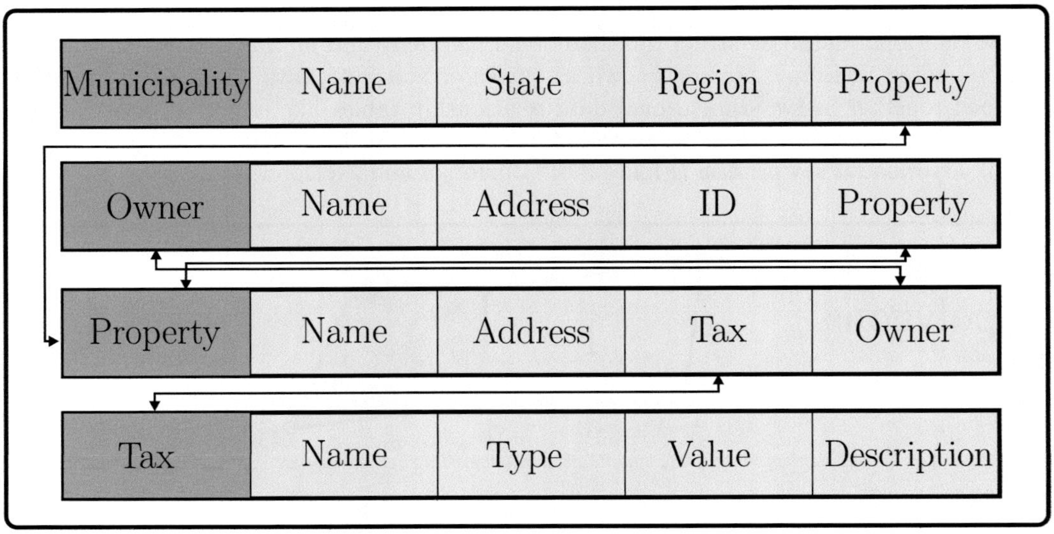

FIGURE 3.6 Example of a relational database applied to rural property tax registration.

The disadvantages of the relational structure are (Burrough and McDonnell, 1998; Longley et al., 2001):

- Complexity in the relationship between tables;
- Considerable computational processing time due to the number of joins needed in searches.

Even so, this method is considered to be one of the best tools for manipulating geospatial entity attributes in commercial geocomputation software (Burrough and McDonnell, 1998).

3.6.4 OBJECT-ORIENTED DATABASE STRUCTURE

The concepts of object-oriented programming structure arose from problems of redundancy and sequential searching in the relational structure. In geocomputation, the use of object-oriented structure is stimulated by the need to manipulate complex geospatial entities more intelligently than a simple point, line or polygon and by problems of database modifications when performing spatial operations between geometries. In this case, the speed of hierarchical and network approaches is used with the flexibility of the relational method when organizing data around entities, data records, and logical relationships between attributes and values. In the object-oriented database, data is defined as single objects organized into similarity groups of object classes according to a natural structure. The relationship between different objects and classes is defined by explicit links (Burrough and McDonnell, 1998).

The characteristics of any object can be described in the database with respect to the attributes of the object and the set of methods used in the description. With respect to all attributes of the object, there must be a unique database identifier regardless of changes, reduction or increase in attribute values. Link segments are used to link the unique identifiers of objects in the database with relationships and hierarchy. After establishing the hierarchy general formations, sub- and super classes can be created to get better efficiency for characterizing attributes and getting answers from the database. In object-oriented approach, "Municipality", "Property" and "Owner" are defined as objects interrelated by set state, such as situates, contains, belongs, uses, among others (Figure 3.7) (Burrough and McDonnell, 1998).

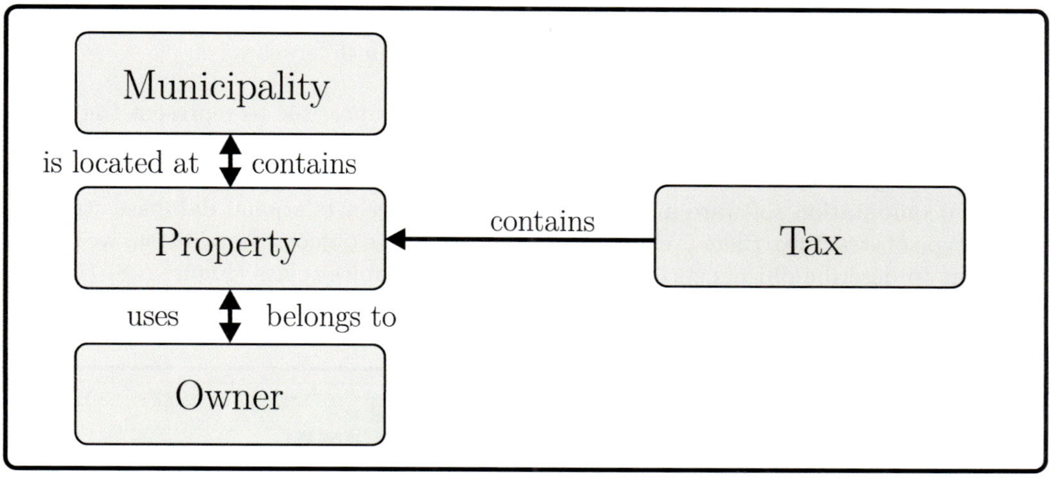

FIGURE 3.7 Object-oriented approach to reduce data storage and improve record linkage.

The types of queries possible to the database can vary according to the operations used to define the objects. The response of the object to a message will depend on the situation, context and polymorphism of the object. The data used in a object-oriented database must be defined as single entities in order to have an efficient hierarchical organizational structure and interrelationships. With this, queries can be referred to specific objects in an efficient structure. Still, the relational database structure may be better for answering questions of an attribute by indexing and clustering methods (Burrough and McDonnell, 1998).

With object-oriented database structures it is possible to unify the attribute with the storage geometry. Geographic data is characterized by a series of attributes with values that define the spatial, graphical, temporal, textual, and numeric dimensions. Part of the attribute definition is performed to describe the nature of the geometry of the object (point, line, polygon, or pixel) at a given spatial scale. Hierarchical relationships can be set up in different classes and sub-classes, and explicit topological links can be defined by direction, intersection, adjacency, overlap, right of/left of operators. Furthermore, with the object-oriented organizational basis, it is possible to model real-world semantics and processes in an integrated and intuitive way. The hierarchical structure and complex relationships can be controlled directly, giving flexibility in updating and modifying the database (Burrough and McDonnell, 1998).

3.6.5 RELATIONAL DATABASE HYBRID STRUCTURES

In hybrid relational database structures, geometric representations are joined to attributes by dividing the spatial problem into two parts. In the first part, the topology of spatial objects is represented, and in the second part are the attributes of spatial objects. The attributes of spatial objects can be stored in a relational database. With the use of the resulting hybrid, object-relational structure, there are some advantages (Burrough and McDonnell, 1998):

- Attributes are not stored with the spatial data geometry;

- Attribute maintenance is independent of the modification of the spatial geometry of the objects;
- You can access the data structure by standard methods and languages, independent of the relational storage system;
- The attributes in the database do not affect the basic principle of geocomputation layers;
- Attributes are linked to the spatial data geometry in different ways.

With the development of hybrid database structures, it is possible to represent the spatial and temporal nature of geographic data flexibly and efficiently (Burrough and McDonnell, 1998). The object-oriented database structure, although little used commercially, is used by some geocomputation software in a version of hybrid object-relational database structure. In this type of structure, there is extensibility to manipulate objects. In addition, we can use functions to describe objects such as by color, size, and object class (Figure 3.8) (Longley et al., 2001).

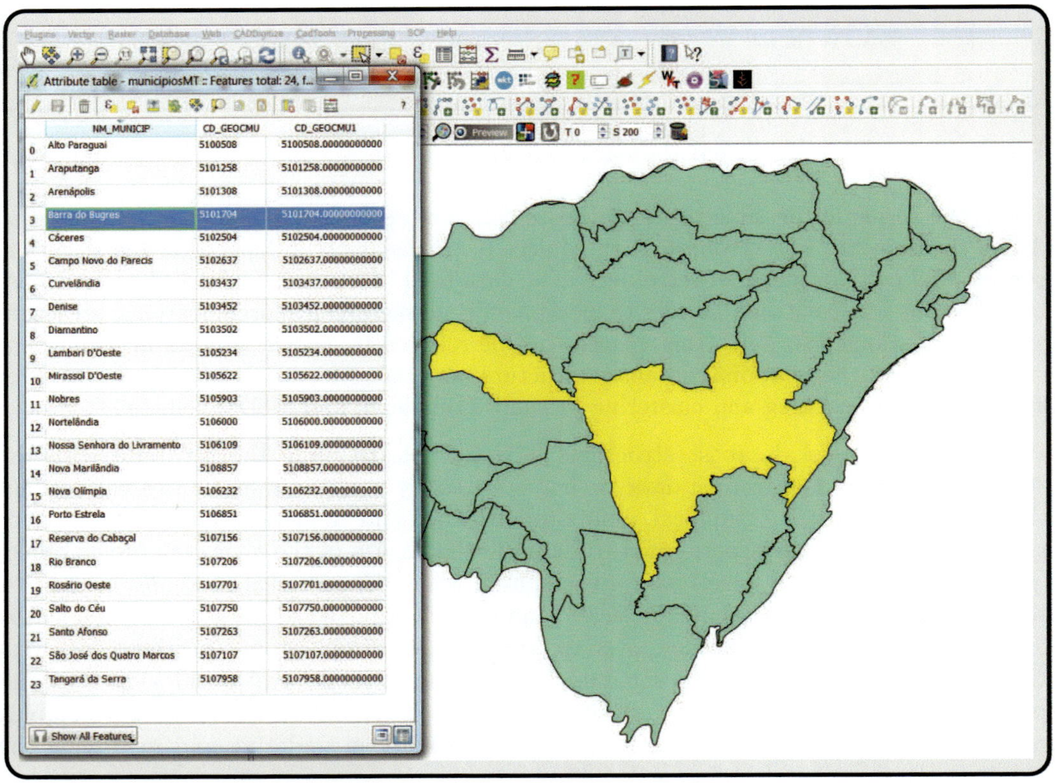

FIGURE 3.8 Object-relational database table from QGIS 2.10.1-Pisa software associated with polygon spatial objects on a map of municipalities' watershed information in the state of Mato Grosso, Brazil.

Another example of a relational object database can be obtained with the `mapview` package (Appelhans et al., 2020) from R. In this code chunk, data from municipalities in Minas Gerais is obtained with the `brazilmaps` package (Siqueira, 2017) and then mapped by `mapview` using RStudio (Figure 3.9).

```r
# Load packages
library(brazilmaps)
library(mapview)
# Get data
mg <- get_brmap(geo = "City", geo.filter = list(State = 31), class = "sf")
# Map
mapview(mg)
```

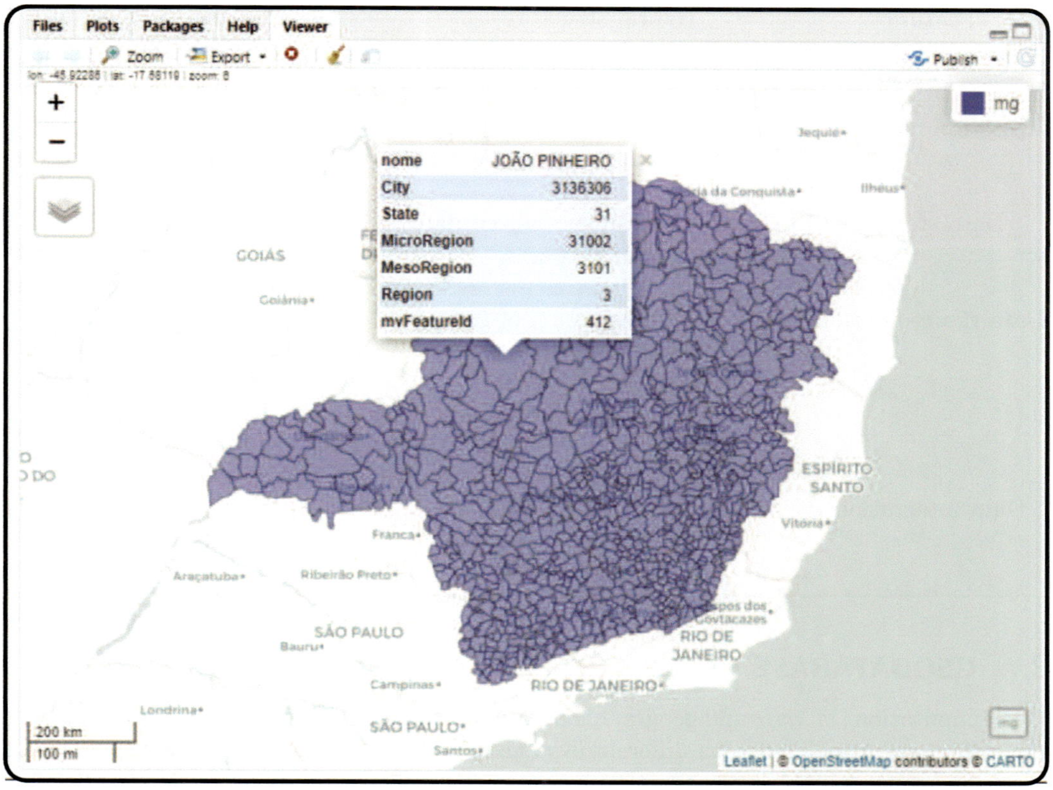

FIGURE 3.9 Object-relational database table from RStudio using mapview package related to a map of municipalities of Minas Gerais state, Brazil.

In an object-relational database the following features applied to geographic objects are highlighted (Longley et al., 2001):

- A query parser mechanism with functions applied to geographic data;
- An efficient optimizer mechanism for geographic queries;
- A query language for geographic data types such as points and polygons and functions such as: select polygons that touch each other;
- Indexing services with multidimensional support for geographic data (x, y, z coordinates);
- Storage management of large volumes of geographic records with different sizes, geometric and topological relationships, through specialized storage structures;
- Short (sub-second) and long transaction services, common in many geographic applications;

- Database replication in case of problems reconciling changes made by users in distributed geocomputation systems.

A debate can be held regarding the use of hybrid and object-oriented database in geocomputation systems with advantages and disadvantages of both methods (Table 3.3) (Burrough and McDonnell, 1998).

TABLE 3.3 Discussion about advantages and disadvantages of geocomputation systems with hybrid, object-oriented database storage structure.

Feature	Hybrid structure	Object-oriented structure
Functionality	Limited by the geocomputation software vendor. Ease of use.	Enables user-defined functions.
Question handling	Slow for complex objects. SQL search language.	Fast for complex objects. Difficult to identify objects on continuous surfaces.
Data storage	Data stored in specialized file structures.	Minimum-maintenance storage. Requires less storage space than entities related by indexed files.
Data representation	Poor representation of time series. Solid theoretical basis of relational database.	Smaller semantic gap in representing real-world objects and concepts than relational database. Limited indexing. Requires definition of objects, functions and topology.
Data integration	Easy integration of attribute data.	Raster and vector structures can be integrated into the same database.

3.7 GEODATABASE TYPES AND FUNCTIONS

Geocomputation software and geographic database management systems work together with multiple tasks associated with different systems (Figure 3.10) (Longley et al., 2001).

International ISO/OGC standards and standards developed for a hierarchy of feature classes of vector type geometries with attribute table, coordinate system and map projections, by the Open Geospatial Consortium are considered a reference for establishing standardizations of data used in geocomputation (Open Geospatial Consortium, 2021; Longley et al., 2001) (Figure 3.11).

According to the ISO/OGC standard, nine methods for evaluating spatial relationships between geometric objects are defined. The complete set of Boolean Operators for testing spatial relationships between geometries is (Longley et al., 2001):

- "Equals", were the geometries equal?
- "Disjoint", in the geometries, was there a common point?
- "Intersects", were there intersections of geometries?
- "Touches", was there intersection of boundaries of geometries?
- "Crosses", were there overlapping geometries? Were there different types of geometries, such as lines and polygons?
- "Within", was one geometry within the other?
- "Contains", was one geometry completely inside the other?

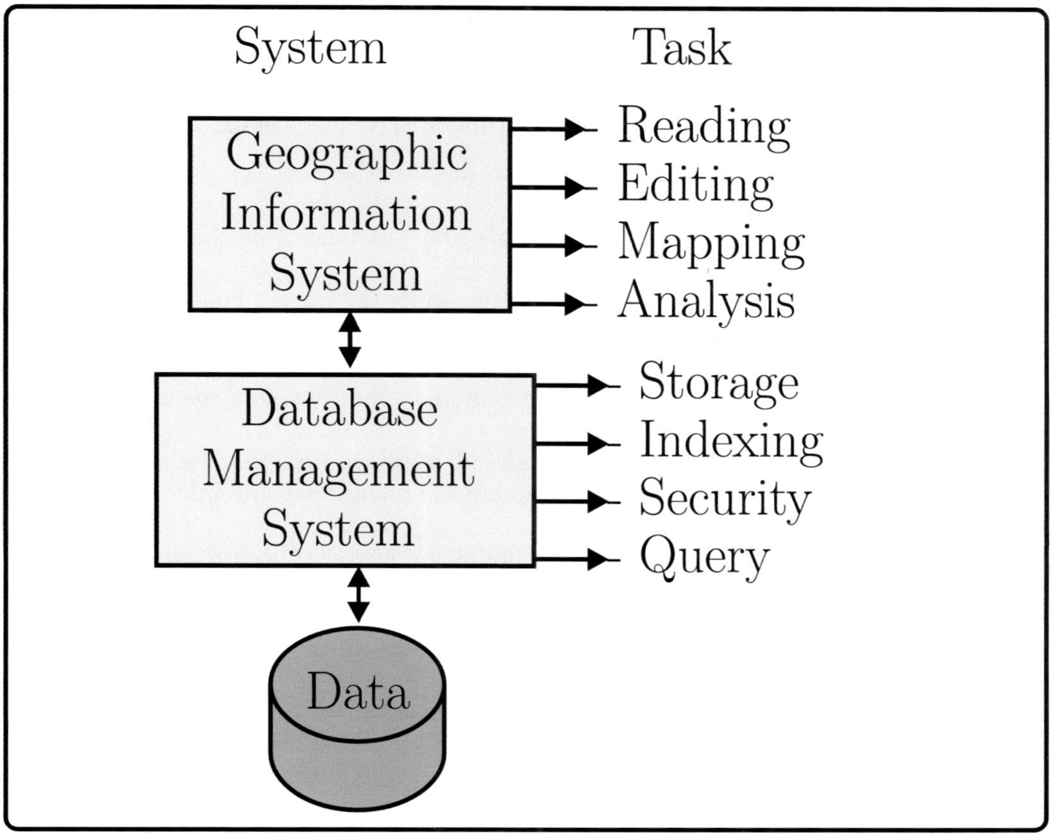

FIGURE 3.10 The functions of geocomputation and geodatabase management software.

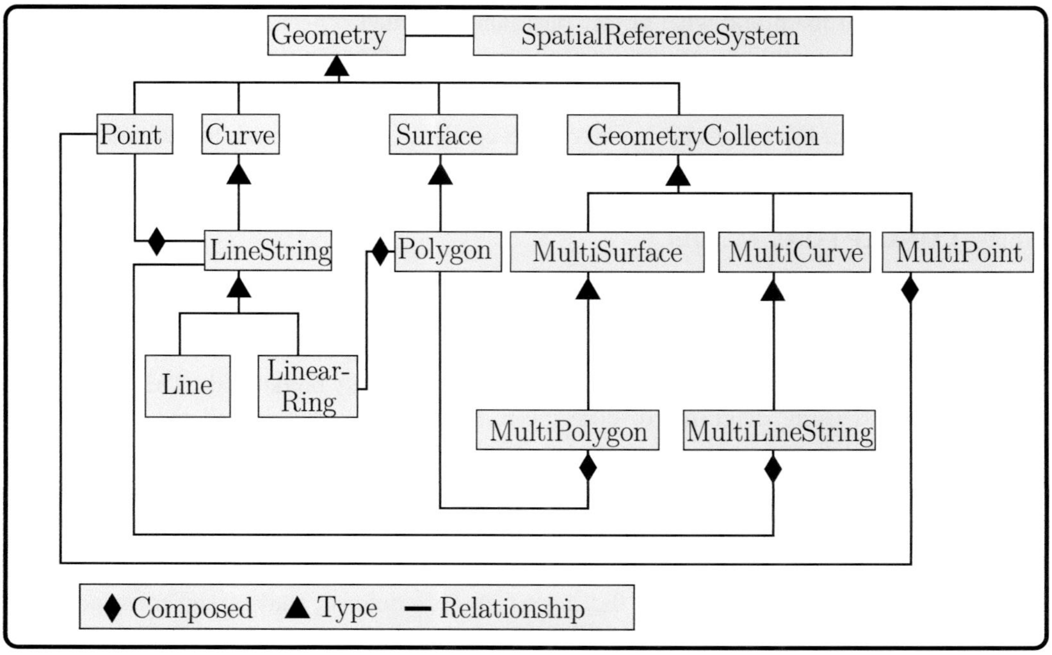

FIGURE 3.11 Geometry class hierarchy.

- "Overlaps", was there overlapping geometry?
- "Relate", was there intersection between the inside, boundary, or outside of the geometries?

Some methods are used in spatial analysis of geometries (Longley et al., 2001):

- "Distance", does it return the shortest distance between two points in two geometries;
- "Buffer", returns a geometry with all points whose distance from the geometry was less than or equal to a user-defined distance;
- "ConvexHull", returns a geometry that represents the convex hull of a geometry. A convex hull is the smallest polygon that can wrap around another geometry without any concave areas;
- "Intersection", returns a geometry that contains only the points common to both input geometries;
- "Union", returns a geometry that contains all points in both input geometries;
- "Difference", returns a geometry containing the points that are different between the two geometries;
- "SymDifference", returns a geometry containing points that occurred in both input geometries.

3.8 GEOGRAPHIC DATABASE DESIGN

In geocomputation and database management system software, there may be its own data model, object types and relationships that can be used. In the DBMS you can define and implement a model for data types and access functions, with a language type. In the DBMS there can also be simple resources, data types and relationships. A geocomputation system can be built with these simple features to create more advanced data types and relationships. Geocomputation software can be combined with domain-specific data models of an object to define classes and relationships of interest. In the project, specific physical fields are created in instance model for filling with object data and specified object classes. For example, a coffee agribusiness planning can use a database for crop irrigation management in summer and drought situations, based on soil moisture monitoring (Figure 3.12).

In database design, key stages are created with concepts of conceptual, logical and physical models, with the use of practical steps (Figure 3.13).

3.8.1 CONCEPTUAL MODEL

In the conceptual model, the modeling of the database from the user's point of view is performed. To do this, the necessary functions and data must be identified and organized into groups to facilitate data management. Then the necessary functions must be specified, and the object and class type relationships must be defined. This process can be accomplished using diagrams to describe a set of object classes and the relationships between them. The geographic representation should be selected with the representation types in discrete objects or continuous fields (Longley et al., 2001).

3.8.2 LOGICAL MODEL

In the logical model, the geographic database types are defined, that is, the correspondence between the types of objects to be studied to specific data types used in geocomputation to create and maintain the database. The data model in geocomputation is usually independent of the actual storage mechanism. In the organization of geographic database

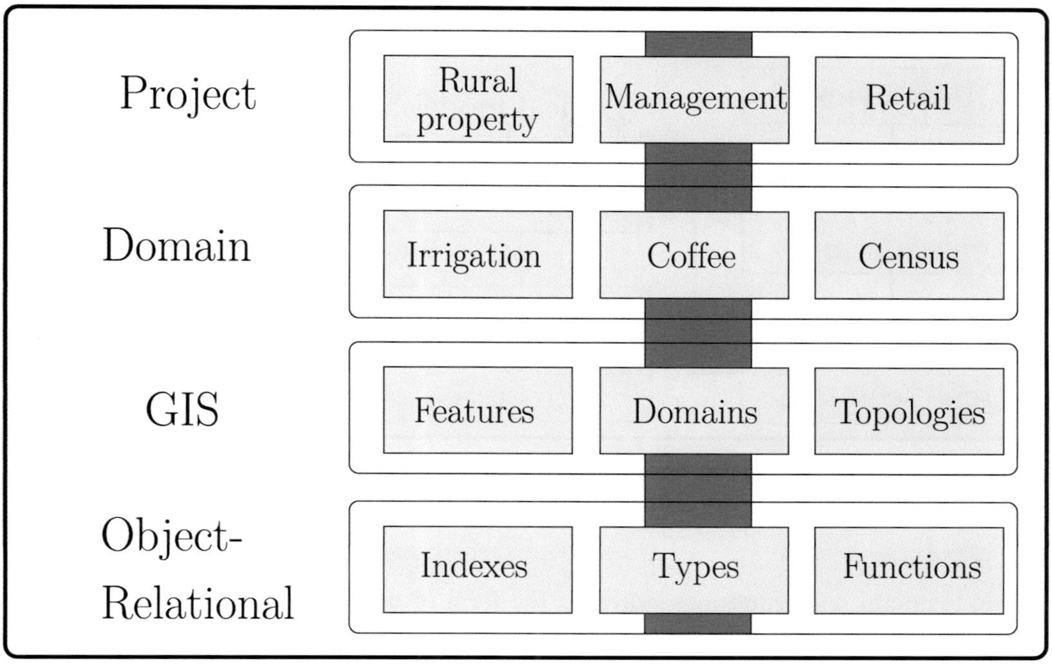

FIGURE 3.12 Levels of geographic database design realization using GIS and geographic database management systems (DBMS) for application in coffee agribusiness.

structure should be defined topological associations, specification of rules, relationships and coordinate system assignment (Longley et al., 2001).

3.8.3 PHYSICAL MODEL

The physical model is the final stage of database modeling, used to define the database schema. In this case, database data values are determined in the physical model with geographically extended database management system data definition language (Longley et al., 2001).

3.9 ORGANIZATION OF GEOGRAPHIC INFORMATION

Major database structures affect how geographic data are discretized and stored in geographic information systems. In many systems methods with feature planes or overlapping layers are used to structure geospatial data intuitively into useful groups (Burrough and McDonnell, 1998).

3.9.1 ORGANIZATION OF DATA IN RASTER-TYPE STRUCTURES

A raster database can be constructed with multiple overlapping layers in the same reference coordinate system in order to optimize access, minimize storage and computational processing (Figure 3.14) (Burrough and McDonnell, 1998).

The structure of file storages is performed from header with information about the projection system of the map, spatial resolution, number of rows and columns, class of the data. Therefore, there is more efficient storage of the coordinate system associated with different types of geometries (Figure 3.15) (Burrough and McDonnell, 1998).

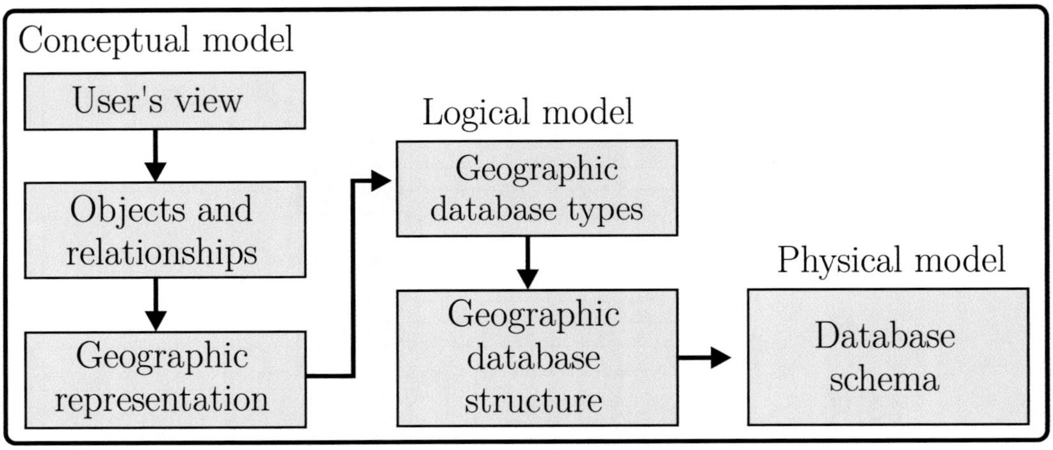

FIGURE 3.13 Key practical steps used in database project development within conceptual, logical, and physical models.

Spatio-temporal data cubes are stored in the form of dense arrays, with space and time being dimensions of the array with applications in different areas, such as (Pebesma and Bivand, 2021b):

- Socioeconomic or demographic data;
- Environment variables monitored at fixed stations;
- Raster maps;
- Time series of satellite images with multiple spectral bands;
- Spatial simulations;
- Climate or weather variables model results.

Data cubes can have two spatial dimensions, as in a raster, and the other dimensions, time (Figure 3.16) (Pebesma and Bivand, 2021b).

Data cubes can be characterized by cubes with higher dimension (hypercubes), such as a five-dimensional cube. In addition to time, there are variations in spectral bandwidth and number of sensors (Figure 3.17) (Pebesma and Bivand, 2021b).

3.9.2 ORGANIZATION OF DATA IN VECTOR-TYPE STRUCTURES

A vector database can be built based on the user's perception of the number of feature planes used to characterize the studied phenomenon with distinction of different classes (Figure 3.18) (Burrough and McDonnell, 1998).

The storage structure of vector type files is also performed by means of header information. In the case of geospatial polygon header of the R package `sp` (Pebesma et al., 2021b), information about attributes of the file with minimum and maximum values, name of data columns, number of variables, coordinate reference system, rectangular extent of the object, number of features and the class of the object are described (Figure 3.19).

In the structure of vector type file stores by the `sf` package (Pebesma et al., 2021a), the header is presented differently than `sp` (Pebesma et al., 2021b). Information about the feature type with the number of features and attribute fields, the geometry type, dimension, rectangular extent of the object, range in Z, coordinate reference system, and the first 10 records in the attribute table are described in the header (Figure 3.20).

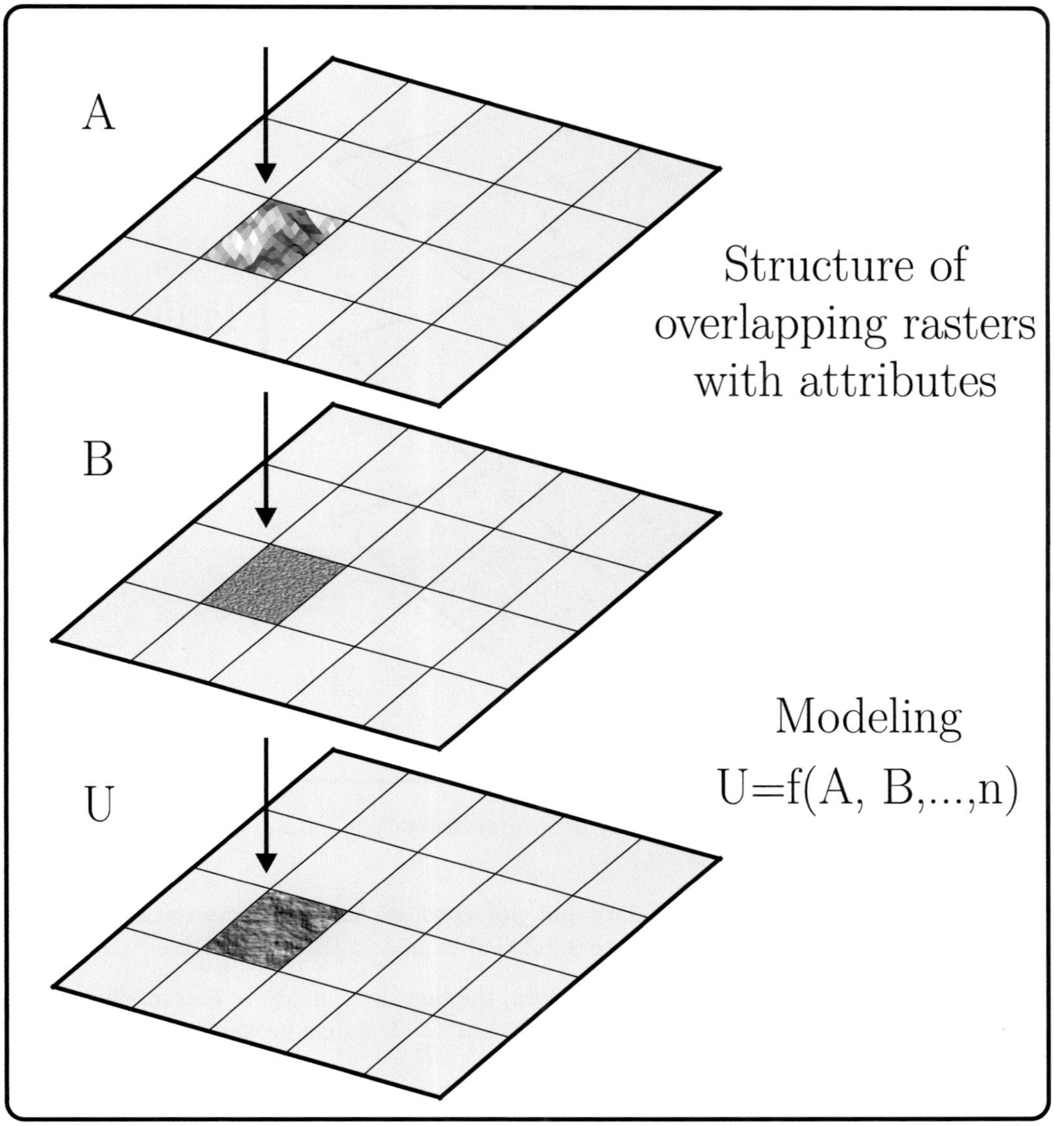

FIGURE 3.14 Simple overlay raster data planes in an overlay layer structure.

```
class       : RasterLayer
dimensions  : 337, 379, 127723  (nrow, ncol, ncell)
resolution  : 0.0008333333, 0.0008333333  (x, y)
extent      : -45.18083, -44.865, -21.40167, -21.12083  (xmin, xmax, ymin, ymax)
crs         : +proj=longlat +datum=WGS84 +no_defs
source      : E:/Aulas/Topografia/Aula11/demLavras_mask.tif
names       : altitude
```

FIGURE 3.15 Example of a digital elevation model header from the R raster package, with the altitude variation of the relief in Lavras, Minas Gerais, Brazil.

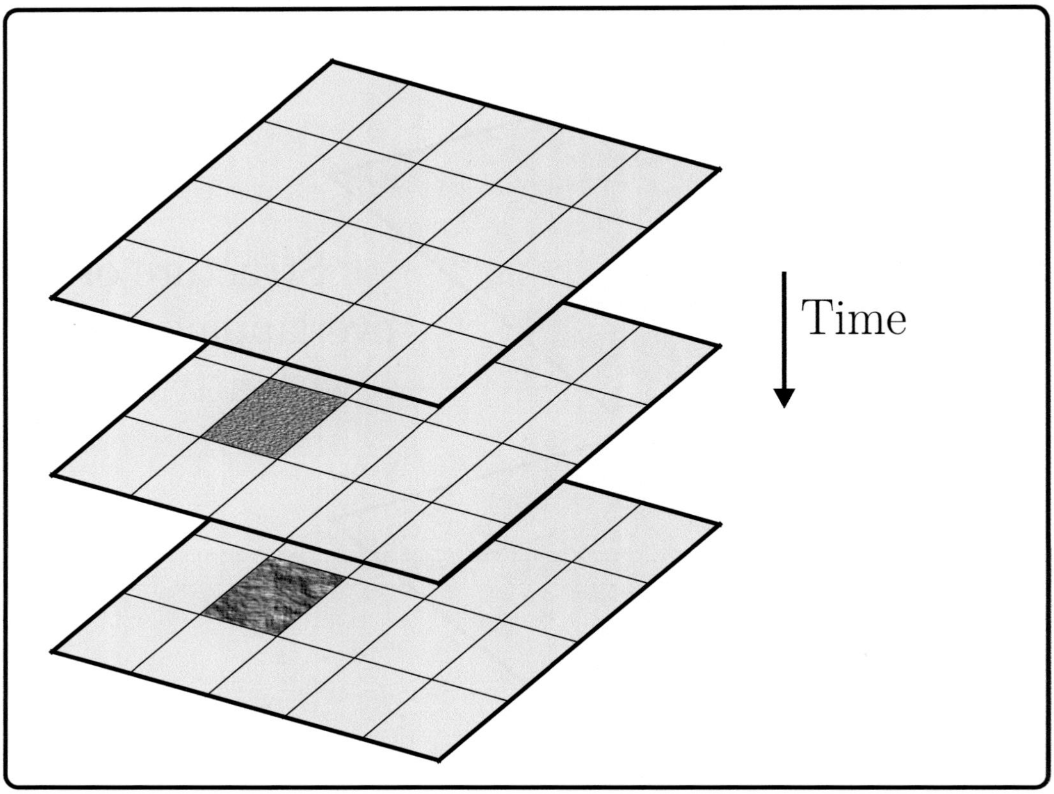

FIGURE 3.16 Spatio-temporal array of several raster-type data.

Arrays of vector data cubes of points and polygons are used to store data of different geographic attributes over time (Figure 3.21) (Pebesma and Bivand, 2021b).

With the structuring and organization of data, the questions, analyses and mappings were carried out efficiently. The main techniques used to structure geographic databases are (Longley et al., 2001):

- Topological creation;
- Indexing.

3.9.3 TOPOLOGY CREATION

Without topology, the database management system cannot distinguish where the sharing of boundaries or the individual separation of two polygons occurs (Figure 3.22) (Burrough and McDonnell, 1998). Topology can be created for vector data sets using either batch or interactive techniques. In batch topology, unstructured vector data is imported from non-topological systems such as CAD, and adaptations must be made to create the topology. In the case of the topology created in an iterative process, the topology creation is performed dynamically at the moment the objects have been added to a geocomputation software database. The connectivity between the objects can be checked, and if there is conformity with any database with pre-established object connectivity rules (Longley et al., 2001).

The normalized and physical database model approaches are used to store and manage the topology. In the normalized model, data is stored in a node and arc structure. This

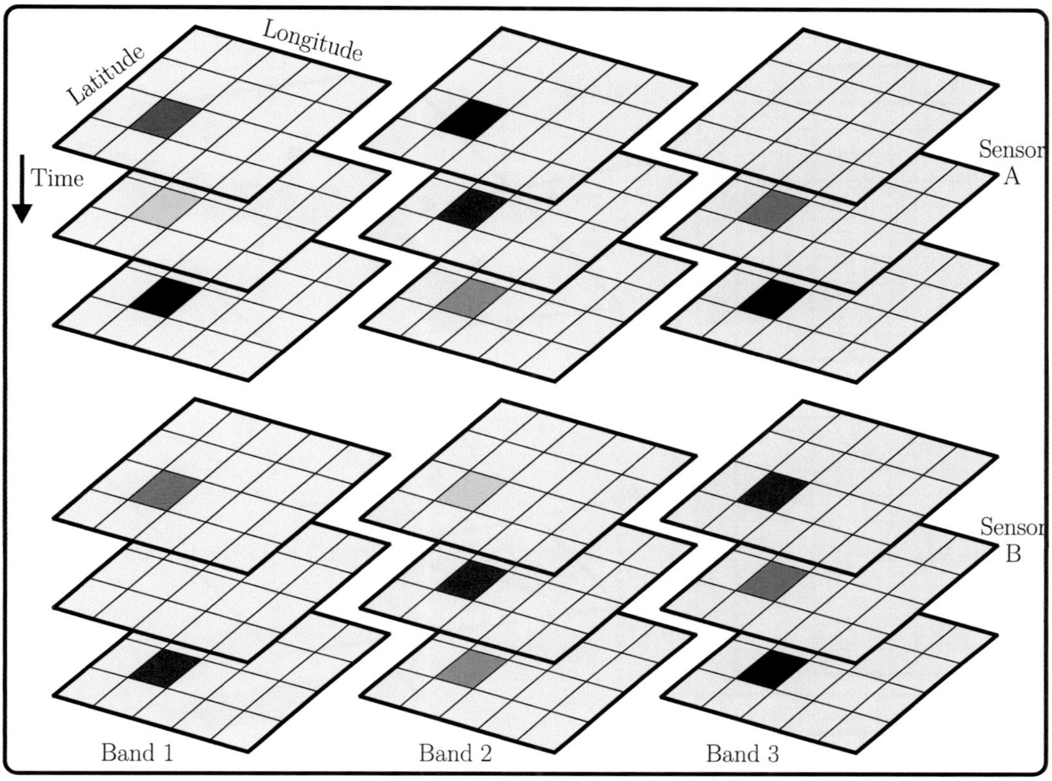

FIGURE 3.17 Array of data hypercubes.

model is called normalized because each object is decomposed into topological primitives for storage in a database, and then the data is reassembled when querying. Polygon objects are assembled at query time by joining tables with line segment geometries and topological primitives with topological relationships. In the normalized approach, the data geometry is stored only once, minimizing database size and avoiding double scanning. However, there are disadvantages of this method in query performance, update and integrity checking (Longley et al., 2001).

In the physical topological model, geometric primitives are not stored in the database, but the entire topology is stored in an object-specific set. Topological relationships are computed on-the-fly whenever required by applications. In the physical model, an external client or middle-tier application server is responsible for validating the topological integrity of the datasets. The topologically correct features are stored in the database with a simpler structure than in the normalized model, with advantages in simplicity and performance over the normalized model. In this case, the geometry of each object is stored in the same column and row value of the table, without the need to store topological primitives and cross-references. This makes it faster to access the object geometry and recalculate the topology outside the database through queries than to retrieve the geometry and topology from the database (Longley et al., 2001).

3.9.4 INDEXING

Considering the size of geographic databases, geographic queries can take a long time. Topological structure can help speed up some types of queries, such as adjacency and

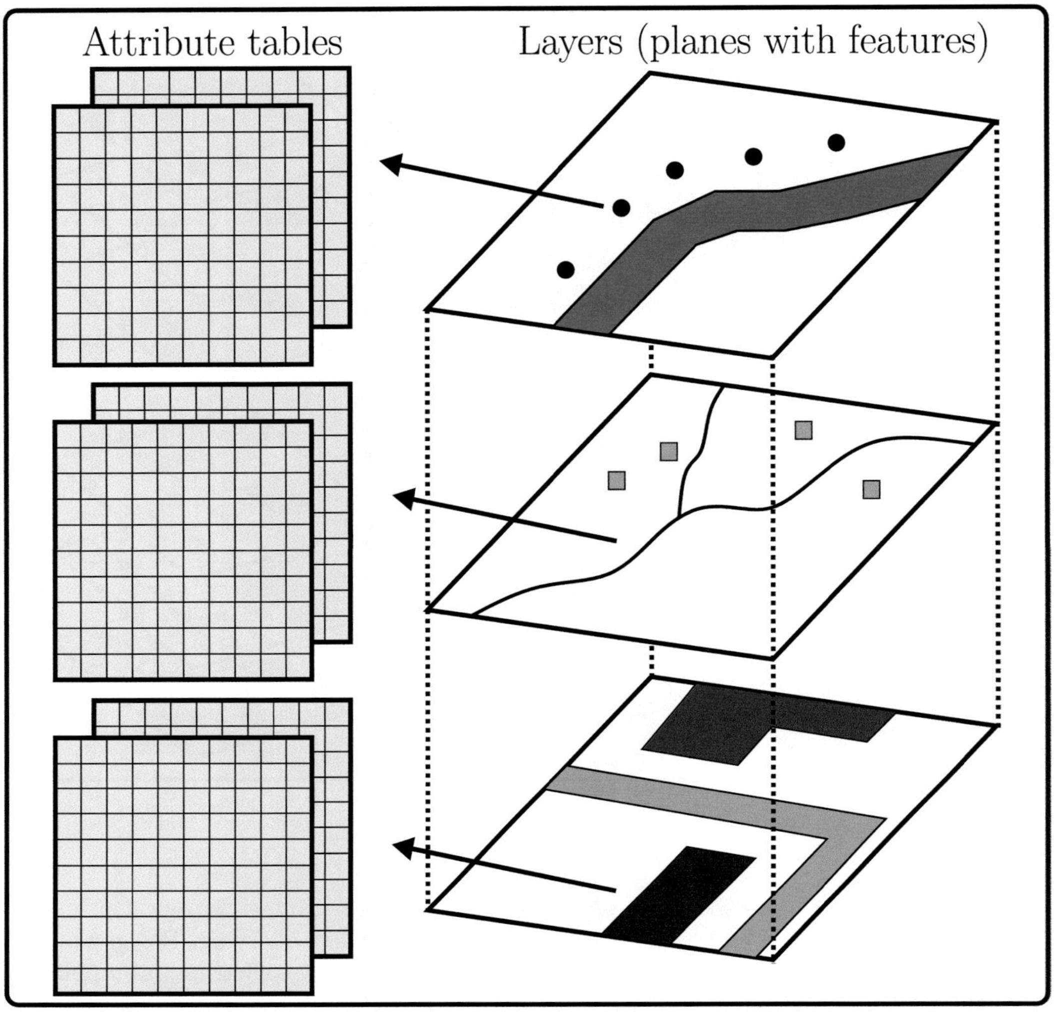

FIGURE 3.18 Plan vector data types in overlay feature layer structure associated with attribute tables.

```
class        : SpatialPolygonsDataFrame
features     : 1214
extent       : -45.17835, -44.9013, -21.39319, -21.1386  (xmin, xmax, ymin, ymax)
crs          : +proj=longlat +datum=WGS84 +no_defs
variables    : 14
names        :  Name, descriptio, timestamp, begin, end, altitudeMo, tessellate,
extrude, visibility, drawOrder, icon, Irrigado, area_ha, NM_MUN
min values   : CafÃ©,           NA,          NA,     NA,  NA,            NA,          -1,
-1,          -1,          NA,    NA,      NÃ£o,    0.087, Lavras
max values   : CafÃ©,           NA,          NA,     NA,  NA,            NA,          -1,
-1,          -1,          NA,    NA,      NÃ£o, 50.0758, Lavras
```

FIGURE 3.19 Example of a geospatial polygon header from the sp package of R, with attributes of coffee crops in Lavras, Minas Gerais, Brazil (note in the attributes that file was not imported with UTF-8 encoding).

```
Simple feature collection with 1214 features and 14 fields
geometry type:  POLYGON
dimension:      XYZ
bbox:           xmin: -45.17835 ymin: -21.39319 xmax: -44.9013 ymax: -
21.1386
z_range:        zmin: 0 zmax: 0
geographic CRS: WGS 84
First 10 features:
   Name descriptio timestamp begin  end altitudeMo tessellate extrude
1  Café       <NA>      <NA>  <NA> <NA>       <NA>         -1      -1
2  Café       <NA>      <NA>  <NA> <NA>       <NA>         -1      -1
3  Café       <NA>      <NA>  <NA> <NA>       <NA>         -1      -1
4  Café       <NA>      <NA>  <NA> <NA>       <NA>         -1      -1
5  Café       <NA>      <NA>  <NA> <NA>       <NA>         -1      -1
6  Café       <NA>      <NA>  <NA> <NA>       <NA>         -1      -1
7  Café       <NA>      <NA>  <NA> <NA>       <NA>         -1      -1
8  Café       <NA>      <NA>  <NA> <NA>       <NA>         -1      -1
9  Café       <NA>      <NA>  <NA> <NA>       <NA>         -1      -1
10 Café       <NA>      <NA>  <NA> <NA>       <NA>         -1      -1
   visibility drawOrder icon Irrigado area_ha NM_MUN
1          -1        NA <NA>      Não  1.6060 Lavras
2          -1        NA <NA>      Não  2.4289 Lavras
3          -1        NA <NA>      Não  6.0715 Lavras
4          -1        NA <NA>      Não  2.4959 Lavras
5          -1        NA <NA>      Não  1.5787 Lavras
6          -1        NA <NA>      Não  1.1142 Lavras
7          -1        NA <NA>      Não  2.6799 Lavras
8          -1        NA <NA>      Não  0.3697 Lavras
9          -1        NA <NA>      Não  2.2400 Lavras
10         -1        NA <NA>      Não  8.0826 Lavras
                        geometry
1  POLYGON Z ((-45.17551 -21.1...
2  POLYGON Z ((-45.16676 -21.1...
3  POLYGON Z ((-45.09658 -21.3...
4  POLYGON Z ((-45.04816 -21.3...
5  POLYGON Z ((-44.92276 -21.2...
6  POLYGON Z ((-44.95828 -21.2...
7  POLYGON Z ((-45.09613 -21.3...
8  POLYGON Z ((-45.1654 -21.33...
9  POLYGON Z ((-45.04659 -21.3...
10 POLYGON Z ((-44.9122 -21.31...
```

FIGURE 3.20 Example of a polygon simple feature header from the R package `sf`, with attributes of coffee crops in Lavras, Minas Gerais, Brazil.

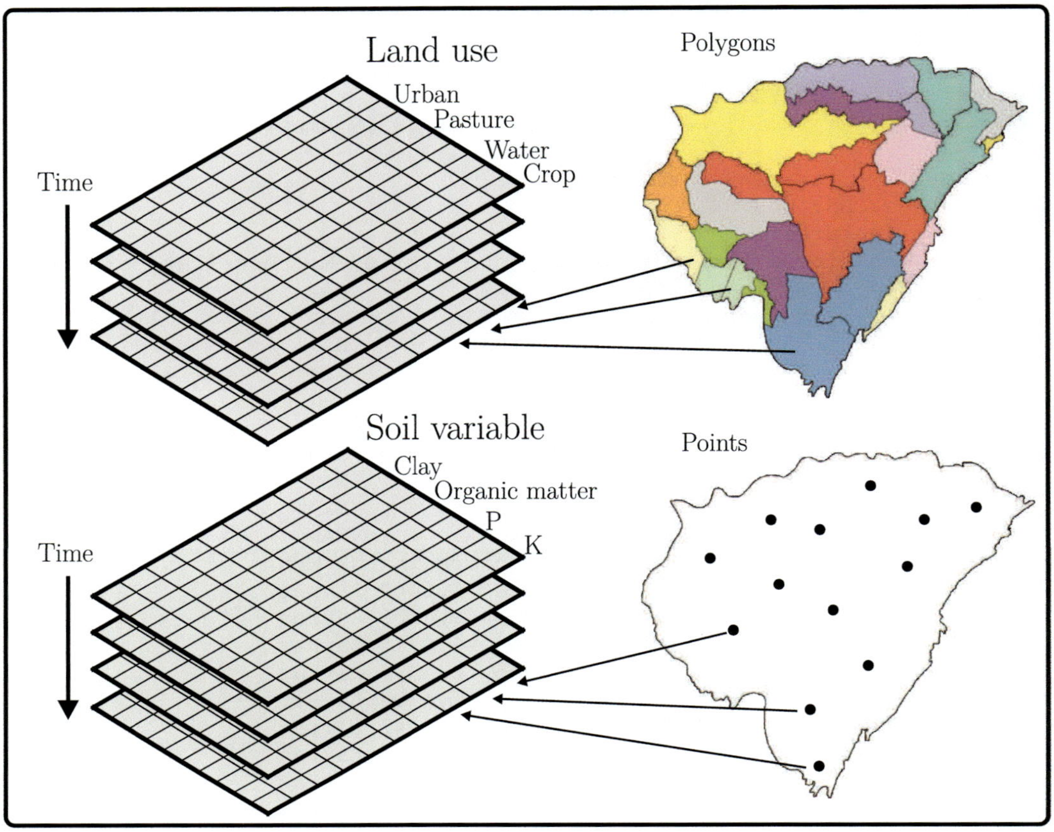

FIGURE 3.21 Spatio-temporal arrays of several vector type data.

network connectivity. A second way to speed up queries was to index a database and use the index to locate data records faster. A database index is defined as an ordered list of data in a table. The number of computational tests to locate a given set of records is reduced by creating an index stored as a column of the table (Longley et al., 2001).

An example of an index can be used to separate the Earth into continental geographic regions and then group countries indexed to those regions. The R package `googleVis` (Gesmann et al., 2021) can be used to demonstrate this type of indexing on a page.

```
library(googleVis)
# Welcome to googleVis version 0.7.3
# package 'googleVis' was built under R version 4.3.3
```

The `gvisOrgChart` function is used to define a flowchart with interactive hierarchical indexing in a tree format of countries in specific regions of the Earth. Interactive mapping is performed in HTML to access the flowchart's trigger options (Figure 3.23).

```
Org <- gvisOrgChart(Regions, options = list(width = 600, height = 250,
                    size = 'large', allowCollapse = TRUE))
plot(Org) # Map flowchart in html
```

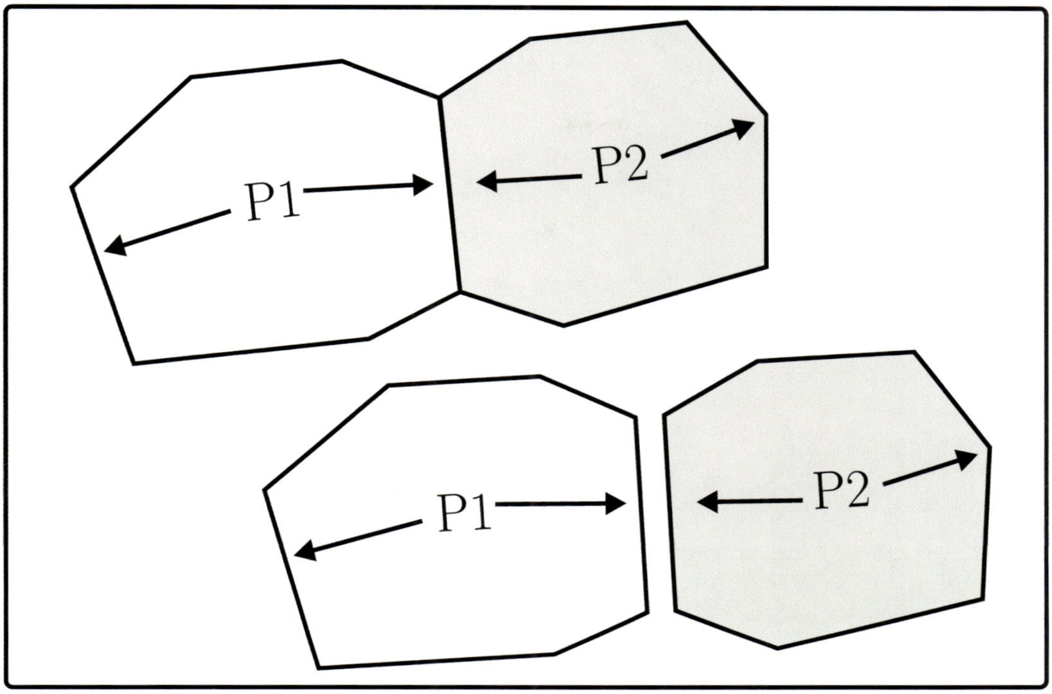

FIGURE 3.22 Without topology, it is not possible to define whether polygon boundaries are shared.

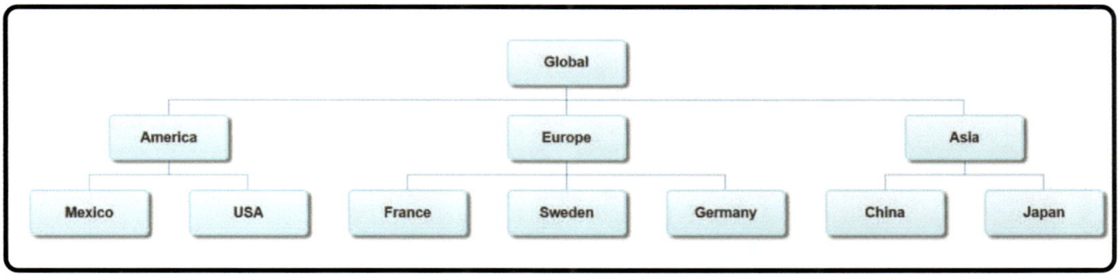

FIGURE 3.23 Hierarchical indexing of countries in specific regions of the Earth by interactive tree map.

The mapping of these regions can also be performed in the form of an interactive tree map by means of the `gvisTreeMap` function (Figure 3.24).

```
Tree <- gvisTreeMap(Regions, "Region", "Parent", "Val", "Fac",
                    options = list(fontSize = 16))
plot(Tree)
```

Another example of tree mapping can be done with the `treemap` package (Tennekes and Ellis, 2023).

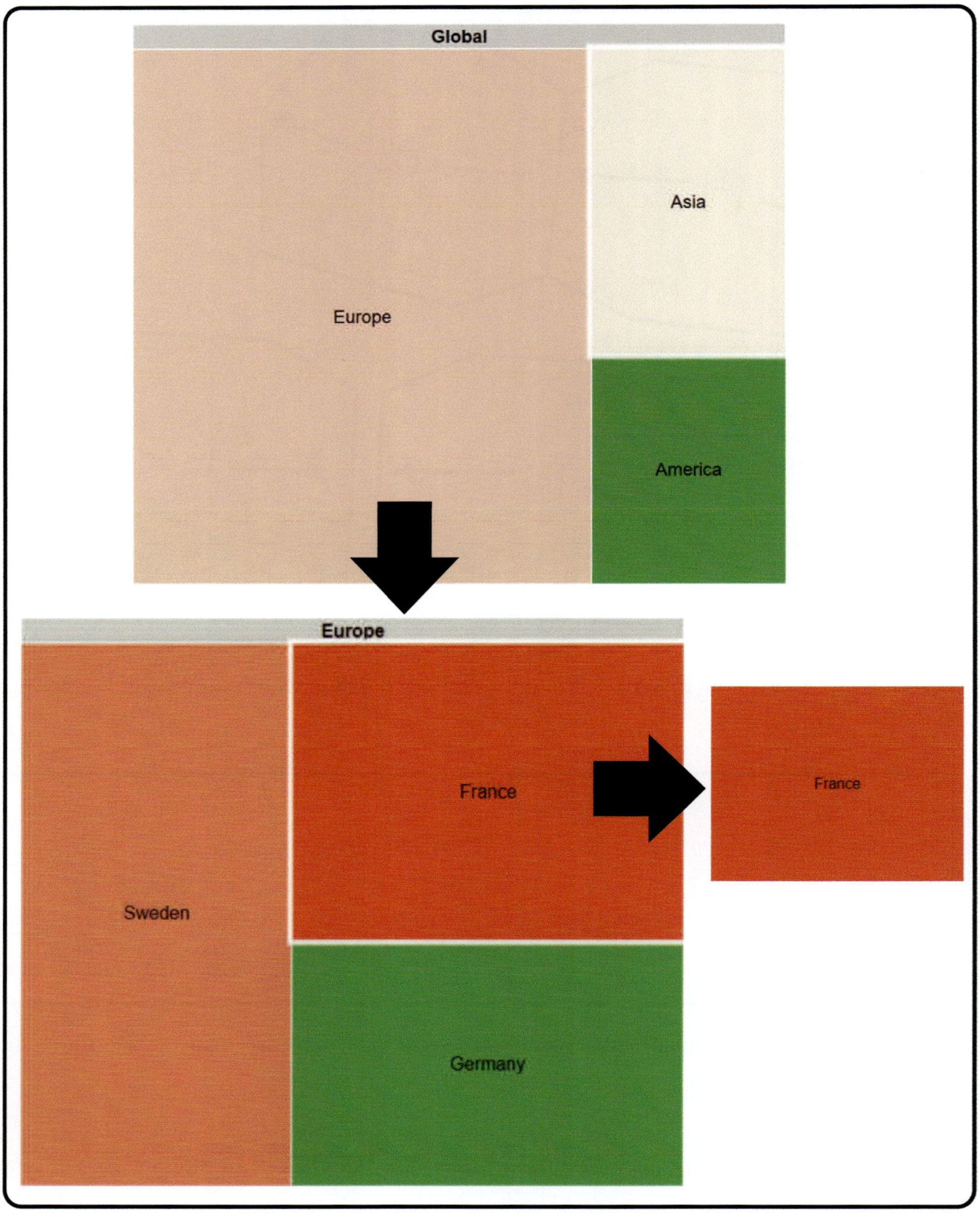

FIGURE 3.24 Interactive indexing of countries in specific regions of the Earth by interactive tree map.

```
library(treemap)
# package 'treemap' was built under R version 4.3.3
```

In the most basic tree map that can be made using the `treemap` function, the input data-set consists of three groups, and in each group an area-mapped value is stored in a `data.frame` (Figure 3.25).

```
# Create database
group <- c("group-1", "group-2", "group-3")
value <- c(12, 18, 26)
data <- data.frame(group, value)
# treemap
treemap(data, index = "group", vSize = "value", type = "index")
```

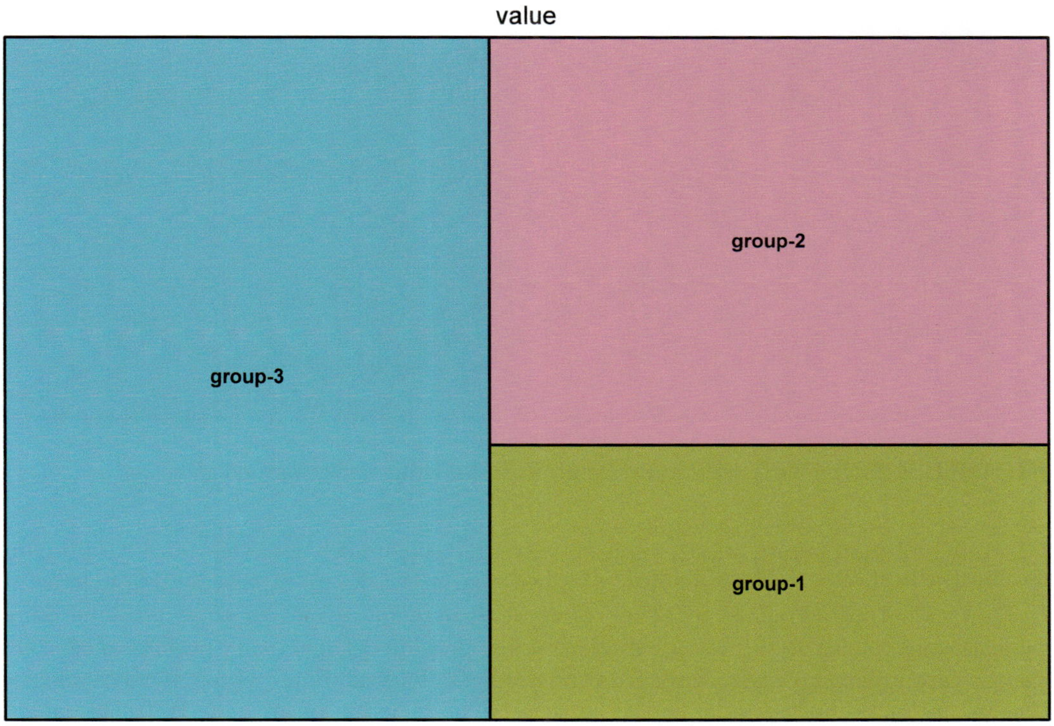

FIGURE 3.25 Simple indexing with three groups per tree map.

The following is an example with a tree map with three groups, each group with several subgroups. In each subgroup, a mapped value in one area is recorded. In the index parameter, it is necessary to specify the levels in the order of importance of group greater than subgroup (Figure 3.26).

```
# Create database
group <- c(rep("group-1", 4), rep("group-2", 2), rep("group-3", 3))
subgroup <- paste("subgroup", c(1, 2, 3, 4, 1, 2, 1, 2, 3), sep = "-")
```

```
value <- c(13, 5, 22, 12, 11, 7, 3, 1, 23)
data <- data.frame(group, subgroup, value)
# treemap
treemap(data, index = c("group", "subgroup"), vSize = "value", type = "index")
```

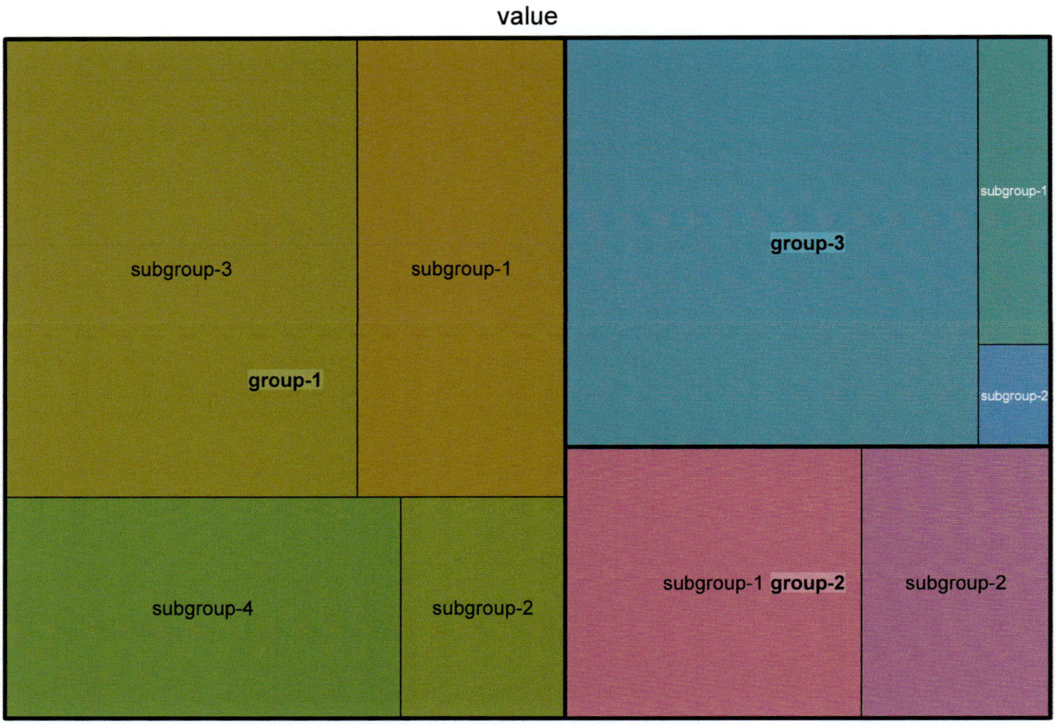

FIGURE 3.26 Indexing with multiple groups and subgroups mapped in a tree.

Creating and maintaining indexes can be very time consuming, and this can be a problem if frequent data updates are required. Indexes can also take up a considerable amount of computational storage space, and the requirements may not be met for large datasets. Some indexes can exploit specific data characteristics to provide optimal query performance, some can be fast updating, and others are robust across different data types. Indexes such as B-trees are one-dimensional and poor at geographic indexing of objects. With this, many indexing techniques have been developed for optimization of geographic data to achieve improvements in simplicity, speed of generation and update. Some indexing methods used in geocomputation are grid indexes, quadtrees and R-trees (Longley et al., 2001).

3.9.4.1 Grid index

An index with grid levels can be used as a regular mesh over a layer of geographic objects. The topmost (coarsest) grid is divided, for example, into four cells of equal sizes. The same process is repeated for the second level index (less coarse grid). A query is used to locate a searched and indexed object in the first list. Then the geometry or attributes of the object are retrieved by overlap, adjacency or contention functions. These are called Primary and Secondary filtering. In Secondary filtering, there is filtering by geometry processing, with higher computational cost. The performance of the index is related to the grid size, the object size and the object density. If the grid size is too large relative to the object size, too many objects are retrieved by the primary filter, and secondary computational processing is required. If the grid size is too small, any large object will be spread over too many grid cells, making it inefficient to perform database queries for the purpose of displaying objects on the computer screen. For data layers with varying object density, multiple levels can be used to optimize computational performance. Three levels of the grid are sufficient for good overall performance of grid indexes. Grid indexes are one of the simplest and most robust indexing methods for creating and updating a wide variety of data types and densities faster (Longley et al., 2001).

3.9.4.2 Quadtree Indexes

Quadtree type indexes are constructed by recursively dividing geographic space into quadrants. In many respects, quadtrees are special types of grid indexes. Quadtrees are data structures used to index and compress geographic database layers. Quadtree types can be classified according to the type of data model indexed, such as points, lines, areas, surfaces, or rasters. The algorithm is used to split the indexed layer by fixed or variable resolution decomposition. In a point quadtree, the space is divided successively into four rectangles based on the location of the points. The root of the tree corresponds to the region as a whole. The rectangular region is divided into four irregular parts based on the X and Y coordinates of the first point. Successive points subdivide each new sub-region into quadrants until all points are indexed. The quadtree index is created to successively divide a layer into quadrants. With layer decomposition, a linear index can be created to reduce the two-dimensional geographic data to a single linear dimension. A standard B-tree can be used to find data quickly (Longley et al., 2001).

3.9.4.3 R-tree Indexes

In R-tree indexes (R-tree) the grouping of objects is accomplished by a rectangular approximation of the location by a minimal bounding rectangle. Groups of point, line, or polygon objects can be indexed based on the minimum delimiter. Objects can be added to the index relative to the rectangular delimiter with minimum expansion to accommodate each new object. If the object has determined expansion of the bounding rectangle, some parameters are predefined, and the rectangle is again split into two parts. In an R-tree with two levels, at the lowest level there are leaf type nodes and at the highest, node with pointers to the bounding rectangle of leaf nodes. The minimal delimiting rectangle is used to reduce the number of objects analyzed in front of a query performed. R-trees are suitable for different geographic objects and for multidimensional data with satisfactory query performance, but with slower update speed than grids and quadtrees (Longley et al., 2001).

3.10 GEODATABASE EDITING AND MAINTENANCE

In the editing process, changes are made at a location in the geographic database in order to add new objects or change existing objects, update the database, and perform maintenance operations. Editing objects can be displayed on the map, with symbology, usually in a reference coordinate system, and there may be background layers such as aerial photographs or satellite imagery to make it easier to see details of interest on the terrain that should be included in the edit. At the end of the edit, the result is an X, Y coordinate layer with optional altitude or attribute Z values. The data entered with the edit must be stored in a system file or database. Access to the database must be managed to ensure continued security and satisfactory quality of use (Longley et al., 2001).

However, despite the simplicity of the polygon editing method, there may be editing problems that must be considered, such as duplicate storage of adjacent polygon lines. Other problems may be related to editing polygons with open edges and internal contours within the polygon geometry itself or gaps between polygon boundaries, probably at the expense of editing errors (Figure 3.27) (Burrough and McDonnell, 1998):

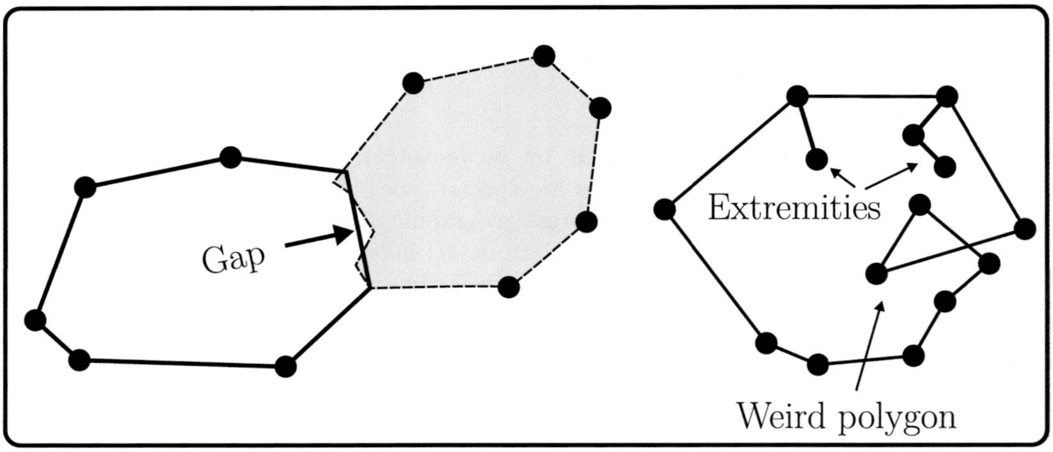

FIGURE 3.27 Topological errors in polygon editing.

3.10.1 MULTIUSER DATABASE EDITING

Multiuser editing in geocomputation is used to allow multiple users to edit the same database at the same time. To do this, solutions are created to enable this task through multiuser networked database transaction models and version control (Longley et al., 2001).

3.10.1.1 Transactions

A group of edits to a database is called a transaction. While some geographic transactions are of short duration, short transactions, many can extend over hours, weeks, and months, called long transactions. In multiuser editing there can be problems updating multiple users, as well as a system crash during editing, resulting in loss of work unless there is a procedure for storing updates in the database (Longley et al., 2001).

3.10.1.2 Version control

Maintenance updates resulting from new service connections or repairs change the database and can affect the use of objects in new versions. The database version control scheme isolates changes made during the editing process. However, depending on the database management system, it may be possible to perform tasks on two different versions of the software installed on the computer in order to solve the problem without leading to major conflicts. Version control is a very useful mechanism to allow multiple users to edit the same geographic database simultaneously. The same version control technology is used to transfer data between systems with synchronization and to integrate changes made to the software (Longley et al., 2001).

3.11 COMPUTATIONAL PRACTICE

Data structures in R are organized by dimensionality (1d, 2d or nd), homogeneity and heterogeneity of content type (Table 3.4) (Wickham, 2019).

TABLE 3.4 Data structures in R according to dimension, homogeneity and heterogeneity.

Dimension	Homogeneous	Heterogeneous
1d	Atomic vector	List
2d	Matrix	Data frame
nd	Array	

The basic types of data are:

- Numerical and integer values;
- Character values;
- Logical values;
- Factors;
- Missing data;
- Temporal variables.

The basic data structure is:

- Matrix and array structure;
- List structure;
- Data frame structure.

Indexing examples are used in relation to data in structure:

- Vector indexing;
- Matrix indexing;
- List indexing;
- Data frame indexing;
- Other types of indexing.

3.11.1 BASIC DATA TYPES

In R there are no 0-dimensional or scalar data structure types. For example, individual numbers are not scalars, but vectors of length 1. Given an object, its structure can be understood by the `str()` function. `str()` is short for structure and provides a compact, readable description of any data structure in R (Wickham, 2019).

3.11.1.1 Vectors

The vector is considered a basic data structure in R subdivided into atomic vectors and lists with the following characteristics in common (Wickham, 2019):

- Type, `typeof()`, what is the object's type?
- Length, `length()`, how many elements are there in the object?
- Attributes, `attributes()`, arbitrary additional metadata.

All elements of an atomic vector must be of the same type, but the elements of a list can be of different types. The `is.vector()` function makes it possible to evaluate whether an object is a vector without attributes. The `is.atomic()` and `is.list()` functions are used to evaluate whether an object is a vector.

3.11.1.1.1 Atomic vectors

There are four types of atomic vectors: logical, integer, double (numeric), and character. Atomic vectors are created with the `c()` function, short for combine. Missing values can be specified with NA, a logical vector of length 1.

3.11.1.1.2 Numeric and integer values

Both the arrow and the equal sign can be used to indicate an assignment action in R. The `show` and `print` functions can be used to display the result 10. The `class` function makes it possible to obtain the class of the created object in this case, which is numerical.

```
a <- 10
a = 10
show(a)
# [1] 10
print(a)
# [1] 10
a
# [1] 10
class(a)
# [1] "numeric"
```

In the numeric class are represented decimal numbers classified as integers through the function `as.integer`.

```
a <- as.integer(10)
a
# [1] 10
class(a)
# [1] "integer"
```

Alternatively, you can use the letter `L` in front of the number to classify it as an integer type.

```
a <- 10L
class(a)
# [1] "integer"
```

The `combine`(c) function is used to create a vector of several numbers with decimal places.

```
b <- c(0.26, 1.3, 5.7, 6.5)
b
# [1] 0.26 1.30 5.70 6.50
```

A colon (:) is used to create data streams in ascending or descending order.

```
d <- 1:10
d
# [1]  1  2  3  4  5  6  7  8  9 10
```

```
10:1
# [1] 10  9  8  7  6  5  4  3  2  1
```

Data sequences by defined intervals can be created with the `seq` function and the `from`, `to` and `by` parameters.

```
e <- seq(from = 1, to = 10, by = 2)
e
# [1] 1 3 5 7 9
seq(from = 10, to = 1, by = -2)
# [1] 10  8  6  4  2
```

The `rev` function creates an inverted sequence.

```
r <- rev(e)
r
# [1] 9 7 5 3 1
```

The `rep` function creates repeated sequences.

```
rep(10, times=5)
# [1] 10 10 10 10 10
rep(1:10, times=3)
# [1]  1  2  3  4  5  6  7  8  9 10  1  2  3  4  5  6  7  8  9 10  1  2  3  4
# 5  6  7  8  9 10
rep(1:10, each=3)
# [1]  1  1  1  2  2  2  3  3  3  4  4  4  5  5  5  6  6  6  7  7  7  8  8  8
# 9  9  9 10 10 10
```

3.11.1.2 Attribute values

All objects can have additional arbitrary attributes used to store metadata about the object. Attributes can be thought of as a list of names. Attributes can be accessed individually with `attr()` or all at once (as a list) with `attributes()` (Wickham, 2019).

A character variable can represent letters, codes, or words.

```
x <- 'Lavras'
y <- 'MG'
class(x)
# [1] "character"
x
# [1] "Lavras"
countries <- c('Brasil', 'Brasil', 'Argentina', 'Paraguai', 'Bolívia')
class(countries)
# [1] "character"
countries
# [1] "Brasil"    "Brasil"    "Argentina" "Paraguai"  "Bolívia"
```

```
length(countries)
# [1] 5
nchar(countries)
# [1] 6 6 9 8 7
```

Constant variables can be created using the example of the word `letters` in lowercase. With the `toupper` function, it is possible to convert them to uppercase.

```
z <- letters
z
# [1] "a" "b" "c" "d" "e" "f" "g" "h" "i" "j" "k" "l" "m" "n" "o" "p" "q" "r"
# "s" "t" "u" "v" "w" "x" "y" "z"
up <- toupper(z)
up
# [1] "A" "B" "C" "D" "E" "F" "G" "H" "I" "J" "K" "L" "M" "N" "O" "P" "Q" "R"
# "S" "T" "U" "V" "W" "X" "Y" "Z"
```

Variables can be concatenated using the `paste` function.

```
Farm <- "Cafua"
Location <- "Ijaci"
paste(Farm, "is in", Location)
# [1] "Cafua is in Ijaci"
```

Manipulation to generate substrings of variables can be performed with the `substr` function.

```
substr('Lavras, Minas Gerais', 1, 6)
# [1] "Lavras"
substr('Lavras, Minas Gerais', 7, 10)
# [1] ", Mi"
```

Variable substitution can be performed using the `gsub` function.

```
gsub('L', '!!', 'Lavras, Minas Gerais')
# [1] "!!avras, Minas Gerais"
gsub('Lavras', 'Ijaci', 'Lavras, Minas Gerais')
# [1] "Ijaci, Minas Gerais"
```

The `grep` function can be used to find elements with a specific pattern.

```
d <- c('a10', 'a11', 'a12', 'b10', 'b11', 'b12')
i <- grep('b', d)
i
# [1] 4 5 6
d[i]
# [1] "b10" "b11" "b12"
grep('2$', d)
```

```
# [1] 3 6
grep('^b', d)
# [1] 4 5 6
```

3.11.1.3 Logical values

Logical or Boolean values can be registered with the terms `TRUE` and `FALSE`. Queries can be performed to obtain logical responses.

```
z <- FALSE
z
# [1] FALSE
class(z)
# [1] "logical"
z <- c(TRUE, TRUE, FALSE)
z
# [1]  TRUE  TRUE FALSE
x <- 10
x > 3
# [1] TRUE
x == 3
# [1] FALSE
x <= 5
# [1] FALSE
```

A true value is equal to "one" and a false value is equal to "zero".

```
y <- TRUE
y + 1
# [1] 2
as.logical(0)
# [1] FALSE
as.logical(1)
# [1] TRUE
as.logical(5.5)
# [1] TRUE
```

3.11.1.4 Factors

Factors are an array of predefined values used to store categorical data with attribute levels. Factors are defined by tiered nominal categories.

```
f1 <- as.factor(countries)
f1
# [1] Brasil    Brasil    Argentina Paraguai  Bolívia
# Levels: Argentina Bolívia Brasil Paraguai
f2 <- c(5:10, 5:10, 5:10)
f2
```

```
# [1]  5  6  7  8  9 10  5  6  7  8  9 10  5  6  7  8  9 10
f2 <- as.factor(f2)
f2
# [1] 5  6  7  8  9  10 5  6  7  8  9  10 5  6  7  8  9  10
# Levels: 5 6 7 8 9 10
```

Factor conversion is not used in numeric expressions. In this case, the factors must be converted to whole numbers (`as.integer`) or to decimal places (`as.numeric`).

```
as.integer(f2)
# [1] 1 2 3 4 5 6 1 2 3 4 5 6 1 2 3 4 5 6
```

The `as.character` function is used to convert factors to characters.

```
fc2 <- as.character(f2)
fc2
# [1] "5"  "6"  "7"  "8"  "9"  "10" "5"  "6"  "7"  "8"  "9"  "10" "5"  "6"  "7"
# "8"  "9"  "10"
```

Missing data is recorded as `NA` .

```
m <- c(1, NA, 3, 2, NA, 7)
m
# [1]  1 NA  3  2 NA  7
```

3.11.1.5 Time variables

There are different calendars and representations with the `as.Date` function.

```
d1 <- as.Date('2019-4-11')
d2 <- as.Date('2019-3-11')
class(d1)
# [1] "Date"
d1 - d2
# Time difference of 31 days
```

Advanced temporal representation classes are obtained with the `as.POSIXlt` and `as.POSIXct` functions.

```
as.POSIXlt(d1)
# [1] "2019-04-11 UTC"
as.POSIXct(d1)
# [1] "2019-04-11 UTC"
```

3.11.2 BASIC DATA STRUCTURE

A vector is a one-dimensional structure previously used to create vectors of numeric values, integers, characters, factors, and Boolean values. Additional multidimensional data structures can store basic data via matrix, table and list.

3.11.2.1 Matrix and array structure

A matrix without data can be created with the `matrix` function.

```
matrix(ncol = 3, nrow = 4)
#      [,1] [,2] [,3]
# [1,]   NA   NA   NA
# [2,]   NA   NA   NA
# [3,]   NA   NA   NA
# [4,]   NA   NA   NA
```

A matrix with data from 1 to 12 can be created with data sorted by column or by row, setting the argument `byrow = TRUE`.

```
matrix(1:12, ncol = 3, nrow = 4)
#     [,1] [,2] [,3]
# [1,]    1    5    9
# [2,]    2    6   10
# [3,]    3    7   11
# [4,]    4    8   12
matrix(1:12, ncol = 3, nrow = 4, byrow = TRUE)
#      [,1] [,2] [,3]
# [1,]    1    2    3
# [2,]    4    5    6
# [3,]    7    8    9
# [4,]   10   11   12
```

The matrix transposition is performed with the `t` function.

```
m <- matrix(1:12, ncol = 3, nrow = 4)
m
#      [,1] [,2] [,3]
# [1,]    1    5    9
# [2,]    2    6   10
# [3,]    3    7   11
# [4,]    4    8   12
t(m)
#      [,1] [,2] [,3] [,4]
# [1,]    1    2    3    4
# [2,]    5    6    7    8
# [3,]    9   10   11   12
```

A matrix can be created by joining vectors with the `cbind` and `rbind` functions.

```
a <- c(1, 2, 3)
b <- 4:6
m1 <- cbind(a, b)
```

```
m1
#      a b
# [1,] 1 4
# [2,] 2 5
# [3,] 3 6
m2 <- rbind(a, b)
m2
#   [,1] [,2] [,3]
# a    1    2    3
# b    4    5    6
m3 <- cbind(b, b, a)
m <- cbind(m1, m3)
m
#      a b b b a
# [1,] 1 4 4 4 1
# [2,] 2 5 5 5 2
# [3,] 3 6 6 6 3
```

You can obtain information about matrix characteristics, such as the number of rows and columns, using the `nrow` and `ncol` functions, respectively.

```
nrow(m)
# [1] 3
ncol(m)
# [1] 5
```

The dimension of the matrix is obtained with the `dim` function.

```
dim(m)
# [1] 3 5
```

The number of cells in the matrix is obtained with the `length` function.

```
length(m)
# [1] 15
```

You can obtain the titles of columns containing variables using the `colnames` function.

```
colnames(m)
# [1] "a" "b" "b" "b" "a"
```

The column name can be reconfigured using the `colnames` function associated with an atomic vector.

```
colnames(m) <- c('ID', 'X', 'Y', 'v1', 'v2')
m
#     ID X Y v1 v2
```

```
# [1,]  1 4 4  4  1
# [2,]  2 5 5  5  2
# [3,]  3 6 6  6  3
```

The name of the lines can be reconfigured with the `paste0` function.

```
rownames(m) <- paste0('row_', 1:nrow(m))
m
#        ID X Y v1 v2
# row_1   1 4 4  4  1
# row_2   2 5 5  5  2
# row_3   3 6 6  6  3
```

The creation of matrix with characters and numerical values can be performed with the `vchar` and `vnumb` functions, respectively.

```
cbind(vchar = c('a', 'b'), vnumb = 1:2)
#      vchar vnumb
# [1,] "a"   "1"
# [2,] "b"   "2"
```

While the matrix is a two-dimensional structure, the array can be multidimensional.

```
# Two scalar arguments to specify rows and columns
a <- matrix(1:12)
a
#       [,1]
# [1,]     1
# [2,]     2
# [3,]     3
# [4,]     4
# [5,]     5
# [6,]     6
# [7,]     7
# [8,]     8
# [9,]     9
# [10,]   10
# [11,]   11
# [12,]   12

# One vector argument to describe all dimensions
b <- array(1:12)
b
# [1]  1  2  3  4  5  6  7  8  9 10 11 12
```

Arrays can be performed in different ways using the `array` function..

```
x1 <- array(1:5, c(1, 1, 5))
x1
# , , 1
#      [,1]
# [1,]    1
# , , 2
#      [,1]
# [1,]    2
# , , 3
#      [,1]
# [1,]    3
#, , 4
#      [,1]
#[1,]    4
# , , 5
#      [,1]
# [1,]    5
x2 <- array(1:5, c(1, 5, 1))
x2
# , , 1
#      [,1] [,2] [,3] [,4] [,5]
# [1,]    1    2    3    4    5
x3 <- array(1:5, c(5, 1, 1))
x3
# , , 1
#      [,1]
# [1,]    1
# [2,]    2
# [3,]    3
# [4,]    4
# [5,]    5
```

3.11.2.2 List structure

The list can be used to store data. Each element of a list can contain objects such as vector, matrix, table, other list, and other complex data types.

```
list(1:10)
e <- list(c(2, 10), 'abc')
e
# [[1]]
# [1]  2 10
# [[2]]
# [1] "abc"
names(e) <- c('first', 'last')
e
# $first
#[1]  2 10
# $last
```

```
# [1] "abc"
m <- matrix(1:9, ncol = 3, nrow = 3)
f <- list(e, m, 'abc')
f
# [[1]]
# [[1]]$first
# [1]  2 10
# [[1]]$last
# [1] "abc"
# [[2]]
#      [,1] [,2] [,3]
# [1,]    1    4    7
# [2,]    2    5    8
# [3,]    3    6    9
# [[3]]
# [1] "abc"
```

3.11.2.3 Data frame structure

In the data frame, the structure is rectangular like the matrix, but it can contain columns with different types of data.

```
ID <- as.integer(1:4)
name <- c('Lavras', 'Búzios', 'Ilhéus', 'Luminárias')
state <- as.factor(c('MG','RJ','BA','MG'))
score <- c(10.2, 9, 13.5, 18)
d <- data.frame(ID, name, state, score, stringsAsFactors = FALSE)
d
#   ID       name state score
# 1  1     Lavras    MG  10.2
# 2  2     Búzios    RJ   9.0
# 3  3     Ilhéus    BA  13.5
# 4  4 Luminárias    MG  18.0
class(d)
# [1] "data.frame"
length(d)
# [1] 4
is.list(d)
# [1] TRUE
names(d)
# [1] "ID"    "name"  "state" "score"
nrow(d)
# [1] 4
dim(d)
# [1] 4 4
colnames(d)
# [1] "ID"    "name"  "state" "score"
```

3.11.3 INDEXING

Database values can be accessed and replaced in various ways in different data structures. The most common indexing approach used is by slicing. In R, square brackets `[ ]` are used for indexing, while parentheses `( )` are used to perform a function.

3.11.3.1 Vector indexing

Some vector elements obtained by indexing are exemplified.

```
v <- 1:10
v
# [1]  1  2  3  4  5  6  7  8  9 10
```

Get the first element of a vector.

```
v[1]
# [1] 1
```

Get the second element of a vector.

```
v[2]
# [1] 2
```

Get the second and third element of a vector.

```
v[2:3]
# [1] 2 3
# or
v[c(2, 3)]
# [1] 2 3
```

A more advanced example, returns all elements except the second.

```
v[c(1,3:10)]
# [1]  1  3  4  5  6  7  8  9 10
# or
v[-2]
# [1]  1  3  4  5  6  7  8  9 10
```

An index can also change values.

```
v[1] <- 11
v
# [1] 11  2  3  4  5  6  7  8  9 10
v[3:10] <- -99
v
# [1]  11   2 -99 -99 -99 -99 -99 -99 -99 -99
```

A single number can be assigned to the first three elements of a vector `b` so that the number is used three times. Then two numbers in a sequence of 3 to 6 are considered, so both numbers are used twice.

```
b[1:3] <- 2
b
# [1]  2  2  2  4  5  6  7  8  9 10 11 12
b[3:6] <- c(10, 20)
b
# [1]  2  2 10 20 10 20  7  8  9 10 11 12
```

3.11.3.2 Matrix indexing

Consider the matrix `m`.

```
m <- matrix(1:9, nrow = 3, ncol = 3, byrow = TRUE)
colnames(m) <- c('a', 'b', 'c')
m
#      a b c
# [1,] 1 2 3
# [2,] 4 5 6
# [3,] 7 8 9
```

Like vectors, matrix values can be accessed through indexing. There are different ways to do this, but it is generally easier to use two numbers in a double index, the first for row numbers and the second for column numbers.

```
# one value
m[2, 2]
# b
# 5
# another one
m[1, 3]
# c
# 3
```

You can also get multiple values at once.

```
m[1:2, 1:2] # 2 columns and rows
#      a b
# [1,] 1 2
# [2,] 4 5
m[2,] # entire row
# a b c
# 4 5 6
m[,2] # entire column
# [1] 2 5 8
```

You can also use column names for subsettings.

```
m[,'b'] # single column
# [1] 2 5 8
m[,c('a', 'c')] # two columns
#      a c
# [1,] 1 3
# [2,] 4 6
# [3,] 7 9
```

Instead of indexing with two numbers, you can also use a single number. You can think of this as a "mobile number". Cells are numbered in columns (i.e., first column rows, then second column, etc.). Therefore,

```
m[2,2]
# b
# 5
m[5]
# [1] 5
```

Notice that,

```
m[,2]
# [1] 2 5 8
```

returns a vector. This is because a single-column matrix can be simplified to a vector. In this case, the matrix structure is "discarded". This is not always desirable, and to prevent this from happening, you can use the `drop = FALSE` argument.

```
m[,2, drop = FALSE]
#      b
# [1,] 2
# [2,] 5
# [3,] 8
```

Defining the values of an array is similar to defining a vector, except that you now need to deal with two dimensions.

```
m[1,1] <- 5 # one value
m
#      a b c
# [1,] 5 2 3
# [2,] 4 5 6
# [3,] 7 8 9
m[3,] <- 10 # a row
m
#       a  b  c
# [1,]  5  2  3
# [2,]  4  5  6
# [3,] 10 10 10
```

```
m[,2:3] <- 3:1 # two columns, with recycling
m
#        a b c
# [1,]   5 3 3
# [2,]   4 2 2
# [3,]  10 1 1
```

The `diag` function sets the values on the diagonal.

```
diag(m)
# [1] 5 2 1
diag(m) <- 0
m
#        a b c
# [1,]   0 3 3
# [2,]   4 0 2
# [3,]  10 1 0
```

3.11.3.3 List indexing

Indexing lists can be a bit confusing, as you can refer to list elements or data elements in one of the list elements.

First create a list `e` of objects.

```
m <- matrix(1:9, nrow = 3, ncol = 3, byrow = TRUE)
colnames(m) <- c('a', 'b', 'c')
e <- list(list(1:3), c('a', 'b', 'c', 'd'), m)
```

Data access within a list element can be done by combining double and single square brackets. Using the double square brackets, the list structure is eliminated. Note the difference that double brackets make. The operation [2] returns a list (of length 1), but [[2]] returns what is inside that element of the list (an array in this case).

```
e[2]
# [[1]]
# [1] "a" "b" "c" "d"
e[[2]]
# [1] "a" "b" "c" "d"
```

List elements can have names.

```
names(e) <- c('zzz', 'xyz', 'abc')
```

Elements can be extracted by their names, like an index or using the `$` operator.

```
e$xyz
# [1] "a" "b" "c" "d"
e[['xyz']]
# [1] "a" "b" "c" "d"
```

Note that `$` can also be used with data frame objects, but not arrays.

3.11.3.4 Data frame indexing

Indexing a data frame can usually be done as arrays and lists. First create a data frame from the `m` array.

```
d <- data.frame(m)
class(d)
# [1] "data.frame"
```

You can extract a column-by-column number.

```
d[,2]
# [1] 2 5 8
```

The following is an alternative way to address the column number in a data frame.

```
d[2]
#   b
# 1 2
# 2 5
# 3 8
```

Note that although `[2]` is the second element in an array, it refers to the second column in a data frame. This is because a data frame is a special type of list and not a special type of array.

```
d[,'b']
# [1] 2 5 8
```

With a data structure, you can also obtain data values using the `$` operator or double brackets, returning the registered vectors.

```
d$b
# [1] 2 5 8
# or
d[['b']]
# [1] 2 5 8
```

The complexity of the data structure is not eliminated if only a bracket or the argument `drop = FALSE` is used.

```
d['b']
#   b
# 1 2
# 2 5
# 3 8
# or
d[,'b', drop = FALSE]
#   b
# 1 2
# 2 5
# 3 8
```

R functions require specific data, such as an array or data frame, and report an error if they get something different. A common situation is that you think you are providing the right type of data, such as a data frame, but you are actually providing a vector because the structure has been broken.

3.11.3.5 Other types of indexing

It is often necessary to perform queries to find data, such as checking which indices of elements in a matrix have values greater than 15. This can be done with the `which` function or bracket operator.

```
x <- 10:20
i <- which(x > 15)
x
# [1] 10 11 12 13 14 15 16 17 18 19 20
i
# [1]  7  8  9 10 11
x[i]
# [1] 16 17 18 19 20
```

It is also possible to use a logical vector for indexing, and values for which the index is `TRUE` are returned.

```
x <- 10:20
b <- x > 15
x
# [1] 10 11 12 13 14 15 16 17 18 19 20
b
# [1] FALSE FALSE FALSE FALSE FALSE FALSE  TRUE  TRUE  TRUE  TRUE  TRUE
x[b]
# [1] 16 17 18 19 20
```

A useful operator allows you to ask whether a set of values is present in a vector with the term `%in%`.

```
x <- 10:20
j <- c(7, 9, 11, 13)
```

```
j %in% x
# [1] FALSE FALSE  TRUE  TRUE
which(j %in% x)
# [1] 3 4
```

The `match` function is similar and useful for matching.

```
match(j, x)
# [1] NA NA  2  4
```

This result means that the third value in `j` is equal to the second value in `x`, and that the fourth value in `j` is equal to the fourth value in `x`. The match is asymmetric, meaning that the match `(j, x)` is not equal to the match `(x, j)`.

```
match(x, j)
# [1] NA  3 NA  4 NA NA NA NA NA NA NA
```

3.11.4 EVALUATE RASTER DATA ARRAYS

Landsat-7 raster imaging data arrays from the municipality of Olinda, Brazil, in six spectral bands, are evaluated and mapped by the `stars` (Pebesma et al., 2021c) package from R. To enable the necessary functions to manipulate the raster file array, the `dplyr` (Wickham et al., 2021a) and `stars` (Pebesma et al., 2021c) packages are installed and then enabled with the commands below.

```
suppressPackageStartupMessages(library(dplyr))
library(stars)
```

Spatial data cubes are arrays with one or more dimensions. Raster data cubes have at least two spatial dimensions that form the raster tessela or set of pot locations. To open an example raster data array, we used the `system.file` function.

```
tif <- system.file("tif/L7_ETMs.tif", package = "stars")
```

Image array mapping is performed after reading the .tif file with the `read_stars` function, followed by slicing with the `slice` function (Figure 3.28).

```
read_stars(tif) |> slice(index = 1:6, along = "band") |> plot(axes = TRUE)
```

Spatio-temporal arrays are stored in objects of class `stars`. The methods developed for the `stars` class are obtained with the `methods` function.

```
methods(class = "stars")
# [1] $<-              %in%              [                 [[<-          ...
# [7] aggregate        aperm             as.data.frame     as.POSIXct    ...
# [13] coerce           contour           cut               dim           ...
```

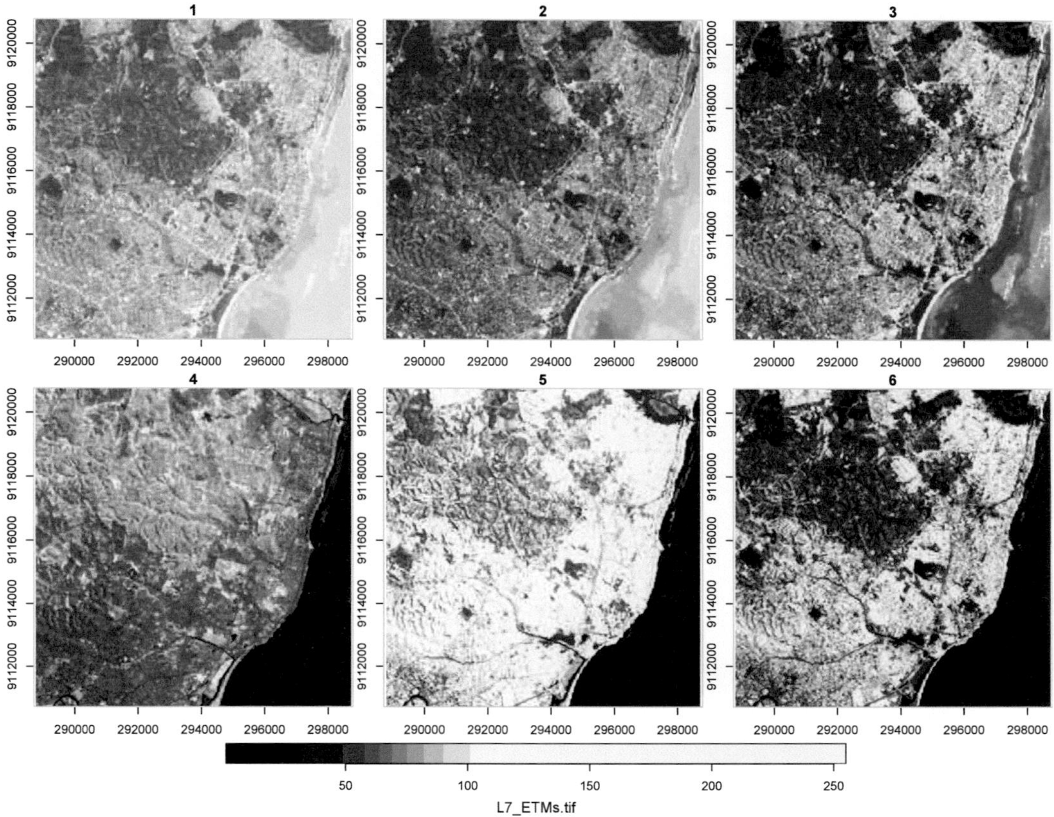

FIGURE 3.28 Landsat-7 image of Olinda as an example of a raster data cube obtained by the stars package.

```
# [19] droplevels           filter              hist                image           ...
# [25] Math                 merge               mutate              Ops             ...
# [31] print                pull                rast                rename          ...
# [37] show                 slice               slotsFromS3         split           ...
# [43] st_as_sf             st_as_sfc           st_as_stars         st_bbox         ...
# [49] st_crs               st_crs<-            st_dimensions       st_dimensions<- ...
# [55] st_geometry          st_interpolate_aw st_intersects         st_join         ...
# [61] st_redimension       st_sample           st_set_bbox         st_transform    ...
# [67] write_stars
```

The dimensions X, Y, and bands of the raster are displayed in a header.

```
x <- read_stars(tif)
x
# stars object with 3 dimensions and 1 attribute
# attribute(s):
#              Min. 1st Qu. Median     Mean 3rd Qu. Max.
# L7_ETMs.tif     1      54     69 68.91242      86  255
# dimension(s):
```

```
#        from  to  offset delta                     refsys point x/y
# x         1 349  288776  28.5 SIRGAS 2000 / UTM zone 25S FALSE [x]
# y         1 352 9120761 -28.5 SIRGAS 2000 / UTM zone 25S FALSE [y]
# band      1   6      NA    NA                         NA    NA
```

4 Space-time Georeferencing with Environmental Geocomputation and R

4.1 LEARNING QUESTIONS

The learning questions answered by reading the chapter are:

- What are the requirements for an efficient georeferencing system?
- How is the Earth measured and modeled for positioning purposes?
- What are the basic principles of map projections?
- How to select a map projection.
- What are the types of reference systems used in geocomputation?

4.2 LEARNING OUTCOMES

The learning outcomes answered by reading the chapter are:

- Know the requirements for an efficient georeferencing system.
- Know how the Earth is measured and modeled for positioning purposes.
- Understand the basic principles of map projections.
- Check how to select a map projection.
- Understand the types of reference systems used in geocomputation.

4.3 INTRODUCTION

Information about location is a basic requirement in geocomputation. With the use of coordinate reference systems, it is possible to (Longley et al., 2001):

- Perform mapping;
- Associate information about the same location;
- Measure distances and areas.

Several terms have been used to describe the act of assigning locations to geographic objects. The verbs georeference, geographic locate, and geocode are used to report that the data used is georeferenced or geocoded. The main requirements of a georeferencing are (Longley et al., 2001):

- The geographic reference must be unique so that there is only one location associated with a given location without confusion about the referenced location;
- The geographic reference should be shared for broad use;
- The geographic reference must persist over time to avoid update problems;
- There is a spatial resolution associated with the georeference;
- It is based on simple nomenclature or metric measurement reference units;
- Latitude, longitude, altitude, and various types of coordinate system are included;
- There is potential for detailed spatial resolution using precise data;

DOI: 10.1201/9781003724377-4

- You can determine the distance between two points and the area within a polygon of at least three points.

4.4 EARTH COORDINATE SYSTEMS

The basic characteristics of the Earth's coordinate system have been defined by a network of imaginary lines called the graticule. The graticule consists of lines of latitude and longitude that cross the surface of the Earth so that the location of points can be described by a sexagesimal system in degrees, minutes, and seconds. The arrangement of this coordinate system is based on the relative position of the Earth in relation to the Sun, as well as the alignment with celestial objects such as the North Star. In this arrangement, a reference line is established at the position of the Equator for latitude values that should start at the Equator and vary by 90° to the North and South Poles, respectively. The first meridian is set as it passes through England at Greenwich and used as a reference line for measuring longitude with a variation of 180° for east and west, respectively (Slocum et al., 2009).

Accurate determination of latitude and longitude is possible after the invention of the chronometer around 1700 by John Harrison. With this, it is possible to use a mechanism other than the pendulum for measuring time. Another important event is the International Meridian Conference, where the Prime Meridian was established as the origin to divide the Earth into Western and Eastern Hemispheres from Greenwich, and into Northern and Southern Hemispheres from the Equator (Slocum et al., 2009).

4.4.1 BASIC CHARACTERISTICS OF THE EARTH'S GRID

The use of addresses, zip codes, and telephone area codes are some ways to obtain the location of places on Earth. However, in these methods there is no uniformity of usage criteria across countries, and it is not possible to realize coverage of the Earth as a whole. To overcome these limitations, a system of latitude and longitude, known as a graticule, has been developed. Lines of latitude are used to divide the Earth into parallel lines that never intersect, called parallels. Lines of longitude, on the other hand, pass through the North to the South, converging at the poles. The term meridian is applied to any line of longitude that passes through the poles dividing the Earth into two equal parts. The 180° meridian is opposite the 0° meridian and coincides with the international date line. The symbols of (+) and (-) before the value in degrees of latitude ($0°, Equator; +90°N; -90°S$) and longitude ($0°, Greenwich; +180°E; -180°W$) are used to define the position of a point in the Northern (N) Hemisphere with positive, Southern (S), negative, Eastern (E), positive, and Western (W), negative values, respectively (Figure 4.1) (Slocum et al., 2009).

Latitude and longitude are used to locate points on the Earth by knowing the angular distance from the Equator to a given place (the latitude) and the angular distance from the Prime Meridian to the same place. As an example, a point is positioned at 60° North latitude and 45° East longitude. The 60° latitude value (ϕ) is the angle formed between the line passing through the point on the surface to the center of the Earth and then to the Equator. The longitude (λ) is the angle formed between a plane passing through a point and the center of the Earth and then to the plane of the Prime Meridian (Figure 4.2) (Slocum et al., 2009).

The array of parallels and meridians on the Earth's surface determines relationships of distance and direction that are used in different types of projections and maps. The shortest distance between two points is a straight line. On the Earth's curved surface, the shortest distance between two points is an arc of a great circle. The great circle is the result

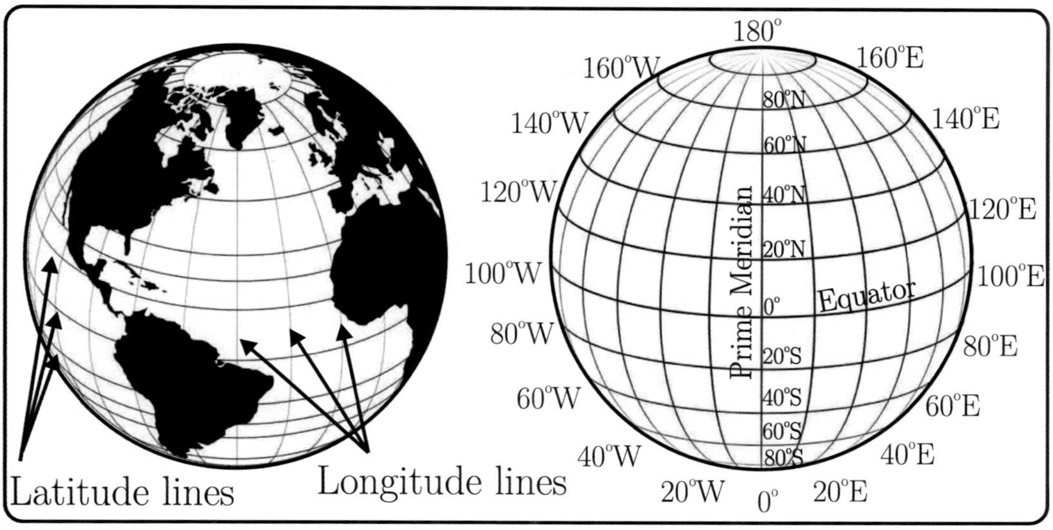

FIGURE 4.1 Latitude and longitude lines combined in the form of a graticule to define positions on the Earth.

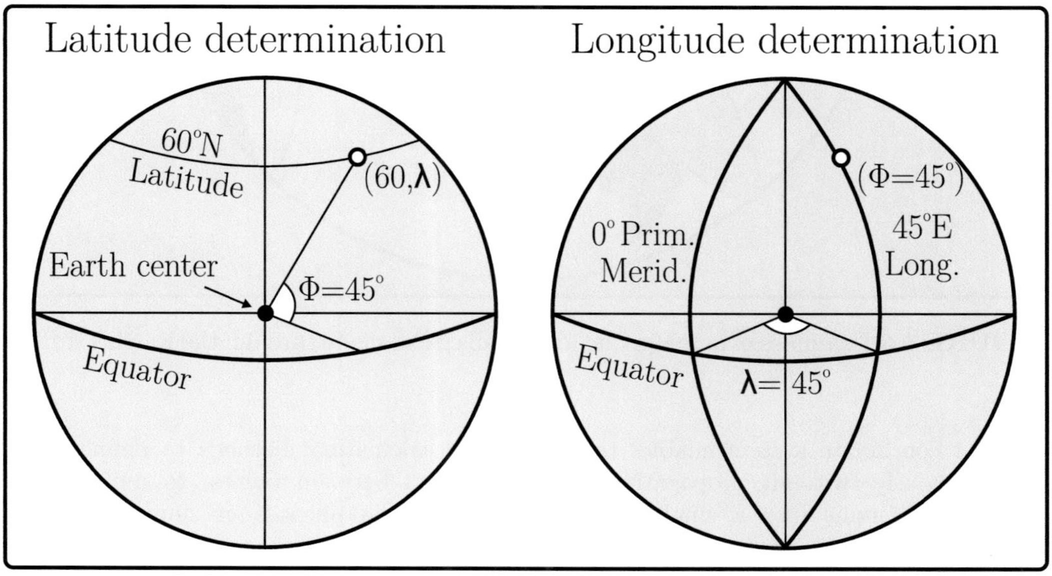

FIGURE 4.2 Determining latitude (left) and longitude (right) angle values on the Earth.

of the intersection of any plane with the center and surface of the Earth. Therefore, all meridians are considered great circles; however, in the case of the parallels, only the Equator is considered a great circle because it passes through the center of the Earth. All other parallels are called small circles. A small circle results from a plane that passes through the curved surface of the Earth without intersecting the Earth's center. Therefore, not all small circles are coincident with parallels (Figure 4.3). The angle formed between each meridian and the great circle is defined as azimuth (or direction). Lines on the Earth's surface with constant direction are called rhumb lines or loxodromes (Slocum et al., 2009).

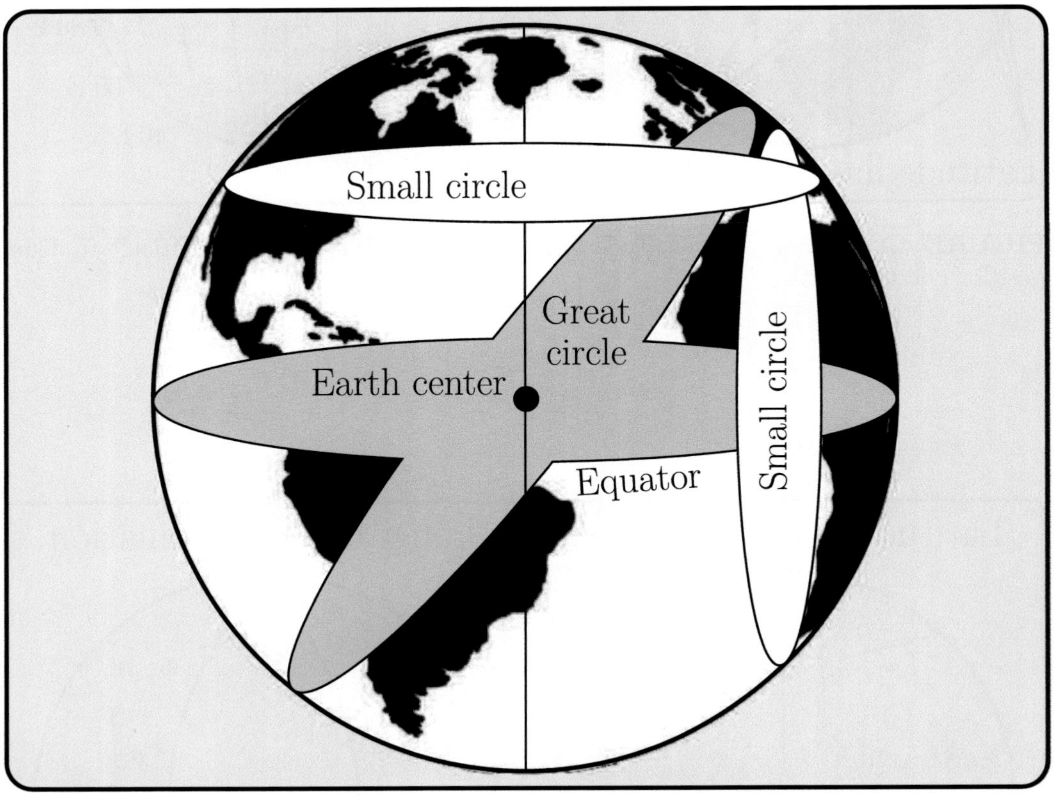

FIGURE 4.3 Examples of large (great) and small circles made through the Earth's surface.

Using a coordinate system enables to standardize calculation methods to define points unambiguously, without propagating graphical errors between points, to unify several individual systems into a single overall system for the purpose of simplifying point identification and management in a project. The main coordinate systems are (Silva and Segantine, 2015; Pebesma and Bivand, 2021a):

- Plane polar coordinate system;
- Cartesian plane coordinate system or rectangular plane system;
- Geocentric spatial Cartesian coordinate system;
- Geodetic geographic coordinate system;
- Time-based coordinate system.

4.4.2 PLANE POLAR COORDINATE SYSTEM

The plane polar coordinate system is defined from an origin point (o) of a surveying survey, a direction (ϕ) related to the reference axis and a distance (r) between the origin and the point at which the coordinates are to be determined. The angle (ϕ) is also called the polar angle and the distance (r) the radius vector. The pair of values (ϕ, r) is called planar polar coordinates (Figure 4.4) (Silva and Segantine, 2015; Pebesma and Bivand, 2021a).

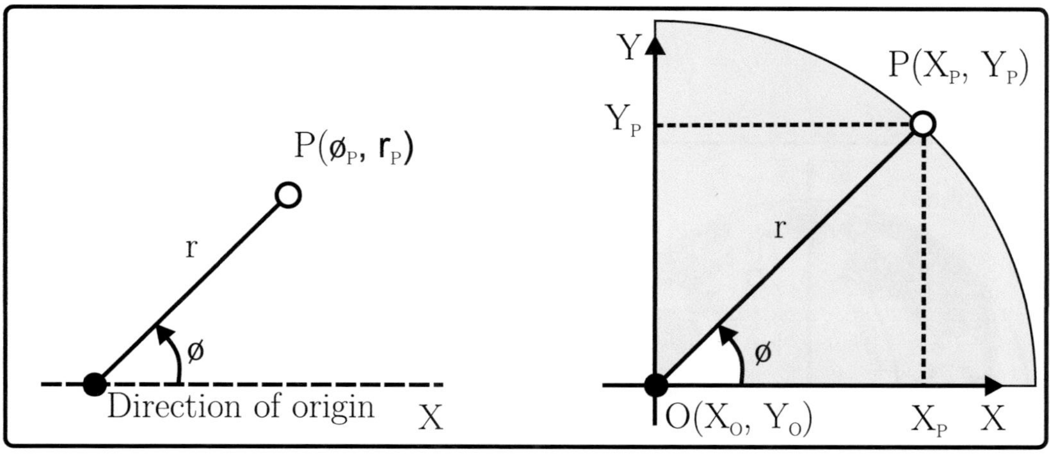

FIGURE 4.4 Plane polar (left) and plane Cartesian (right) coordinate systems.

4.4.3 RECTANGULAR PLANE COORDINATE SYSTEM

The rectangular plane coordinate system is based on the Cartesian plane coordinate system. The system consists of two geometric axes in the same plane and perpendicular to each other forming four quadrants. The intersection of the axes defines the origin of the system. The primary axis, horizontal, is called the abscissa, X, and the secondary axis, vertical, perpendicular to the abscissa axis is called the ordinate, Y. The axes are graduated according to a scale defined for the system. The X and Y axes are positive to the right and positive to the top, respectively. The pair of X, Y values is called the Cartesian plane or rectangular plane coordinates. A ray is defined by two points or by a point and a geometric direction in this system. The direction is indicated by the angle Az (Figure 4) (Silva and Segantine, 2015).

The Euclidean distance d_{ij} between two points p_i and p_j in a Cartesian coordinate system, considering two dimensions is (Pebesma and Bivand, 2021a):

$$d_{ij} = \sqrt{(x_i - x_j)^2 + (y_i - y_j)^2} \tag{4.1}$$

where the distance is referring to a straight line between points i and j.

4.4.4 GEOCENTRIC SPATIAL CARTESIAN COORDINATE SYSTEM

When positioning a point in a Cartesian coordinate system, a third Z-axis is added to the Cartesian plane coordinate system. The Z-axis is added perpendicular to the plane established by the (X, Y)-axes. This system is suitable for the satellite positioning signal with its origin at the center of the Earth, with the (X, Y)-axes at the Equator and the Z-axis coincident with the Earth's mean rotational axis. The X-axis is directed to intercept a meridian adopted as reference, defining a geocentric spatial Cartesian coordinate system (Figure 4.5) (Silva and Segantine, 2015).

The Euclidean distance d_{ij} between two points p_i and p_j in a Cartesian coordinate system, considering three dimensions is (Pebesma and Bivand, 2021a):

$$d_{ij} = \sqrt{(x_i - x_j)^2 + (y_i - y_j)^2 + (z_i - z_j)^2} \tag{4.2}$$

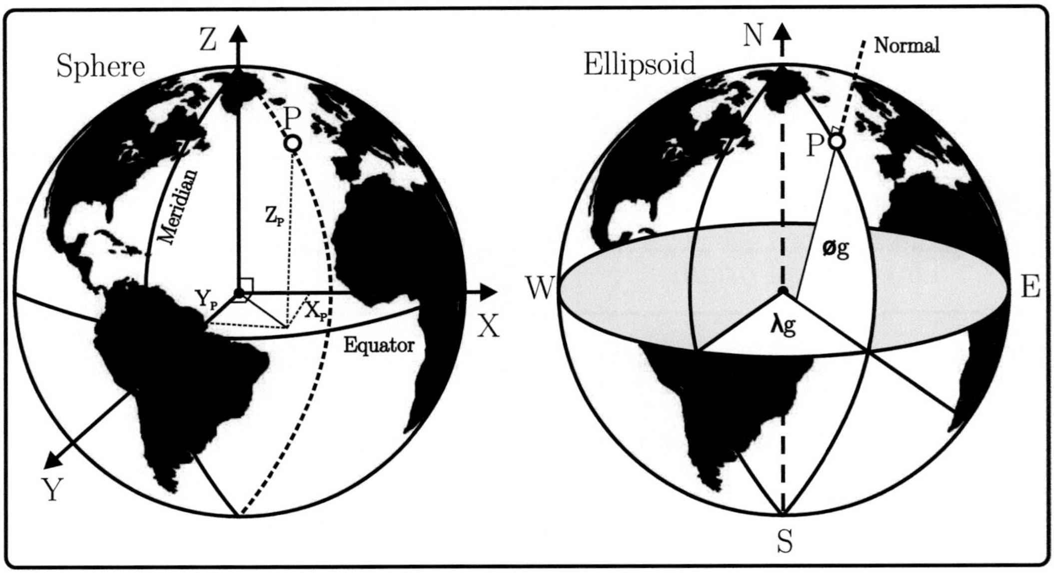

FIGURE 4.5 Geocentric spatial Cartesian (left) and geodetic geographic (right) coordinate systems.

Considering two points on a circle, the arc length (s_{ij}) between the two points $c_1 = (r, \phi_i)$ and $c_2 = (r, \phi_j)$ is (Pebesma and Bivand, 2021a):

$$s_{ij} = r|\phi_i - \phi_j| = r\theta \tag{4.3}$$

where θ is the angle between ϕ_i and ϕ_j in radians. For small values of θ, s_{ij} is approximately dij.

Considering two points, $P_i = (\lambda_i, \phi_i), P_j = (\lambda_j, \phi_j)$ on a sphere with radius r, the great circle distance is the arc length between P_i, P_j on the circle passing through these points with the center equal to that of the sphere is (Pebesma and Bivand, 2021a):

$$s_{ij} = r\theta_{ij} \tag{4.4}$$

where

$$\theta ij = arccos[sen\phi_i sen\phi_j + cos\phi_i cos\phi_j cos(|\lambda_i - \lambda_j|)] \quad (4.5)$$

where the angles between P_i and P_j are in radians.

Spherical geometry can be produced in R with data, spatial analysis functions, orthographic projection, and coordinate transformations from the `s2`, `sf` and `stars` packages with a code chunk (Figure 4.6) (Pebesma and Bivand, 2023).

```
par(mfrow = c(1, 1), mar = c(2, 2, 0.3, 0.3), mgp = c(1.5, 0.6, 0))
library(sf) # Load packages
library(s2) # Load packages
library(stars) # Load packages
g <- as_s2_geography(TRUE) # Earth
co <- s2_data_countries() # Countries
oc <- s2_difference(g, s2_union_agg(co)) # Oceans
b <- s2_buffer_cells(as_s2_geography("POINT(-30 -10)"), 9800000) # Visible half
i <- s2_intersection(b, oc) # Visible ocean
co <- s2_intersection(b, co)
ortho <- st_crs("+proj=ortho +lat_0=-10 +lon_0=-30")
st_transform(st_as_sfc(i), ortho) |> plot(col = "lightblue")
st_transform(st_as_sfc(co), ortho) |> plot(col = "antiquewhite",
                    add = TRUE, border = "grey")
# Sampling randomly from globe
sf_use_s2(FALSE) |> suppressMessages()
if (packageVersion("sf") >= "1.0.16") {
  globe = st_as_stars() |> st_bbox() # st_sample only needs bbox
} else {globe = st_as_stars() |> st_bbox() |> st_as_sfc()}
# Fibonacci
sf_use_s2(FALSE) |> suppressMessages()
st_sample(globe, 1000, type = "Fibonacci", exact = FALSE) |>
  suppressMessages() -> pts
sf_use_s2(TRUE) |> suppressMessages()
pts |> s2_intersection(i) |> st_as_sfc() -> pts
st_transform(pts, ortho) |> plot(add = TRUE, pch = 3, cex = .5)
```

4.4.5 GEODETIC GEOGRAPHIC COORDINATE SYSTEM

In the geodetic geographic coordinate system, the reference surface of the system is a reference ellipsoid. In this case, arcs determined on the reference surface with origin at a meridian and at the Equator plane are called geodetic latitude and longitude, or ϕ_g and λ_g, respectively. Geodetic latitudes are referenced from the Equator from $0°$ *to* $90°$ in the Northern Hemisphere (N), and from $0°$ *to* $-90°$ in the Southern Hemisphere (S). Geodetic longitudes are referenced relative to the Greenwich meridian with a range of $0°$ *a* $360°$ in the East (E) direction, or $0°$ *a* $180°$ in the East direction, and $0°$ *a* $-180°$ in the West (W) direction. Due to the flattened shape of the reference ellipsoid, the normal line that defines the geodetic latitude of point P does not pass through the center of the ellipsoid (Silva and Segantine, 2015).

In addition to the normal line that defined the geodetic latitude (ϕ_g) and longitude (λ_g) at a point on the Earth's surface, geocentric (ϕ_s) latitude and (λ_s) longitude are defined

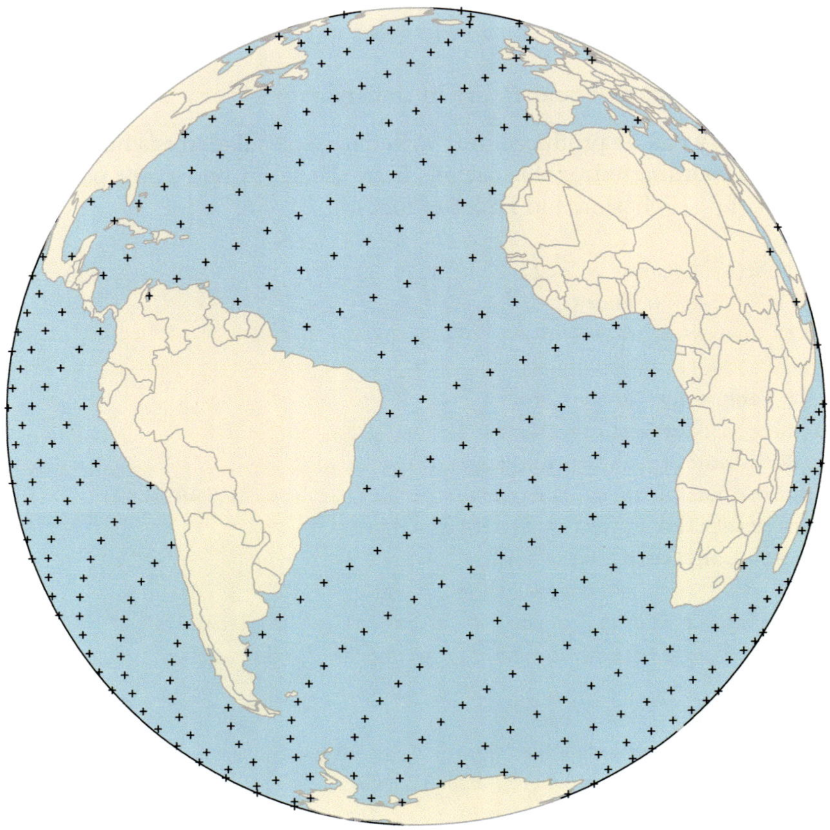

FIGURE 4.6 Orthographic projection in the context of the R programming language used to obtain projections of points from a sphere onto a plane, such as for globe or map visualizations.

by a line passing through the center of the Earth, and astronomical (ϕ_a) latitude and (λ_a) longitude are established by a vertical. Therefore, with geodetic coordinates, it is possible to define a geographical point univocally, without two points having the same coordinate; whereas the astronomical geographical coordinates can vary depending on the variation of gravity on the Earth's surface (Figure 4.7) (Silva and Segantine, 2015).

Details on calculating the distance between two points on the ellipsoid is more complex than on the sphere, so more details can be obtained from Karney (2013).

4.4.6 TIME-BASED COORDINATE SYSTEM

When compared to geographic space, time is linear and one-dimensional. Some objects and processes can vary throughout the day or year, so time can be represented by a tuple with year and day of the year or by day and hour of the day. Time can be represented in cyclical form when the phenomenon studied has a variation of start and end over hours, days, weeks or years (Pebesma and Bivand, 2021a). R's date system uses a standardized date system

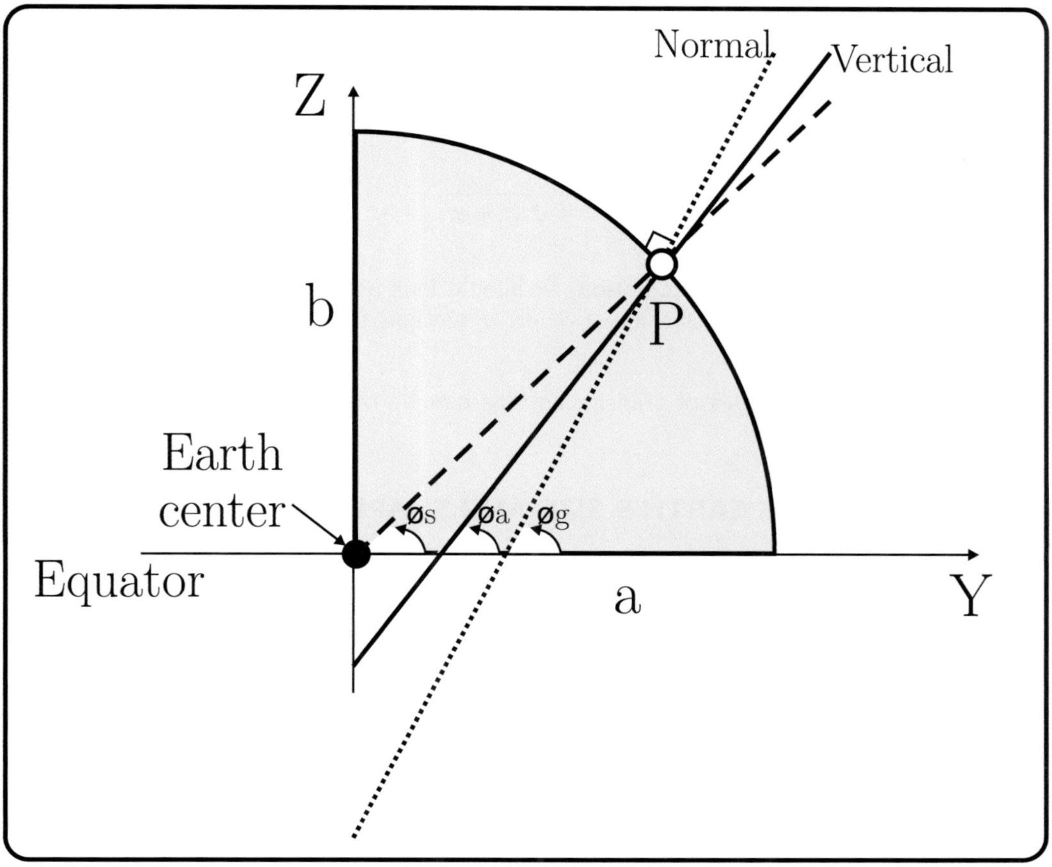

FIGURE 4.7 Representation of geodetic, geocentric and astronomical latitudes related to a point on the surface.

for reading and writing dates (ISO, 2000) and time zone of the universal time coordinate system (UTC) (Pebesma and Bivand, 2021a).

4.4.7 COORDINATE TRANSFORMATION

In many engineering applications, it is necessary to convert the coordinates of points from one system to another or to use different coordinate systems to represent a point. In coordinate conversion, data transformation is performed which consists of a mathematical operation to relate two coordinate systems (Silva and Segantine, 2015).

The direct and inverse conversion between Cartesian and polar coordinates is obtained by (Pebesma and Bivand, 2021a):

$$x = rcos\phi \tag{4.6}$$

$$y = rsen\phi \tag{4.7}$$

$$\phi = arctan(\frac{y}{x}) \tag{4.8}$$

$$r = \sqrt{x^2 + y^2} \tag{4.9}$$

In the case of tangent arc determination, we should look at the corresponding quadrant in the calculation, and the function `atan2` or other method to obtain correct results can be useful in this case.

More details on other types of transformations can be obtained in Silva and Segantine (2015).

4.5 DETERMINING EARTH'S SIZE AND SHAPE

Geodesy is the science used to understand and explain the size and shape of the Earth (Slocum et al., 2009).

4.5.1 DETERMINING EARTH SIZE

Based on historical accounts, the Greek mathematician and scholar Eratosthenes (276-195 BCE) was working in the library of Alexandria in Egypt and was able to determine the circumference of the Earth according to the following equation (Slocum et al., 2009):

$$C = 360° \frac{d}{\phi} \tag{4.10}$$

where d is the distance between two locations on the Earth's surface and, ϕ, the angular latitude difference between the two locations considered.

In the Eratosthenes method, the size of the Earth is estimated by means of a Line (L) intersected by another Line (P_1), creating the angle ϕ. With the projection of L to the center of the Earth, there is intersection of another line (P_2), parallel to (P_1), to form the same angle ϕ according to the properties of geometry of triangles (Figure 4.8).

The cities of Syene (present-day Aswan, Egypt) and Alexandria, in northern Egypt, were the two locations considered in Eratosthenes' work, since they are located on the same meridian. The distance between Alexandria and Syene was estimated based on the number of days of travel between the two cities at 810 km. The angular difference in latitude was determined based on the angular variation of a shadow in Alexandria of $7°12'$. The geometric relationship of triangles was used to determine the angular variation between the two locations relative to the center of the Earth from the angular measurement taken in Alexandria (Slocum et al., 2009).

Carl Sagan explains in detail how Eratosthenes discovered that the Earth is spherical and measured its circumference from a representation of the region on a map of the time when the work was done. By folding the map subjected to front lighting, simulating the curvature of the Earth, it is possible to see the variation of the shadow of two towers represented vertically above the map (Figure 4.9) (Sagan, 2013).

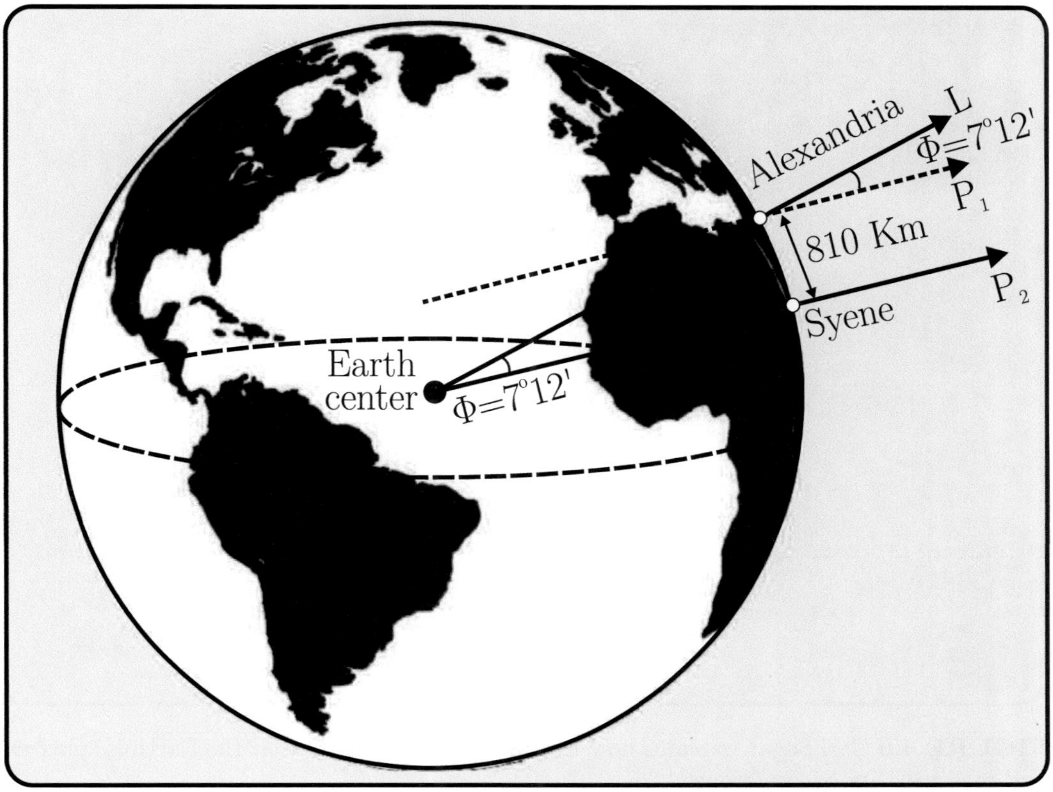

FIGURE 4.8 Method used by Eratosthenes to estimate the size of the Earth.

4.5.2 DETERMINING EARTH SHAPE

The idea that the Earth was a perfect sphere proposed by Pythagoras, Aristotle, and Archimedes predominated for over 1000 years. However, around 1600, the Cassini family conducted a precise triangulation survey in France in which a difference of one degree in latitude was verified by comparing measurements taken in the North and South of the country. From this, it was concluded that the Earth is not a sphere, but a prolate spheroid, flattened in an east-west direction. Astronomer and Physicist Jean Richer (1630-1696) observed periods of motion of a pendulum at different locations on Earth, determining the hypothesis that gravity is weaker in Cayenne, French Guiana, when compared to Paris, France. Another researcher of gravity, Sir Isaac Newton (1642-1727) applied the gravitational concept to the Earth with the hypothesis that with the centrifugal rotation of the Earth, mass is pushed off-center, determining the formation of an oblate spheroid. Newton's hypothesis was later proved by measurements taken at the equator and in a region near Finland. Thus, the Earth is described as an oblate spheroid, characterized by an ellipse with major (a) and minor (b) semi-axis. The degree of deviation of the ellipse from the circle is called the flattening constant (f) (Figure 4.10) (Slocum et al., 2009).

$$f = \frac{(a-b)}{a} \tag{4.11}$$

A two-dimensional ellipsoid can be rotated about its minor semi-axis to determine a 3-D

FIGURE 4.9 Carl Sagan explains how Eratosthenes discovered that the Earth is spherical and measured its circumference.

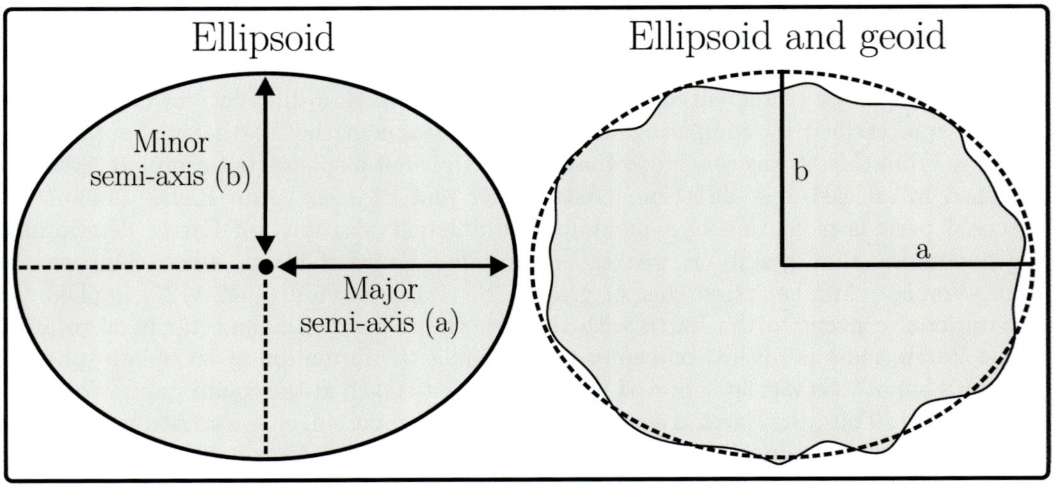

FIGURE 4.10 Parameters defined in a reference ellipsoid used to describe the shape of the Earth.

figure, called the reference ellipsoid, used to describe the size and shape of the Earth (Slocum et al., 2009).

4.5.3 REFERENCE ELLIPSOID

With the correct determination of the Earth's shape, it is possible to verify Sir Isaac Newton's hypothesis that the Earth was a bulging spheroid or ellipsoid with a bulge at the equator as a result of centrifugal force. This concept of the ellipsoid is extended to the reference ellipsoid, a solid body that more accurately defines the shape of the Earth than the hypothesized spherical shape of the Earth. In the reference ellipsoid, the major semi-axis coincides with the equatorial radius, corresponding to the distance from the Earth's center to a point at the Equator, and the minor semi-axis coincides with the polar radius, referring to the distance from the Earth's center to the North Pole (Slocum et al., 2009).

Using the reference ellipsoid, topographic surveys are performed in several countries to create ellipsoids with better fit for specific regions of the Earth and for the whole Earth (Table 4.1) (Slocum et al., 2009).

TABLE 4.1 Some ellipsoids and parameters used in different countries and in the world.

Ellipsoid	Date	a	b	f	Country
Airy	1830	6377563.44	6356256.91	1/299.33	Great Britain
Everest	1830	6377276.35	6356075.42	1/300.80	India
Clarke	1866	6378206.40	6356583.80	1/294.98	North America
International	1924	6378388.00	6356911.95	1/297.00	Selected areas
Krassovsky	1940	6378245.00	6356863.03	1/298.30	Soviet Union
WGS	1972	6378135.00	6356750.50	1/298.26	US Dept. of Defense
GRS	1980	6378137.00	6356752.31	1/298.257	World
WGS	1984	6378137.00	6356752.31	1/298.257	World

Latitude can have different values depending on the surface shape and the type of surface used to represent the Earth. Geodetic latitude is obtained in relation to the center of the reference ellipsoid. In geocentric latitude, the latitude is obtained in relation to the center of the sphere. On the spherical surface, geodetic and geocentric latitudes have the same value when using a spherical model of the Earth (Figure 4.11).

On the sphere, the distance between degrees of latitude is constant and can be calculated as a function of the measurement of the Earth's circumference divided by 360°.

4.5.4 GEOID

With advances in satellite science and technology, it has been found that the Earth is not as smooth as in the mathematical definition of the reference ellipsoid, but roughened as a result of different gravitational forces that define it as a geoid. The geoid is defined as a smooth, undulating surface of the Earth under the effect of gravity. To determine the geoidal surface, it is necessary to specify the nature of the gravitational forces on Earth. The perpendicular line of the geoid does not coincide with the perpendicular line to the reference ellipsoid. It should be noted that gravity varies with the type and density of rocks with minerals that can affect gravity. The geoid should be considered as a reference surface representation model if there is a variation of land flooding by the oceans, and it is possible to adjust for gravitational differences between land and water bodies. The geoid is also used

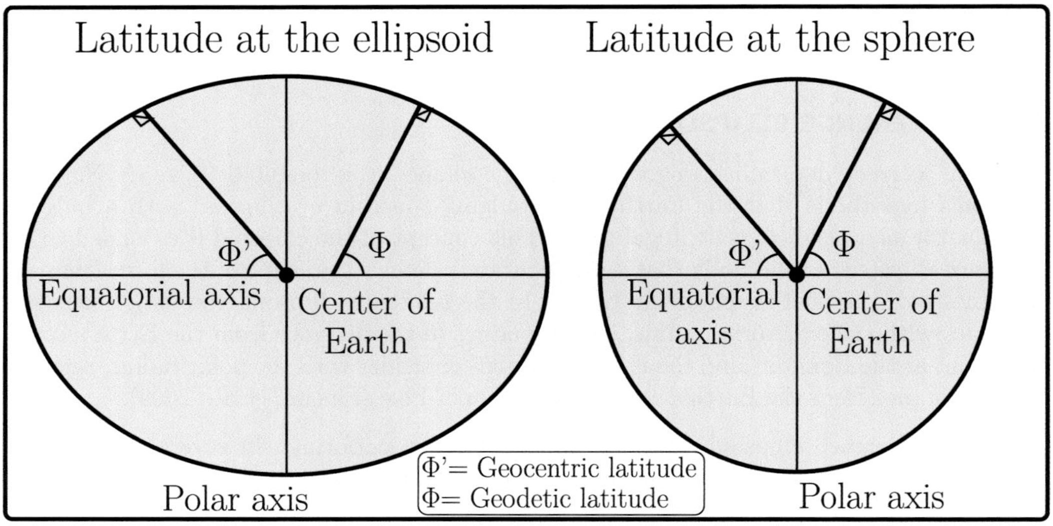

FIGURE 4.11 Comparison between geodetic and geocentric latitudes on a reference ellipsoid and on the sphere.

as a reference in defining a geodetic datum (Figure 4.12) (Slocum et al., 2009; European Space Agency, 2021).

4.5.5 GEODETIC DATUM

The relationship between the geoid and the ellipsoid at a given location is useful for a number of mapping applications. The coincidence between the geoid and the ellipsoid at a point is used as a reference to define a datum. A datum is a quantity used as a reference base for determining other values. A geodetic datum is a model used to describe places, directions, and scale relationships according to the datum's origin on the surface. With respect to geodesy, there are two types of datum (Slocum et al., 2009):

- Horizontal;
- Vertical.

In the horizontal datum, geodetic latitude and longitude locations are specified with respect to a reference ellipsoid where the origin of the latitude and longitude values are coincident with the Equator and Prime Meridian, respectively. The accuracy in determining latitude and longitude from a horizontal datum can be determined based on the reference parameters of semi-major axis (equatorial radius), minor axis and flattening of the ellipsoid. Reference values for determining elevations are obtained from a vertical datum, and the concept of mean sea level must be considered as the base value (Slocum et al., 2009).

The reference ellipsoid, geoid, geodetic datum and geodesy are applied in mapping with a requirement for accuracy in distance, direction and area measurements. However, the selection of an accurate Earth representation model such as the GRS80 reference ellipsoid may present little difference in visual appearance of the map when compared to the spherical model. Thus, the spherical model can be used for thematic applications, as the mathematics involved in map creation is simplified by performing map projections (Slocum et al., 2009).

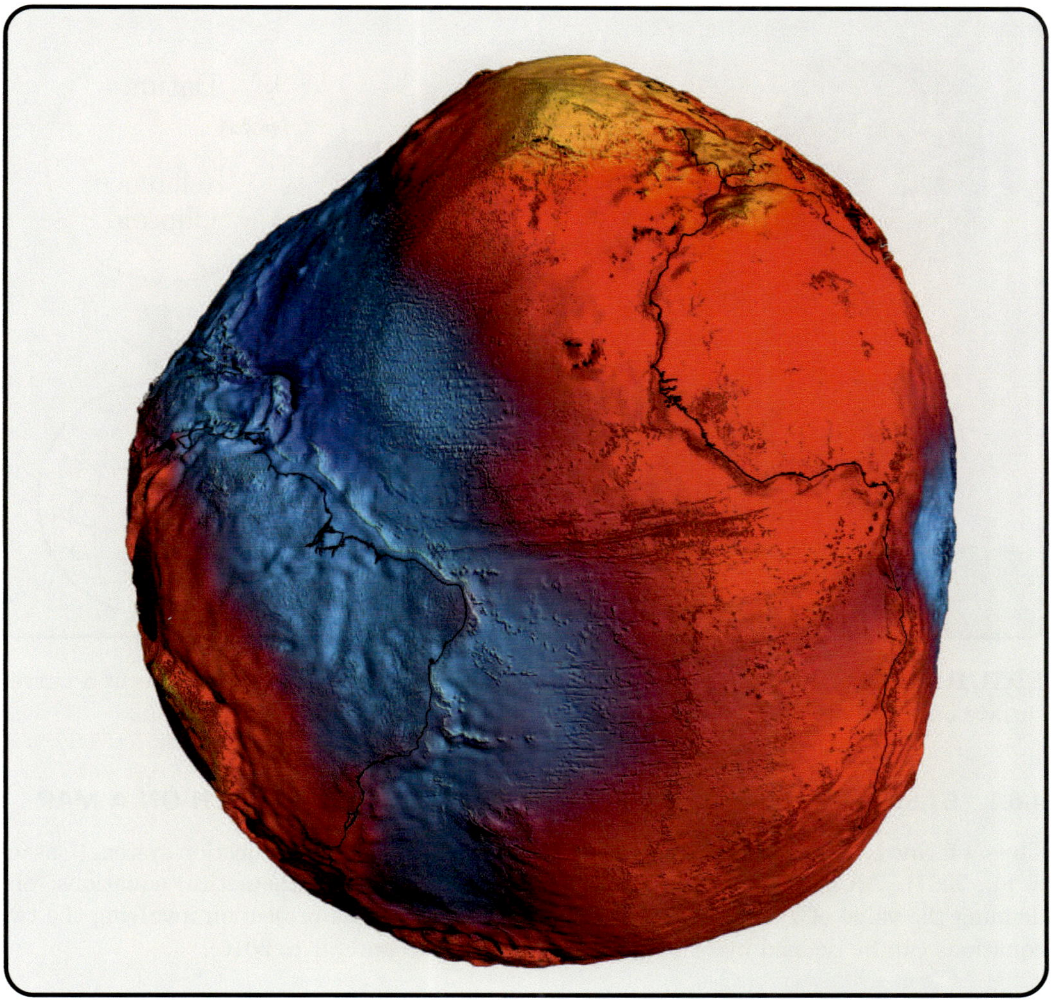

FIGURE 4.12 Definition of the Earth's geoid based on measurements from the European Space Agency (ESA) GOCE mission.

4.6 ELEMENTS OF MAP PROJECTIONS

Geographic referencing or georeferencing is defined by fixing places and real-world features within a spatial network of a given coordinate system. The goal of georeferencing is to realize a spatial network rigid enough to measure, calculate, record, and analyze positions of Earth features. Georeferencing can be considered in practice as a series of techniques and concepts to transform measurements made on irregular surfaces of the Earth into a plane surface represented on a map by means of a coordinate system. In georeferencing, the physical shape of the Earth is characterized by the concepts of geoid and ellipsoid. Therefore, the concept of georeferencing involves a series of progressive transformations to transform the curved surface of the Earth into a plane surface of a map (Figure 4.13) (Lo and Yeung, 2007).

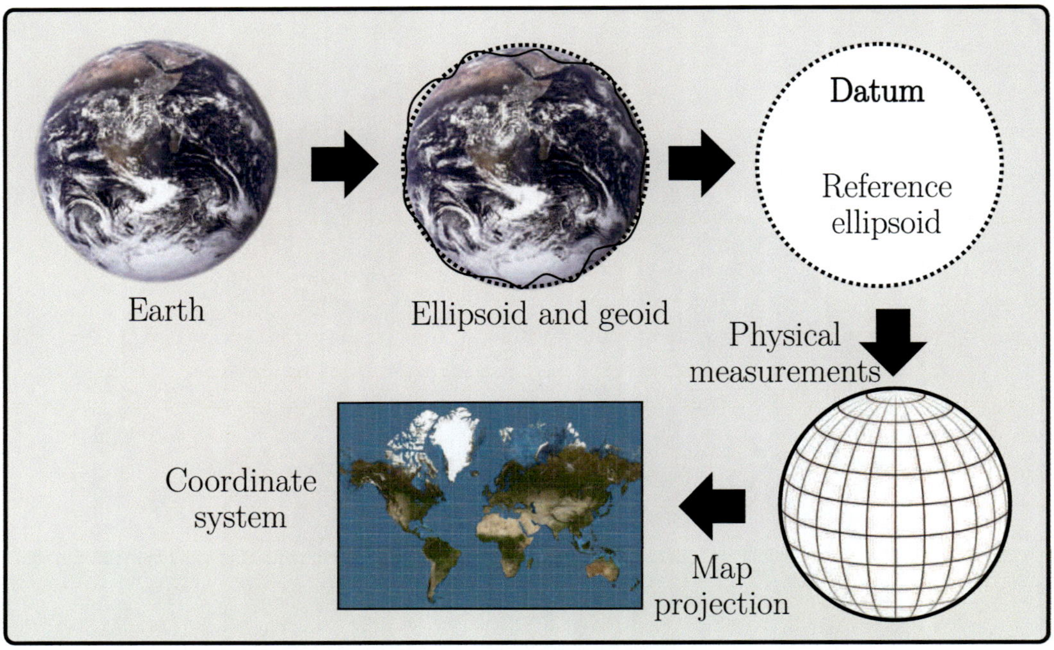

FIGURE 4.13 Georeferencing the Earth and transformations used to represent a curved surface on a flat map.

4.6.1 BASIC MATHEMATICS OF PROJECTIONS OF THE EARTH ON A MAP

The set formed by the projection and the parameters is called map projection system (Casaca et al., 2007). All map projections must contain at least two mathematical equations: one defining the value of x and the other the value of y. A simple projection involving the two equations can be defined by (Slocum et al., 2009; Alves and Silva, 2016):

$$x = r(\lambda - \lambda_0) \tag{4.12}$$

$$y = r\phi \tag{4.13}$$

where λ refers to the longitude value, ϕ, to the latitude, λ_0, to the central meridian value, and, r, to the radius of the reference globe.

To calculate the x and y values based on these equations, the longitude value of the central meridian must be obtained. If the longitude value is 0°, the central meridian value coincides with the prime meridian. All latitude and longitude values must be converted to radians. This conversion is necessary to perform computational calculations (Alves and Silva, 2016).

The R packages `sf` (Pebesma et al., 2021a), `rnaturalearth` (Massicotte and South, 2023), and `tmap` (Tennekes et al., 2020) are used as examples to determine some global map projections. The `library` function typed in the R console is used to load the packages.

```r
library(sf)
# Linking to GEOS 3.11.2, GDAL 3.7.2, PROJ 9.3.0; sf_use_s2() is TRUE
library(rnaturalearth)
# package 'rnaturalearth' was built under R version 4.3.3
library(tmap)
# package 'tmap' was built under R version 4.3.3
```

Country polygons, class `sf`, are obtained from the Internet and then converted to the multipolygon geometry type `sfc`.

```r
Earth <- countries110 |>  st_geometry() # Get data and convert to sfc
Earth # See header
# Geometry set for 177 features
# Geometry type: MULTIPOLYGON
# Dimension:     XY
# Bounding box:  xmin: -180 ymin: -90 xmax: 180 ymax: 83.64513
# Geodetic CRS:  WGS 84
# First 5 geometries:
# MULTIPOLYGON (((180 -16.06713, 180 -16.55522, 1...
# MULTIPOLYGON (((33.90371 -0.95, 34.07262 -1.059...
# MULTIPOLYGON (((-8.66559 27.65643, -8.665124 27...
# MULTIPOLYGON (((-122.84 49, -122.9742 49.00254,...
# MULTIPOLYGON (((-122.84 49, -120 49, -117.0312 ...
```

The Plate Carrée projection is one of the oldest projections developed by the Greeks. We observed that in the longitude and latitude lines of the Plate Carrée projection, there is spacing at equal intervals. The longitude lines are equal in size, but there is no convergence at the poles, as there should be in conditions of an ideal projection (Slocum et al., 2009; Alves and Silva, 2016). The Earth is reprojected and mapped in the Plate Carrée projection using the `qtm` function (Figure 4.14).

```r
qtm(Earth, fill = "black", projection = "+proj=eqc") + tm_grid()
```

We can expand the projected map based on the simple equations just by adding trigonometric functions. For example, by adding sine to the longitude, the equation is modified by (Slocum et al., 2009; Alves and Silva, 2016):

$$x = r(\lambda - \lambda_0) \tag{4.14}$$

$$y = rsen\phi \tag{4.15}$$

With this mathematical transformation, the Lambert cylindrical projection of equal area is determined, developed in 1772 by Johann H. Lambert. The meridians are equally spaced similarly to the Plate Carrée projection, but the spacing between the parallels decreases with increasing distance from the Equator (Figure 4.15).

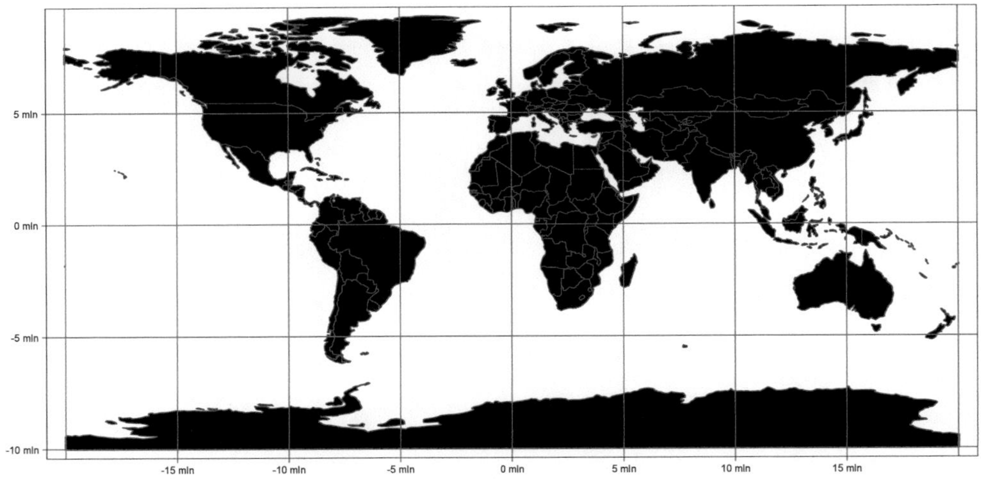

FIGURE 4.14 Plate Carrée mapping projection applied on the Earth's surface.

```
qtm(Earth, fill = "black", projection = "+proj=cea") + tm_grid()
```

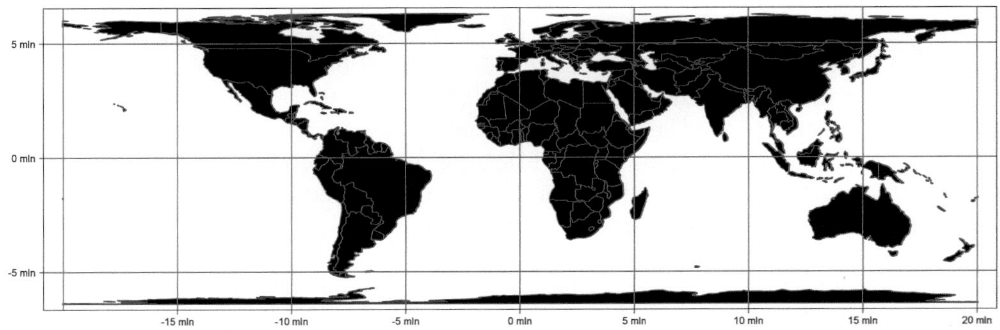

FIGURE 4.15 Lambert's cylindrical map projection applied on the Earth's surface.

An alternative can be realized by introducing a cosine function into the simple equations, changing the way the longitude is determined (Slocum et al., 2009; Alves and Silva, 2016):

$$x = r(\lambda - \lambda_0)cos\phi \tag{4.16}$$

$$y = r\phi \tag{4.17}$$

The modification in longitude results in the pseudo-cylindrical sinusoidal projection. In this case, the lines of latitude are parallel and equally spaced. However, the meridians are curved converging to the North and South Poles (Figure 4.16).

```
qtm(Earth, fill = "black", projection = "+proj=sinu") + tm_grid()
```

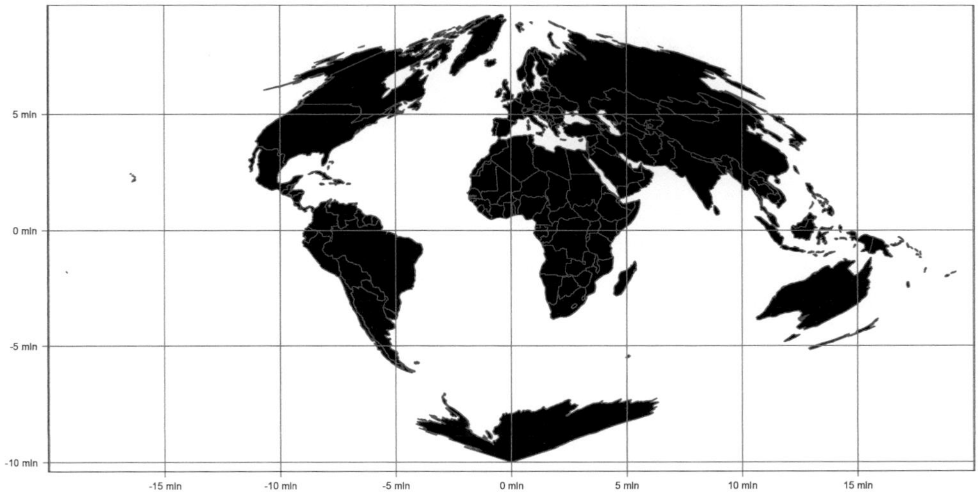

FIGURE 4.16 Pseudo-cylindrical sinusoidal map projection applied on the Earth's surface.

By combining the sine and cosine functions into simple equations, the polycylindrical or Mollweide projection is obtained (Slocum et al., 2009; Alves and Silva, 2016):

$$x = r(\lambda - \lambda_0)cos\phi \tag{4.18}$$

$$y = rsen\phi \tag{4.19}$$

With this, the meridians are curved as in the sinusoidal projection and the parallels are similar to that of the other two cylindrical projections shown (Figure 4.17).

```
qtm(Earth, fill = "black", projection = "+proj=moll") + tm_grid()
```

The parameters of the projections can be modified through coordinate reference system (crs) definitions (Lovelace et al., 2019). For example, by modifying the center of the Lambert azimuthal projection of equal area to longitude -47.8919 and latitude -15.7975, the map is centered on Brasilia, Distrito Federal, Brazil (Figure 4.18).

```
qtm(Earth, fill = "black", projection="+proj=laea +x_0=0 +y_0=0 +lon_0=-47.8919
                                        +lat_0=-15.7975") + tm_grid()
```

The projection concepts described earlier apply equally to data with raster geometry. However, there are important differences in vector and raster projections (Lovelace et al., 2019):

- In transforming a vector object, there is a change in the coordinates of each vertex;

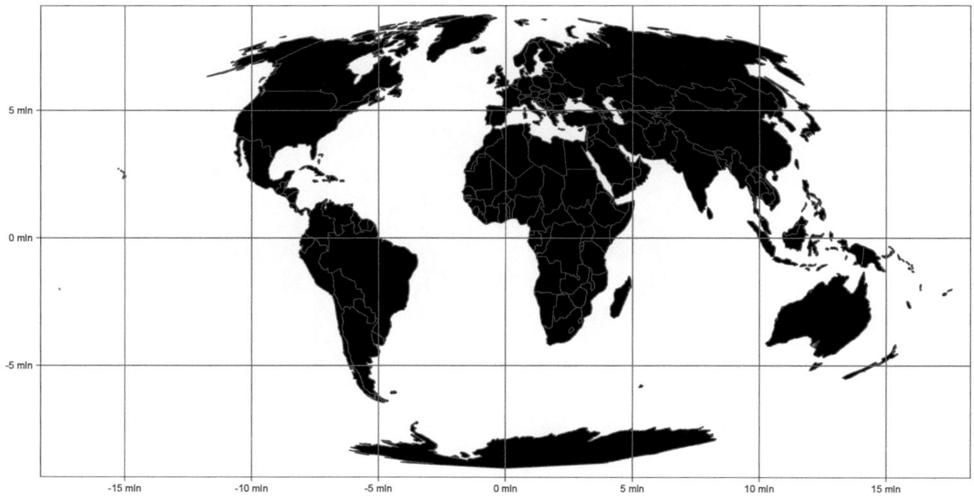

FIGURE 4.17 Mollweide map projection applied on the Earth's surface.

- A raster is composed of rectangular cells of the same size, and it is not possible to transform pixel coordinates separately;
- In raster projection, a new raster object is created, usually with a different number of columns and rows than the original;
- In raster projection, the attributes must be subsequently reestimated, so that new pixels can be filled with other values.

In raster data projection, two separate spatial operations are considered (Lovelace et al., 2019):

- A vector projection of pixel centroids to another crs;
- Calculation of new pixel values by resampling.

4.6.2 CHARACTERISTICS OF MAP PROJECTIONS

Since the shape of the Earth is not flat, constructing a map with all desired conditions for all situations is a scientific challenge. In an operational map projection, the parameters geodetic datum, center point, scale factor, and a new origin for map coordinates are specified (Casaca et al., 2007). Map projections are classified by different methodologies in order to determine the best fit for the represented surface (Fitz, 2008; Instituto Brasileiro de Geografia e Estatística - IBGE, 1995).

4.6.2.1 Classification according to method

Map projections are classified according to geometric and analytical methods. In geometric projections, the projection is based on a plane according to a point of view. They can be obtained by intersecting over the projection surface the bundle of straight lines that pass through points on the reference surface starting from a perspective center. In analytical projections, mathematical formulations are used to calculate the projections (Casaca et al., 2007).

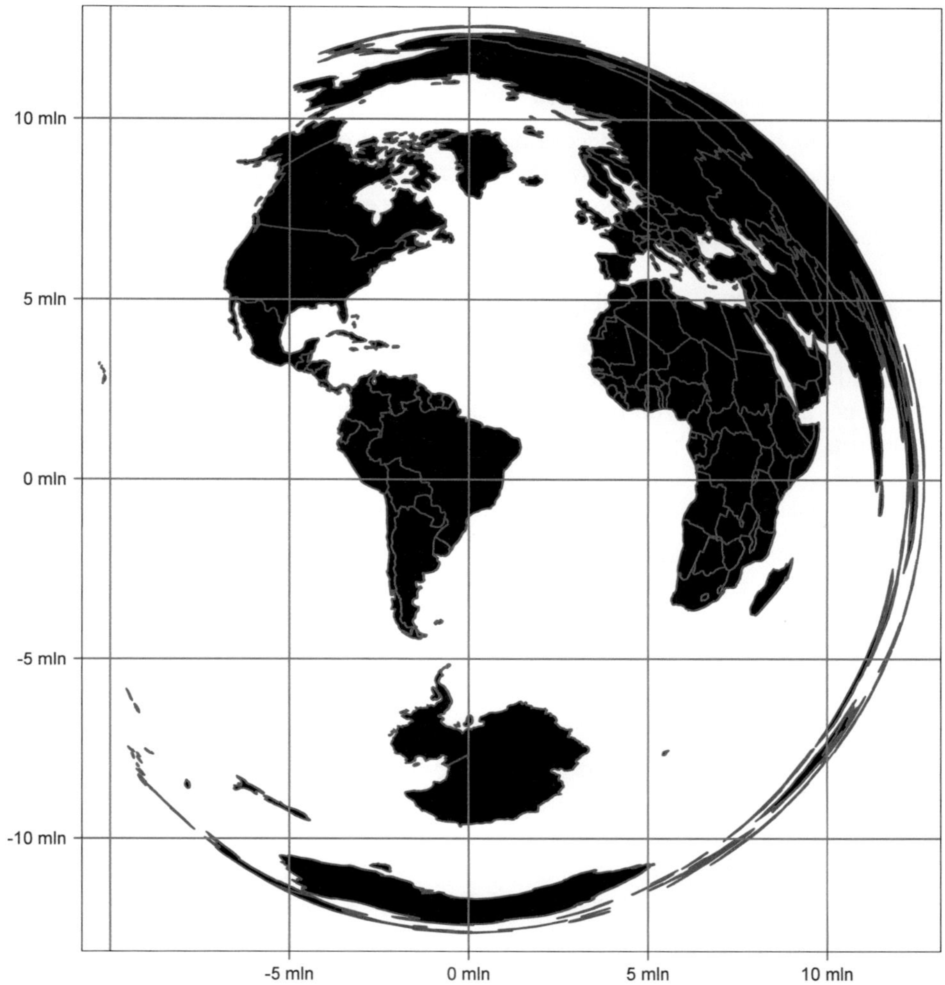

FIGURE 4.18 Lambert azimuthal equal area map projection applied on the Earth's surface.

4.6.2.2 Classification according to deformations

Considering the impossibility of developing a spherical or ellipsoidal surface on a plane without deformations, in practice, projections are sought to reduce or eliminate part of the deformations according to the desired application, according to the following properties (Fitz, 2008):

- Conformal or similar <- The true shape of the areas represented is maintained by not deforming the angles on the map;
- Equidistant <- There is constant variation of the distance represented, without linear deformation;
- Equivalent <- There is constant variation of the relative dimensions of the areas represented;
- Azimuthal <- The directions and azimuths of all lines arising from the central point of the projection are equal to those of the corresponding lines on the Earth's

sphere;
- Aphilatic or arbitrary <- Does not preserve areas, angles, distances or azimuths.

4.6.2.3 Classification according to the location of the point of view

Map projections are classified according to the point of view into gnomic, stereographic and orthographic. In gnomic or central projections, the point of view is located in the center of the ellipsoid, in the stereographic, at the opposite end of the projection surface, and in the orthographic the point of view is at infinity (Alves and Silva, 2016).

4.6.2.4 Classification according to the type of projection surface

Map projections are classified according to the type of projection surface into plane, conical, cylindrical and polyhedral. In plane projections, the projection surface is a plane, in conical projections, a cone, in cylindrical projections, a cylinder, and in polyhedral projections, a polyhedron (Alves and Silva, 2016).

4.6.2.5 Classification according to the position of the projection surface

Map projections are classified according to the position of the projection surface into equatorial, polar, transverse and oblique (Figure 4.19). In equatorial type projections, the axis of the projection surface is located at the Earth's equator; in the polar type, at a pole; in the transverse type, perpendicular to the Earth's axis of rotation; and in the oblique type, at any other position (Lo and Yeung, 2007; Alves and Silva, 2016).

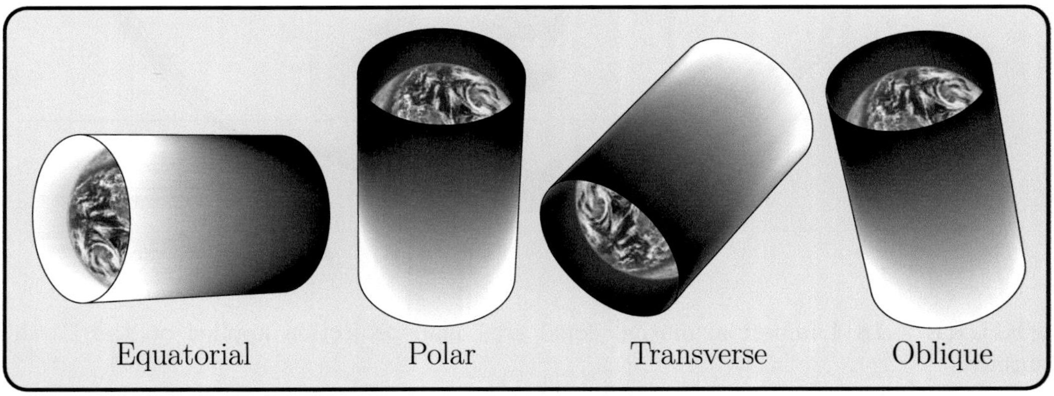

FIGURE 4.19 Different aspects of map projection regarding the position of the projection surface.

4.6.2.6 Classification according to the situation of the projection surface

Map projections are classified according to the situation of the projection surface into tangent and secant (Figure 4.20). In tangent situations, the projection surface tangents the ellipsoid in a point (plane) or in a line (cylinder or cone); in secant situations, it cuts the ellipsoid in two points (plane) or in two lines (cylinder or cone) of secance (Alves and Silva, 2016).

4.7 MAP PROJECTION SELECTION

The selection of map projections is based on the projection selection guide by John Snyder. In this guide, it is recommended to separate projections based on the geographic extent to

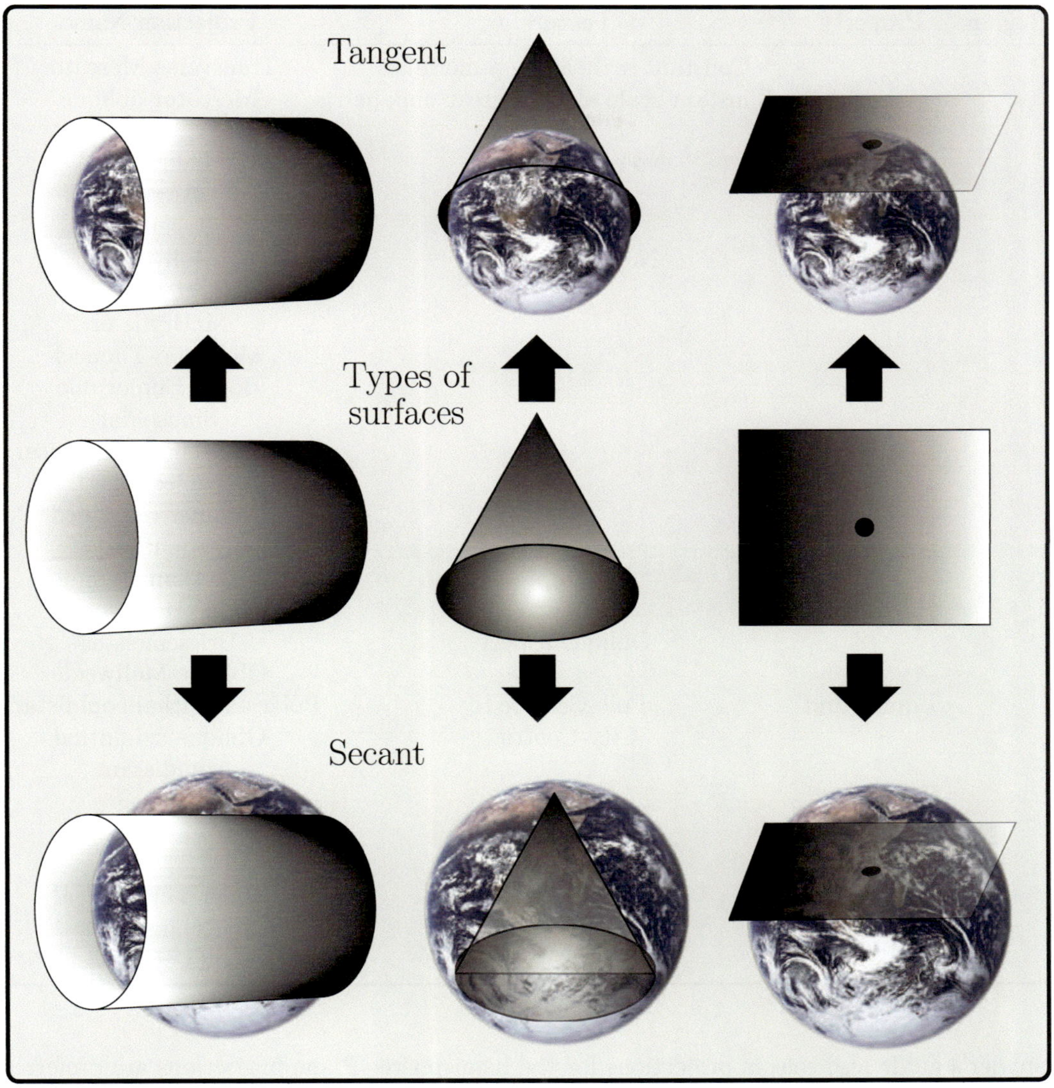

FIGURE 4.20 Types of surfaces of cylindrical, conical and plane projections in situations of tangency and secance to surfaces.

be mapped: (1) Earth, (2) Hemisphere, (3) Continent, ocean, or smaller region. Global projections are arranged according to the desired feature, such as conformal, and, special features, such as the center of the projection (Slocum et al., 2009).

According to the map projections guide developed by Snyder to create a map projection of the Earth as a whole, we can choose the projection based on the preserved property, feature and projection name (Table 4.2) (Slocum et al., 2009).

TABLE 4.2 Snyder Map Projection Guide for selecting a projection of the globe.

Region	Property	Feature	Projection Name
Globe	Conformal	Constant scale along the Equator	Mercator

Region	Property	Feature	Projection Name
		Constant scale along a meridian	Transverse Mercator
		Constant scale along a great oblique circle	Mercator oblique
		No constant scale	Lagrange August Eisenlohr
	Equivalent	Non-interrupted	Mollweide Eckert IV & VI McBryde or McBryde-Thomas Boggs Eumorphic Sinusoidal Other pseudo-cylindrical projections Hammer (modified azimuthal)
		Interrupted	Other than Hammer Goode Homolosine
		Oblique aspect	Briesemeister Oblique Mollweide
	Equidistant	Pole-Centered	Polar azimuthal equidistant
		City-Centric	Oblique azimuthal equidistant
	Straight bearing lines		Mercator
	Transigent Distorted		Miller's cylindrical Robinson's pseudo-cylindrical

Snyder's guide also selects projections for the hemisphere. Plane projections are preferred for mapping hemispheres because the standard point is placed at the center of the projection to enable recentralization in areas of interest (Table 4.3) (Slocum et al., 2009).

TABLE 4.3 Snyder map projection guide for selecting a plane projection of the Earth's hemisphere.

Region	Property	Projection Name
Hemisphere	Conformal	Conformal stereographic
	Equivalent	Lambert azimuthal
	Equidistant	Azimuthal equidistant
	Global view	Orthographic

The Snyder guide is used for mapping a continent, ocean, or small region by means of hierarchy based on the directional predominance in area extent of the mapped terrain (Table 4.4) (Slocum et al., 2009).

TABLE 4.4 Snyder's map projection guide for selecting a projection on a continent, ocean or small region.

Region	Directional Extension	Location	Property	Projection Name
Continent, ocean or small region	East_West	Along Equator	Conformal	Mercator
			Equivalent	Equivalent cylindrical
		From Equator	Conformal	Lambert's conformal conic
			Equivalent	Albers Equal-Area Conic
	North-South	Along a meridian	Conformal	Transverse Mercator
			Equivalent	Cylindrical transverse equivalent
	Oblique	In any place	Conformal	Mercator Oblique
			Equivalent	Oblique cylindrical equivalent
	Equal extension	Polar, equatorial or oblique	Conformal	Stereographic
			Equivalent	Lambert azimuthal

The criteria used to select a map projection are (Slocum et al., 2009):

1. Define the mapping objective;
2. Define the projection fitted to the data with minimum distortion;
3. Prioritize the analysis of the effect of distortion on mapping as the area to be mapped increases;
4. A well-known and widely used projection is not necessarily the best projection for a specific application;
5. Observe the appearance of the projection and its effect as a whole on the map drawn.

4.8 COMPUTATIONAL PRACTICE

4.8.1 REFERENCE SYSTEMS

The location of observations is characterized by coordinates in a coordinate system and applied to geospatial data. Different coordinate systems can be used, with the coordinates defined in a two-dimensional, three-dimensional space, referenced to orthogonal axes (Cartesian coordinates), or by means of distance, directions (polar coordinates, spherical coordinates), time, in a given system of units of measurement.

4.8.1.1 Units of measurement

A property of an object or phenomenon can have magnitude expressed as a number and a reference measurement unit, according to a measurement procedure, a reference material, or a combination of these. For proper data handling, numbers are accompanied by meaning and context information (Pebesma and Bivand, 2021a).

With Package `units` (Pebesma et al., 2016), it is possible to get support for units of measurement interfaced with the units database udunits[1] and the C library for defining and converting units.

The `install.packages` function is used to install the `units` package.

```
install.packages("units")
```

The `library` function is used to enable the `units` package.

```
library(units)
```

The metric unit setting of values between 1 and 2 is done with the `set_units` function.

```
(a <- set_units(1:2, m))
```

Similarly, we can set up the same values in kilometer units.

```
(a_km <- set_units(1:2, km))
```

By adding the units together, the final result is shown in meters.

```
(a + a_km)
```

Time units are set from 1 to 2 seconds.

```
(b <- set_units(2:3, s))
```

We can also perform operations in different units of measurement, as in the case of distance per unit of time.

```
(a/b)
```

Errors are obtained in the case of meaningless operations.

```
a - b
#Error: cannot convert s into m
```

4.8.1.2 Time reference systems

4.8.1.2.1 Date

In R there are two native classes for time-varying data, `POSIXt` and `Date`, used to specify dates and times, respectively. Objects of type `Date` are vectors of class `Date` with numbers of

[1] https://www.unidata.ucar.edu/software/udunits/

days since January 1, 1970. For example, the date 1970-02-01 is specified with the `as.Date` function.

```
(d <- as.Date("1970-02-01"))
```

When converting the date to a numerical value with the function `as.numeric`, 31 days are observed with reference to January 1970.

```
as.numeric(d)
```

By specifying a second date, on 200-02-01, 10,988 days have passed.

```
(d1 <- as.Date("2000-02-01")) # Specify date d1
as.numeric(d1) # Convert to a numerical value
```

ISO 8601 is used as a reference for specifying dates. If the date record is not in the standard format, we must use the `format` parameter to specify the sequence of day (d.), month (m.), and year (Y).

```
(d <- as.Date("01.02.2000", format = "%d.%m.%Y"))
```

The `format` function is used to change the date stamp from the default format. The default format may vary according to the local settings of the computer being used.

```
format(d, format = "%d.%m.%Y")
```

4.8.1.2.2 POSIXt

The `POSIXt` is a native R class used in temporal specification. There are two subclasses in `POSIXt` (Pebesma and Bivand, 2021a):

- POSIXct: It represents time as a numeric value, with decimal seconds since 01/01/1970 00:00 UTC;
- POSIXlt: There is time component information in a list of second, minute, hour, day of month, month, and year.

```
(t <- as.POSIXct("1970-01-01 02:00:00", tz = "UTC"))
```

With the `as.numeric` function, the time of 2 h is converted into 7,200 seconds.

```
as.numeric(t)
```

The components of the time list are listed with the `names` function.

```
names(unclass(as.POSIXlt(t)))
```

The number of logged hours is accessed with the `unclass` function.

```
unclass(as.POSIXlt(t))$hour
```

4.8.1.2.3 *Time zone*

The time zone is a local time modifier, in which time is stored in relation to Coordinated Universal Time (UTC). A local time zone can be specified according to Pacific Standard Time (PST).

```
(t = as.POSIXct("2000-01-01 00:00:00", tz = "PST"))
```

With this, a time zone is added to `t`, redefining the origin of time according to the PST time zone.

```
as.numeric(t)
```

We can use the `as.Date` function to convert `POSIXt` to `Date`, that is, to convert the date to the local date. When converting back to POSIXct, the time of day and time zone information is lost.

```
(t = as.POSIXct("2000-01-01 23:00:00", tz = "PST"))
as.Date(t)
format(as.POSIXct(as.Date(t)), tz = "UTC")
```

When working with local time zones, there may be confusion under the situation of different time references. In this case, the `Sys.setenv` function can help to set the time zone.

```
Sys.setenv(TZ = "UTC")
```

4.8.1.3 Coordinate reference systems

PROJ is a type of code used by coordinate reference systems (datums) and map projections called `proj4string`. Some of these codes have been compiled by the International Association of Oil & Gas Producers with a code called epsg. In the `sf` package there is a `crs` class initialized by an `EPSG` code (Pebesma and Bivand, 2021a).

```
library(sf)
```

```
st_crs(4326)
```

or by a PROJ parameter, `+proj`, as:

```
st_crs("+proj=longlat")
```

and may not have a value, i.e., NA.

```r
st_crs()
```

Several methods are available for crs objects:

```r
methods(class = "crs")
```

A `crs` object can be converted to a well-known text (WKT) object and displayed with the `cat` function.

```r
cat(st_as_text(st_crs(4326), pretty = TRUE))
```

The `methods` function accesses the generic methods used by `st_crs`.

```r
methods(st_crs)
```

On objects of type `sfc` reference ellipsoid parameters can be obtained, as in the case of the WGS-84 ellipsoid, `crs` 4326, according to the following code.

```r
st_crs(st_sfc(crs = 4326), parameters = TRUE)[c(1:4, 8)]
```

The semi-major axis lengths and flattening are used to calculate distances of great circles on the ellipsoid by means of the Karney (2013) algorithm. This algorithm has been implemented in `PROJ`, and can be accessed in the packages and functions `lwgeom::st_geod_distance` and `sf::st_distance` on objects with geodesic coordinates.

4.8.1.4 Vector and raster geometry projection

The `raster` (Hijmans et al., 2020), `rgdal` (Bivand et al., 2021) and `sf` (Pebesma et al., 2021a) packages can be used to perform the map projection transformation of a raster file and vector feature. The packages are installed with the `install.packages` function and enabled with the `library` function. If the package is already installed on the computer, just enable the package to perform the practice.

```r
library(raster)
library(rgdal)
library(sf)
```

4.8.1.5 Vector geometry projection

In some geocomputation situations, the map projection of an exported file may be lost. For a simple vector file of class `sf`, the `st_set_crs(st_crs())` function can be combined with the `st_crs` function to obtain the `crs` of one file and then apply the `+proj` parameter to another file to set the same map projection configuration. This may be necessary if the `crs` status is unknown when importing the file. The `crs` can be obtained either from a vector file or from a raster file by accessing the file header. In the case of files of class `sf`, the `st_transform` function can be used to transform geodetic map projections to plane projections and vice-versa. For more precise map projection, it is recommended to use the `+proj` parameter set according to the projection of interest or to use the `st_crs` function to obtain the `+proj` of a file of interest. Next, the `st_read` function is used to obtain municipalities of Minas Gerais in the WGS-84 geodetic projection for the plane Albers Equal-Area Conic projection, using the `+proj` parameter as a reference for the coordinate transformation of the vertices of the polygon.

The polygon with the political municipal boundaries of some municipalities in Minas Gerais, Brazil obtained from a subset of the `rnaturalearth` package data (Massicotte and South, 2023) is available for download[2] for the reproduction of the practical activity.

The spatial polygons with administrative boundary attributes of municipalities are imported into R by means of the `st_read` function.

```
munsf <- st_read("C:/livros/geo/shape/municipiosMG2007.shp")
```

The coordinates of the municipal boundaries are transformed to the Albers Plane coordinate system of Equal Area using the `st_transform` function.

```
mg <- st_transform(munsf, "+proj=aea +lat_1=-5 +lat_2=-42 +lat_0=-32
                    +lon_0=-60 +x_0=0 +y_0=0 +ellps=aust_SA +units=m +no_defs")
```

In the case of class `sp` data, the `spTransform` function is used for coordinate transformation of simple vector features with attributes.

4.8.1.6 Projection of raster geometries

The moderate resolution climate surface of annual mean air temperature for the period 1970 to 2000 obtained from the WorldClim project (Fick and Hijmans, 2017; Harris et al., 2014), clipped and masked into the polygons of municipalities surrounding Lavras is available for download[3].

The annual mean air temperature data are imported into R using the `raster` function, and the climate variable is renamed with the `names` function.

```
temp <- raster("C:/geo/c2/ty1970_2000.tif") # Import
names(temp) <- c("ty1970_2000") # Rename
```

The `projectRaster` function is used for the transformation from the original WGS-84 geodetic projection to the plane Albers Equal-Area Conic projection, by setting the `+proj` parameter as a reference for the transformation of raster pixel coordinates.

```
t <- projectRaster(temp, crs="+proj=aea +lat_1=-5 +lat_2=-42 +lat_0=-32
                    +lon_0=-60 +x_0=0 +y_0=0 +ellps=aust_SA +units=m +no_defs")
```

The annual mean air temperature variation is mapped in layers by adding the boundary polygons of municipalities. The `plot` function is an option used in mapping polygon features with attributes and the temperature raster surface. The `add=TRUE` parameter is set to

[2]http://www.sergeo.deg.ufla.br/geo/downloads/municipiosMG2007.zip

[3]http://www.sergeo.deg.ufla.br/geo/downloads/ty1970_2000.zip

perform hybrid mapping of the raster and vector data models on the same map and the transparency of polygons is obtained with the parameter `col=NA` (Figure 4.21).

```
plot(t, axes=T, col=topo.colors(20)) # temperature
plot(st_geometry(mg), col=NA, add = TRUE) # municipalities
```

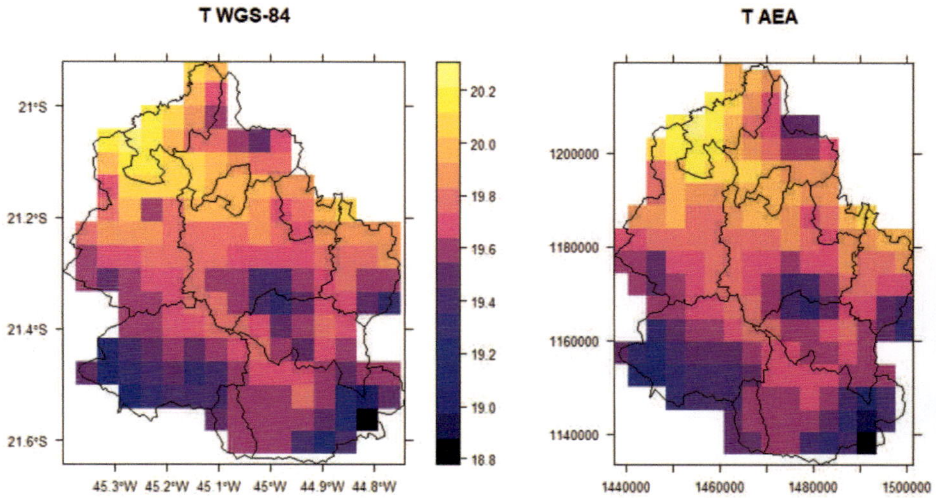

FIGURE 4.21 Comparative mapping of the mean annual air temperature (T) (Celsius) in municipalities of Minas Gerais, in the period from 1970 to 2000 in the WGS-84 (left) and Albers Equal-Area Conic (AEA) projections (right).

Note that some functions can be used together to open and reproject vector data based on the map projection of an existing raster layer. For example, the municipalities of Minas Gerais can be imported with the `st_read` function and the `%>%` operator enables the vector feature with attributes to also be transformed according to the raster projection using the `st_transform` and `st_crs` functions.

```
mg <- st_read("C:/livros/geo/shape/municipiosMG2007.shp") |> st_transform(st_crs(t))
```

5 Spatial Data Models with Environmental Geocomputation and R

5.1 LEARNING QUESTIONS

The learning questions answered by reading the chapter are:

- What is the definition of geographic data model?
- What is the importance of spatial data models in geocomputation?
- How to create spatial data models in geocomputation.
- How geographic data models are applied in geocomputation.
- How to define topology and its usefulness in data analysis and validation.
- What language is used in data models?
- How to perform terrestrial environment modeling to create useful geographic database.

5.2 LEARNING OUTCOMES

The learning outcomes answered by reading the chapter are:

- Define geographic data model and its importance in geocomputation.
- Understand how to create spatial data models in geocomputation.
- Characterize the main types of geographic data models used in geocomputation.
- Understand about topology and its usefulness in data analysis and validation.
- Understand data model language and notation.
- Perform modeling of terrestrial environments to create useful geographic databases.

5.3 INTRODUCTION

When observing the environment, the inherent complexity of the place is simplified by major features in order to develop a model of the observed region. This cognitive exercise is influenced by cultural notions of the observer and the purpose of the study. Different stages of model development produce geographic data to be used and shared in digital form (Figure 5.1) (Burrough and McDonnell, 1998).

A satisfactory model in geocomputation should incorporate various applications from environmental studies, agriculture, geology, and human utilities such as water supply, sanitation, and electric power (Table 5.1) (Câmara and Medeiros, 1998).

DOI: 10.1201/9781003724377-5

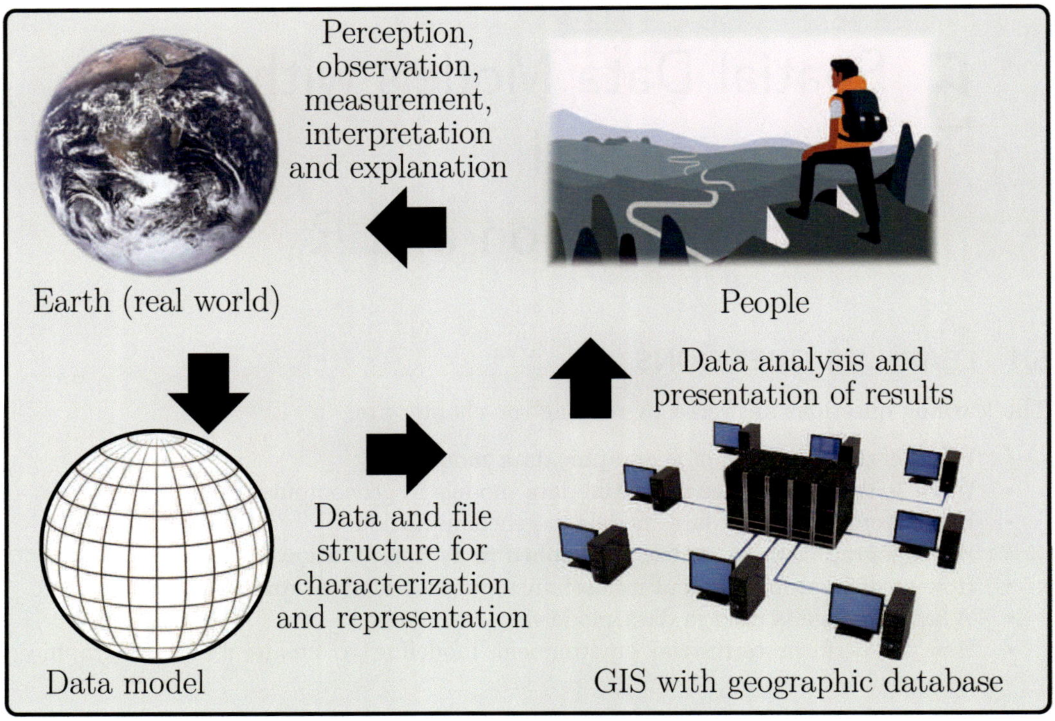

FIGURE 5.1 The context of data modeling in geocomputation.

TABLE 5.1 Example of data model applications in geocomputation.

Application	Data type	Graphical representation	Operation
Agricultural management	Thematic remote sensing data, digital elevation model	Matrix, vector, grid, TIN	Spatial analysis, slope
Rural planning	Rural registry	Matrix, vector	Image classification, spatial query
Geology	Digital elevation model, imagery, thematic data	Matrix, grid	IHS transformation, 3-D visualization
Utility network (electricity, sanitation, supply)	Linear networks	Vector	Spatial query, calculations
Urban planning	Networks, urban cadastre	Vector	Spatial query

In an approach with two extremes, in the first, the geospace can be occupied by different entities described and mapped, in a coordinate system, according to specific characteristics. In the second, a field of continuous variation, without an exactly defined boundary, can be used. Data modeling allows to characterize, to standardize these concepts, from basic units that can be registered and mapped. In the main approaches, sets of points, lines, polygons, and tessellation units are used to describe the different landscape features. The use of a particular model influences the type of data used to describe a phenomenon and perform a particular type of geospatial analysis (Burrough and McDonnell, 1998).

Digital terrain modeling has been used in different areas of science and technology, with mathematical approaches and algorithm development, for terrain quantification, contour creation, interpolation, and three-dimensional continuous surface visualization. Terrain data samples taken at regular or irregular distance intervals, after processing and analysis, generate a continuous digital elevation model that can be stored in an arrangement or network of irregular triangles in Triangulated Irregular Network (TIN) (Lo and Yeung, 2007) (Figure 5.2).

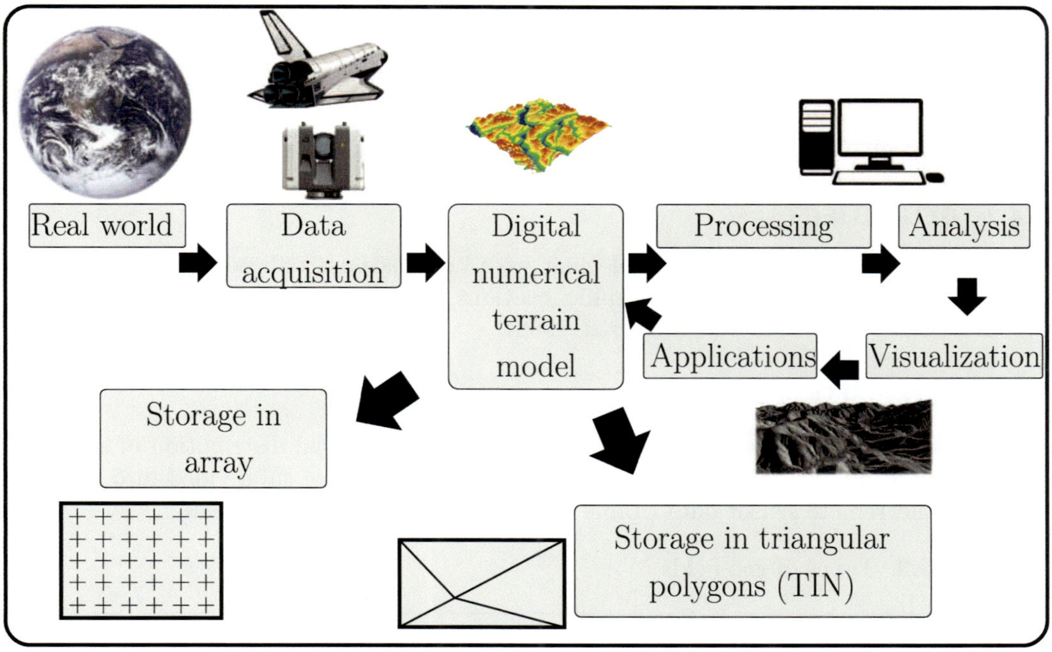

FIGURE 5.2 Main steps performed in digital terrain elevation modeling in geocomputation.

5.4 DEFINITIONS OF DATA MODELS

The heart of any geographic information system is the data model (Longley et al., 2001). A data model is defined by the conceptual toolkit used to describe how geographic reality will be represented by objects and processes in the digital environment of computing systems (Câmara and Medeiros (1998); Longley et al. (2001)).

Data model is critical in determining the usability and learning speed of the system. In the object-oriented data model, for example, the most relevant contributions are (Câmara and Medeiros, 1998):

- Unified approach to modeling, combining ideas of objects and fields;
- Integration of remote sensing images, digital terrain model, thematic maps, cadastral maps and networks;

- Support for multiple geometric representations of the same real-world entity;
- Coexistence of vector, matrix and grid representations in the same system.

Geographic data modeling is developed based on art, science, and technology, and it requires scientific understanding of specific real-world geographic features, including the state and behavior of objects, and the relationships between them, for modeling. With this, it is possible to control the types of data represented and the operations performed from the adopted data model (Longley et al., 2001).

Digital terrain modeling or digital terrain modeling is defined as a statistical representation of a continuous terrain surface by a large number of points with known coordinates X, Y, Z (Lo and Yeung, 2007).

5.4.1 GEOGRAPHICAL REGION

Starting from a continuous region of the Earth's surface, a geographic region R is defined as a surface belonging to a geographic space, and it can be represented in a plane or grid, with a cartographic projection. The geographic region should serve as a geometric support to locate geographic entities using vector or matrix geometric representations associated with the object (Câmara and Medeiros, 1998).

5.4.2 GEO-OBJECT

A geo-object is defined as a single element with non-spatial attributes, also called entity, associated with multiple exact geographic locations, and with distinction of the object from its surroundings (Câmara and Medeiros, 1998).

5.4.3 GEO-FIELD

A geo-field, also called field data model, is defined as the spatial distribution of a variable with values at all points in a geographic region. Geo-fields are subdivided into thematic, numeric, and remote sensor data (Table 5.2) (Câmara and Medeiros, 1998).

TABLE 5.2 Types of geo-fields.

Geo-field type	Feature	Example
Thematic	Each point in space is associated with a theme on the map, in a R geographic region	Maps of soils, climate and cultivars
Numeric	Each point in space is associated with a real value on the map, in a R geographic region	Map of magnetic field and altimetry
Remote sensor data	Specialization of the numerical geo-field by discretization of the sensor signal over a R geographic region	Landsat image of Lavras, Brazil

A thematic layer of polygons with the boundary of Brazilian states above the geo-field with the digital terrain model with soil organic carbon data can be seen in Figure 5.3 (Santos et al., 2018; Vasques et al., 2017).

5.4.4 NON-SPATIAL OBJECT

A non-spatial object has no directly associated spatial locations. In many situations, it may be convenient to associate non-spatial information to a geographic database through links

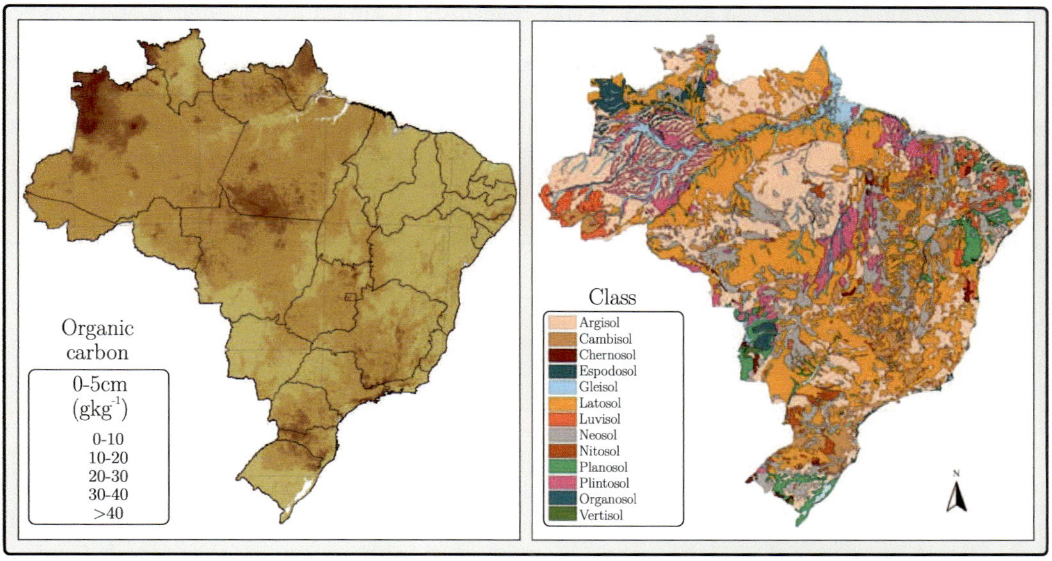

FIGURE 5.3 Numerical and thematic geo-fields of soil organic carbon and soil class in the Brazilian soil classification system.

between geo-objects and existing alphanumeric information, such as in a rural property cadastre with key codes between different property cadastral tables and tax information are matched to link the geo-object, with associated spatial geometry, and the non-spatial object, with additional rural property information (Figure 5.4) (Câmara and Medeiros, 1998).

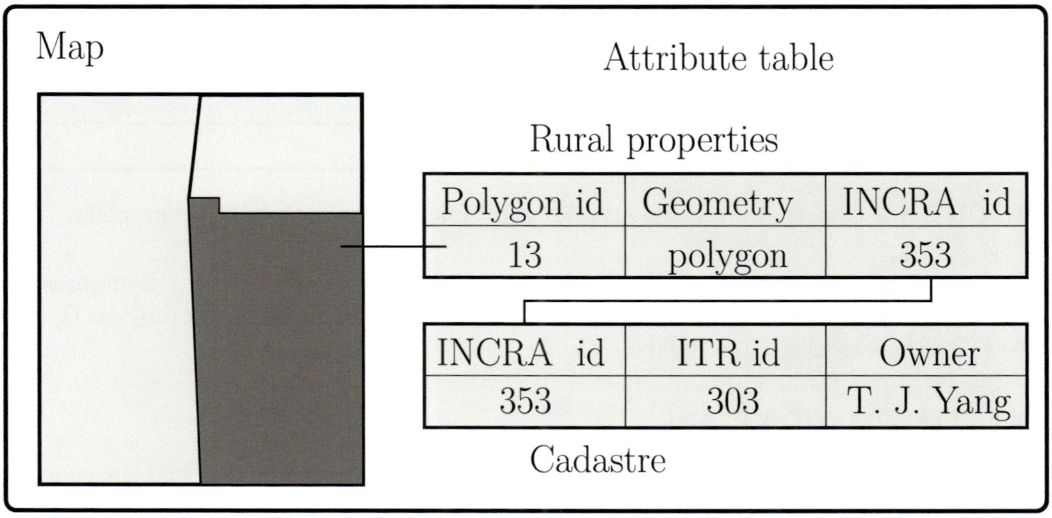

FIGURE 5.4 Example of link between geo-object and non-spatial object.

5.4.5 CADASTRAL MAP

In geocomputation, the geometric representation of geo-objects can be stored together with the neighborhood, maintaining the topology relationships, according to the idea of a cadastral map. A cadastral map is defined as a complex object that groups geo-objects in a cartographic projection and geographic region, R (Câmara and Medeiros, 1998).

An urban cadastral map, for example, may contain geo-object mappings of the classes lots, hospitals, and schools. The class of lots, containing objects from the classes built lots and land, and the class of schools, containing primary and secondary schools. The geo-objects of the class lots include all the lots in the city with the necessary cadastral information (Figure 5.5) (Câmara and Medeiros, 1998).

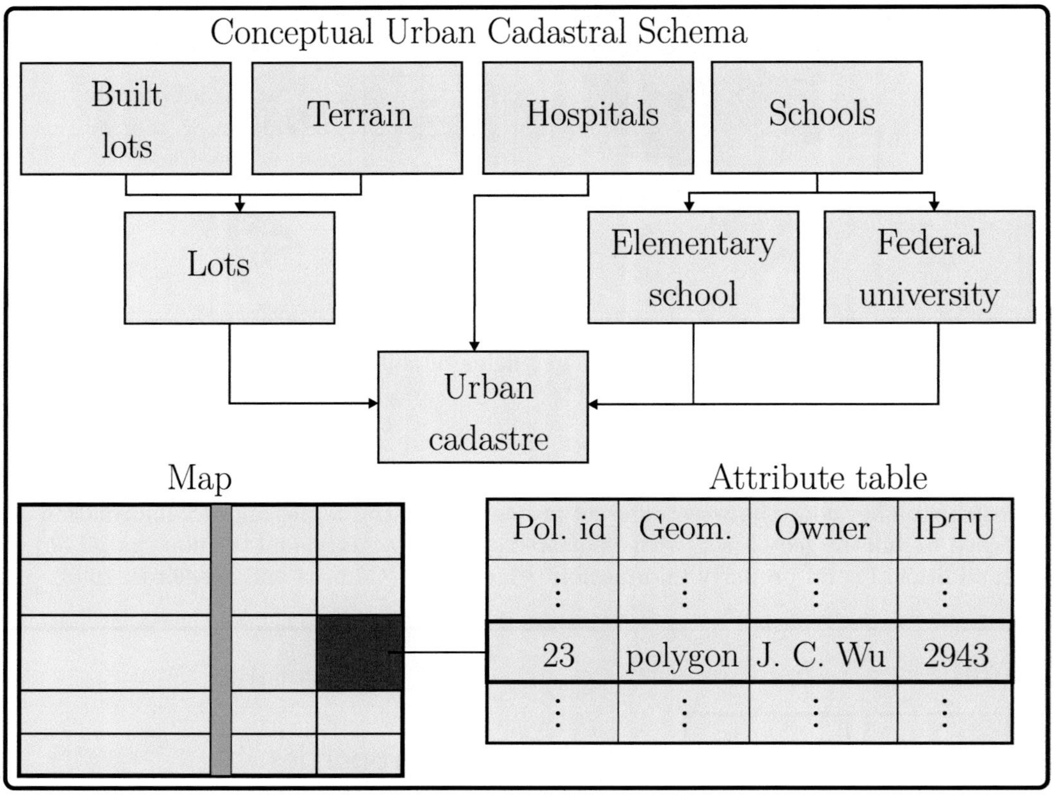

FIGURE 5.5 Conceptual scheme and cadastral map representation of urban plots.

In a power grid cadastral map, location information and connected linear structures are characterized by arc-node topology, with mapping of different classes of geo-objects (Figure 5.6) (Câmara and Medeiros, 1998).

5.4.6 INFORMATION PLANE

The idea of an information plane is useful as an interface between the user and the concepts of geo-field and cadastral map. An information plane is defined as the generalization of geo-fields and geo-objects, in a geographic region R (Câmara and Medeiros, 1998). The information plane is also called `layer` in geographic information systems.

5.4.7 GEOGRAPHIC DATABASE

A geographic database is defined as a set of information planes, geo-objects and non-spatial objects, with no restrictions on data scale and continuity of the information planes that make up the database (Câmara and Medeiros, 1998).

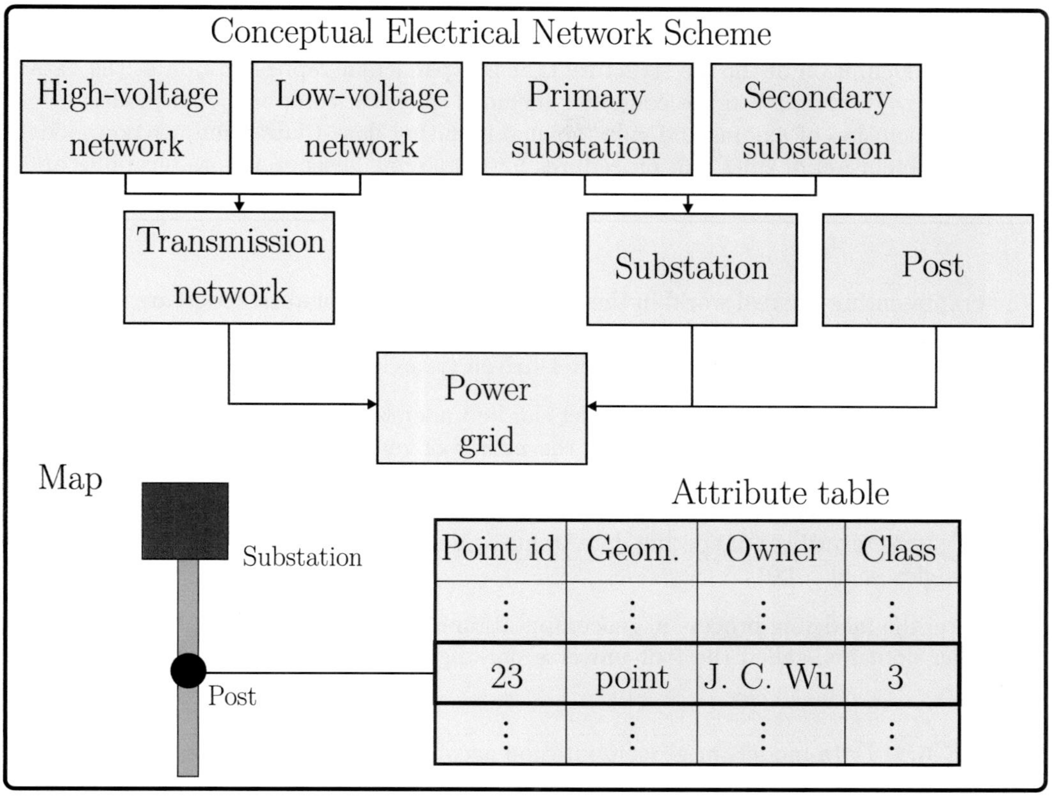

FIGURE 5.6 Conceptual scheme and representation of electrical network cadastral map.

In summary, the representation of a data model has as its basic entity a geographic database composed of information planes, geo-objects, geo-fields and non-spatial objects (Figure 5.7).

In the geographic data model definition, the information planes can be geo-fields or geo-objects. Other specialized classes can also be part of the geographic database.

5.5 LEVELS OF DATA MODEL ABSTRACTION

The main steps for representing standardized models from human observations of the Earth's environment, either directly or by remote sensing data or by sample statistics from a locality are presented in (Table 5.3) (Burrough and McDonnell, 1998).

TABLE 5.3 Levels of development and abstraction used to create analog digital models of discrete entities and continuous variables in the environment.

Level	Description
1	A view of reality (conceptual model)
2	Human conceptualization leading to an analogous abstraction (analog model)
3	Formalization of the analog abstraction without implementation conventions or constraints (geospatial data model)
4	Representation of the data model for computer record (database model)

Level	Description
5	Definition of the file structure that is a particular representation of the data structure in the computer memory (physical computational model)
6	Inclusion of axioms and rules for manipulating data (data manipulation model)
7	Inclusion of rules and procedures for displaying geospatial data, graphics and tables (graphic model)

When representing the real world in the computer, four levels of abstraction are used, based on generalization and simplification of the real world. As abstraction increases through the stages, human influence becomes computer-driven (Longley et al., 2001).

In geocomputation systems, a data model can be understood differently by developers and users. For developers, a data model is the means of representing an application domain in a design that can be implemented in a computer system for controlling, analyzing, and performing operations. For users, the data model is considered a description of the structure of the system, regardless of specific data items or details of particular applications (Figure 5.8) (Longley et al., 2001).

Therefore, the modeling process in geocomputation is performed to translate the real world into other domains, called the four-universe paradigm (Table 5.4) (Câmara and Medeiros, 1998).

TABLE 5.4 Data models in geocomputation according to the four-universe paradigm.

Universe	Description
Real world	Entities and processes modeled in the system, such as: soil types, urban and rural cadastre, geophysical and topographical data.
Mathematical	Conceptual universe with formal mathematical definition of the entities that will be modeled, according to the characteristics of continuous data and object entities. In this universe, geographic data can be divided into thematic and cadastral data, digital terrain model and remote sensing data.
Representation	Mapping of entities of the mathematical universe based on geometrical, matrix and vector representations, which vary according to scale, cartographic projection and time of data acquisition.
Implementation	Data structures and algorithms chosen according to performance characteristics, equipment capacity, and data volume. Programming language is used to implement data structures, such as quaternary trees and R-trees.

In the conceptual universe, geographic space is modeled with the complementary views of field and object model. In the field model, geographic space is considered a continuous surface where phenomena occur in a variety of ways, with different distributions. For example, at each point on a vegetation map, there is an association of the point on the map with a type of vegetation cover. In a geochemical map, each point is associated with the content and anomalies of a given element, and so on. In the object model, the geographic space is represented by a set of distinct and identifiable entities, as in a cadastre of lots in a

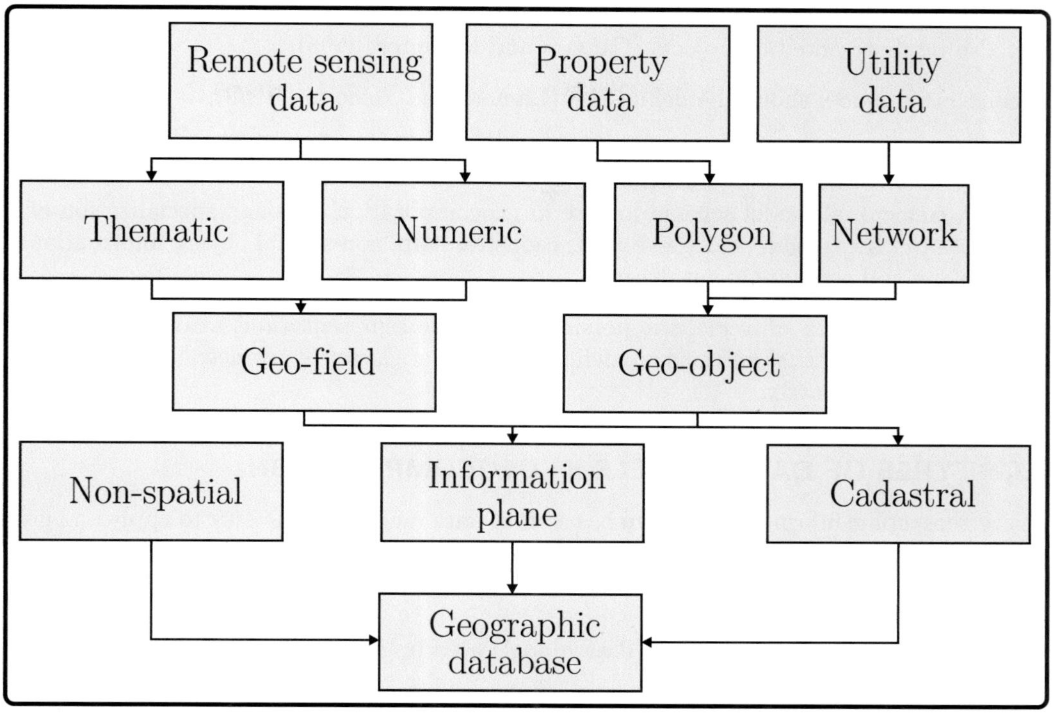

FIGURE 5.7 Conceptual representation of a geographic data model with a geographic database, information plane, geo-field, geo-object cadastral for raster and vector representation of rural and urban data.

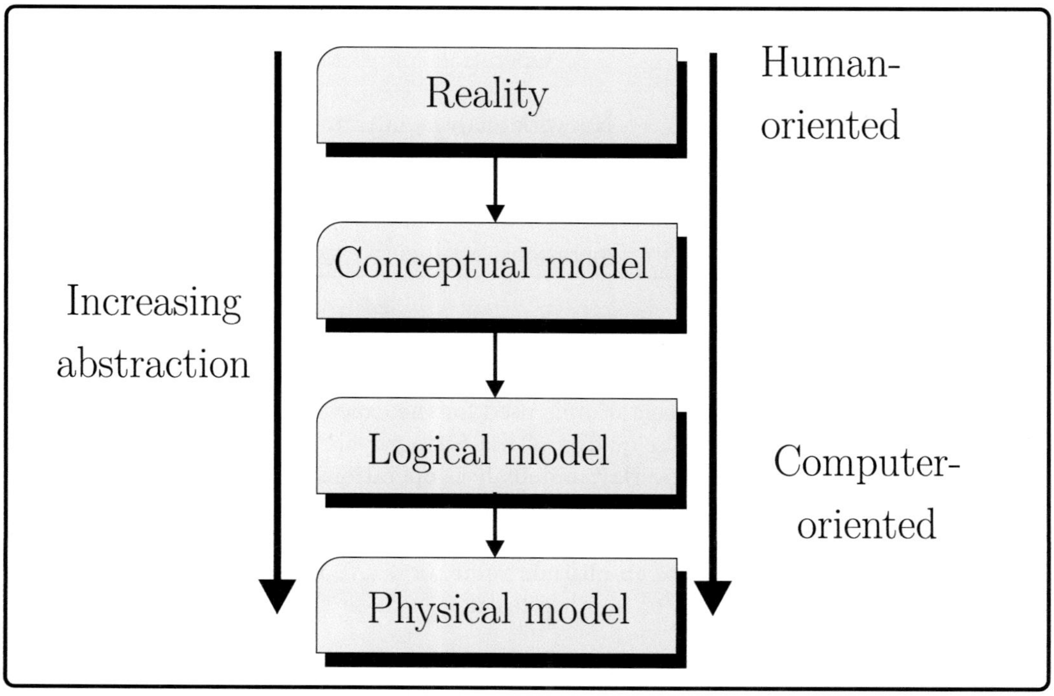

FIGURE 5.8 Levels of data model abstraction in geocomputation.

municipality with attributes about each lot. Rivers in a watershed and airports in a state can also be represented by objects (Câmara and Medeiros, 1998).

A conceptual model should be defined by (Câmara and Medeiros, 1998):

- Basic classes of model with relationships based on principles of specialization, generalization, and aggregation;
- A conceptual model schema for use in geocomputation, through specialization of basic classes of geo-field, geo-object, cadastral map, non-spatial object, information plane and geographic database.

In the representation universe, the possible geometric representations associated with the classes of the conceptual universe are defined, the major classes of geometric representations being vector and matrix.

5.6 TYPES OF DATA MODELS IN GEOCOMPUTATION

Every geographic information system has a set of data models that is easy to apply and gives greater work productivity. The set of entities with the same type of geometric dimensionality is defined as a representation layer (Table 5.5) (Longley et al., 2001).

TABLE 5.5 Types of geographic data models used in geocomputation.

Data Model	Application Example
Computer-Aided Design (CAD)	Automated engineering design
Non-topological Graph	Simple mapping
Image	Image processing
Raster and grid	Spatial analysis and modeling in environmental, natural resources and agriculture
Vector with topological relationships	Operations on vector geometric features
Network	Network analysis in transportation, hydrology and utilities
Triangulated Irregular Network (TIN)	Terrain modeling, analysis and visualization
Object	Operations on all entity types in various applications

5.7 VECTOR REPRESENTATION HIERARCHY

In geocomputation, vector representation is used for the exact nature of the representation method, the storage efficiency, the quality of the cartographic output, and the availability of functions for map projection, overlay, and analysis operations. The vector model presents coordinates that define the geometry of objects of type point, line, or polygon (Figure 5.9) with the number of dimensions equal to 2 (X, Y), column and line, or longitude and latitude; 3 (X, Y, Z), with the addition of an altitude value; or 4 (X, Y, Z, M), with the addition of a M value to represent time or another feature, such as the depth of the soil sample (Longley et al., 2001).

In general, the hierarchy of representations of vector geometric primitives is defined with coordinates 2-D and 3-D, point 2-D, line 2-D, isoline, oriented arc 2-D node, grid node, 3-D node, and polygon. After defining the geometric vector primitives, a hierarchy is established by specializations of 2-D point set, isoline set, planar subdivision, oriented graph, triangle grid, and 3-D point map (Figure 5.10) (Câmara and Medeiros, 1998).

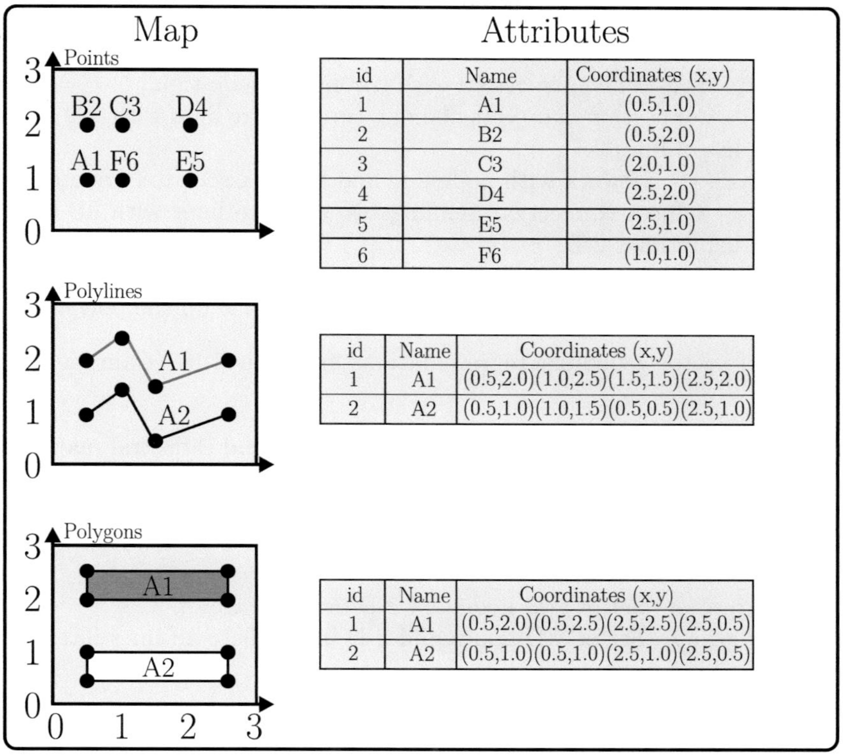

FIGURE 5.9 Representation of point, line and polygon in the vector data model.

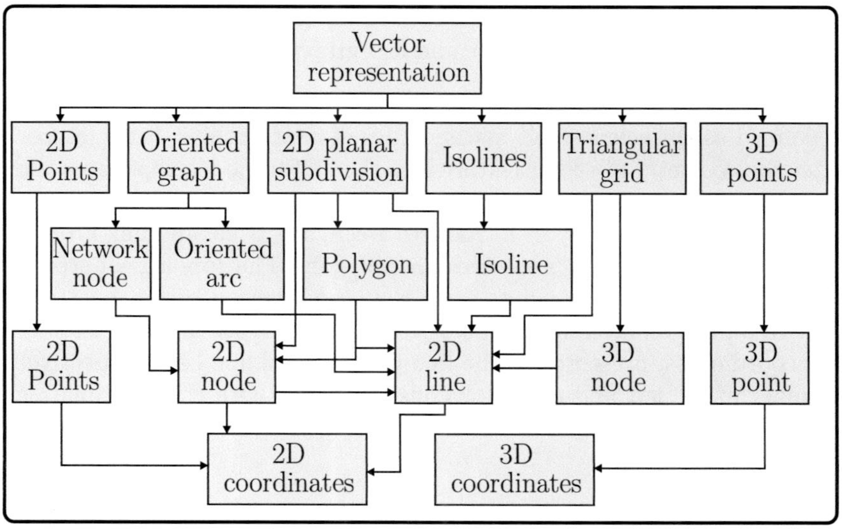

FIGURE 5.10 Class hierarchy for vector representation.

The geometric primitives are defined by (Câmara and Medeiros, 1998):

- Coordinates 2-D <- Object with singular location $(X, Y) \in R$;

- Coordinates 3-D <- Object with singular location $(X,\ Y,\ Z)$ where $(X,\ Y) \in R$;
- Point 2-D <- Object with descriptive attributes and coordinates 2-D;
- Point 3-D <- Object with descriptive attributes and coordinates 3-D;
- Line 2-D <- 2D line with attributes and coordinate set 2-D;
- Isoline <- Line with an associated real value (e.g., elevation);
- Oriented arc <- 2-D line associated with traversal orientation;
- Node 2-D <- Connection of objects defining two or more lines with 2-D coordinates $\in R$ and a list of lines 2-D;
- Node network <- Network with node 2-D and list of associated oriented arcs;
- Node 3-D <- Object connection defining two or more lines with 3D coordinates $\in R$ and a list of lines 2-D;
- Polygon <- Polygon with a list of 2-D lines and 2-D nodes representing the coordinates of the outer area and outer areas that make up the polygon.

Specializations of vector geometric representations are defined by (Câmara and Medeiros, 1998):

- 2-D point set <- Set of 2-D points used for location and cadastral mapping, along space;
- Set of isolines <- Set of lines with a coordinate and that do not intersect;
- Planar subdivision <- Set of polygons, 2-D lines and 2-D nodes in a geographic region R;
- Oriented graph <- Set of grid node and 2-D oriented arc;
- Triangular grid <- Set of 3-D nodes and 2-D lines where all lines intersect at the start and end points of each line;
- Point map 3-D <- Set of 3-D points.

5.7.1 VECTOR FEATURE

The vector representation is encoded by the vector feature model with cadastral tables with simple point, line and polygon geometry and topology. This type of dataset is used in geocomputation for ease creation, storage and fast graphical display (Câmara and Medeiros, 1998).

Topology, defined as the science and mathematics of relationships, is structured with rules and validates the geometry of vector features. In the vector model, topology validates vector dataset geometry and for network tracing operations and polygon adjacency testing. The topological structure of vector layers introduces 1-cell, arc, edge, and link properties for line, as well as 2-cell, area, and face properties for polygon. The topological structure of lines provides editing options at exact points using the snap function within a certain distance of the line end. In a polygon layer with topology, the polygons are made of lines, and the lines are made of coordinates, presented in the line-polygon and line lists, respectively. For each line, the number of the left and right polygons are stored with the coordinate geometry of the data, defining the contiguity of a structured polygon layer (Figure 5.11) (Longley et al., 2001).

The geographic relationships between vector object classes are based on geographic operators such as overlap, adjacency, within and between, which determine the interaction between the objects. Some rules are defined for attributes, as connectivity, relationship and geography. The rules are used to define what happens to the characteristics of objects when there is, for example, merging editing and splitting objects. In the hypothetical case of geometric division or merger of a subdivision in an urban region, the area and tax values

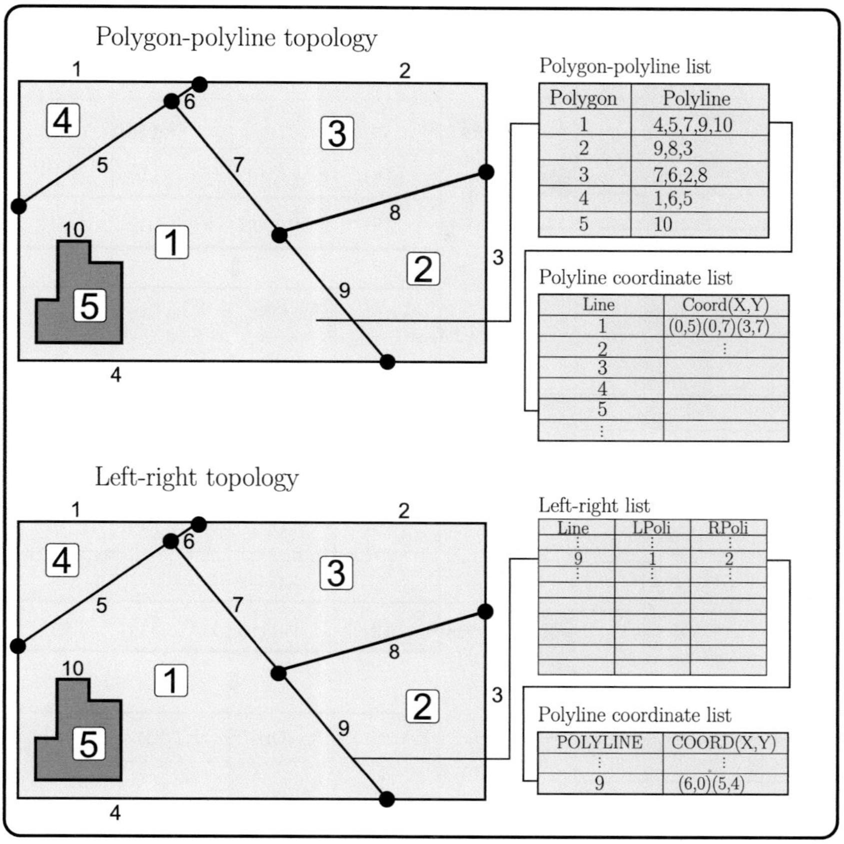

FIGURE 5.11 Topological structure and contiguity of a structured topology of a polygon-type feature layer.

should be updated according to the rule for defining the tax value based on the property area (Figure 5.12) (Longley et al., 2001).

5.7.2 NETWORK MODEL

The network model is a special type of topology that can define network flows in radial and loop form. In the radial, or tree, the flow modeling has an up or down direction. In the loop network, intersections can occur and interrupt the network flow (Longley et al., 2001).

5.7.3 TRIANGLE NETWORK MODEL

Triangulated Irregular Networks (TIN) are irregular triangles used to create surfaces in geocomputation. In the TIN surface, contiguous triangular elements define the surface. A TIN network can be created from points with coordinates X, Y, Z, after interpolation by Delaunay triangulation, in order to define areas with faces that touch neighbors at all edges. The advantages of the TIN representation are the incorporation of the original sampled points to check the modeling accuracy, efficient storage of terrain data, and simple evaluation of the difference in elevation, slope, aspect and line of sight between points. In the TIN topology structure, triangles always have three nodes and usually three neighboring triangles. In the TIN periphery, triangles may have one or two neighbors (Figure 5.13) (Longley et al., 2001).

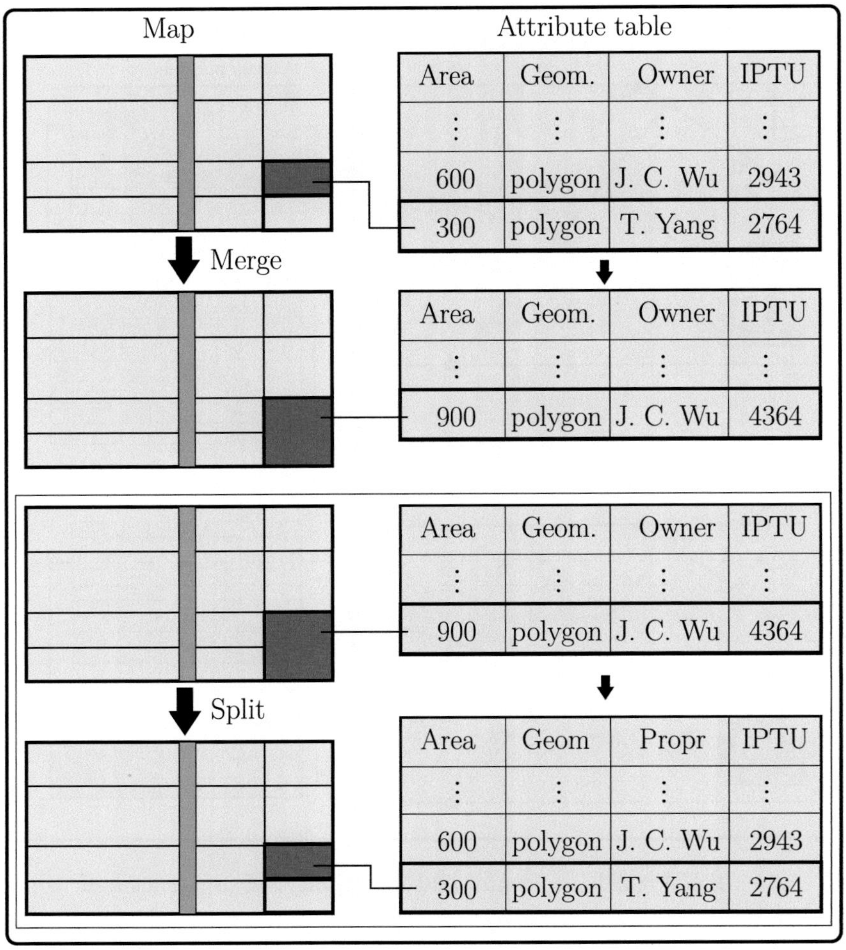

FIGURE 5.12 Example of using merge and split rules for objects in a subdivision in an urban region.

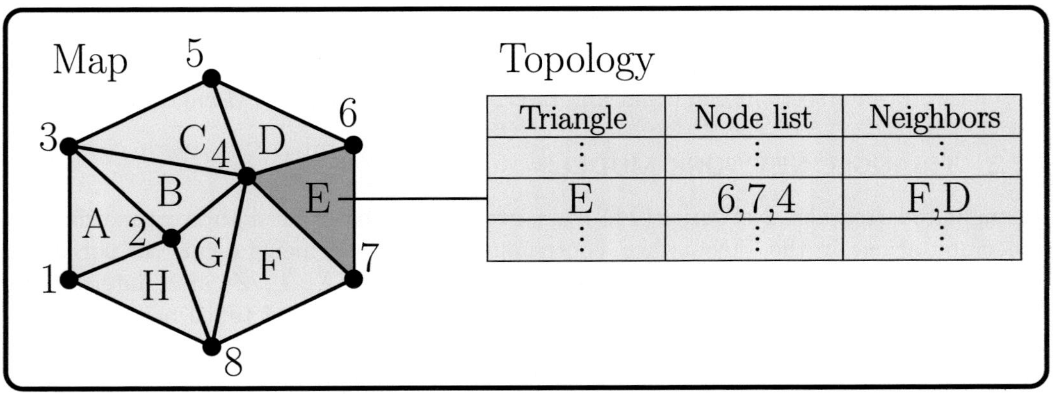

FIGURE 5.13 Topology of the TIN triangles network model.

Object data models have desirable features in geocomputation modeling, such as (Longley et al., 2001):

- Encapsulation;
- Inheritance;
- Polymorphism.

Encapsulation refers to the fact that each individual object has a state and behavior description. The state of the object can be related to attribute characteristics, and the behavior can be related to the operations that can be applied on the object, such as create, delete, draw, split, merge and query. Inheritance is the possibility of using existing characteristics of one object in another object with modifications, additions or removals of previously existing characteristics. Polymorphism describes the process in which each object is developed, drawn, created, or deleted according to its implementation (Longley et al., 2001).

5.8 MATRIX REPRESENTATION HIERARCHY

The matrix representation hierarchy can be represented by raster model with arrangement of cells or pixels to represent real-world objects and processes. In the cells there can be an attribute value according to coding schemes in integer and floating point value categories. In the simplest case, a binary representation can be used. Rasters can be stacked to represent multiple attributes at a single pixel location. In rasters there is usually a header and geographic coordinate system associated with the pixels. Due to the large amount of data stored in rasters, compression techniques are developed for indexing, such as run-length encoding, block encoding, wavelet and quadtrees (Longley et al., 2001).

The matrix representation hierarchy can be specialized into the following classes: regular grid, image band, thematic image, and synthetic image (Figure 5.14) (Câmara and Medeiros, 1998):

- Regular grid <- Matrix of real numbers;
- Image band <- Image with matrix with values of digital numbers;
- Thematic image <- Image with a numerical value associated with an element of the matrix;
- Synthetic image <- Image in false-color composition.

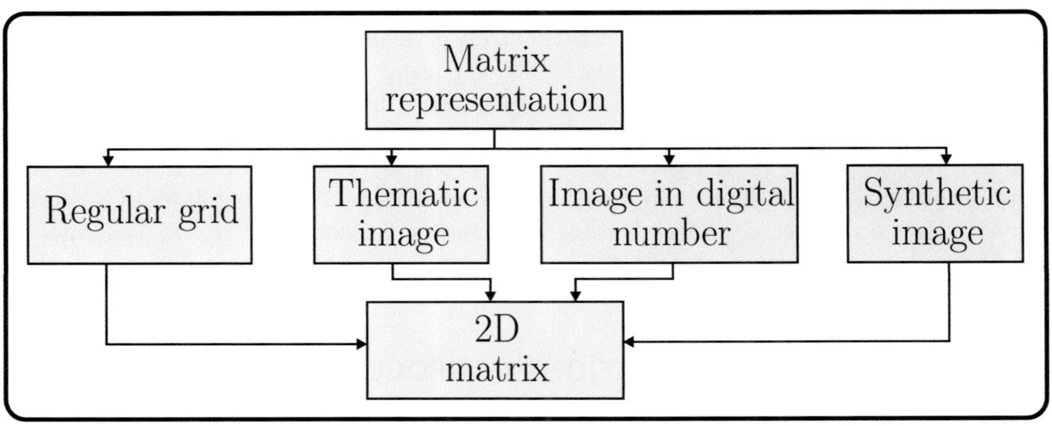

FIGURE 5.14 Class hierarchy for matrix representation.

5.9 DATA MODEL IMPLEMENTATION

In data model implementation, decisions about the programming used in a geocomputation system vary according to the availability of algorithms for geographic data analysis and the performance of the hardware used. Spatial indexing frameworks are used to access spatial data with data structures and search and retrieval algorithms. These methods are developed on multidimensional keys that vary according to the associated data (Table 5.6) (Câmara and Medeiros, 1998).

TABLE 5.6 Multidimensional key subdivision for geographic data access by spatial indexing.

Data type	Spatial indexing
Points	K-D trees
Lines and polygons	R and R+ trees
Images	Quaternary trees
Multiresolution data	V-trees

As an example, instances of the remote sensing data class can be stored as gray-level images and synthetic images. A numeric geo-field can be represented as a matrix (regular grid) or as vectors (isolines, triangular grid, and 3-D point set). A thematic geo-field can be represented as vectors with topological structure (planar subdivision), or as matrix (thematic image). Cadastral instances can be represented by planar subdivision instances or by 2-D point sets. A network can be represented by oriented graph (Câmara and Medeiros, 1998).

5.10 RASTER COMPRESSION TECHNIQUES

One of the main operational problems associated with raster is the amount of raw data that must be stored. Many types of raster compression techniques have been developed to improve storage efficiency, such as run-length encoding, block encoding, wavelet compression, and quadtrees. Simple run-length encoding techniques can compress the file size by a factor of 5. Wavelet compression techniques result in a compression rate of almost 40. Run-length encoding is perhaps the simplest compression method involving encoding row cells that have the same value, with a pair of values indicating the number of cells with the same value and the actual value. Block encoding is like a two-dimensional version of run-length encoding in which the array is defined as a series of square blocks of the largest size possible. The array is divided using blocks of a smaller and smaller size. It is sometimes described as a quadtree data sctructure. Wavelet compression techniques remove information by recursively examining patterns in datasets at different scales. Wavelet compressed raster layers can be quickly viewed at different scales with appropriate amounts of detail, especially for compressing aerial photographs. This results in very high compression ratios, because information is lost, and its use is limited to applications that do not need to use the raw digital numbers for processing or analysis (Longley et al., 2001).

5.11 DATA MODEL APPLICATIONS IN GEOCOMPUTATION

In the process of defining the conceptual schema, the specialization hierarchy defined by modeling enables the creation of classes for use in geographic databases, with correspondence between the different universes used in modeling (Table 5.7).

TABLE 5.7 Example of correspondence between the modeling universes for representing the environment.

Reality	Concept	Representation	Implementation	Product
Vegetation	Thematic geo-field	Thematic image	2-D matrix	Vegetation map
Altitude	Numerical geo-field	Set of 3-D points	3-D points with KD-tree	Elevation map
Urban	Cadastral geo-object	Planar subdivision with polygons	2-D polygons with R-Tree	Urban plots map
Rural	Cadastral geo-object	Planar subdivision with polygons	2-D polygons with R-Tree	Property map
Utilities	Cadastral geo-object	Network with oriented graph	2-D lines with R-Tree	Power grid map

As an example, the definition of a rural cadastral modeling scheme is considered. The class rural properties is specialized into geo-objects subdivided into large farm and small farm for cadastral mapping. The class, soil map, specialized as planar thematic subdivision, stores instances of soil types on the properties. The classes' elevation and slope, specializations of the digital terrain model, store the altitude and slope on the properties. Land use monitoring is performed with Landsat remote sensor data over the studied region (Figure 5.15).

5.12 COMPUTATIONAL PRACTICE

The R package `sp` (Pebesma et al., 2021b) is used in spatial data analysis in R. In `sp` sets of classes are defined to represent spatial data. A class defines a particular type of data, such as the table-like class type called `data.frame`. The definition of classes creates a standard representation of a given data model making it easy to apply functions. The R package `sp` has many functions for modifying or analyzing spatial data. In the R `sp` package there are several classes defined with names starting with the name `Spatial`. For vector data, the basic types are `SpatialPoints`, `SpatialLines` and `SpatialPolygons`. These classes represent only geometries. A `data.frame` is used to store attributes associated with geometries, defining hybrid data model structures of `SpatialPointsDataFrame`, `SpatialLinesDataFrame` and `SpatialPolygonsDataFrame`.

The R package `raster` (Hijmans et al., 2020) is developed with classes of type `RasterLayer`, `RasterBrick` and `RasterStack`, called `raster` objects. In the R `raster` package there are functions for creating, reading, manipulating and writing raster data and general raster data manipulation functions with specific functions such as reading a block of raster values from a file, converting pixel numbers to coordinates and vice versa, and performing algebra of rasters.

The `install.packages` function installs the `raster` R package.

```
install.packages("raster")
```

The `library` function enables the `raster` and load `sp` package.

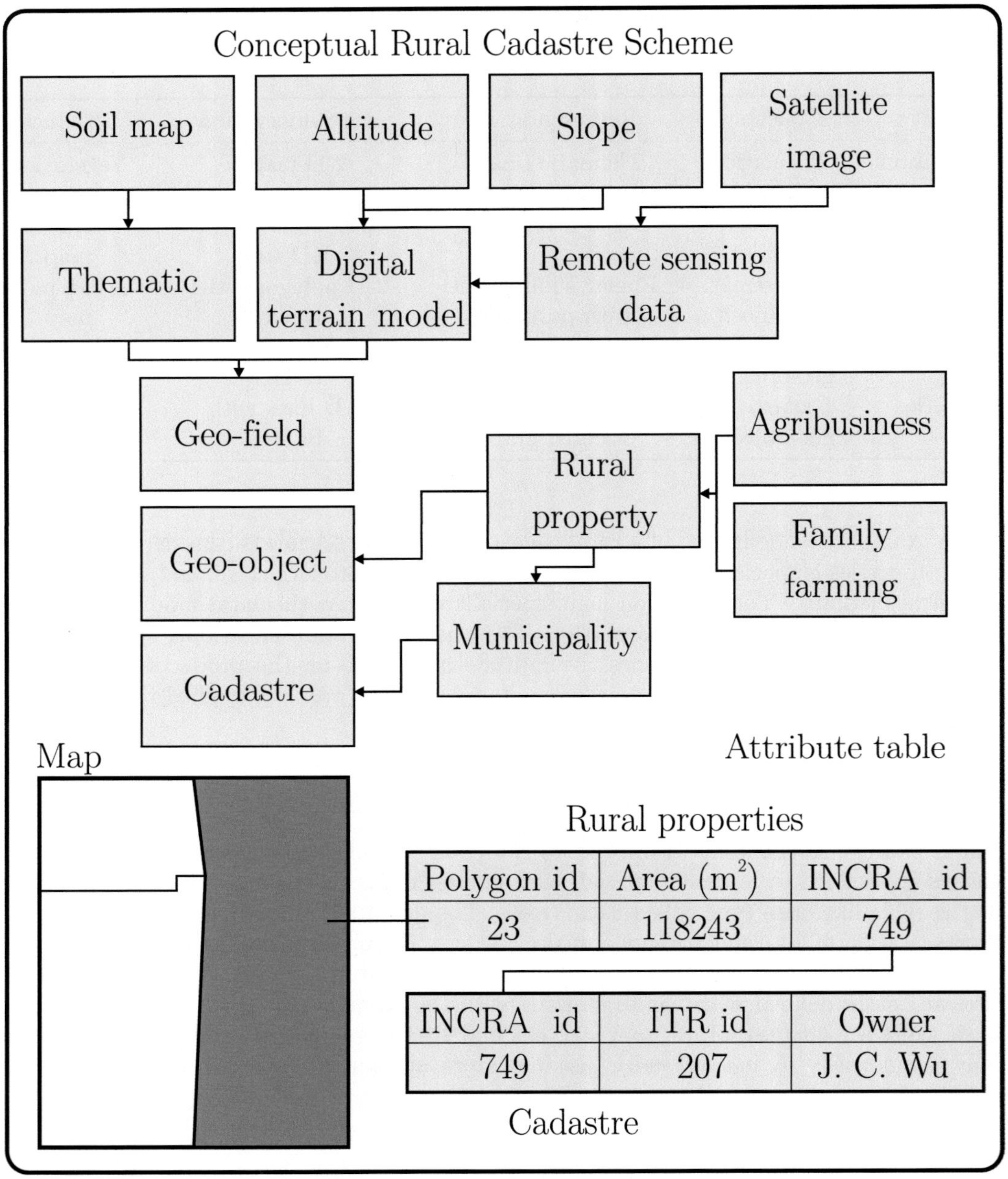

FIGURE 5.15 Data model with conceptual scheme and rural cadastre representation of ecological and economic variables.

```
library(raster)
# Loading required packages: sp
```

5.12.1 VECTOR MODEL

5.12.1.1 Creating a spatial point with attributes

Geographical coordinates (degrees) of latitude and longitude are used to create geospatial points, defined in the WGS-84 coordinate reference system, on the campus of the Universidade Federal de Lavras (UFLA), Brazil.

```
longitude <- c(-44.975860, -44.973541)
latitude <- c(-21.227613, -21.227719)
lonlat <- cbind(longitude, latitude)
```

The longitude and latitude values of the two points are converted to `SpatialPoints`, using the `SpatialPoints` function.

```
pts <- SpatialPoints(lonlat)
```

The `class` function enables to look at the class of the object created.

```
class(pts)
# [1] "SpatialPoints"
# attr(,"package")
# [1] "sp"
```

The contents of the class are taken by the `showDefault` function.

```
showDefault(pts)
# An object of class "SpatialPoints"
# Slot "coords":
#      longitude  latitude
# [1,] -44.97586 -21.22761
# [2,] -44.97354 -21.22772
# Slot "bbox":
#                 min       max
# longitude -44.97586 -44.97354
# latitude  -21.22772 -21.22761
# Slot "proj4string":
# Coordinate Reference System:
# Deprecated Proj.4 representation: NA
```

A coordinate reference system (CRS) is assigned to the points by setting the `proj4string` parameter.

```
crs <- CRS('+proj=longlat +datum=WGS84')
pts <- SpatialPoints(lonlat, proj4string = crs)
```

A database is created with an identifier code (id) and name associated with the points using the `data.frame` function.

```
df <- data.frame(ID = 1:nrow(lonlat), name = c("p1", "p2"))
df
#   ID name
# 1  1   p1
# 2  2   p2
```

The database is associated with the vector of spatial points by means of the `SpatialPointsDataFrame` function.

```
ptsdf <- SpatialPointsDataFrame(pts, data = df)
ptsdf
# class       : SpatialPointsDataFrame
# features    : 2
# extent    : -44.97586, -44.97354, -21.22772, -21.22761  (xmin, xmax, ymin, ymax)
# crs         : +proj=longlat +datum=WGS84 +no_defs
# variables   : 2
# names       : ID, name
# min values  :  1,   p1
# max values  :  2,   p2
```

The file structure is checked with the `str` function.

```
str(ptsdf)
# Formal class 'SpatialPointsDataFrame' [package "sp"] with 5 slots
# ..@ data       :'data.frame': 2 obs. of  2 variables:
# .. ..$ ID  : int [1:2] 1 2
# .. ..$ name: chr [1:2] "p1" "p2"
# ..@ coords.nrs : num(0)
# ..@ coords     : num [1:2, 1:2] -45 -45 -21.2 -21.2
# .. ..- attr(*, "dimnames")=List of 2
# .. .. ..$ : NULL
# .. .. ..$ : chr [1:2] "longitude" "latitude"
# ..@ bbox       : num [1:2, 1:2] -45 -21.2 -45 -21.2
# .. ..- attr(*, "dimnames")=List of 2
# .. .. ..$ : chr [1:2] "longitude" "latitude"
# .. .. ..$ : chr [1:2] "min" "max"
# ..@ proj4string:Formal class 'CRS' [package "sp"] with 1 slot
# .. .. ..@ projargs: chr "+proj=longlat +datum=WGS84 +no_defs"
# .. .. ..$ comment: chr "GEOGCRS[\"unknown\",\n
# DATUM[\"World Geodetic System 1984\",\n
# ELLIPSOID[\"WGS 84\",6378137,298.25722"| __truncated__
```

Another way to check the created file is with the `showDefault` function.

```
showDefault(ptsdf)
# An object of class "SpatialPointsDataFrame"
```

```
# Slot "data":
#   ID name
# 1  1   p1
# 2  2   p2
# Slot "coords.nrs":
# numeric(0)
# Slot "coords":
#      longitude  latitude
# [1,] -44.97586 -21.22761
# [2,] -44.97354 -21.22772
# Slot "bbox":
#                 min       max
# longitude -44.97586 -44.97354
# latitude  -21.22772 -21.22761
# Slot "proj4string":
# Coordinate Reference System:
# Deprecated Proj.4 representation: +proj=longlat +datum=WGS84 +no_defs
# WKT2 2019 representation:
# GEOGCRS["unknown",
#     DATUM["World Geodetic System 1984",
#         ELLIPSOID["WGS 84",6378137,298.257223563,
#             LENGTHUNIT["metre",1]],
#         ID["EPSG",6326]],
#     PRIMEM["Greenwich",0,
#         ANGLEUNIT["degree",0.0174532925199433],
#         ID["EPSG",8901]],
#     CS[ellipsoidal,2],
#         AXIS["longitude",east,
#             ORDER[1],
#             ANGLEUNIT["degree",0.0174532925199433,
#                 ID["EPSG",9122]]],
#         AXIS["latitude",north,
#             ORDER[2],
#             ANGLEUNIT["degree",0.0174532925199433,
#                 ID["EPSG",9122]]]]
```

5.12.1.2 Create spatial line

The `spLines` function is used to create a spatial line between previously defined points with the same coordinate reference system.

```
lns <- spLines(lonlat, crs = crs)
```

5.12.1.3 Create spatial polygon

Similarly, a spatial polygon is created with the `spPolygons` function. In this case, you need four vertices to define the polygon.

```
# Set longitude (x) and latitude (y) coordinates
x <- c(-44.97537, -44.97537, -44.97417, -44.97417)
y <- c(-21.22805, -21.22722, -21.22722, -21.22805)
# Join longitude (x) and latitude (y) coordinates
xy <- cbind(x, y)
# Create spatial polygon
pols <- spPolygons(xy, crs = crs)
```

5.12.1.4 Mapping objects

The `plot` function maps the created objects (Figure 5.16).

```
plot(pts, axes = TRUE, col = "black")
plot(pols, col = "grey90", add = TRUE)
plot(lns, col = "black", add = TRUE)
```

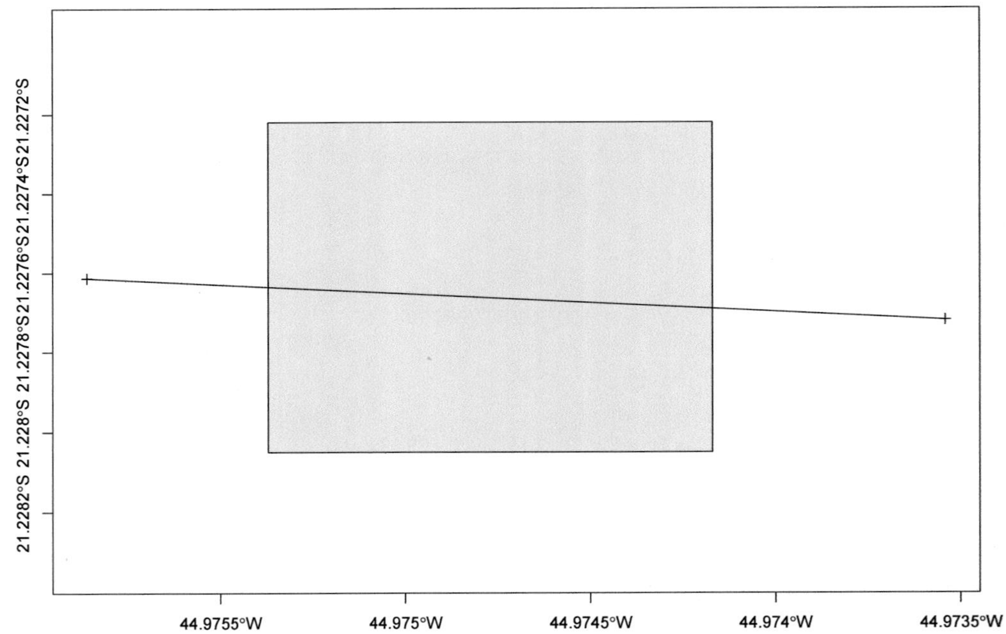

FIGURE 5.16 Vector data model mapping.

5.12.2 RASTER MODEL

5.12.2.1 Raster layer

A `RasterLayer` object is used to represent single or variable layer raster data with fundamental parameters including number of columns and rows, spatial extent, and coordinate reference system. A `RasterLayer` can store pixel value information in a file or in computer memory. The `raster` function is used to set the number of columns (ncol), rows (nrow), minimum and maximum extent in x and y of the raster.

```
r <- raster(ncol = 10, nrow = 10, xmn = -44.97537, xmx = -44.97417,
            ymn = -21.22805, ymx = -21.22722)
```

The header of the created raster is checked as follows.

```
r
# class      : RasterLayer
# dimensions : 10, 10, 100  (nrow, ncol, ncell)
# resolution : 0.00012, 8.3e-05  (x, y)
# extent   : -44.97537, -44.97417, -21.22805, -21.22722  (xmin, xmax, ymin, ymax)
# crs        : +proj=longlat +datum=WGS84 +no_defs
```

A vector of random numbers is associated with the pixels of the raster with the following function.

```
values(r) <- runif(ncell(r))
r
# class      : RasterLayer
# dimensions : 10, 10, 100  (nrow, ncol, ncell)
# resolution : 0.00012, 8.3e-05  (x, y)
# extent : -44.97537, -44.97417, -21.22805, -21.22722  (xmin, xmax, ymin, ymax)
# crs        : +proj=longlat +datum=WGS84 +no_defs
# source     : memory
# names      : layer
# values     : 0.01410259, 0.9944192  (min, max)
```

A new raster is created from the previous raster and called `r1`. A vector of numbers can also be associated with the pixels of the raster with the following function.

```
r1 <- r
values(r1) <- 1:ncell(r1)
r1
# class      : RasterLayer
# dimensions : 10, 10, 100  (nrow, ncol, ncell)
# resolution : 0.00012, 8.3e-05  (x, y)
# extent     : -44.97537, -44.97417, -21.22805, -21.22722  (xmin, xmax, ymin, ymax)
# crs        : +proj=longlat +datum=WGS84 +no_defs
# source     : memory
# names      : layer
# values     : 1, 100  (min, max)
```

The raster `r1` is mapped together with the points, line and polygon created earlier into different overlay layers (Figure 5.17).

```
plot(ptsdf, axes = TRUE)
plot(r1, legend = FALSE, add = TRUE)
plot(pols, col = "transparent", add = TRUE)
plot(lns, add = TRUE)
```

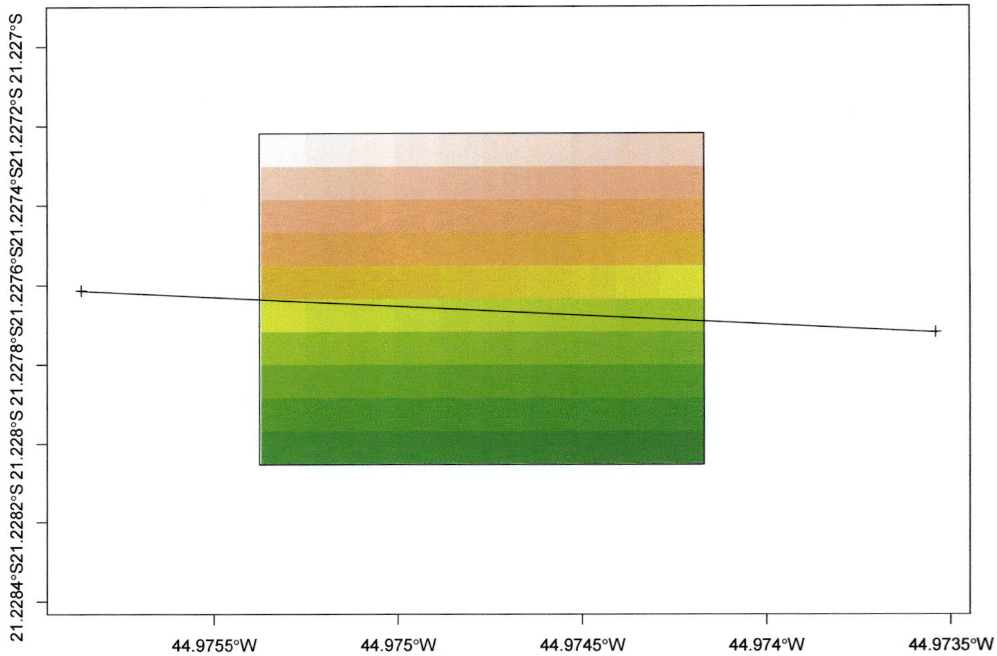

FIGURE 5.17 Mapping raster and vector data model.

5.12.2.2 Raster stack and brick

Raster data can have multiple variables and hence multilayers. Multilayer rasters can be represented by the classes, `RasterStack` and `RasterBrick`. In a `RasterStack`, collections of `RasterLayer` objects with the same extent and spatial resolution are gathered together, defined by a list of `RasterLayer` objects. A `RasterBrick` also assembles multiple layers of objects, but can be more processing-efficient than the `RasterStack`. The main difference between a `RasterStack` and a `RasterBrick` is the more flexible ability of the `RasterStack` to bundle different `RasterLayer` objects with the same length and resolution, while a `RasterBrick` can bundle the structured objects into just a single file.

A `RasterStack` is performed on the two previously created and renamed raster layers, r and r1.

```
rstack <- stack(r, r1) # stack
names(rstack) <- c("r", "r1") # rename
```

The variables from the two previously created raster files are included in the header of the stacked raster.

```
rstack
# class       : RasterStack
# dimensions : 10, 10, 100, 2  (nrow, ncol, ncell, nlayers)
# resolution : 0.00012, 8.3e-05  (x, y)
# extent      : -44.97537, -44.97417, -21.22805, -21.22722  (xmin, xmax, ymin, ymax)
# crs         : +proj=longlat +datum=WGS84 +no_defs
# names       :          r,          r1
# min values :  0.01410259,  1.00000000
# max values :   0.9944192, 100.0000000
```

The mapping of the stacked raster is performed with the `plot` function (Figure 5.18).

```
plot(rstack)
```

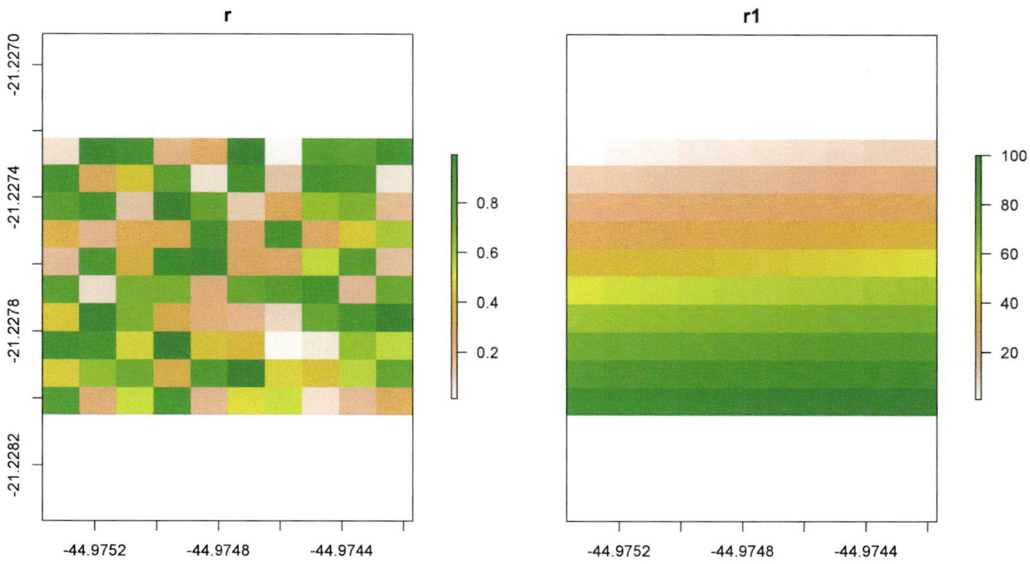

FIGURE 5.18 Stacked raster data model mapping.

In the R `sp` package, there is also support for using raster data in a grid with the `SpatialGridDataFrame` and `SpatialPixelsDataFrame` classes (Pebesma and Bivand, 2005).

6 Space-time Geovisualization with Environmental Geocomputation and R

6.1 LEARNING QUESTIONS

The learning questions answered by reading the chapter are:

- What is the nature of maps and cartography?
- What are the key principles for map design?
- What choices are used to make maps?
- What are the types of map sympathology?
- How can geocomputation be used for visual communication?
- What types of interfaces are used for the user to access geospatial information and conduct research?
- How to use existing data representation models in geocomputation.
- How do 3D geovisualization and virtual reality enhance the ability to understand the world?

6.2 LEARNING OUTCOMES

The learning outcomes answered by reading the chapter are:

- Know about the nature of maps and cartography;
- Know the key principles for map design;
- Verify the options used to make maps;
- Check the types of map symmetry;
- Using geocomputation for visual communication;
- Understand types of interfaces used to access geospatial information and conduct research;
- Check how to use existing models of data representations in geocomputation;
- Verify how 3D geovisualization and virtual reality can improve the ability to understand the world.

6.3 INTRODUCTION

Decision support through geocomputation is performed based on messages of readily interpretable representations in the minds of decision-makers, for different applications in the environment. The main goal of geocomputation is to utilize the huge resource of available files, avoiding information overload. Digital fragments of evidence scattered in space and time are gathered on different applications and types of maps, but with similarities regarding the use of geocomputation (Longley et al., 2001).

The meaning of a spatial real-world representation is communicated publicly and commercially through mapping. Historically, the paper map was the only means of interface available between the map creator and the user. The characteristics of this type of

DOI: 10.1201/9781003724377-6

representation are permanent, with a fixed set of attributes, predetermined and invariant scale, but they rarely provide quantitative or qualitative indications about the uncertainty of the attributes represented, and it is expensive and labor intensive to convert paper maps by digitization into digital format (Longley et al., 2001; Lo and Yeung, 2007). These attributes limit the use of more data, continuously updated and spread across the Internet. Paper and digital maps are important for economic, environmental aspects and social activities as a support tool to communicate and make important decisions (Longley et al., 2001).

Cartography is defined as the art, science and technique of making maps and charts. The term map is used as a reference for mapping in terrestrial areas and, the chart, as a reference for mapping in maritime regions. However, the term map is used for both situations. Maps are effective for summarizing and communicating the results of geocomputation operations to the public. The importance of producing a map is that many consumers of geographic information can interact with geocomputation through the use of map products. Maps can be produced as a formal map, according to cartographic conventions as a referential communication product, or, as a transient map, used only for visualization to display, analyze, edit, and query geographic information (Longley et al., 2001).

Geospatial data of different aspects of biophysical and socioeconomic environments are made available by public and commercial mapping agencies. The wide availability of geospatial data diminishes the importance of the existence of maps; however, the relationship of geographic information systems to maps cannot be ignored. Geocomputation users and scientists must have a certain level of competence about maps to understand critical questions about geocomputation data, implementation, and applications, such as whether in a given application it is necessary to evaluate different scenarios or perform geospatial analysis and modeling (Lo and Yeung, 2007).

Geovisualization is based on established principles of map production and display. The creation and use of visual representations for geovisualization facilitates thinking about, understanding, and building knowledge about environments at measurable geographic scales. Geovisualization is used in research that integrates visualization approaches in scientific computing, cartography, image analysis, information visualization, exploratory data analysis, with geocomputation. With this, it is possible to develop theories, methods and tools for visualization, analysis, synthesis and presentation of geospatial data with different topics and subtopics, such as those defined in the research agenda of the International Cartographic Association (Table 6.1) (Longley et al., 2001).

TABLE 6.1 Geovisualization research topics and subtopics as defined by the International Cartographic Association's research agenda.

Topics and subtopics
Cognitive and usability aspects in geovisualization:
- Virtual geographic environments and 3-D geovisualization;
- Dynamic representations (including animated and interactive mapping);
- Improved interface design;
- Understanding individual and group interaction using geovisualization;
- Collaborative viewing;
- Better evaluation of existing methods.
Representation for cartographic visualization:
- Existing and potential functions of map displays;
- Techniques for producing maps and visualizing spatial data;

Topics and subtopics
- Technology to improve cartographic practice;
- Relationships between user cognition, map design, and the nature of spatial data.
Geovisualization applied to knowledge:
- Visual approaches to gaining knowledge;
- Visual support for obtaining knowledge through geocomputation;
- Use of databases and data models to support knowledge gathering.
User interface aspects for geovisualization:
- Interface development and use;
- Usability and navigability in geovisualization;
- Resources for user collaboration.

Geovisualization is much more than conventional map drawing, becoming an application that uses geographic data resources to meet various scientific and social needs, as well as a line of research for the development of new methods and visualization tools for better user interaction with data. This is possible by performing queries, geographic data transformation, and user immersion in this context (Longley et al., 2001).

6.4 CHARACTERISTICS OF MAPS

Maps are used as mechanisms for storing and communicating geographic information. Maps can present the results of analyses, as well as communicate spatial relationships between phenomena on the same map, or between maps of the same or different areas. The use of maps assists in identifying spatial order and differentiation with effective decision support and ease of interpretation. Maps not only organize and convey known information about the environment, but they also create or reinforce a particular message (Longley et al., 2001).

The disadvantages of using maps are (Longley et al., 2001):

- Maps can accidentally or purposefully impair communication due to incorrect use of symbols, conveying the wrong message to users;
- A map is the result of a single realization of a spatial process, so that each map instance represents the result of a sampling trial performed, and thus a single occurrence generated from all possible maps of the same subject for the same area;
- Maps are in general created by complex rules, symbology and conventions, making it difficult for an untrained user to understand and interpret a map, especially in the case of multivariate and statistical thematic mapping, where the idiosyncrasies of classification schemes and color symbology can be complex.

Map design is considered a creative process in which the cartographer conveys the message of the map's purpose to share information, highlight patterns, processes, and illustrate outcomes of interest. A secondary objective is to create a pleasing and interesting image without detracting from the primary objectives. Designing a map is a very complex procedure that requires simultaneous optimization of many variables and harmonization of multiple methods (Longley et al., 2001).

Considering a map with a standardized geographic coordinate system or plane coordinate system conventions, the mapping process varies according to the purpose of the map, characteristics of the area to be mapped, available data, audience, conditions of use, and technical boundaries (Longley et al., 2001), as well as other map characteristics in the mapping process, such as (Lo and Yeung, 2007):

- Abstraction;
- Simplification;
- Focus;
- Symbolization;
- Scale;
- Location;
- Purpose;
- Classification.

6.4.1 SCALE

Geographic representation at a given scale is a primary characteristic of a geospatial datum. Map scale refers to the rate or ratio between distances measured on the map and the corresponding distance on the ground. The ways generally used to define scale in maps are (Lo and Yeung, 2007):

- Representative fraction <- Representation of scale in rate form such as 1:100 or 1/100, where 1 is the distance on paper and 100 the corresponding distance on the terrain. In this case, 1 cm on paper is equal to 1 m on the ground;
- Nominal scale <- Representation of scale in descriptive form, i.e., 1 cm on paper is equivalent to 1 m on the terrain;
- Bar scale <- Representation of scale in graphical form drawn on the map to facilitate estimating distances on the terrain from measurements made on the map.

One option for using vector data from countries at different levels of detail is to obtain it from websites such as GADM[1]. In this case, level 1 detail data is obtained for Brazil in order to get vectors for the country and states. The `sf` (Pebesma et al., 2021a) and `tmap` (Tennekes, 2018) packages are loaded for mapping.

```
library(sf)
# Linking to GEOS 3.11.2, GDAL 3.7.2, PROJ 9.3.0; sf_use_s2() is TRUE
library(tmap)
# package 'tmap' was built under R version 4.3.3
```

The `read_sf` function is used to import the GADM database for Brazil into R. The `read_sf` function is used to import the GADM database for Brazil into R. The state of Minas Gerais is obtained from a subset of Brazil using bracket operators and attribute names.

```
br1 <- read_sf("C:/geo/GADM/Brazil/gadm36_BRA_1.shp")
# Simple feature collection with 27 features and 10 fields
# Geometry type: MULTIPOLYGON
# Dimension:     XY
# Bounding box:  xmin: -73.98971 ymin: -33.74708 xmax: -28.84694 ymax: 5.264878
# Geodetic CRS:  WGS 84
# A tibble: 27 × 11
brmg1 <- br1[br1$NAME_1 == "Minas Gerais",]
# Simple feature collection with 1 feature and 10 fields
# Geometry type: MULTIPOLYGON
```

[1]https://gadm.org/data.html

```
# Dimension:     XY
# Bounding box:  xmin: -51.04588 ymin: -22.92275 xmax: -39.85676 ymax: -14.23343
# Geodetic CRS:  WGS 84
# A tibble: 1 × 11
```

The functions tm_shape, tm_polygons, tm_compass, and tm_scale_bar are used to map the polygons with a graphic scale and wind rose. The function tmap_arrange is used to map the country and state side by side (Figure 6.1).

```
br <- tm_shape(br1) +
    tm_polygons() +
    tm_compass(type = "8star", position = c("right", "top")) +
    tm_scale_bar(breaks = c(0, 200, 400), text.size = 1)
mg <- tm_shape(brmg1) +
    tm_polygons() +
    tm_compass(type = "8star", position = c("right", "top")) +
    tm_scale_bar(breaks = c(0, 200, 400), text.size = 1)
tmap_arrange(br, mg, ncol = 2)
```

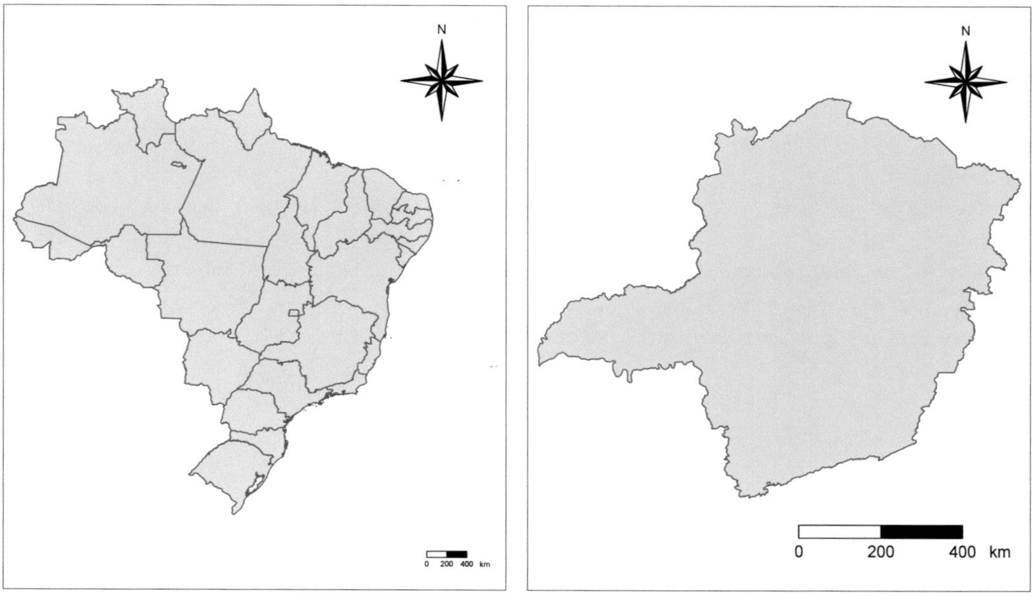

FIGURE 6.1 Graphic bar scale and wind rose used in mapping conventions for Brazilian states and the state of Minas Gerais.

Scale is one of the main factors that determine the design, production, and use of maps. The choice of scale for a given application is basically determined by the purpose of the map. Maps can be made at large scales, such as 1:5000, as in the case of engineering studies, plot registration, and land planning, as well as at small scales, as in the case of an atlas map or large area map, usually smaller than 1:100000. Medium scale maps between 1:5000 and 1:100000 represent large areas with a reasonable amount of detail, such as urban areas, minor roads and bridges, typically used in topographic mapping at the national or state level and provided by mapping agencies (Lo and Yeung, 2007).

The scale can also refer to the detail of the database used. As an example, the mapping of rivers and lakes from the Natural Earth website database to assess the scales at which the data is constituted is represented in the scales 1:110000000 (110), 1:50000000 (50) and 1:10000000 (10). In Minas Gerais, rivers are found only at scales 50 and 10, while for lakes, only at scale 10 is the largest number of lakes found in the state.

The `rnaturalearth` (Massicotte and South, 2023) and `sf` packages are loaded with the `library` function for mapping global rivers and lakes around the state of Minas Gerais, Brazil.

```
library(rnaturalearth)
# package 'rnaturalearth' was built under R version 4.3.3
library(sf)
```

The `ne_download` function is used to download data on rivers and lakes at scales 110, 50, and 10. Note that the size of the files increases as the scale number decreases due to the increased detail in the database.

```
lakes110 <- ne_download(scale = 110, type = 'lakes', category = 'physical')
# 'https://naturalearth.s3.amazonaws.com/110m_physical/ne_110m_lakes.zip'
# Content type 'application/zip' length 23622 bytes (23 KB)
# downloaded 23 KB
# Simple feature collection with 24 features and 37 fields
# Geometry type: POLYGON
# Dimension:     XY
# Bounding box:  xmin: -124.9536 ymin: -16.53641 xmax: 109.9298 ymax: 66.9693
# Geodetic CRS:  WGS 84
lakes50 <- ne_download(scale = 50, type = 'lakes', category = 'physical')
# 'https://naturalearth.s3.amazonaws.com/50m_physical/ne_50m_lakes.zip'
# Content type 'application/zip' length 252615 bytes (246 KB)
# downloaded 246 KB
# Simple feature collection with 412 features and 39 fields
# Geometry type: POLYGON
# Dimension:     XY
# Bounding box:  xmin: -165.8985 ymin: -50.62002 xmax: 176.0827 ymax: 81.94033
# Geodetic CRS:  WGS 84
lakes10 <- ne_download(scale = 10, type = 'lakes', category = 'physical')
# 'https://naturalearth.s3.amazonaws.com/10m_physical/ne_10m_lakes.zip'
# Content type 'application/zip' length 2349685 bytes (2.2 MB)
# downloaded 2.2 MB
# Simple feature collection with 1355 features and 41 fields
# Geometry type: MULTIPOLYGON
# Dimension:     XY
# Bounding box:  xmin: -165.9656 ymin: -50.66967 xmax: 177.1544 ymax: 81.95521
# Geodetic CRS:  WGS 84
rivers110 <- ne_download(scale = 110, type = 'rivers_lake_centerlines',
                         category = 'physical')
# 'https://naturalearth.s3.amazonaws.com/110m_physical/
```

```
# ne_110m_rivers_lake_centerlines.zip'
# Content type 'application/zip' length 30879 bytes (30 KB)
# downloaded 30 KB
# Simple feature collection with 13 features and 35 fields
# Geometry type: LINESTRING
# Dimension:     XY
# Bounding box:  xmin: -135.3134 ymin: -33.99358 xmax: 129.956 ymax: 72.90651
# Geodetic CRS:  WGS 84
rivers50 <- ne_download(scale = 50, type = 'rivers_lake_centerlines',
                        category = 'physical')
# 'https://naturalearth.s3.amazonaws.com/50m_physical/
# ne_50m_rivers_lake_centerlines.zip'
# Content type 'application/zip' length 504454 bytes (492 KB)
# downloaded 492 KB
# Simple feature collection with 478 features and 36 fields
# (with 1 geometry empty)
# Geometry type: MULTILINESTRING
# Dimension:     XY
# Bounding box:  xmin: -165.2439 ymin: -50.24014 xmax: 176.3258 ymax: 73.3349
# Geodetic CRS:  WGS 84
rivers10 <- ne_download(scale = 10, type = 'rivers_lake_centerlines',
                        category = 'physical')
# 'https://naturalearth.s3.amazonaws.com/10m_physical/
# ne_10m_rivers_lake_centerlines.zip'
# Content type 'application/zip' length 2079507 bytes (2.0 MB)
# downloaded 2.0 MB
# Simple feature collection with 1473 features and 38 fields
# Geometry type: MULTILINESTRING
# Dimension:     XY
# Bounding box:  xmin: -164.9035 ymin: -52.15775 xmax: 177.5204 ymax: 75.79348
# Geodetic CRS:  WGS 84
```

The geometry results of river and lake vectors are mapped side by side with functions from the `sf` package (Figure 6.2).

```
par(mfrow = c(2, 3))
# Rivers
plot(st_geometry(brmg1), main = "rivers110", col = sf.colors(n = 1, alpha = 0.3),
     bgc = 'gray92', axes = TRUE)
plot(st_geometry(rivers110), col = 'blue', add = TRUE)
plot(st_geometry(brmg1), main = "rivers50", col = sf.colors(n = 1, alpha = 0.3),
     bgc = 'gray92', axes = TRUE)
plot(st_geometry(rivers50), col = 'blue', add = TRUE)
plot(st_geometry(brmg1), main = "rivers10", col = sf.colors(n = 1, alpha = 0.3),
     bgc = 'gray92', axes = TRUE)
plot(st_geometry(rivers10), col = 'blue', add = TRUE)
# Lakes
plot(st_geometry(brmg1), main = "lakes110", col = sf.colors(n = 1, alpha = 0.3),
     bgc = 'gray92', axes = TRUE)
```

```
plot(st_geometry(lakes110), col = 'lightblue', border = "blue", add = TRUE)
plot(st_geometry(brmg1), main = "lakes50", col = sf.colors(n = 1, alpha = 0.3),
     bgc = 'gray92', axes = TRUE)
plot(st_geometry(lakes50), col = 'lightblue', border = "blue", add = TRUE)
plot(st_geometry(brmg1), main = "lakes10", col = sf.colors(n = 1, alpha = 0.3),
     bgc = 'gray92', axes = TRUE)
plot(st_geometry(lakes10), col = 'lightblue', border = "blue", add = TRUE)
```

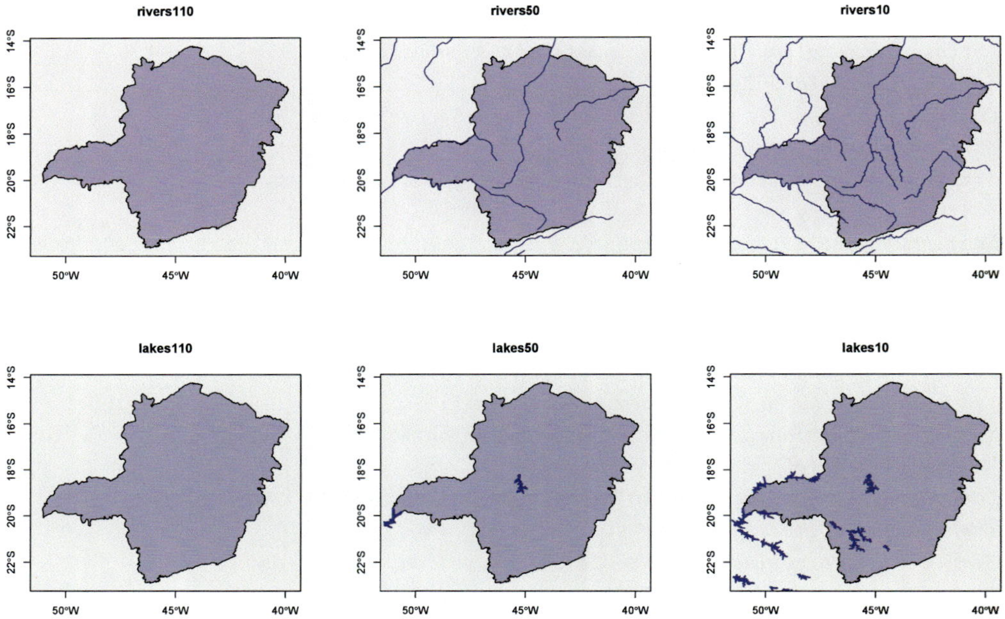

FIGURE 6.2 Mapping rivers and lakes in the state of Minas Gerais, obtained from the Natural Earth website, at scales 1:110 million, 1:50 million and 1:10 million.

6.4.2 CLASSIFICATION

Mapping classes vary according to the purpose of use in (Lo and Yeung, 2007):

- Topographic or general purpose maps;
- Thematic or specific purpose maps.

Topographic maps are useful for different purposes characterized by standards of accuracy and content. The focus of these maps is the location and representation of physical and cultural features, such as the landscape, drainage network, roads, highways, airports, built-up areas, agricultural cultivation areas, and conservation areas. These maps are the reference base for making thematic maps and integrating geospatial data from different sources. In geocomputation, topographic maps in digital format are called digital topographic base and used as reference for georeferencing other data or for specific thematic applications (Lo and Yeung, 2007).

Topographic maps with thematic purpose are used in specific measurement and classification situations. These maps are produced at different scales using a base reference map and applied in thematic mapping of biophysical and socioeconomic data. In geocomputation, these maps, in digital format, are called overlay plots, to provide geospatial data for applications to specific objects and processes (Lo and Yeung, 2007).

6.4.3 THE MAPPING PROCESS

The mapping process is associated with different technologies and consists of the following working steps (Figure 6.3) (Lo and Yeung, 2007):

- Planning;
- Data acquisition;
- Production;
- Product delivery.

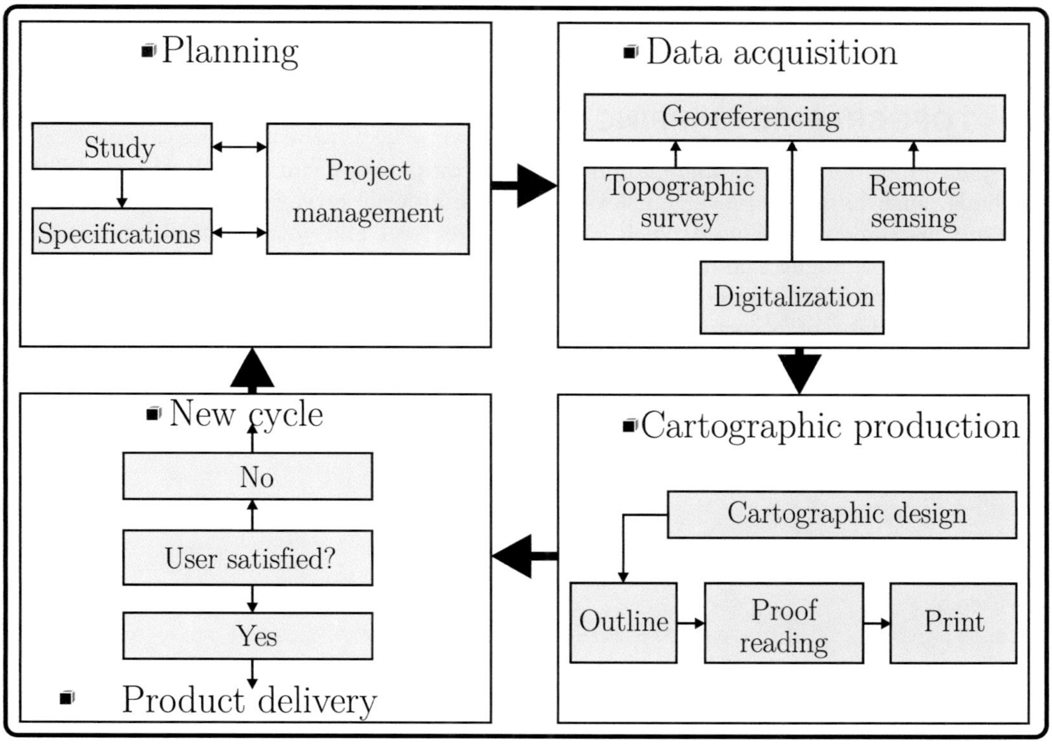

FIGURE 6.3 The mapping process with different work steps.

The planning phase is initiated according to user-defined study and analysis needs. This phase depends on the nature of the mapping project according to the objective of the work. The necessary costs and requirements are justified according to the resources and mapping agencies, according to existing specifications and standards. The data acquisition phase includes selecting and establishing geographic reference network and obtaining data by geodetic survey methods with topographic and remote sensing equipment. The major work in the data acquisition phase is the evaluation of available data sources. In the cartographic production phase, the raw data are converted into maps ensuring accuracy, correctness and compilation according to the required specifications. In the product delivery phase, maps are stored and managed for printing and distribution. Geospatial data can be distributed through data catalogs, digital geospatial libraries, and Internet websites of public and private agencies (Lo and Yeung, 2007).

6.4.4 MAP COMPOSITION

A fundamental requirement for a good map is a layout with good visual balance. The map composition defined in the process of creating a map comprises several closely related elements (Longley et al., 2001):

- Map body <- It should contain the main part of the map;
- Overview map <- Map with broad location to set the context of the main map;
- Title <- Inform the reader about the content of the map;
- Legend <- List of items with description about the symbology used in the map;
- Scale <- Indication about the size of objects and distance relationship between them;
- Direction indicator <- Direction can be evaluated based on wind rose symbol and grid lines;
- Map metadata <- Information about map projection, creation date, data sources and map authorship.

6.5 TOPOGRAPHIC MAPPING

Geospatial data for topographic mapping and geocomputation can be collected by different methods, such as ground-level, air-level, and orbital-level surveys. At the ground level, sighting instruments are used, such as theodolite, level and total station for obtaining measurements of angles and distances in polygons, as well as the use of satellite positioning systems (such as GNSS) for geodetic surveys. Remote sensing data, digital elevation models, and spectral data of objects on the surface are used for characterization of terrestrial environments, and in thematic mapping at air and orbital levels (Lo and Yeung, 2007).

The GADM, level 2 data for Brazil are loaded into R using the `read_sf` function.

```
br2 <- read_sf("C:/geo/GADM/Brazil/gadm36_BRA_2.shp")
br2
# Simple feature collection with 5504 features and 13 fields
# Geometry type: MULTIPOLYGON
# Dimension:     XY
# Bounding box:  xmin: -73.98971 ymin: -33.74708 xmax: -28.84694 ymax: 5.264878
# Geodetic CRS:  WGS 84
# A tibble: 5,504 × 14
```

A subset of the state of Minas Gerais is performed with bracket operator and attribute name.

```
brmg2 <- br2[br2$NAME_1 == "Minas Gerais",]
brmg2
# Simple feature collection with 854 features and 13 fields
# Geometry type: MULTIPOLYGON
# Dimension:     XY
# Bounding box:  xmin: -51.04588 ymin: -22.92275 xmax: -39.85676 ymax: -14.23343
# Geodetic CRS:  WGS 84
# A tibble: 854 × 14
```

The geometry results for 854 municipalities in Minas Gerais are mapped using functions from the `sf` package (Figure 6.4).

```
plot(st_geometry(brmg2), col = sf.colors(n = 1, alpha = 0.3), bgc = 'gray92',
    axes = TRUE)
```

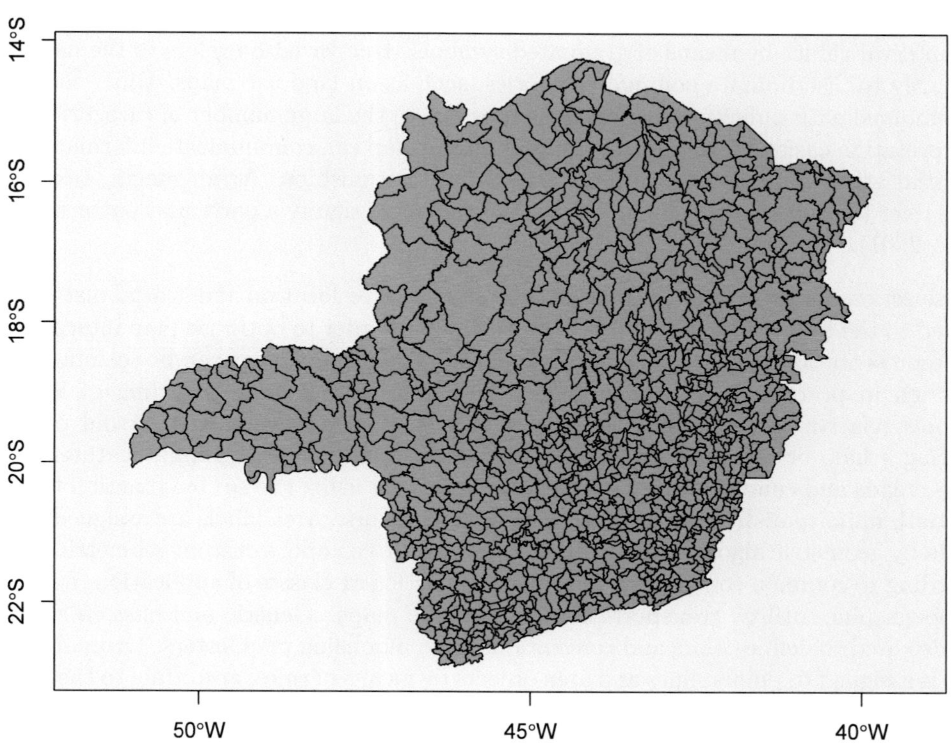

FIGURE 6.4 Base mapping of administrative boundaries of the state of Minas Gerais, Brazil using the `sf` package.

6.6 THEMATIC MAPPING

Many of the conventions used in mapping are related to the use of symbols and colors. Attribute mapping is represented by graphical symbols, by point, line, or area symbols, modified in different ways to communicate different kinds of information. Symbol modifications vary according to cognitive principles and the accumulated experience of application implementations (Longley et al., 2001).

Attribute data are used to describe features related to the terrestrial environment. Thematic mapping is defined as the process of obtaining attributes from data and compiling them into maps for specific purposes. The attributes presented in a thematic mapping are recorded at different measurement scales, such as (Lo and Yeung, 2007):

- Nominal <- Scale represented by names, numbers and codes to identify objects on the map;

- Ordinal <- Scale represented by order or ranking, such as soil drainage condition;
- Interval <- Scale represented by magnitude intervals of values, such as land elevation;
- Rate or ratio <- Scale with support for arithmetic operations of addition, subtraction, multiplication, and division, with the use of an absolute zero or true origin as a reference.

The size and orientation of point and line symbols are varied to distinguish changes in ordinal and interval values by means of graduated symbols. Hue variation refers to the use of color, primarily to discriminate nominal categories, such as on land use maps. Different hues can be combined with different textures or shapes due to the large number of categories making interpretation easier. The shape of symbols on the map can communicate information about a spatial attribute, spatial location, and spatial relationships. Arrangement, texture, and focus refer to properties within and between symbols to signify a particular pattern (Longley et al., 2001).

The selection of appropriate graphic variables to describe location and spatial distribution is related to the best position of symbols on the map in order to optimize map interpretability. The representation of nominal data by graphical symbols and icons can be seemingly trivial, although in practice there are problems with automating the positioning of labels and symbols relative to geographic objects. Point labels are positioned to avoid overlap by creating a halo or rectangle around the text or symbols. Linear graphic features, such as rivers, roads and contours, can be labeled with text by using the `spline` function to provide a smooth uniform distribution, or by using distinct colors. Area labels are assigned to center points by geometric algorithms similar to those used to compute custom geometric centroids according to common conventions and rules for particular classes of application, for example in topographic, utility, transportation and seismic maps. Generic and custom algorithms are also used to define colors and conventions for symbolizing map letters. Ordinal attribute data is assigned to points, lines and area objects by means of rules, according to the hierarchy of graphic variables (symbols and letters, sizes, types, colors, intensities). However, in general, the typical user cannot differentiate between the hierarchy of more than seven ordinal categories. Proportional circles and bar charts are used to assign the interval or ratio scale to point location data. Choropleth mapping can generate questions about the uniformity of an attribute within an area or by a density map of points with different diameters with proportional circles. The variation in color intensity or attribute shading is associated with the magnitude of attributes on a map (Longley et al., 2001).

To facilitate user interpretation in thematic mapping, some methods are used to classify interval and rate data into the following categories (Longley et al., 2001):

- Natural breaks (Jenks);
- Quantiles;
- Equal intervals;
- Standard deviation.

The R package `sf` (Pebesma et al., 2021a) is used to map the production of processed coffee beans (ton) by the method of natural breaks (Jenks) in the period from 2014 to 2019, in Minas Gerais, Brazil, configuring a common legend for all maps and with the border of the municipal boundaries transparent (Figure 6.5).

The `sf`, `readr` (Wickham et al., 2021c), and `brazilmaps` (Siqueira, 2017) packages are loaded into R using the `library` function.

```
library(sf)
# Linking to GEOS 3.11.2, GDAL 3.7.2, PROJ 9.3.0; sf_use_s2() is TRUE
library(readr)
# package 'readr' was built under R version 4.3.3
library(brazilmaps)
```

The `read_csv2` function is used to import a file with the `.csv` extension from a directory on the computer. These data contain information on coffee production by municipality collected by the Instituto Brasileiro de Geografia e Estatística (IBGE). Note that since the file contains special characters with missing data and Latin origin, it is necessary to configure the `na` and `locale` arguments so that the data is imported correctly.

```
tab <- read_csv2("C:/geo/SIDRA_IBGE/cafe.csv", col_names = TRUE,
                 na = c("...", "-"), locale = locale(encoding = 'ISO-8859-1'))
# Rows: 5533 Columns: 48
# Delimiter: ";"
# chr  (1): municipio
# dbl (47): cod_ibge, 1974, 1975, 1976, 1977, 1978, 1979, 1980, 1981, 1982,
# 1983, 1984, 1985, 1986, 1987, 1988, 1989, 1990, 1991,...
# ☒ Use `spec()` to retrieve the full column specification for this data.
# ☒ Specify the column types or set `show_col_types = FALSE` to quiet this message.
```

The `get_brmap` function is used to import a database containing Brazilian municipalities. Next, brackets are used to create a subset of data for the state of Minas Gerais using the state code.

```
brasil <- get_brmap(geo = "City", class = "sf") # Get Brazilian municipalities
brasil
# Simple feature collection with 5570 features and 6 fields
# Geometry type: MULTIPOLYGON
# Dimension:     XY
# Bounding box:  xmin: -73.99045 ymin: -33.74391 xmax: -32.39494 ymax: 5.271841
# Geodetic CRS:  +proj=longlat +ellps=GRS80 +no_defs
mg1 <- brasil[brasil$State == "31",] # Create subset of Minas Gerais
mg1
# Simple feature collection with 853 features and 6 fields
# Geometry type: MULTIPOLYGON
# Dimension:     XY
# Bounding box:  xmin: -51.04608 ymin: -22.92276 xmax: -39.85683 ymax: -14.24667
# Geodetic CRS:  +proj=longlat +ellps=GRS80 +no_defs
```

The `merge` function is used to merge the vector database with the database in the spreadsheet with the `.csv` extension from data columns that match the IBGE code for each municipality.

```
tab_mg <- merge(mg1, tab, by.x = c('City'), by.y = c('cod_ibge'), all.x = TRUE)
```

The results are mapped using functions from the `sf` package in order to generate a comparative choropleth map between different years of coffee yield (Figure 6.5).

```
par(mfrow = c(3, 2))
plot(tab_mg[48:53], key.pos = 4, key.width = lcm(1.3), key.length = 1.0,
     axes = TRUE, graticule = TRUE, border = 'transparent', breaks = "jenks")
```

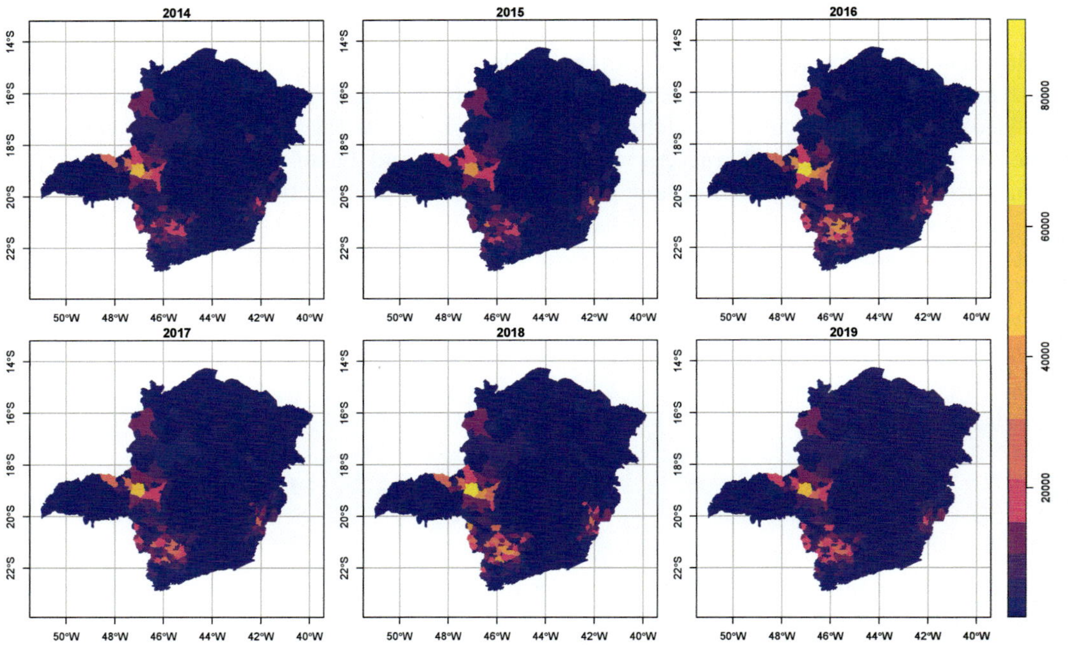

FIGURE 6.5 Mapping processed coffee production (ton) by natural breaks method from 2014 to 2019, in Minas Gerais, Brazil, using `sf` package.

The same mapping is performed comparatively with the `tmap` package (Tennekes et al., 2020), but without adding the coordinate grid and including the border with the municipal boundaries and a legend for each of the years for the period from 2014 to 2019 (Figure 6.6).

The choice of the best classification method is a subject of research in geocomputation based on convenience and the accumulated experience of the cartographer. The frequency distribution of the data helps to evaluate the most appropriate choice for each situation.

In addition to various mapping styles, color palette options from the `tmap` package can be obtained in a `shiny` application (Chang et al., 2021) where you can evaluate different color possibilities and commands to use them. Using the `tmaptools` package function (Tennekes, 2021), `palette_explorer`, you can access the `shiny` application to view various types of color palettes (Figure 6.7).

```
tmaptools::palette_explorer()
# Loading required namespace: shinyjs
# Loading required package: shiny
# Warning: package 'shiny' was built under R version 4.3.3
# Listening on http://127.0.0.1:7111
```

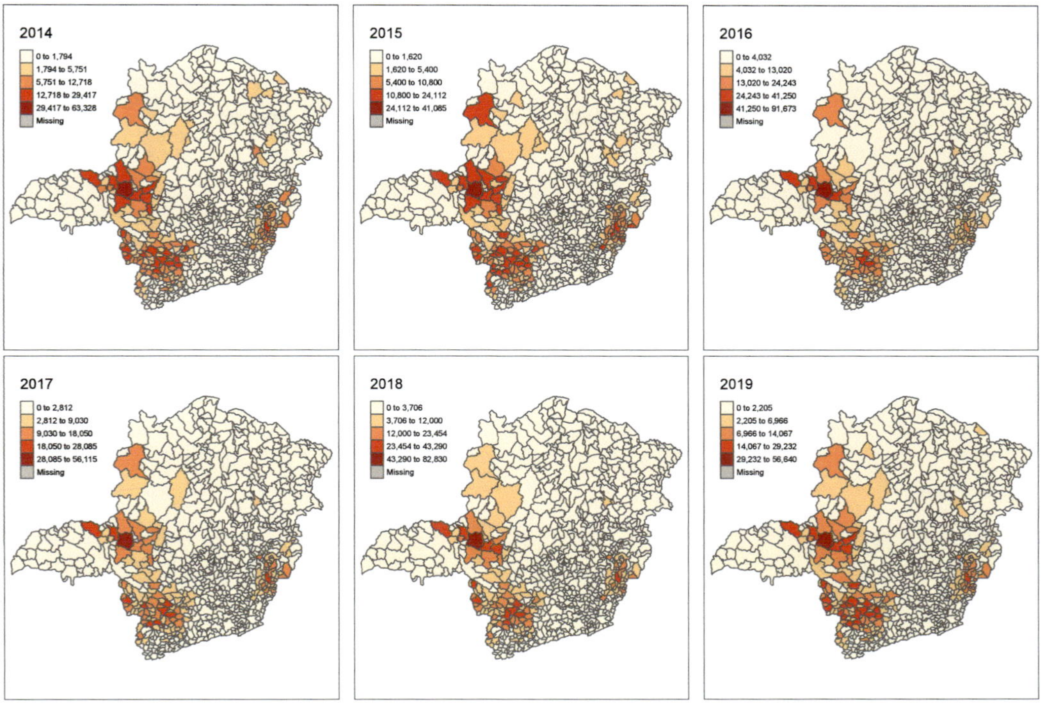

FIGURE 6.6 Mapping processed coffee (ton) production by natural breaks method from 2014 to 2019, in Minas Gerais, Brazil, using `tmap` package.

6.7 GEOVISUALIZATION

With the advancement of geoprocessing technology, the traditional map-oriented approach is modified with the use of digital mapping and increased analytical capacity of geospatial data, extending the decision-making capacity, resulting in the development of new concepts, techniques and applications of cartography. With the use of computational techniques it is possible to create immersive and interactive cartographic environments for dynamic visualization and analysis of geospatial information. Geoinformation products include not only maps, but also statistical graphics, tabulated reports, and multimedia graphics associated with geospatial information in modern geocomputation environments. With this, human-machine interaction and visualization imply the terms scientific visualization and visual information obtained through a new conceptual network of geospatial information presentation dependent on computer graphics, image processing, high performance computing, and other areas of computer technology. Data modeling makes it possible to structure and store geospatial data, and human-machine interaction makes it possible to work in an optimized environment to communicate, inspect, and present geoinformation for scientific interpretation (Figure 6.8) (Lo and Yeung, 2007).

The interaction between the human visual system and computer visualization in geocomputation enables to explain the relationship between methods of representation, analysis, spatial understanding and the use of digital technologies in modern cartography. The emergence of interest from geocomputation users in using the term geographic visualization is referred to as geovisualization. Geovisualization is the result of the synthesis of methods and trends in scientific visualization and geoinformation visualization. Geovisualization is defined as the integration of scientific and geoinformation visualization

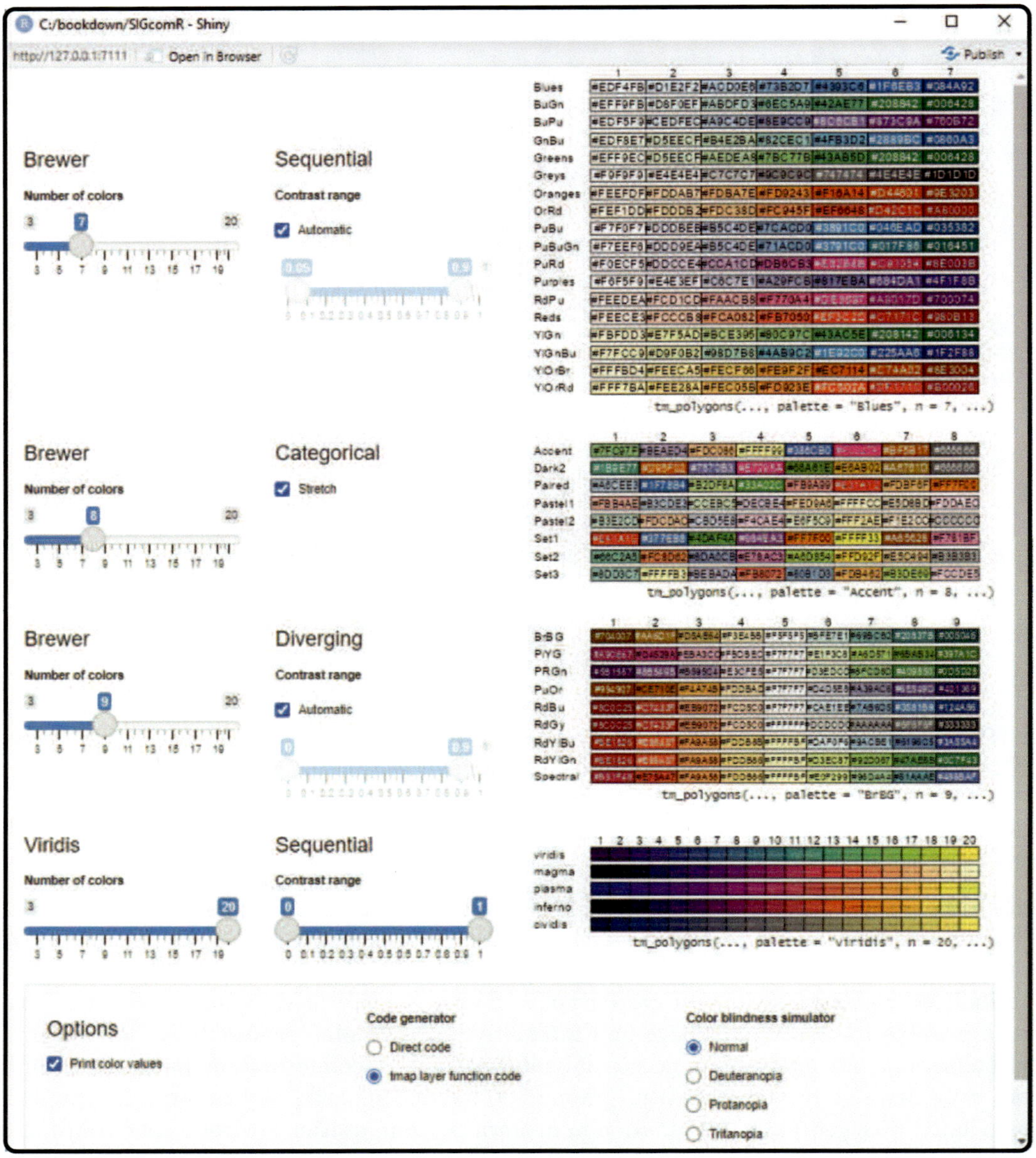

FIGURE 6.7 Example of color palettes that can be used in mapping with the `tmap` package.

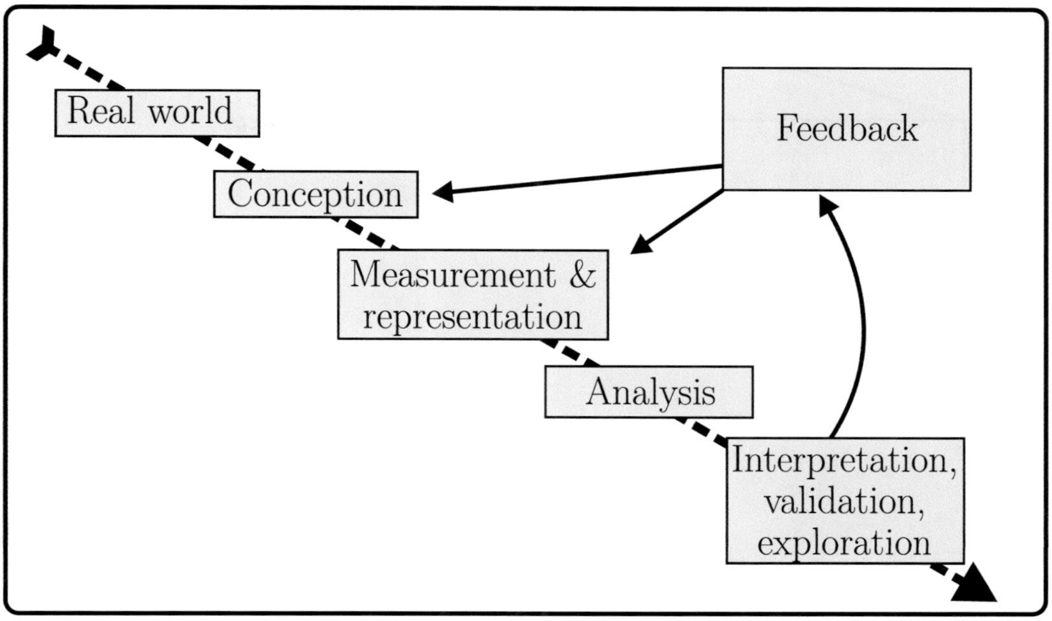

FIGURE 6.8 Design, measurement, analysis, and visualization for scientific interpretation.

approaches with the use of cartography, image analysis, exploratory data analysis, and geocomputation. The overall goal of geovisualization is to articulate and integrate approaches from these domains by providing theories, methods, and tools for visual inspection, analysis, synthesis, and presentation of geospatial information (Figure 6.9). In the geovisualization environment, maps and graphs are tools that simulate visual thinking about graphical patterns, relationships, and trends, as the purpose of formulating hypotheses and developing problem solving by building knowledge (Lo and Yeung, 2007).

The changing nature of maps in geocomputation allows for a more sophisticated and interactive geospatial information interface. Thus, the map is used as a spatial index to access geospatial data and as an intuitive tool to search geospatial information and related resources. Tools for spatial search are developed to make the map a gateway to obtain geospatial information on a daily basis, such as (Figure 6.10) (Lo and Yeung, 2007):

- Google Maps[2];
- Google Earth[3];
- Yahoo Maps[4];
- TerraServer[5];
- MapQuest[6].

Geovisualization techniques are realized by different ways, according to categories (Lo and Yeung, 2007):

[2]http://maps.google.com

[3]http://earth.google.com

[4]http://maps.yahoo.com

[5]http://www.terraserver.com

[6]http://www.mapquest.com

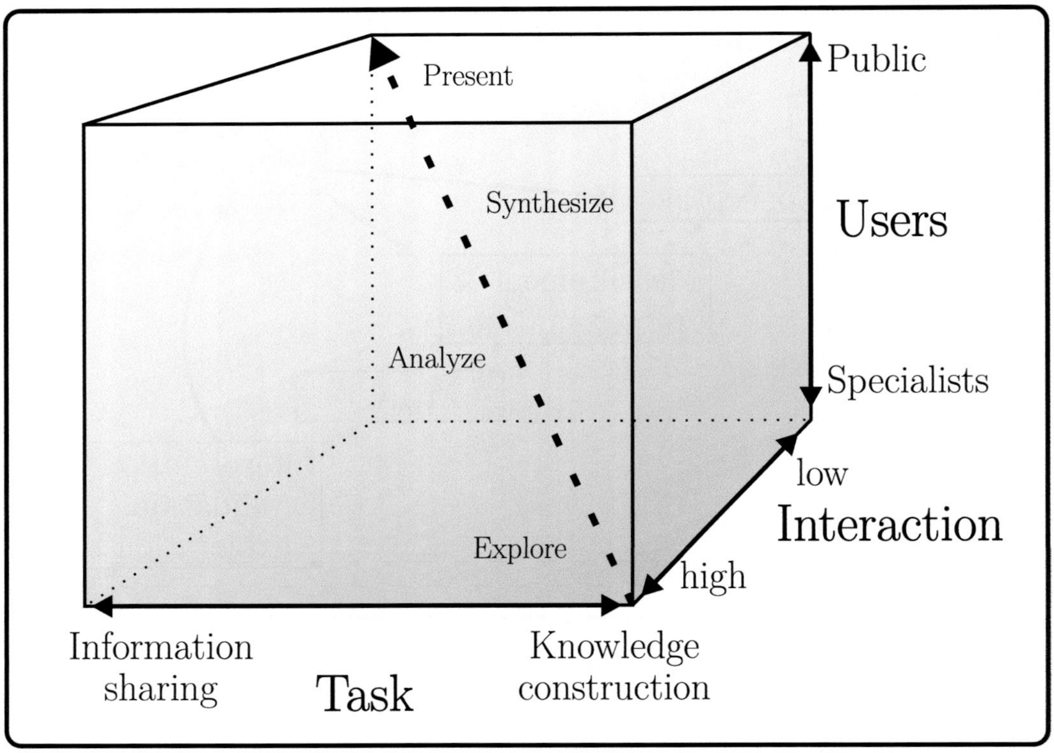

FIGURE 6.9 Geovisualization dimensions along a diagonal in space defined by task types, user and interaction level according to the interface used.

- Two-dimensional graphics;
- Three-dimensional graphics;
- Planimetric visualization;
- Three-dimensional perspective visualization;
- Animation.

Three-dimensional perspective is created by different forms and techniques depending on data volume, level of realism, operational and complexity representation (Table 6.2) (Lo and Yeung, 2007):

TABLE 6.2 Description of visualization characteristics by different techniques.

Technical	Data volume	Level of realism	Operational complexity	Data integrity
Geometric modeling	High	Low to moderate	Moderate to high	High
Video imaging	Low	High	Moderate	Low to moderate
Geometric video imaging	Moderate to high	High	Moderate to high	Moderate to high
Image draping	Low to moderate	Moderate	Moderate to high	Moderate

FIGURE 6.10 Interactive photo-realistic geovisualization of the line represented by the Greenwich meridian at the Royal Observatory, Greenwich in England, taken in Google Earth's virtual reality system, street view mode.

The animation visualization can determine a flight over a three-dimensional area according to a sequence of produced images. The sequence of animations can also be realized in a virtual reality environment being able to move and view any part of the animated scene, as well as by hyperlinks of individual objects related to geospatial and non-geospatial data sources (Lo and Yeung, 2007).

6.8 GEOGRAPHIC DATA SOURCES FOR MAPPING

Various types of geographic data are obtained from the Internet, whether basic data with administrative or thematic boundaries (Table 6.3).

TABLE 6.3 Geospatial data infrastructure on the Internet.

Name	Link	Data Type
GEOSS portal	http://www.geoportal.org/	Global raster
Copernicus Open Access Hub	https://scihub.copernicus.eu/	Global raster
SEDAC	http://sedac.ciesin.columbia.edu/	Global and regional vectors
INSPIRE geoportal	http://inspire-geoportal.ec.europa.eu/	Global and regional vectors
EarthExplorer	https://earthexplorer.usgs.gov/	Global raster
Sentinel API	https://scihub.copernicus.eu/twiki/do/view/SciHubWebPortal/APIHubDescription	Global raster

Name	Link	Data Type
GADM	https://gadm.org/data.html	Global and regional vectors of administrative boundaries
SRTM Data	https://srtm.csi.cgiar.org/srtmdata/	SRTM data from around the globe with variable spatial resolution
Sidra IBGE	https://sidra.ibge.gov.br/home/pmc/brasil	Brazil's socioeconomic and agricultural production geospatial data at the administrative boundaries level

The R packages `raster` (Hijmans et al., 2020) and `sf`, enable mapping of states in Brazil with vector and raster data obtained from GADM and SRTM data.

The `raster` (Hijmans et al., 2020), `sf`, and `geodata` (Hijmans et al., 2024a) packages are loaded into R with the `library` function.

```
library(raster)
# Loading required packages: sp
library(sf)
library(geodata)
# Loading required packages: terra
# terra 1.7.60
```

The SRTM altitude data for Brazil is obtained using the `elevation_30s` function from the `geodata` package.

```
srtm <- elevation_30s(country = "BRA", path=tempdir())
# Trying the URL 'https://geodata.ucdavis.edu/geodata/elv/BRA_elv_msk.zip'
# Content type 'application/zip' length 20394349 bytes (19.4 MB)
# downloaded 19.4 MB
srtm
# class       : SpatRaster
# dimensions  : 4716, 5448, 1  (nrow, ncol, nlyr)
# resolution  : 0.008333333, 0.008333333  (x, y)
# extent      : -74.1, -28.7, -33.9, 5.4  (xmin, xmax, ymin, ymax)
# coord. ref. : lon/lat WGS 84 (EPSG:4326)
# source      : BRA_elv_msk.tif
# name        : BRA_elv_msk
# min value   :        -155
# max value   :        2703
```

The `raster` function is used to transform `SpatRaster` class data into `RasterLayer`.

```
srtm_raster <- raster(srtm)
# class      : RasterLayer
```

```
# dimensions : 4716, 5448, 25692768  (nrow, ncol, ncell)
# resolution : 0.008333333, 0.008333333  (x, y)
# extent     : -74.1, -28.7, -33.9, 5.4  (xmin, xmax, ymin, ymax)
# crs        : +proj=longlat +datum=WGS84 +no_defs
# source     : BRA_elv_msk.tif
# names      : BRA_elv_msk
# values     : -155, 2703  (min, max)
```

The crop and mask functions are used to crop and mask the data within Brazil's borders.

```
mgcrop <- crop(srtm_raster, brmg1) # Crop
mgsrtm <- mask(mgcrop, brmg1) # Mask
```

The writeRaster function is used to store the results in a directory of interest on the computer.

```
writeRaster(srtm_raster, filename="G:/geo/SRTM/srtmBrasil.tif", format="GTiff",
            overwrite = TRUE)
writeRaster(mgsrtm, filename="G:/geo/SRTM/srtmMG.tif", format="GTiff",
            overwrite = TRUE)
```

The files can be reimported into R using the raster function if the data analysis is to be continued at a later date.

```
srtm_raster <- raster("C:/geo/SRTM/srtmBrasil.tif") # Brazil
names(srtm_raster) <- c("Altitude") # Rename layer
mgsrtm <- raster("C:/geo/SRTM/srtmMG.tif") # MG
names(mgsrtm) <- c("Altitude") # Rename layer
```

The functions of the sf package are used to map the results obtained in Brazil and Minas Gerais (Figure 6.11).

```
par(mfrow = c(1, 2), mar = c(2, 2, 0.3, 0.3), mgp = c(1.5, 0.6, 0))
plot(srtm_raster, col = RColorBrewer::brewer.pal(n = 9, name = "Greys"),
     horizontal = TRUE)
plot(st_geometry(br1), add = TRUE)
plot(mgsrtm, col = RColorBrewer::brewer.pal(n = 9, name = "Greys"),
     horizontal = TRUE)
plot(st_geometry(brmg1), add = TRUE)
```

The functions in the tmap package can also be used to perform similar mapping (Figure 6.12).

```
br_srtm <- tm_shape(br1) + tm_polygons() + tm_shape(srtm_raster) +
  tm_raster(alpha = 0.7, title = "Brazil SRTM elevation (m)") +
  tm_compass(type = "8star", position = c("right", "top")) +
  tm_scale_bar(breaks = c(0, 200, 400), text.size = 1)
```

```
mg_srtm <- tm_shape(brmgl) + tm_polygons() + tm_shape(mgsrtm) +
  tm_raster(alpha = 0.7, title = "Minas Gerais SRTM elevation (m)") +
  tm_compass(type = "8star", position = c("right", "top")) +
  tm_scale_bar(breaks = c(0, 200, 400), text.size = 1)
tmap_arrange(br_srtm, mg_srtm, ncol = 2)
```

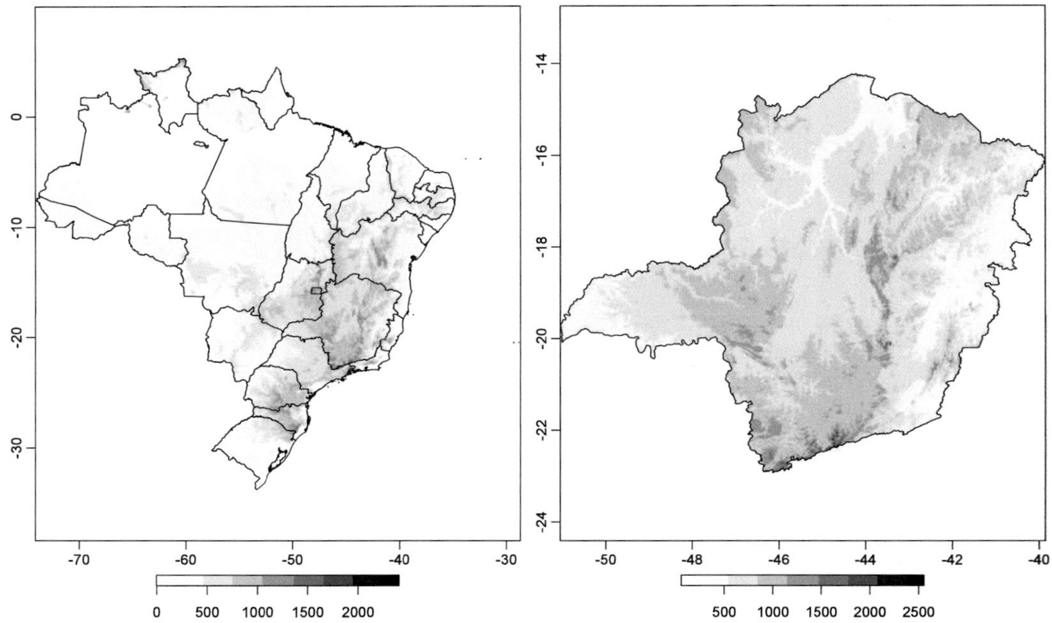

FIGURE 6.11 Geovisualization of altitude variation in the Brazilian states (left) and in the state of Minas Gerais (right) using `sf` package.

Geographic data can also be obtained via R packages (Table 6.4).

TABLE 6.4 Some R packages and data sources used to obtain geographic data.

Package	Description	Source
`getlandsat`	Access to Landsat 8 data	Chamberlain (2018)
`osmdata`	OpenStreetMap data access	Padgham et al. (2021)
`raster`	Access to administrative, elevation, and WorldClim data through the `getData` function	Hijmans et al. (2020)
`rnaturalearth`	Access to physical and cultural data of vector and raster type from the terrestrial environment	Massicotte and South (2023)
`rnoaa`	Access to climate data from the National Oceanic and Atmospheric Administration (NOAA)	Chamberlain et al. (2021)
`rWBclimate`	Access to World Bank climate data	Hart (2014)
`spData`	Data used by the `sp` package	Bivand et al. (2020b)

Package	Description	Source
`spDataLarge`	Large geographic data used by the `sp` package	
`brazilmaps`	Geometric data source and cadastral database of political-administrative boundaries of Brazil with IBGE code	Siqueira (2017)
`sidrar`	Access to the IBGE automatic retrieval system (SIDRA)	Siqueira (2021)
`tigris`	Load Census TIGER/Line Shapefiles from the United States Census Bureau	Walker (2025)
`USAboundaries`	Historical and Contemporary Boundaries of the United States of America	Mullen and Bratt (2025)
`rnaturalearthdata`	World Vector Map Data from Natural Earth Used in rnaturalearth	South et al. (2024)
`GSODR`	Global Surface Summary of the Day (GSOD) Weather Data Client	Sparks et al. (2025)

Using administrative data obtained from the portal GADM maps and data (GADM database of Global Administrative Areas, 2021), we can see different levels of administrative boundaries in the state of Minas Gerais, Brazil (Figure 6.13).

With the advent of the Open Geospatial Consortium (OGC), web services are created with OWS specifications: Web Feature Service (WFS), Web Map Service (WMS), Web Map Tile Service (WMTS), Web Coverage Service (WCS), and Wep Processing Service (WPS). Some requirements can be made to an OWS service, such as `getCapabilities` to discover the functionality of a FAO service.

6.9 FILE AND DATA FORMAT USED IN MAPPING

6.9.1 FILE FORMATS IN VECTOR AND RASTER STRUCTURES

Different files in vector and raster structures are created with specific extensions and usage and sharing modalities (Table 6.5).

TABLE 6.5 Types of vector and raster file structures with extensions, usage modalities, and sharing.

Name	Extension	Information	Structure	Usage
ESRI Shapefile	.shp (main file)	Popular format	Vector	Partially free
GeoJSON	.geojson	JSON format for representing simple features	Vector	Free
KML	.geojson	XML format for geospatial visualization for use in Google Earth with compressed version in KMZ format	Vector	Free

Name	Extension	Information	Structure	Usage
GPX	.gpx	XML created for data obtained by GPS	Vector	Free
GeoTIFF	.tif/.tiff	Popular format with spatial metadata files	Raster	Free
Arc ASCII	.asc	Text format with header and raster cell values in rows and columns	Raster	Free
R-raster	.gpx	Native raster format	Raster	Open
SQLite/SpatiaLite	.gpx	Relational database	Vector and raster	Open
ESRI file GDB	.gdb	Spatial and non-spatial objects created by ArcGIS with multiple feature classes, topology, and GDAL support	Vector and raster	Proprietary
GeoPackage	.gpkg	XML created for GPS acquired data	Vector and raster	Open

6.9.2 RASTER ATTRIBUTES REGISTER FIELDS

The values of raster file record fields vary according to the type of data used, with specific minimum and maximum values (Table 6.6).

TABLE 6.6 Types of raster values according to minimum and maximum possible value.

Variable	Code	Minimum value	Maximum value
Logical values	LOG1S	FALSE (0)	TRUE (1)
Integer values	INT1S	-127	127
Integer values	INT1U	0	255
Integer values	INT2S	-32,767	32,767
Integer values	INT2U	0	65,534
Integer values	INT4S	-2,147,483,647	2,147,483,647
Integer values	INT4U	0	4,294,967,296
Decimal values	FLT4S	-3.4e+38	3.4e+38
Decimal values	FLT8S	-1.7e+308	1.7e+308

6.9.3 FORMATS USED FOR EXPORTING VISUAL GRAPHICS FILES

The formats commonly used to export analysis and mapping results for standard printing and sharing among different applications are `.jpeg`, `.bmp`, `.png`, `.tif`, `.pdf`, `.html`, `.doc`, `.docx`, among others.

6.10 PACKAGES USED IN MAPPING

Several R packages are used with general and specific flexibility for mapping static variables (Table 6.7) (Lovelace et al., 2019).

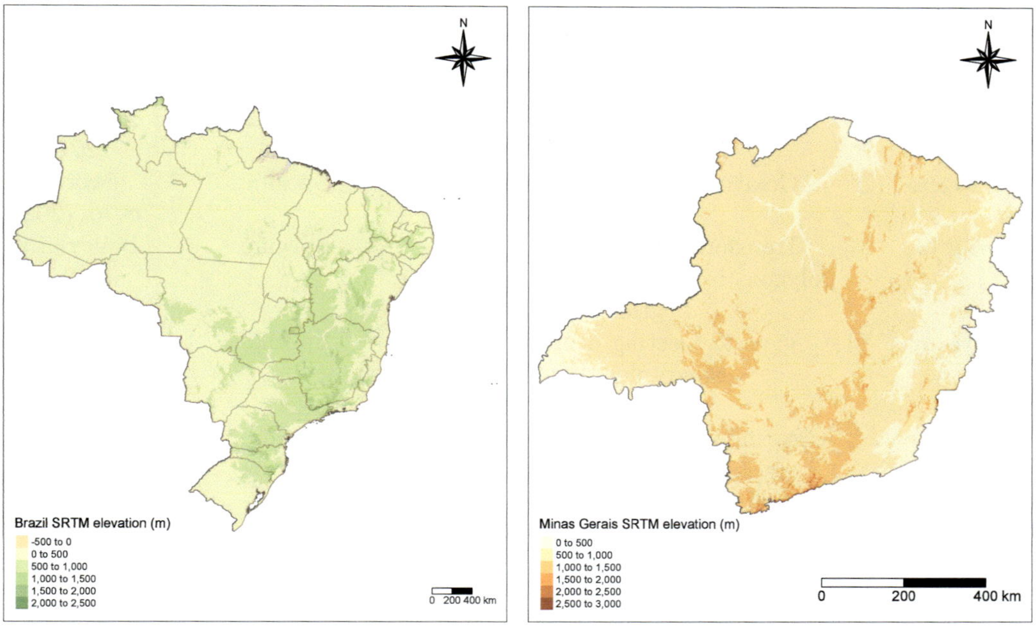

FIGURE 6.12 Geovisualization of altitude variation in the Brazilian states (left) and in the state of Minas Gerais (right) using `tmap` package.

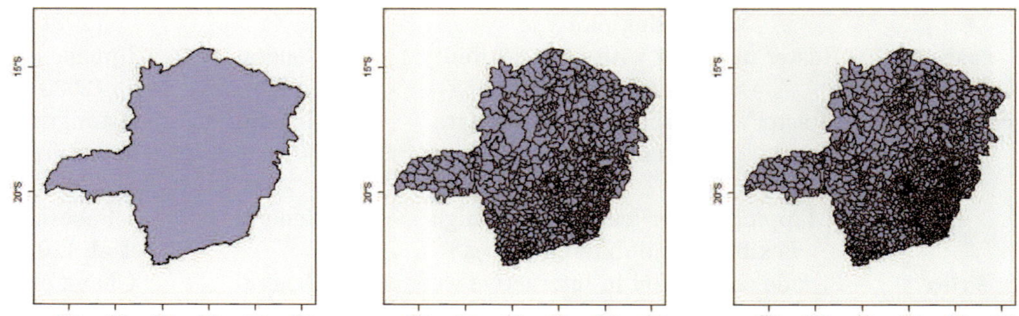

FIGURE 6.13 Different GADM administrative levels, level 1 (left), level 2 (middle) and level 3 (right), representing the administrative boundaries of the state of Minas Gerais, Brazil using the `sf` package.

TABLE 6.7 R packages for mapping with different purposes and flexibilities of map types.

Package	Objective	Flexibility	Source
`cartography`	Thematic cartography	General	Giraud et al. (2021)
`cartogram`	Creating cartograms with R	Specific	Jeworutzki et al. (2020)
`geogrid`	Transformation of geospatial polygons into regular or hexagonal grids	Specific	Bailey et al. (2018)
`geofacet`	ggplot2 faceting utilities for geographic data	Specific	Hafen and Schloerke (2020)
`ggplot2`	Creating data visualizations with graphics grammar with high flexibility and different types	General	Wickham et al. (2020)
`globe`	2-D and 3-D mapping of the Earth, including the main coastline	Specific	Baddeley et al. (2017)
`googleway`	Access to Google Maps APIs to retrieve data and plot maps	General	Cooley et al. (2020)
`ggspatial`	Spatial data structure for ggplot2	General	Dunnington and Thorne (2021)
`leaflet`	Creation of interactive web maps	General	Cheng et al. (2021)
`linemap`	Line mapping	Specific	Giraud (2023)
`mapsf`	Thematic cartography	General	Giraud (2025)
`mapview`	Interactive visualization of geospatial data in R	General	Appelhans et al. (2020)
`plotly`	Creating interactive graphics on the Web using plotly.js	General	Sievert et al. (2021)
`raster`	Raster mapping with high flexibility and different types	General	Hijmans et al. (2020)
`rasterVis`	RasterVis methods for raster data visualization	General	Lamigueiro and Hijmans (2020)
`sf`	Mapping vector features with high flexibility and different types	General	Pebesma et al. (2021a)
`shiny`	R-code convention in interactive web applications	General	Chang et al. (2021)
`tmap`	Thematic mapping	General	Tennekes et al. (2020)
`usmap`	US maps including Alaska and Hawaii	General	Lorenzo (2025)

An example of mapping the `usmap` package (Lorenzo, 2025) can be performed in conjunction with the `ggplot2` package (Wickham et al., 2020). Mapping of US counties is performed with the `plot_usmap` function (Figure 6.14).

```
library(usmap) # Load packages
library(ggplot2) # Load packages
plot_usmap(regions = "counties") + labs(title = "US Counties", # Map
       subtitle = "Counties of the United States.") +
  theme(panel.background = element_rect(color = "black", fill = "lightblue"))
```

The mapping of the main rivers in the US is performed using the `usmap_transform` function to obtain the rivers in the `sf` class, and functions from the `ggplot2` package are used to map the results (Figure 6.15).

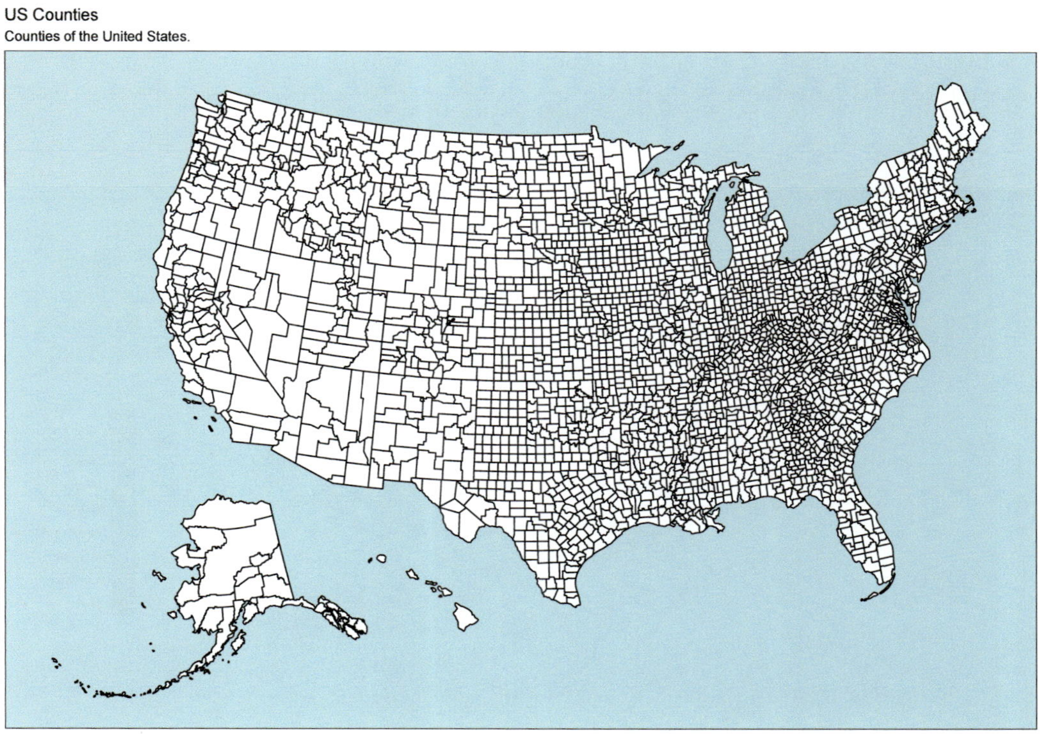

FIGURE 6.14 Mapping US counties with the `usmap` package.

```
rivers_t <- usmap_transform(usrivers) # Get rivers
# Simple feature collection with 55 features and 4 fields
# Geometry type: GEOMETRY
# Dimension:     XY
# Bounding box:  xmin: -1966923 ymin: -2184789 xmax: 2395130 ymax: 732140.8
# Projected CRS: NAD27 / US National Atlas Equal Area
# First 10 features:
#         NAME    SYSTEM      MILES    Shape_Length                          Shape
# 1  Alabama                604.4729     9.629580 LINESTRING (1443627 -102127...
# 2  Arkansas Mississippi 1215.1264    20.520755 LINESTRING (-548393.1 -6056...
# 3  Bear                   274.1188     4.324926 LINESTRING (-912836 -411458...
# 4  Brazos                 847.9190    13.480166 LINESTRING (-317314.3 -1146...
```

```
# 5  Canadian Mississippi  827.8695    13.763149 MULTILINESTRING ((-401564.6...
# 6  Chattahoochee         441.2986     6.783564 LINESTRING (1475503 -992711...
# 7  Clark Fork  Columbia  329.0775     6.087929 LINESTRING (-1209897 493876...
# 8  Columbia    Columbia  656.0648    11.889121 LINESTRING (-1273123 588612...
# 9  Connecticut           326.4272     5.216401 LINESTRING (2194427 426418....
# 10 Flathead    Columbia  178.6006     3.098627 LINESTRING (-1046196 539999...
plot_usmap("counties", color = "gray80") +  # Map
  geom_sf(data = rivers_t, aes(linewidth = Shape_Length,
  color = SYSTEM, fill = SYSTEM)) +
  scale_linewidth_continuous(range = c(0.3, 1.8), guide = "none") +
  scale_color_discrete(guide = "none") +
  labs(title = "Major Rivers in the United States",
  subtitle = "Source: ESRI 2010", fill = "River") +
  theme(legend.position = "right")
```

FIGURE 6.15 Mapping major rivers in the United States with the `usmap` package.

6.11 COMPUTATIONAL PRACTICE

Different mapping types can be performed in R, such as static, animated, and interactive mappings. However, this practice has the main focus on static mapping, using the R packages `sf` and `tmap`. The data for Arabica coffee production in processed beans are obtained from the IBGE's SIDRA system. The spatial simple feature vector base with the administrative boundaries of the municipalities are obtained by the R package `brazilmaps`. The third-party software `ImageMagick-7.0.10-59-Q16-HDRI-x64-dll` is needed as support for performing the animated mapping. This software is installed on the computer and accessed in connection with R for animated mapping of annual coffee production in municipalities of Minas Gerais. Other packages used are described below.

The `install.packages` function installs the major packages, `sf`, `tmap`, `mapview`, `readr`, `brazilmaps`, `dplyr` (Wickham et al., 2021a) and `gifski` (Ooms et al., 2025), used in computational practice.

```
install.packages("sf")
install.packages("tmap")
install.packages("mapview")
install.packages("readr")
devtools::install_github("rpradosiqueira/brazilmaps")
install.packages("dplyr")
install.packages("gifski")
```

The `library` function is used to enable the `sf`, `tmap`, `mapview`, `readr`, `brazilmaps`, `dplyr` and `gifski` packages.

```
library(sf)
library(tmap)
library(mapview)
library(readr)
library(brazilmaps)
library(dplyr)
library(gifski)
```

6.11.1 GET VECTOR FILE FOR MAPPING

The shapefile used in the practice can be obtained by download[7] on the Internet. Then unzip the file into a working directory to perform the computational practice.

6.11.2 IMPORT SHAPEFILE

```
tab_mg <- st_read("C:/geo/Croplands/cafeMG.shp")
st_crs(tab_mg) <- 4326
```

In this vector file, the production data of processed coffee beans between 2014 and 2019 are organized in relational form, in columns.

6.11.3 BASIC STATIC MAPPING

Static mapping is the most common type of visualization in geocomputation (Figure 6.16). The `plot` function is the most common for making maps of raster and vector objects. This type of map evolved with the creation of the `sp`, `sf`, `tmap`, `ggplot2`, `raster`, and `stars` packages.

[7] http://www.sergeo.deg.ufla.br/geo/downloads/cafeMG.zip

```
plot(st_geometry(tab_mg), graticule = TRUE, col="grey90", axes = TRUE)
```

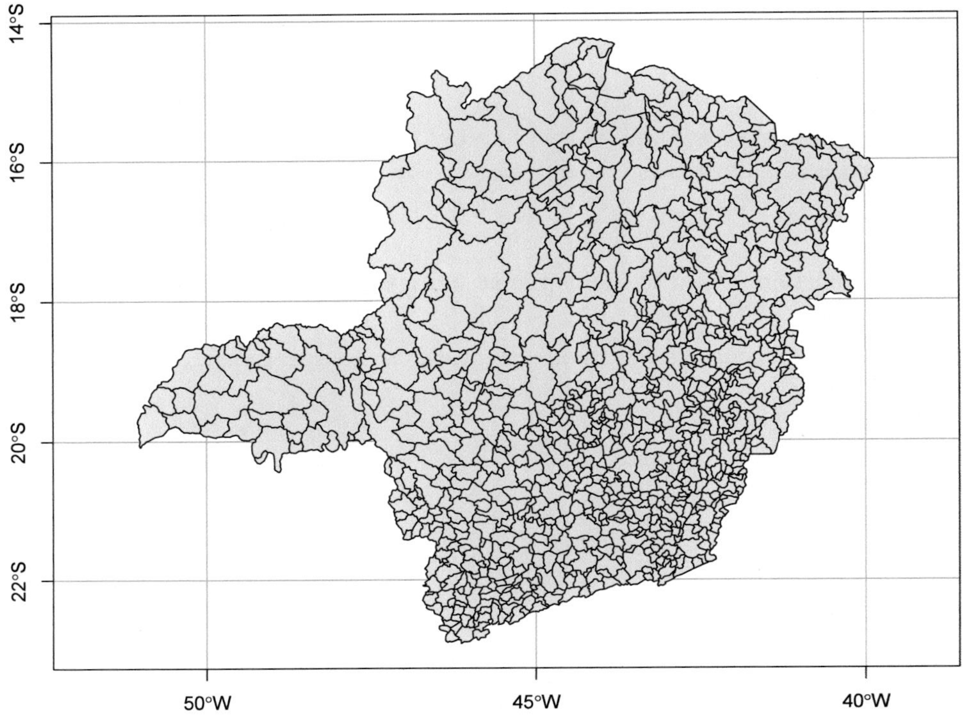

FIGURE 6.16 Static mapping of administrative boundaries at the municipal level in the state of Minas Gerais, Brazil.

6.11.4 CHOROPLETH MAPPING

The choropleth thematic mapping of the 2019 production of processed Arabica coffee in Minas Gerais is performed in relation to production with the `tmap` package (Figure 6.17).

```
coffee2019 <- tm_shape(tab_mg) + tm_polygons(col = "X2019",
              title = "Coffee Production, 2019 (ton)", style = "jenks",
              palette = RColorBrewer::brewer.pal(n = 9, name = "OrRd")) +
  tm_text("nome", size = 1/2, remove.overlap=TRUE, col = "black",
          shadow = TRUE, bg.color = "white", bg.alpha = 0.2)+
  tm_layout(legend.text.size = 1.0, bg.color = "lightgrey") +
   tm_compass(type = "8star", position = c("right", "bottom")) +
  tm_scale_bar(breaks = c(0, 100, 200), text.size = 1)
coffee2019
```

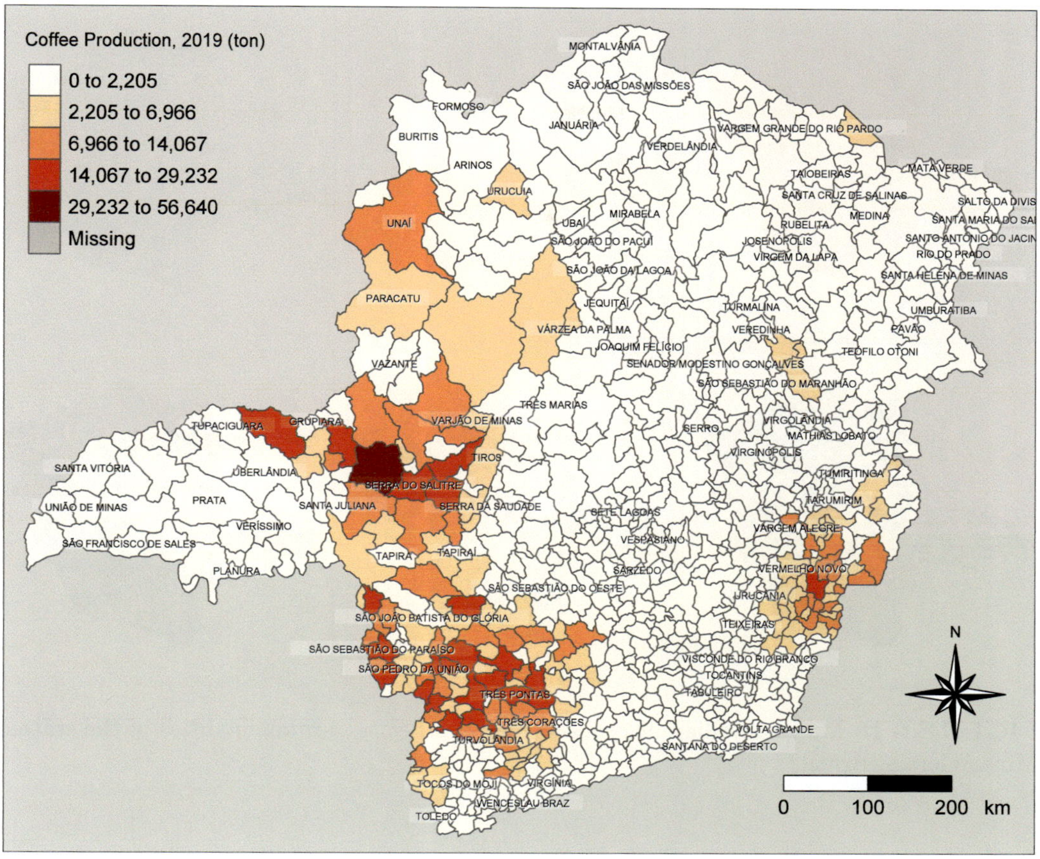

FIGURE 6.17 Coropleth mapping of the production of Arabica coffee in municipalities of Minas Gerais, in 2019, with the `tmap` package.

6.11.5 AESTHETICS

The `tm_shape` function of the R package `tmap` is performed to map different types of colors and the filling of edges with different aesthetics (Figure 6.18).

```
ma1 <- tm_shape(tab_mg) + tm_fill(col = "red")
ma2 <- tm_shape(tab_mg) + tm_fill(col = "red", alpha = 0.3)
ma3 <- tm_shape(tab_mg) + tm_borders(col = "blue")
ma4 <- tm_shape(tab_mg) + tm_borders(lwd = 3)
ma5 <- tm_shape(tab_mg) + tm_borders(lty = 2)
ma6 <- tm_shape(tab_mg) + tm_fill(col = "red", alpha = 0.3) +
  tm_borders(col = "blue", lwd = 3, lty = 2)
tmap_arrange(ma1, ma2, ma3, ma4, ma5, ma6)
```

6.11.6 FACETED MAPS WITH POLYGONS

A faceted mapping defined by multiple variables of Arabica coffee production by municipality in Minas Gerais. In this case, the definition of production classes is performed

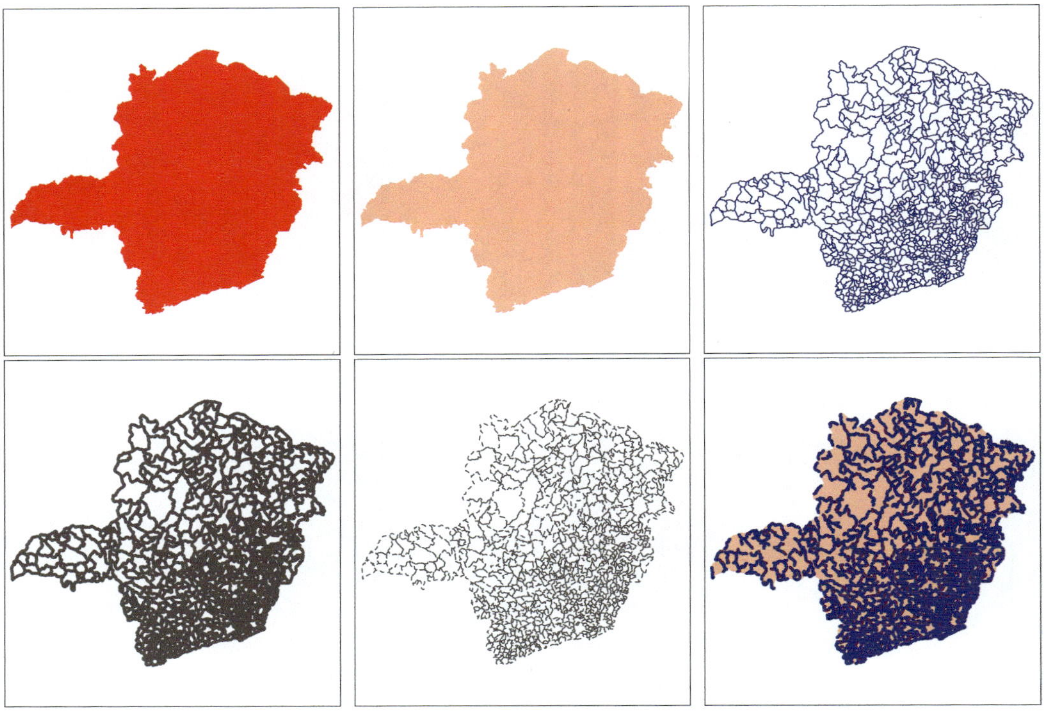

FIGURE 6.18 Mapping different types of map aesthetics in municipalities of the state of Minas Gerais, Brazil.

by the `kmeans` classification method. Coffee production between 2014 and 2019 is mapped using the `tmap` package and the `tm_shape`, `tm_fill`, `tm_borders`, and `tm_layout` functions (Figure 6.19).

```
coffee <- tm_shape(tab_mg) + tm_fill(c("X2014", "X2015",
    "X2016", "X2017", "X2018", "X2019"), style = "kmeans",
    palette = list("OrRd", "OrRd", "OrRd", "OrRd", "OrRd", "OrRd"),
    title = c("2014", "2015", "2016", "2017", "2018", "2019")) +
    tm_borders() + tm_layout(bg.color = "lightgrey")
coffee
```

6.11.7 FACETED MAPS WITH SYMBOLS

The file with the coffee production database is reorganized with the columns `cod_ibge`, `municipio`, `Ano` and `Prod` and exported in `.csv` format to perform the practice of faceted and animated maps. The database file used in the practice can be obtained by download[8] on the Internet. Then unzip the file into a working directory to perform the computational practice.

[8]http://www.sergeo.deg.ufla.br/geo/downloads/cafeAnimado.zip

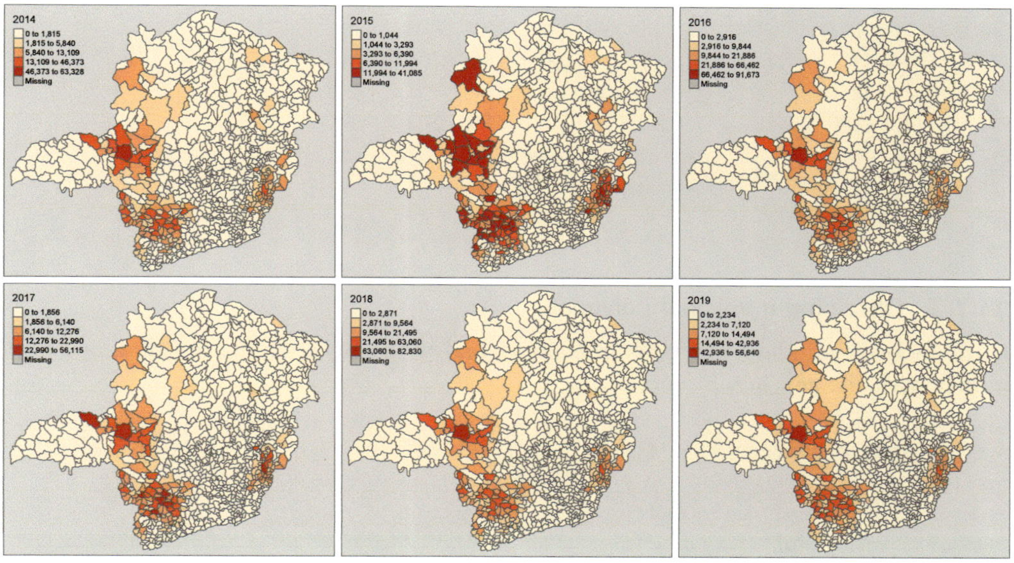

FIGURE 6.19 Mapping the production of coffee in the municipalities of Minas Gerais classified by the `kmeans` method, in the period from 2014 to 2019.

6.11.7.1 Import spreadsheet with data

The file `cafeAnimado.csv` is imported into R with the `read_csv2` function. Missing data values represented by `...` and `-` are considered `NA`. The parameter `encoding = 'ISO-8859-1'` is set to properly represent the Latin terms.

```
tab1 <- read_csv2("C:/geo/SIDRA_IBGE/cafeAnimado2.csv", col_names = TRUE,
            na = c("...", "-"), locale = locale(encoding = 'ISO-8859-1'))
# Rows: 44272 Columns: 4
# ── Column specification ────────────
# Delimiter: ";"
# chr (1): municipio
# dbl (3): cod_ibge, Ano, Prod
# ☒ Use `spec()` to retrieve the full column specification for this data.
# ☒ Specify the column types or set `show_col_types = FALSE`
# to quiet this message.
```

6.11.7.2 Convert data to data.frame

The data is converted into `data.frame` with the `as.data.frame` function.

```
tab1 <- as.data.frame(tab1)
#     cod_ibge                     municipio  Ano Prod
# 1    1100015    Alta Floresta D'Oeste (RO) 2012   NA
# 2    1100023               Ariquemes (RO) 2012   NA
# 3    1100031                  Cabixi (RO) 2012   NA
# 4    1100049                  Cacoal (RO) 2012   NA
```

```
# 5     1100056              Cerejeiras (RO) 2012    NA
# 6     1100064       Colorado do Oeste (RO) 2012    NA
# 7     1100072              Corumbiara (RO) 2012    NA
# 8     1100080           Costa Marques (RO) 2012    NA
# 9     1100098         Espigão D'Oeste (RO) 2012    NA
# 10    1100106           Guajará-Mirim (RO) 2012    NA
# ...
```

6.11.7.3 Obtaining municipal polygons

The municipal boundary polygons in class `sf` and with IBGE coding are obtained on the Internet with the `get_brmap` function.

```
brazil <- get_brmap(geo = "City", class = "sf")
brazil
# Simple feature collection with 5570 features and 6 fields
# Geometry type: MULTIPOLYGON
# Dimension:     XY
# Bounding box:  xmin: -73.99045 ymin: -33.74391 xmax: -32.39494 ymax: 5.271841
# Geodetic CRS:  +proj=longlat +ellps=GRS80 +no_defs
```

6.11.7.4 Making a subset in Minas Gerais

A subset is made in the state of Minas Gerais, code 31, using brackets.

```
mg1 <- brazil[brazil$State == "31",]
mg1
# Simple feature collection with 853 features and 6 fields
# Geometry type: MULTIPOLYGON
# Dimension:     XY
# Bounding box:  xmin: -51.04608 ymin: -22.92276 xmax: -39.85683 ymax: -14.24667
# Geodetic CRS:  +proj=longlat +ellps=GRS80 +no_defs
```

6.11.7.5 Join the tables

The attribute tables are joined using IBGE codes for the municipalities.

```
tab1_mg <- merge(mg1, tab1, by.x = c('City'), by.y = c('cod_ibge'), all.x = TRUE)
tab1_mg
# Simple feature collection with 6824 features and 9 fields
# Geometry type: MULTIPOLYGON
# Dimension:     XY
# Bounding box:  xmin: -51.04608 ymin: -22.92276 xmax: -39.85683 ymax: -14.24667
# Geodetic CRS:  +proj=longlat +ellps=GRS80 +no_defs
```

6.11.7.6 Perform faceted mapping with symbols

Faceted mapping with symbols of coffee production by municipality in the years from 2014 to 2019 is performed first with the filtering of the attribute columns with the years between

2014 and 2019.

```
coffee_2014_2019 <- tab1_mg |> filter(Ano %in% c(2014, 2015, 2016, 2017, 2018,
                                                   2019))
coffee_2014_2019
# Simple feature collection with 5118 features and 9 fields
# Geometry type: MULTIPOLYGON
# Dimension:      XY
# Bounding box:  xmin: -51.04608 ymin: -22.92276 xmax: -39.85683 ymax: -14.24667
# Geodetic CRS:  +proj=longlat +ellps=GRS80 +no_defs
```

Then the functions `tm_shape`, `tm_symbols`, `tm_facets` and `tm_layout` are used in the mapping (Figure 6.20).

```
tm_shape(tab1_mg) + tm_polygons() +
  tm_shape(coffee_2014_2019) + tm_symbols(col = "coral", border.col = "red",
  size = "Prod", breaks = TRUE, style = "kmeans") +
  tm_facets(by = "Ano", nrow = 2, free.coords = FALSE, scale.factor = 4) +
  tm_layout(legend.outside = TRUE, legend.outside.position = "bottom",
  legend.position = c(0.4, 0.6), main.title = "", main.title.position = "centre",
  title = "Coffee Production (ton)", legend.title.color = "white",
  outer.margins = c(-0.1, 0, 0.05, 0))
```

6.11.8 ANIMATED MAPPING

The `ImageMagick` software is installed to perform animated mapping, and then set the `PATH` location of where the software is installed. The `Sys.setenv` function is used to set the directory where `ImageMagick` is installed on the computer.

```
Sys.setenv(PATH = paste("c:/Arquivos de Programas/ImageMagick-7.0.10-Q16-HDRI",
                        Sys.getenv("PATH"), sep = ";"))
```

The `tm_facets` function is configured to display only one year of coffee production at a time.

```
coffee_anim1 <- tm_shape(tab1_mg) + tm_polygons() +
  tm_shape(coffee_2014_2019) + tm_dots(size = "Prod", breaks = TRUE,
                                       style = "kmeans") +
  tm_facets(along = "Ano", ncol = 1, nrow = 1, free.coords = FALSE)
coffee_anim1
```

The `tmap_animation` function creates an animated map file in `.gif` extension (Figure 6.21).

```
tmap_animation(coffee_anim1, filename = "C:/assets/img7/cafe_anim2.gif",
               delay = 200)
```

6.11.9 INTERACTIVE MAPPING

Interactive mapping with the `tmap` package is useful to visualize the state of Minas Gerais in a context of the existence of other cartographic bases on the Internet, from the `view` mode

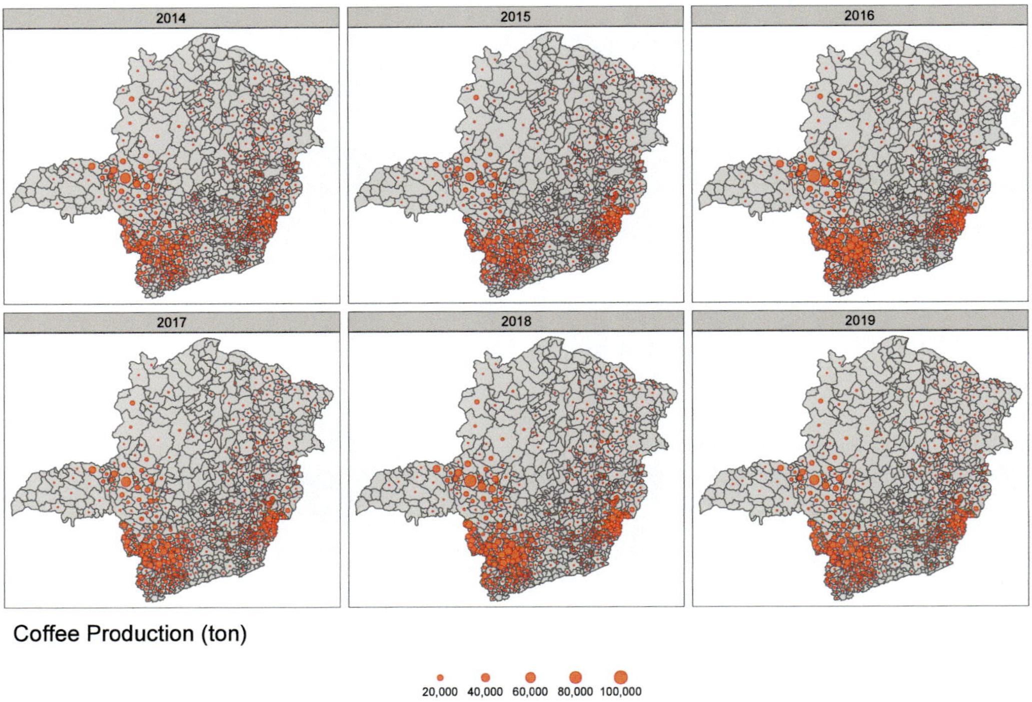

FIGURE 6.20 Mapping facets with symbols of the production of processed coffee, classified by the `kmeans` method, in the municipalities of Minas Gerais, in the period from 2014 to 2019.

of the `tmap_mode` function (Figure 6.22).

```
tmap_mode("view")
ma5
```

Interactive mapping can also be performed with the `mapview` package using the `mapview` function (Figure 6.23).

```
mapview(tab_mg)
```

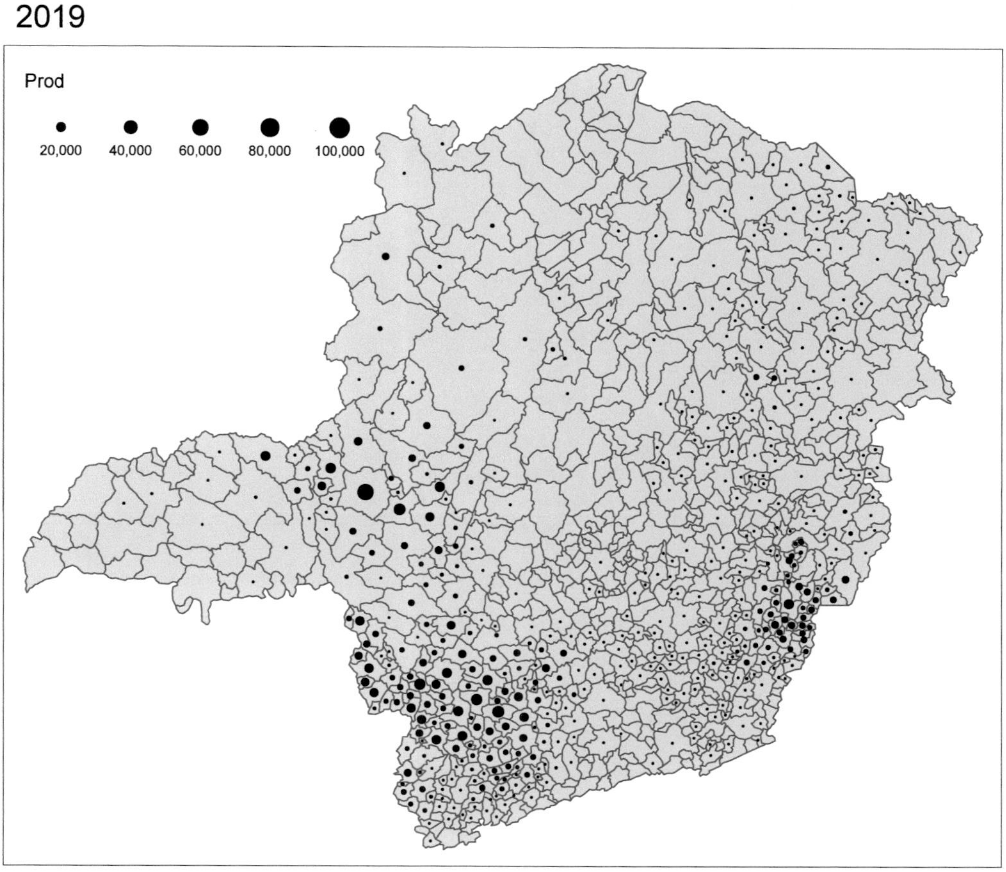

FIGURE 6.21 Animated mapping with symbols of the production of processed coffee, classified by the `kmeans` method, in the municipalities of Minas Gerais, in the period from 2014 to 2019.

FIGURE 6.22 Interactive web mapping of the state of Minas Gerais with the `tmap` package.

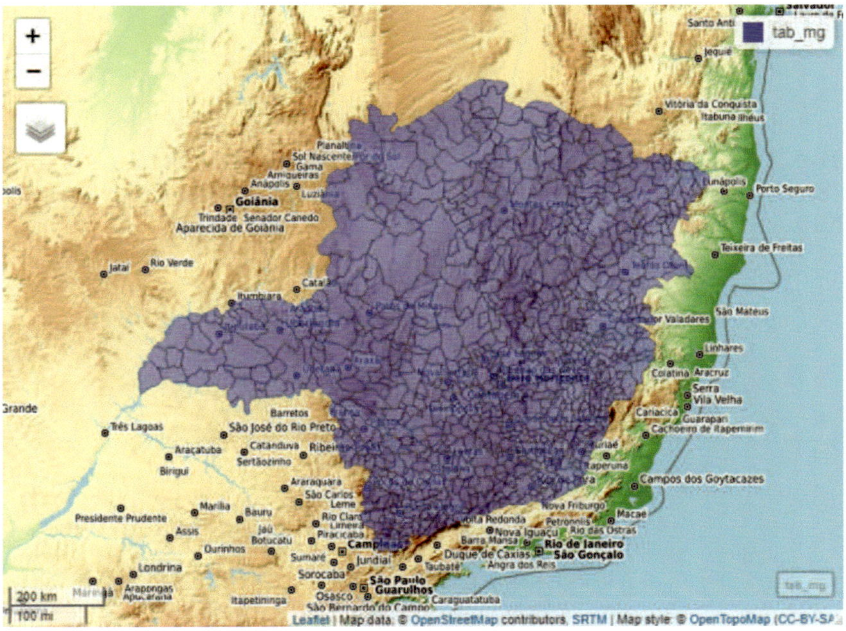

FIGURE 6.23 Interactive web mapping of the state of Minas Gerais with the `mapview` package.

7 Project Elaboration with Environmental Geocomputation and R

7.1 LEARNING QUESTIONS

The learning questions answered by reading the chapter are:

- How to perform science planning with geocomputation;
- How geocomputation applications relate to scientific practice;
- How to solve geographic problems objectively;
- How to perform science research planning in geocomputation.

7.2 LEARNING OUTCOMES

The learning outcomes answered by reading the chapter are:

- Understand about the use of geocomputation in everyday life;
- Appreciate the variety and diversity of applications of geocomputation as a science;
- Identify hypotheses in scientific studies of geographic problems;
- Verify an example of elaboration of a scientific project in geocomputation.

7.3 INTRODUCTION

Geospatial data and analysis have become so common that society often takes them for granted. For example, most people are not concerned about how the fastest route to a location is determined when one of the Internet mapping sites (Bing Maps, Google Maps, Yahoo! Maps) or personal cell phone navigation device is used. The only concern is with the correct functioning of the device. Therefore, it has become essential to worry about improving the scientific applications of geocomputation so that there is the highest possible accuracy and efficiency in the decision-making process (Jensen and Jensen, 2012).

Geocomputation is used in everyday activities of working and living on Earth. The everyday uses of geocomputation are highlighted below (Longley et al., 2001):

- Effectiveness in decision-making;
- Application in various areas related to socioeconomics and environment;
- Use in mapping, measuring, managing, monitoring and modeling operations;
- Measurable economic benefits;
- Combinations with other technologies.

The factors driving the everyday use of geocomputation in modern society are (Longley et al., 2001):

- Increased availability of geographic information systems on the Internet and local enterprise networks;
- Relative reductions in the price of geocomputation hardware and software;

DOI: 10.1201/9781003724377-7

- Increased awareness of the population to include the geographic dimension in decision-making;
- Possibility of user interaction with windowed environments;
- Technology supporting applications to visualize, manage data and analysis, as well as linking with other software;
- Availability of satellite positioning system;
- Availability of user-friendly applications;
- Accumulated experience in using geocomputation applications.

7.4 CONTEXT OF GEOSPATIAL INFORMATION

The context in which geospatial information is used consists of the set of knowledge acquired in scientific research and geospatial applications, technologies used to develop and manage computer systems and, the institutions created to facilitate the acquisition and development of geospatial information. The context of geospatial information has affected sectors of the economy in different application areas (Figure 7.1) (Lo and Yeung, 2007).

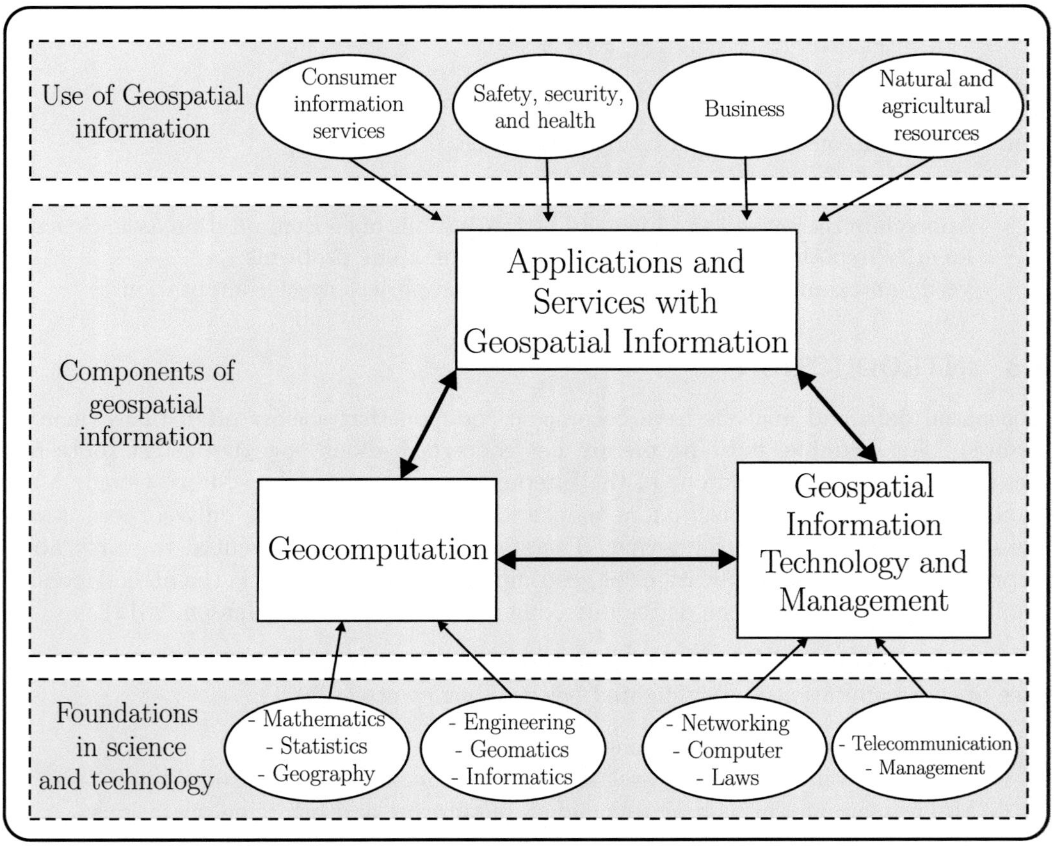

FIGURE 7.1 The context of geospatial information.

The main sources of geospatial information production are (Burrough and McDonnell, 1998):

- Topographic mapping;
- Property registration;
- Hydrography mapping;

- Military organizations;
- Remote sensing and satellite agencies and companies;
- Survey of natural resources such as geology, hydrology, soil, ecology, biogeography, meteorology, climatology, and oceanography.

The main types of geographic data available are (Burrough and McDonnell, 1998):

- Topographic maps at different scales;
- Imagery at different data collection altitudes;
- Administrative boundaries, census data, zip code, statistics, people, land cover and land use at different resolutions;
- Marketing survey data;
- Utility data, such as gas, water, sewage, power lines, and Internet network and their location;
- Data on rocks, water, soil, atmosphere, biological activity, natural disasters, and other types.

The main applications of geographic data are (Neves et al., 1998; Silva and Assad, 1998; Burrough and McDonnell, 1998):

- Agricultural sciences, for monitoring, management, description and scenario building;
- Archaeology, for description and study of the past;
- Environmental monitoring and management;
- Health and epidemiology, for locating diseases in relation to the environment;
- Emergency services, for optimizing routes for ambulance, police, fire escape, investigation and location of crimes;
- Navigation, for air, sea, and land;
- Marketing, for locating places and target groups, delivery optimization;
- Regional and local cost planning, maintenance, and site management;
- Planning and management of highways, railroads, and airways;
- Property and inventory evaluation, calculation of cut, fill, and volume of materials;
- Social studies, to analyze population movement, local and regional development;
- Tourism for locating and managing attractions and facilities;
- Everyday utilities for locating, managing, and planning water, drainage, gas, electricity, telephone, and cabling services.

In the necessity of improving the quality of products, services, and processes, geocomputation is used at different structural levels of government organizations for the following reasons (Figure 7.2) (Longley et al., 2001):

- Inventory resources and infrastructure, plan transportation routing, improve public service delivery, manage land development, and generate revenue by increased economic activity;
- Use of geocomputation in long-term geographic problems with health, safety and welfare of citizens, incorporating public values in decision-making, providing services in a fair and equitable manner, and representing citizen opinions by democratic work;
- Applications of geocomputation in public health risk monitoring, housing stock management, social welfare fund allocation, crime tracking, geodemographic analysis, operational, tactical and strategic decision-making in enforcement, health planning and education management;
- Asset inventory, policy analysis, modeling and strategic planning.

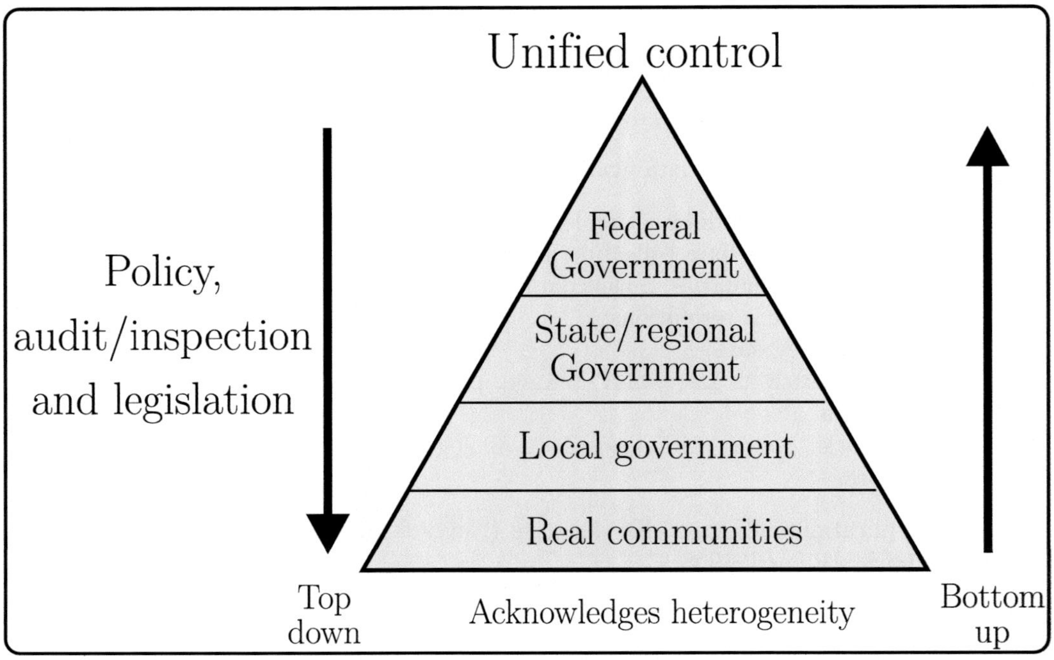

FIGURE 7.2 Use of geocomputation at different structural levels of governmental decision-making.

7.5 SCIENTIFIC APPLICATIONS OF GEOCOMPUTATION

In developing scientific applications of geocomputation, some fundamental concepts are presented in order to clarify any conceptual doubts about science, scientist, research, experiment, hypothesis, and methodology (Table 7.1).

TABLE 7.1 Fundamental concepts used in scientific applications of geocomputation.

Term	Meaning
Science	Knowing and solving geographical problems.
Scientist	Every individual who generates knowledge.
Research	Set of activities oriented toward the search for a particular knowledge.
Experiment	A planned activity to obtain new facts, confirm results from previous experiments, generate or validate technologies.
Hypothesis	Testable proposition in the experiment that involves the solution to the problem.
Methodology	Use of experimental procedures.

In the deductive scientific method, a chain of descending reasoning of analysis from the general to the particular is used to reach a conclusion. In the inductive scientific method, inferences about a general or universal truth not contained in observations are made from particular observed data.

In basic research, useful knowledge is generated for the advancement of science with no foreseen practical application. In applied research, the generation of knowledge for practical application is directed to the solution of specific problems.

Since in science the objective of knowing and solving geographic problems varies according to the complexity of the scientific principles and techniques available, some project objectives used in geocomputation are (Longley et al., 2001):

- Rational, effective and efficient allocation of resources according to criteria;
- Monitoring and understanding geospatial context;
- Understanding regional differences of an object or process;
- Understanding processes in natural and anthropic environments;
- Prescribing strategies for environmental maintenance, air quality, soil and water conservation.

Geocomputation applications should be based on sound concepts and theory to solve different types of geographic problems (Longley et al., 2001).

7.6 GEOCOMPUTATION SCIENTIFIC PROJECT DEVELOPMENT

In the geographic information system development process, the phases of hypothesis testing, *in situ* data collection, and geospatial infrastructure data collection are addressed in a systematic way, with evaluation and interpretation methodology, to obtain the necessary knowledge to make decisions about the solution of a geographic problem (Figure 7.3).

A complete content of known digital libraries can be accessed through keyword searches used to search the subject of interest relative to the selected geographic problem (Volpato, 2007; Arvanitou et al., 2021):

- web of science[1];
- ScienceDirect[2];
- IEEExplore[3];
- ACM[4];
- Scopus[5];
- Google Scholar[6]

In the preparation of the scientific project in the area of geocomputation, the topics should be written with the verb in the future tense, because the text is prepared in the context of presenting a proposition to study and generate relevant scientific conclusions about a particular geographic problem. The following topics used in the preparation of a scientific research proposal (Table 7.2) are contemplated in the preparation of the research project.

TABLE 7.2 Meaning of topics used in scientific project development in geocomputation.

Topic	Meaning
Title	The title should be succinct and try to contain in a few words what you intend to accomplish with the project.

[1]https://www.periodicos.capes.gov.br/?option=com_pcollection&mn=70&smn=79&cid=81

[2]https://www.sciencedirect.com/

[3]https://ieeexplore.ieee.org/Xplore/home.jsp

[4]https://dl.acm.org/

[5]https://www.scopus.com/home.uri

[6]https://scholar.google.com.br/

Topic	Meaning
Author	Project proponent, usually the project coordinator. Provide information on the call for proposals to which the project will be submitted.
Abstract and Keywords	Include index words for the abstract that have not already been put in the title.
Introduction	Go straight to the point, pointing out the problem, the generating demand, and show that you have not only been able to diagnose the problem, but are also able to solve it. Get information about what has already been done in other work on a similar subject.
Hypothesis	Hypothesis should be written in the affirmative in order to elucidate the testable proposition of the project.
Objectives	Objectives should be written in a way that leaves no doubt about what you want to achieve in the project.
Literature Review	Recent scientific articles published in high-impact journals on the researched subject.
Materials and Methods	It is necessary to put as many details as possible, but without exaggerating, e.g., variable studied, date when the radiometric data was obtained, analysis performed on the data.
Expected Results	Present the results obtained based on the data analysis.
Execution Schedule	Schedule of time in which the project will be carried out.
Diffusion of Technology	Teaching, research, and extension associated with the project.
Budget	Consider the request for permanent goods, consumables, bibliographic material, and researcher grants.
Team	Team of researchers, with information on branch data, institution, how they work on the project.
References	Include the bibliographical references used.

7.7 PRACTICE TO PREPARE PROJECT

In the practice of project elaboration, a summary of a scientific project in the area of geocomputation and related areas with application of climate change in agriculture is performed.

To implement the project, it is necessary to obtain a database to develop the proposal, which will later be converted into a scientific article. As a suggestion for obtaining data, you can use the AppEEARS platform (Krehbiel et al., 2024) as an example, from which data was obtained to develop the article contained in the laboratory manual of this book.

In this case, data on population count and density in South America were obtained by geocomputation of nighttime illumination products from remote sensing (Figure 7.4).

Data from the Google Earth Engine platform can also be accessed in R using the `RGEE` package (Aybar et al., 2022).

The data used in this example was obtained from the Forest Data Partnership[7] with Google Earth Engine, where raster datasets with a resolution of 10 meters were created, mapping

[7]https://github.com/google/forest-data-partnership/tree/main/models

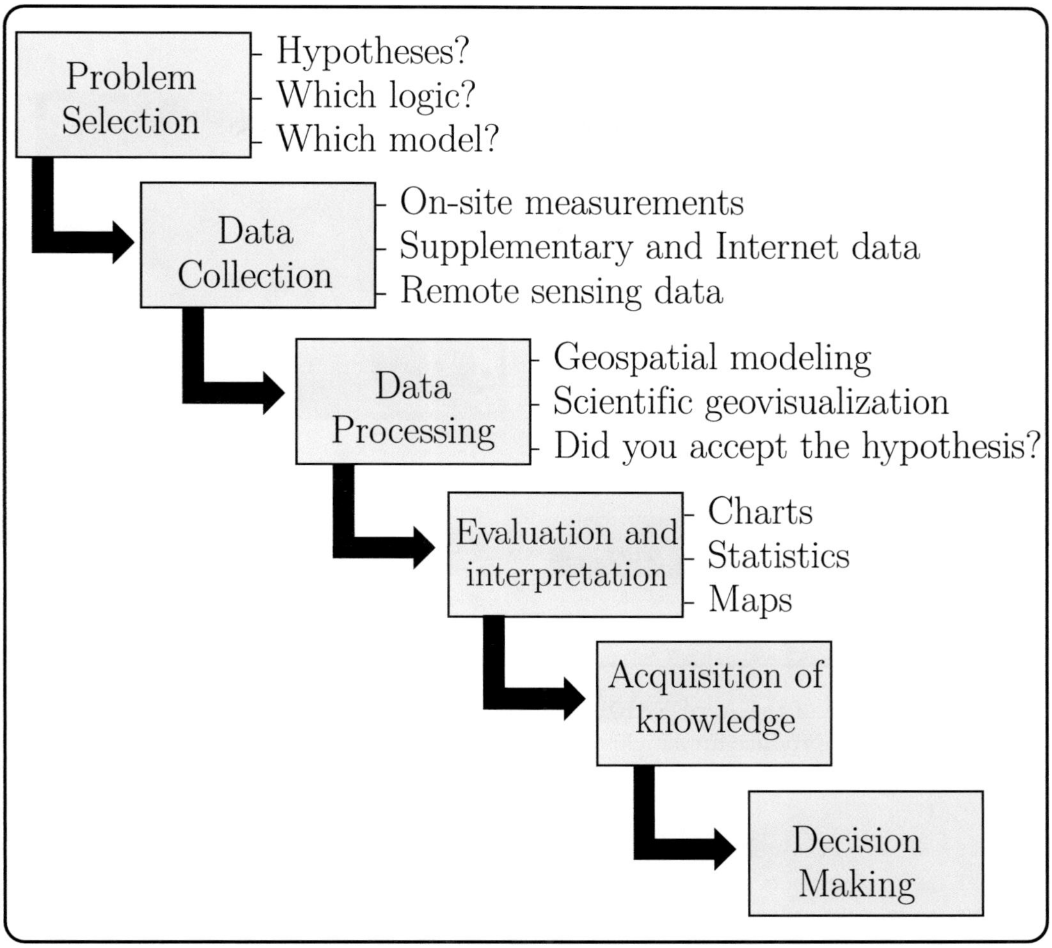

FIGURE 7.3 Stages of the decision-making process in geographic information science.

the main commercial crops: palm, rubber, cocoa, coffee, natural forest, and other land cover types. This was done by combining terrain data (slope) and annual compositions of Sentinel-1, Sentinel-2, and Palsar images to generate training data. These training datasets were fed into a machine learning model, which was then applied to generate semi-global commodity probability maps for the years from 2017 to 2023.

The `library` function is used to load the `rgee` package.

```
library(rgee)
# package ‘rgee’ was built under R version 4.3.3
```

The `ee_Initialize` function is used to initialize the `Earth Engine`.

```
ee_Initialize(email = "****@gmail.com", drive = TRUE)
# ---------------- rgee 1.1.7 ------------------------- earthengine-api 0.1.323 --
#  [OK] user: not_defined
#  [OK] Google Drive credentials:  FOUND
```

FIGURE 7.4 Use of the AppEEARS platform to get remote sensing data used in project implementation to produce an article on environmental geocomputation.

```
#  [OK] Initializing Google Earth Engine:  DONE!
#  [OK] Earth Engine account: users/***
#  [OK] Python Path: C:/Users/UFLA/AppData/Local/r-miniconda/envs/rgee/python.exe
# --------------------------------------------------------------------------------
```

Coffee crop data is obtained using the `ee$ImageCollection` and `collection$filterDate` functions.

```
collection <- ee$ImageCollection(
  'projects/forestdatapartnership/assets/coffee/model_2025a')
coffee2020 <- collection$filterDate('2020-01-01', '2020-12-31')$
  mosaic()$gt(0.95) # Select date
```

The functions `Map$setCenter` and `Map$addLayer` are used to map coffee fields in the region of Patrocínio, Minas Gerais, 2020 (Figure 7.5).

```
Map$setCenter(-46.84, -18.91, 9) # Set center
Map$addLayer(coffee2020$selfMask(), list(min = 0.5, max = 1,  # Map
                palette = c("#FFFFFF", "#A52A2A")), 'coffee 2020')
```

7.7.1 PREPARATION OF THE PROJECT ABSTRACT

The title, abstract and keywords of the project are prepared as follows:

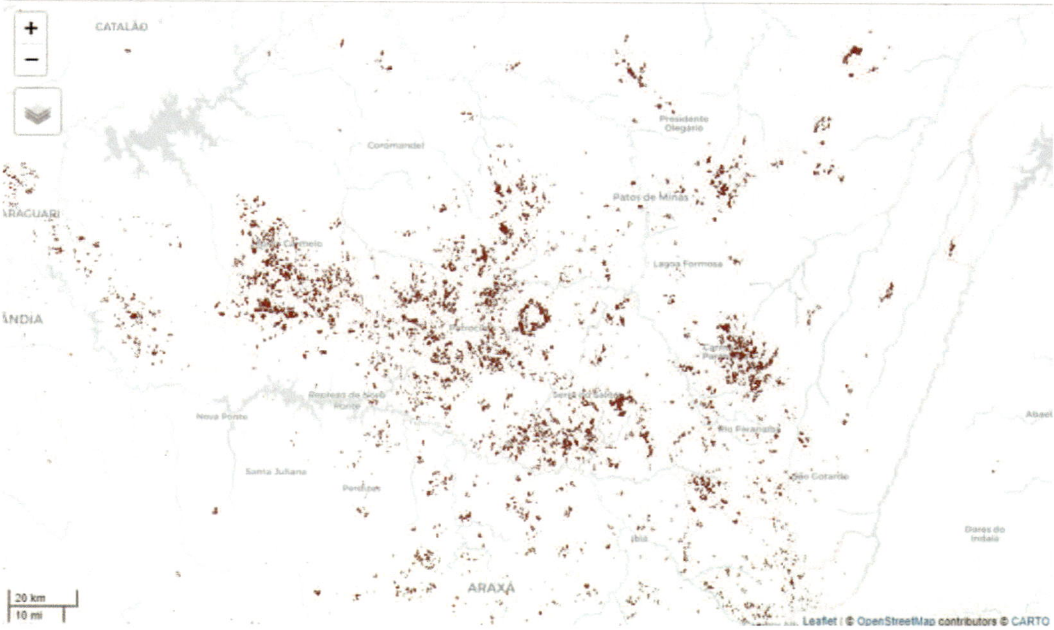

FIGURE 7.5 Using the RGEE package to access probability of coffee crop occurrence (2020) mapped with remote sensing and machine learning modeling from the Forest Data Partnership Project, with mapping results centered in the Patrocínio region, Minas Gerais, Brazil.

Geostatistical Surfaces of Climatological Normals of Mean Air Temperature in Minas Gerais

Marcelo de Carvalho Alves - Federal University of Lavras, Department of Agricultural Engineering, email: *marcelo.alves@ufla.br*

Abstract

Air temperature, a vital component for the terrestrial environment sustainability, can be used as an indicator and an important factor used in short- and long-term meteorological modeling at different scales. Temperature must be monitored on a spatial and temporal scale with high precision. Terrain elevation can be used as the main influence factor depending on the measurement scale. In small and medium scales, factors related to local relief were modeled with geostatistics including external variables in temperature modeling. We aimed to evaluate the use of universal kriging in the modeling of air temperature in order to create temperature surfaces at each km^2 in Minas Gerais State, Brazil using altitude, longitude and latitude covariates. The organized mean air temperature data of climatological normals of the National Institute of Meteorology were submitted to summary statistics, statistical regression and geostatistical analysis. Monthly and annual normals of the mean air temperature compensated for the period from 1981 to 2010 were modeled using temperature as a dependent variable and altitude, longitude and latitude as co-variables. Multiple regression modeling performed on temperature using altitude, longitude and latitude covariates determined significant parameters for monthly and annual mean air temperature global prediction. Relief and coordinates were used as external drift on variography and universal kriging with block for local temperature interpolation and prediction in order to generate 1 km moderate resolution surfaces of monthly and annual mean air temperature. Universal kriging determined smoothing effect of standard deviation of geospatial variation with prediction errors varying between $0.6° - 1.5°C$. Higher prediction error values were observed between June and August. Mean air temperature local prediction presented greater errors mainly in the lower altitude regions and in the colder months. In both monthly and annual temperature predictions, universal kriging with external drift enabled to circumvent the problem of performing spatial prediction from sparse punctual attribute data, conferring a temperature downscaling effect in Minas Gerais.

Keywords: Variogram with external drift; Topography; Universal kriging; Moderate resolution surface; Regression kriging; Uncertainty.

8 Vector Spatial Analysis with Environmental Geocomputation and R

8.1 LEARNING QUESTIONS

The learning questions answered by reading the chapter are:

- How to define spatial analysis and how to evaluate whether a method is spatial;
- How to manipulate and analyze attribute data;
- How to use geographic queries in geocomputation;
- What is the concept of reasoning in geocomputation;
- How to determine spatial subsets, topological relationships, spatial joins, non-overlapping joins, spatial data aggregation, and distance relationships in geocomputation.

8.2 LEARNING OBJECTIVES

The learning objectives answered by reading the chapter are:

- Understand the definition of spatial analysis and evaluate whether a specific method is spatial;
- Verify how to manipulate and analyze attribute data;
- Verify the use of operations on attribute data in geocomputation;
- Understand the concept of reasoning in geocomputation;
- Know the methods used to determine spatial subsets, topological relationships, spatial joins, non-overlapping joins, aggregation of spatial data and distance relationships in geocomputation.

8.3 INTRODUCTION

A geographic information system is distinguished from other types of information systems by spatial analysis functions performed on spatial and non-spatial attributes of entities stored in a geographic database intended for queries about the characteristics of stored objects (Câmara and Medeiros, 1998). Spatial analysis techniques can be applied to any kind of data arranged in any space, not just geographic space, such as analysis in outer space by astronomers or brain scan analysis by neuroscientists. Therefore, the term geospatial analysis is used, configuring the use of spatial analysis in the geographic environment of planet Earth. Geospatial analysis or analytical mapping is crucial in geocomputation, as it includes all the transformations, manipulations, and methods that can be applied to data to add value to it, to support decisions, and to reveal patterns and anomalies that are not immediately obvious (Longley et al., 2001).

Geospatial analysis is defined as the process by which raw data is transformed into useful information in pursuit of scientific discovery, or more effective decision-making about phenomena occurring on planet Earth. However, many procedures are not really spatial

DOI: 10.1201/9781003724377-8

because they affect only the attribute of the data, but not the size, feature, or shape of spatial entities such as point, line, polygon, and pixel (Burrough and McDonnell, 1998; Longley et al., 2001).

Spatial analysis methods can be very sophisticated, but also very simple. Spatial analysis methods may involve mathematical calculations, and the mathematical complexity may indicate the importance of the technique used. However, the human eye and brain also visually analyze geographic data and detect patterns and anomalies in maps and images. Therefore, spatial analysis should be considered a continuum of sophistication, from the simplest types of analysis, quickly intuitive under human visualization, to software types with complex and sophisticated mathematical application. The best spatial analysis is that performed with collaboration between computer and human, both playing vital roles in achieving satisfactory results (Longley et al., 2001).

Applications of geospatial analysis for scientific discovery and decision are performed in different forms (Longley et al., 2001):

- Inductive <- Using spatial analysis to examine empirical evidence in search of patterns that might support new theories or general principles;
- Deductive <- Focusing on testing theories and evaluating known principles from existing data and mappings performed;
- Normative <- Using spatial analysis to develop or prescribe new or better designs to the solution of practical problems enabling satisfactory decision-making.

Geospatial analysis can be used to perform geographic queries, distance measurements, summary statistics, optimizing object locations using established criteria, and hypothesis testing focusing on the reasoning used to evaluate a dataset and make generalizations from that data. Geospatial analysis can begin with the inclusion or exclusion of objects in the analysis and the intersection of lines and areas of different types that generate new entities. Spatial interactions are not limited to entity boundaries, but they can also include functions involving distances and topology, such as distance to river networks or roads, and in other different areas such as meteorology, archaeology, geodemography, rural and urban property and real estate valuation, and land planning (Burrough and McDonnell, 1998; Longley et al., 2001).

The initial problem of geographic data analysis can be a user-specific problem, with a geographic database containing information of interest that can solve a problem or provide access to get answers about queries through maps, tables and figures. Answering a query requires a database configured to perform a set of operations to determine and present results. The types of analysis methods that can be used on geospatial data can vary according to the data model and the type of representation adopted. In this case, for each data model and representation type, different approaches can be taken. The key issue is to determine whether the model used refers to entities in space or the continuous variation of an attribute over space. In the case of entities, the data analysis and query is concerned with the attributes and the connectivity between the entities and measurements made in space. In the case of continuous fields, data analysis is performed based on the spatial characteristics of the field. The analysis can be made more complex by the fact that continuous fields are discretized into a set of triangles or a regular grid and individual triangles or grid cells analyzed as individual entities (Burrough and McDonnell, 1998).

The nature of geographic analysis operations is defined from transformation and mixed operators between geo-objects and geo-fields (Câmara and Medeiros, 1998). However, in this chapter, the topics are centered on geospatial vector entity analysis in object-oriented relational data model. The fundamental axioms of geospatial analysis are presented including computational language used to solve practical applications.

8.4 BASIC CLASSES OF OPERATIONS FOR GEOSPATIAL ANALYSIS

In the entity model of objects in space, the basic information is: what is it, where is it, and what is the relationship to other entities? The nature of the entity is obtained based on the attributes, location by coordinates, and spatial relationships with different entities in terms of topology (proximity and connectivity). The aspects of location, proximity and topology distinguish geographic data from other types of data used in information systems (Burrough and McDonnell, 1998).

The basic classics of entity data analysis by attribute operations, distance, location, and topology construction result in new attributes attached to the original entities as the database grows. Data obtained by query or spatial analysis can be recorded in computer files for further processing or mapping for digital or printed visualization (Table 8.1) (Burrough and McDonnell, 1998).

TABLE 8.1 Basic classics of entity data analysis through attribute operations, distance, location, and topology construction.

Operation class	Description
Attribute	Operations on one or more attributes of an entity
Attribute	Operations on one or more attributes of several entities overlapped in space
Attribute	Operations on one or more linked attributes
Attribute	Operations on attributes of entities contained in other entities
Distance or location	Operations to locate entities with distance or location criteria
Distance or location	Operations to create contour zones around entities
Topology	Operations for modeling spatial interactions in a network of connected entities

8.5 OPERATIONS ON ATTRIBUTE DATA

Attributes refer to the geographic location of longitude, latitude, and altitude, the geospatial characteristics of boundary, area, shape, and contiguity of the entity itself, and other qualitative and quantitative information that defines the non-spatial characteristics of the object. New attributes can be added to entities as a result of logical and mathematical operations performed on the database or simply by modifying the symbol to represent an entity on the map. Performing a query with a search criterion can be performed and obtaining as a response a new temporary attribute created after being selected by the defined criterion (Burrough and McDonnell, 1998). When modifying attributes of entities of municipalities, the shape and units of the map are not modified; only the color and shading characteristics are modified, as in the case of the name of municipalities of the state of Minas Gerais, Brazil. The same occurs when performing a query of objects within polygons and a calculation performed, such as the sum of the area of coffee crop within each municipality and the mean altitude above sea level in the areas of coffee fields, within each municipality of Minas Gerais (Figure 8.1).

To perform this example, it is necessary to have a combination of spatial and geometric analysis of vector and raster data, although the focus of this chapter is on spatial analysis

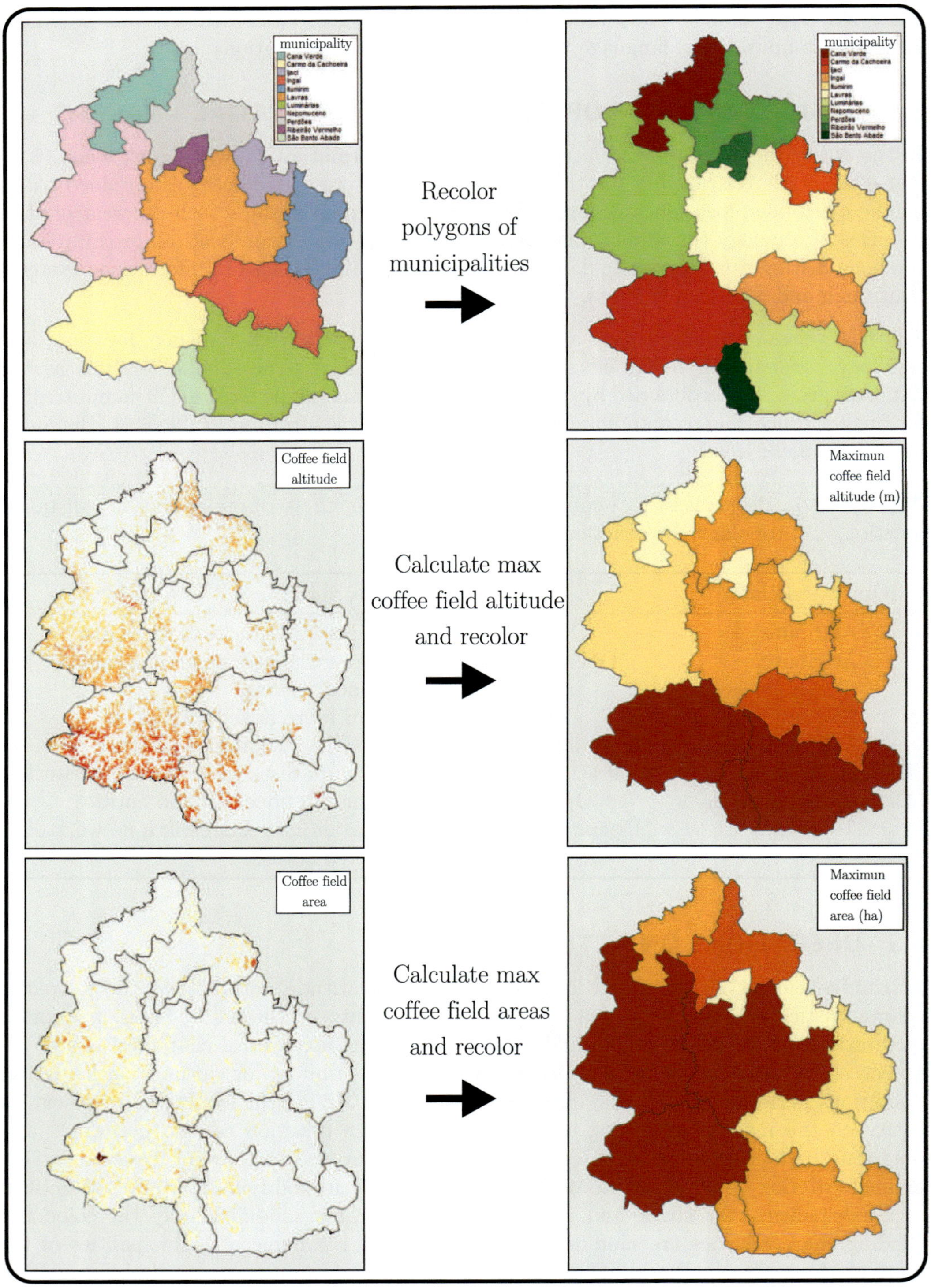

FIGURE 8.1 Color change of entities of municipalities in Minas Gerais (top), max altitude above sea level in coffee fields (middle), and max area of coffee fields within each municipality in Minas Gerais, Brazil (bottom).

of vector data. The packages `sf` (Pebesma et al., 2021a), `tmap` (Tennekes et al., 2020), `exactextractr`(Daniel Baston, 2023), `rgdal` (Bivand et al., 2021), `raster` (Hijmans et al., 2020) and `dplyr` (Wickham et al., 2021a) needed to detail what was done in Figure 8.1 are loaded into R with the `library` function.

```
library(sf)
library(tmap)
library(exactextractr)
library(rgdal)
library(raster)
library(dplyr)
```

The data from municipalities in Minas Gerais considered in the analysis are loaded into R using the `read_sf` function.

```
munMG2007sf <- read_sf(
  "C:/Aulas/2021_1/Geoprocessamento/shape/municipiosMG2007.shp",
  options = "ENCODING=WINDOWS-1252")
munMG2007sf
# Simple feature collection with 11 features and 9 fields
# Geometry type: POLYGON
# Dimension:     XY
# Bounding box:  xmin: -45.39973 ymin: -21.64273 xmax: -44.74138 ymax: -20.92158
# Geodetic CRS:  WGS 84
# A tibble: 11 × 10
```

The mapping of municipalities with different colors is performed with functions from the `tmap` package, coloring the municipality polygon with the `tm_fill` function and recoloring with the `palette` argument of the `tm_polygons` function (Figure 8.2).

```
color1 <- tm_shape(munMG2007sf[4:4]) + tm_polygons("NOME_MUNIC",
  title = "Municipalities") + tm_legend(position = c("right", "top"),
                                        frame = TRUE)
color2 <- tm_shape(munMG2007sf[4:4]) + tm_polygons("NOME_MUNIC",
                                                   palette = "RdYlGn",
  title = "Municipalities") + tm_legend(position = c("right", "top"),
                                        frame = TRUE)
tmap_arrange(color1, color2, ncol = 2)
```

The coffee polygons in the municipalities considered in the analysis are loaded into R using the `read_sf` function.

```
cana_verde_coffee <- read_sf(
  "C:/Aulas/2021_1/Geoprocessamento/shape/cana_verde_cafe.shp")
carmo_da_cachoeira_coffee <-read_sf(
  "C:/Aulas/2021_1/Geoprocessamento/shape/carmo_da_cachoeira_cafe.shp")
ijaci_coffee <- read_sf("C:/Aulas/2021_1/Geoprocessamento/shape/ijaci_cafe.shp")
ingai_coffee <- read_sf("C:/Aulas/2021_1/Geoprocessamento/shape/ingai_cafe.shp")
itumirim_coffee <- read_sf(
```

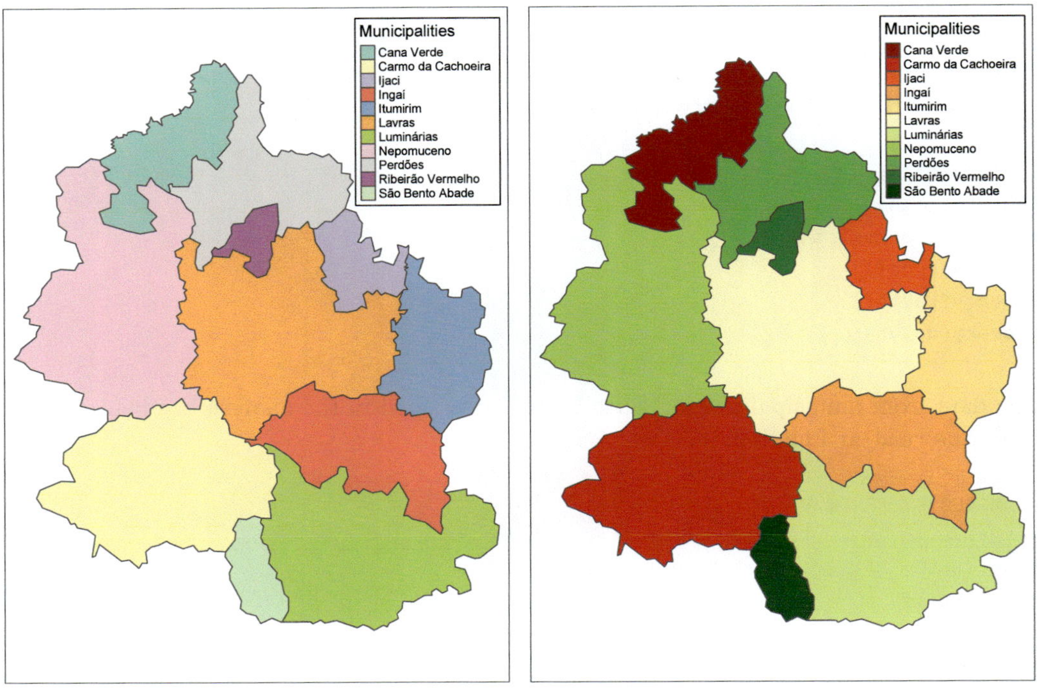

FIGURE 8.2 Color change of municipalities in Minas Gerais, Brazil.

```
  "C:/Aulas/2021_1/Geoprocessamento/shape/itumirim_cafe.shp")
lavras_coffee <- read_sf(
  "C:/Aulas/2021_1/Geoprocessamento/shape/lavras_cafe.shp")
luminarias_coffee <- read_sf(
  "C:/Aulas/2021_1/Geoprocessamento/shape/luminarias_cafe.shp")
nepomuceno_coffee <- read_sf(
  "C:/Aulas/2021_1/Geoprocessamento/shape/nepomuceno_cafe.shp")
perdoes_coffee <- read_sf(
  "C:/Aulas/2021_1/Geoprocessamento/shape/perdoes_cafe.shp")
ribeirao_vermelho_coffee <- read_sf(
  "C:/Aulas/2021_1/Geoprocessamento/shape/ribeirao_vermelho_cafe.shp")
sao_bento_do_abade_coffee <- read_sf(
  "C:/Aulas/2021_1/Geoprocessamento/shape/sao_bento_do_abade_cafe.shp")
```

The `rbind` function is used to join the polygons into a single geo-object of class `sf data.frame`.

```
coffeesf <- do.call(rbind, list(cana_verde_coffee, carmo_da_cachoeira_coffee,
  ijaci_coffee, ingai_coffee, itumirim_coffee, lavras_coffee, luminarias_coffee,
  nepomuceno_coffee, perdoes_coffee, ribeirao_vermelho_coffee,
  sao_bento_do_abade_coffee))
```

You can inspect your data to identify the problematic polygons and attempt to correct their geometry using GIS software or by manually editing the coordinates. A more programmatic approach in R is to use the `st_make_valid()` function from the `sf` package. This function is used to repair invalid geometries.

```
coffeesf <- st_make_valid(coffeesf)
coffeesf
# Simple feature collection with 8005 features and 14 fields
# Geometry type: POLYGON
# Dimension:     XY
# Bounding box:  xmin: -45.39288 ymin: -21.63094 xmax: -44.77784 ymax: -20.94797
# Geodetic CRS:  WGS 84
# A tibble: 8,005 × 15
names(coffeesf)
# [1] "Name" "descriptio" "timestamp" "begin" "end" "altitudeMo" "tessellate"
# "extrude"    "visibility" "drawOrder"
# [11] "icon"       "Irrigado"    "area_ha"     "NM_MUN"      "geometry"
```

Altitude data from the Shuttle Radar Topography Mission (SRTM) in the region of interest were obtained from the Earth Explorer platform (https://earthexplorer.usgs.gov/). In this case, two tiles were needed to cover the entire study area. The data are imported into R using the raster function.

```
dem1 <- raster("C:/sr/C6/srtm_27_17.tif")
dem1
# class      : RasterLayer
# dimensions : 6000, 6000, 3.6e+07  (nrow, ncol, ncell)
# resolution : 0.0008333333, 0.0008333333  (x, y)
# extent     : -50, -45, -25, -20  (xmin, xmax, ymin, ymax)
# crs        : +proj=longlat +datum=WGS84 +no_defs
# source     : srtm_27_17.tif
# names      : srtm_27_17
dem2 <- raster("C:/sr/C6/srtm_28_17.tif")
dem2
# class      : RasterLayer
# dimensions : 6000, 6000, 3.6e+07  (nrow, ncol, ncell)
# resolution : 0.0008333333, 0.0008333333  (x, y)
# extent     : -45, -40, -25, -20  (xmin, xmax, ymin, ymax)
# crs        : +proj=longlat +datum=WGS84 +no_defs
# source     : srtm_28_17.tif
# names      : srtm_28_17
```

The mosaic of the two tiles is created using the `mosaic` function, which calculates the mean of the intersection between the two tiles using the `fun = mean` operator.

```
dem <- mosaic(dem1, dem2, fun = mean)
dem
# class      : RasterLayer
# dimensions : 6000, 12000, 7.2e+07  (nrow, ncol, ncell)
# resolution : 0.0008333333, 0.0008333333  (x, y)
# extent     : -50, -40, -25, -20  (xmin, xmax, ymin, ymax)
# crs        : +proj=longlat +datum=WGS84 +no_defs
```

```
# source      : r_tmp_2025-08-19_181022.624276_15956_07975.grd
# names       : layer
# values      : -165, 2854  (min, max)
```

The vector file of class `sf` is transformed into class `sp` with the `as_Spatial` function.

```
coffeesp <- as_Spatial(coffeesf)
coffeesp
# class       : SpatialPolygonsDataFrame
# features    : 8005
# extent     : -45.39288, -44.77784, -21.63094, -20.94797  (xmin, xmax, ymin, ymax)
# crs         : +proj=longlat +datum=WGS84 +no_defs
# variables   : 14
```

A subset of the digital elevation model around the municipalities of interest is performed with the `crop` function, and the `mask` function is used to maintain the brightness of the pixels only in the region of interest. With this masking operation, the region around the area has brightness values equal to `NA`. This provides information on the altitude of the terrain only in regions with coffee fields. As a result, coffee is grown in this region at altitudes between 765 and 1,240 meters above sea level.

```
srtm_crop <- crop(dem, coffeesp)
srtm_coffee <- mask(srtm_crop, coffeesp)
srtm_coffee
# class      : RasterLayer
# dimensions : 819, 738, 604422  (nrow, ncol, ncell)
# resolution : 0.0008333333, 0.0008333333  (x, y)
# extent     : -45.3925, -44.7775, -21.63083, -20.94833  (xmin, xmax, ymin, ymax)
# crs        : +proj=longlat +datum=WGS84 +no_defs
# source     : memory
# names      : layer
# values     : 765, 1240  (min, max)
```

The `rasterize` function is used to convert the areas registered in each polygon into a raster with values of the maximum area of each coffee field per pixel. Note that coffee fields ranged in size from 0.0135 to 170.6717 ha.

```
coffee_area <- rasterize(coffeesp, srtm_coffee, field = "area_ha", fun = "max")
coffee_area
# class      : RasterLayer
# dimensions : 819, 738, 604422  (nrow, ncol, ncell)
# resolution : 0.0008333333, 0.0008333333  (x, y)
# extent     : -45.3925, -44.7775, -21.63083, -20.94833  (xmin, xmax, ymin, ymax)
# crs        : +proj=longlat +datum=WGS84 +no_defs
# source     : memory
# names      : layer
# values     : 0.0135, 170.6717  (min, max)
```

The `exact_extract` function is used to obtain the maximum values for altitude and area cultivated with coffee for each municipality considered in the analysis.

```
alt_coffee_max <- exact_extract(srtm_coffee, munMG2007sf, 'max')
# [1] 1013 1215  990 1064 1032 1039 1240 1110 1119  947 1161
area_coffee_max <- exact_extract(coffee_area, munMG2007sf, 'max')
# [1]  17.1836 170.6717  25.9855  37.4560  12.7177  50.0758  44.6074  83.1584
# 112.0619  15.7800  35.8402
```

The `st_sf` function is used to record the maximum values obtained in new data columns in the municipality attribute table.

```
coffeeMGsf <- sf::st_sf(data.frame(munMG2007sf, alt_coffee_max, area_coffee_max))
names(coffeeMGsf)
# [1] "GEOCODIG_M" "UF" "SIGLA" "NOME_MUNIC" "REGIão" "MESORREGIã"  "NOME_MESO"
# [8] "MICRORREGI" "NOME_MICRO" "alt_coffee_max"  "area_coffee_max" "geometry"
```

The elevation and area of coffee fields are mapped side by side using the `plot` function. Note that the geometry of the municipalities is mapped in a layer below the raster layer (Figure 8.3).

```
par(mfrow = c(1, 2), mar = c(2, 2, 0.3, 0.3), mgp = c(1.5, 0.6, 0))
plot(st_geometry(munMG2007sf), col = "grey92")
plot(srtm_coffee, col = RColorBrewer::brewer.pal(n = 9, name = "YlOrRd"),
  horizontal = TRUE, legend.args = list(text ='Altitude (m)'), add = TRUE)
plot(st_geometry(munMG2007sf), col = "grey92")
plot(coffee_area, col = RColorBrewer::brewer.pal(n = 9, name = "YlOrRd"),
  horizontal = TRUE, legend.args = list(text ='Area (ha)'), add = TRUE)
```

The municipalities with the highest altitude and cultivated area in coffee fields are mapped using `tmap` package functions. The mapping is classified into quantile intervals based on the `style` argument (Figure 8.4).

```
c_alt <- tm_shape(coffeeMGsf[10:10]) + tm_polygons("alt_coffee_max",
    title = "Maximum altitude (m)", style="quantile") +
    tm_legend(position = c("right", "top"), frame = TRUE)
c_area <- tm_shape(coffeeMGsf[11:11]) + tm_polygons("area_coffee_max",
    title = "Maximum cultivated area (ha)", style = "quantile") +
    tm_legend(position = c("right", "top"), frame = TRUE)
tmap_arrange(c_alt, c_area, ncol = 2)
```

For a given location, x, the value of an attribute U_i created by the query-based selection process is (Burrough and McDonnell, 1998):

$$U_i = f(A_{i1}, A_{i2}, ..., A_{in}) \tag{8.1}$$

where $A_{i1}, A_{i2}, ..., A_{in}$ are attribute values used to estimate U_i.

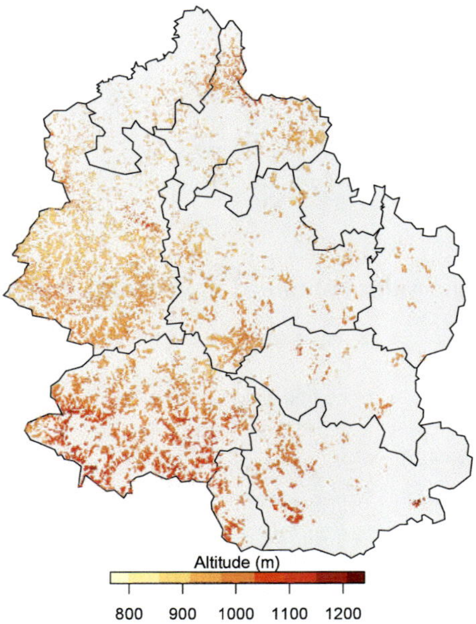

FIGURE 8.3 Mapping elevation and area of coffee fields in municipalities of Minas Gerais, Brazil.

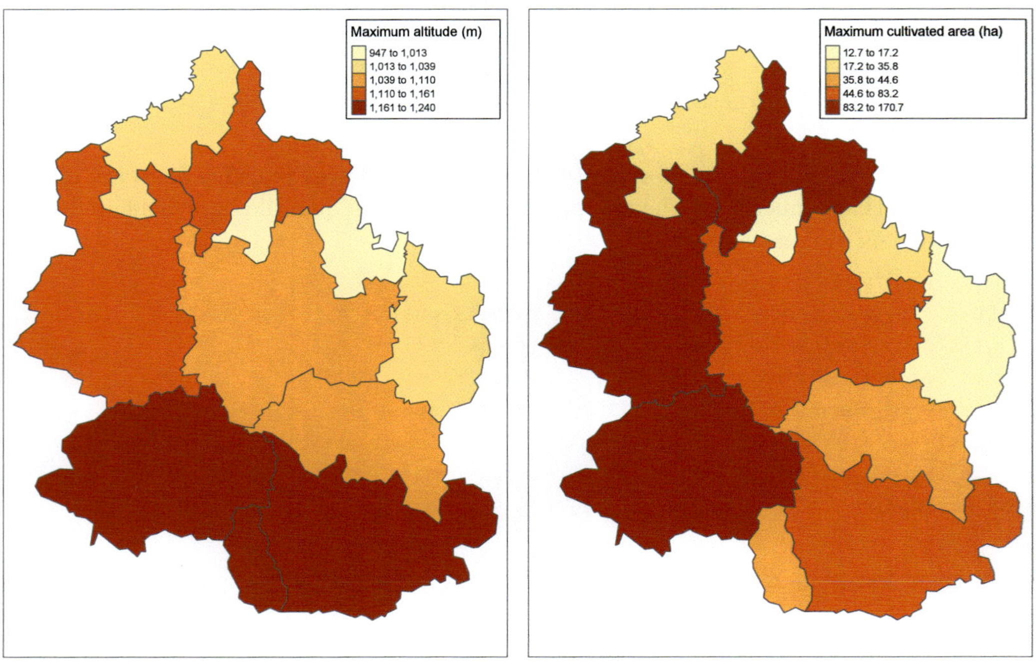

FIGURE 8.4 Choropleth mapping of maximum altitude and cultivated area in coffee fields in municipalities of Minas Gerais, Brazil.

The $f()$ function can be any of the following simple operations or in combination (Table 8.2) (Burrough and McDonnell, 1998):

- Logical (Boolean) operations;
- Simple and complex arithmetic operations and numerical models;
- Univariate statistical analysis;
- Multivariate and Bayesian statistical methods for classification and discrimination;
- Multicriteria methods and methods based on computational intelligence.

TABLE 8.2 Mathematical operations and transformation of attribute data.

Operation	Description
Logic	True or false, 0 or 1, resulting from the union, intersection, negation, and exclusion of two or more sets
Arithmetic	Result of addition, subtraction, multiplication, division, raised to power, exponential, logarithm, truncation, square root operations that created new attributes from the original attribute
Trigonometric	sine, cosine, tangent, arc-tangent operations to create new attributes from the original attribute
From data type	The new attribute originates from the original attribute on a different data type (Boolean, nominal, ordinal, directional, integer, real, topological)
Statistics	Statistical operations on mean, mode, median, standard deviation, variance, minimum, mean, maximum values, range, skewness, kurtosis, performed on quantitative values of a given attribute
Multivariate	New attributes obtained after using regression model, by numerical model of a physical process, principal component, factor, correspondence, multivariate, probability, discriminant, maximum likelihood, Bayesian, fuzzy and neural network analyses applied to original objects

Boolean logic operations, for example, can be realized in practice by using symbol operators (Table 8.3) and are illustrated by the Venn diagram (Figure 8.5) (Burrough and McDonnell, 1998; Lovelace et al., 2019).

TABLE 8.3 Symbol operators applied in Boolean logic in geocomputation.

Symbol	Name
==	Equal to
!=	Not equal to
>, <	Major, minor than
>=, <=	Major or equal, minor or equal
&, |, !	Logical operators, and, or, not

8.5.1 MANIPULATION OF VECTOR DATA ATTRIBUTES

Vector geographic data consists of one column with the attribute variable and one row per observation or feature. In some objects, geometry data is stored in a column, called

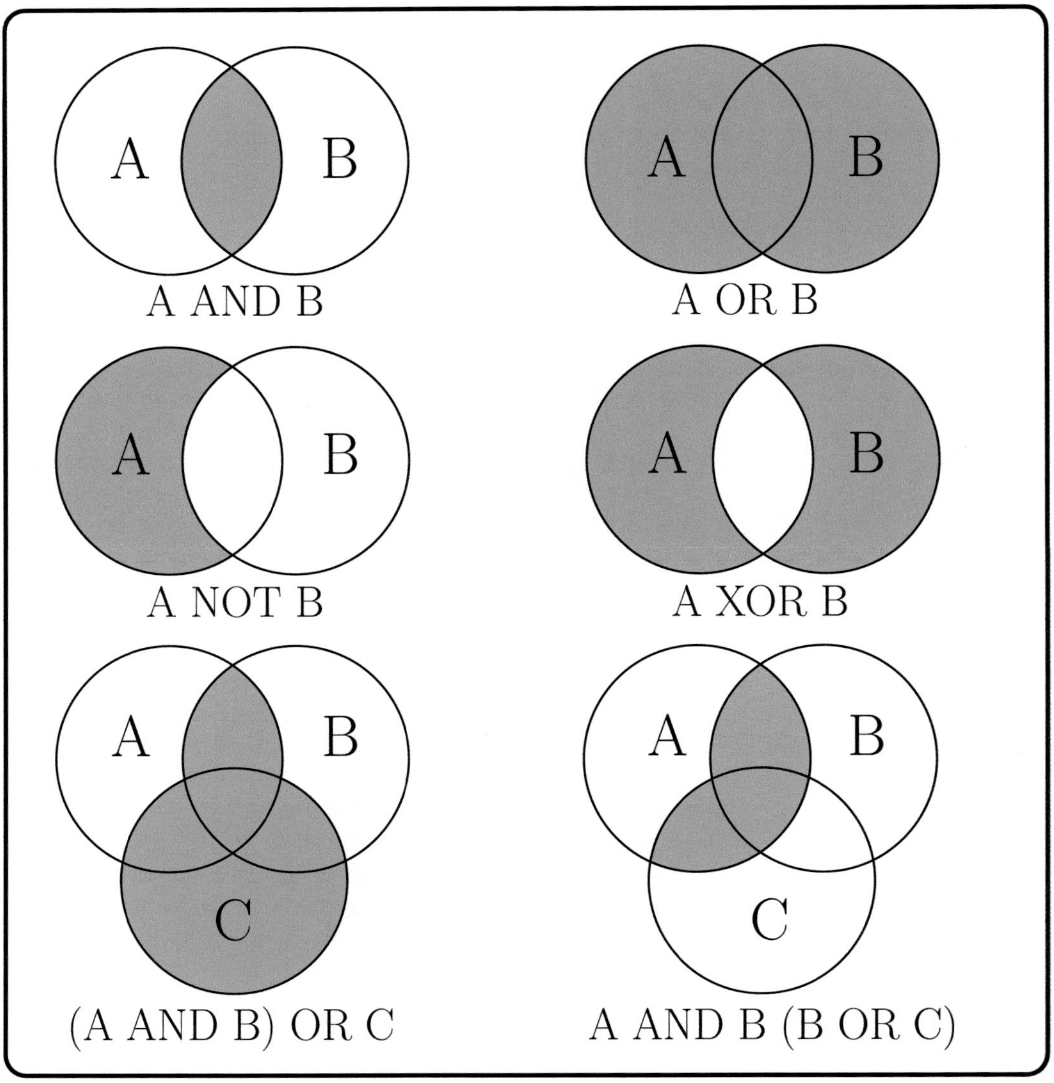

FIGURE 8.5 Venn diagram with the results of applying Boolean logic to join and intersect one or more datasets, where shaded graphs mean true.

geometry. The geometry column is special in that it is a list column, which can contain various geographic entities such as points, lines, and polygons per line. Methods are used for objects to function as a regular data frame and store spatial and non-spatial data in the same way, as columns in a `data.frame` (Lovelace et al., 2019). Attribute manipulation operations are used to perform queries, calculations, and obtain new files of interest, according to the use of different attribute manipulation operators for geographic objects. For example, an attribute selection operator over a set of geo-objects, given a constraint based on descriptive attributes determines a subset whose members satisfy the constraint as a result (Câmara and Medeiros, 1998).

The main operations for manipulating vector geo-object attributes are (Lovelace et al., 2019):

- Obtaining subsets of attribute data;
- Aggregation of vector attributes;
- Joining of vector attributes;
- Creating attributes and removing geospatial information.

Geospatial simple feature vector data consists of a column with the attribute variable and a row per observation or feature. Geometry data are stored in a column, called geometry, in list form, which can contain one or several geographic entities of points, lines, and polygons per data record line. Objects are characterized by a regular data frame used to store geospatial and non-geospatial data in columns as in a database table (Lovelace et al., 2019).

8.5.1.1 Obtaining subsets of attribute data

The basic functions to perform subsetting of data preserve the spatial component of attributes for the following purposes (Lovelace et al., 2019):

- Perform subsetting of rows and columns of the data;
- Obtain a result based on query criteria;
- Extract a vector from an attribute variable;
- Select columns and rows by name and position.

The subsets can be obtained through logical vectors creating new objects; as an example, from a cadastral map of municipalities of Minas Gerais, we can extract a cadastral subset based on the search criteria by name of the municipality (Figure 8.6).

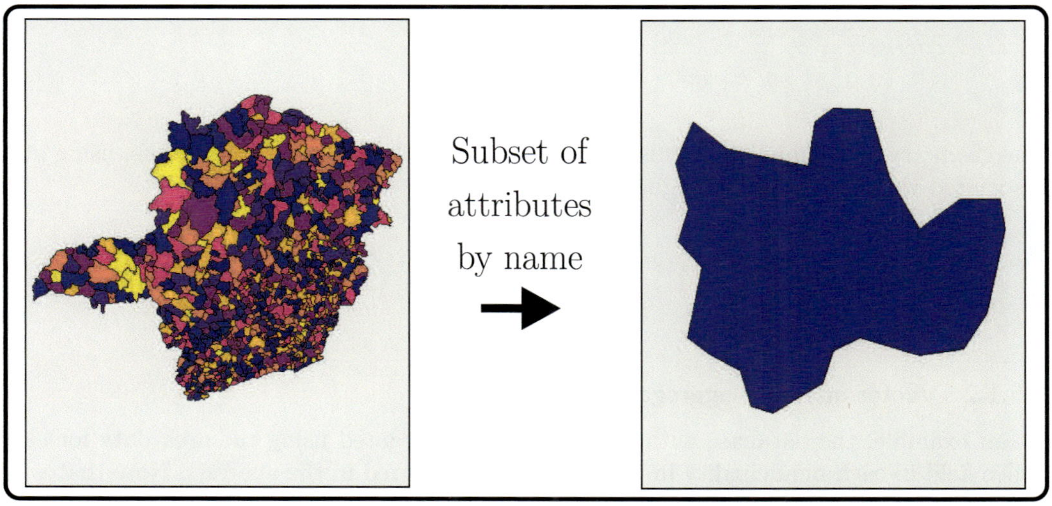

FIGURE 8.6 Subset of attributes per municipality name, determining the mapping of the municipality of Lavras, Minas Gerais, Brazil.

In this case, the location of the Lavras municipality implies nine records in the database because of the different sampling data collection times determined by the census conducted on municipal coffee production (Figure 8.7).

The subset of the municipality Nepomuceno is created from the data of municipalities in Minas Gerais used for coffee mapping in the previous example. The subset is created based on the name of the municipality registered in the attribute table database using square brackets and `$` to select the column of attribute data of interest.

```
   Mode    FALSE     TRUE
logical     7660        9
Simple feature collection with 9 features and 9 fields
geometry type:  POLYGON
dimension:      XY
bbox:           xmin: -45.1768 ymin: -21.39891 xmax: -44.86478 ymax: -21.13287
geographic CRS: GCS_unknown
        City   nome State MicrRgn MesoRgn Region      municip  Ano Prod                       geometry
3925 3138203 LAVRAS    31   31057    3111      3 Lavras (MG) 2012 5280 POLYGON ((-44.99302 -21.134...
3926 3138203 LAVRAS    31   31057    3111      3 Lavras (MG) 2013 5889 POLYGON ((-44.99302 -21.134...
3927 3138203 LAVRAS    31   31057    3111      3 Lavras (MG) 2018 7392 POLYGON ((-44.99302 -21.134...
3928 3138203 LAVRAS    31   31057    3111      3 Lavras (MG) 2015 3981 POLYGON ((-44.99302 -21.134...
3929 3138203 LAVRAS    31   31057    3111      3 Lavras (MG)   NA   NA POLYGON ((-44.99302 -21.134...
3930 3138203 LAVRAS    31   31057    3111      3 Lavras (MG) 2014 4979 POLYGON ((-44.99302 -21.134...
3931 3138203 LAVRAS    31   31057    3111      3 Lavras (MG) 2019 6750 POLYGON ((-44.99302 -21.134...
3932 3138203 LAVRAS    31   31057    3111      3 Lavras (MG) 2017 7275 POLYGON ((-44.99302 -21.134...
3933 3138203 LAVRAS    31   31057    3111      3 Lavras (MG) 2016 7140 POLYGON ((-44.99302 -21.134...
```

FIGURE 8.7 Result of query operation by municipality name.

```
nepomuceno <- coffeeMGsf[coffeeMGsf$NOME_MUNIC == "Nepomuceno",]
nepomuceno
# Simple feature collection with 1 feature and 11 fields
# Geometry type: POLYGON
# Dimension:     XY
# Bounding box:  xmin: -45.39973 ymin: -21.37724 xmax: -45.14439 ymax: -21.0458
# Geodetic CRS:  WGS 84
#   GEOCODIG_M UF SIGLA NOME_MUNIC  REGIão MESORREGIã NOME_MESO MICRORREGI ...
# 8    3144607 31    MG Nepomuceno Sudeste         3111 Campo das Vertentes   ...
#  area_coffee_max                        geometry
# 8          83.1584 POLYGON ((-45.28033 -21.053...
```

The geometry of the municipality is mapped with functions from the `sf` package, using `plot` associated with `st_geometry` (Figure 8.8).

```
plot(st_geometry(nepomuceno), graticule = TRUE, col = "white", bgc = "grey92",
     border = 'black', axes = TRUE)
```

8.5.1.2 Vector attribute aggregation

As an example, the database with coffee fields is aggregated using the area data for each coffee field in each municipality in Minas Gerais considered in the analysis. Note that with the `names` function, it is possible to check the names of each data column in the attribute table. The columns of interest in this case are those for area (`area_ha`) and municipality name (`NM_MUN`).

```
names(coffeesf)
# [1] "Name" "descriptio" "timestamp" "begin" "end" "altitudeMo" "tessellate"...
# [9] "visibility" "drawOrder" "icon" "Irrigado" "area_ha" "NM_MUN" "geometry"
```

The `aggregate` function is used to sum the areas of coffee fields by municipality. However, the geometry results remain with the same geometry as the crops and not with the geometry of the municipalities.

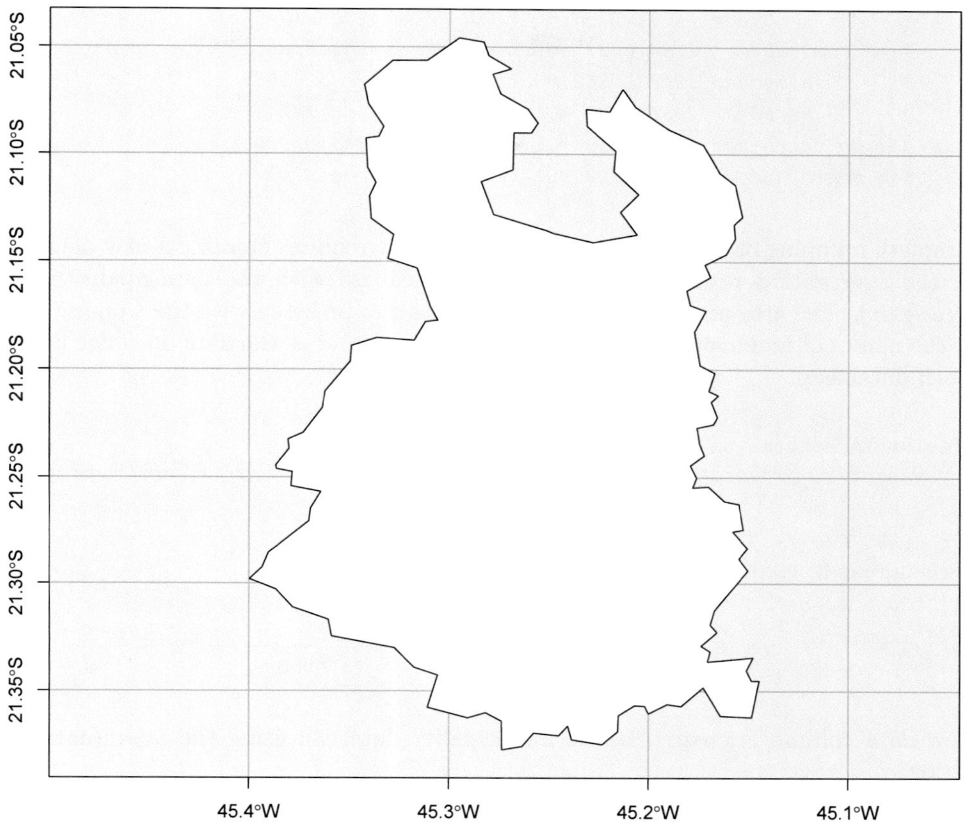

FIGURE 8.8 Mapping Nepomuceno based on a subset from the municipal database of Minas Gerais, Brazil.

```
coffeeTotalArea <- aggregate(coffeesf["area_ha"], by = list(coffeesf$NM_MUN),
                             FUN = sum, na.rm = TRUE)
print(coffeeTotalArea, n = 11)
# Simple feature collection with 11 features and 2 fields
# Attribute-geometry relationships: aggregate (1), identity (1)
# Geometry type: MULTIPOLYGON
# Dimension:     XY
# Bounding box:  xmin: -45.39288 ymin: -21.63094 xmax: -44.77784 ymax: -20.94797
# Geodetic CRS:  WGS 84
# First 10 features:
#               Group.1    area_ha                       geometry
# 1          Cana Verde   850.0477 MULTIPOLYGON (((-45.24484 -...
# 2  Carmo da Cachoeira 13270.2751 MULTIPOLYGON (((-45.33587 -...
# 3               Ijaci   161.8435 MULTIPOLYGON (((-44.9604 -2...
```

```
# 4              Ingaí    563.0123 MULTIPOLYGON (((-45.02132 -...
# 5           Itumirim    294.9510 MULTIPOLYGON (((-44.86633 -...
# 6             Lavras  4911.1740 MULTIPOLYGON (((-45.1654 -2...
# 7         Luminárias  2234.8637 MULTIPOLYGON (((-45.03861 -...
# 8         Nepomuceno 13823.0962 MULTIPOLYGON (((-45.35485 -...
# 9            Perdões  2988.7973 MULTIPOLYGON (((-45.1424 -2...
# 10 Ribeirão Vermelho   154.0259 MULTIPOLYGON (((-45.09929 -...
# 11   São Bento Abade   979.6204 MULTIPOLYGON (((-45.09937 -...
```

Choropleth mapping of crop areas in each municipality requires creating a new data column with the aggregation results calculated in the database with the municipality polygons (`coffeeMGsf`). The area and municipality name data can be listed with the `$` operator. Note that the names of municipalities are listed in the same order as the data on coffee field areas in both databases.

```
coffeeTotalArea$area_ha
# [1] 850.0477  13270.2751  161.8435  563.0123  294.9510  4911.1740  2234.8637
# 13823.0962
# [9] 2988.7973  154.0259  979.6204
coffeeMGsf$NOME_MUNIC
# [1] "Cana Verde"    "Carmo da Cachoeira" "Ijaci"          "Ingaí"
# [5] "Itumirim"      "Lavras"             "Luminárias"     "Nepomuceno"
# [9] "Perdões"       "Ribeirão Vermelho"  "São Bento Abade"
```

A new data column is created in the municipality database using the `st_sf(data.frame)` function.

```
coffeeMGsf <- sf::st_sf(data.frame(coffeeMGsf, coffeeTotalArea$area_ha))
names(coffeeMGsf)
# [1] "GEOCODIG_M" "UF"               "SIGLA"             "NOME_MUNIC"
# [5] "REGIão"     "MESORREGIã"       "NOME_MESO"         "MICRORREGI"
# [9] "NOME_MICRO" "alt_coffee_max" "area_coffee_max" "coffeeTotalArea.area_ha"
# [13] "geometry"
```

The column name `coffeeTotalArea.area_ha` is too long and can be renamed using the `rename` function to `field_area_ha`.

```
coffeeMGsf <- coffeeMGsf |> rename(field_area_ha = coffeeTotalArea.area_ha)
names(coffeeMGsf)
# [1] "GEOCODIG_M" "UF" "SIGLA" "NOME_MUNIC"  "REGIão"  "MESORREGIã"
# [7] "NOME_MESO" "MICRORREGI" "NOME_MICRO" "alt_coffee_max" "area_coffee_max"
# "field_area_ha"
# [13] "geometry"
```

The choropleth mapping of the attribute aggregation of coffee-growing areas by municipality is performed with quantile classification and ranking by count values. Note that the argument `style = cat` is used to map the numerical variable as a categorical variable and not by intervals as in the case of `style = quantile` (Figure 8.9).

```
field_quant <- tm_shape(coffeeMGsf[12:12]) + tm_polygons("field_area_ha",
    title = "Coffee field area (ha)", style = "quantile", palette = "YlOrRd",) +
    tm_legend(position = c("right", "top"), frame = TRUE) + tm_text("field_area_ha")
field_cont <- tm_shape(coffeeMGsf[12:12]) + tm_polygons("field_area_ha",
    title = "Coffee field area (ha)", style = "cat", palette = "YlOrRd",) +
    tm_legend(position = c("right", "top"),  frame = TRUE)
tmap_arrange(field_quant, field_cont, ncol = 2)
```

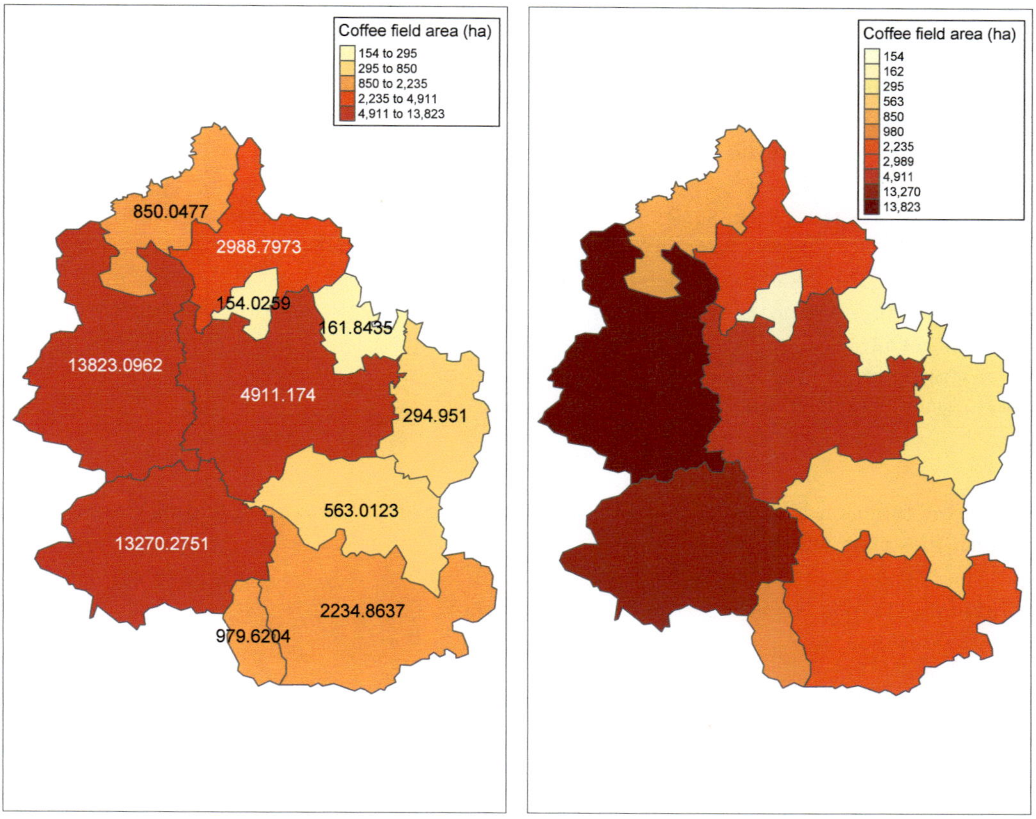

FIGURE 8.9 Choropleth mapping of attribute aggregated coffee fields at the municipal level using the quantile method and ranking by numerical values as a categorical variable.

8.5.1.3 Joining vector attributes

The joining of vector attributes performed with the functions applied against two databases, b1 and b2, enables the following results to be obtained:

- Inclusion of all rows in b1 and b2;
- Inclusion of all rows in b1;
- Inclusion of all rows in b2;
- Inclusion of all rows in b1 or b2.

A table with columns of Arabica coffee production data per year in Brazil and polygons with cadastral data of municipalities in Minas Gerais are used to perform the joining of vectorial attributes. The data can be joined based on a column in common between the databases. After joining, the mapping of data is performed, for example, filling the color of the polygon with values of coffee production 2012 attribute (Figure 8.10).

To join attributes, two separate databases are used, but with one column of data in common, usually a column with numeric key identifier values. The `sf` and `readr` packages are required to perform this analysis and are loaded into R with the `library` function.

```
library(sf)
library(readr)
```

The `read_csv2` function is used to import the database edited in a spreadsheet, and the `st_read` function imports simple polygon feature data. With the `names` function, you can check the names of the data columns in both files. The `set_crs` function is used to assign a coordinate system to the geo-object.

```
tab <- read_csv2("C:/geo/SIDRA_IBGE/tabelaCafeNew.csv", col_names = TRUE,
    na = c("...", "-"), locale = locale(encoding = 'ISO-8859-1')) # Load data
names(tab) # See database 1 names
# [1] "cod_ibge" "municipio" "2012" "2013" "2014" "2015" "2016" "2017" "2018"...
mg <- st_read('C:/geo/c7/mg1.shp') # Load MG data
st_crs(mg)<-st_crs(4326) # Set crs # Set CRS
names(mg) # See database 2 names
# [1] "nome" "City" "State" "MicrRgn" "MesoRgn" "Region" "geometry"
```

The `print` function is used to check the first five records of the corresponding data columns in both files that will be joined.

```
print(tab[1], n = 5)
# A tibble: 5,533 × 1
#   cod_ibge
#      <dbl>
# 1  1100015
# 2  1100023
# 3  1100031
# 4  1100049
# 5  1100056
print(mg[2], n = 5)
# Simple feature collection with 853 features and 1 field
# Geometry type: POLYGON
# Dimension:     XY
# Bounding box:  xmin: -51.04608 ymin: -22.92276 xmax: -39.85683 ymax: -14.24667
# Geodetic CRS:  WGS 84
# First 5 features:
#      City                       geometry
# 1 3100104 POLYGON ((-47.42967 -18.165...
# 2 3100203 POLYGON ((-45.1393 -19.2104...
# 3 3100302 POLYGON ((-42.31071 -20.142...
# 4 3100401 POLYGON ((-43.02387 -20.410...
# 5 3100500 POLYGON ((-42.36208 -18.925...
```

The left_join function is used to join the spreadsheet data with the geo-object database. Using the names function, we can see that the geo-object data columns also contain coffee production data from 2012 to 2019.

```
tab_mg <- left_join(mg, tab, by = c(City = "cod_ibge"))
names(tab_mg)
# [1] "nome" "City" "State" "MicrRgn" "MesoRgn" "Region" "municipio" "2012"
# "2013" "2014"
# [11] "2015" "2016" "2017" "2018" "2019" "geometry"
```

With the plot function, coffee production data is mapped in the municipalities of Minas Gerais for the year 2012 (Figure 8.10).

```
plot(tab_mg["2012"])
```

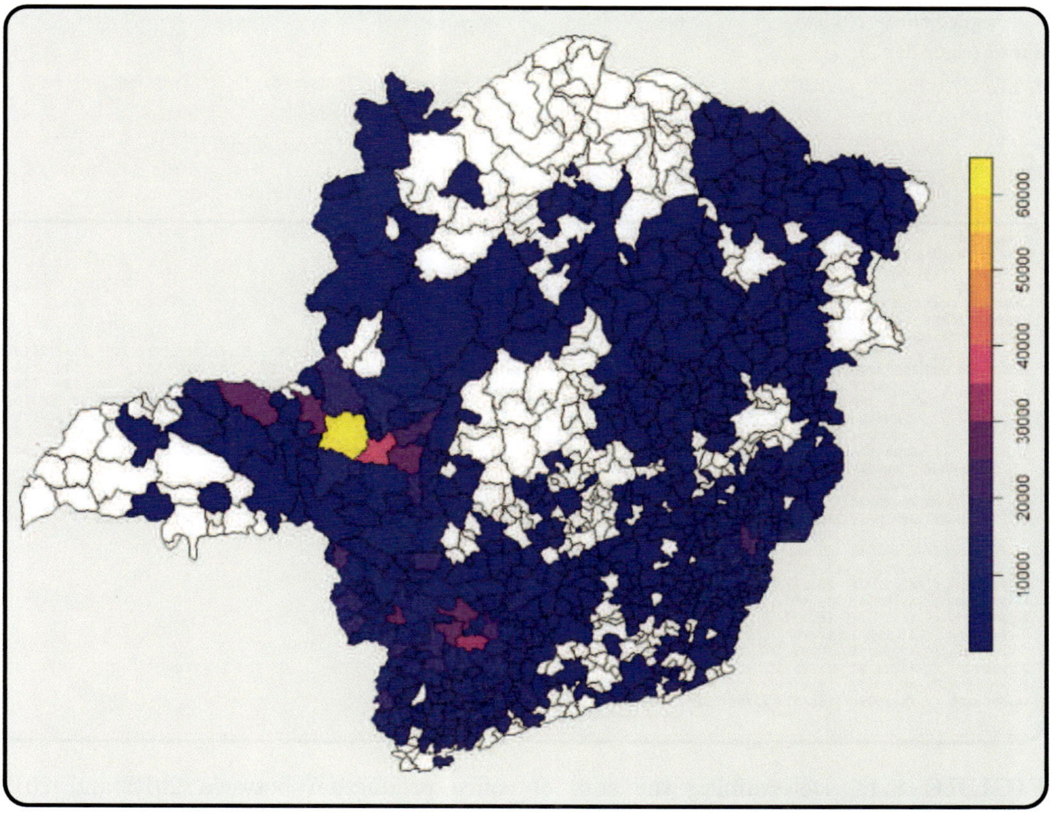

FIGURE 8.10 Joining performed between municipal coffee production database and Minas Gerais political boundary of municipalities for mapping the production of processed coffee in 2012.

According to the map, municipalities with the white color have no Arabica coffee production records specified as no data (NA).

8.5.2 CREATE AND REMOVE ATTRIBUTES

New attribute columns can be created based on calculations performed on existing columns. The sum of Arabica coffee production can be determined, for example, between 2012 and 2019 by creating a new data column (Figure 8.11).

In order to create a new data column based on the other coffee production columns, a new geo-object is created in relation to the data from the previous example. Using the `$` operator, a new data column is created based on the sum of the other coffee production columns between 2012 and 2019. With the `names` function, you can verify that a new data column `prod_sum` was created at the end of the database.

```
mgCoffee <- tab_mg # Create a new geo-object
names(mgCoffee)
# [1] "nome"      "City"      "State"     "MicrRgn"   "MesoRgn"   "Region"
# [7] "municipio" "2012"      "2013"      "2014"      "2015"      "2016"
# [13] "2017"      "2018"      "2019"      "geometry"
mgCoffee$prod_sum <- (mgCoffee$`2012` + mgCoffee$`2013` + mgCoffee$`2014` +
  mgCoffee$`2015` + mgCoffee$`2016` + mgCoffee$`2017`+ mgCoffee$`2018`+
    mgCoffee$`2019`)
names(mgCoffee)
# [1] "nome"      "City"      "State"     "MicrRgn"   "MesoRgn"   "Region"
# [7] "municipio" "2012"      "2013"      "2014"      "2015"      "2016"
# [13] "2017"      "2018"      "2019"      "geometry"  "prod_sum"
```

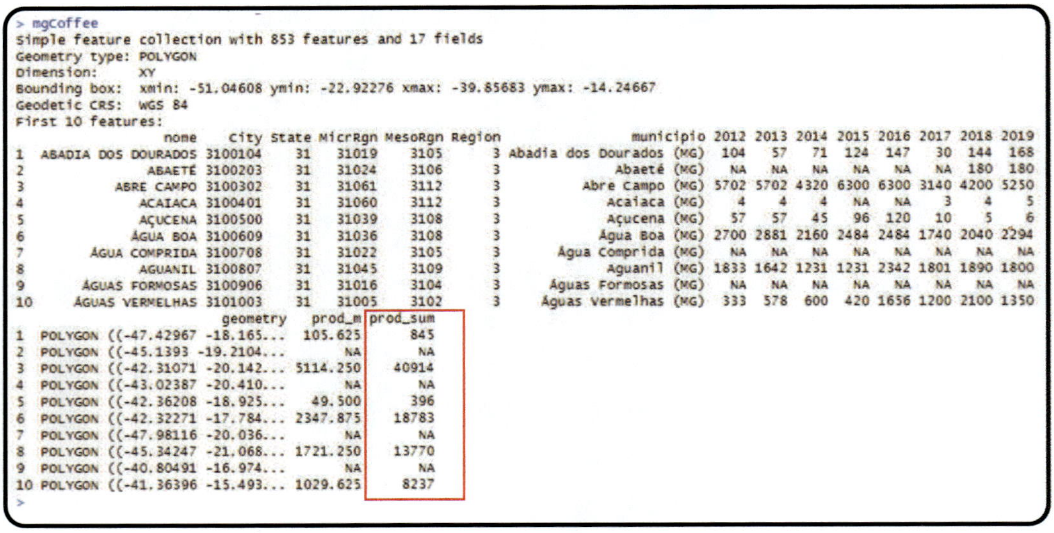

```
> mgCoffee
Simple feature collection with 853 features and 17 fields
Geometry type: POLYGON
Dimension:     XY
Bounding box:  xmin: -51.04608 ymin: -22.92276 xmax: -39.85683 ymax: -14.24667
Geodetic CRS:  WGS 84
First 10 features:
                 nome    City State MicrRgn MesoRgn Region                  municipio 2012 2013 2014 2015 2016 2017 2018 2019
1  ABADIA DOS DOURADOS 3100104    31   31019    3105      3 Abadia dos Dourados (MG)  104   57   71  124  147   30  144  168
2               ABAETÉ 3100203    31   31024    3106      3              Abaeté (MG)   NA   NA   NA   NA   NA   NA  180  180
3           ABRE CAMPO 3100302    31   31061    3112      3          Abre Campo (MG) 5702 5702 4320 6300 6300 3140 4200 5250
4             ACAIACA 3100401    31   31060    3112      3             Acaiaca (MG)    4    4    4   NA   NA    3    4    5
5             AÇUCENA 3100500    31   31039    3108      3             Açucena (MG)   57   57   45   96  120   10    5    6
6            ÁGUA BOA 3100609    31   31036    3108      3            Água Boa (MG) 2700 2881 2160 2484 2484 1740 2040 2294
7       ÁGUA COMPRIDA 3100708    31   31022    3105      3       Água Comprida (MG)   NA   NA   NA   NA   NA   NA   NA   NA
8             AGUANIL 3100807    31   31045    3109      3             Aguanil (MG) 1833 1642 1231 1231 2342 1801 1890 1800
9      ÁGUAS FORMOSAS 3100906    31   31016    3104      3      Águas Formosas (MG)   NA   NA   NA   NA   NA   NA   NA   NA
10    ÁGUAS VERMELHAS 3101003    31   31005    3102      3     Águas Vermelhas (MG)  333  578  600  420 1656 1200 2100 1350
                         geometry   prod_m prod_sum
1  POLYGON ((-47.42967 -18.165...  105.625      845
2  POLYGON ((-45.1393 -19.2104...       NA       NA
3  POLYGON ((-42.31071 -20.142... 5114.250    40914
4  POLYGON ((-43.02387 -20.410...       NA       NA
5  POLYGON ((-42.36208 -18.925...   49.500      396
6  POLYGON ((-42.32271 -17.784... 2347.875    18783
7  POLYGON ((-47.98116 -20.036...       NA       NA
8  POLYGON ((-45.34247 -21.068... 1721.250    13770
9  POLYGON ((-40.80491 -16.974...       NA       NA
10 POLYGON ((-41.36396 -15.493... 1029.625     8237
>
```

FIGURE 8.11 Determining the sum of coffee production between 2012 and 2019, generating a new cadastral column in the geographic database.

Data columns can be removed in different ways, such as using bracket operators. In this case, columns 8 to 15 and column 17 are removed. The results can be observed using the `names` function. It should be noted that the `subset` and `dplyr::select` functions can also be used to remove columns.

```
mg <- mgCoffee[,-c(8:15, 17)]
names(mg)
[1] "nome" "City" "State" "MicrRgn" "MesoRgn" "Region"  "municipio" "geometry"
```

8.6 SPATIAL ANALYSIS OPERATIONS ON VECTOR DATA

Given a geographic region with a set of geo-objects and a cadastral map that maps the objects in the region, spatial query operations can be defined by means of a spatial predicate that consists of a spatial constraint defined by a topological or metric relationship. The result of this operation is a subset of the original set composed of all the geo-objects that define the response according to the spatial predicate used (Câmara and Medeiros, 1998).

The main spatial analysis operations used on vector data features of cadastral type are (Lovelace et al., 2019):

- Spatial subset determination;
- Topological relations;
- Spatial join;
- Non-overlapping join;
- Aggregation of spatial data;
- Distance relations.

8.6.1 SPATIAL SUBSET

Spatial subsetting is the process of selecting features of an object based on its spatial relationship with another object. A subset can be performed with an object location operator by name, in a similar way to the one performed earlier to locate the municipality of Nepomuceno, keeping the geometry of the located geo-object.

The intersection of spatial objects with attributes and coordinate systems, such as polygons, can be intersected using the `st_intersection()` function from the `sf` package. With this function, it is possible to obtain the geometric intersection between objects by creating a new spatial object representing the overlapping regions. Note that when working with spatial objects, it is essential to ensure that both objects have the same coordinate reference system for accurate intersection calculations. If the coordinate systems are different, the `st_transform()` function can be used to convert one or both objects to a common coordinate system before performing the intersection.

The `st_intersection` function is used as an example to obtain the intersection between the `mg` and `munMG2007sf` geo-objects from previous examples. Using the `names` function, it is possible to see that the database resulting from the intersection contains the fields common to both databases, which means that a spatial join of the database was also performed.

```
mg_int <- st_intersection(mg, munMG2007sf)
names(mg_int)
# [1] "nome" "City" "State" "MicrRgn" "MesoRgn" "Region" "municipio"
# "GEOCODIG_M" "UF" "SIGLA"
# [11] "NOME_MUNIC" "REGIão" "MESORREGIã" "NOME_MESO" "MICRORREGI"
# "NOME_MICRO" "geometry"
names(mg)
# [1] "nome" "City" "State" "MicrRgn" "MesoRgn" "Region" "municipio" "geometry"
```

```
names(munMG2007sf)
# [1] "GEOCODIG_M" "UF" "SIGLA" "NOME_MUNIC" "REGIão" "MESORREGIã" "NOME_MESO"
# "MICRORREGI" "NOME_MICRO" "geometry"
```

The results are mapped using the `plot` function (Figure 8.12). Note that since the databases have different numbers of vertices that provide different levels of polygon detail, the municipal boundaries appear unusual. It is recommended to use databases with the same level of detail in order to use the resulting intersection data without the need for editing.

```
plot(mg_int[1], key.pos = 1, axes = TRUE, bgc = "gray92", main = "")
```

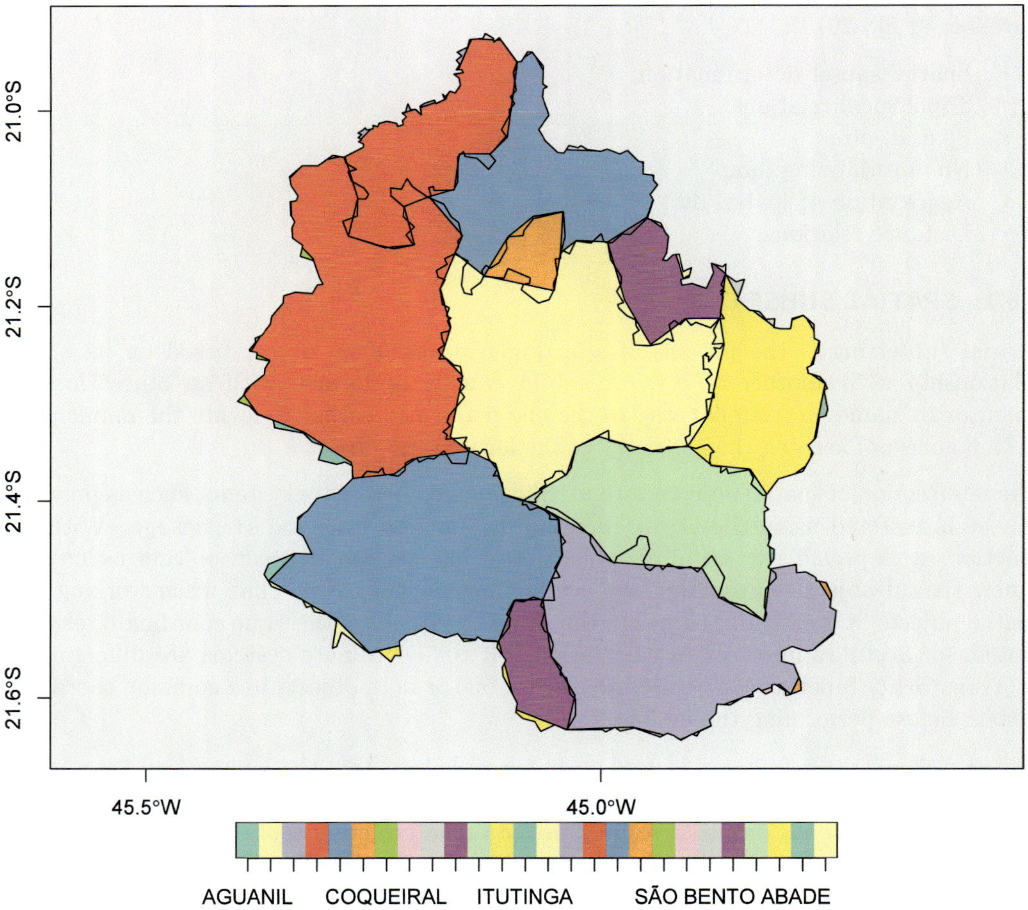

FIGURE 8.12 Determination of spatial subsets using the geo-object intersection method.

8.6.2 TOPOLOGICAL RELATIONS

In topological spatial relationships, geo-objects are associated with graphical representations of points, lines and polygons divided into (Câmara and Medeiros, 1998):

- Topological relationships invariant to biunivocal and bicontinuous transformations

such as those of scale, translation and rotation, such as, "within" and "adjacent to";
- Directional relationships, such as "above" and "next to";
- Metric relationships, derived from distance and direction operations.

Topological relationships can be used to obtain geospatial subsets of objects according to the situation of one in relation to another, as in the case of objects mapped in an area of the Federal University of Lavras, by means of points, lines and polygons. We can observe that a point is located inside the polygon, that the line intersects the polygon and that two points are touched by the line (Figure 8.13).

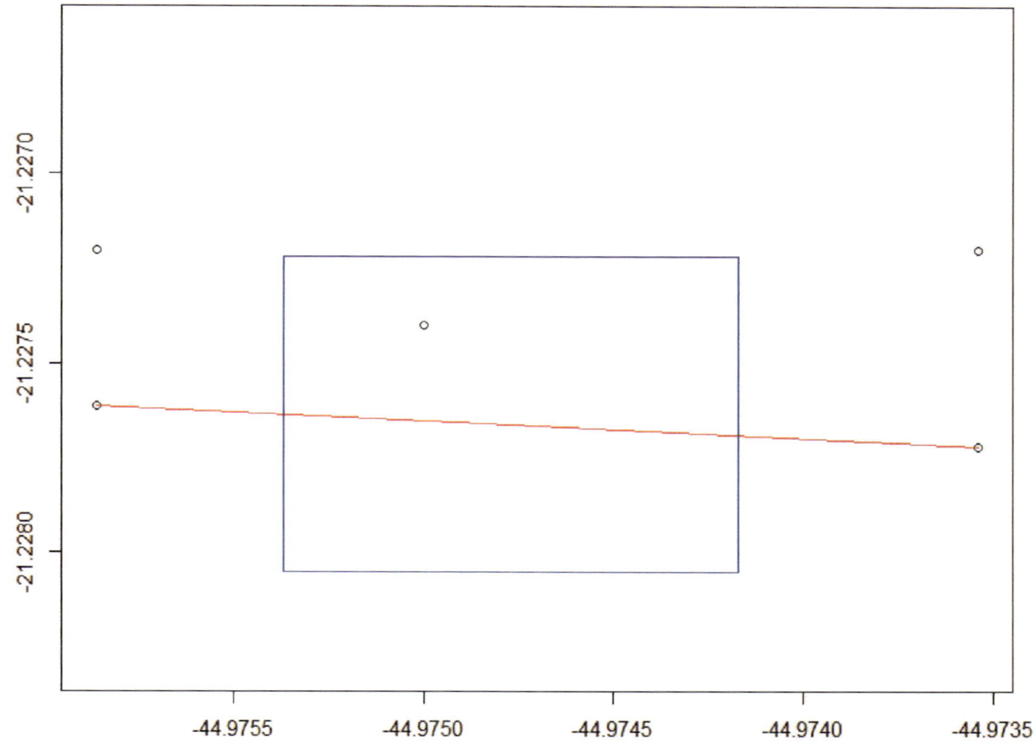

FIGURE 8.13 Topological situation relations of objects mapped on the campus of the Federal University of Lavras, such as points, line and polygon.

A simple query allows to know which of the points is intercepted by the polygon, or to get a logical result from using the function with the answer values TRUE or FALSE. More advanced queries can be performed to evaluate if an object is within a specified distance from another object. In this case, the distance can be obtained with respect to the geographic coordinate system, with the distance measure in degrees, or by converting the geographic projection into a plane map projection, in metric units of the coordinate system (Figure 8.14).

8.6.3 SPATIAL JOIN

The spatial join operation produces a set of objects and values as a result of using a spatial constraint. The term spatial join is used by analogy to the join operation in a conventional database (Câmara and Medeiros, 1998).

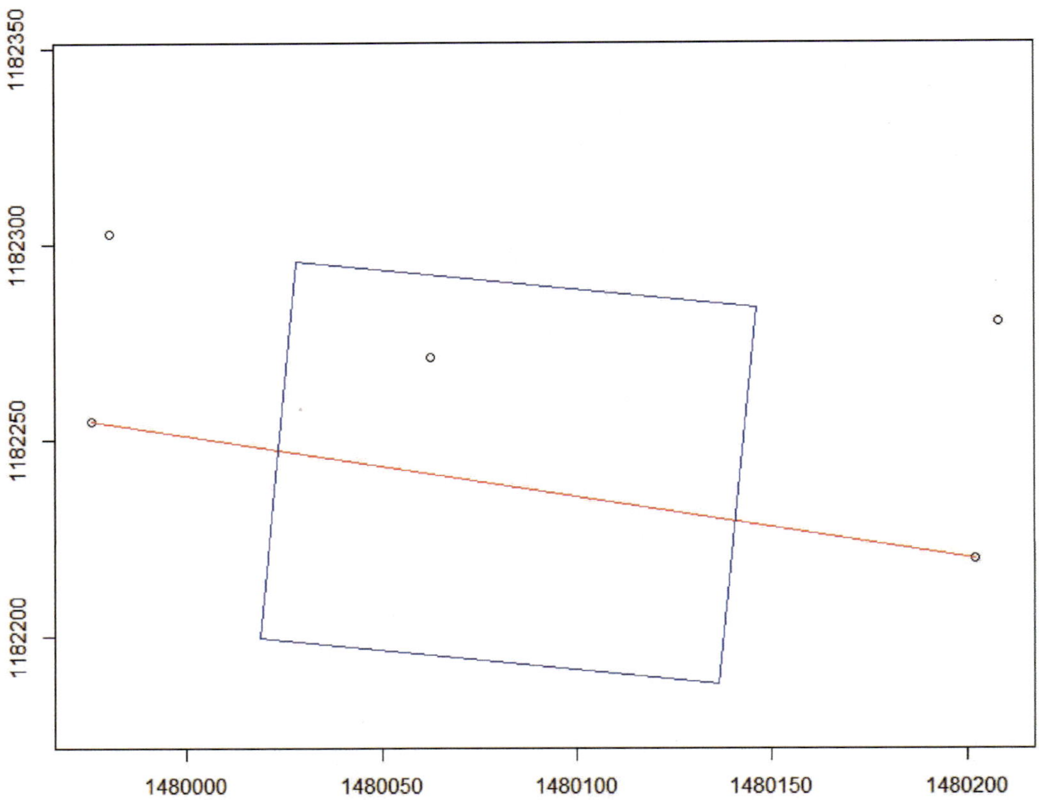

FIGURE 8.14 Topological relations of points, line and polygon in conic Albers Equal Area map projection.

To illustrate a type of geospatial join applied in coffee growing, four random points are created to perform the geospatial join in the region comprising the state of Minas Gerais. The created points and the polygons with municipal boundaries are mapped to facilitate the visualization of the points in relation to the polygons (Figure 8.15) (See computational practice section for code details).

With this, four data records are detected where there was a match in the join, considering the temporal information of coffee production registered in each municipality. The join with the database preserves the geometry column and the name of the municipalities on the left. There may be occasions when more than one point is inserted within a polygon. In this case, the records will be duplicated together with the polygon feature creating a new record line for each combination (Figure 8.16) (See computational practice section for code details).

8.6.3.1 Packages used

An example of spatial join can be performed with climate data in the United States. The packages `sf`, `dplyr`, `viridis` (Garnier et al., 2024), `ggplot2` (Wickham et al., 2020), `rnaturalearth` (Massicotte and South, 2023), `GSODR` (Sparks et al., 2025; Sparks, 2018), `ggrepel` (Slowikowski, 2024), `cowplot` (Wilke, 2024), `nngeo` (Dorman, 2024), `tigris` (Walker, 2025), and `purrr` (Wickham and Henry, 2023) are loaded into R with the `library` function.

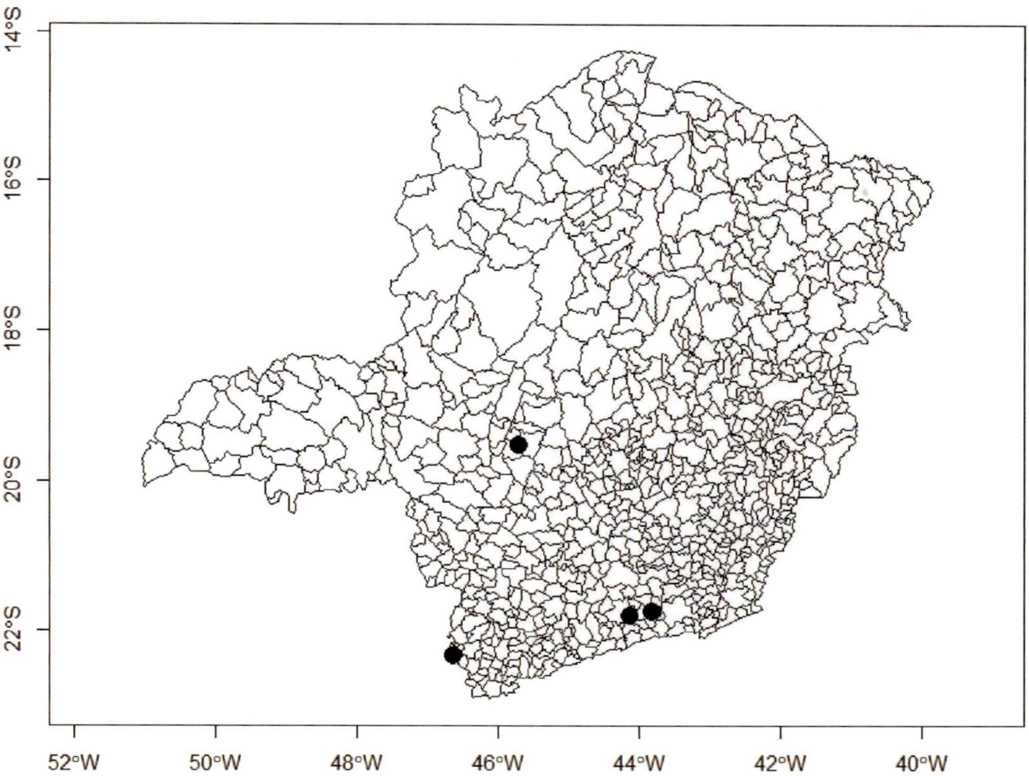

FIGURE 8.15 Random points created in the state of Minas Gerais to perform geospatial join with the polygons of corresponding municipalities.

```
library(sf)
library(dplyr)
library(viridis)
library(ggplot2)
library(rnaturalearth)
library(GSODR)
library(ggrepel)
library(cowplot)
library(nngeo)
library(tigris)
library(purrr)
```

8.6.3.2 Polygon data

Data for US states is obtained in low spatial resolution, excluding Puerto Rico, Alaska, and Hawaii, using the `us_states` function. The `filter` function is used to exclude unwanted polygons. The boundaries of the state of Montana are obtained using the `filter` function. A rectangle is drawn around Montana for illustrative purposes when mapping the region with the `st_make_grid` function. Montana counties are obtained using the `counties` function. The `NAME` of the counties database is replaced by `name` to differentiate it from the country

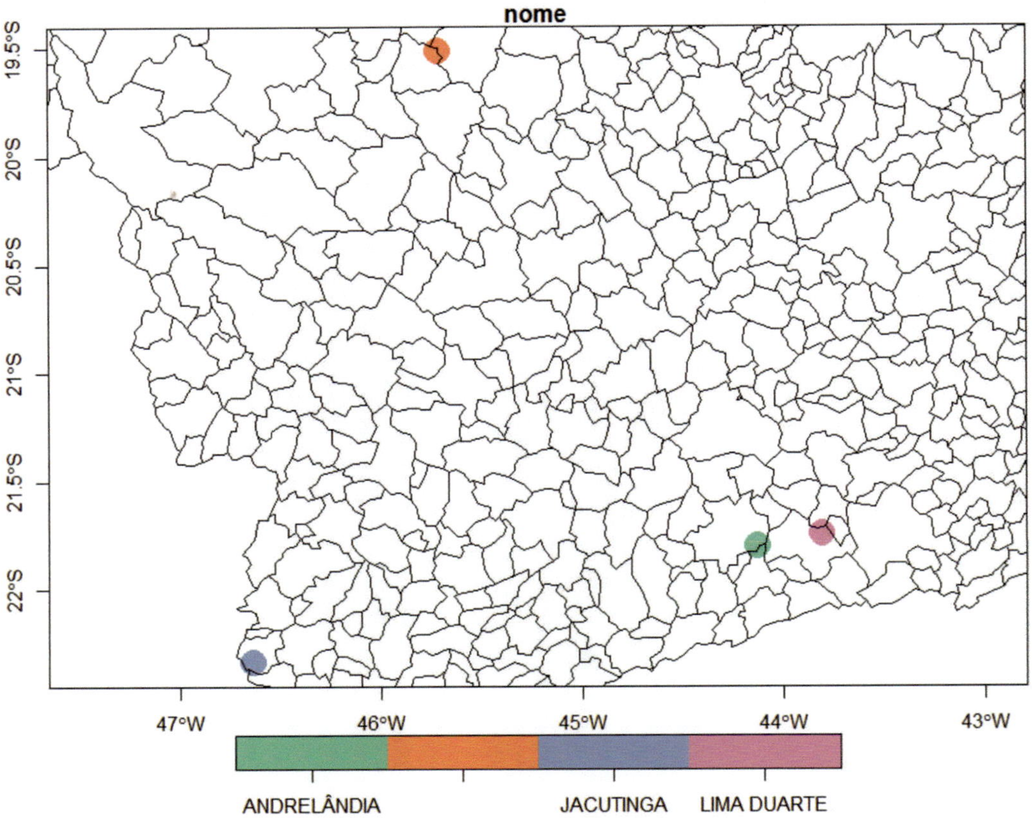

FIGURE 8.16 Municipalities of Andrelandia, Dores do Indaia, Jacutinga and Lima Duarte randomly selected in Minas Gerais by the geospatial joining method.

database. Note that you can use the `glimpse` function to check the database of interest.

```
us <- states() |> st_transform(4326) # Get US and transform projection
# Retrieving data for the year 2024
us <- us |> filter(!STUSPS %in% c("PR", "AK",
                "HI", "MP", "AS","VI", "GU")) # Filter out unwanted polygons
mt <- us |> filter(STUSPS == "MT") # Get Montana only
mt_box <- st_make_grid(mt, n = 1) # Make a bok
mt_co <- counties(state = "MT") |> st_transform(4326) |> rename(name = NAME)
# Retrieving data for the year 2024
# Using FIPS code '30' for state 'MT'
```

A preliminary mapping of the analyzed region is performed using functions from the `sf` package (Figure 8.17).

```
plot(us$geometry, axes = TRUE, col = "white", bgc = "gray92")
plot(mt$geometry, add = TRUE, col = "gray92", border = "black")
plot(mt_co$geometry, add = TRUE, border = "blue", col = NA)
plot(mt_box, add = TRUE, border = "red3", col = NA, lwd = 2)
```

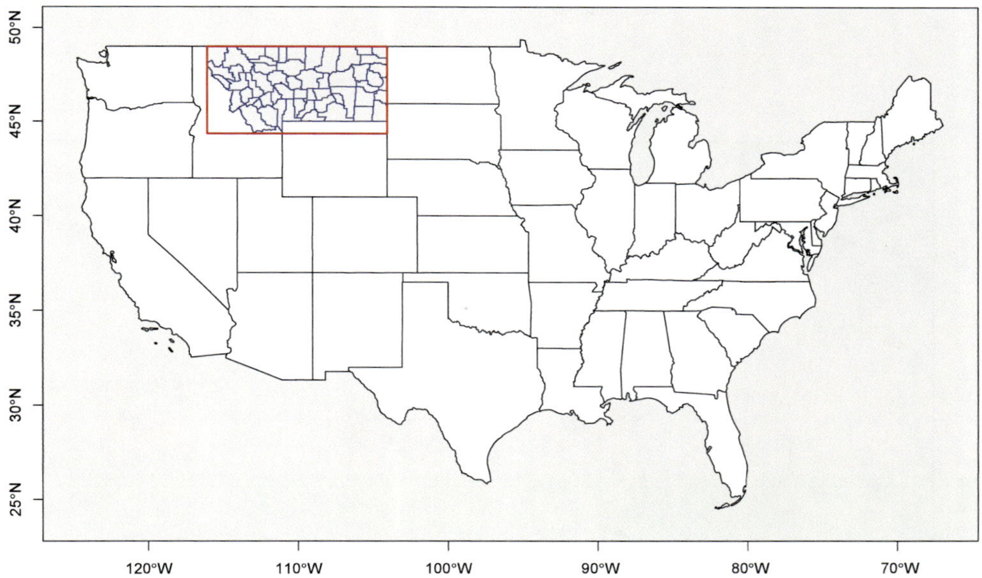

FIGURE 8.17 Mapping Montana counties in the United States.

8.6.3.3 Point data

Climate data from the `GSODR`[1] package is used to obtain simple point features representing global climate stations. The `isd_history.rda` file containing station data is obtained using the `load(system.file)` function. The file containing station data of class `data.frame` is converted into a point geo-object using `st_as_sf` function, considering the data column with the longitude and latitude coordinates and a coordinate reference system. Station data is filtered only for the US and Montana using `dplyr::filter` function.

```
load(system.file("extdata", "isd_history.rda", package = "GSODR"))
isd_history <- as.data.frame(isd_history) |>
  st_as_sf(coords=c("LON", "LAT"), crs = 4326, remove = FALSE) # Make spatial
isd_history_mt <- dplyr::filter(isd_history, CTRY == "US", STATE == "MT")
```

A simple feature collection of country multipolygons is obtained from the `ne_countries` function in the `rnaturalearth` package.

```
world <- ne_countries(scale = "medium", returnclass = "sf")
```

The global mapping of `GSODR` weather stations is performed with the `plot` function (Figure 8.18).

[1] https://docs.ropensci.org/GSODR/

```
plot(world$geometry, bgc = "#D8F4FF", col = "antiquewhite", axes = TRUE)
plot(isd_history$geometry, pch = 16, cex = 0.2, col = "gray50", add = TRUE)
```

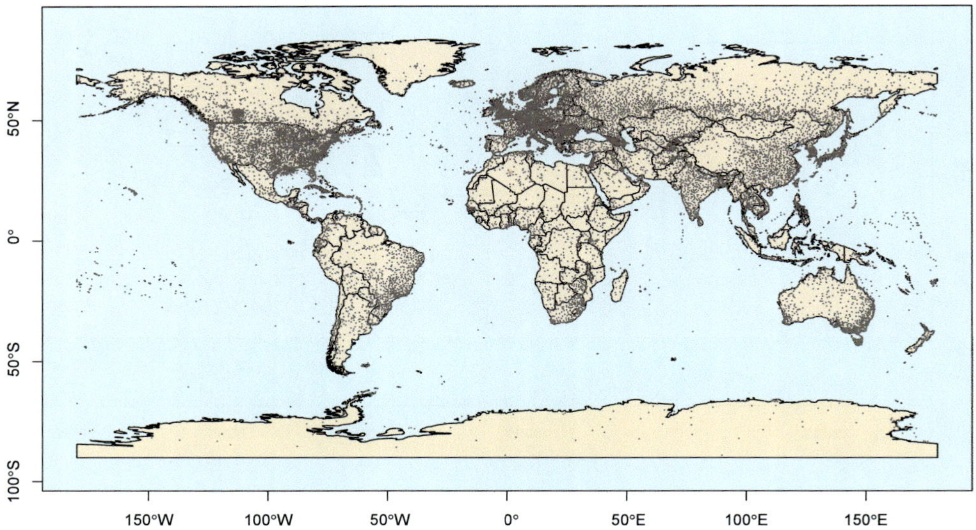

FIGURE 8.18 Mapping weather stations around the world obtained with the `GSODR` package.

The result of the subset of weather stations in Montana is mapped with the `plot` function (Figure 8.19).

```
plot(mt$geometry, col = "white", border = "black", lwd = 1.5, bgc = "gray92",
     axes = TRUE)
plot(isd_history_mt$geometry, add = T, pch = 21, bg = "white", cex = 0.7,
     col = "black")
```

Note that now we have a dataframe in our environment that has both global climate station locations, and only stations associated with Montana, USA.

8.6.3.4 Spatial joins

Using the bracket operator, it is possible to explore topological relationships and determine polygons in Montana counties that have points intersected by weather stations, thereby determining a subset of polygons. The reverse situation can also be performed, creating a subset of polygons that do not have weather stations. In this case, the `!lengths` function returns a logical vector of all objects that do not intersect with each other.

```
mt_co_isd_poly <- mt_co[isd_history,] # Find counties that contain points
mt_co_poly_anti <- isd_history_mt[!lengths(
```

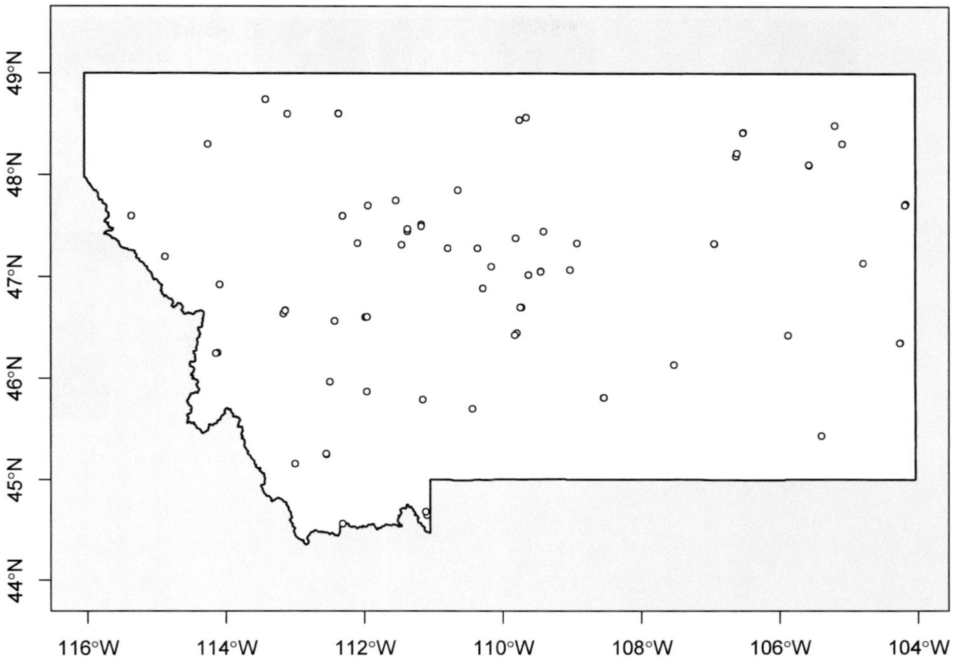

FIGURE 8.19 Mapping weather stations obtained with the `GSODR` package in the state of Montana, USA.

```
  st_intersects(isd_history_mt, mt_co)),] # Find points not within counties
mt_co_poly_anti2 <- mt_co[!lengths(st_intersects(
  mt_co, isd_history_mt)),] # Find counties that don't contain points
```

The results of these topological relations are mapped using the `plot` function (Figure 8.20).

```
plot(mt_co$geometry, col = alpha("pink", 0.3))
plot(mt_co_poly_anti$geometry, pch = 15, add = TRUE, col = "maroon")
plot(mt_co_poly_anti2$geometry, add = TRUE, col = "darkgreen")
```

We can add the county name attribute to the point dataset. The `st_join` function is used to spatially join the county name attribute with the point locations. The argument `left = TRUE` will retain all the points in the dataset rather than just the spatial overlaps where points fall inside polygons. We used `left = FALSE` in this case, that is equivalent to a `dplyr::inner_join`.

```
isd_mt_co_pts <- st_join(isd_history, left = FALSE, mt_co["name"])
```

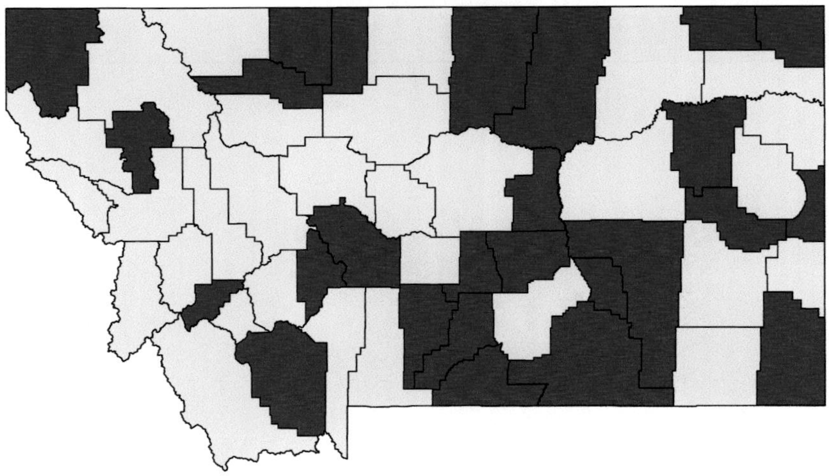

FIGURE 8.20 Anti-joins with Montana and GSOD climate stations where counties with no stations are green. There are no station points outside county boundaries in Montana, USA.

The results of these topological relations are mapped using the `plot` function (Figure 8.21).

```
plot(isd_mt_co_pts$geometry, pch = 21, cex = 0.7, col = "purple", bg = "gray80")
plot(mt_co$geometry, border = "gray20", col = NA, add = TRUE)
```

Now we have only points that fall inside Montana county, and the new database now has a new column attribute called `name`, being all climate station points with a named Montana county associated with them. We can easily specify additional columns inside the `st_join` function, and if we don't specify any columns, then all columns from the polygon `data.frame` that spatially joined the points data will be added to the points `data.frame` using the following function.

```
isd_mt_co_pts <- st_join(isd_history, left = FALSE, mt_co)
```

8.6.3.5 Practice with climate data example

Spatial joins can be used to select, obtain climate data from the `GSODR` package and create geovisualizations. Data from stations between 1980 and 2018 are used as an example, although some stations have data from the 1930s, enabling other time periods to be used. The availability of stations is checked for a date range as follows.

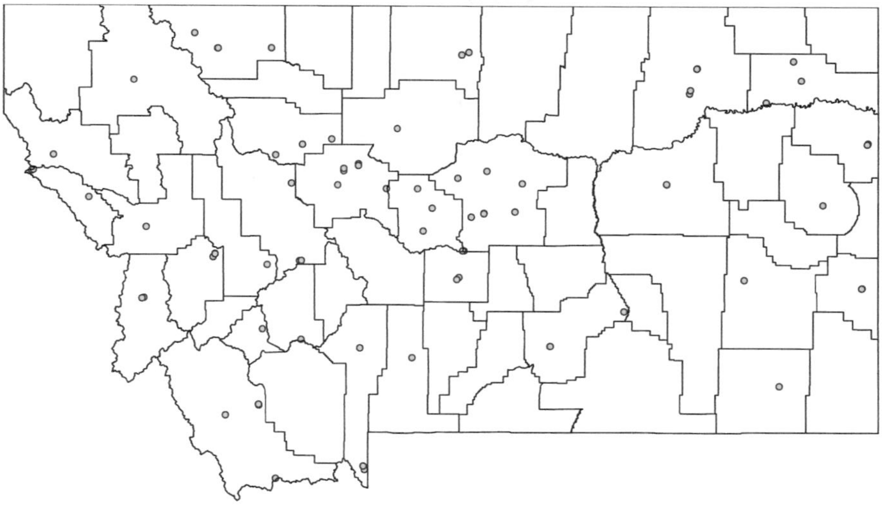

FIGURE 8.21 Spatial join with Montana and GSOD climate stations for points that fall within Montana counties, adding attributes and retaining just the spatial overlaps.

```
stations <-
  isd_mt_co_pts[isd_mt_co_pts$BEGIN <= 19800101 & isd_mt_co_pts$END >= 20181231,]
```

GSOD stations with data between 1980 and 2018 are mapped with `ggplot` functions (Figure 8.22).

```
ggplot() +
  geom_sf(data = mt, fill = "steelblue", alpha = 0.4) +
  geom_sf(data = mt_co, color = "gray50", lwd = 0.4, alpha = 0.4) +
  geom_sf(data = stations, size = 2, pch = 21, color = "thistle1",
          fill = "black") +
  theme_bw()
```

You can use basic functions from the `dplyr` package to calculate the number of stations per county. Station data is grouped by name to count the number of stations per county. Stations are joined back to county layer with a spatial join using the `left = FALSE` argument.

```
# Calculate stations per county
stations_n <- stations |> rename(cnty_name = name) |> group_by(cnty_name) |>
  mutate(n = n()) |> ungroup()
# Join back to county layer
mt_co_stations <- st_join(mt_co, stations_n, left = FALSE)
```

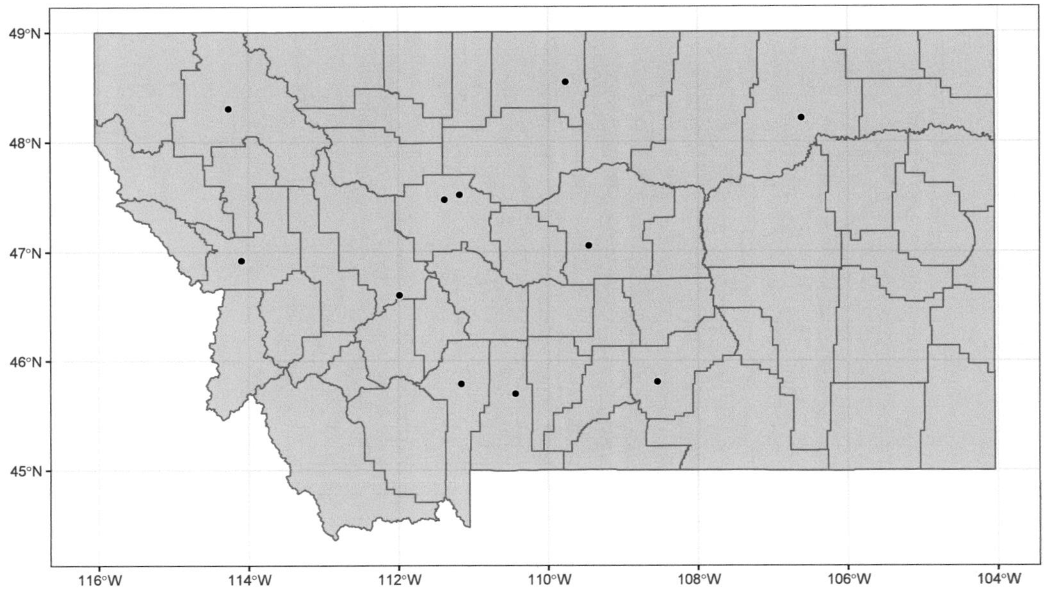

FIGURE 8.22 GSOD stations with data between 1980 and 2018 within Montana counties.

The number of GSOD stations with data between 1980 and 2018 per county are mapped with `ggplot` functions (Figure 8.23).

```
ggplot() +  geom_sf(data = mt) +  geom_sf(data = mt_co_stations,
 aes(fill = as.factor(n)), lwd = 0.4, alpha = 0.4) +
  theme(legend.position = "bottom") +
 scale_fill_viridis_d("Number of Stations")
```

If there is more than one station per county, it may be useful to select only stations that are closest to the county's centroid. In step 1, it is necessary to select only counties with more than one station. It is also necessary to add the centroid of each county and then select the station closest to the centroid. In step 2, the `purrr` package is used to apply the `st_centroid` function to each county in the database.

```
# Step 1: filter to stations with n > 1
stations_n2 <- stations_n |> filter(n > 1)
# Step 2: Get Centroid points for counties with n > 1
mt_co_centroid_pts <- mt_co_stations |> filter(n > 1) |>
  mutate(cent_lon = map_dbl(geometry, ~st_centroid(.x)[[1]]),
         cent_lat = map_dbl(geometry, ~st_centroid(.x)[[2]])) |> as.data.frame()
# Make sf object using the centroid
```

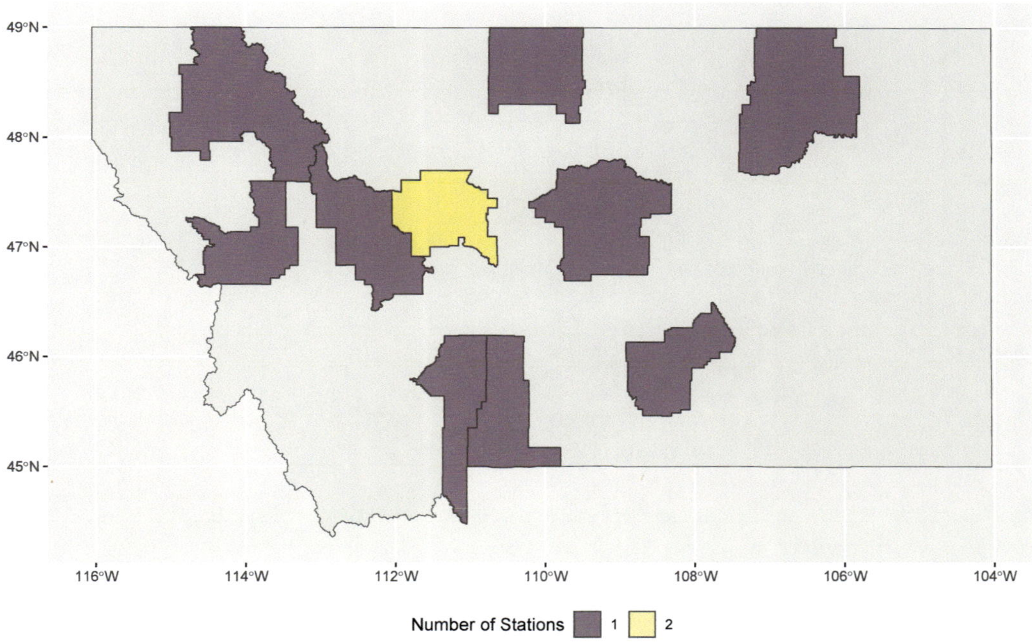

FIGURE 8.23 Number of GSOD stations with data between 1980 and 2018 by county in Montana.

```
mt_co_centroid_pts <- st_as_sf(mt_co_centroid_pts,
        coords = c("cent_lon", "cent_lat"), crs = 4326, remove = FALSE)
```

The centroids of counties with more than one GSOD station are mapped using `ggplot` functions (Figure 8.24).

```
ggplot() +  geom_sf(data = mt_co) + geom_sf(data = mt_co_centroid_pts,
  aes(fill = as.factor(n)), pch = 21, size = 4) +
  theme(legend.position = "bottom") +
  scale_fill_viridis_d("Number of Stations")
```

Using the `st_nn` package, it is possible to determine the nearest neighbors of points, lines, and polygons and provide the distance. This enables to obtain the nearest station to the centroid in the county according to the following codes.

```
# Find the first nearest neighbor GSOD station for each centroid
nearest_stations <- st_nn(mt_co_centroid_pts, stations_n2,
                    returnDist = TRUE, k = 1, progress = FALSE)
# Add distances and centroid ID and make a dataframe with distinct IDs
nearest_stations_df <- tibble(dist = nearest_stations$dist, centrID =
  do.call(rbind, nearest_stations$nn)) |> dplyr::distinct(.keep_all = TRUE)
# Make a data.frame of stations in counties where n = 1
```

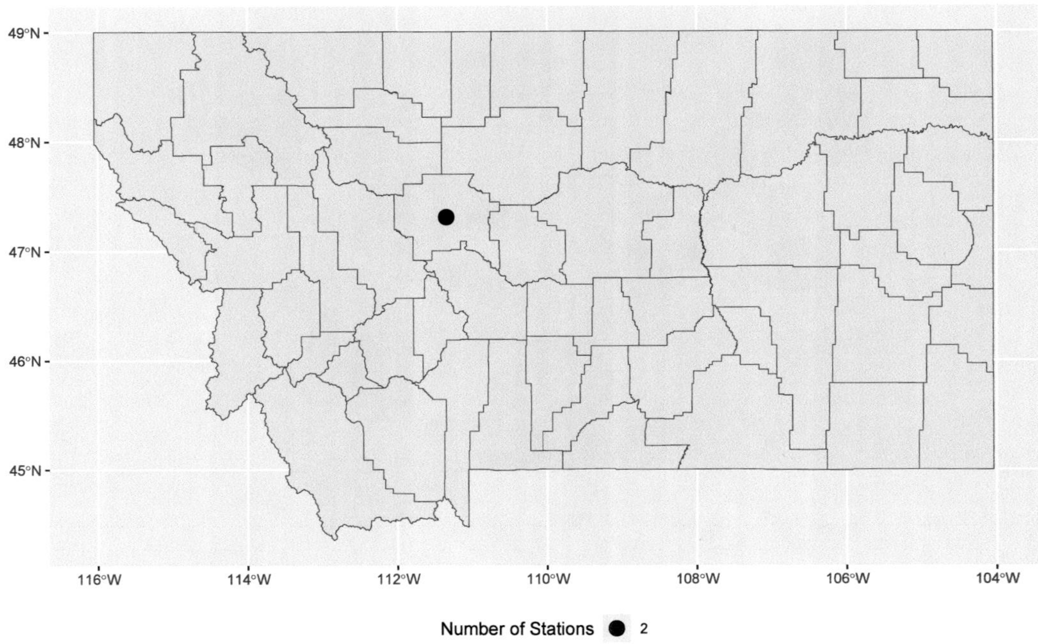

FIGURE 8.24 Centroids of GSOD stations with more than one station in Montana, USA.

```
stations_n1 <- stations_n |> filter(n == 1)
# Select the nearest to centroid stations and bind to the n1 dataset
stations_final <- stations_n2[nearest_stations_df$centrID,] |> rbind(stations_n1)
```

The county centroids, final stations selected and counties with more than one GSOD station are mapped using `ggplot` functions (Figure 8.25).

```
ggplot() + geom_sf(data = mt_co) + geom_sf(data = mt_co_centroid_pts,
 pch = 21, size = 2) + geom_sf(data = stations_n2, col = "orange",
 alpha = 0.9, pch = 17, size = 1.8) +
 geom_sf(data = stations_final, col = "blue3", alpha = 0.9) +
 theme_bw() + labs(x = "Lon", y = "Lat") +
 geom_label_repel(data = stations_final, aes(x = st_coordinates(geometry)[,1],
           y = st_coordinates(geometry)[,2], label = cnty_name), size = 2.5) +
  labs(caption = "County Centroids (open circle),
  Final Stations Selected (blue circle),
       Stations n>1 (orange)")
```

The final set of stations obtained can be used to get climate data. In this example, stations with the lowest and highest altitude will be obtained. The list of stations is used to get daily data from each station for a period of interest. This selects stations with the code `STNID` and `cnty_name` in Park and Valley.

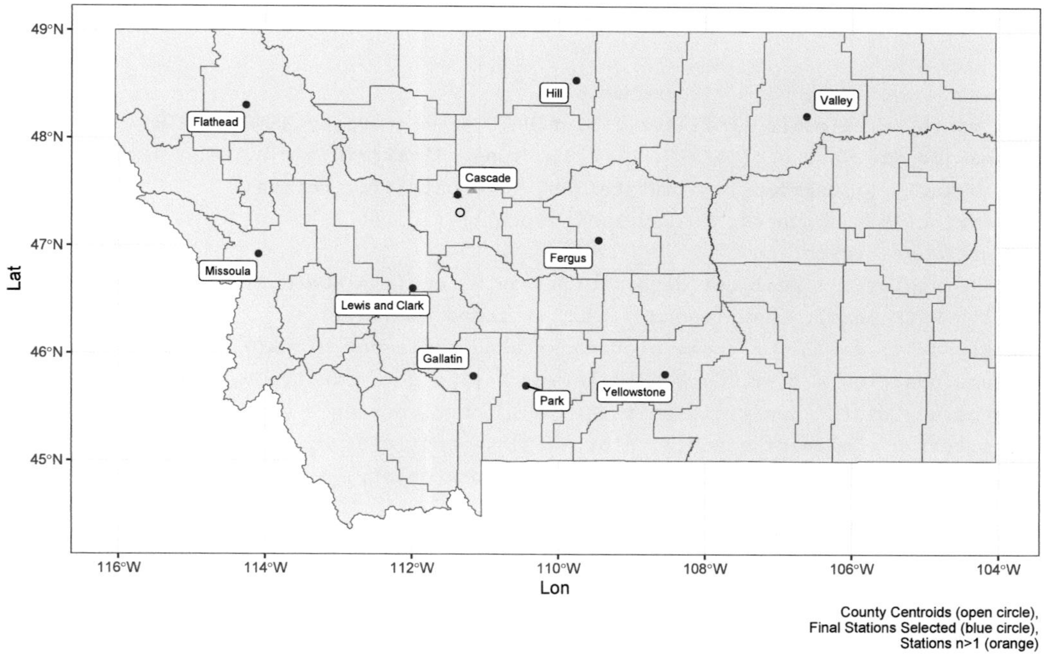

FIGURE 8.25 County centroids, final stations selected and counties with more than one GSOD station in Montana, USA.

```
# Get highest and lowest altitude station IDs
(stationIDs <- stations_final |> filter(min(`ELEV(M)`) ==`ELEV(M)` |
    max(`ELEV(M)`) ==`ELEV(M)` | cnty_name == "cnty_name") |>
    select(STNID, cnty_name, -geometry) |> as.data.frame() |> select(-geometry))
#          STNID cnty_name
# 1 726798-24150      Park
# 2 727680-94008    Valley
```

Climate data is obtained using the `get_GSOD` function, and the `distinct` function is used to obtain only distinct names from the data frame. The names of the stations were modified using the `mutate` function to improve visualization on the map.

```
# Get climate data
climdata <- get_GSOD(station = stationIDs$STNID, years = c(1980:2018))
# Recode stations name
climdata <- climdata |> mutate(NAME=recode(NAME,
        "MISSION FIELD AIRPORT" = "M.F.Airport",
        "WOKAL FIELD/GLASGOW INTL AIRPORT" = "W.F.G.I.Airport"))
# Select unique rows from dataset
climdata_stations <- dplyr::distinct(climdata, NAME, .keep_all = TRUE) |>
    st_as_sf(coords = c("LONGITUDE", "LATITUDE"), crs = 4326, remove = FALSE)
```

Monthly mean rainfall are calculated for the selected distinct stations and then mapped (Figure 8.26).

```
# Filter month data with missing data out
clim_month <- climdata |> filter(!is.na(PRCP)) |> group_by(NAME, MONTH) |>
 summarize_at(.vars = c("TEMP","PRCP"), .funs = list(min = min, mean = mean,
 max = max)) |> ungroup() |> mutate(NAME = forcats::fct_relevel(
   NAME, c("M.F.Airport", "W.F.G.I.Airport")))
# Monthly precipitation
mPPT <- ggplot() + geom_col(data = clim_month, aes(x = MONTH, y = PRCP_mean,
 fill = PRCP_mean), show.legend = TRUE) + theme_minimal() +
 labs(y = "", x = "") + theme(plot.background = element_blank(),
 legend.position = c(-0.25, 0.55), legend.key.height = unit(.15, units = "in"),
 legend.key.width = unit(.1, units = "in"), panel.border = element_blank(),
 axis.text.y = element_blank(), plot.margin = unit(c(0, 0, 0 ,0), "mm")) +
 scale_fill_viridis_c("Mean \nPPT(in)") + coord_polar() +
 facet_wrap(clim_month$NAME~., nrow = 3)
# Make map
siteMap <- ggplot() + geom_sf(data = mt_co, fill = "sienna", alpha = 0.3) +
 geom_sf(data = climdata_stations, fill = "dodgerblue", alpha = 0.9, pch = 21,
 size = 4) + geom_label_repel(data = climdata_stations, aes(x = LONGITUDE,
 y = LATITUDE, label = NAME), size = 3.7) + theme_bw() + theme(panel.border =
 element_blank(), plot.caption = element_text(hjust = 0)) + coord_sf(datum = NA) +
 labs(subtitle = "Monthly Precipitation for GSOD Stations, 1980-2018", y = "",
 x = "", caption = "Monthly precipitation polar plots for each station")
# Plot it together using cowplot
ggdraw(siteMap) + draw_plot(mPPT, width = 0.6, height = .53, x = -0.18, y = 0.01)
```

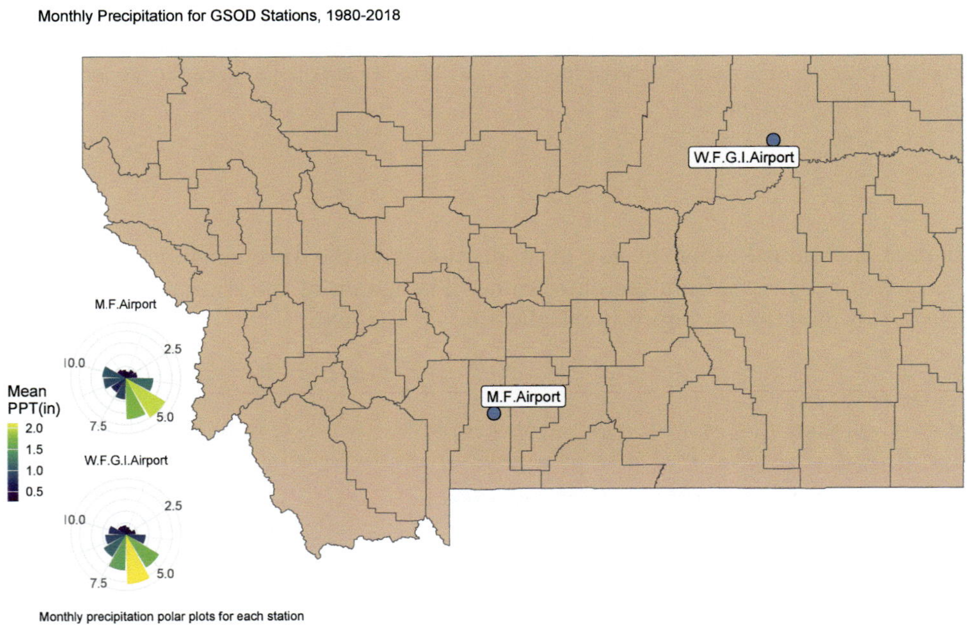

FIGURE 8.26 Monthly mean rainfall for selected GSOD stations, 1980-2018.

A similar procedure is performed for daily mean air temperature data (Figure 8.27).

```
# YDAY filter
clim_day <- climdata |> filter(!is.na(PRCP)) |> group_by(NAME, YDAY) |>
 summarize_at(.vars=vars(TEMP, PRCP), .funs = list(min = min, mean = mean,
                                                  max = max))
# Daily mean temp
(dTEMP <- ggplot() + geom_col(data = clim_day, aes(x = YDAY, y = TEMP_mean,
fill = TEMP_mean), show.legend = TRUE) + theme_minimal() + labs(y = "", x = "") +
 scale_x_continuous(breaks=c(1, 90, 180, 270), labels = c("Jan","Apr","Jul",
                                                         "Oct")) +
 theme(plot.background = element_blank(), legend.position = c(-0.25, 0.55),
 legend.key.height = unit(.15,units = "in"), legend.key.width = unit(.1,
                                                             units = "in"),
 panel.border = element_blank(), axis.text.y = element_blank(),
 plot.margin = unit(c(0, 0, 0 ,0), "mm")) +
 scale_fill_viridis_c("Mean \nTemp(C)") + coord_polar() + facet_wrap(NAME~.,
                                                                  nrow = 3))
# Make map
siteMap2 <- ggplot() + geom_sf(data = mt_co, fill = "sienna",alpha = 0.3) +
geom_sf(data = climdata_stations, fill = "dodgerblue", alpha = 0.9, pch = 21,
size = 4) + geom_label_repel(data = climdata_stations,
aes(x = LONGITUDE, y = LATITUDE, label = NAME), size = 3.7) +
theme_bw() + theme(panel.border = element_blank(),
plot.caption = element_text(hjust = 0)) + coord_sf(datum = NA) +
labs(subtitle = "Daily Mean Temperature for GSOD Stations, 1980-2018",
y = "", x = "", caption = "Polar plots of daily temp for each station")
# Plot it together using cowplot
ggdraw(siteMap2) + draw_plot(dTEMP, width = 0.63, height = .53, x = -0.19,
                             y = 0.01)
```

8.6.4 NON-OVERLAPPING JOINS

In some situations, although two geographic datasets do not touch, there may be strong geographic relationships allowing associations. In this case, a distance threshold can be set to perform the spatial join by including non-overlapping objects. Identification attributes, point name, measured *in situ* temperature value, and sampling date are defined for two distinct point sets to illustrate this method. The cadastral point maps are transformed into the Albers Equal Area conical projection of South America to perform the join using the distance criterion between each point in one dataset and the neighboring point in the other dataset. Then, a distance of 20 m is defined to perform the non-overlapping join. It was observed that due to the defined distance, it is not possible to perform the join of points 1 and 2, being necessary to increase the distance criterion to contemplate all the points in the non-overlapping join. By modifying the distance to 30 m, all points are contemplated in the join (Figure 8.28) (see computational practice section for code details).

In the case of the Montana, USA, the climate data example given above, for joins that are not overlapped by the distance criterion, note that as the distance tolerance for the join increases from 20 to 200 to 2000, the number of rows included in the join increases from 98 to 99 to 107, respectively.

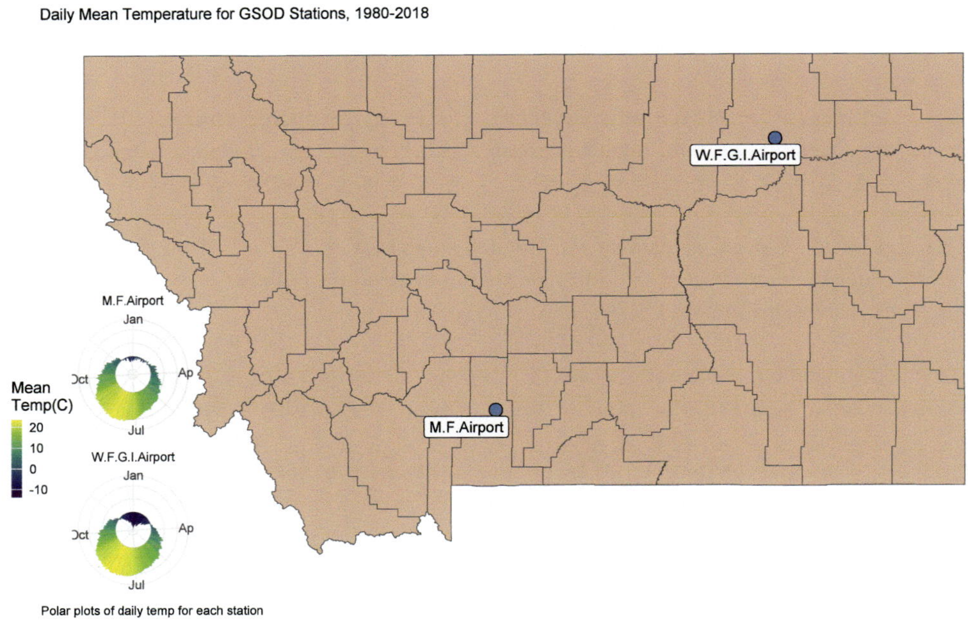

FIGURE 8.27 Daily mean air temperature for GSOD stations, 1980-2018.

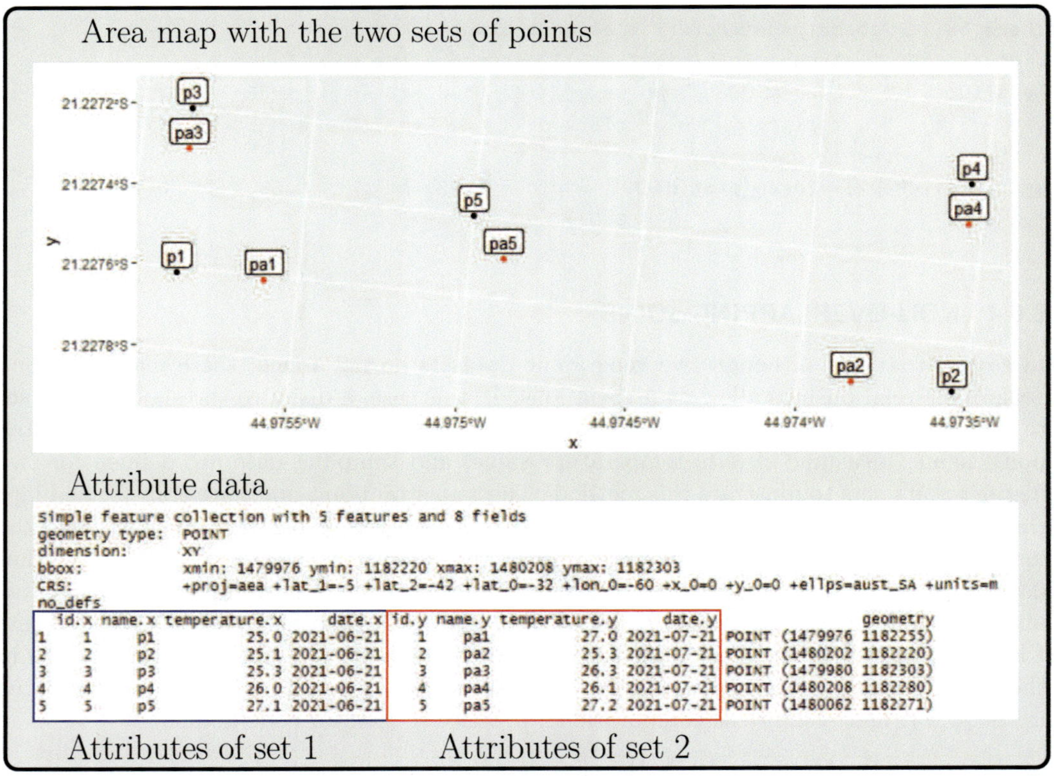

FIGURE 8.28 Non-overlapping multiple point feature datasets used in a join performed with distance condition in the neighborhood of the points.

```
isd_non_overlap_20 <- st_join(isd_history, left = FALSE,
    mt_co["name"], st_is_within_distance, dist = 20) # distance 20
nrow(isd_non_overlap_20)
# [1] 98
isd_non_overlap_200 <- st_join(isd_history, left = FALSE,
          mt_co["name"], st_is_within_distance, dist = 200) # distance 200
nrow(isd_non_overlap_200)
# [1] 99
isd_non_overlap_2000 <- st_join(isd_history, left = FALSE,
        mt_co["name"], st_is_within_distance, dist = 2000) # distance 2000
nrow(isd_non_overlap_2000)
# [1] 107
```

8.6.5 SPATIAL DATA AGGREGATION

Spatial data aggregation can be performed changing the geometry of the data grouped (union) in the state of Minas Gerais, containing in addition to the results of the sum of production, the spatial object, using the `summarize` function.

Attribute aggregation operations for geographic data can summarize the dataset based on a grouping variable in an attribute column. As an example, one can determine the aggregation of the municipal production of Arabica coffee from 2012 to 2019, in tons of processed coffee, in the state of Minas Gerais, from the production data per municipality, by using a sum operation (Table 8.4).

TABLE 8.4 Municipal Arabica coffee production from 2012 to 2019, in tons of processed coffee, in the state of Minas Gerais, Brazil.

Year	Production (ton)
2012	1578355
2013	1584487
2014	1346517
2015	1325471
2016	1816203
2017	1437330
2018	1881219
2019	1477975

With this, it is possible to obtain a non-spatial result of the sum and observe the bienniality of production, mainly from 2016 to 2017 and from 2018 to 2019, being that, mainly in the last even years there was higher production and the bienniality was more evident.

A query can be performed to select the years with the highest Arabica coffee production and the number of municipalities used in the sum of production in each year by means of ranking. Ranking only the two largest production records, in the 853 municipalities surveyed, the highest production values are in 2018 and 2016, with 1881219 and 1816203 tons of coffee produced, respectively (see computational practice section for code details).

This operation can be performed by loading the `sf` package with the `library` function. Although some examples like this have been classified as aggregation by attributes in computational practice, considering that the results obtained alter the geometry of the input geo-objects, we consider it more appropriate to classify this operation as spatial and geometric. However, more details on geometric analysis of vector data will be provided in subsequent chapters.

```
library(sf)
library(dplyr)
```

The coffee production data is imported using the `st_read` function, and the coordinate reference system is assigned using the `st_crs` function.

```
coffeeProd <- st_read('C:/geo/c7/tab2_mg.shp') # Import data
st_crs(coffeeProd) <- st_crs(4326) # Assign geographic projection
coffeeProd
# Simple feature collection with 6824 features and 9 fields
# Geometry type: POLYGON
# Dimension:     XY
# Bounding box:  xmin: -51.04608 ymin: -22.92276 xmax: -39.85683 ymax: -14.24667
# Geodetic CRS:  WGS 84
# ...
```

The `summarize` function is used to determine the sum of coffee production and the union of the original polygons. The total coffee production in all evaluated years and municipalities indicated a total production of 12447557 tons of coffee in 6824 database records (Figure 8.29).

```
coffeeUnion <- coffeeProd |> summarize(Prod = sum(Prod, na.rm = TRUE), n = n())
coffeeUnion
# Simple feature collection with 1 feature and 2 fields
# Geometry type: POLYGON
# Dimension:     XY
# Bounding box:  xmin: -51.04608 ymin: -22.92276 xmax: -39.85683 ymax: -14.24667
# Geodetic CRS:  WGS 84
#       Prod    n                       geometry
# 1 12447557 6824 POLYGON ((-49.8893 -18.6099...
```

Spatial aggregation calculates statistics in areas where an input layer overlaps a boundary layer. The cadastral map of coffee fields mapped in municipalities of Minas Gerais is used as an example of the use of geospatial aggregation.

The polygons with geopolitical information of 11 municipalities of Minas Gerais are used as the delimiting area to perform the aggregation. The coffee polygons with the data column with area values of each polygon are mapped within the municipalities of Minas Gerais before performing the geospatial aggregation function.

Spatial aggregation differs from aggregation by attributes. Aggregation using only the database can generate different results, since crops may occur between municipal boundaries, and this fact may not have been considered when creating the database. In this case, the `aggregate` function is used to perform aggregation between two different databases, one containing coffee fields and the other containing the polygons of municipalities in Minas

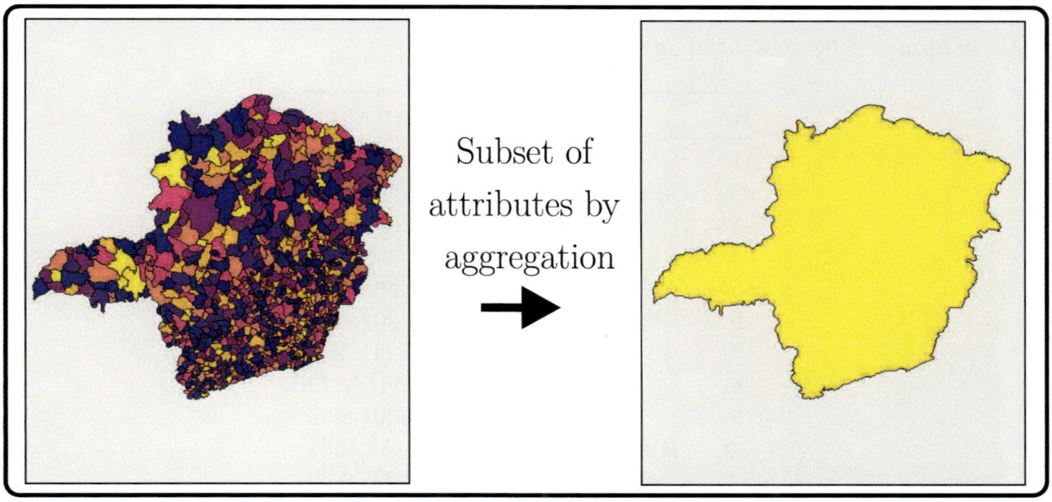

FIGURE 8.29 Spatial data aggregation of the municipal production of coffee in Minas Gerais state, Brazil.

Gerais. Therefore, the resolution of boundaries according to the number of vertices can also determine different results.

```
coffeeSpatialAgg <- aggregate(x = coffeesf["area_ha"], by = munMG2007sf,
                              FUN = sum)
print(coffeeSpatialAgg, n = 11)
# Simple feature collection with 11 features and 1 field
# Geometry type: POLYGON
# Dimension:     XY
# Bounding box:  xmin: -45.39973 ymin: -21.64273 xmax: -44.74138 ymax: -20.92158
# Geodetic CRS:  WGS 84
#       area_ha                       geometry
# 1    869.0856 POLYGON ((-45.11883 -20.926...
# 2  13266.9534 POLYGON ((-45.17252 -21.348...
# 3    263.8040 POLYGON ((-44.93474 -21.114...
# 4    566.5135 POLYGON ((-44.9751 -21.3412...
# 5    290.9689 POLYGON ((-44.86342 -21.167...
# 6   4901.4056 POLYGON ((-44.99714 -21.133...
# 7   2231.3625 POLYGON ((-45.00372 -21.454...
# 8  13818.8130 POLYGON ((-45.28033 -21.053...
# 9   3007.8268 POLYGON ((-45.04777 -21.043...
# 10   152.1581 POLYGON ((-45.05039 -21.151...
# 11  1029.4486 POLYGON ((-45.06405 -21.500...
```

After performing the aggregation to add up the areas of each coffee polygon within the municipalities, the largest Arabica coffee producing municipalities were Nepomuceno and Carmo da Cachoeira, with areas of 13818.8130 and 13266.1646 ha, respectively (Table 8.5).

TABLE 8.5 Area cultivated with Arabica coffee in Municipalities of Minas Gerais, in 2018, determined by geospatial aggregation.

ID	Name	Area (ha)
1	Cana Verde	869.0856
2	Carmo da Cachoeira	13266.1646
3	Ijaci	263.8040
4	Ingai	566.5135
5	Itumirim	290.9689
6	Lavras	4901.4056
7	Luminárias	2231.3625
8	Nepomuceno	13818.8130
9	Perdoes	3007.8268
10	Ribeirão Vermelho	152.1581
11	Sao Bento Abade	1029.4486

Note that the results of spatial aggregation differed from the attribute aggregation performed previously. In addition, it can be observed that the results for coffee fields are already aggregated by municipality, so there is no need to add them to the database as was previously done in the aggregation by attributes.

The results are mapped using the `plot` function from the `sf` package, where you can see the union of the aggregated coffee fields as municipalities in Minas Gerais (Figure 8.30). Note that this is also a geometric operation of union that will be discussed in more detail in Chapter 10.

```
plot(coffeeSpatialAgg[1])
```

Using the `summarize` function, it is possible to determine the spatial aggregation by sum of the polygons with coffee areas, also generating a geometric result of the union of polygons with the total area of coffee planted in the municipalities of Minas Gerais.

```
mgSpatialAgg <- coffeeSpatialAgg |> summarize(area_ha = sum(area_ha, na.rm=T),
                                              n = n())
# Simple feature collection with 1 feature and 2 fields
# Geometry type: POLYGON
# Dimension:     XY
# Bounding box:  xmin: -45.39973 ymin: -21.64273 xmax: -44.74138 ymax: -20.92158
# Geodetic CRS:  WGS 84
#     area_ha n                       geometry
# 1 40398.34 11 POLYGON ((-45.09272 -20.950...
```

The results are mapped using the `plot` function from the `sf` package, where you can see the union of the polygons (Figure 8.31).

```
plot(mgSpatialAgg[1])
```

Considering the sum of the total coffee area produced in 2018 in the assessed muncipalities, 40,398.34 ha is grown in the surveyed region.

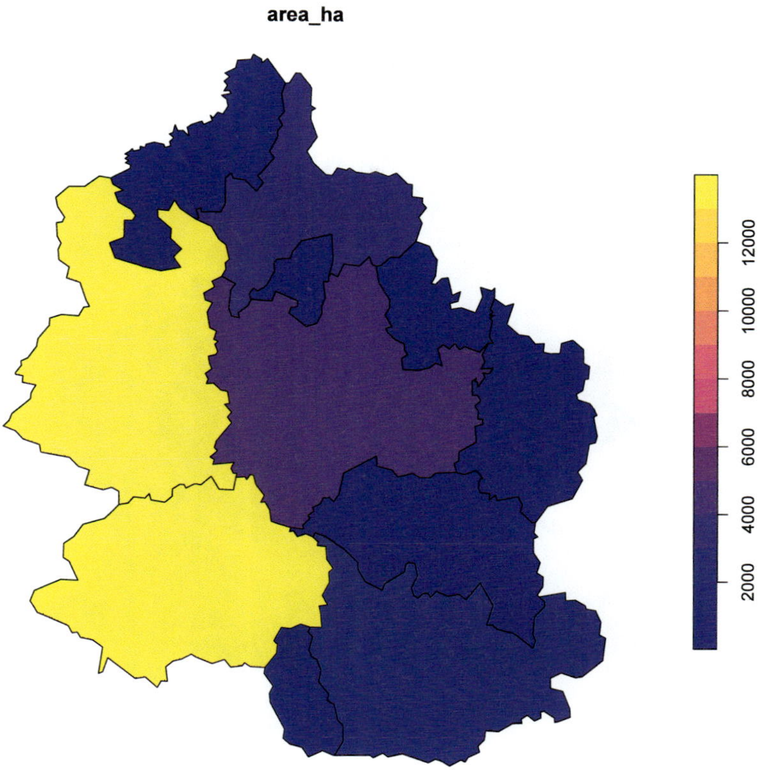

FIGURE 8.30 Spatial aggregation of coffee fields in municipalities of Minas Gerais using the `aggregate` function.

8.6.6 DISTANCE RELATIONSHIPS

Distance relationships can be performed using geodetic, spherical calculations, as well as considering the surface in a given cartographic projection.

As a way to illustrate the determination of distance between two points, the data from the previous example is used where point set cadastral maps are created to determine the ellipsoidal distance between point 1, from point set `1`, to point 4, from point set `2` (Figure 8.13). The ellipsoidal distance between the two points is 243.3245 m. In the case of Figure 8.14, the plane Euclidian distance between points 1 and 4 joined by a red line is 243.2272 m. The spherical distance between these same two points is 242.9749 m (see computational practice section below for code details).

8.7 COMPUTATIONAL PRACTICE

The `install. packages` installs the `sf` (Pebesma et al., 2021a), `dplyr` (Wickham et al., 2021a), `ggplot2` (Wickham et al., 2020), `rgdal` (Bivand et al., 2021), `readr` (Wickham et al., 2021c), `stringr` (Wickham and RStudio, 2019), `tidyerse` (Wickham and RStudio, 2021), `s2` (Dunnington et al., 2021), `geobr` (Pereira et al., 2021) and `lwgeom` (Pebesma et al., 2020) packages to perform different types of spatial analysis of vector geographic object with attributes in computational practice. The `s2` package is already used by `sf` in newer versions. The `s2` package is already used by `sf` in newer versions but is needed to illustrate

FIGURE 8.31 Calculation of total coffee production in simple feature polygons, resulting in a geometric union using the `summarize` function.

a specific spherical distance calculation

```
install.packages("sf")
install.packages("dplyr")
install.packages("ggplot2")
install.packages("rgdal")
install.packages("readr")
install.packages("stringr")
install.packages("tidyverse")
install.packages("s2")
install.packages("geobr")
install.packages("lwgeom")
```

The `library` function is used to enable the R packages used in computational practice.

```
library(sf)
library(dplyr)
library(ggplot2)
library(rgdal)
```

```
library(readr)
library(stringr)
library(tidyverse)
library(s2)
library(geobr)
library(lwgeom)
```

8.7.1 MANIPULATION OF VECTOR GEOGRAPHIC OBJECT WITH ATTRIBUTES

Vector geographic data, in the R package `sf`, consists of one column with the attribute variable and one row per observation or feature. In some objects, geometry data is stored in a column, called "geometry". The geometry column is special in that it is a list column and can contain several geographic entities, such as points, lines and polygons per line. The methods are used as a regular data frame and store both spatial and non-spatial data in the same way, as columns in a `data.frame`. The vector attribute manipulation performed on `data.frame` objects is used in the R package `sf` with support for the `tibble` and `tbl` classes in the `tidyverse`, enabling geospatial data analysis workflows. Many functions used to manipulate objects of class `sf` can be checked with the `methods` function (Lovelace et al., 2019).

```
methods(class = "sf")
#[1] $<-              [                 [[<-           aggregate      anti_join
#[6] arrange          as.data.frame    cbind          coerce         dbDataType
#[11] dbWriteTable     distinct          dplyr_reconstruct filter      full_join
#[16] gather           group_by          group_split identify          initialize
#[21] inner_join       left_join         merge        mutate           nest
#[26] pivot_longer     plot              print        rbind            rename
#[31] right_join       rowwise           sample_frac sample_n          select
#[36] semi_join        separate          separate_rows show            slice
#[41] slotsFromS3      spread            st_agr      st_agr<-          st_area
#[46] st_as_s2         st_as_sf          st_bbox     st_boundary       st_buffer
#[51] st_cast          st_centroid       st_collection_extract  st_convex_hull
#st_coordinates
#[56] st_crop          st_crs            st_crs<-     st_difference    st_filter
#[61] st_geometry      st_geometry<-     st_inscribed_circle    st_interpolate_aw
#st_intersection
#[66] st_intersects    st_is             st_is_valid    st_join        st_line_merge
#[71] st_m_range       st_make_valid     st_nearest_points  st_node    st_normalize
#[76] st_point_on_surface  st_polygonize   st_precision    st_reverse  st_sample
#[81] st_segmentize   st_set_precision  st_shift_longitude st_simplify st_snap
#[86] st_sym_difference  st_transform   st_triangulate    st_union     st_voronoi
#[91] st_wrap_dateline  st_write        st_z_range        st_zm        summarise
#[96] transform          transmute       ungroup           unite        unnest
```

The `st_read` function is used to import a file of polygon features with a database of coffee production in the state of Minas Gerais. The `st_crs` function is used to set the map projection of the polygons in WGS-84. The polygons used are available for download[2].

[2]http://www.sergeo.deg.ufla.br/geo/downloads/tab2_mg.zip

```
coffeeProd <- st_read('C:/geo/c7/tab2_mg.shp') # Import data
st_crs(coffeeProd) <- st_crs(4326) # Assign geographic projection
```

The `dim`, `nrow` and `ncol` functions are used to obtain a measure of the dataset of municipal coffee production in Minas Gerais.

```
# Number of dimensions
dim(coffeeProd)
# [1] 6824 10
# Number of rows
nrow(coffeeProd)
# [1] 6824
# Number of columns
ncol(coffeeProd)
# [1] 10
```

The `st_drop_geometry` function is used to extract attribute data from an object `sf` with geometric features.

```
coffee_df <- st_drop_geometry(coffeeProd)
class(coffee_df)
# [1] "data.frame"
```

The R packages `tidyverse` and `dplyr` can be used for data frame manipulation in many urban planning and transportation applications (Lovelace et al., 2019).

8.7.1.1 Obtain subsets of attribute data

The basic functions to perform subsetting of data in R and `dplyr` are used, preserving the spatial component of attributes, such as (Lovelace et al., 2019):

- `[` <- to perform subsetting of rows and columns;
- `subset()` <- to obtain a result based on query criteria;
- `$` <- to extract a vector from an attribute variable;
- `select()` <- to select columns by name and position;
- `slice()` <- to perform subsetting of rows based on position;
- `filter()` <- to get a result based on query criteria;
- `pull()` <- to extract a vector from an attribute variable.

A subset of attribute features is realized in the file `coffeeProd` according to the data position in the row.

```
coffeeProd[11:16,]
# Simple feature collection with 6 features and 9 fields
# Geometry type: POLYGON
# Dimension:     XY
# Bounding box:  xmin: -45.65972 ymin: -19.31677 xmax: -45.08112 ymax: -18.87596
# Geodetic CRS:  WGS 84
#     City    nome State MicrRgn MesoRgn Region   municipio  Ano Prod   geometry
```

```
# 11 3100203 ABAETÉ  31 31024  3106  3 Abaeté (MG) 2018 180 POLYGON ((-45.1393...
# 12 3100203 ABAETÉ  31 31024  3106  3 Abaeté (MG) 2016  NA POLYGON ((-45.1393...
# 13 3100203 ABAETÉ  31 31024  3106  3 Abaeté (MG) 2014  NA POLYGON ((-45.1393...
# 14 3100203 ABAETÉ  31 31024  3106  3 Abaeté (MG) 2012  NA POLYGON ((-45.1393...
# 15 3100203 ABAETÉ  31 31024  3106  3 Abaeté (MG) 2017  NA POLYGON ((-45.1393...
# 16 3100203 ABAETÉ  31 31024  3106  3 Abaeté (MG) 2019 180 POLYGON ((-45.1393...
```

A subset of attribute features is held in the `coffeeProd` file according to the data position in the column.

```
coffeeProd[,7:9]
# Simple feature collection with 6817 features and 3 fields
# Geometry type: POLYGON
# Dimension:     XY
# Bounding box:  xmin: -51.04608 ymin: -22.92276 xmax: -39.85683 ymax: -14.24667
# Geodetic CRS:  WGS 84
# First 10 features:
#                   municipio  Ano Prod                       geometry
# 1  Abadia dos Dourados (MG) 2015  124 POLYGON ((-47.42967 -18.165...
# 2  Abadia dos Dourados (MG) 2016  147 POLYGON ((-47.42967 -18.165...
# 3  Abadia dos Dourados (MG) 2017   30 POLYGON ((-47.42967 -18.165...
# 4  Abadia dos Dourados (MG) 2018  144 POLYGON ((-47.42967 -18.165...
# 5  Abadia dos Dourados (MG) 2013   57 POLYGON ((-47.42967 -18.165...
# 6  Abadia dos Dourados (MG) 2014   71 POLYGON ((-47.42967 -18.165...
# 7  Abadia dos Dourados (MG) 2012  104 POLYGON ((-47.42967 -18.165...
# 8  Abadia dos Dourados (MG) 2019  168 POLYGON ((-47.42967 -18.165...
# 9                Abaeté (MG) 2015   NA POLYGON ((-45.1393 -19.2104...
# 10               Abaeté (MG) 2013   NA POLYGON ((-45.1393 -19.2104...
```

A subset of attribute features is made in the file `coffeeProd` separating data by name, city and state.

```
coffeeProd[,c("nome", "City", "State")]
# Simple feature collection with 6817 features and 3 fields
# Geometry type: POLYGON
# Dimension:     XY
# Bounding box:  xmin: -51.04608 ymin: -22.92276 xmax: -39.85683 ymax: -14.24667
# Geodetic CRS:  WGS 84
# First 10 features:
#                   nome    City State                       geometry
# 1  ABADIA DOS DOURADOS 3100104    31 POLYGON ((-47.42967 -18.165...
# 2  ABADIA DOS DOURADOS 3100104    31 POLYGON ((-47.42967 -18.165...
# 3  ABADIA DOS DOURADOS 3100104    31 POLYGON ((-47.42967 -18.165...
# 4  ABADIA DOS DOURADOS 3100104    31 POLYGON ((-47.42967 -18.165...
# 5  ABADIA DOS DOURADOS 3100104    31 POLYGON ((-47.42967 -18.165...
# 6  ABADIA DOS DOURADOS 3100104    31 POLYGON ((-47.42967 -18.165...
# 7  ABADIA DOS DOURADOS 3100104    31 POLYGON ((-47.42967 -18.165...
# 8  ABADIA DOS DOURADOS 3100104    31 POLYGON ((-47.42967 -18.165...
```

```
# 9              ABAETÉ 3100203     31 POLYGON ((-45.1393 -19.2104...
# 10             ABAETÉ 3100203     31 POLYGON ((-45.1393 -19.2104...
```

The subsets can be obtained by means of logical vectors in order to create new geo-objects, as an example of the location of the municipality Lavras, which has eight occurrences in the database due to the different sampling data collection times determined by the census performed.

```
coffeeLavras <- coffeeProd$nome == "LAVRAS" # Define subset
summary(coffeeLavras)
# Mode    FALSE     TRUE
# logical    6809        8
```

A subset using brackets is performed to create a new geo-object referring to the municipality of Lavras.

```
lavras <- coffeeProd[coffeeLavras,] # Create new geo-object
lavras
# Simple feature collection with 8 features and 9 fields
# Geometry type: POLYGON
# Dimension:     XY
# Bounding box:  xmin: -45.1768 ymin: -21.39891 xmax: -44.86478 ymax: -21.13287
# Geodetic CRS:  WGS 84
#  City    nome State MicrRgn MesoRgn Region    municipio  Ano Prod      geometry
# 3489 3138203 LAVRAS 31 31057 3111 3 Lavras(MG) 2013 5889 POLYGON ((-44.99302...
# 3490 3138203 LAVRAS 31 31057 3111 3 Lavras(MG) 2019 6750 POLYGON ((-44.99302...
# 3491 3138203 LAVRAS 31 31057 3111 3 Lavras(MG) 2016 7140 POLYGON ((-44.99302...
# 3492 3138203 LAVRAS 31 31057 3111 3 Lavras(MG) 2012 5280 POLYGON ((-44.99302...
# 3493 3138203 LAVRAS 31 31057 3111 3 Lavras(MG) 2014 4979 POLYGON ((-44.99302...
# 3494 3138203 LAVRAS 31 31057 3111 3 Lavras(MG) 2015 3981 POLYGON ((-44.99302...
# 3495 3138203 LAVRAS 31 31057 3111 3 Lavras(MG) 2018 7392 POLYGON ((-44.99302...
# 3496 3138203 LAVRAS 31 31057 3111 3 Lavras(MG) 2017 7275 POLYGON ((-44.99302...
```

Using the ‘subset’ function, we can get the same result by making a subset of the Lavras municipality.

```
coffeeLavras <- subset(coffeeProd, nome == "LAVRAS")
coffeeLavras
# Simple feature collection with 8 features and 9 fields
# Geometry type: POLYGON
# Dimension:     XY
# Bounding box:  xmin: -45.1768 ymin: -21.39891 xmax: -44.86478 ymax: -21.13287
# Geodetic CRS:  WGS 84
#...
```

The `select()` function of `dplyr` is used to perform subsets by position.

```
coffeeProdAno <- dplyr::select(coffeeProd, nome, Ano, Prod)
names(coffeeProdAno)
# [1] "nome"      "Ano"       "Prod"      "geometry"
```

The `select()` function can also use the :, -, and rename variables operators at the same time.

```
# Operator :
coffeeProdAno1 <- dplyr::select(coffeeProd, nome, municip:Prod)
names(coffeeProdAno1)
# [1] "nome"      "municip"  "Ano"       "Prod"      "geometry"
# Operator -
coffeeAno <- dplyr::select(coffeeProdAno1, -Prod)
names(coffeeAno)
# [1] "nome"      "municip"  "Ano"       "geometry"
# Rename
coffeeAno1 <- dplyr::select(coffeeAno, mun = municip, year = Ano, name = nome)
names(coffeeAno1)
# [1] "mun"       "year"      "name"      "geometry"
```

The same subsets can also be made and renamed using bracket and equal operators.

```
# Select by column name
coffeeAno2 <- coffeeProd[,c("municip", "Ano")]
names(coffeeAno2)
# [1] "municip"  "Ano"       "geometry"
# Rename the column
names(coffeeAno2)[names(coffeeAno2) == "municip"] = "mun"
names(coffeeAno2)
# [1] "mun"       "Ano"       "geometry"
```

In `dplyr` the pipe operator `|>` is also used to perform expressive chaining code, so that the output of a previous function becomes the first argument of the next function, enabling chaining. Thus, we filter the county `Lavras` in the `coffeeProd` database, and subset the `nome`, `Ano` and `Prod` columns and five rows.

```
coffeeAnoProd <- coffeeProd |> filter(nome == "LAVRAS") |>
  dplyr::select(nome, Ano, Prod) |> slice(1:5)
names(coffeeAnoProd)
# [1] "nome"      "Ano"       "Prod"      "geometry"
```

Without using the pipe operator, the same function is performed as follows:

```
coffeeAnoProd1 <- slice(dplyr::select(filter(coffeeProd, nome == "LAVRAS"),
    nome, Ano, Prod), 1:5)
names(coffeeAnoProd1)
# [1] "nome"      "Ano"       "Prod"      "geometry"
```

8.7.1.2 Vector attribute aggregation

Attribute data aggregation operations are used to summarize datasets based on a grouping variable, into an attribute column. An example of attribute aggregation is to calculate Arabica coffee production from 2012 to 2019 in the state of Minas Gerais, from production data by municipality, using sum operation and the `aggregate` function.

```
coffeeProdAgg <- aggregate(Prod ~ Ano, FUN = sum, data = coffeeProd, na.rm = TRUE)
coffeeProdAgg
#    Ano     Prod
# 1 2012 1578355
# 2 2013 1584487
# 3 2014 1346517
# 4 2015 1325471
# 5 2016 1816203
# 6 2017 1437330
# 7 2018 1881219
# 8 2019 1477975
```

With this, a non-spatial result of the sum is obtained, indicating the bienniality of production, mainly from 2016 to 2017 and from 2018 to 2019, mainly in the last even years, with higher production and the bienniality more evident.

The same previous function can be used keeping the geometry of the data grouped in the state of Minas Gerais, containing in addition to the production sum results the spatial object.

```
coffeeProdAgg1 <- aggregate(coffeeProd["Prod"], by = list(coffeeProd$Ano),
                            FUN = sum, na.rm = TRUE)
names(coffeeProdAgg1)
# [1] "Group.1"  "Prod"      "geometry"
```

8.7.1.3 Joining vector attributes

The joining of vector attributes with `dplyr` can be performed with the functions applied against two databases, b1 and b2.

- inner_join() <- Inclusion of all rows in b1 and b2;
- left_join() <- Inclusion of all rows in b1;
- right_join() <- Inclusion of all rows in b2;
- full_join() <- Inclusion of all rows in b1 or b2.

It should be noted that if a row in b1 corresponds to multiple rows in b2, all rows in b2 will be returned once for each corresponding row in b1.

The joining of vector attributes is performed with data in a table with columns of Arabica coffee production data per year in Brazil and polygons with cadastral data of municipalities in Minas Gerais. The data are joined based on a common column between the databases.

The spreadsheet with production data is obtained using the `read_csv2` function from the R package `readr`. The data used are available for download[3].

[3]http://www.sergeo.deg.ufla.br/geo/downloads/tabelaCafeNew.zip

```
tab <- read_csv2("C:/geo/SIDRA_IBGE/tabelaCafeNew.csv", col_names = T,
                 na = c("...", "-"), locale = locale(encoding = 'ISO-8859-1'))
# Rows: 5533 Columns: 10
# Delimiter: ";"
# chr (1): municipio
# dbl (9): cod_ibge, 2012, 2013, 2014, 2015, 2016, 2017, 2018, 2019
# ⊠ Use `spec()` to retrieve the full column specification for this data.
# ⊠ Specify the column types or set `show_col_types = FALSE` to quiet this message.
```

The objects of polygons with attributes of the municipalities of the state of Minas Gerais are obtained with the `st_read` function. The data used are available for download[4].

```
mg <- st_read('C:/geo/c7/mg1.shp')
# Reading layer `mg1' from data source `C:\geo\c7\mg1.shp' using driver
# `ESRI Shapefile'
# Simple feature collection with 853 features and 6 fields
# Geometry type: POLYGON
# Dimension:     XY
# Bounding box:  xmin: -51.04608 ymin: -22.92276 xmax: -39.85683 ymax: -14.24667
# Geodetic CRS:  GCS_unknown
st_crs(mg)<-st_crs(4326) # Set crs
```

The column in common between both databases, with the code of the municipalities, is renamed to perform the join.

```
tab_r <- rename(tab, cod = cod_ibge)
tab_mg_r <- left_join(mg, tab_r, by = c(City = "cod"))
```

Next, the mapping of Arabica coffee production in 2018 is performed (Figure 8.32).

```
plot(tab_mg_r["2018"])
```

The `setdiff` function makes it possible to evaluate rows where there is no match between join codes. Comparing the `city` and `cod` columns in the `coffeeProd` and `tab_r` files created earlier, there is a complete match between the codes used for Brazilian municipalities.

```
setdiff(coffeeProd$city, tab_r$cod)
NULL
```

The R package `stringr` (Wickham and RStudio, 2019) with the `str_subset` function allows you to locate a name in the database and evaluate how it is written.

```
str_subset(tab_r$municipio, "Lavra*")
#[1] "Lavras da Mangabeira (CE)" "Pedra Lavrada (PB)"
```

[4]http://www.sergeo.deg.ufla.br/geo/downloads/mg1.zip

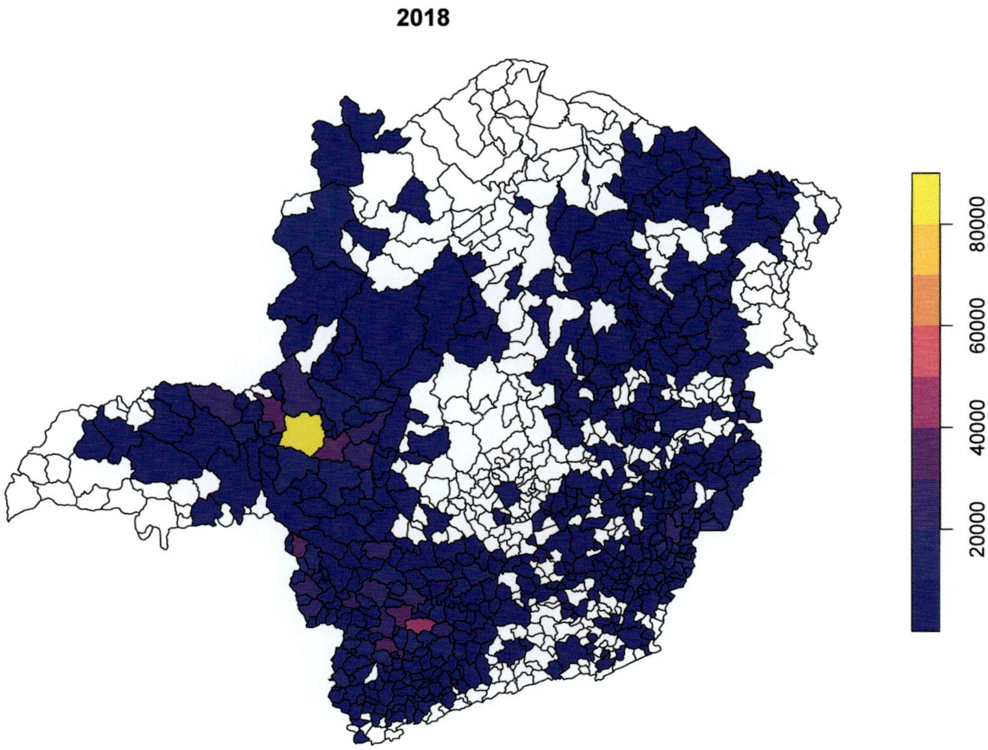

FIGURE 8.32 Mapping Arabica coffee production in 2018 in the state of Minas Gerais.

```
#[3] "Lavras (MG)"              "Lavrinhas (SP)"
#[5] "Lavras do Sul (RS)"
```

8.7.2 CREATE ATTRIBUTES AND REMOVE SPATIAL INFORMATION

New attribute columns can be created based on calculations performed on existing columns. The average Arabica coffee production is determined between 2012 and 2019 using basic R commands by creating a new data column called `prod_m`.

```
mgcoffee <- tab_mg_r # Create a new dataset
names(mgcoffee)
# [1] "nome"       "City"       "State"      "MicrRgn"    "MesoRgn"    "Region"
# [7] "municipio" "2012"       "2013"       "2014"       "2015"       "2016"
# [13] "2017"       "2018"       "2019"       "geometry"
mgcoffee$prod_m <- (mgcoffee$`2012` + mgcoffee$`2013` + mgcoffee$`2014` +
  mgcoffee$`2015`+ mgcoffee$`2016` + mgcoffee$`2017`+ mgcoffee$`2018`+
  mgcoffee$`2019`) / 8
names(mgcoffee)
# [1] "nome"       "City"       "State"      "MicrRgn"    "MesoRgn"    "Region"
# [7] "municipio" "2012"       "2013"       "2014"       "2015"       "2016"
```

```
# [13] "2017"       "2018"       "2019"       "geometry"  "prod_m"
```

The same operation is performed with the `mutate` function from the R package `dplyr`.

```
mgcoffeeMutate <- mgcoffee |>
  mutate(prod_m = (mgcoffee$`2012` + mgcoffee$`2013` + mgcoffee$`2014` +
    mgcoffee$`2015` + mgcoffee$`2016` + mgcoffee$`2017`+ mgcoffee$`2018`+
    mgcoffee$`2019`) / 8 )
names(mgcoffeeMutate)
# [1] "nome"       "City"       "State"      "MicrRgn"    "MesoRgn"    "Region"
# [7] "municipio" "2012"       "2013"       "2014"       "2015"       "2016"
# [13] "2017"       "2018"       "2019"       "geometry"  "prod_m"
```

When using the `transmute` function, the `prod_m` column overrides all other existing columns to generate only one attribute column with the geometry column.

```
coffeeTransmute <- mgcoffee |> transmute(prod_m = (mgcoffee$`2012` +
  mgcoffee$`2013` + mgcoffee$`2014` + mgcoffee$`2015` + mgcoffee$`2016` +
    mgcoffee$`2017`+ mgcoffee$`2018`+ mgcoffee$`2019`) / 8 )
names(coffeeTransmute)
# [1] "prod_m"    "geometry"
```

The `unite` function can be used to join two columns of data, including a separator. In this case, the columns `nome` and `City` are joined with the `:` separator.

```
coffee_unite <- mgcoffee |> unite("nome_city", nome:City, sep = ":",
                                  remove = TRUE)
coffee_unite
# Simple feature collection with 853 features and 15 fields
# Geometry type: POLYGON
# Dimension:     XY
# Bounding box:  xmin: -51.04608 ymin: -22.92276 xmax: -39.85683 ymax: -14.24667
# Geodetic CRS:  WGS 84
# First 10 features:
#                        nome_city State MicrRgn MesoRgn Region
# 1   ABADIA DOS DOURADOS:3100104    31   31019    3105      3
# 2               ABAETÉ:3100203    31   31024    3106      3
# 3           ABRE CAMPO:3100302    31   31061    3112      3
# 4              ACAIACA:3100401    31   31060    3112      3
# 5              AÇUCENA:3100500    31   31039    3108      3
# 6            ÁGUA BOA:3100609    31   31036    3108      3
# 7       ÁGUA COMPRIDA:3100708    31   31022    3105      3
# 8             AGUANIL:3100807    31   31045    3109      3
# 9     ÁGUAS FORMOSAS:3100906    31   31016    3104      3
# 10  ÁGUAS VERMELHAS:3101003    31   31005    3102      3
```

With the `separate` function, the opposite of the `unite` function is performed, in order to separate the previously joined data.

```
coffee_separate <- coffee_unite |> separate("nome_city", c("nome", "City"),
                                              sep = ":")
coffee_separate
# Simple feature collection with 853 features and 16 fields
# Geometry type: POLYGON
# Dimension:     XY
# Bounding box:  xmin: -51.04608 ymin: -22.92276 xmax: -39.85683 ymax: -14.24667
# Geodetic CRS:  WGS 84
# First 10 features:
#                    nome    City State MicrRgn MesoRgn Region
# 1  ABADIA DOS DOURADOS 3100104    31   31019    3105      3
# 2               ABAETÉ 3100203    31   31024    3106      3
# 3           ABRE CAMPO 3100302    31   31061    3112      3
# 4              ACAIACA 3100401    31   31060    3112      3
# 5              AÇUCENA 3100500    31   31039    3108      3
# 6            ÁGUA BOA 3100609    31   31036    3108      3
# 7        ÁGUA COMPRIDA 3100708    31   31022    3105      3
# 8             AGUANIL 3100807    31   31045    3109      3
# 9       ÁGUAS FORMOSAS 3100906    31   31016    3104      3
# 10     ÁGUAS VERMELHAS 3101003    31   31005    3102      3
```

The rename and set_names functions are used to rename columns. In the first function, the old nome of the coffee_separate file is replaced with a new name, name.

```
coffeeSeparete <- coffee_separate |> rename(name = nome)
names(coffeeSeparete)
# [1] "name"       "City"       "State"      "MicrRgn"    "MesoRgn"    "Region"
# [7] "municipio"  "2012"       "2013"       "2014"       "2015"       "2016"
# [13] "2017"       "2018"       "2019"       "prod_m"     "geometry"
```

The set_names function is used to rename all columns at once, as in the following example.

```
new_name <- c("n", "c", "s", "mi", "me", "r", "mu", "a12", "a13", "a14",
              "a15","a16","a17","a18","a19","pm","geometry")
newColumns <- coffee_separate |> set_names(new_name)
names(newColumns)
# [1] "n"         "c"         "s"         "mi"        "me"        "r"
# [7] "mu"        "a12"       "a13"       "a14"       "a15"       "a16"
# [13] "a17"       "a18"       "a19"       "pm"        "geometry"
```

8.7.3 SPATIAL ANALYSIS OPERATIONS ON VECTOR DATA

8.7.3.1 Topological relations

Various topological relations can be used to obtain geospatial subsets, according to the situation of one geo-object in relation to another. Simple vector features of points, lines and polygons can be used to analyze how the spatial analysis process works by considering topological relations between geo-objects. The functions st_polygon, st_linestring and st_multipoint are used to create polygon, line and points used in computational practice.

```r
# Create a polygon
a_poly <- st_polygon(list(rbind(c(-44.97537, -21.22805), c(-44.97537, -21.22722),
    c(-44.97417, -21.22722),c(-44.97417, -21.22805), c(-44.97537, -21.22805))))
a <- st_sfc(a_poly)
# Create a line
l_line <- st_linestring(x = matrix(c(-44.975860, -44.973541, -21.227613,
                                      -21.227719), ncol = 2))
l <- st_sfc(l_line)
# Create points
p_matrix <- matrix(c(-44.975860, -44.973541, -44.975860, -44.973541, -44.9750,
        -21.227613, -21.227719, -21.2272, -21.2272, -21.2274), ncol = 2)
p_multi <- st_multipoint(x = p_matrix)
p <- st_cast(st_sfc(p_multi), "POINT")
```

The area mapping facilitates the topological visualization of the objects created from WGS-84 geographical coordinates in decimal degrees (Figure 8.33).

```r
plot(p[], col = "black", pch = 21, bg = "black", axes = TRUE)
plot(l[], col = "black", add = TRUE)
plot(a[], col = "transparent", border = "black", add = TRUE)
```

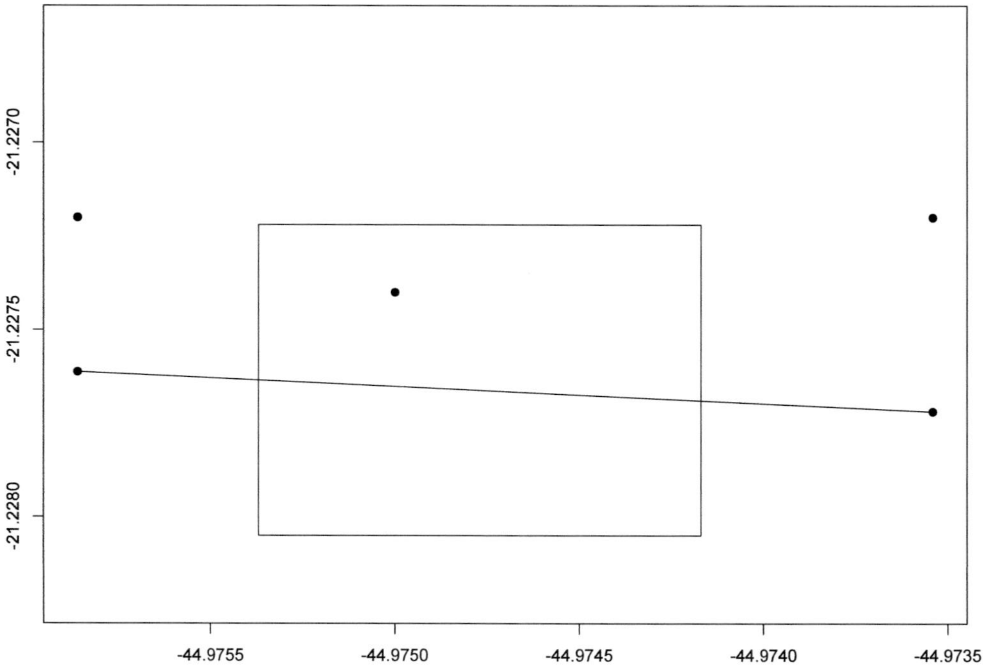

FIGURE 8.33 Topological geovisualization of point, line and polygon objects.

The `st_intersects` function enables to know which point is intersected by the polygon.

```
st_intersects(p, a)
# Sparse geometry binary predicate list of length 5, where the predicate
# was `intersects'
# 1: (empty)
# 2: (empty)
# 3: (empty)
# 4: (empty)
# 5: 1
```

The `sparse = FALSE` parameter is set in the code to obtain a logical result from the use of the `st_intersects` function, with response values of TRUE or FALSE according to the query performed. In this case, the point located inside the polygon gives the result TRUE.

```
st_intersects(p, a, sparse = FALSE)
#       [,1]
# [1,] FALSE
# [2,] FALSE
# [3,] FALSE
# [4,] FALSE
# [5,]  TRUE
```

The opposite of the `st_intersects` function is `st_disjoint`, which returns only objects that are not spatially related to the selected object.

```
st_disjoint(p, a, sparse = FALSE)
#       [,1]
# [1,]  TRUE
# [2,]  TRUE
# [3,]  TRUE
# [4,]  TRUE
# [5,] FALSE
```

The parameter [, 1] can be used to convert the result into a vector.

```
st_disjoint(p, a, sparse = FALSE)[,1]
# [1]  TRUE  TRUE  TRUE  TRUE FALSE
```

With the use of the `st_within` function, only objects completely within the selected object got a TRUE response.

```
st_within(p, a, sparse = FALSE)[,1]
# [1] FALSE FALSE FALSE FALSE  TRUE
```

Notice that although the fifth point is inside the rectangle, it does not touch any part of its border. For this reason, using the `st_touches` function determines all FALSE returns for the same query.

```
st_touches(p, a, sparse = FALSE)[,1]
# [1] FALSE FALSE FALSE FALSE FALSE
```

In the case of using the same function for a query of points touching the line, points 1 and 2 touch the ends of the line.

```
st_touches(p, l)
# Sparse geometry binary predicate list of length 5, where the predicate
# was `touches'
# 1: 1
# 2: 1
# 3: (empty)
# 4: (empty)
# 5: (empty)
```

Objects can also be selected using the `st_is_within_distance` function. This function uses a spatial index for geographic coordinates when `sf` uses s2 geometry. Since in this case the coordinates are geographic, the `s2` package is enabled with the `library` function.

```
library(s2)
```

The distance between point 1 and the left edge of the rectangle is 0.00049 degrees, based on the longitude values of the objects.

```
-44.975860 - -44.97537
# [1] -0.00049
```

The `st_touches` function is used to check whether the line touches any of the evaluated points.

```
st_touches(p, l)
# Sparse geometry binary predicate list of length 5, where the predicate
# was `touches'
# 1: 1
# 2: 1
# 3: (empty)
# 4: (empty)
# 5: (empty)
```

The `st_crosses` function is used to check if the line crosses the polygon.

```
st_crosses(a, l)
# Sparse geometry binary predicate list of length 1, where the predicate
# was `crosses'
# 1: 1
```

Using the `st_is_within_distance` function, we observe that the distance of 0.00048 is not enough to reach both objects, but only the distance 0.00049 is enough to achieve the desired result.

```
d <- st_is_within_distance(p[1], a, dist = 0.00048)
lengths(d) > 0
# [1] FALSE
d1 <- st_is_within_distance(p[1], a, dist = 0.00049)
lengths(d1) > 0
# [1] TRUE
```

The points, line and polygon are assigned the WGS-84 geographic projection.

```
st_crs(p) <- 4326
st_crs(l) <- 4326
st_crs(a) <- 4326
```

The points, line and polygon are reprojected to the Albers Equal Area conic projection of South America.

```
p_p <- st_transform(p, crs = "+proj=aea +lat_1=-5 +lat_2=-42 +lat_0=-32
   +lon_0=-60 +x_0=0 +y_0=0 +ellps=aust_SA +units=m no_defs")
l_p <- st_transform(l, crs = "+proj=aea +lat_1=-5 +lat_2=-42 +lat_0=-32
   +lon_0=-60 +x_0=0 +y_0=0 +ellps=aust_SA +units=m no_defs")
a_p <- st_transform(a, crs = "+proj=aea +lat_1=-5 +lat_2=-42 +lat_0=-32
   +lon_0=-60 +x_0=0 +y_0=0 +ellps=aust_SA +units=m no_defs")
```

With the geographic objects in cartographic projection, it is possible to query the distance between point 1 and the polygon using the `st_is_within_distance` function. Note that at 49 m it is possible to reach both objects with the query, but at 48 m it is not.

```
dm <- st_is_within_distance(p_p[1], a_p, dist = 49)
lengths(dm) > 0
# [1] TRUE
d1m <- st_is_within_distance(p_p[1], a_p, dist = 48)
lengths(d1m) > 0
# [1] FALSE
```

The area mapping is performed to facilitate the topological visualization of the objects created in Albers Equal Area cartographic projection (Figure 8.34).

```
plot(p_p[], col = "black", pch = 21, bg = "black", axes = TRUE)
plot(l_p[], col = "black", add = TRUE)
plot(a_p[], col = "transparent", border = "black", add = TRUE)
```

8.7.3.2 Spatial subset

Geospatial subsetting is the process of selecting features of an object based on its geospatial relationship with another object.

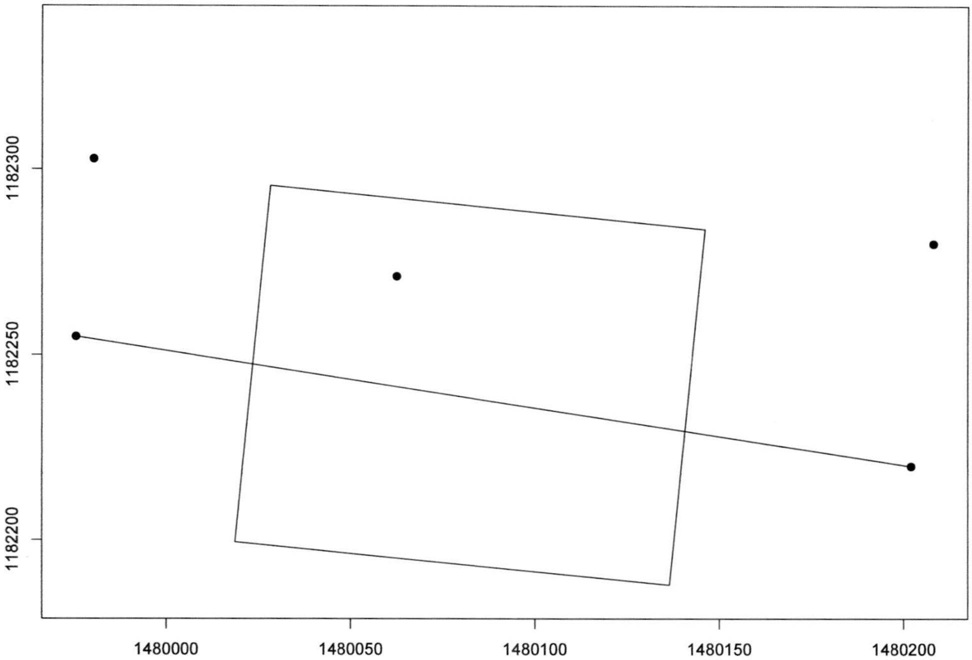

FIGURE 8.34 Topological geovisualization of point, line and polygon geo-objects in the Albers Equal Area plane map projection.

The `geobr` package is used in this case to obtain the geometry of municipalities at an enhanced resolution to perform more accurate aggregation calculations with this data in the practice on spatial data aggregation.

The `read_municipality` function is used to obtain the municipalities of Minas Gerais, referring to the year 2007.

```
muni <- read_municipality(code_muni = "MG", year = 2007)
# Using year 2007
# Downloading: 1.3 MB
muni
# Simple feature collection with 853 features and 4 fields
# Geometry type: MULTIPOLYGON
# Dimension:     XY
# Bounding box:  xmin: -51.0459 ymin: -22.92272 xmax: -39.85685 ymax: -14.23321
# Geodetic CRS:  GRS 1980(IUGG, 1980)
# First 10 features:
#    code_muni            name_muni code_state abbrev_state                geom
# 1    3100104 Abadia Dos Dourados   31         MG MULTIPOLYGON (((-47.44451 -...
# 2    3100203               Abaeté   31         MG MULTIPOLYGON (((-45.18241 -...
# 3    3100302           Abre Campo   31         MG MULTIPOLYGON (((-42.36996 -...
```

```
# 4     3100401             Acaiaca   31          MG MULTIPOLYGON (((-43.11432 -...
# 5     3100500             Açucena   31          MG MULTIPOLYGON (((-42.38632 -...
# 6     3100609            Água Boa   31          MG MULTIPOLYGON (((-42.35442 -...
# 7     3100708       Água Comprida   31          MG MULTIPOLYGON (((-48.10599 -...
# 8     3100807             Aguanil   31          MG MULTIPOLYGON (((-45.41611 -...
# 9     3100906      Águas Formosas   31          MG MULTIPOLYGON (((-40.96316 -...
# 10    3101003     Águas Vermelhas   31          MG MULTIPOLYGON (((-41.40315 -...
```

Coordinates with CRS, GRS 1980, are transformed to WGS-84 with the `st_transform` function.

```
muniwgs <- st_transform(muni, crs = 4326)
```

Subsetting can be performed with the bracket operator of R or with the `filter` function of the `tidyverse`. A target variable is created with the name of the counties of interest in the filtering, and the `%in%` operator is used to apply the filter.

```
target <- c("Cana Verde", "Carmo Da Cachoeira", "Ijaci", "Ingaí", "Itumirim",
          "Lavras", "Luminárias", "Nepomuceno", "Perdões", "Ribeirão Vermelho",
          "São Bento Abade")
munMG <- muniwgs |> filter(name_muni %in% target)
```

8.7.3.3 Spatial data aggregation

The cadastral map of coffee fields mapped in municipalities of Minas Gerais is used as an example of the use of spatial aggregation. The data used are available for download[5] and were obtained and preprocessed from the project source GeoPortal do Café[6].

The `st_read` function is used to import the geometry of coffee crops in a region of the state of Minas Gerais, Brazil.

```
coffee <- st_read('C:/geo/c8/cafesf.shp')
# Reading layer `cafesf' from data source `C:\geo\c8\cafesf.shp' using driver
# `ESRI Shapefile'
# Simple feature collection with 8005 features and 14 fields
# Geometry type: POLYGON
# Dimension:     XY
# Bounding box:  xmin: -45.39288 ymin: -21.63094 xmax: -44.77784 ymax: -20.94797
# Geodetic CRS:  WGS 84
```

The 'plot' function is used in mapping the crops and state boundaries (Figure 8.35).

```
plot(st_geometry(munMG), col = "white", axes = TRUE, bgc = "grey92")
plot(st_geometry(coffee), col = "brown", border = "transparent", add = TRUE)
```

[5]http://www.sergeo.deg.ufla.br/geo/downloads/cafesf.zip

[6]http://portaldocafedeminas.emater.mg.gov.br/

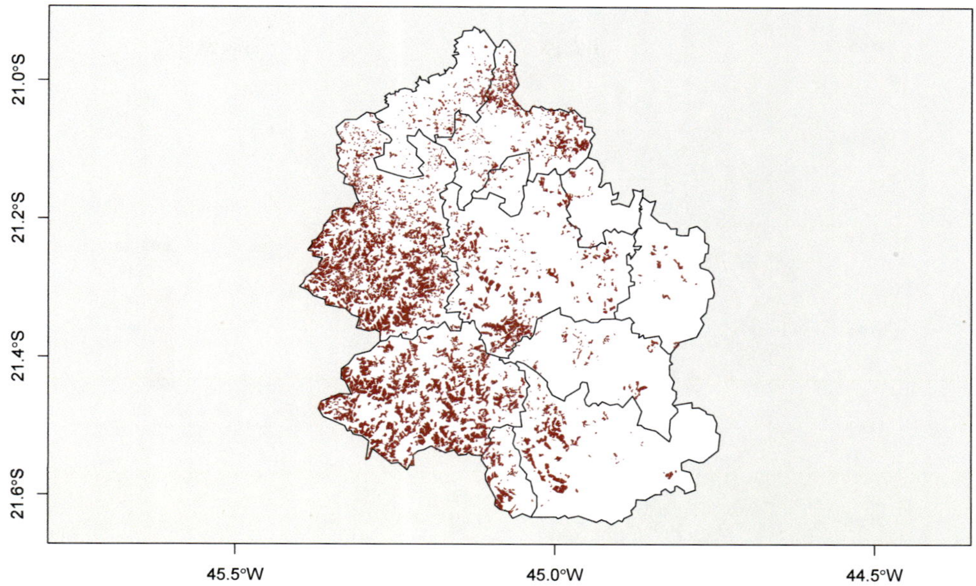

FIGURE 8.35 Geometry of coffee crops in municipalities of Minas Gerais for the 2018 harvest.

The polygons with geopolitical information of 11 municipalities of Minas Gerais are used as the delimiting area to perform the aggregation. The coffee polygons with area attribute (ha) of each polygon are mapped within the municipalities of Minas Gerais (Figure 8.36).

```
ggplot() + geom_sf(data = munMG, fill = "white", size = 0.1) +
  geom_sf(data = coffee, aes(fill = area_ha), colour = NA)  +
  scale_fill_distiller(palette = "Reds")
```

The `s2` package is disabled in this case to avoid problems with invalid geometries when using the `sf_use_s2(FALSE)` function.

```
sf_use_s2(FALSE)
```

The function `aggregate` is used to sum the areas of each coffee polygon within the municipalities. The municipality Nepomuceno is the largest producer with an area of 13818.8130 ha, followed by Carmo da Cachoeira, with 13266.1646 ha.

```
mgagg <- aggregate(x = coffee[,13], by = munMG, FUN = sum)
print(mgagg, n = 11)
#Simple feature collection with 11 features and 1 field
#Geometry type: MULTIPOLYGON
#Dimension:     XY
```

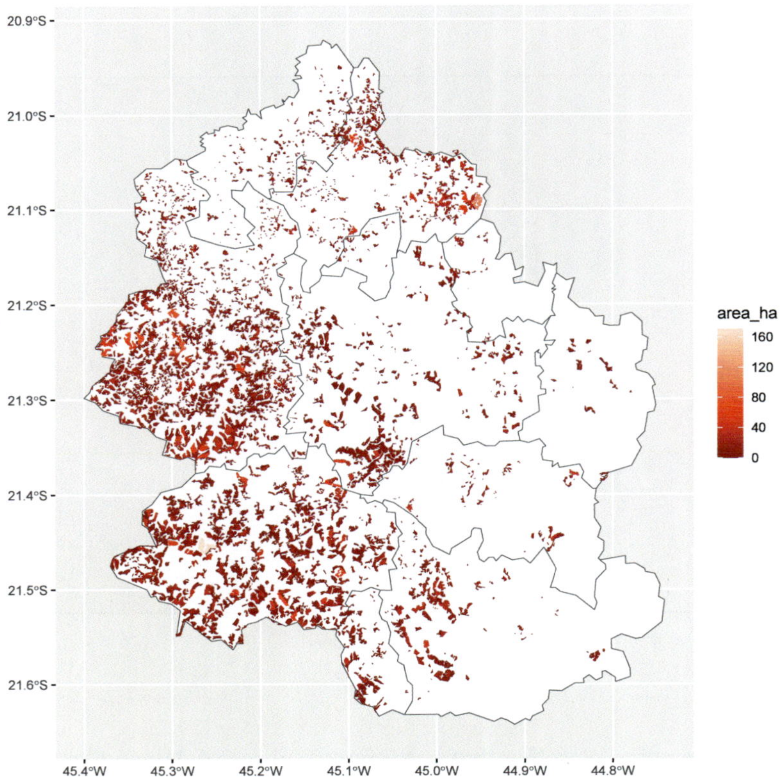

FIGURE 8.36 Attribute mapping of coffee plantation areas in municipalities of Minas Gerais referring to the 2018 crop.

```
#Bounding box:  xmin: -45.39973 ymin: -21.64273 xmax: -44.74138 ymax: -20.92158
#Geodetic CRS:  WGS 84
#      area_ha                       geometry
#1    869.0856 MULTIPOLYGON (((-45.11883 -...
#2  13266.1646 MULTIPOLYGON (((-45.17252 -...
#3    263.8040 MULTIPOLYGON (((-44.93474 -...
#4    566.5135 MULTIPOLYGON (((-44.9751 -2...
#5    290.9689 MULTIPOLYGON (((-44.86342 -...
#6   4901.4056 MULTIPOLYGON (((-44.99714 -...
#7   2231.3625 MULTIPOLYGON (((-45.00372 -...
#8  13818.8130 MULTIPOLYGON (((-45.28033 -...
#9   3007.8268 MULTIPOLYGON (((-45.04777 -...
#10   152.1581 MULTIPOLYGON (((-45.05039 -...
#11  1029.4486 MULTIPOLYGON (((-45.06405 -...
```

The total area produced in the municipalities with the sum of the attribute column is 40,397.55 ha. Note that the sum of attributes determined a result different from 40,398.34 ha, obtained with spatial aggregation by the `summarize` function performed previously. This is probably due to the calculation of decimal places in the sum by attributes, being the result of spatial aggregation apparently more reliable.

```
sum(mgagg$area_ha)
#[1] 40397.55
```

It also should be noted that the resolution of polygons of administrative boundaries of municipalities used in the analyses can alter the results obtained.

8.7.3.3.1 Spatial aggregation for data from the entire state of Minas Gerais

The `summarize` function is used in the `dplyr` package in a manner equivalent to `aggregate`, and the grouping of variables is specified by the `group_by` function.

```
coffeeProdAgg2 <- coffeeProd |>  group_by(Ano) |>
  summarize(Prod = sum(Prod, na.rm = T))
names(coffeeProdAgg2)
# [1] "Ano"       "Prod"      "geometry"
coffeeProdAgg2
#    Ano     Prod         geometry
#   <dbl>   <dbl>         <POLYGON [degrees]>
# 1   2012 1578355 ((...
# 2   2013 1584487 ((...
# 3   2014 1346517 ((...
# 4   2015 1325471 ((...
# 5   2016 1816203 ((...
# 6   2017 1437330 ((...
# 7   2018 1881219 ((...
# 8   2019 1477975 ((...
```

In this case, the dataset is transformed into `tibble` format, and the column of geometries is described as a vector of type `sfc`.

The coffee production in all evaluated years can be obtained by including the number of values considered in the summation, using the `summarize` function.

```
coffeeProd |> summarize(Prod = sum(Prod, na.rm = TRUE), n = n())
# Simple feature collection with 1 feature and 2 fields
# Geometry type: POLYGON
# Dimension:     XY
# Bounding box:  xmin: -51.04608 ymin: -22.92276 xmax: -39.85683 ymax: -14.24667
# Geodetic CRS:  WGS 84
#       Prod    n                            geometry
# 1 12447557 6824 POLYGON ((-45.42035 -22.621...
```

The years with the highest Arabica coffee production and the number of municipalities used in the sum of production in each year can be consulted by ranking.

```
coffeeProd |> dplyr::select(Prod, Ano) |> group_by(Ano) |>
  summarize(Prod = sum(Prod, na.rm = TRUE), nome = n()) |>
  top_n(n = 2, wt = Prod) |> st_drop_geometry()
#  A tibble: 2 x 3
```

```
#      Ano     Prod  nome
# *  <dbl>    <dbl> <int>
# 1   2016 1816203   853
# 2   2018 1881219   853
```

After the query, it is concluded that a total of 852 surveyed municipalities with production, being verified higher production in 2018, followed by 2016, with 1881219 and 1816203 tons of processed coffee beans, respectively.

8.7.3.4 Spatial join

An example of spatial join is performed using four random points created to perform the spatial join in the region encompassed by the state of Minas Gerais. Random seeds are created using the `set.seed` function within the boundary of the state of Minas Gerais, defined with the `st_bbox` function.

```
set.seed(2020) # set seed for reproducibility
(bb_coffeeProd = st_bbox(muniwgs)) # the MG's bounds
```

The geometry of the points is defined with the `tibble` function according to the previous definitions of minimum and maximum data boundaries in the x,y directions.

```
random_df <- tibble(x = runif(n = 4, min = bb_coffeeProd[1], max =
 bb_coffeeProd[3]), y = runif(n = 4, min = bb_coffeeProd[2],
                             max = bb_coffeeProd[4]))
```

Coordinate and geographic CRS settings are assigned to the points based on the Minas Gerais municipalities file.

```
random_points <- random_df |> st_as_sf(coords = c("x", "y")) |> # set coordinates
  st_set_crs(st_crs(muniwgs)) # set geographic CRS
```

The points created and the polygons with municipality boundaries are mapped to make it easier to see the points in relation to the polygons (Figure 8.37).

```
plot(st_geometry(muniwgs), col = 'white', axes = TRUE, bg="gray92")
plot(st_geometry(random_points), add = TRUE, pch = 21, lwd = 2, cex = 2,
     col = 'red')
```

Before creating the joined dataset, the spatial subset of points is used to create a new dataset with only the municipalities with random points. The `left=FALSE` parameter is set so that only results at the intersection between the two geographic databases used are considered.

```
random_joined <- st_join(random_points, left = FALSE, muniwgs["name_muni"])
nrow(random_joined)
#[1] 4
random_joined$name_muni
#[1] Lima Duarte      Jacutinga        Andrelândia      Dores Do Indaiá
```

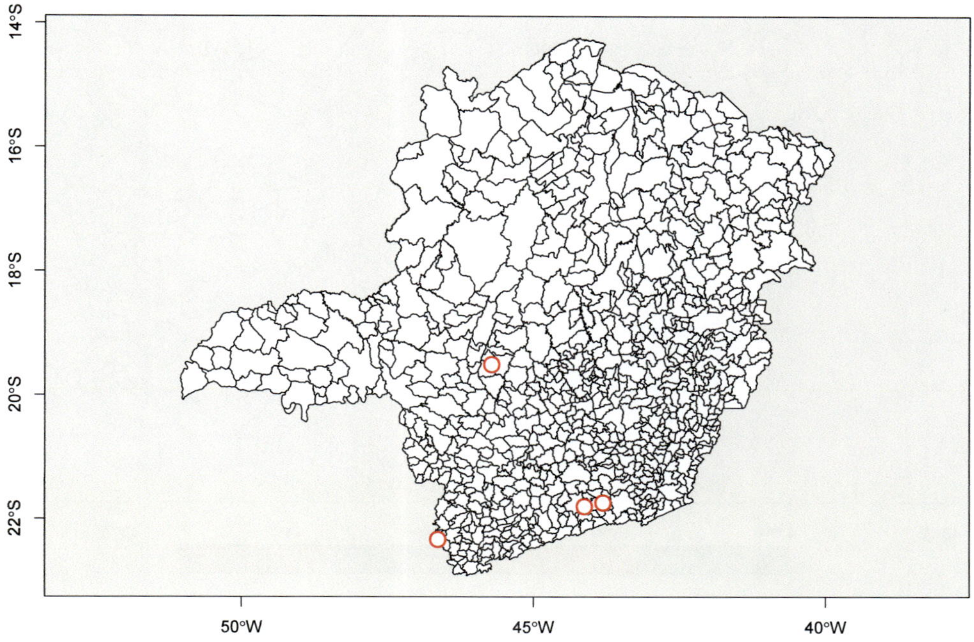

FIGURE 8.37 Mapping random points defined in Minas Gerais to perform spatial join.

With this, 32 data records are detected where there was a match in the join.

The results are exported for archiving and then imported back into R for mapping (Figure 8.38).

```
st_write(random_joined, dsn = "C:/geo/c8/random_joined.shp") #Export
random_j <- st_read("C:/geo/c8/random_joined.shp") #Import
# Reading layer `random_joined' from data source `C:\geo\c8\random_joined.shp'
# using driver `ESRI Shapefile'
# Simple feature collection with 4 features and 1 field
# Geometry type: POINT
# Dimension:     XY
# Bounding box:  xmin: -46.63489 ymin: -22.33718 xmax: -43.80767 ymax: -19.50672
# Geodetic CRS:  WGS 84
```

```
plot(random_j[,1], reset = FALSE, axes = TRUE, pch = 16, cex = 3, lwd = 5,
     bgc = 'gray92', key.pos = 1)
plot(st_geometry(muniwgs), add = TRUE)
```

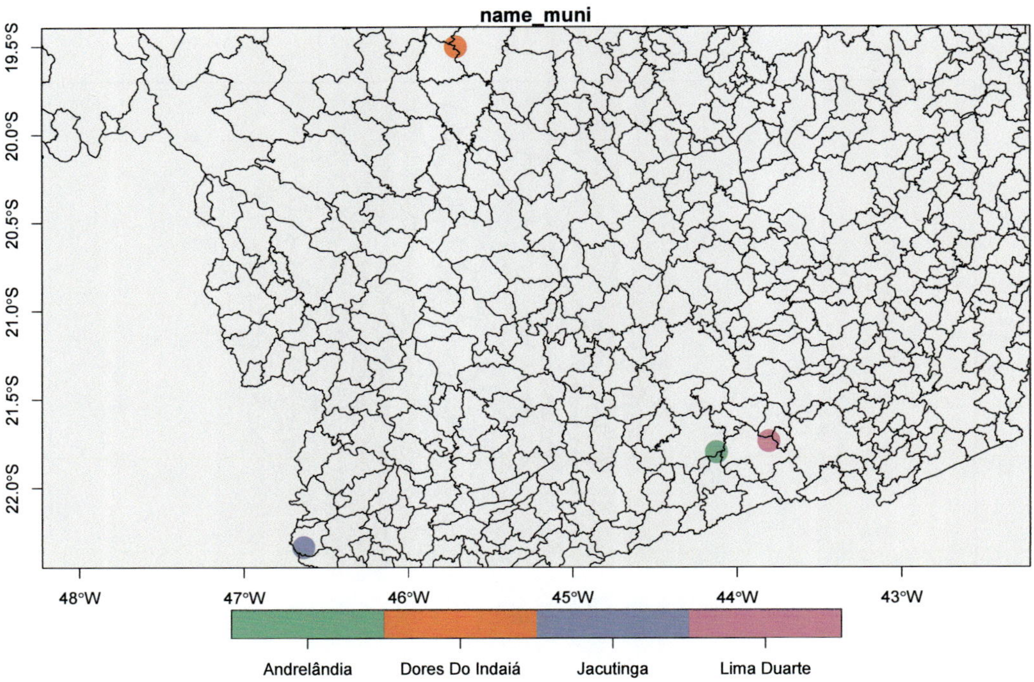

FIGURE 8.38 Mapping randomly located municipalities in Minas Gerais using points and joined with the database of municipalities by spatial join.

8.7.3.5 Conditional non-overlapping joins

In some situations, although two geographic datasets do not touch, there may be strong geographic relationships allowing associations. In this case, a distance threshold can be established to perform the spatial join including non-overlapping objects. Two sets of point geometries, `mp1` and `mp2`, are created to evaluate this type of join. An matrix of longitude and latitude coordinates of points is created with the `matrix` function. The point geometry is defined with the `st_multipoint` function and in class file `sfc` with the `st_cast(st_sfc())` function.

```
# Create point1
p_matrix1 <- matrix(c(-44.975860, -44.973541, -44.975860, -44.973541, -44.9750,
          -21.227613, -21.227719, -21.2272, -21.2272, -21.2274), ncol = 2)
p_multi1 <- st_multipoint(x = p_matrix1)
mp1 <- st_cast(st_sfc(p_multi1), "POINT")
# Create point2
p_matrix2 <- matrix(c(-44.9756, -44.973842, -44.975859, -44.973540, -44.9749,
          -21.227611, -21.227718, -21.2273, -21.2273, -21.2275), ncol = 2)
p_multi2 <- st_multipoint(x = p_matrix2)
mp2 <- st_cast(st_sfc(p_multi2), "POINT")
```

Attributes of identification (id), point name (name), temperature value (temperature) and date (date) are set for each of the points in the two evaluated datasets.

```
# Attributes 1
lnd_attrib1 <- data.frame(id = c(1, 2, 3, 4, 5), name = c("p1","p2","p3",
  "p4","p5"), temperature = c(25, 25.1, 25.3, 26, 27.1), date =
  as.Date("2021-06-21")) # data.frame object
lnd_sf1 <- st_sf(lnd_attrib1, geometry = mp1)    # sf object
lnd_sf1 <- st_set_crs(lnd_sf1,  "+proj=longlat +datum=WGS84 +no_defs") # set CRS
lnd_sf1
# Simple feature collection with 5 features and 4 fields
# Geometry type: POINT
# Dimension:     XY
# Bounding box:  xmin: -44.97586 ymin: -21.22772 xmax: -44.97354 ymax: -21.2272
# CRS:           +proj=longlat +datum=WGS84 +no_defs
#   id name temperature       date                   geometry
# 1  1   p1        25.0 2021-06-21 POINT (-44.97586 -21.22761)
# 2  2   p2        25.1 2021-06-21 POINT (-44.97354 -21.22772)
# 3  3   p3        25.3 2021-06-21  POINT (-44.97586 -21.2272)
# 4  4   p4        26.0 2021-06-21  POINT (-44.97354 -21.2272)
# 5  5   p5        27.1 2021-06-21    POINT (-44.975 -21.2274)

# Attributes 2
lnd_attrib2 <- data.frame(id = c(1, 2, 3, 4, 5), name = c("pa1","pa2","pa3",
  "pa4","pa5"), temperature = c(27, 25.3, 26.3, 26.1, 27.2), date =
  as.Date("2021-07-21")) # data.frame object
lnd_sf2 <- st_sf(lnd_attrib2, geometry = mp2)    # sf object
lnd_sf2 <- st_set_crs(lnd_sf2,  "+proj=longlat +datum=WGS84 +no_defs") # set CRS
lnd_sf2
# Simple feature collection with 5 features and 4 fields
# Geometry type: POINT
# Dimension:     XY
# Bounding box:  xmin: -44.97586 ymin: -21.22772 xmax: -44.97354 ymax: -21.2273
# CRS:           +proj=longlat +datum=WGS84 +no_defs
#   id name temperature       date                   geometry
# 1  1  pa1        27.0 2021-07-21  POINT (-44.9756 -21.22761)
# 2  2  pa2        25.3 2021-07-21 POINT (-44.97384 -21.22772)
# 3  3  pa3        26.3 2021-07-21  POINT (-44.97586 -21.2273)
# 4  4  pa4        26.1 2021-07-21  POINT (-44.97354 -21.2273)
# 5  5  pa5        27.2 2021-07-21   POINT (-44.9749 -21.2275)
```

The point data with attributes are transformed into the Universal Transverse Mercator projection, zone 23 South.

```
lnd_sf1_p <- st_transform(lnd_sf1, crs = 32723)
lnd_sf2_p <- st_transform(lnd_sf2, crs = 32723)
```

Next, a distance of 20 m is defined to perform the non-overlapping junction. It was observed that due to the defined distance, it is not possible to perform the junction of points 1 and 2, being necessary to increase the distance criteria to contemplate all the points in the non-overlapping junction.

```
njoined <- st_join(lnd_sf1_p, lnd_sf2_p, join = st_is_within_distance, dist = 20)
njoined
# Simple feature collection with 5 features and 8 fields
# Geometry type: POINT
# Dimension:     XY
# Bounding box:  xmin: 502505 ymin: 7652650 xmax: 502745.7 ymax: 7652707
# Projected CRS: WGS 84 / UTM zone 23S
#  id.x name.x temperature.x  date.x id.y name.y temperature.y date.y geometry
# 1 1  p1   25.0 2021-06-21 NA   <NA>   NA          <NA>    POINT (502505 7652661)
# 2 2  p2   25.1 2021-06-21 NA   <NA>   NA          <NA> POINT (502745.7 7652650)
# 3 3  p3   25.3 2021-06-21 3    pa3    26.3 2021-07-21    POINT (502505 7652707)
# 4 4  p4   26.0 2021-06-21 4    pa4    26.1 2021-07-21 POINT (502745.7 7652707)
# 5 5  p5   27.1 2021-06-21 5    pa5    27.2 2021-07-21 POINT (502594.3 7652685)
```

By changing the distance to 32 m, all points are covered at the junction.

```
njoined <- st_join(lnd_sf1_p, lnd_sf2_p, join = st_is_within_distance, dist = 32)
njoined
# Simple feature collection with 5 features and 8 fields
# Geometry type: POINT
# Dimension:     XY
# Bounding box:  xmin: 502505 ymin: 7652650 xmax: 502745.7 ymax: 7652707
# Projected CRS: WGS 84 / UTM zone 23S
#  id.x name.x temperature.x  date.x id.y name.y temperature.y  date.y  geometry
# 1 1  p1  25.0 2021-06-21  1  pa1       27.0 2021-07-21    POINT (502505 7652661)
# 2 2  p2  25.1 2021-06-21  2  pa2       25.3 2021-07-21 POINT (502745.7 7652650)
# 3 3  p3  25.3 2021-06-21  3  pa3       26.3 2021-07-21    POINT (502505 7652707)
# 4 4  p4  26.0 2021-06-21  4  pa4       26.1 2021-07-21 POINT (502745.7 7652707)
# 5 5  p5  27.1 2021-06-21  5  pa5       27.2 2021-07-21 POINT (502594.3 7652685)
```

The datasets are mapped with the name attributes of each point (Figure 8.39).

```
ggplot() + geom_sf(data = lnd_sf1_p, color = "black") +
  geom_sf_label(data = lnd_sf1_p, aes(label = name), nudge_x = 0, nudge_y = 5) +
  geom_sf(data = lnd_sf2_p, color = "red") +
  geom_sf_label(data = lnd_sf2_p, aes(label = name), nudge_x = 0, nudge_y = 5)
```

8.7.3.6 Distance relationships

Distance relationships can be performed by using geodetic, spherical calculations, as well as considering the surface in a given map projection, as a way to illustrate the determination of distance between two points.

Two sets of point geometries are created, `mp1` and `mp2`. A matrix of longitude and latitude coordinates of points is created with the `matrix` function. The point geometry is defined with the `st_multipoint` function in a `sfc` class file with the `st_cast(st_sfc())` function.

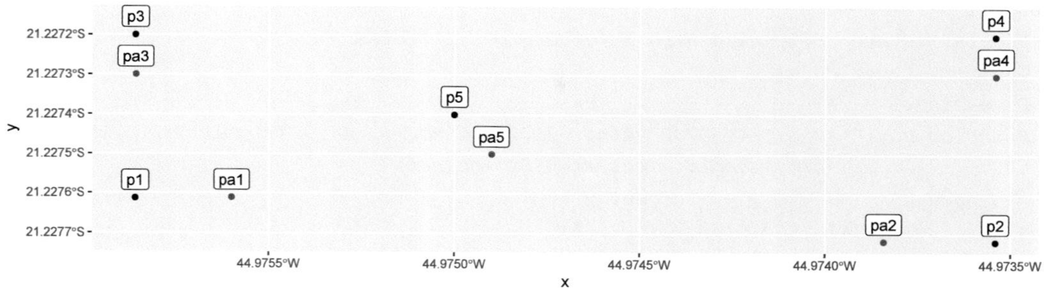

FIGURE 8.39 Mapping points with attributes used in conditional spatial join based on distance threshold.

```
# Create point 1
p_matrix1 <- matrix(c(-44.974899, -21.229064), ncol = 2)
p_multi1 <- st_multipoint(x = p_matrix1)
mp1 <- st_cast(st_sfc(p_multi1), "POINT")
# Geometry set for 1 feature
# Geometry type: POINT
# Dimension:     XY
# Bounding box:  xmin: -44.9749 ymin: -21.22906 xmax: -44.9749 ymax: -21.22906
# CRS:           NA
# POINT (-44.9749 -21.22906)
```

```
# Create point 2
p_matrix2 <- matrix(c(-44.980719, -21.198968), ncol = 2)
p_multi2 <- st_multipoint(x = p_matrix2)
mp2 <- st_cast(st_sfc(p_multi2), "POINT")
# Geometry set for 1 feature
# Geometry type: POINT
# Dimension:     XY
# Bounding box:  xmin: -44.98072 ymin: -21.19897 xmax: -44.98072 ymax: -21.19897
# CRS:           NA
# POINT (-44.98072 -21.19897)
```

A geographic database with `id` and point name is created with the `data.frame` function. The geometry created earlier is used as the basis for the attributes of points 1 (pa1) and 2 (pb1) created with the `st_sf` function. The `st_set_crs` function is used to assign a reference coordinate system to the points, in this case the SIRGAS 2000 system (GRS80 ellipsoid).

```
# Attributes p1
lnd_attrib1 <- data.frame(id = c(1), name = c("pa1")) # data.frame object
lnd_sf1 <- st_sf(lnd_attrib1, geometry = mp1)    # sf object
lnd_sf1 <- st_set_crs(lnd_sf1, "+proj=longlat +ellps=GRS80
                      +towgs84=0,0,0 +no_defs") # set CRS
```

```
lnd_sf1
# Simple feature collection with 1 feature and 2 fields
# Geometry type: POINT
# Dimension:     XY
# Bounding box:  xmin: -44.9749 ymin: -21.22906 xmax: -44.9749 ymax: -21.22906
# CRS:           +proj=longlat +datum=WGS84 +no_defs
#   id name                  geometry
# 1  1   pa1 POINT (-44.9749 -21.22906)

# Attributes p2
lnd_attrib2 <- data.frame(id=c(1), name = c("pb1")) # data.frame object
lnd_sf2 = st_sf(lnd_attrib2, geometry = mp2)    # sf object
lnd_sf2 = st_set_crs(lnd_sf2, "+proj=longlat +ellps=GRS80
                        +towgs84=0,0,0 +no_defs") # set CRS
lnd_sf2
# Simple feature collection with 1 feature and 2 fields
# Geometry type: POINT
# Dimension:     XY
# Bounding box:  xmin: -44.98072 ymin: -21.19897 xmax: -44.98072 ymax: -21.19897
# CRS:           +proj=longlat +ellps=GRS80 +towgs84=0,0,0 +no_defs
#   id name                    geometry
# 1  1  pb1 POINT (-44.98072 -21.19897)
```

The point data with attributes are transformed into the Universal Transverse Mercator projection, zone 23 South (SIRGAS 2000/UTM zone 23S).

```
lnd_sf1_p <- st_transform(lnd_sf1, crs = 31983)
lnd_sf2_p <- st_transform(lnd_sf2, crs = 31983)
```

The plane distance between points 1 and 2 can be determined by the `st_distance` function, giving a distance of 3385.211 m.

```
st_distance(lnd_sf1_p[1,], lnd_sf2_p[1,])
# Units: [m]
#          [,1]
# [1,] 3385.211
```

The `st_coordinates` function is used to obtain the X and Y coordinates of points 1 and 2 in order to be able to check the results obtained with the `st_distance` function.

```
p1prj <- st_coordinates(lnd_sf1_p)
p1prj
#          X       Y
# 1 502604.7 7652501
p2prj <- st_coordinates(lnd_sf2_p)
p2prj
#          X       Y
# 1 502001.2 7655832
```

The same results can be obtained when applying the Pythagorean theorem, obtaining the Euclidean distance of 3385.211 m.

```
d <- sqrt((p1prj[,1] - p2prj[,1])^2 + (p1prj[,2] - p2prj[,2])^2)
d
#[1] 3385.211
```

The spherical distance between point 1 (`lnd_sf1`) and point 2 (`lnd_sf2`) can be determined with the `st_distance` function. This gives a distance of 3400.474 m.

```
st_distance(lnd_sf1[1,], lnd_sf2[1,])
# Units: [m]
#          [,1]
# [1,] 3400.474
```

The spherical distance can also be calculated by the `s2` package, with the `s2_distance` function, giving the same value of 3400.474 m.

```
s2_distance(lnd_sf1[1,], lnd_sf2[1,])
# [1] 3400.474
```

The geodetic (ellipsoidal) distance can be determined using the `lwgeom` package. In this case, the `st_geod_distance` function is used to calculate the distance between the two points, yielding a value of 3386.565 m.

```
st_geod_distance(lnd_sf1[1,], lnd_sf2[1,])
# Units: [m]
#          [,1]
# [1,] 3386.565
```

It can be seen that the plane distance between the two points considered is smaller than the ellipsoidal distance, followed by the spherical distance.

9 Raster Spatial Analysis with Environmental Geocomputation and R

9.1 LEARNING QUESTIONS

The learning questions answered by reading the chapter are:

- How do you define and manipulate raster layer attributes?
- What spatial analysis operations are used on raster data?
- How to analyze the spatial relationship of a cell with a neighborhood of cells.
- How to obtain morphometry, hydrology, illumination and terrain shadow variables from raster spatial analysis.
- How to use the concept of variable synthesis in mathematical modeling applied to spatial analysis of raster data.

9.2 LEARNING OUTCOMES

The learning outcomes answered by reading the chapter are:

- Define and verify how to manipulate raster layer attributes;
- Verify the use of spatial analysis operations on raster data;
- Understand how to analyze the spatial relationship of a cell with respect to its neighbors;
- Verify how to obtain variables of morphometry, hydrology, lighting and terrain shadow from the raster spatial analysis;
- Understand the concept of variable synthesis in mathematical modeling applied to spatial analysis of raster data.

9.3 INTRODUCTION

The term geocomputation, used in the literature on geographic information systems, can create questions about what geographic data processing is. Some people have considered geocomputation as a set of spatial analysis functions. Others define geocomputation more generally as a set of techniques and procedures used to create and modify spatial data with one or more parameters. In a geographic information system, there may be hundreds of geocomputation tools developed to implement spatial techniques using geographic information, including data digitization, database creation and maintenance, spatial data analysis, modeling, and generation of products with statistical and cartographic information (Lo and Yeung, 2007).

The geocomputation characteristics of raster data are dependent on tools and procedures used in processing the raster data model, and it is important to understand the characteristics of raster data as well as applications used in solving geographic problems. In the raster data model, the geographic space of interest is divided into regular cells with specific dimensions and the attribute value of each cell is represented by an alphanumeric

DOI: 10.1201/9781003724377-9

code. Therefore, raster type data is best used to represent continuous spatial phenomena such as: elevation, temperature, and air pollution concentration (Lo and Yeung, 2007).

The modeling approach used on raster type data enables to analyze the terrain as a continuous surface, where it is possible to determine new attributes, such as a slope map, or extract the topology of the steepest slope paths. The state of a S cell on the terrain is determined by attributes, local functions, lateral interactions, neighborhood, and topology (Figure 9.1).

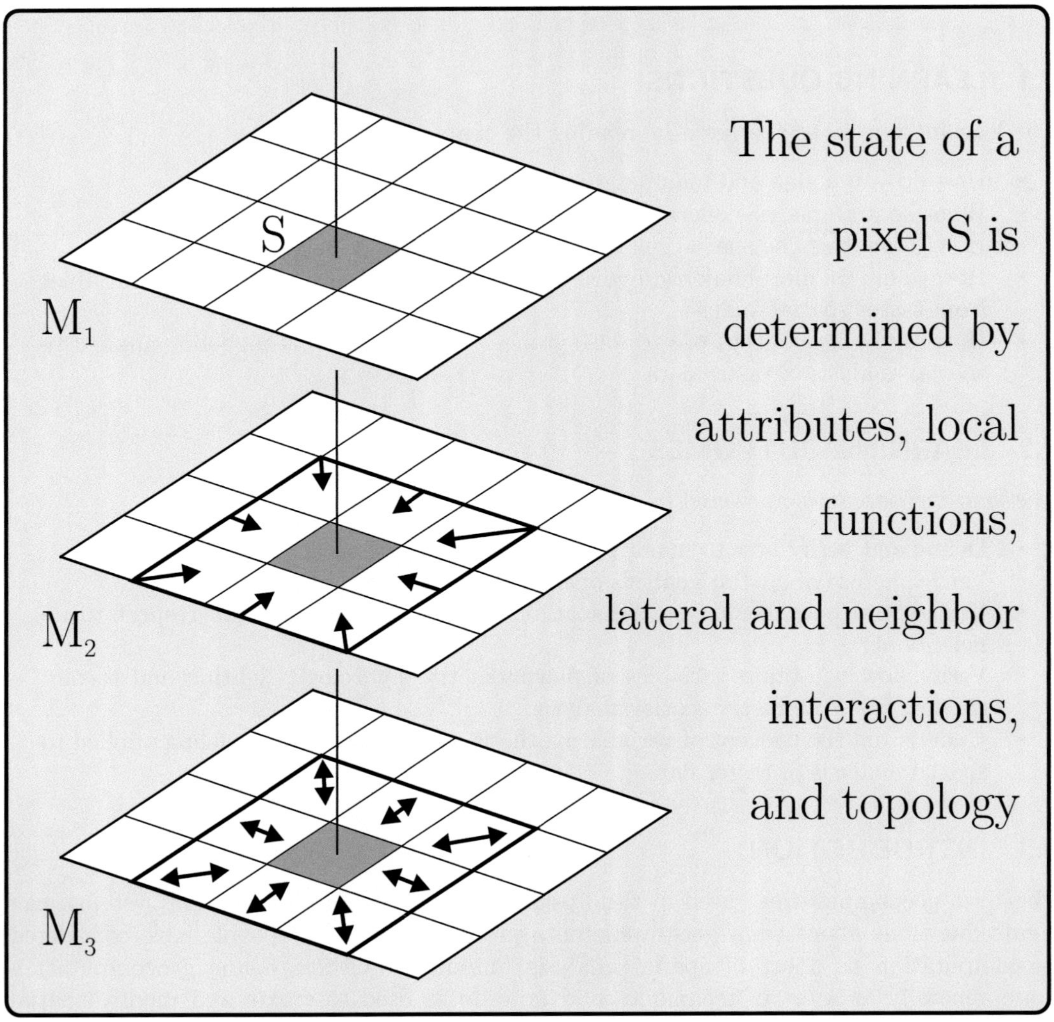

FIGURE 9.1 The state of a cell as a function of local operations and lateral interactions with the neighborhood.

Spatial analysis operations on thematic, numerical, and remote sensor data geo-fields are classified as (Câmara and Medeiros, 1998; Longley et al., 2001; Lovelace et al., 2019):

- Local or per cell <- Operations to analyze the raster cell-by-cell, comparing the value of each cell in one layer with the values of the same cell in other layers;
- Focal or neighborhood operations <- Operations to compare the value of each cell with the values of cells in regions neighboring the cell, such as, eight neighbors;

- Zonal <- Operations to calculate the result of blocks of contiguous cells that share the same value, such as calculating the shape of contiguous areas with the same land use type. These are similar to focal operations, but the surrounding cell grid on which the new values will be calculated may have irregular sizes and shapes;
- Global or raster scanning operations<- Operations that produce results over the entire raster layer, such as, the average value.

9.4 RASTER ATTRIBUTE MANIPULATION

Raster attribute data operations are related to creating continuous and categorical raster layers and extracting cell values from single and multilayer raster subsets. Global raster operations can be used to characterize raster datasets as a whole. In contrast to the vector data model where simple features were used to represent points, lines, and polygons as discrete entities in space, raster data represented continuous surfaces (Lovelace et al., 2019).

Rasters geo-fields can contain numeric, integer, logical, or factor class values. To use character values in raster, it is necessary to create coded classes to record data in the raster, for example, coffee canopy morphometry variables that varied according to average plant foliage percentage via visual rating scale. Categorical values can be composed of logical class or factor variables. For example, a raster can be created to represent categories with levels of plant foliage at: Very High, High, Medium, Low, and Very Low.

9.4.1 DEFINING RASTER SUBSETS

A raster subset can be made with operators for row and column indexing, cell IDs, coordinates, or other raster. Rasters can be stacked into multilayered raster blocks, with cell values in layer order. Cell values can be modified by overwriting existing values in conjunction with a subset operation (Lovelace et al., 2019).

9.4.2 SUMMARY STATISTICS OF RASTER

Summary statistics can be obtained from a raster layer, such as minimum, mean, maximum values, interquartile range, number of NAs/no data, standard deviation (sd), among others. In case of stacked raster, summary statistics can be obtained for both stacked layers (Lovelace et al., 2019).

9.4.3 MAP ALGEBRA

Many generic functions are implemented by raster algebra methods, via algebraic and logical operators. Map algebra is processed quickly on raster by facilities inherent to the data model used (Lovelace et al., 2019).

Map algebra, or cartographic modeling, raster operations are divided into subclasses, and it is possible to work with one or multiple grids simultaneously using different types of operation (Lovelace et al., 2019).

Raster map algebra operations can also be classified by disciplines in multiple areas of knowledge beyond geocomputation, such as:

- Topography;
- Hydrology;
- Climatology;
- Remote Sensing;
- Digital image processing.

9.4.4 LOCAL OPERATIONS

Local operations are used in all cell-by-cell operations on one or multiple layers, such as in the method of reclassifying values by means of numerical intervals. Map algebra is considered a type of local operation (Lovelace et al., 2019). The calculation of the normalized difference vegetation index (NDVI) is also an example of a cell-by-cell raster operation:

$$NDVI = \frac{\rho NIR - \rho Red}{\rho NIR + \rho Red} \tag{9.1}$$

where ρNIR and ρRed are the terrain reflectance values recorded in the sensor bands in the near infrared and red wavelengths, respectively.

9.4.5 FOCAL OPERATIONS

Focal operations are performed on a central cell and its neighbors. The neighborhood may be called a kernel, filter, or moving window, in 3-by-3 cell regions in the raster, for example. A focal operation can be used to apply a mathematical function to all cells within a specified neighborhood (Burrough and McDonnell, 1998; Lovelace et al., 2019).

Focal functions or filters have played a dominant role in image processing. In low-pass or smoothing filters, the average function is used to remove extreme values. High-pass filters, on the other hand, accentuate the extreme values, such as the Laplace and Sobel line detection filters (Lovelace et al., 2019).

Geocomputation of terrain variables to calculate topographic variables, such as slope, aspect, and flow directions, are performed using focal functions. The function `terrain`, can be used to calculate these variables. These and other variables can be determined by different methods, such as that of Zevenbergen and Thorne, used to calculate slope, but not implemented in this function of the `terrain` package. Other algorithms for determining terrain curvature, recharge area, and moisture indices can be obtained using the `RSAGA` package and `SAGA GIS` software, for example (Lovelace et al., 2019).

9.4.6 ZONAL OPERATIONS

Zonal operations are similar to focal operations, except that zonal filters can take any shape rather than a predefined rectangular window.

9.4.7 GLOBAL AND DISTANCE OPERATIONS

Global operations are a special case of zonal operations with the entire raster dataset representing a single zone. The most common global operations are summary statistics applied to the entire raster dataset, such as minimum, mean, and maximum values. In addition, global operations are also useful for calculating distance and weighting rasters. For example, we can calculate the distance from each cell to a specific target cell, such as to the nearest coast, and add an elevation raster to weight the distance with elevation so that with each additional meter of elevation, the Euclidean distance is extended. The visibility and `viewshed` calculations are also relative to the global operations family (Lovelace et al., 2019).

The smallest distance between two objects in a plane space is determined by a straight line. If the distance between two points is defined by the coordinates (x_1, y_1) and (x_2, y_2), the distance, d between the two points is the length of the hypotenuse in a right triangle, given by (Longley et al., 2001):

$$d = \sqrt{(x_2 - x_1)^2 + (y_2 - y_1)^2} \tag{9.2}$$

The determination of two points on a curved surface of the Earth, represented by spherical distance, for example, is determined as a function of latitude (ϕ) and longitude (λ) coordinates of the two points to calculate the arc length (D) (Longley et al., 2001):

$$D = Rcos^{-1}[sen\phi_1 sen\phi_2 + cos\phi_1 cos\phi_2 cos(\lambda_1 - \lambda_2)] \tag{9.3}$$

where R is the Earth's radius of 6378 km.

For geodetic distance determination situations, the calculations are more complex, and an ellipsoid model must be used.

9.4.8 RASTER MOSAIC

Suppose that the objective of a work is to calculate NDVI and, additionally, to determine terrain attributes from elevation data, for observations within a study area. Given that these calculations are dependent on remote sensing information, the imagery of locations can be divided into scenes that cover a specific spatial extent. Since a study area can comprise the extent of more than one scene, it is necessary to merge the scenes covered at each locality to obtain a raster mosaic of the entire region studied. The mosaic operation will occur at the intersection between the raster objects; however, the final result changes the size of the original raster, which is also a geometric operation.

9.5 SPATIAL ANALYSIS IN CELLS

All operations performed at a location are called a cell-by-cell operation. The advantage of the cell-based approach is related to the use of variance and smoothing concepts. In addition, determinations can be made on attributes, with zonal or neighborhood spatial analysis functions, around a cell. Since the raster model approach is useful for characterizing continuous surfaces, it is possible to perform analysis on how to visualize the surface from a location, determine the steepest region in a given area, as well as extract site topology based on hydrological processes and define ways to optimize access in different terrain types (Burrough and McDonnell, 1998).

All the numerical operations used in vector type data can also be used on cells of raster type data. Thus, cell values in different layers can be selected, added, subtracted, or manipulated by any mathematical method of interest (Figure 9.2) (Burrough and McDonnell, 1998).

9.6 SPATIAL FILTERING

The simplest and most used form of spatial filtering is performed by passing a square moving window over a continuous surface in order to compute a new value of the central cell in the window $C_{i,j}$ as a function of cell values within the window. This operation is called convolution and the window used can be cells in arrangements, 3x3, 5x5, 7x7, or in distance measure. The most common types of operation are low-pass and high-pass filtering, as

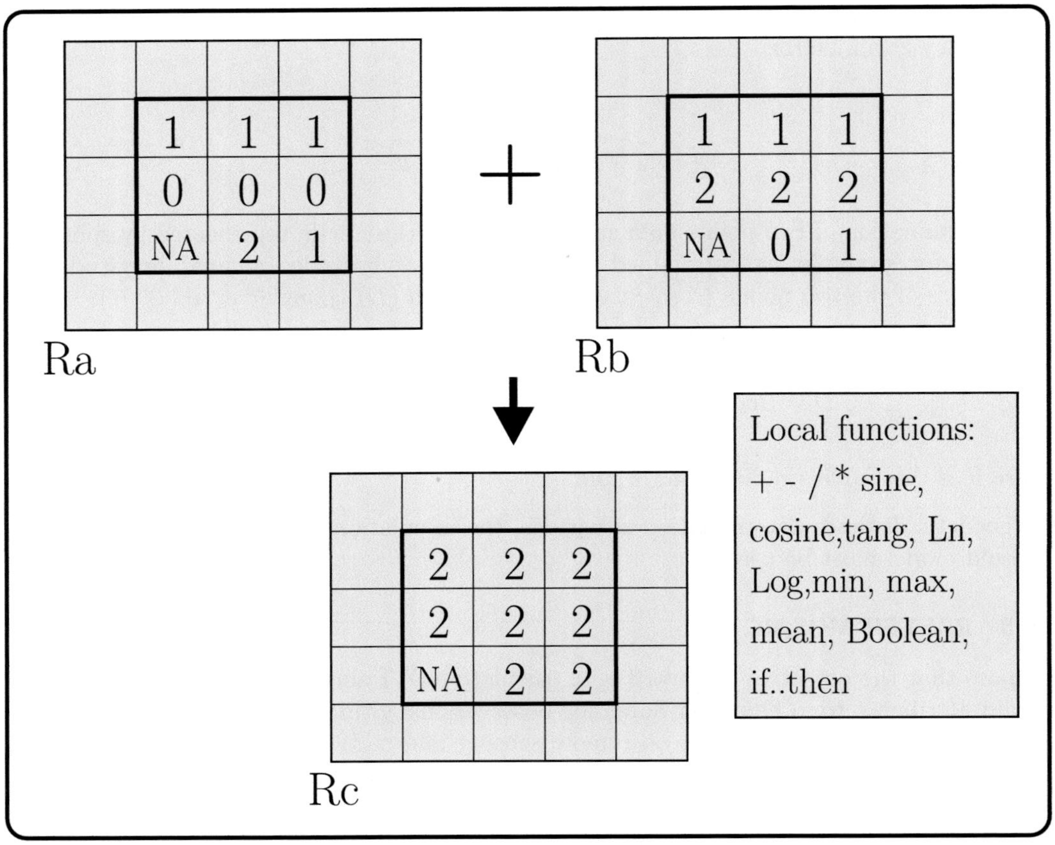

FIGURE 9.2 Cell-by-cell operations on the Ra and Rb rasters to generate the environment variable Rc.

well as any other mathematical operation or function. In filtering, the value of the cell in the center of the window is calculated as an arithmetic function of neighboring cell values (Figure 9.3) (Burrough and McDonnell, 1998).

9.7 SPATIAL ANALYSIS OF DIGITAL ELEVATION MODEL

A large number of products can be derived from continuous surfaces and discretized into regular grids. The most common continuous variable is the digital elevation model and the variables that can be obtained from spatial analysis of this data. Data such as slope, aspect, line of sight, hillshade, and irradiance can be obtained and, from this, determine useful information such as the topology of the surface and how material flow will travel through the drainage networks of the area (Table 9.1) (Burrough and McDonnell, 1998).

TABLE 9.1 Attributes that can be determined from a digital elevation model.

Attribute	Definition	Applications
Elevation	Altitude above sea level or a local reference.	Climate variables, soil-relief relationship, vegetation, volume of material for cut and fill, energy potential

Attribute	Definition	Applications
Slope	Rate of elevation change	Terrain slope, flow, environment classification, remote sensing image correction
Aspect	Horizontal angle in the direction of the steepest slope	Solar irradiance, evapotranspiration, vegetation attributes, remote sensing imagery correction
Profile curvature	Rate of change of slope	Stream acceleration, erosion and deposition zones, soil and terrain evaluation indices, vegetation
Plane curvature	Rate of aspect change	soil and water characteristics, convergence and divergence of flow
Direction of drainage	Direction of flow on steeper slope	Calculation of watering attritutes as a function of flow topology. Access to local transport of materials over drainage network
Specific recharge area	Upslope area	Upslope recharge area of a location
Length of flow	Length of longest flow path	Acceleration of flow, erosion rate, sediment yield
Channel flow	Cells with water flow	Flow intensity, flow location, erosion and sedimentation
Ridge	Cells without upstream contributing area	Drainage divide, vegetation, soil, erosion, geological analysis, connectivity
Moisture index	ln(specific recharge area)/tan(slope))	Moisture retention index
Index of flow power	Specific recharge area * tan(slope)	Overland flow erosive power
Sediment transport index (LS factor)	$(n+1)(A_S/22.13)^n(sen\beta/0.0896)^m$	Characterization of erosion and deposition processes in the USLE model
Length of recharge	Distance from the highest point to the outflow	Overland flow attenuation
Intervisible points	Intervisibility zones	Microwave transmission towers, towers for fire visualization, hotels, military applications
Irradiance	Amount of solar energy received per unit area	Studies of vegetation, soil, evapotranspiration, hillshade, location of areas for solar plant construction

Since the raster surface is in principle mathematically continuous, mathematical derivative calculations can be performed in relation to the altitude raster.

A simple statistical model based on linear regression can be used to obtain the variation of temperature in the Swiss Alps as a function of elevation, according to the equation (Burrough and McDonnell, 1998):

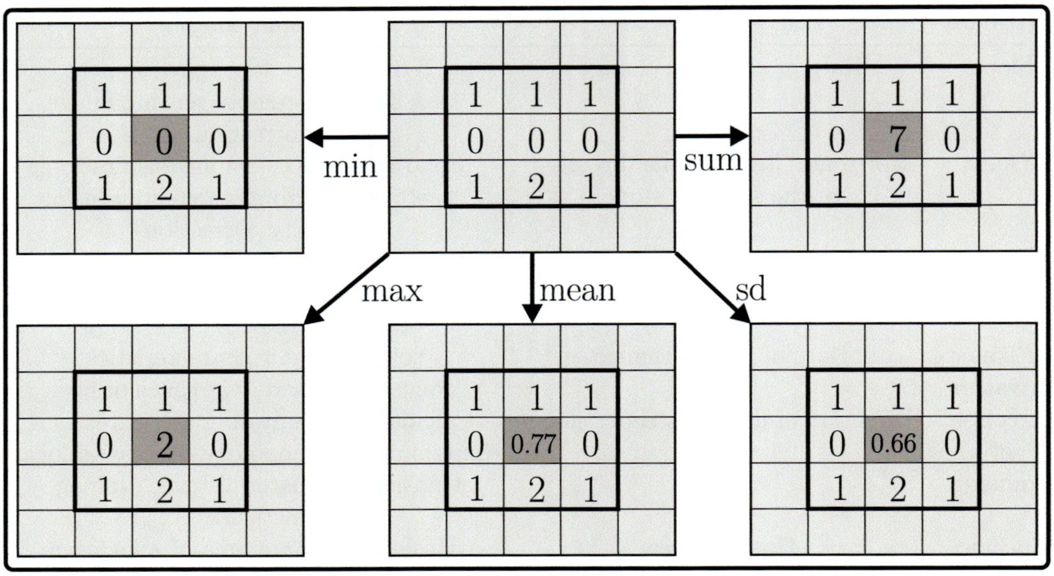

FIGURE 9.3 3x3 window operations in spatial filtering based on summary statistics functions, minimum (min), mean (mean), maximum (max), sum (sum) and, standard deviation (sd) values.

$$T = 5.697 - 0.00443 * E \tag{9.4}$$

Where T is the temperature in degrees Celsius and E is the altitude in meters.

A similar equation is obtained in municipalities of Minas Gerais as a function of climatological normals of mean air temperature, monitored in conventional climatological stations of INMET. A linear regression fit is performed between the mean annual mean air temperature obtained over a 30-year period, between 1961 and 1990, in 32 municipalities in the state. The regression model obtained is:

$$T = 25 - 0.0054 * E \tag{9.5}$$

The packages `readr` (Wickham et al., 2021c), `ggpubr` (Kassambara, 2025), `gridExtra` (Auguie and Antonov, 2017), `RStoolbox` (Leutner et al., 2019), and `raster` (Hijmans et al., 2020) needed to perform this example are loaded into R with the `library` function.

```
library(readr)
# package 'readr' was built under R version 4.3.3
library(ggpubr)
# Loading required packages: ggplot2
library(gridExtra)
library(RStoolbox)
# This is version 1.0.0 of RStoolbox
```

```
# package 'RStoolbox' was built under R version 4.3.3
library(raster)
# Loading required packages: sp
```

The mean sea level altitude and mean annual air temperature (°C) (1961-1990) raster data from the Instituto Nacional de Meteorologia[1] main climatological stations are imported into R using the `read_csv2` function.

```
mgAT <- read_csv2("C:/geo/c9/mgAT.csv", col_names = TRUE,
                  locale = locale(encoding = 'ISO-8859-1'))
# Rows: 32 Columns: 3
# -- Column specification ---------------------------------------------------
# Delimiter: ";"
# chr (1): Cidade
# dbl (2): Altitude, T-ano
# [i] Use spec() to retrieve the full column specification for this data.
names(mgAT)
# [1] "Cidade"   "Altitude" "T-ano"
names(mgAT)
[1] "Cidade"   "Altitude" "T-ano"
names(mgAT) <- c("City", "Altitude", "T_year")
mgAT
# A tibble: 32 x 3
#   City           Altitude T_year
#   <chr>             <dbl>  <dbl>
# 1 Aimores              83   24.6
# 2 Araxa              1004   20.4
# 3 Aracuai             284   24.4
# 4 Bambui              661   20.7
# 5 Barbacena          1126   18.0
# 6 Belo Horizonte      850   21.1
# 7 C. Mato Dentro      652   20.8
# 8 Caparao             843   18.8
# 9 Capinopolis         620   23.0
#10 Caratinga           609   21.2
# [i] 22 more rows
# [i] Use print(n = ...) to see more rows
```

A regression is used to fit the mean annual air temperature as a function of temperature by a first-order polynomial model using a formula. The plot with equations including parameters and model fit quality coefficients are obtained with the `ggscatter` function.

```
formula <- y ~ poly(x, 1, raw = TRUE) # Define the polynomial equation order
reg <- ggscatter(mgAT, x = "Altitude", y = "T_year", add = "reg.line") + # Plot
  stat_cor(label.x = 1000, label.y = 25) +
  geom_text(aes(label = City), hjust = 0, vjust = 0, check_overlap = TRUE) +
```

[1]https://portal.inmet.gov.br/normais

```
  stat_regline_equation(label.x = 900, label.y = 24.5, aes(label =
  paste(..eq.label.., ..adj.rr.label.., sep = "~~~~")),
  formula = formula) + labs(x = "Altitude (m)", y = "Temperature (°C)")
```

A digital elevation model SRTM[2] of Minas Gerais is used as a reference for altitude data to implement the regression model created, in order to estimate annual mean air temperature values in Minas Gerais according to the variation in the region's relief. The SRTM data are imported into R using the `raster` function and renamed using the `names` function.

```
srtmMG <- raster("C:/geo/SRTM/srtmMG.tif")
names(srtmMG) <- c("Altitude")
srtmMG
# class      : RasterLayer
# dimensions : 1043, 1343, 1400749  (nrow, ncol, ncell)
# resolution : 0.008333333, 0.008333333  (x, y)
# extent     : -51.05, -39.85833, -22.925, -14.23333  (xmin, xmax, ymin, ymax)
# crs        : +proj=longlat +datum=WGS84 +no_defs
# source     : srtmMG.tif
# names      : Altitude
# values     : 63, 2599  (min, max)
```

The regression equation is used to estimate temperature data based on SRTM altitude values using the R calculator.

```
mgT <- 25 - 0.0054 * srtmMG$Altitude
mgT
# class      : RasterLayer
# dimensions : 1043, 1343, 1400749  (nrow, ncol, ncell)
# resolution : 0.008333333, 0.008333333  (x, y)
# extent     : -51.05, -39.85833, -22.925, -14.23333  (xmin, xmax, ymin, ymax)
# crs        : +proj=longlat +datum=WGS84 +no_defs
# source     : memory
# names      : Altitude
# values     : 10.9654, 24.6598  (min, max)
```

The results are mapped using the `ggscatter`, `ggR`, `arrangeGrob`, and `grid.arrange` functions (Figure 9.4).

```
m1 <- reg
m2 <- ggR(srtmMG, geom_raster = TRUE) +
 scale_fill_gradientn(colours = rainbow(100), name = "Altitude (m)",
 breaks = c(500, 1500, 2500)) + theme(legend.position = "bottom")
m3 <- ggR(mgT, geom_raster = TRUE) +
  scale_fill_gradientn(colours = rev(rainbow(100)),
name = expression(paste("Temperature (",degree~C,")"))) +
```

[2]https://www.earthdata.nasa.gov/data/instruments/srtm

```
  theme(legend.position="bottom")
bottom_row <- arrangeGrob(m2, m3, ncol = 2)
grid.arrange(m1, bottom_row, ncol = 1)
```

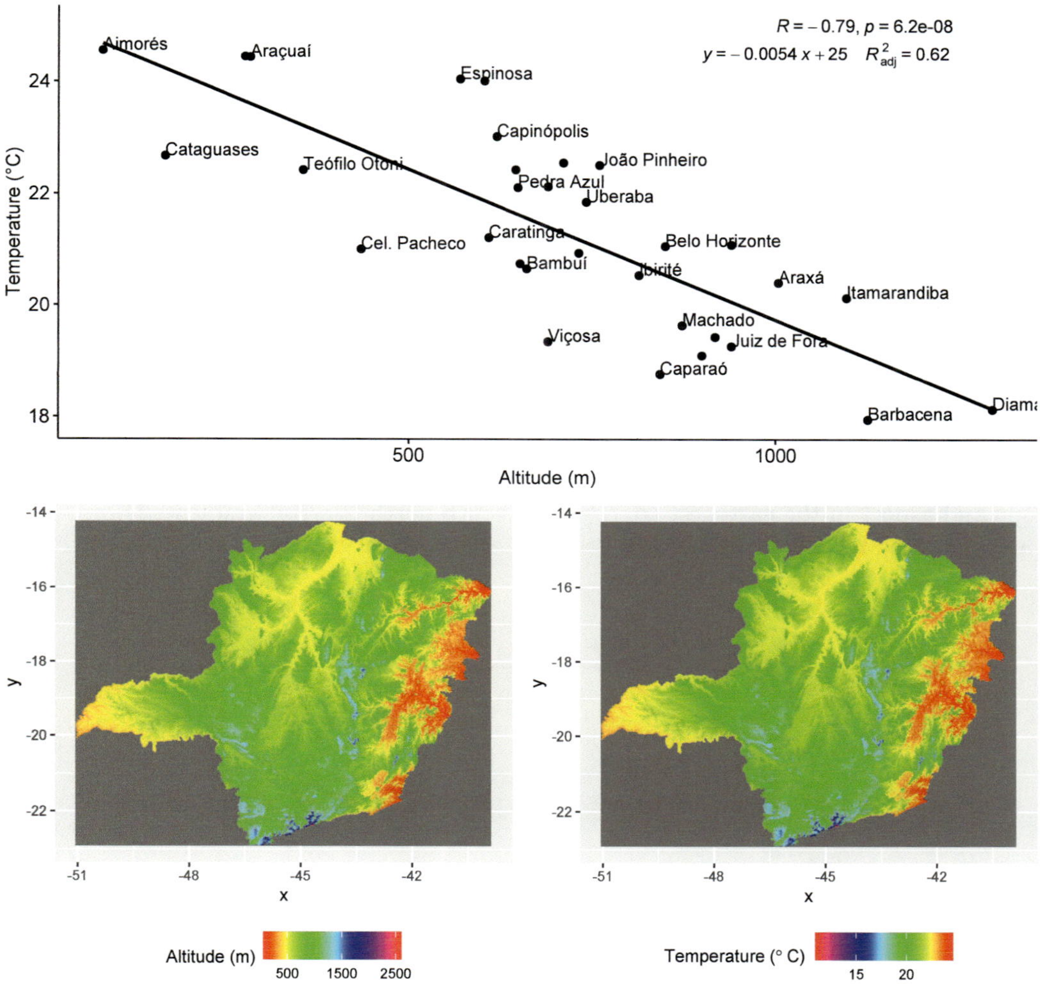

FIGURE 9.4 Use of spatial analysis with linear regression model (top) to determine the variation of climatological normals of mean air temperature in Minas Gerais as a function of altitude of the data collection station (bottom).

Similar studies can be conducted to determine the relationship between heavy metal pollution from mining dam rupture in relation to distance from a river or floodplain.

9.7.1 OBTAINING MORPHOMETRIC TERRAIN VARIABLES

The results of first-order derivatives of an elevation raster are the slope and aspect of the terrain. The second-order derivatives of the raster are the convex and plane curvatures of the profile. Slope can be defined by the plane tangent to the surface of the digital elevation model at any point, comprised of the gradient in relation to the maximum altitude, and aspect, defined by the direction at a horizontal angle, measured clockwise, in the direction of maximum slope change. The aspect may also be called the terrain exposure (Burrough and McDonnell, 1998).

The derivatives of the hypsometric curve are calculated cell-by-cell in a 3x3 window that has moved successively over the raster. The slope is calculated by the second order finite difference method adjusted to the four nearest neighbors of the moving window (Burrough and McDonnell, 1998):

$$tanS = [(\partial z/\partial x)^2 + (\partial z/\partial y)^2]^{0.5} \tag{9.6}$$

where z is the altitude, and x, y are the coordinates of the x and y axes, respectively.

The aspect is determined by (Burrough and McDonnell, 1998):

$$tanA = -(\partial z/\partial y)/(\partial z/\partial x) \qquad (-p < A < p) \tag{9.7}$$

A third-order finite difference estimator is used to compute the derivatives used to calculate slope and aspect in the east-west ($\partial z/\partial x$) and south-north ($\partial z/\partial y$) gradients (Horn, 1981; Burrough and McDonnell, 1998):

$$[\partial z/\partial x] = \frac{[(Z_{i+1,j+1} + 2z_{i+1,j} + z_{i+1,j-1}) - (Z_{i+1,j-1} + 2z_{i,j-1} + z_{i-1,j-1})]}{8\partial x} \tag{9.8}$$

$$[\partial z/\partial y] = \frac{[(Z_{i+1,j+1} + 2z_{i,j+1} + z_{i-1,j+1}) - (Z_{i+1,j-1} + 2z_{i,j-1} + z_{i-1,j-1})]}{8\partial x} \tag{9.9}$$

These same attributes are calculated as (Zevenbergen and Thorne, 1987) (Figure 9.5):

$$A = [(Z1 + Z3 + Z7 + Z9)/4 - (Z2 + Z4 + Z6 + Z8)/2 + Z5]/d^4 \tag{9.10}$$

$$B = [(Z1 + Z3 - Z7 - Z9)/4 - (Z2 - Z8)/2]/d^3 \tag{9.11}$$

$$C = [(-Z1 + Z3 - Z7 + Z9)/4 + (Z4 - Z6)/2]/d^3 \tag{9.12}$$

$$D = [(Z4 + Z6)/2 - Z5)]/d^2 \tag{9.13}$$

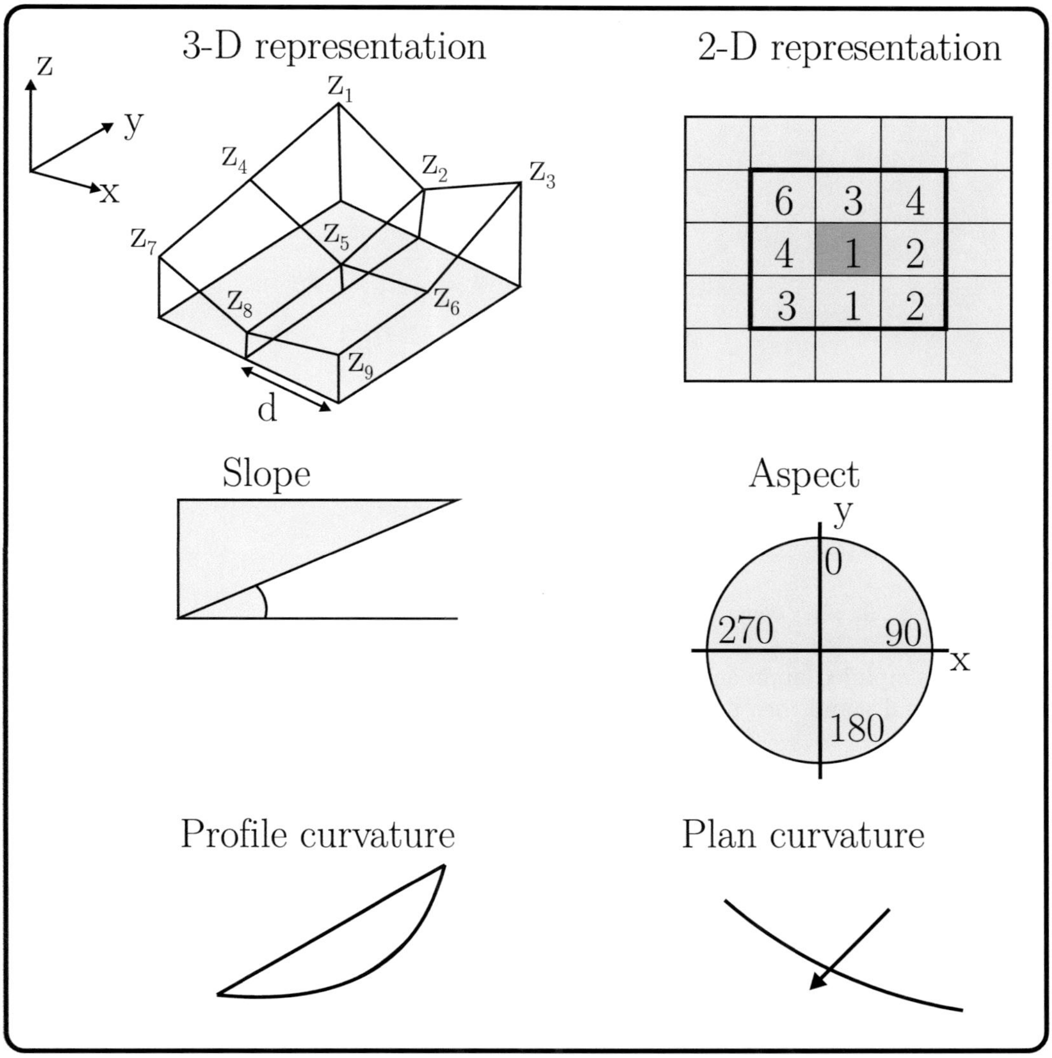

FIGURE 9.5 Determination of morphometric terrain variables.

$$E = [(Z2 + Z8)/2 - Z5)]/d^2 \tag{9.14}$$

$$F = [(-Z1 + Z3 + Z7 - Z9)/4d^2 \tag{9.15}$$

$$G = [(-Z4 + Z6)/2d \tag{9.16}$$

$$H = [(Z2 - Z8)/2d \tag{9.17}$$

$$I = Z5 \tag{9.18}$$

$$Slope = sqrt(G^2 + H^2) \tag{9.19}$$

$$Aspecto = arctan(-H/-G) \tag{9.20}$$

$$C_{Perfil} = 2(DG^2 + EH^2 + FGH)/(G^2 + H^2) \tag{9.21}$$

$$C_{Plano} = -2(DH^2 + EG^2 - FGH)/(G^2 + H^2) \tag{9.22}$$

In a more simplified approach, the equations are summarized in determining two initial variables x and y and then determining the slope tangent by (Longley et al., 2001):

$$tan(slope) = \sqrt{y^2 + x^2} \tag{9.23}$$

The slope in degrees is obtained by:

$$slope = \frac{1}{tan(slope)} \tag{9.24}$$

The slope in percent is obtained by the tangent of the slope in degrees multiplied by 100:

$$slopep = tan(slope) * 100 \tag{9.25}$$

The aspect, referring to the variation of the horizontal angle, measured clockwise, from zero to the direction of the greatest slope is obtained by (Longley et al., 2001):

$$tan(aspect) = \frac{b}{c} \tag{9.26}$$

$$aspect = tan(aspect) \pm C \tag{9.27}$$

where C is 0, if $x>0$, $y>0$; 180, if $y<0$; and 360, if $x<0$ and $y>0$.

The values of x and y are determined relative to the 8-neighbor raster, per (Longley et al., 2001):

$$y = \frac{(z_4 + 2z_2 + z_2 - z_6 - 2z_4 - z_3)}{8d} \tag{9.28}$$

$$x = \frac{(z_6 + 2z_3 + z_4 - z_3 - 2z_1 - z_2)}{8d} \tag{9.29}$$

where d is the grid distance and z is the raster brightness value, with cells on a regular grid with altitude values across the terrain.

Alternative methods can be used to fit a regression model to the nine altitude values in the 3x3 window and derive the slope and aspect from there (Burrough and McDonnell, 1998).

SRTM digital elevation model data is used to remove sinks to enhance the detection of drainage routes in a watershed area of Mato Grosso, Brazil. Morphometric and hydrological variables are then determined in the evaluated region of the basin. The morphometric variables determined by geospatial analysis and mapped are: altitude (m), slope (slope, degrees) and aspect (aspect, degrees). The hydrological variable determined is the topographic wetness index (TWI) (Figure 9.6). The analyses are performed in grid format file (`.sgrd`) of SAGA GIS (Conrad et al., 2015).

It is recommended to use a spatial resolution of 30 m for the raster cell to calculate morphometric and hydrological variables of the environment, such as in the study of watersheds, to obtain more accurate results (Chang et al., 2019). Cells with a more detailed spatial resolution than this can determine erroneous features in the landscape, such as rivers where there are no rivers.

The `RSAGA` package (Brenning et al., 2018a) is used for the analyses; however, another more current option is to use the `Rsagacmd` package (Pawley, 2024). In both cases, SAGA[3] must be installed on the computer so that the command line version of `SAGA` can be used in R as third-party software. `RSAGA` is loaded into R with the `library` function.

```
library(RSAGA)
# Loading required packages: gstat
# Loading required packages: plyr
# Loading required packages: shapefiles
# Loading required packages: foreign
# package 'RSAGA' was built under R version 4.3.3
```

The `SRTM` altitude data at a spatial resolution of 30 m is imported into R in geotiff format using the raster function.

[3]https://saga-gis.sourceforge.io/en/index.html

```
srtm30 <- raster("C:/Projetos/Uniselva/raster/dem/srtm30.tif")
srtm30
# class       : RasterLayer
# dimensions : 7952, 8884, 70645568  (nrow, ncol, ncell)
# resolution : 30, 30  (x, y)
# extent      : 321451, 587971, 8200370, 8438930  (xmin, xmax, ymin, ymax)
# crs         : +proj=utm +zone=21 +south +ellps=GRS80 +units=m +no_defs
# source      : srtm30.tif
# names       : srtm30
```

The `writeRaster` function is used to export `SRTM` data in `SAGA` format. The `NAflag` argument is configured so that the data remains masked in cells with `NA` values. This creates files with the extensions `.sdat` and `.sgrd` in the directory of interest.

```
writeRaster(srtm30, "C:/uniselva/raster/dem/srtm30/srtm30.sdat",
            format = "SAGA", NAflag = -9999, overwrite = TRUE)
```

Sinks are filled using the `rsaga.sink.removal` function.

```
rsaga.sink.removal(in.dem = "C:/Projetos/Uniselva/raster/dem/srtm30/srtm30.sgrd",
 out.dem = "C:/Projetos/Uniselva/raster/dem/srtm30/srtm30fill.sgrd",
 method = 1, env = env)
```

The slope and aspect in degrees (Zevenbergen and Thorne, 1987) are determined using the `rsaga.slope.asp.curv` function. Other terrain morphometric variables can also be determined using this function.

```
rsaga.slope.asp.curv(
  in.dem = "C:/Projetos/Uniselva/raster/dem/srtm30/srtm30fill.sgrd",
  out.slope = "C:/Projetos/Uniselva/raster/dem/srtm30/slope30.sgrd",
  out.aspect = "C:/Projetos/Uniselva/raster/dem/srtm30/aspect30.sgrd",
  method = "poly2zevenbergen", unit.slope = "degrees", unit.aspect = "degrees",
  env = env)
```

The topographic wetness index is determined using the `rsaga.wetness.index` function.

```
rsaga.wetness.index(in.dem = "C:/Projetos/Uniselva/raster/dem/srtm30/srtm30.sgrd",
out.wetness.index = "C:/Projetos/Uniselva/raster/dem/srtm30/twi30.sgrd")
```

The filled digital elevation model, slope, aspect, and calculated topographic wetness index are imported back into R using the `raster` function.

```
dem_fill <- raster("C:/Projetos/Uniselva/raster/dem/srtm30/srtm30fill.sdat")
crs(dem_fill) <- " +proj=utm +zone=21 +south +ellps=GRS80 +units=m +no_defs"
slope <- raster("C:/Projetos/Uniselva/raster/dem/srtm30/slope30.sdat")
crs(slope) <- " +proj=utm +zone=21 +south +ellps=GRS80 +units=m +no_defs"
aspect <- raster("C:/Projetos/Uniselva/raster/dem/srtm30/aspect30.sdat")
```

```
crs(aspect) <- " +proj=utm +zone=21 +south +ellps=GRS80 +units=m +no_defs"
swi <- raster("C:/Projetos/Uniselva/raster/dem/srtm30/twi30.sdat")
crs(swi) <- " +proj=utm +zone=21 +south +ellps=GRS80 +units=m +no_defs"
```

The calculated data is stacked into a single file using the `stack` function.

```
mterrain <- stack(dem_fill, slope, aspect, swi)
mterrain
# class      : RasterStack
# dimensions : 7952, 8884, 70645568, 4  (nrow, ncol, ncell, nlayers)
# resolution : 30, 30  (x, y)
# extent     : 321451, 587971, 8200370, 8438930  (xmin, xmax, ymin, ymax)
# crs        : +proj=utm +zone=21 +south +ellps=GRS80 +units=m +no_defs
# names      : srtm30fill, slope30, aspect30, twi30
```

The results are mapped using the `ggR` and `grid.arrange` functions.

```
m1 <- ggR(mterrain$srtm30fill, geom_raster = TRUE) +
   scale_fill_gradientn(colours = rainbow(100), name = "Altitude (m)")
m2 <- ggR(mterrain$slope30, geom_raster = TRUE) +
   scale_fill_gradientn(colours = rainbow(100), name =
                        expression(paste("Slope (",~degree,")")))
m3 <- ggR(mterrain$aspect30, geom_raster = TRUE) +
   scale_fill_gradientn(colours = rev(rainbow(100)),
                    name = expression(paste("Aspect (",~degree,")")))
m4 <- ggR(mterrain$twi30, geom_raster = TRUE) +
   scale_fill_gradientn(colours = rev(rainbow(100)), name = "TWI")
grid.arrange(m1, m2, m3, m4, ncol = 2)
```

In general, geomorphologic response times are longer than hydrologic system response times. Consequently, the morphology recorded on topographic maps may be more a reflection of past hydrologic regimes than a product of the interdependence of present-day water flows. Water catchment morphology and hydrologic processes are inextricably linked through the geomorphic development of a basin. However, momentary hydrologic response can vary as a function of spatial heterogeneity leading to complex patterns of response. The morphology of the basin recharge area can be used to simulate the dynamics of the contributing areas and the effect of the channel network on the runoff path in the watershed (Beven et al., 1988).

9.7.2 OBTAINING SURFACE TOPOLOGY AND DRAINAGE NETWORK

Drainage network, flow, recharge areas, watershed, and ridges are landscape features that have contributed to the understanding of material flow. This type of variable can be constructed from an elevation raster. Automated ways of creating drainage networks are useful tools for hydrologists to estimate the flow of water and sediment through the terrain and linked to dynamic hydrological processes in geographic information systems.

The flow of material on a grid surface is determined by considering the steepest downward direction on the terrain. Several algorithms are used in this calculation, such as the deterministic $D8$ algorithm. In this case, the approximate direction of flow is determined

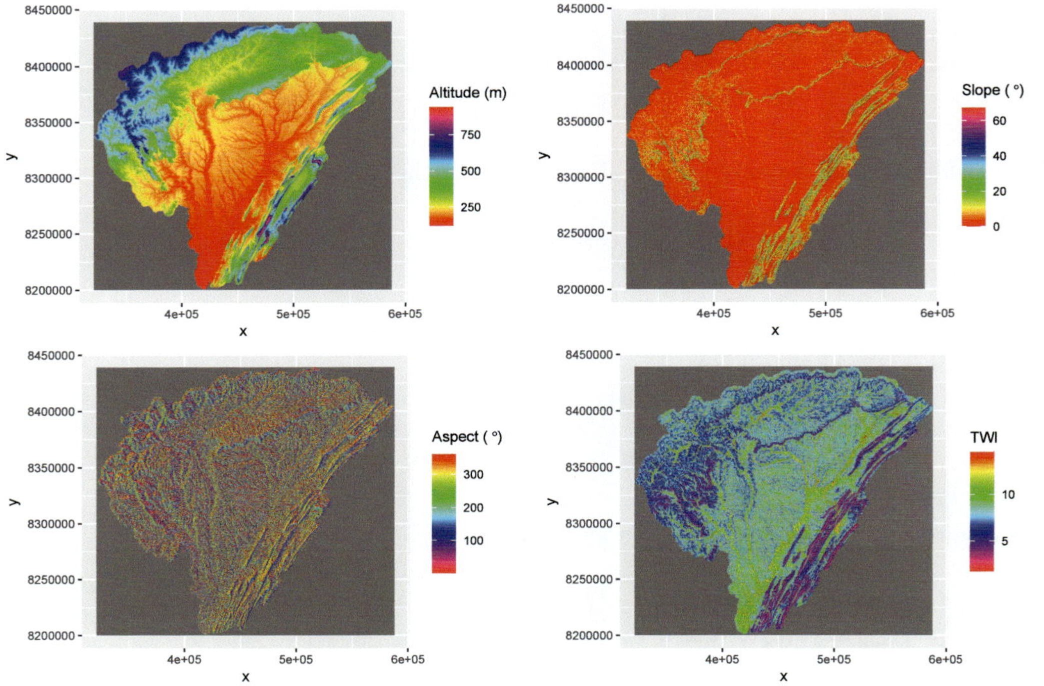

FIGURE 9.6 Geospatial analysis and mapping of morphometric (altitude, slope and aspect) and hydrological (TWI) variables determined in the upper-middle and upper Paraguay planning and management units, Mato Grosso, Brazil.

using a 3x3 moving window. The $Rho8$ algorithm is a stochastic version of $D8$ to represent aspects of the terrain. The $FD8$ and $FRho8$ algorithms are modifications of the original algorithm to model flow dispersion and recharge dispersion (Burrough and McDonnell, 1998).

Sinks can be discrete relief depressions or just artifacts generated with digital elevation model processing. In elevation raster modeling with sinkholes, there can be disruption of drainage topology, so removing these objects from the raster by spatial analysis is a way to obtain a terrain network with better continuous structure (Figure 9.7) (Burrough and McDonnell, 1998). Several methods for filling depressions in digital elevation model are developed to improve hydrologic modeling results. More details on the methodology can be obtained from Planchon and Darboux (2002) and Wang and Liu (2006).

A mass balance can be determined for each cell according to (Burrough and McDonnell, 1998):

$$S = P - I - F - E \tag{9.30}$$

where S is the water surplus per cell, P, the precipitation, I, the interception, F, the infiltration, and, E, the evaporation. The accumulated flow in a drainage network on the terrain is obtained by S in connected cells.

A drainage network map with the flow direction from upstream to downstream (outflow) can be useful for calculating other terrain indices, such as the topographic wetness index

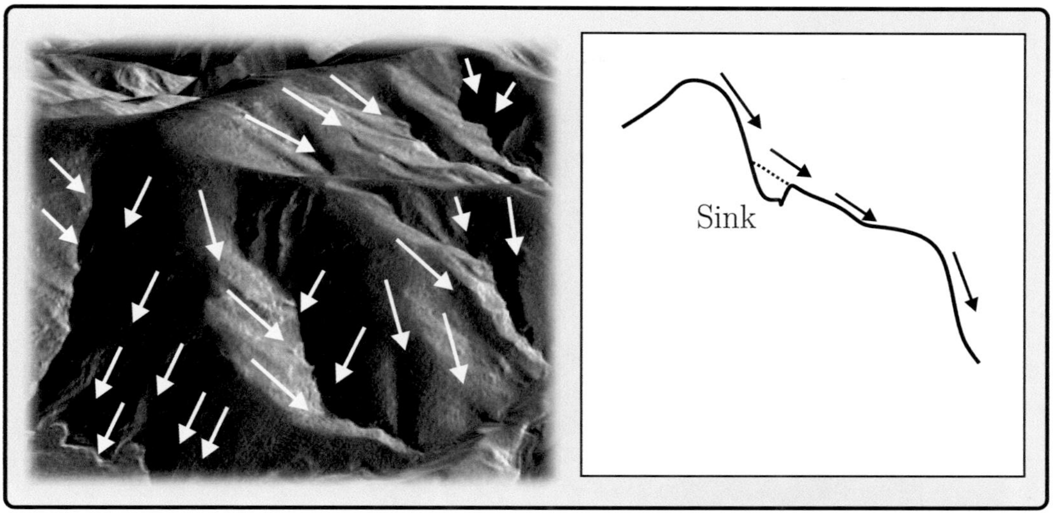

FIGURE 9.7 Drainage direction indicated by arrows in the direction of the steepest slope and the presence of sinks in the digital elevation model removed by spatial analysis method.

(Burrough and McDonnell, 1998):

$$wetness\ index = ln(\frac{A_s}{tan\beta}) \tag{9.31}$$

where A_s is the recharge area in m^2 and β, the slope in degrees.

The sediment transport index is defined by (Burrough and McDonnell, 1998):

$$\tau = (\frac{A_s}{22.13})^{0.6}(\frac{sen\beta}{0.0896})^{1.3} \tag{9.32}$$

This index is used to characterize the process of erosion and deposition and the effect of varying land topography on soil loss.

9.7.3 HILL SHADING, IRRADIANCE AND INTERVISIBLE POINTS

9.7.3.1 Hillshading

The use of hillshade maps can be useful for visualizing the three-dimensional appearance of the terrain. By using this type of information, determined as a function of slope and terrain aspect, it is possible to increase the realism of the final map obtained with image overlay and use of raster layer transparency functions (Figure 9.8).

The `hillShade` function is used to determine the shaded relief of the Mato Grosso watershed used in the previous example.

The results are mapped using transparency and layer overlay with the `ggR` and `scale_fill_gradientn` functions.

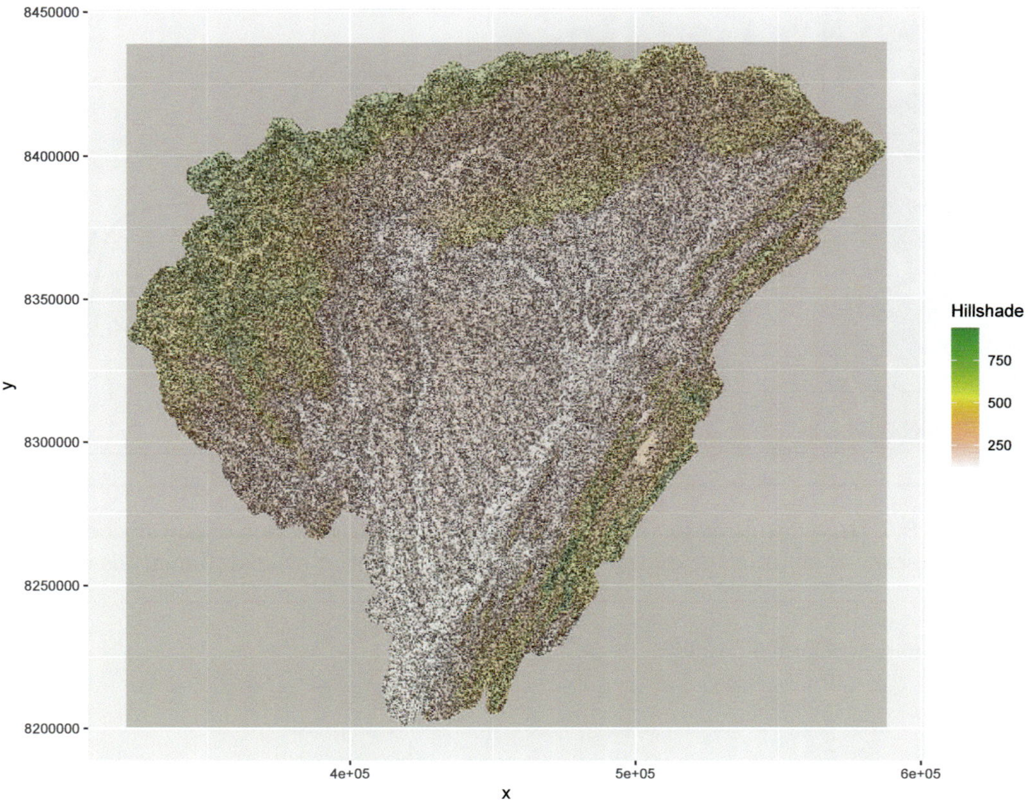

FIGURE 9.8 Hillshade map overlayed on sink corrected altitude layer under transparency and color gradient as application of the multiconcept used in remote sensing.

The region with green coloration in the image is related to the higher ground, while lighter shades define lower ground and complex drainage patterns. Dichotomous drainage patterns are observed in most of the watershed. In the northern part of the basin, a dentritic drainage pattern is observed. Tangará da Serra is located in a region with light green tones in the central-northern portion of the watershed, defining two large drainage areas in a region of higher altitude, transversal in the west direction, defining distinct drainage regions in the southern and northern portions of the ridge.

9.7.3.2 Solar irradiance

The irradiance map is, in principle, an extension of the hill shading fundamentals used to determine the amount of solar energy reaching the land surface. In this case, it is necessary to consider the solar position according to latitude and time of day throughout the year. In addition, it is necessary to include the atmospheric light attenuation effect in the model, as well as the shading of the terrain as a function of terrain variation (Burrough and McDonnell, 1998).

For an atmospheric transmittance, T_0, the direct irradiance I on a slope is obtained by (Burrough and McDonnell, 1998) (Figure 9.9):

$$I = cosiS_0exp(\frac{-T_o}{cos\theta_0}) = [cos\theta_0cos\beta + sin\theta_0sin\beta cos(\phi_0 - A)]S_0exp(\frac{-T_0}{cos\theta_0}) \quad (9.33)$$

where $cosi$ is the cosine of the solar illuminance on the slope, S_0, the angle of the solar extraterrestrial flux, θ_0, the solar zenith angle, ϕ_0, solar azimuth, A, azimuth of the slope, and β, the angle of the slope. As β and A are obtained from the digital elevation model data, the spatial variation of the component in the above equation is the solar radiation incident on an area, defined as solar irradiance (Wm^{-2}). Since $cosi$ varies with day and year, it is possible, for example, to determine the spatio-temporal variation of irradiance over a day on a sunlit terrain (Burrough and McDonnell, 1998).

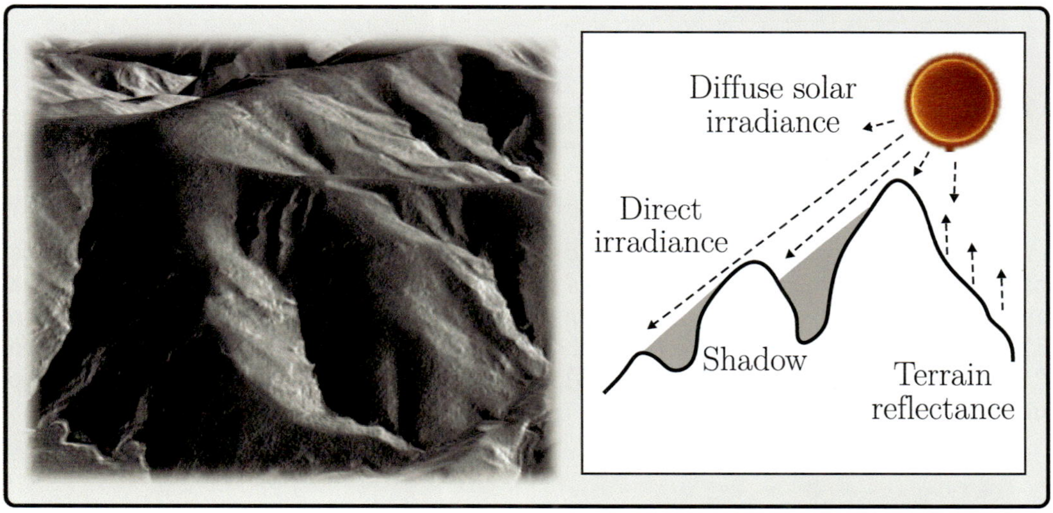

FIGURE 9.9 Determining solar irradiance and shading on sloping terrain.

9.7.3.3 Line of sight on maps

New attributes can be obtained from the three-dimensional knowledge of the relief on a continuous surface. The method used for line of sight calculation is a simple operation to determine places in the landscape that can be viewed from a given point. Intervisibility can be coded as a binary variable, where 0 is invisible and 1, visible. A raster map with values 1 is called a `viewshed`. For large distance situations, it may be necessary to include the curvature of the Earth in the analysis, as well as to consider the transparency of the atmosphere in the analysis (Burrough and McDonnell, 1998).

Determining intervisibility can be complicated to perform by *in situ* survey due to the large number of profiles that must be extracted and compared (Burrough and McDonnell, 1998). Intervisualization of landscapes using `viewshed` techniques associated with photographs taken of the landscape at specific locations *in situ* may be promising and cost-effective for environmental planning and management. The application of `viewshed` helped conceptualize visual aspects of landscapes in a region with lakes near an urban area in Van, eastern Turkey. The visibility information is useful for increasing the efficiency of development plans for urban areas around the city of Van (Alphan and Aşur, 2021).

9.8 OBTAINING SYNTHESIS VARIABLES WITH SPATIAL ANALYSIS

One of the major advantages of raster representation is to record each attribute on a different layer so that any mathematical operation can be used on one or more attributes of the same cell with ease. Thus, any mathematical operation used on simple numbers can also be applied to raster data. Map algebra is used to apply algebraic techniques to build models for geospatial analysis in cartography. A spatial analysis operation with raster is performed choosing layers and operators, as in this example used to generate sum (M_S) and average ($\bar{M_S}$) output synthesis maps from the input raster datasets $Map_i, Map_{i1}, ..., Map_n$, under sum and mean functions, respectively:

$$M_S = Map_i + Map_{i1} + ... + Map_n \tag{9.34}$$

$$\bar{M_S} = \sum_{i=1}^{n} \frac{Map_i}{n} \tag{9.35}$$

As an example, we can consider the need to determine the agricultural suitability of Arabica coffee cultivation on family farms. In this case, agricultural suitability is determined as a function of nutrient availability (N), oxygen (O), water (W), and susceptibility to soil erosion (E). In addition, rainfall erosivity (R), soil erodibility (K), cropping (C) and soil protection (P) parameters, slope length (L), slope (S), and soil class are considered a function of crop suitability (Figure 9.10):

$$Suitability = f(N, O, W, E) \tag{9.36}$$

$$Erosion_{(USLE)} = f(RKLSCP) \tag{9.37}$$

Soil loss by erosion is determined, in this example, by the universal soil loss equation (USLE) model. Further details of model application can be obtained from Smith and Wischmeier (1957), Wischmeier and Smith (1978), and Burrough and McDonnell (1998).

With the complex nature of modeling and error propagation, it is necessary to define practical ways to extend the ways of solving geospatial problems, linked to the design and methodology of data analysis. The sensitivity analysis approach is used to calibrate spatial analysis results to evaluate new data or ways of preparing input variables, model parameters, and selection criteria to generate results and evaluate how adjusting procedures changed the final outcome of the generated product (Lo and Yeung, 2007) (Figure 9.11). Therefore, sensitivity analysis is considered a dynamic modeling technique to deal with the subjectivity and variability of parameters to solve geospatial problems. Sensitivity analysis is useful for assessing uncertainties about the modeling performed and will be covered in more detail later.

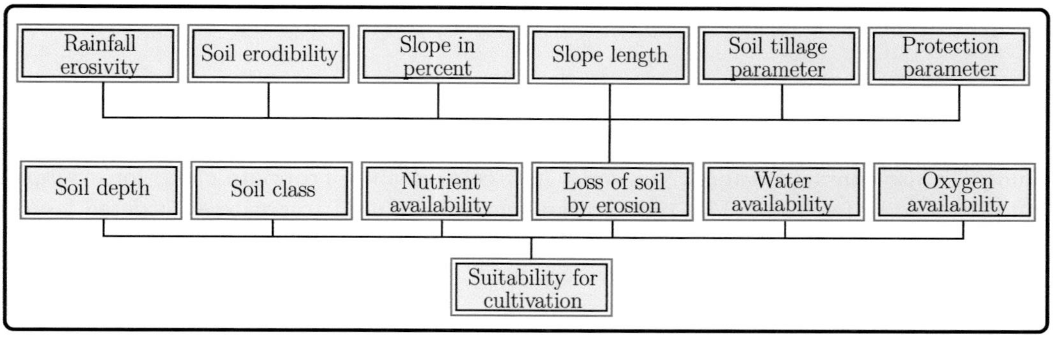

FIGURE 9.10 Determination of the crop suitability map as a function of land use, cover, class, depth, nutrients, water availability, soil oxygen, and rainfall erosion characteristics.

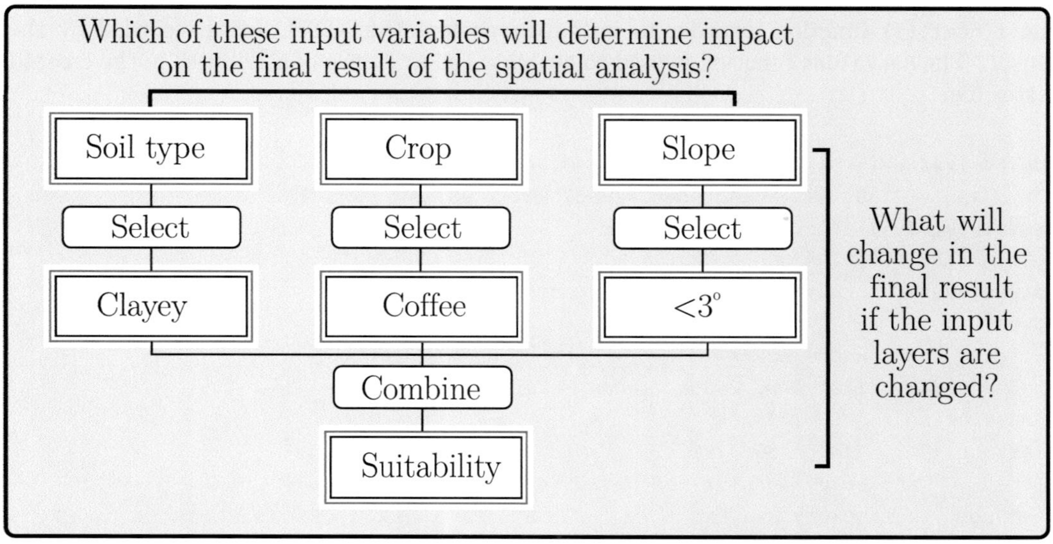

FIGURE 9.11 Sensitivity analysis to obtain a synthesis variable with satisfactory meaning.

9.9 COMPUTATIONAL PRACTICE

The `install.packages` function is used to install the packages needed to perform the computational practice activities.

```
install.packages("raster")
install.packages("dplyr")
```

The `library` function is used to enable the `raster` and `dplyr` packages needed to perform the tasks.

```
library(raster)
library(dplyr)
```

Raster can be created from scratch, considering their unique structure, whereas subset operations and other geospatial analysis operations are performed differently in R (Lovelace et al., 2019).

9.9.1 RASTER ATTRIBUTE MANIPULATION

A raster layer can contain numeric, integer, logical, or factor class values, but not character values. To use character values, you must first convert an appropriate class, for example using the `factor` function. The `levels` argument can be used to create an ordered factor with different variables (Lovelace et al., 2019), such as the coffee canopy morphometry variables that varied according to the average plant clumping percentage via a visual rating scale.

The `raster` function is used to create a raster dataset to represent a variable, such as the normalized difference vegetation index (NDVI) obtained by the GreenSeeker sensor at 30 cm from the coffee plant canopy at 1.5 m spatial resolution. A vector of random numbers is assigned with a length equal to the number of cells in the raster layer with the `runif(ncell())` function in order to initially represent the NDVI signal recorded by the sensor. The `hasValues` function is used to evaluate the existence of attributes in the created raster file.

```
ndvi <- raster(nrows = 120, ncols = 240, res = 1.5,
      xmn = -180, xmx = 180, ymn = -90, ymx = 90) # Create the raster
hasValues(ndvi) #Evaluate attributes
# [1] FALSE
values(ndvi) <- runif(ncell(ndvi)) #Create attributes
ndvi
# class      : RasterLayer
# dimensions : 120, 240, 28800  (nrow, ncol, ncell)
# resolution : 1.5, 1.5  (x, y)
# extent     : -180, 180, -90, 90  (xmin, xmax, ymin, ymax)
# crs        : +proj=longlat +datum=WGS84 +ellps=WGS84 +towgs84=0,0,0
# source     : memory
# names      : layer
# values     : 7.857336e-06, 0.9999862  (min, max)
```

The result is a raster with 120 rows, 240 columns and 28800 cells at 1.5 by 1.5 m spatial resolution and a minimum and maximum spatial extent in the x and y directions (`xmn, xmx, ymn, ymax`). The `vals` argument can also be set to contain numerical values from 0 to 28799 if you want to constitute a numbered raster layer of the same raster created earlier.

```
r <- raster(nrows = 120, ncols = 240, res = 1.5, xmn = -180, xmx = 180, ymn = -90,
            ymx = 90, vals = 0:28799)
r
# class      : RasterLayer
# dimensions : 120, 240, 28800  (nrow, ncol, ncell)
# resolution : 1.5, 1.5  (x, y)
# extent     : -180, 180, -90, 90  (xmin, xmax, ymin, ymax)
# crs        : +proj=longlat +datum=WGS84 +ellps=WGS84 +towgs84=0,0,0
# source     : memory
# names      : layer
# values     : 0, 28799  (min, max)
```

Raster can also contain categorical values of logical variables of class or factor in R. For example, you can create a raster to represent categories with levels of tree foliage in: Very High, High, Medium, Low, and Very Low. The `vals` argument defines values of each cell, totaling from 1 to 28800 cells.

```
leafCoffee <- c("Very High", "High", "Medium", "Low", "Very Low")
leafChar <- sample(leafCoffee, 28800, replace = TRUE)
leafFact <- factor(leafChar, levels = leafCoffee)
coffee <- raster(nrows = 120, ncols = 240, res = 1.5,
     xmn = -180, xmx = 180, ymn = -90, ymx = 90, vals = leafFact)
coffee
# class      : RasterLayer
# dimensions : 120, 240, 28800  (nrow, ncol, ncell)
# resolution : 1.5, 1.5  (x, y)
# extent     : -180, 180, -90, 90  (xmin, xmax, ymin, ymax)
# crs        : +proj=longlat +datum=WGS84 +ellps=WGS84 +towgs84=0,0,0
# source     : memory
# names      : layer
# values     : 1, 5  (min, max)
# attributes :
# ID      VALUE
#  1 Very High
#  2      High
#  3    Medium
#  4       Low
#  5  Very Low
```

The results are mapped using the `ggR`, `scale_fill_gradientn`, and `grid.arrange` functions (Figure 9.12).

```
m1 <- ggR(ndvi$layer, geom_raster = TRUE) +
  scale_fill_gradientn(colours = rainbow(100), name = "NDVI")
m2 <- ggR(coffee, geom_raster = TRUE)
grid.arrange(m1, m2, ncol = 1)
```

In the raster created above, categorical variables are represented as integer identifiers, so the `coffee[5, 5]` code chunk returns a number that represents a unique identifier, but not the name of the identifier. A table of attributes is stored in the raster in a slot called `attributes`, which can be viewed with the `ratify` function.

```
coffee[5, 5]
# [1] 5
ratify(coffee)
# class      : RasterLayer
# dimensions : 120, 240, 28800  (nrow, ncol, ncell)
# resolution : 1.5, 1.5  (x, y)
# extent     : -180, 180, -90, 90  (xmin, xmax, ymin, ymax)
# crs        : +proj=longlat +datum=WGS84 +ellps=WGS84 +towgs84=0,0,0
# source     : memory
```

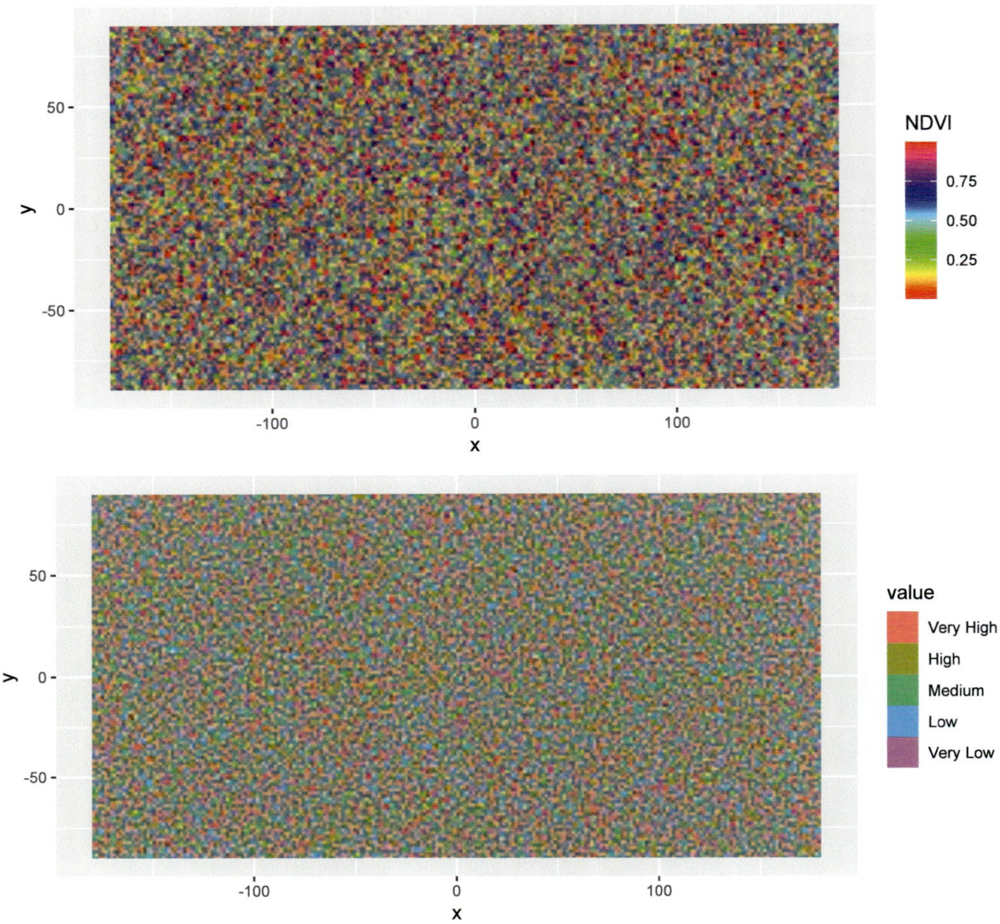

FIGURE 9.12 Rasters with numerical (top) and categorical (bottom) values.

```
# names      : layer
# values     : 1, 5  (min, max)
# attributes :
# ID
#  1
#  2
#  3
#  4
#  5
```

The `levels` function is used to add new factor levels to the attribute table, for example, soil texture levels.

```
levels(coffee)[[1]] <- cbind(levels(coffee)[[1]],
  texture = c("Very clayey", "Clayey", "Medium", "Sandy", "Silty"))
levels(coffee)
```

```
# [[1]]
#   ID      VALUE      texture
# 1  1 Very High Very clayey
# 2  2      High      Clayey
# 3  3    Medium      Medium
# 4  4       Low       Sandy
# 5  5  Very Low       Silty
```

Using the `factorValues` function, it is possible to query the value of the raster created with attributes on cell identifiers (IDs) 1, 10 and 100.

```
factorValues(coffee, coffee[c(1, 10, 100)])
# VALUE texture
# 1  High  Clayey
# 2  High  Clayey
# 3   Low   Sandy
```

The attribute mapping can be done with the raster name followed by the attribute with quotes, as in the code below (Figure 9.13).

```
plot(coffee, "texture")
```

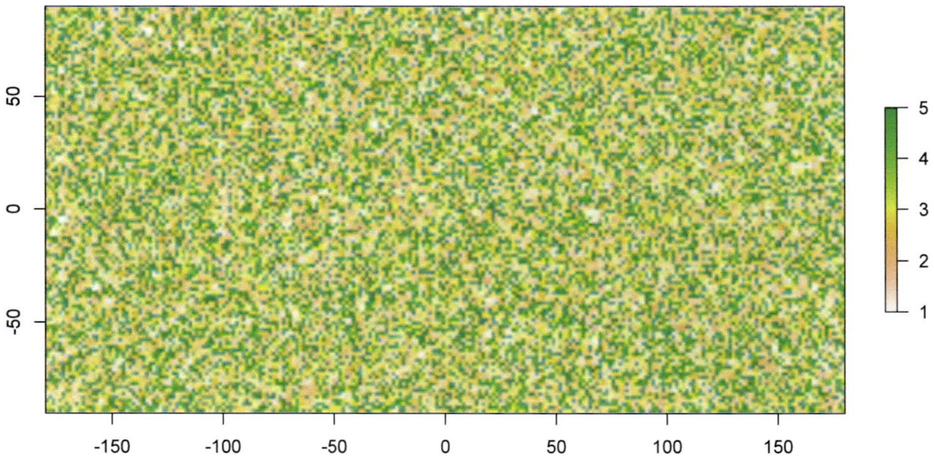

FIGURE 9.13 Mapping a raster with attributes.

9.9.1.1 Defining raster subsets

A raster subset can be made with the `[` operator, with the following types of entries (Lovelace et al., 2019):

- Row and column indexing;
- Cell IDs;
- Coordinates;

- Another raster.

The first two options can be considered non-spatial operations. Using the first two subset options, both returned the value of the upper-left cell in the `ndvi` raster.

```
ndvi[1, 1] # row 1, column 1
#[1] 0.5180495
ndvi[1] # cell ID 1
#[1] 0.5180495
```

To extract all values or complete rows, you can use the `values` and `getValues` functions. The results are rounded to one decimal place with the `round` function.

```
ndviValues <- getValues(ndvi, row = 10)
round(ndviValues, digits = 1)
# [1] 0.1 0.0 0.3 0.5 0.8 0.4 0.1 0.2 0.9 0.3 0.3 0.9 0.3 0.9 0.1 0.5 0.8 0.4
# ...
```

The `stack` function can be used to stack or make blocks of multilayered raster, with cell values in layer order. With the `names` argument, you can rename the stacked raster.

```
coffeeStack <- stack(ndvi, coffee)
names(coffeeStack) <- c("ndvi", "coffee")
coffeeStack
# class       : RasterStack
# dimensions : 120, 240, 28800, 2  (nrow, ncol, ncell, nlayers)
# resolution : 1.5, 1.5  (x, y)
# extent      : -180, 180, -90, 90  (xmin, xmax, ymin, ymax)
# crs         : +proj=longlat +datum=WGS84 +ellps=WGS84 +towgs84=0,0,0
# names       :          ndvi,          coffee
# min values : 7.857336e-06, 1.000000e+00
# max values :    0.9999862,     5.0000000
```

For multilayer raster, you can perform raster subsets with the `subset` function, used to extract layers from a stacked raster. As another option, the `[[` and `$` operators can also be used to perform subsets.

```
raster::subset(coffeeStack, "ndvi")
coffeeStack[["ndvi"]]
coffeeStack$ndvi
# class       : RasterLayer
# dimensions : 120, 240, 28800  (nrow, ncol, ncell)
# resolution : 1.5, 1.5  (x, y)
# extent      : -180, 180, -90, 90  (xmin, xmax, ymin, ymax)
# crs         : +proj=longlat +datum=WGS84 +ellps=WGS84 +towgs84=0,0,0
# source      : memory
# names       : ndvi
# values      : 7.857336e-06, 0.9999862  (min, max)
```

Cell values can be modified by overwriting existing values in conjunction with a subset operation. As an example, the upper-left cell of the `ndvi` raster is replaced with a 0.

```
ndvi[1, 1] <- 0
ndvi[1, 1]
# [1] 0
```

In the case of modifying several raster cells with the same value, we can use the following command:

```
ndvi[1,1:10] <- 0
ndvi[1:15]
# [1] 0.0000000 0.0000000 0.0000000 0.0000000 0.0000000 0.0000000 0.0000000
# 0.0000000 0.0000000 0.0000000
# [11] 0.7995349 0.7864938 0.4564505 0.6473681 0.1971507
```

9.9.1.2 Summary statistics for raster

Summary statistics can be obtained from raster. Printing a raster in the R console by typing the name returns minimum and maximum values for a raster. Using the `summary` function, common summary statistics are obtained, such as minimum, maximum, interquartile range, and number of NAs/no data. Other statistics, such as standard deviation (sd) or custom summary statistics, can be calculated with the `cellStats` function.

```
cellStats(ndvi, sd)
# [1] 0.2876624
```

Using the `summary` and `cellStats` functions on stacked raster, summary statistics are obtained for both stacked layers, with the use of the `brick` function.

```
summary(brick(ndvi, coffee))
#           layer.1 layer.2
# Min.    0.0000000       1
# 1st Qu. 0.2556412       2
# Median  0.5012948       3
# 3rd Qu. 0.7505820       4
# Max.    0.9999584       5
# NA's    0.0000000       0
```

Other functions used to obtain summary statistics are `boxplot`, `density` and `hist` (Figure 9.14).

```
boxplot(ndvi)
density(ndvi)
hist(ndvi)
```

```
par(mfrow = c(2, 2), mar = c(2.5, 2.5, 0.3, 0.3), mgp = c(1.5, 0.6, 0))
boxplot(ndvi)
density(ndvi)
hist(ndvi, main="")
```

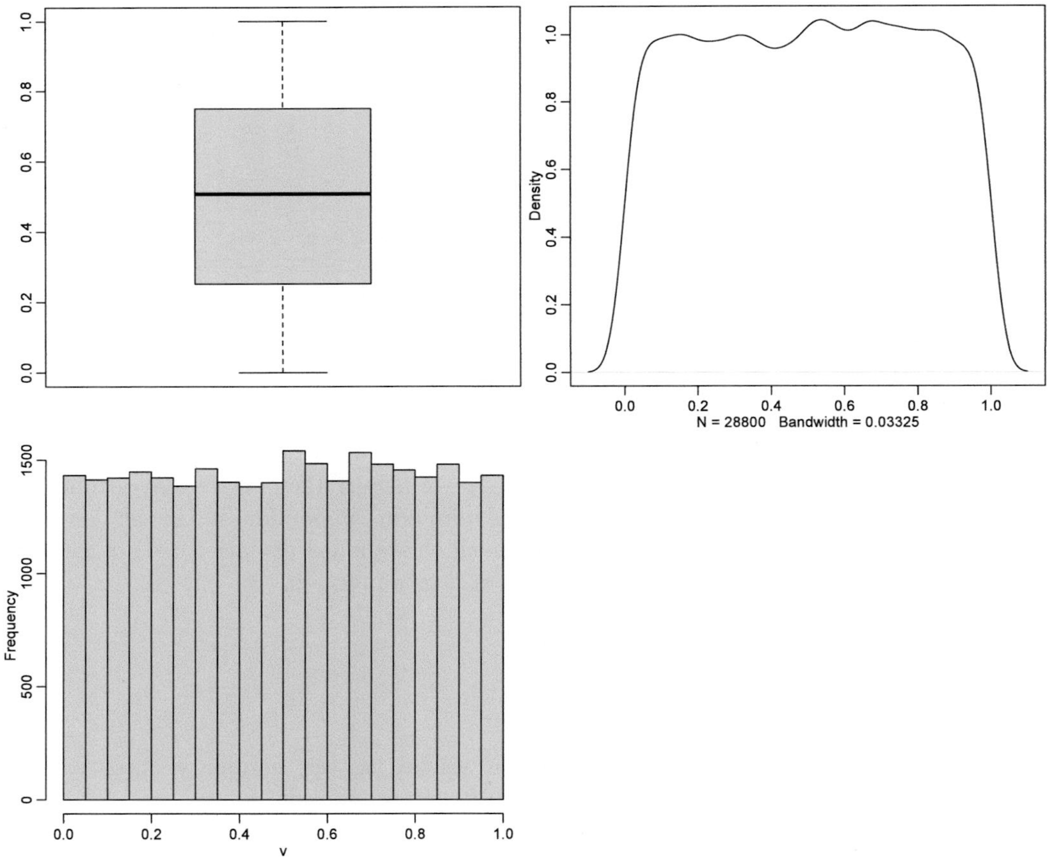

FIGURE 9.14 Boxplot (top-left), density (top-right) and histogram (bottom) summary statistics applied to raster layers.

There can be conflict between some R package function names, for example the `select` function used in the `dplyr` and `raster` packages. In this case, the `detach` function can be used to disable the unwanted package and perform the operation.

```
detach("package:raster", unload = TRUE, force = TRUE)
```

9.9.2 SPATIAL OPERATION ON A RASTER

Spatial operations also apply to raster. The spatial subset of raster can be used, as well as stacking raster blocks into a single raster. The most important spatial operation may be map algebra for many applications. Map algebra is also a prerequisite for raster distance calculations. It should be noted that in spatial operations two raster layers must have the same coordinate reference system (Lovelace et al., 2019).

9.9.2.1 Spatial subset

A spatial subset can be obtained by specifying the location of the x,y coordinates of the cell. Then, the cell value can be obtained by the `cellFromXY` and `extract` functions.

```
id <- cellFromXY(ndvi, xy = c(0.1, 0.1))
ndvi[id]
# [1] 0.8965427
raster::extract(ndvi, data.frame(x = 0.1, y = 0.1))
# [1] 0.8965427
```

After mapping the `ndvi` raster, the values of an enclosing rectangle in the image are obtained to make a subset in the raster. The drawing of the rectangle is performed interactively by defining two points on the image with the left mouse button, one in the upper-left corner of the rectangle, and the other in the lower-right corner, defining the geometric figure at the size of interest. The interactive drawing process is initiated with the `drawExtent` function (Figure 9.15).

```
plot(ndvi)
drawExtent() # Interactive drawing on the raster
# xmin <- -39.87603
# xmax <- 35.02066
# ymin <- -19.93802
# ymax <- 19.31818
# or manual with known coordinates,
e <- extent(-39.87603, 35.02066, -19.93802, 19.31818)
plot(ndvi)
plot(e, col = "red", lwd = 2, add= TRUE)
```

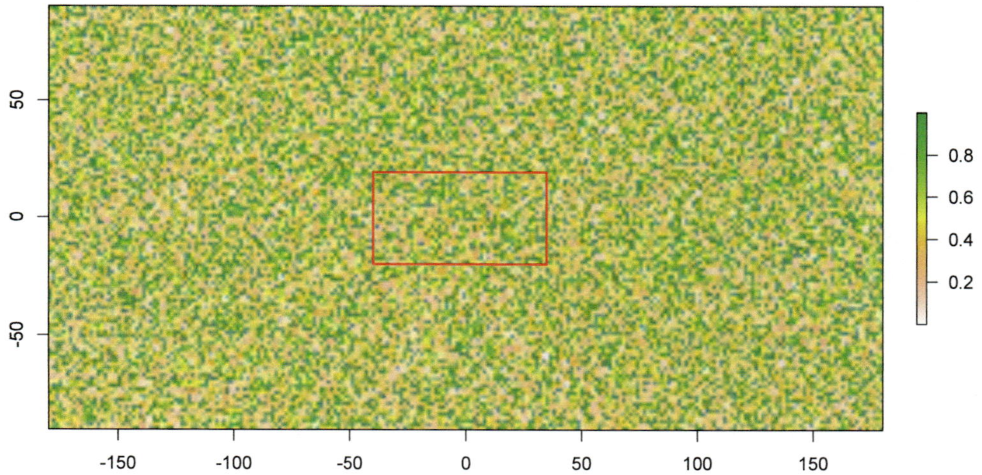

FIGURE 9.15 Obtaining coordinates to make a subset in the raster with the `drawExtent` function.

A raster subset called `clip` is created to perform a subset of `ndvi` from a raster subset, according to the code below.

```
clip <- raster(xmn = -39.87603, xmx = 35.02066, ymn = -19.93802, ymx = 19.31818,
               res = 1.5, vals = rep(0.99, 1))
ndvi[clip]
# [1] 0.321270042 0.452781357 0.321557307 0.631252910 0.366205655 0.841701852
# 0.168788964 0.827878981
# [9] 0.341318787 0.358925330 0.546224207 0.807697638 0.038189800 0.093945083
# 0.988478887 0.850167925
# ...
```

Basically, the values of the first raster, `ndvi` are obtained within the extent of the second raster, `clip`.

A subset can also be performed to obtain a spatial object as output. For this, the `drop` argument is set to `FALSE`. A subset of the `ndvi` raster is obtained to get the first two cells of the raster. A subset can also be obtained by row and column indexing of the raster, determining the same result.

```
ndvi[1:2, drop = FALSE] # spatial subset with cell ID
# class       : RasterLayer
# dimensions : 1, 2, 2  (nrow, ncol, ncell)
# resolution : 1.5, 1.5  (x, y)
# extent      : -180, -177, 88.5, 90  (xmin, xmax, ymin, ymax)
# crs         : +proj=longlat +datum=WGS84 +ellps=WGS84 +towgs84=0,0,0
# source      : memory
# names       : layer
# values      : 0.7873853, 0.905303  (min, max)
ndvi[1, 1:2, drop = FALSE] # spatial subset by row and column index
# class       : RasterLayer
# dimensions : 1, 2, 2  (nrow, ncol, ncell)
# resolution : 1.5, 1.5  (x, y)
# extent      : -180, -177, 88.5, 90  (xmin, xmax, ymin, ymax)
# crs         : +proj=longlat +datum=WGS84 +no_defs
# source      : memory
# names       : layer
# values      : 0.7873853, 0.905303  (min, max)
```

A raster with logical or NA values can be used to make a subset in another raster. A raster is created with NA values with NDVI less than 0.8. The subset on the ndvi raster is performed from the mask file with the `[` operator, and the functions, `mask()` and `overlay`, generating the same result with these operations.

```
rmask <- ndvi
rmask[rmask < 0.8] <- NA # create the mask raster with NA
# create spatial subset
ndvi[rmask, drop = FALSE]           # with [ operator
mask(ndvi, rmask)                   # with mask()
overlay(ndvi, rmask, fun = "max")   # with overlay
```

```
# class       : RasterLayer
# dimensions : 120, 240, 28800  (nrow, ncol, ncell)
# resolution : 1.5, 1.5  (x, y)
# extent      : -180, 180, -90, 90  (xmin, xmax, ymin, ymax)
# crs         : +proj=longlat +datum=WGS84 +ellps=WGS84 +towgs84=0,0,0
# source      : memory
# names       : layer
# values      : 0.8000077, 0.9999245  (min, max)
```

These operations can be considered as local Boolean, because cells of rasters with the same size are compared (Figure 9.16).

```
m1 <- ggR(ndvi$layer, geom_raster = TRUE) +
   scale_fill_gradientn(colours = rainbow(100), name = "NDVI")
m2 <- ggR(rmask$layer, geom_raster = TRUE) +
   scale_fill_gradientn(colours = rainbow(100), name = "NDVI - NA")
grid.arrange(m1, m2, ncol = 1)
```

9.9.2.2 Map algebra

Many generic functions that allow simple and elegant raster algebra are implemented for raster, including the normal algebraic operators, such as `+`, `-`, `*`, `/`, logical operators, such as `>`,`>=`, `<`, `==`,`!` and, functions like `abs`, `round`, `ceiling`, `floor`, `trunc`, `sqrt`, `log`, `log10`, `exp`, `cos`, `sin`, `atan`, `tan`, `max`, `min`, `range`, `prod`, `sum`, `any`, `all`. In these functions, you can mix raster layers with numbers, as long as the first argument is a raster object. A raster layer, `r`, is created with 10 rows and 10 columns to illustrate map algebra operations. A sum operation is performed on the raster `r` to obtain a raster `s`, defined by $s = r + 100$ (Figure 9.17).

```
r <- raster(ncol = 10, nrow = 10)
values(r) <- 1:ncell(r)
r
# class       : RasterLayer
# dimensions : 10, 10, 100  (nrow, ncol, ncell)
# resolution : 36, 18  (x, y)
# extent      : -180, 180, -90, 90  (xmin, xmax, ymin, ymax)
# crs         : +proj=longlat +datum=WGS84 +ellps=WGS84 +towgs84=0,0,0
# source      : memory
# names       : layer
# values      : 1, 100  (min, max)
s <- r + 100
s
# class       : RasterLayer
# dimensions : 10, 10, 100  (nrow, ncol, ncell)
# resolution : 36, 18  (x, y)
# extent      : -180, 180, -90, 90  (xmin, xmax, ymin, ymax)
# crs         : +proj=longlat +datum=WGS84 +ellps=WGS84 +towgs84=0,0,0
# source      : memory
```

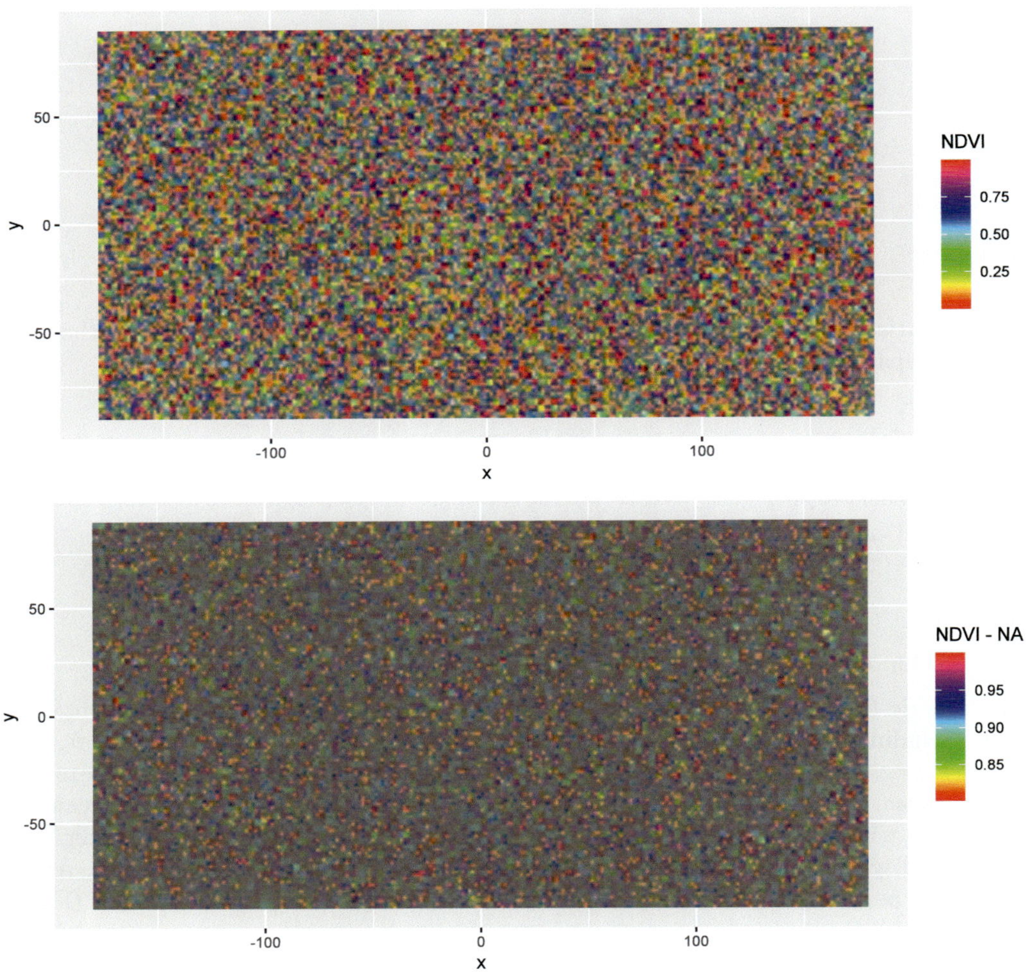

FIGURE 9.16 Rasters with numerical values of simulated NDVI (top) and subset with NDVI mask obtained from another raster of the same size and with the same coordinate system (bottom).

```
# names       : layer
# values      : 101, 200  (min, max)
```

The results are mapped with the `ggR` and `scale_fill_gradientn` functions.

```
m1 <- ggR(r$layer, geom_raster = TRUE) +
  scale_fill_gradientn(colours = topo.colors(100), name = "r")
m2 <- ggR(s$layer, geom_raster = TRUE) +
  scale_fill_gradientn(colours = topo.colors(100), name = "s")
grid.arrange(m1, m2, ncol = 1)
```

Map algebra is processed quickly on raster because raster datasets only store coordinates implicitly. When two or more raster datasets share the same extent, projection, and

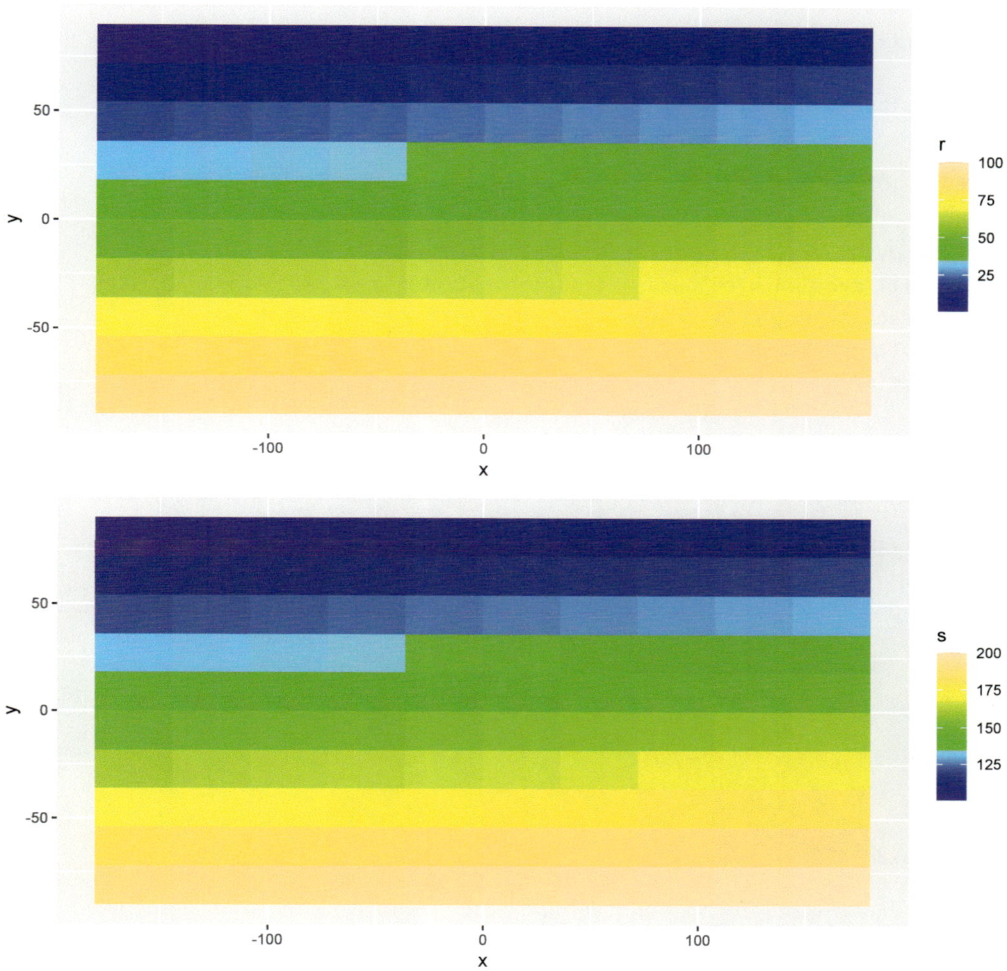

FIGURE 9.17 Original raster, r, created, and output raster, s, after applying 100 unit sum function to the original raster.

resolution, it is possible to treat them as matrices for geocomputation. Operationally, first, the raster package checks the file headers in which any algebraic operation is to be performed, and only if there is a match, processing proceeds. Second, location matching occurs cell-by-cell, without changing positions and making it easy to perform calculations quickly (Lovelace et al., 2019).

9.9.2.3 Local operations

Local operations refer to all cell-by-cell operations on one or multiple layers. For example, sorting ranges of numerical values into groups can be used to group ndvi values into low (class 1), medium (class 2), and high (class 3) ndvi classes. A matrix with intervals and reclassification codes is constructed to reclassify the ndvi raster into the intervals 6.653927e-05-0.1, 0.1-0.8 and 0.8 to 0.9999859 to assume values 1, 2 and, 3, respectively. To do this, in the first two columns of the matrix the minimum and maximum class intervals are placed, and in the third column the new class value, and so on until all classes of interest are filled. The reclassification is performed using the `reclassify` function (Figure 9.18).

```
rcl <- matrix(c(6.653927e-05, 0.1, 1, 0.1, 0.8, 2, 0.8, 0.9999859, 3), ncol = 3,
              byrow = TRUE)
recl <- reclassify(ndvi, rcl = rcl)
```

The results are mapped with the `ggR` and `scale_fill_gradientn` functions.

```
m1 <- ggR(ndvi$layer, geom_raster = TRUE) +
   scale_fill_gradientn(colours = topo.colors(100), name = "ndvi")
m2 <- ggR(recl$layer, geom_raster = TRUE) +
   scale_fill_gradientn(colours = topo.colors(100), name = "ndvi rec")
grid.arrange(m1, m2, ncol = 1)
```

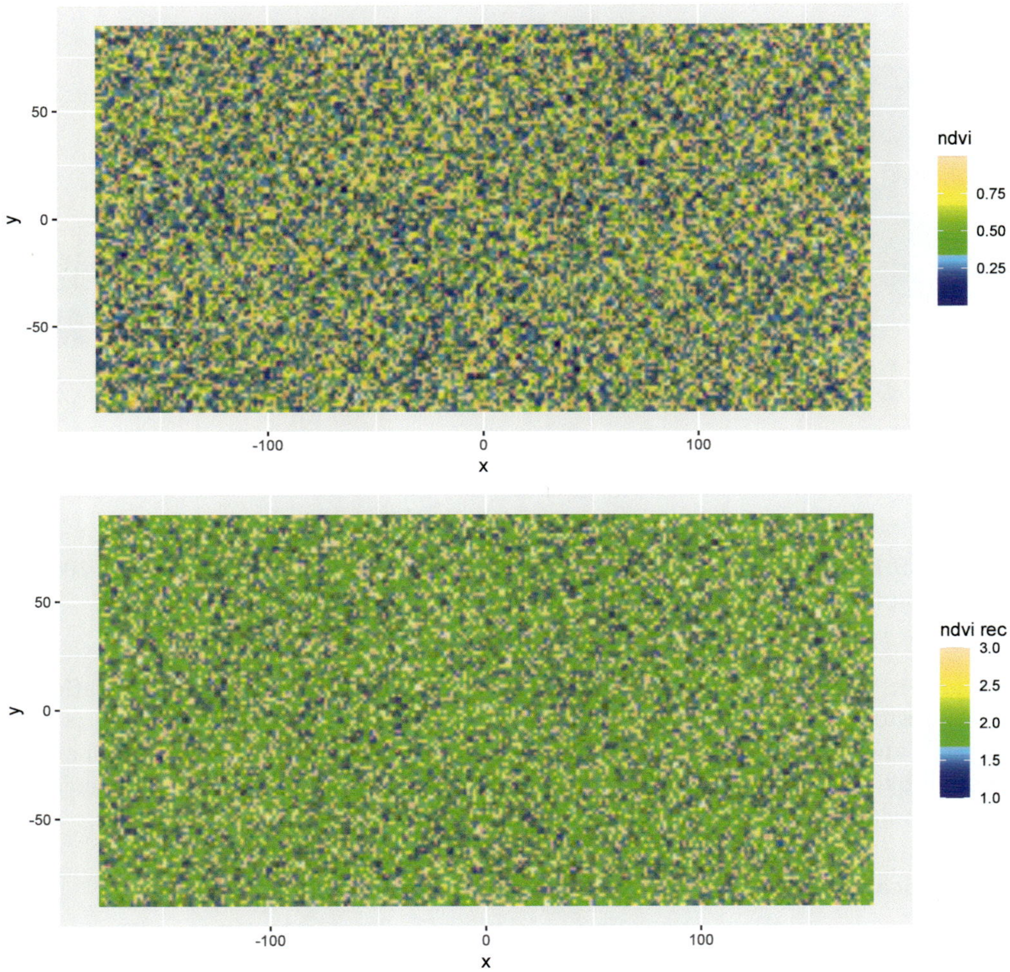

FIGURE 9.18 Original simulated ndvi raster and output raster with reclassified ndvi with three specific classes.

Map algebra is another type of local operation as used on a raster created for this purpose. The calculation of the normalized difference vegetation index is a well-known example of a cell-by-cell operation on a raster. Instead of mathematical operators, the functions `calc` and `overlay` can be used to get more efficient results when facing raster with a large database (Lovelace et al., 2019).

9.9.2.4 Focal operations

While local functions operate on a cell of one or several layers, focal operations consider a central cell and its neighbors. The neighborhood is also called a kernel, filter, or moving window. Usually 3-by-3 cells are used to apply a focal function because it has a central cell and eight surrounding neighbors, but it can take any other shape, not necessarily rectangular, according to the design goal. A focal operation can be used to apply an aggregation function to all the cells within a specified neighborhood, use the output value as a new value for the central cell, and move on to the next central cell. Other names for this type of operation are spatial filtering and convolution (Burrough and McDonnell, 1998; Lovelace et al., 2019).

In R, the `focal` function can be used to perform spatial filtering. In this case, the shape of the moving window is defined as a matrix whose values correspond to `w` weights. The `fun` parameter is used to specify the type of neighborhood function to be used, such as the maximum value, `max`. However, other values can be used, such as `min`, `sum`, `mean`, or `var` (Figure 9.19).

```
ndvi_focal <- focal(ndvi, w = matrix(1, nrow = 3, ncol = 3), fun = max)
```

The results are mapped with the `ggR` and `scale_fill_gradientn` functions.

```
m1 <- ggR(ndvi$layer, geom_raster = TRUE) +
   scale_fill_gradientn(colours = topo.colors(100), name = "ndvi")
m2 <- ggR(ndvi_focal$layer, geom_raster = TRUE) +
   scale_fill_gradientn(colours = topo.colors(100), name = "ndvi max")
grid.arrange(m1, m2, ncol = 1)
```

9.9.2.5 Zonal operations

Zonal operations are similar to focal operations, except that zonal filters can take any shape rather than a predefined rectangular window. For example, a zonal operation is performed to average the simulated `ndvi` raster within the raster levels with plant foliage attribute, `coffee`.

```
z = zonal(ndvi, coffee, fun = "mean") |> as.data.frame()
z
#   zone      mean
# 1    1 0.5054312
# 2    2 0.5003323
# 3    3 0.5004861
# 4    4 0.5053535
# 5    5 0.4995033
```

The average ndvi values obtained per zone can later be added to the raster attribute table.

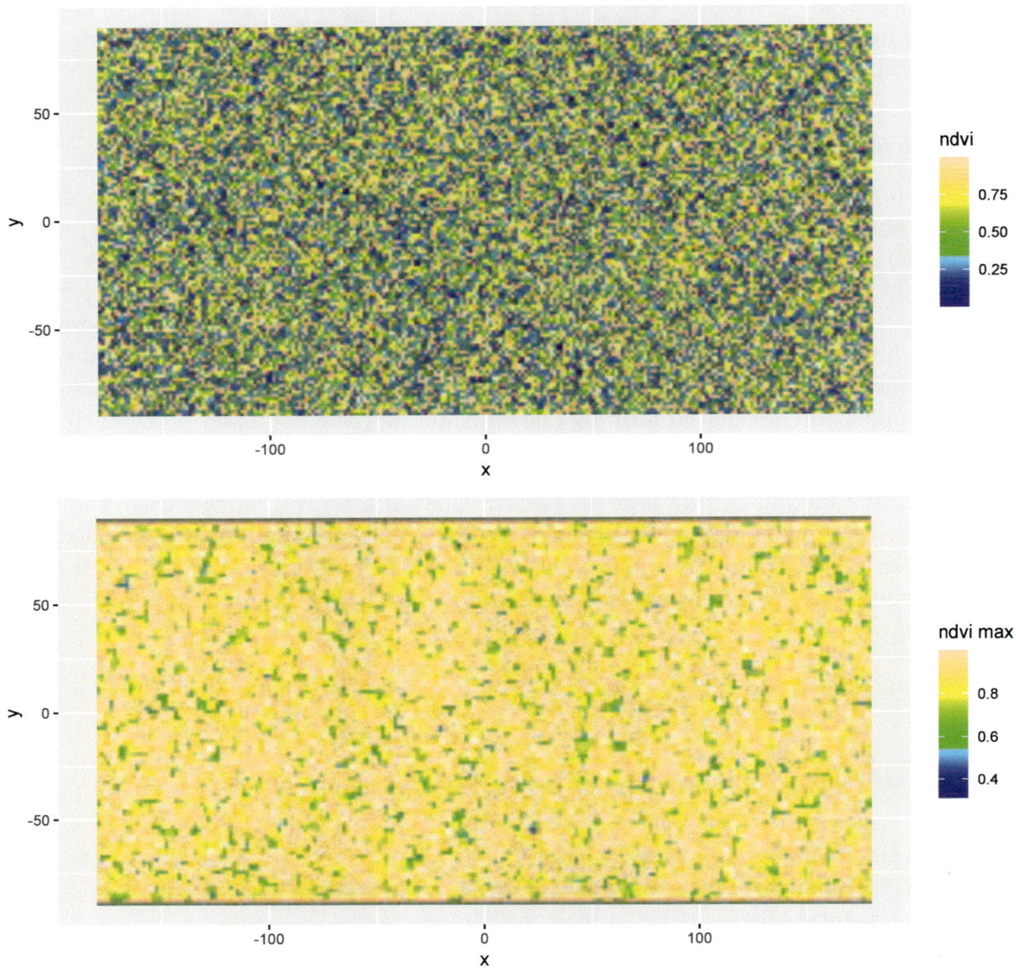

FIGURE 9.19 Original simulated ndvi raster and output raster with ndvi under spatial filtering with 3x3 moving window and maximum value operation.

9.9.2.6 Mosaic operations on raster layers

As an example of raster mosaic application, the `getData` function is used to obtain SRTM altitude data for Brazil and Paraguay. The `merge` function is used later to merge the digital elevation model of the two countries into just one scene, i.e., placing them side-by-side (Figure 9.20).

```
bra <- getData("alt", country = "BRA", mask = TRUE)
pry <- getData("alt", country = "PRY", mask = TRUE)
bra_pry <- merge(bra, pry)
```

Another option for the same procedure can be performed with the `geoData` and `terra` packages. In this case, the `elevation_30s` function is used to obtain digital elevation model data for countries.

```
library(geodata)
# Loading required package: terra 1.7.60
bra <- elevation_30s(country = "BRA", path = tempdir()) # Get elevation data
pry <- elevation_30s(country = "PRY", path = tempdir()) # Get elevation data
bra_pry <- terra::mosaic(bra, pry) # Mosaic
```

The data is exported to file with the `write.raster` function.

```
writeRaster(bra, filename = "C:/geo/C9/bra.tif", overwrite = TRUE)
writeRaster(pry, filename = "C:/geo/C9/pry.tif", overwrite = TRUE)
writeRaster(bra_pry, filename = "C:/geo/C9/bra_pry.tif", overwrite = TRUE)
```

The results are mapped with the `ggR` and `scale_fill_gradientn` functions. Note that the mosaic operation, in addition to being spatial at the intersection between the two layers, is also a geometric operation because it changes the final geometry of the data after the operation is performed.

```
# Plot for raster
m1 <- ggR(bra$BRA_msk_alt, geom_raster = TRUE) +
   scale_fill_gradientn(colours = topo.colors(100), name = "BRA")
m2 <- ggR(pry$PRY_msk_alt, geom_raster = TRUE) +
   scale_fill_gradientn(colours = topo.colors(100), name = "PRY")
m3 <- ggR(bra_pry$layer, geom_raster = TRUE) +
   scale_fill_gradientn(colours = topo.colors(100), name = "Merge")
grid.arrange(m1, m2, m3, ncol = 2)
# Plot for terra
m1 <- ggR(bra, geom_raster = TRUE) +
   scale_fill_gradientn(colours = topo.colors(100), name = "BRA")
m2 <- ggR(pry, geom_raster = TRUE) +
   scale_fill_gradientn(colours = topo.colors(100), name = "PRY")
m3 <- ggR(bra_pry, geom_raster = TRUE) +
   scale_fill_gradientn(colours = topo.colors(100), name = "Merge")
grid.arrange(m1, m2, m3, ncol = 2)
```

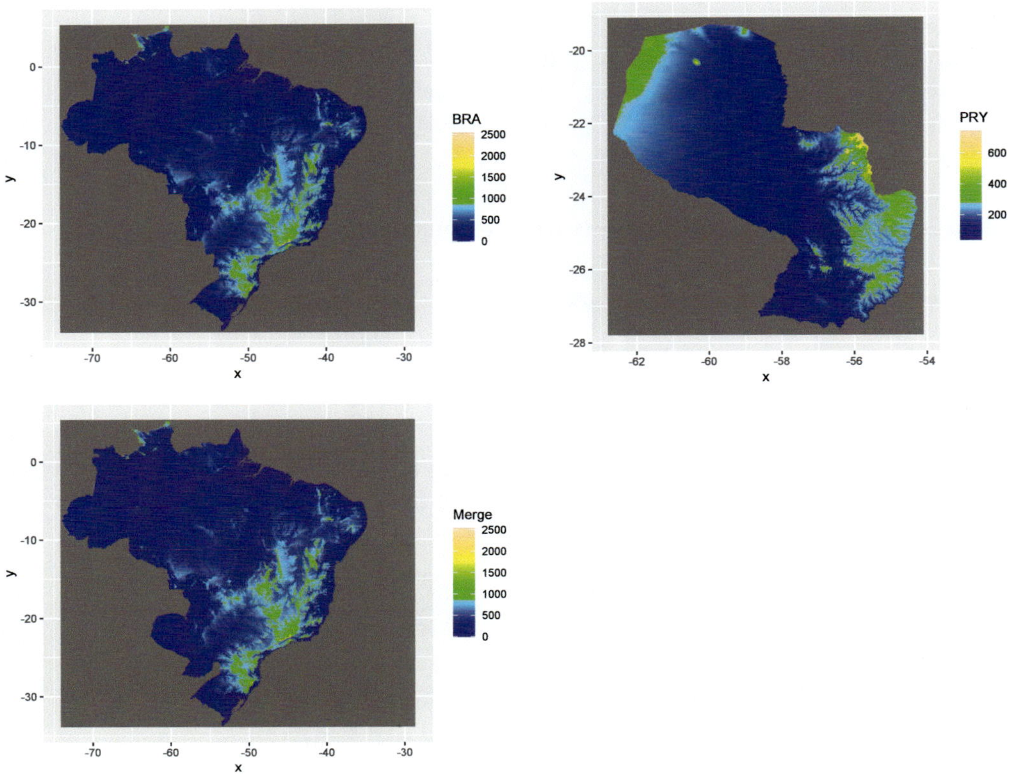

FIGURE 9.20 Performing a spatial merge operation to mosaic the digital elevation model of Brazil (BRA) (top-left) with that of Paraguay (PRY) (top-right) creating a raster mosaic (Merge) (bottom).

10 Vector Geometric Analysis with Environmental Geocomputation and R

10.1 LEARNING QUESTIONS

The learning questions answered by reading the chapter are:

- What types of geometry operations can be used on vector features?
- How to manipulate the geometry column of geo-objects.
- How to determine the compactness of geometric primitives.
- In what situations are geometric operations used in geocomputation?
- How to perform a buffer.
- How to transform a geometric primitive into another.
- What is the advantage of simplifying a simple geometric feature?

10.2 LEARNING OUTCOMES

The learning outcomes answered by reading the chapter are:

- Understand types of geometry operations used on vector features;
- Manipulate the geometry column of geo-objects;
- Determine the compactness of geometric primitives;
- Verify applications of geometric operations in geocomputation;
- Learn to perform a buffer;
- Learn to transform a geometric primitive in another;
- Check the advantage of simplifying a simple geometric feature.

10.3 INTRODUCTION

Geographic datasets can be structured and manipulated based on their non-geographic attributes and geospatial features. In geographic data with vector features, it is possible to have control over the geometry column in geo-objects (Lovelace et al., 2019).

Unary and binary operations can be used to transform vector geometries. In unary operations, geometries are treated in isolation, in operations with points, lines, and polygons to simplify features, create `buffer`, `centroid`, and perform `affine` transformations for changing, scaling, and rotating vectors. With binary transformations, the geometry of one object is modified based on the shape of another. This makes it possible to perform clipping and union of geometries. It is also possible to perform transformations between geometry types, such as from points to lines and polygons (Lovelace et al., 2019).

10.4 GEOMETRIC OPERATIONS ON VECTOR FEATURES

In geocomputation systems, the shape of geo-objects can be represented by objective measurements such as the area of the geo-object. A simple way to compare the shape

DOI: 10.1201/9781003724377-10

of geographic objects is based on the relationship between the object's perimeter and area. A measure of compactness of a geo-object can be obtained by (Longley et al., 2001):

$$S = \frac{P}{3.54\sqrt{A}} \tag{10.1}$$

where P is the length of the perimeter and A is the area. The parameter 3.54 (twice the square root of π) is added to the equation to ensure that the most compact geometry, that of a circle, will return a value of S = 1. Other geometry primitives, such as triangle, square and rectangle, will return values of greater compactness.

Geometric operations are characterized by changing the geometry of objects with vector features (Lovelace et al., 2019). Geometric operations can be performed on vector features with overlapping, intersecting, and varying distance between them. Overlapping and intersecting operations are performed to obtain new entities using Boolean logic operations. In some situations of spurious features or features with faulty edges, simplification operations can be performed to smooth the original entity, using, for example, a simplification algorithm (Douglas and Peucker, 1973; Burrough and McDonnell, 1998).

10.4.1 SIMPLIFICATION

Simplification is a process of generalizing objects with vector features of lines and polygons for use in smaller scale maps to reduce the amount of memory, disk space, and network bandwidth consumed by the object when published as an interactive map on the Internet or in other situations where it is necessary to simplify the geo-object (Lovelace et al., 2019). An algorithm developed by Douglas and Peucker (1973) is implemented in the `GEOS` package `sf` to reduce the number of vertices of the geo-object. As an example, tolerance options of 500 m, 1000 m, and 2000 m are defined to control the generalization level of a cadastral map with 48 vector features and 7 cadastral fields, with multilinestring type geometry, in the Upper-Middle and Upper Paraguay Planning and Management Units, Mato Grosso, Brazil.

The packages `sf` (Pebesma et al., 2021a), `ggplot2` (Wickham et al., 2020), `gridExtra` (Auguie and Antonov, 2017), `ggspatial` (Dunnington and Thorne, 2021), and `rmapshaper` (Teucher and Russell, 2023) packages needed to perform the vector geometric analysis operations examples are loaded into R with the `library` function.

```
library(sf)
# Linking to GEOS 3.11.2, GDAL 3.7.2, PROJ 9.3.0; sf_use_s2() is TRUE
library(ggplot2)
library(gridExtra)
library(ggspatial)
library(rmapshaper)
# package 'rmapshaper' was built under R version 4.3.3
```

The `st_read` function is used to import the drainage network of the watershed and the municipalities of the watershed's geographical region into R. The `st_transform` function is used to transform the data into the UTM 21 South plane map projection.

```
hidrosf <- st_read("C:/Aulas/2021_1/Geoprocessamento/shape/hidrografiaMT.shp",
                   options = "ENCODING=WINDOWS-1252")
hidro <- st_transform(hidrosf,
                      "+proj=utm +zone=21 +south +datum=WGS84 +units=m +no_defs")
hidro
# Simple feature collection with 48 features and 7 fields
# Geometry type: MULTILINESTRING
# Dimension:     XY
# Bounding box:  xmin: 331838.1 ymin: 8212782 xmax: 568639.7 ymax: 8432843
# Projected CRS: +proj=utm +zone=21 +south +datum=WGS84 +units=m +no_defs
# ...
munsf <- st_read("C:/Aulas/2021_1/Geoprocessamento/shape/municipiosMT.shp",
                 options = "ENCODING=WINDOWS-1252")
mun <- st_transform(munsf,
                    "+proj=utm +zone=21 +south +datum=WGS84 +units=m +no_defs")
mun
# Simple feature collection with 24 features and 3 fields
# Geometry type: MULTIPOLYGON
# Dimension:     XY
# Bounding box:  xmin: 326931.5 ymin: 8208648 xmax: 587652.9 ymax: 8436246
# Projected CRS: +proj=utm +zone=21 +south +datum=WGS84 +units=m +no_defs
# ...
```

The function of the sf package, sf_use_s2, is disabled to perform the geometric union of municipalities in the watershed under analysis, thereby creating a geometric object of class sfc. A database is created with an identifier code and name using the data.frame function. The st_as_sf function is used to create a geo-object with the boundary of the sf data.frame class basin.

```
sf::sf_use_s2(FALSE) # Disable s2
border_sfc <- st_union(mun) # Perform union
d <- data.frame(id = 1, name = "basin") # Create data frame
bor <- st_as_sf(d, border_sfc)  # Create sf
bor
# Simple feature collection with 1 feature and 2 fields
# Geometry type: POLYGON
# Dimension:     XY
# Bounding box:  xmin: 326931.5 ymin: 8208648 xmax: 587652.9 ymax: 8436246
# Projected CRS: +proj=utm +zone=21 +south +datum=WGS84 +units=m +no_defs
#   id  name                       border_sfc
# 1  1 basin POLYGON ((470485.4 8247284,...
```

The st_simplify function is used to create simplified versions of the drainage network of the watershed with tolerances of 500, 1000, and 2000 m.

```
hidro2000 <- st_simplify(hidro, dTolerance = 2000)  # 2000 m
hidro1000 <- st_simplify(hidro, dTolerance = 1000)  # 1000 m
hidro500 <- st_simplify(hidro, dTolerance = 500)  # 500 m
```

The `object.size` function is used to check the size of simplified geo-objects when compared to the original object.

```
object.size(hidro)
# 173736 bytes
object.size(hidro500)
# 112784 bytes
object.size(hidro1000)
# 106000 bytes
object.size(hidro2000)
# 102096 bytes
```

The results are mapped using functions from the `ggplot2` and `ggspatial` packages (Figure 10.1).

```
a1 <- ggplot(data = bor) + geom_sf(fill = "antiquewhite") +
 annotate(geom = "text", x = 530000, y = 8237000,
 label = "Simplification = 0 m", fontface = "italic",
 color = "grey22", size = 4.5) + geom_sf(data = hidro, colour = "blue") +
 annotation_scale(location = "bl", width_hint = 0.3) +
 annotation_north_arrow(location = "bl", which_north = "true",
 pad_x = unit(0.75, "in"), pad_y = unit(0.3, "in"),
 style = north_arrow_fancy_orienteering)
a2 <- ggplot(data = bor) + geom_sf(fill = "antiquewhite") +
 annotate(geom = "text", x = 530000, y = 8237000,
 label = "Simplification = 500 m", fontface = "italic",
 color = "grey22", size = 4.5) + geom_sf(data = hidro500, colour = "blue") +
 annotation_scale(location = "bl", width_hint = 0.3) +
 annotation_north_arrow(location = "bl", which_north = "true",
 pad_x = unit(0.75, "in"), pad_y = unit(0.3, "in"),
 style = north_arrow_fancy_orienteering)
a3 <- ggplot(data = bor) + geom_sf(fill = "antiquewhite") +
 annotate(geom = "text", x = 530000, y = 8237000,
 label = "Simplification = 1000 m", fontface = "italic",
 color = "grey22", size = 4.5) + geom_sf(data = hidro1000, colour = "blue") +
 annotation_scale(location = "bl", width_hint = 0.3) +
 annotation_north_arrow(location = "bl", which_north = "true",
 pad_x = unit(0.75, "in"), pad_y = unit(0.3, "in"),
 style = north_arrow_fancy_orienteering)
a4 <- ggplot(data = bor) + geom_sf(fill = "antiquewhite") +
 annotate(geom = "text", x = 530000, y = 8237000,
 label = "Simplification = 2000 m", fontface = "italic",
 color = "grey22", size = 4.5) + geom_sf(data = hidro2000, colour = "blue") +
 annotation_scale(location = "bl", width_hint = 0.3) +
 annotation_north_arrow(location = "bl", which_north = "true",
 pad_x = unit(0.75, "in"), pad_y = unit(0.3, "in"),
 style = north_arrow_fancy_orienteering)
grid.arrange(a1, a2, a3, a4, ncol = 2)
```

The size of the simplified objects is actually reduced in terms of computational memory

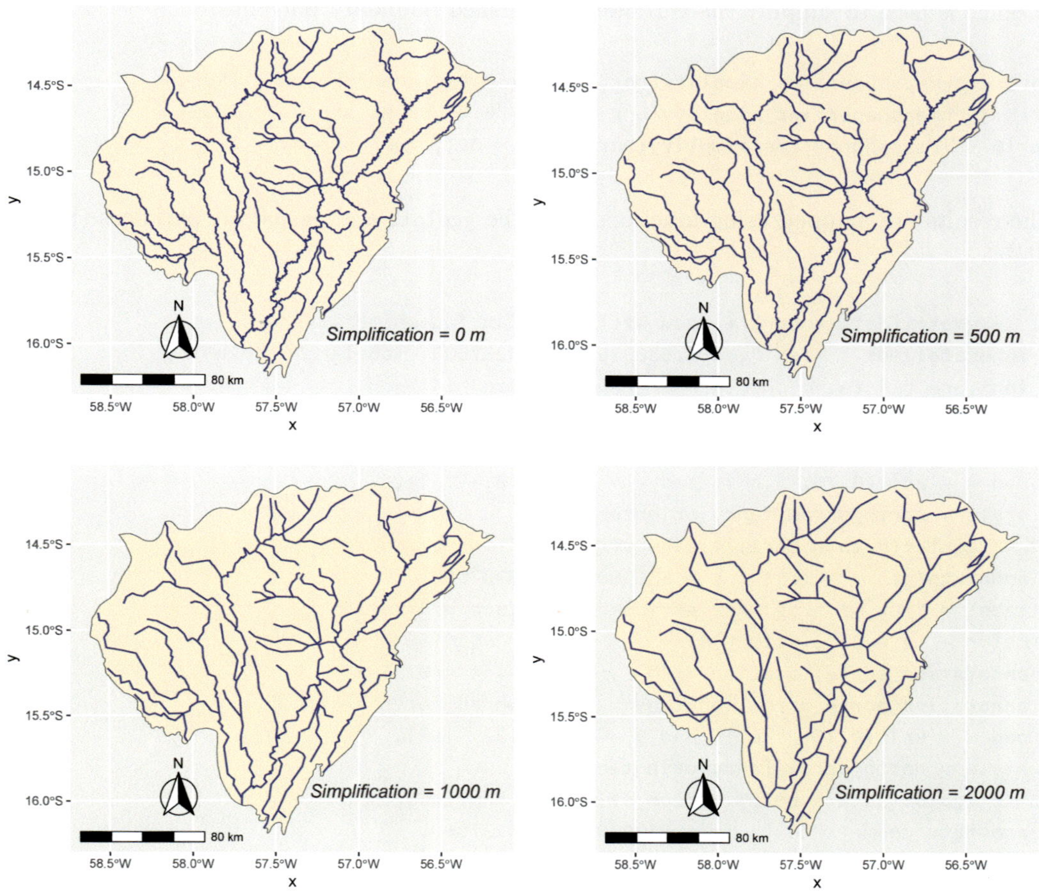

FIGURE 10.1 Drainage network simplification, with tolerance options of 500 m, 1000 m and 2000 m, in the Planning and Management Units of the Upper-Middle and Upper Paraguay River, Mato Grosso, Brazil.

consumption (Table 10.1).

TABLE 10.1 Computational memory consumption of simplified drainage network.

Vector feature	Memory consumption (bytes)
Original drainage network	173736
Simplified drainage network with 500 m tolerance	112784
Simplified drainage network with 1000 m tolerance	106000
Simplified drainage network with 2000 m tolerance	102096

Simplification can also be used on polygon vector features (Lovelace et al., 2019). In this case, simplification is applied to the boundaries of the Upper-Middle and Upper Paraguay Planning and Management Units, Mato Grosso, Brazil. The Visvalingam and Whyatt (1993) algorithm is used to simplify the number of vertices of the watershed boundaries by 1, 5, and 10% of the original number of vertices. The `ms_simplify` function from the `rmapshaper` package is used to simplify the vertices of watershed boundary polygons.

```
bor1 <- rmapshaper::ms_simplify(bor, keep = 0.01, keep_shapes = TRUE)
bor5 <- rmapshaper::ms_simplify(bor, keep = 0.05, keep_shapes = TRUE)
bor10 <- rmapshaper::ms_simplify(bor, keep = 0.1, keep_shapes = TRUE)
```

The results are mapped using functions from the `ggplot2` and `ggspatial` packages (Figure 10.2).

```
a1 <- ggplot(data = bor) + geom_sf(fill = "antiquewhite") +
  annotate(geom = "text", x = 530000, y = 8237000, label = "Original",
  fontface = "italic", color = "grey22", size = 4) +
  annotation_scale(location = "bl", width_hint = 0.3) +
  annotation_north_arrow(location = "bl", which_north = "true",
  pad_x = unit(0.75, "in"), pad_y = unit(0.3, "in"),
  style = north_arrow_fancy_orienteering)
a2 <- ggplot(data = bor1) + geom_sf(fill = "antiquewhite") +
  annotate(geom = "text", x = 530000, y = 8237000,
  label = "10% of original vertices", fontface = "italic",
  color = "grey22", size = 4) +
  annotation_scale(location = "bl", width_hint = 0.3) +
  annotation_north_arrow(location = "bl", which_north = "true",
  pad_x = unit(0.75, "in"), pad_y = unit(0.3, "in"),
  style = north_arrow_fancy_orienteering)
a3 <- ggplot(data = bor5) + geom_sf(fill = "antiquewhite") +
  annotate(geom = "text", x = 530000, y = 8237000,
  label = "5% of original vertices", fontface = "italic",
  color = "grey22", size = 4) +
  annotation_scale(location = "bl", width_hint = 0.3) +
  annotation_north_arrow(location = "bl", which_north = "true",
  pad_x = unit(0.75, "in"), pad_y = unit(0.3, "in"),
  style = north_arrow_fancy_orienteering)
a4 <- ggplot(data = bor10) + geom_sf(fill = "antiquewhite") +
  annotate(geom = "text", x = 530000, y = 8237000,
  label = "1% of original vertices", fontface = "italic",
  color = "grey22", size = 4) +
  annotation_scale(location = "bl", width_hint = 0.3) +
  annotation_north_arrow(location = "bl", which_north = "true",
  pad_x = unit(0.75, "in"), pad_y = unit(0.3, "in"),
  style = north_arrow_fancy_orienteering)
grid.arrange(a1, a2, a3, a4, ncol = 2)
```

10.4.2 CENTROID

Centroid operations are used to identify the center of geographic objects in order to define the geographic center of an object and create representations of more complex vector objects by a single point (Lovelace et al., 2019).

The most commonly used centroid operation is the geographic centroid operation. In this type of operation, a geospatial feature is represented by the center of mass of the object.

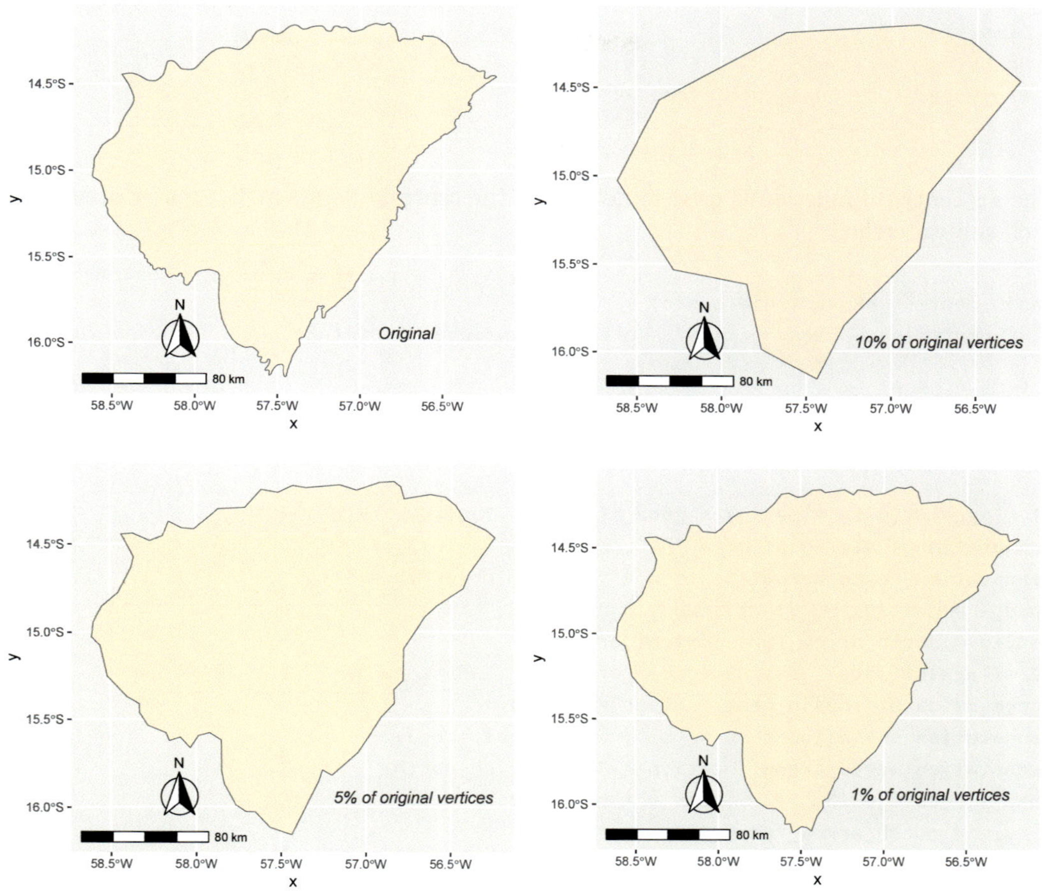

FIGURE 10.2 Simplification of watershed polygons by 1, 5 and 10% of the original number of vertices with the boundary of Planning and Management Units of the Upper-Middle and Upper Paraguay River, Mato Grosso, Brazil.

Geographic centroids are used to create a simplified representation of complex geometries and to estimate distances between polygons (Lovelace et al., 2019). A centroid operation is used to determine the center of mass of sub-basin polygons in Mato Grosso.

The sub-watershed data is used as an example to determine the centroid of each sub-watershed. The `st_read` function is used to import the sub-watershed data, and the `st_transform` function is used to transform the data into the UTM 21 South cartographic projection.

```
basinsf <- st_read("C:/Aulas/2021_1/Geoprocessamento/shape/baciasMT.shp",
                   options = "ENCODING=WINDOWS-1252")
basin <- st_transform(basinsf, "+proj=utm +zone=21 +south +datum=WGS84
                               +units=m +no_defs")
basin
# Simple feature collection with 182 features and 1 field
# Geometry type: MULTIPOLYGON
```

```
# Dimension:      XY
# Bounding box:  xmin: 324536.4 ymin: 8203510 xmax: 585199.8 ymax: 8436151
# Projected CRS: +proj=utm +zone=21 +south
# ...
```

The `st_centroid` function is used to determine the centroid points at the center of mass of each sub-watershed.

```
basin_cent <- st_centroid(basin)
# st_centroid assumes attributes are constant over geometries
```

The results are mapped using functions from the `ggplot2` and `ggspatial` packages (Figure 10.3).

```
a1 <- ggplot(data = basin) + geom_sf(fill = "antiquewhite") +
 annotation_scale(location = "bl", width_hint = 0.3) +
 annotation_north_arrow(location = "bl", which_north = "true",
 pad_x = unit(0.75, "in"), pad_y = unit(0.3, "in"),
 style = north_arrow_fancy_orienteering)
a2 <- ggplot(data = basin) + geom_sf(fill = "antiquewhite") +
 geom_sf(data = basin_cent, colour = "red") +
 annotation_scale(location = "bl", width_hint = 0.3) +
 annotation_north_arrow(location = "bl", which_north = "true",
 pad_x = unit(0.75, "in"), pad_y = unit(0.3, "in"),
 style = north_arrow_fancy_orienteering)
grid.arrange(a1, a2, ncol = 2)
```

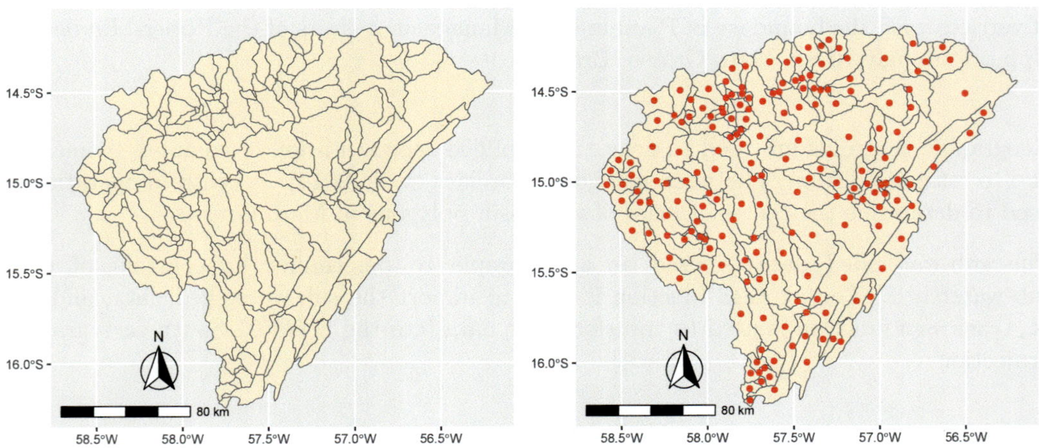

FIGURE 10.3 Determination of centroid (right) of sub-basins (left) in the Upper-Middle and Upper Paraguay Planning and Management Units, Mato Grosso, Brazil.

In some situations, depending on the geometry of the feature, the centroid may fall outside the boundaries of its original objects. Thus, the analyst must choose to use the points created in a geospatial analysis of interest or perform additional geospatial operations to check which objects were in unwanted positions relative to the original object and edit these points.

10.4.3 BUFFER

The buffer is represented by a polygon referring to an area within a certain prespecified distance from an original geometric element of type point, line or polygon (Longley et al., 2001; Lovelace et al., 2019). The buffer operation is used to define geographically bounded zones around the initial entity (Burrough and McDonnell, 1998). Regardless of the input type, the output of a buffer operation will always be a polygon. Unlike simplification, which is used in geovisualization and file size reduction applications, buffering is used for geospatial analysis of geographic data (Lovelace et al., 2019). A buffer can be used either to increase the number of pixels sampled within irregular edges of a polygon (Figure 10.4), or to determine how many points are included within a certain distance of an alignment or the permanent preservation area with riparian forest around a river according to the width of the main watercourse (Figure 10.5).

The geographical boundary of the coffee field is imported into R using the `st_read` function.

```
borM <- st_read("C:/geo/artigoML/shape/poligono.shp")
borM
# Simple feature collection with 1 feature and 1 field
# Geometry type: POLYGON
# Dimension:     XY
# Bounding box:  xmin: 512063.1 ymin: 7689925 xmax: 512345.8 ymax: 7690382
# Projected CRS: WGS 84 / UTM zone 23S
#   Id                       geometry
# 1  0 POLYGON ((512230.6 7689972,...
```

Buffer zones are created with the `st_buffer` function around the coffee field in Minas Gerais at a geographical extension of 10, 30, and 50 m, and around the drainage network of the watershed in Mato Grosso at a geographical extension of 500, 1000, and 2000 m.

```
bor10 <- st_buffer(borM, dist = 10) # Coffe field 10 m
bor30 <- st_buffer(borM, dist = 30) # Coffe field 30 m
bor50 <- st_buffer(borM, dist = 50) # Coffe field 50 m
hidro10 <- st_buffer(hidro, dist = 500) # Basin 500 m
hidro30 <- st_buffer(hidro, dist = 1000) # Basin 1000 m
hidro50 <- st_buffer(hidro, dist = 2000) # Basin 2000 m
```

The results of the buffer around the coffee field (Figure 10.4) and the watershed (Figure 10.5) are mapped using functions from the `ggplot2` and `ggspatial` packages.

```
a1 <- ggplot(data = borM) + geom_sf(fill= "grey92", colour="red") +
  annotation_scale(location = "br", width_hint = 0.5) +
  annotation_north_arrow(location = "br", which_north = "true",
  pad_x = unit(0.2, "in"), pad_y = unit(0.2, "in"),
  style = north_arrow_fancy_orienteering)
a2 <- ggplot(data = bor10) + geom_sf(fill = "antiquewhite") +
```

```
  geom_sf(data = borM, colour = "red") +
  annotation_scale(location = "br", width_hint = 0.5) +
  annotation_north_arrow(location = "br", which_north = "true",
  pad_x = unit(0.2, "in"), pad_y = unit(0.2, "in"),
  style = north_arrow_fancy_orienteering)
a3 <- ggplot(data = bor30) + geom_sf(fill = "antiquewhite") +
  geom_sf(data = borM, colour = "red") +
  annotation_scale(location = "br", width_hint = 0.5) +
  annotation_north_arrow(location = "br", which_north = "true",
  pad_x = unit(0.2, "in"), pad_y = unit(0.2, "in"),
  style = north_arrow_fancy_orienteering)
a4 <- ggplot(data = bor50) + geom_sf(fill = "antiquewhite") +
  geom_sf(data = borM, colour = "red") +
  annotation_scale(location = "br", width_hint = 0.5) +
  annotation_north_arrow(location = "br", which_north = "true",
  pad_x = unit(0.2, "in"), pad_y = unit(0.2, "in"),
  style = north_arrow_fancy_orienteering)
grid.arrange(a1, a2, a3, ncol = 3)
```

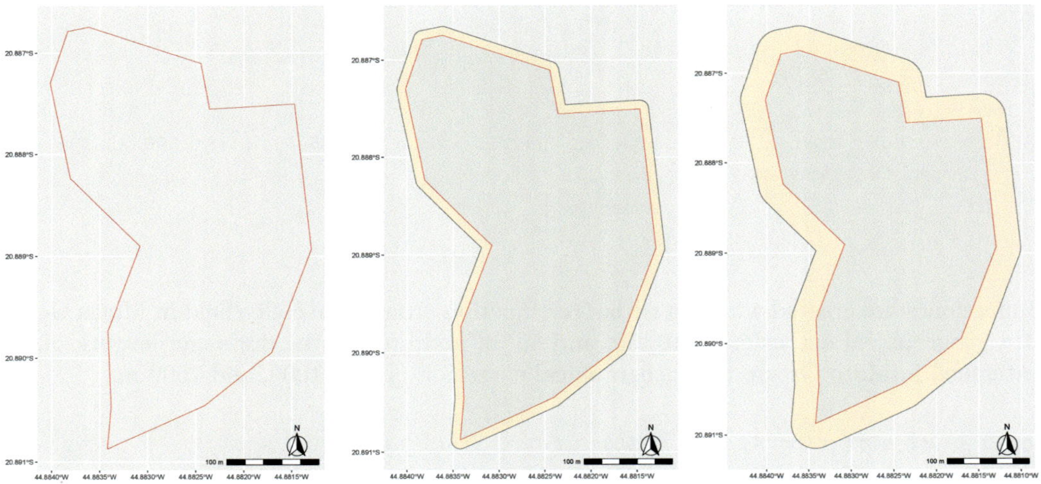

FIGURE 10.4 Polygon feature of coffee plantation in Santo Antônio do Amparo, Minas Gerais, Brazil (left), where 30 m (center) and 50 m (right) buffers were applied to the original polygon to sample pixels from remote sensing products used to study geospatial variation of bacterial blight epidemic in coffee plants.

```
a1 <- ggplot(data = bor) + geom_sf(fill = "antiquewhite") +
 annotate(geom = "text", x = 530000, y = 8237000,
 label = "Drainage network", fontface = "italic", color = "grey22", size = 5) +
 geom_sf(data = hidro, colour = "blue") +
 annotation_scale(location = "bl", width_hint = 0.3) +
 annotation_north_arrow(location = "bl", which_north = "true",
 pad_x = unit(0.75, "in"), pad_y = unit(0.3, "in"),
 style = north_arrow_fancy_orienteering)
```

```
a2 <- ggplot(data = bor) + geom_sf(fill = "antiquewhite") +
 annotate(geom = "text", x = 530000, y = 8237000,
 label = "Buffer 500", fontface = "italic", color = "grey22", size = 5) +
 geom_sf(data = hidro10, colour = "blue") +
 annotation_scale(location = "bl", width_hint = 0.3) +
 annotation_north_arrow(location = "bl", which_north = "true",
 pad_x = unit(0.75, "in"), pad_y = unit(0.3, "in"),
 style = north_arrow_fancy_orienteering)
a3 <- ggplot(data = bor) + geom_sf(fill = "antiquewhite") +
 annotate(geom = "text", x = 530000, y = 8237000,
 label = "Buffer 1000", fontface = "italic", color = "grey22", size = 5) +
 geom_sf(data = hidro30, colour = "blue") +
 annotation_scale(location = "bl", width_hint = 0.3) +
 annotation_north_arrow(location = "bl", which_north = "true",
 pad_x = unit(0.75, "in"), pad_y = unit(0.3, "in"),
 style = north_arrow_fancy_orienteering)
a4 <- ggplot(data = bor) + geom_sf(fill = "antiquewhite") +
 annotate(geom = "text", x = 530000, y = 8237000,
 label = "Buffer 2000", fontface = "italic", color = "grey22", size = 5) +
 geom_sf(data = hidro50, colour = "blue") +
 annotation_scale(location = "bl", width_hint = 0.3) +
 annotation_north_arrow(location = "bl", which_north = "true",
 pad_x = unit(0.75, "in"), pad_y = unit(0.3, "in"),
 style = north_arrow_fancy_orienteering)
grid.arrange(a1, a2, a3, a4, ncol = 2)
```

The number of buffer segments per quadrant can be set when it is necessary to reduce the memory footprint of the buffer or when high precision of the polygon contours is required. In this case you should increase the number of buffer segments per quadrant.

In addition to these applications, the buffer can be used in estimating the layout of roads, highways, railroads, high voltage power lines and their impact on a given area along the development undertaken. Some typical examples of buffer application are:

- Determine the number of restaurants within 2 km of the work region;
- Assess water pollution in the vicinity of dairy industries;
- Verify the world population density within 100 km of the sea;
- Verify the number of rural properties around the nitisol soil class;
- Determine the number of roads for production flow around a rural property;
- Delimit the permanent preservation area around a watercourse.

10.4.4 AFFINE

Any transformation which preserves lines and parallelism is a type of affine or affine transformation. However, angles or lengths are not necessarily preserved with this type of transformation. The types of affine transformations are (Lovelace et al., 2019):

- Shift (translation);
- Scale;
- Rotation;
- Combination between types.

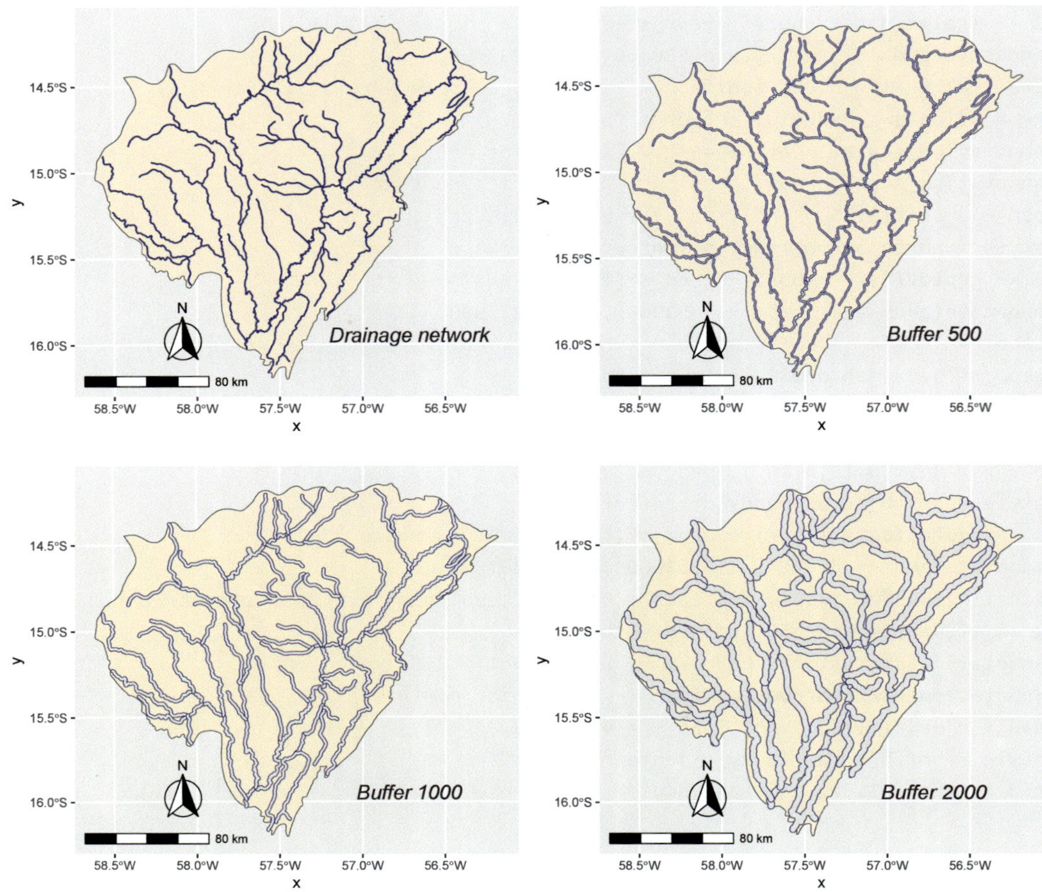

FIGURE 10.5 Multilinestring geometry feature where 500 m, 1000 m and 2000 m buffers are applied around drainage network in the Upper-Middle and Upper Paraguay Planning and Management Units, Mato Grosso, Brazil.

Shift transformations are needed, for example, to place labels on maps. Scale is used on non-contiguous topographic maps to obtain different representation situations on maps of a locality. Many affine transformations are used to redesign or improve a map geometry by transforming the vector feature into a more accurate map projection (Lovelace et al., 2019).

In the shift transformation, we can move points based on units of measurement on a map. For example, a distance of 30,000 m north is added to the geographic boundaries of the Upper-Middle Paraguay and Upper Paraguay Planning and Management Units. Scale changes can alter the geo-object dimension by increasing or shrinking the objects relative to the source coordinates, while keeping all topological relationship geometries intact. The scale change can be performed on the geo-object as a whole or on specific regions of interest. In this case, the geographic boundaries of the watershed are halved by rerunning operations based on the position of the watershed centroid.

Rotation of two-dimensional coordinates is performed using a rotation matrix to rotate vector features clockwise:

$$R = \begin{pmatrix} cos\theta & -sen\theta \\ sen\theta & cos\theta \end{pmatrix} \tag{10.2}$$

Using the rotation matrix, the geographic points bordering the watershed are rotated by 30° clockwise.

The shift operation is performed with the `sfc` class watershed polygon. A coordinate reference system is then assigned, and the object is converted to `sf` with the `st_set_crs` and `st_as_sf` functions, respectively.

```
bord_shift <- border_sfc + c(0, 30000) # shift
bord_shift <- st_as_sf(st_set_crs(bord_shift,
        "+proj=utm +zone=21 +south +datum=WGS84 +units=m +no_defs"))
bord_shift
# Simple feature collection with 1 feature and 0 fields
# Geometry type: POLYGON
# Dimension:     XY
# Bounding box:  xmin: 326931.5 ymin: 8238648 xmax: 587652.9 ymax: 8466246
# Projected CRS: +proj=utm +zone=21 +south +datum=WGS84 +units=m +no_defs
#                                 x
# 1 POLYGON ((470485.4 8277284,...
```

The scale operation is performed from the `sfc` class watershed polygon, where the centroid is calculated using the `st_centroid` function. From the calculated centroid, the scale change operation is performed to create a geo-object half the size of the original object. The new scale parameters are assigned to the watershed geometry using the `st_set_geometry` function. The cartographic projection is assigned to the results using the `st_set_crs` function.

```
bord_cent_sfc <- st_centroid(border_sfc) # Centroid
bord_cent_sfc
# Geometry set for 1 feature
# Geometry type: POINT
# Dimension:     XY
# Bounding box:  xmin: 448083.9 ymin: 8343593 xmax: 448083.9 ymax: 8343593
# Projected CRS: +proj=utm +zone=21 +south
# POINT (448083.9 8343593)
bord_scale <- (border_sfc - bord_cent_sfc) * 0.5 + bord_cent_sfc # Scale calc.
bord_scale
# Geometry set for 1 feature
# Geometry type: POLYGON
# Dimension:     XY
# Bounding box:  xmin: 387507.7 ymin: 8276120 xmax: 517868.4 ymax: 8389919
# CRS:           NA
# POLYGON ((459284.7 8295439, 459242.5 8295289, 4...
bord_scale_sf <- st_set_geometry(bor, bord_scale) # Apply scale to sf
bord_scale_sf <- st_set_crs(bord_scale_sf, # Set CRS
          "+proj=utm +zone=21 +south +datum=WGS84 +units=m +no_defs")
bord_scale_sf
```

```
# Simple feature collection with 1 feature and 2 fields
# Geometry type: POLYGON
# Dimension:     XY
# Bounding box:  xmin: 387507.7 ymin: 8276120 xmax: 517868.4 ymax: 8389919
# Projected CRS: +proj=utm +zone=21 +south +datum=WGS84 +units=m +no_defs
#   id  name                     border_sfc
# 1  1 basin POLYGON ((459284.7 8295439,...
```

A rotation function is created to define the rotation of the watershed with the name `rotation`. The rotation function is applied to the geographical boundaries of the watershed, and then the coordinate reference system is assigned to the results.

```
rotation = function(a){r = a * pi / 180 # Degrees to radians
  matrix(c(cos(r), sin(r), -sin(r), cos(r)), nrow = 2, ncol = 2)} # Function
bord_rotate <- (border_sfc - bord_cent_sfc) * rotation(30) + bord_cent_sfc # Rot.
bord_rotate <- st_as_sf(st_set_crs(bord_rotate, # Set CRS and convert to sf
        "+proj=utm +zone=21 +south +datum=WGS84 +units=m +no_defs"))
bord_rotate
# Simple feature collection with 1 feature and 0 fields
# Geometry type: POLYGON
# Dimension:     XY
# Bounding box:  xmin: 336679.5 ymin: 8224759 xmax: 598085.7 ymax: 8445925
# Projected CRS: +proj=utm +zone=21 +south +datum=WGS84 +units=m +no_defs
#                                 x
# 1 POLYGON ((419329.6 8248986,...
```

The results of affine operations in the watershed are mapped using functions from the `ggplot2` and `ggspatial` packages (Figure 10.6).

```
a1 <- ggplot(data = bor) + geom_sf(fill= "antiquewhite") +
  annotate(geom = "text", x = 350000, y = 8237000, label = "Original",
  fontface = "italic", color = "grey22", size = 5) +
  annotation_scale(location = "br", width_hint = 0.5) +
  annotation_north_arrow(location = "br", which_north = "true",
  pad_x = unit(0.75, "in"), pad_y = unit(0.5, "in"),
  style = north_arrow_fancy_orienteering)
a2 <- ggplot(data = bor) + geom_sf(fill= "antiquewhite") +
  annotate(geom = "text", x = 350000, y = 8237000, label = "Shifting",
  fontface = "italic", color = "grey22", size = 5) +
  geom_sf(data = bord_shift, fill= "NA", linetype = "dashed", size=0.7) +
  annotation_scale(location = "br", width_hint = 0.5) +
  annotation_north_arrow(location = "br", which_north = "true",
  pad_x = unit(0.75, "in"), pad_y = unit(0.5, "in"),
  style = north_arrow_fancy_orienteering)
a3 <- ggplot(data = bor) + geom_sf(fill= "antiquewhite") +
  annotate(geom = "text", x = 350000, y = 8237000, label = "Scaling",
  fontface = "italic", color = "grey22", size = 5) +
  geom_sf(data = bord_scale_sf, fill= "NA", linetype="dashed", size=0.7) +
```

```
  annotation_scale(location = "br", width_hint = 0.5) +
  annotation_north_arrow(location = "br", which_north = "true",
  pad_x = unit(0.75, "in"), pad_y = unit(0.5, "in"),
  style = north_arrow_fancy_orienteering)
a4 <- ggplot(data = bor) + geom_sf(fill= "antiquewhite") +
  annotate(geom = "text", x = 350000, y = 8237000,
  label = "Rotation", fontface = "italic", color = "grey22", size = 5) +
  geom_sf(data = bord_rotate, fill= "NA", linetype="dashed", size=0.7) +
  annotation_scale(location = "br", width_hint = 0.5) +
  annotation_north_arrow(location = "br", which_north = "true",
  pad_x = unit(0.75, "in"), pad_y = unit(0.5, "in"),
  style = north_arrow_fancy_orienteering)
grid.arrange(a1, a2, a3, a4, ncol = 2)
```

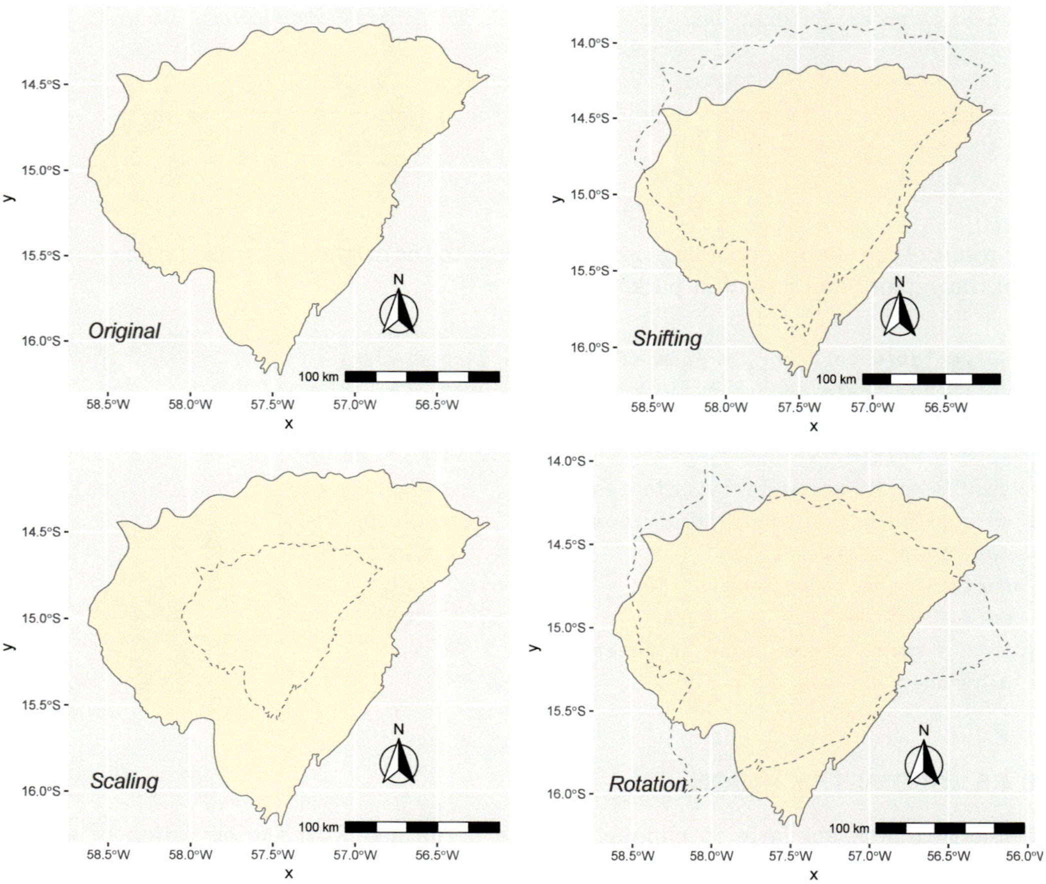

FIGURE 10.6 Affine operations of shifting, scaling and rotation applied to the geographic boundaries of the Upper-Middle and Upper Paraguay Planning and Management Units, Mato Grosso, Brazil.

10.4.5 CLIPPING

Geospatial clipping is a way of performing a spatial subsetting in a way that changes the original geometry of the geo-object by modifying the attribute representation feature. Clipping can be applied to lines and polygons. Clipping of geometries with vector type features can be performed by operations between possible combinations of the Venn diagram (Lovelace et al., 2019). As an example, the difference between the watershed polygon and the rescaled area in the center of the watershed with the same geometry as the edges, represented in the previous figure in the scale operation. With this, the delimited area in the center of the basin is clipped. According to the situation, it may be of interest to perform the clipping using the intersection operation of two figures.

The `st_difference` function is used to determine the geometric difference between the geographic boundaries of the original watershed and the boundaries of the watershed changed in scale, determining a clipping at the center of the original geographic region.

```
bord_dif <- st_difference(bor, bord_scale_sf)
bord_dif
# Simple feature collection with 1 feature and 4 fields
# Geometry type: POLYGON
# Dimension:     XY
# Bounding box:  xmin: 326931.5 ymin: 8208648 xmax: 587652.9 ymax: 8436246
# Projected CRS: +proj=utm +zone=21 +south +datum=WGS84 +units=m +no_defs
#   id  name id.1 name.1                   border_sfc
# 1  1 basin    1  basin POLYGON ((470401.1 8246985,...
```

The result of the clipping operation is mapped next to the original watershed using functions from the ggplot2 and ggspatial packages (Figure 10.7).

```
a1 <- ggplot(data = bor) + geom_sf(fill= "antiquewhite") +
  annotation_scale(location = "br", width_hint = 0.5) +
  annotation_north_arrow(location = "br", which_north = "true",
  pad_x = unit(0.75, "in"), pad_y = unit(0.5, "in"),
  style = north_arrow_fancy_orienteering)
a2 <- ggplot(data = bord_dif) + geom_sf(fill= "antiquewhite") +
  annotation_scale(location = "br", width_hint = 0.5) +
  annotation_north_arrow(location = "br", which_north = "true",
  pad_x = unit(0.75, "in"), pad_y = unit(0.5, "in"),
  style = north_arrow_fancy_orienteering)
grid.arrange(a1, a2, ncol = 2)
```

10.4.6 GEOMETRY UNION

Geometry union, similarly to clipping, can be performed with the operation of joining geo-objects with shared boundaries in a cadastral map (Figure 10.8). However, with the geometric join operation, no calculations are performed on the attribute table, as performed in the `aggregate` and `summarize` operations performed in the chapter on geospatial analysis of vector features. In geospatial analysis methods, the analysis of vector features is performed between attribute tables and geometric features simultaneously to summarize and aggregate the features.

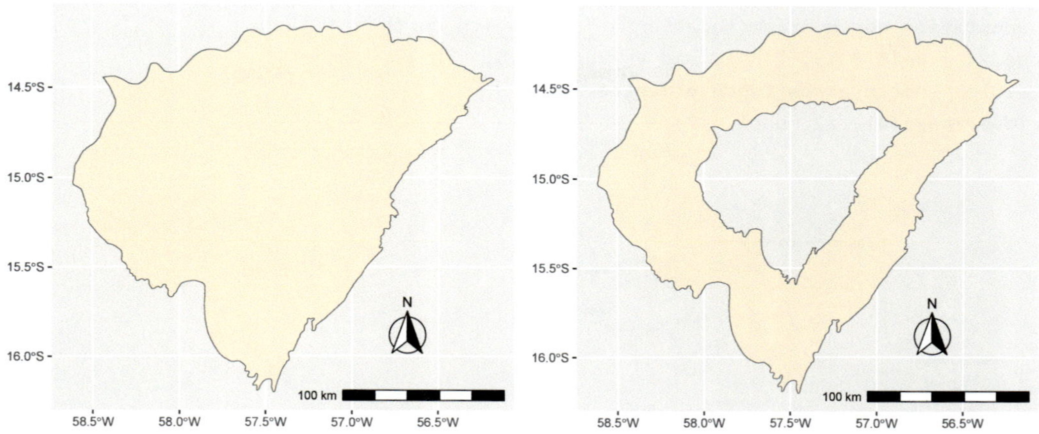

FIGURE 10.7 Applying clipping to the central region of a watershed in order to illustrate the concept of clipping by difference of geometries.

The geometric union operation was demonstrated in the code at the beginning of this chapter. The difference between the geo-objects can be observed by the geometric difference between the objects `mun` and `bor`.

```
mun
# Simple feature collection with 24 features and 3 fields
# Geometry type: MULTIPOLYGON
# Dimension:     XY
# Bounding box:  xmin: 326931.5 ymin: 8208648 xmax: 587652.9 ymax: 8436246
# Projected CRS: +proj=utm +zone=21 +south +datum=WGS84 +units=m +no_defs
# ...
bor
# Simple feature collection with 1 feature and 2 fields
# Geometry type: POLYGON
# Dimension:     XY
# Bounding box:  xmin: 326931.5 ymin: 8208648 xmax: 587652.9 ymax: 8436246
# Projected CRS: +proj=utm +zone=21 +south +datum=WGS84 +units=m +no_defs
#   id  name                     border_sfc
# 1  1 basin POLYGON ((470485.4 8247284,...
```

The results of the original municipalities in the watershed and the geometric union are mapped side-by-side with functions from the `ggplot2` and `ggspatial` packages (Figure 10.8).

```
a1 <- ggplot(data = mun) + geom_sf(fill= "antiquewhite") +
  geom_sf_text(aes(label=NM_MUNICIP), check_overlap = TRUE) +
  annotation_scale(location = "br", width_hint = 0.5) +
  annotation_north_arrow(location = "br", which_north = "true",
  pad_x = unit(0.75, "in"), pad_y = unit(0.5, "in"),
  style = north_arrow_fancy_orienteering)
a2 <- ggplot(data = bor) + geom_sf(fill= "antiquewhite") +
  annotation_scale(location = "br", width_hint = 0.5) +
```

```
  annotation_north_arrow(location = "br", which_north = "true",
  pad_x = unit(0.75, "in"), pad_y = unit(0.5, "in"),
  style = north_arrow_fancy_orienteering)
grid.arrange(a1, a2, ncol = 2)
```

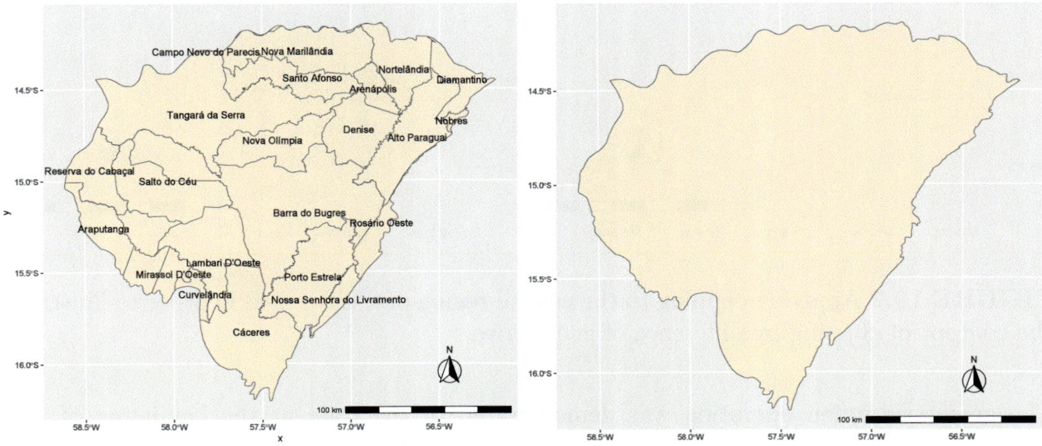

FIGURE 10.8 Geometric union of geo-objects that share the same boundary on a cadastral map used to define the geographic boundaries of the Upper-Middle Paraguay and Upper Paraguay Planning and Management Units, Mato Grosso, Brazil.

10.4.7 TRANSFORMATION BETWEEN GEOMETRY TYPES

The transformation between geometry types, e.g., from points to line or polygon can be performed according to the purpose of the analysis. Converting points to lines can be performed to measure the distance traveled using line segments in a GNSS survey. The conversion of points into polygon can be used for area calculation (Lovelace et al., 2019).

The drainage network of the watershed in Mato Grosso, originally with `multilinestring` type geometry is transformed into `multipoint` type geometry in order to illustrate transformation between geometry types.

The `st_cast` function is used to transform the drainage network of the watershed from `multilinestring` to `multipoint`.

```
hidro_points = st_cast(hidro, "MULTIPOINT")
hidro_points
# Simple feature collection with 242 features and 7 fields
# Geometry type: MULTIPOINT
# Dimension:     XY
# Bounding box:  xmin: 331838.1 ymin: 8212782 xmax: 568639.7 ymax: 8432843
# Projected CRS: +proj=utm +zone=21 +south +datum=WGS84 +units=m +no_defs
# ...
```

The results of the original drainage network with line geometry and with point geometry are mapped side-by-side using functions from the `ggplot2` and `ggspatial` packages (Figure 10.9).

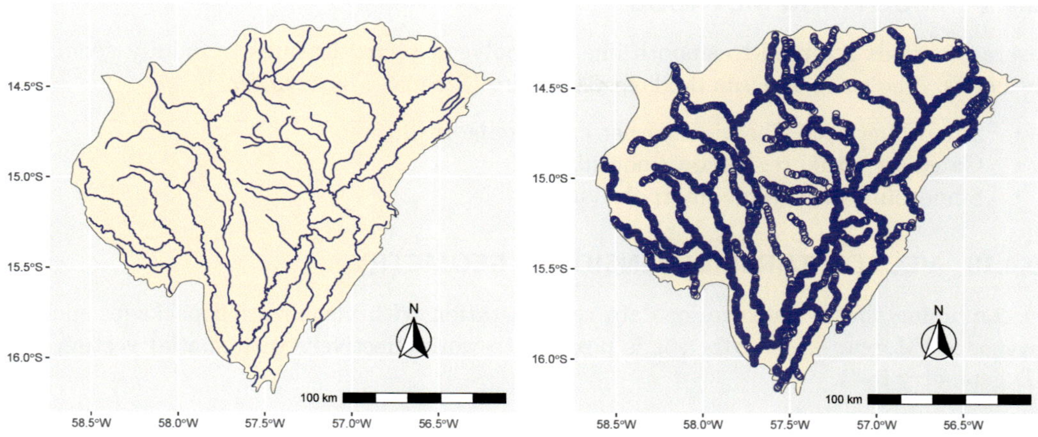

FIGURE 10.9 Transformation of drainage network vector features with multilinestring geometry to multipoint, in the Planning and Management Units of the Upper-Middle and Upper Paraguay River, Mato Grosso, Brazil.

```
a1 <- ggplot(data = bor) + geom_sf(fill= "antiquewhite") +
  geom_sf(data = hidro,  colour = "blue") +
  annotation_scale(location = "br", width_hint = 0.5) +
  annotation_north_arrow(location = "br", which_north = "true",
  pad_x = unit(0.75, "in"), pad_y = unit(0.5, "in"),
  style = north_arrow_fancy_orienteering)
a2 <- ggplot(data = bor) + geom_sf(fill= "antiquewhite") +
  geom_sf(data = hidro_points,  colour = "blue", size=2, shape=1) +
  annotation_scale(location = "br", width_hint = 0.5) +
  annotation_north_arrow(location = "br", which_north = "true",
  pad_x = unit(0.75, "in"), pad_y = unit(0.5, "in"),
  style = north_arrow_fancy_orienteering)
grid.arrange(a1, a2, ncol = 2)
```

10.4.8 DENSIFICATION

Densification is the process of adding additional points to the line segments that define polylines or polygons. Each line segment is divided into several subsegments of equal length, and the original vertices are always included in the densified feature. In the densification algorithm, the Euclidean distance is considered in vertex determinations, along a straight line (Strimas-Mackey, 2021).

Chaikin's corner cutting algorithm is used to iteratively replace each point on a line with two new points with the following criteria (Strimas-Mackey, 2021):

- Define a new point 1/4 of the way to the next point and 1/4 of the way from the previous point.

10.4.9 SMOOTHING METHODS

Several methods are used to smooth line and polygon vector features to modify geometric aesthetics. The smoothing methods available are (Strimas-Mackey, 2021):

- Smoothing with Chaikin's corner cutting algorithm;
- Gaussian kernel regression smoothing;
- Smoothing with spline interpolation.

10.4.10 VECTOR THREE-DIMENSIONAL GEOMETRY

By combining the `sf` package for data representation with visualization packages such as `rayshader` (Morgan-Wall, 2024), it is possible to work effectively with spatial vectors and 3-D geometry in R.

A 3-D vector requires an x, y, and z component to define its position in space. You can create an `sf` object by adding a geometry column to a standard R data frame, where each row represents a feature with corresponding 3-D coordinates. Vectors with three components (x, y, z) are naturally extended from 2-D vectors, representing points in a 3-D coordinate system.

Other packages such as `rgl` (Adler and Murdoch, 2025) can also be used to work with three-dimensional vector geometry and were applied in the interpolation of vector data to create continuous surfaces in Chapter 12.

Functions such as `st_point`, `st_linestring`, and `st_polygon` can be used to create simple three-dimensional `sfg` feature geometries with Z coordinates. The `sf` package can be used to create a 3-D point with the following chunk code.

```
library(sf) # Load pakage
point_3d <- st_point(c(1, 2, 3)) # Create 3-D point
point_3d  # See header
# POINT Z (1 2 3)
```

The `st_sfc` function can be used to create a simple feature geometry list column, `sfc`, and it is possible to combine multiple `sfg` objects into one list column.

```
sfc_3d <- st_sfc(point_3d) # Create sfc object
sfc_3d # See header
# Geometry set for 1 feature
# Geometry type: POINT
# Dimension:     XYZ
# Bounding box:  xmin: 1 ymin: 2 xmax: 1 ymax: 2
# z_range:       zmin: 3 zmax: 3
# CRS:           NA
# POINT Z (1 2 3)
```

The `sf` object can combine an attribute data structure with the `sfc` object using the `st_sf` function, as an example of an `sf` object with a 3-D point and an attribute.

```
attributes <- data.frame(id = 1, name = "3-D Feature") # Create data frame
attributes # See header
```

```
#   id        name
# 1  1    3-D Feature
sf_object_3d <- st_sf(data_attributes, geometry = sfc_3d)
sf_object_3d
# Simple feature collection with 1 feature and 2 fields
# Geometry type: POINT
# Dimension:     XYZ
# Bounding box:  xmin: 1 ymin: 2 xmax: 1 ymax: 2
# z_range:       zmin: 3 zmax: 3
# CRS:           NA
#   id        name        geometry
# 1  1 3-D Feature POINT Z (1 2 3)
```

10.5 COMPUTATIONAL PRACTICE

A computational practice is performed on objects with simple vector features in R. Functions to change the geometry of geo-objects are used on objects of the simple feature classes with the table (`data.frame`) with feature attributes and feature geometries (`sf`) and other geometric derivations needed in some situations, such as the list-column with the geometries for each feature (record) (`sfc`) and the feature geometry of an individual simple feature (`sfg`).

10.5.1 PACKAGES USED

The `install.packages` function is used to install the packages `sf`, `raster` (Hijmans et al., 2020), `dplyr` (Wickham et al., 2021a), `smoothr` (Strimas-Mackey, 2021), `ggplot2`, `gridExtra`, `ggspatial` and `scales` (Wickham et al., 2023) needed to perform the practical computational activities. If the packages are already installed, it is not necessary to install them again for each computational practice.

```
install.packages("sf")
install.packages("raster")
install.packages("dplyr")
install.packages("smoothr")
install.packages("ggplot2")
install.packages("gridExtra")
install.packages("ggspatial")
install.packages("scales")
```

The `library` function is used to load the packages necessary to perform the computational practice.

```
library(sf)
# Linking to GEOS 3.11.2, GDAL 3.7.2, PROJ 9.3.0; sf_use_s2() is TRUE
library(raster)
# Loading required packages: sp
library(dplyr)
library(smoothr)
library(ggplot2)
```

```
library(gridExtra)
library(ggspatial)
# package 'ggspatial' was built under R version 4.3.3
library(scales)
```

10.5.2 SIMPLIFICATION

A drainage network with four lines is created using the `st_linestring` function in order to simulate a drainage network. The `st_sfc` function is used to group the linestrings into a single file. The `st_cast` function is used to convert `linestring` to `multilinestring`. The `st_sf` function is used to convert the drainage network into an `sf` feature. The UTM map projection, zone 23S, datum WGS-84, is applied to the lines using the `st_crs` function.

```
l1 <- st_linestring(rbind(c(502500,7652620),c(502510,7652700),
  c(502650,7652720), c(502700,7652710), c(502740,7652730))) # Create row 1
l2 <- st_linestring(rbind(c(502510,7652700),c(502520,7652680),
  c(502630,7652630), c(502700,7652540)))  # Create row 2
l3 <- st_linestring(rbind(c(502520,7652680),c(502530,7652650),
  c(502625,7652600), c(502650,7652530), c(502720,7652505))) # Create row 3
l4 <- st_linestring(rbind(c(502630,7652630),c(502680,7652660),
  c(502700,7652700), c(502730,7652710))) # Create row 4
sfc14 <- st_sfc(l1, l2, l3, l4) # Convert to sfc
l14 <- st_cast(sfc14, "MULTILINESTRING") # Convert to multilinestring
l14sf <- st_sf(l14) # Convert to sf
# Assign projection
st_crs(l14sf) <- "+proj=utm +zone=23 +south +datum=WGS84 +units=m +no_defs"
l14sf
# Simple feature collection with 4 features and 0 fields
# Geometry type: MULTILINESTRING
# Dimension:     XY
# Bounding box:  xmin: 502500 ymin: 7652505 xmax: 502740 ymax: 7652730
# Projected CRS: +proj=utm +zone=23 +south +datum=WGS84 +units=m +no_defs
#                              l14
# 1 MULTILINESTRING ((502500 76...
# 2 MULTILINESTRING ((502510 76...
# 3 MULTILINESTRING ((502520 76...
# 4 MULTILINESTRING ((502630 76...
```

A simplification function with tolerances of 11, 13 and 16m is applied to the lines using the `st_simplify` function in order to evaluate the implementation of the Douglas-Peucker algorithm in the `sf` package.

```
lsimp11 <- st_simplify(l14sf, dTolerance = 11)  # 11 m
lsimp13 <- st_simplify(l14sf, dTolerance = 13)  # 13 m
lsimp16 <- st_simplify(l14sf, dTolerance = 16)  # 16 m
```

The results are mapped using functions from the `ggplot2` and `ggspatial` packages (Figure 10.10).

```
a1 <- ggplot(data = l14sf) + geom_sf(colour= "blue") +
  annotate(geom = "text", x = 502540, y = 7652570, label = "Original",
  fontface = "italic", color = "grey22", size = 5) +
  annotation_scale(location = "bl", width_hint = 0.5) +
  annotation_north_arrow(location = "bl", which_north = "true",
  pad_x = unit(0.75, "in"), pad_y = unit(0.3, "in"),
  style = north_arrow_fancy_orienteering)
a2 <- ggplot(data = lsimp11) + geom_sf(colour= "blue") +
  annotate(geom = "text", x = 502540, y = 7652570, label = "Simp.11 m",
  fontface = "italic", color = "grey22", size = 5) +
  annotation_scale(location = "bl", width_hint = 0.5) +
  annotation_north_arrow(location = "bl", which_north = "true",
  pad_x = unit(0.75, "in"), pad_y = unit(0.3, "in"),
  style = north_arrow_fancy_orienteering)
a3 <- ggplot(data = lsimp13) + geom_sf(colour= "blue") +
  annotate(geom = "text", x = 502540, y = 7652570, label = "Simp.13 m",
  fontface = "italic", color = "grey22", size = 5) +
  annotation_scale(location = "bl", width_hint = 0.5) +
  annotation_north_arrow(location = "bl", which_north = "true",
  pad_x = unit(0.75, "in"), pad_y = unit(0.3, "in"),
  style = north_arrow_fancy_orienteering)
a4 <- ggplot(data = lsimp16) + geom_sf(colour= "blue") +
  annotate(geom = "text", x = 502540, y = 7652570, label = "Simp.16 m",
  fontface = "italic", color = "grey22", size = 5) +
  annotation_scale(location = "bl", width_hint = 0.5) +
  annotation_north_arrow(location = "bl", which_north = "true",
  pad_x = unit(0.75, "in"), pad_y = unit(0.3, "in"),
  style = north_arrow_fancy_orienteering)
grid.arrange(a1, a2, a3, a4, ncol = 2)
```

The size of the file with the geographic simple features after simplification is evaluated with the `object.size` function.

```
object.size(l14sf) # Original
# 18024 bytes
object.size(lsimp11) # 11m
# 8408 bytes
object.size(lsimp13) # 13m
# 8328 bytes
object.size(lsimp16) # 16m
# 8312 bytes
```

We can see that starting at the 11 m tolerance threshold the drainage network is simplified, and thereafter there is an increase in the simplification of the features with increasing tolerance threshold.

10.5.3 CENTROID

A simple polygon feature is created with the `st_polygon` function to illustrate determining the centroid. Next, two more polygons are created by adding +0.001° and -0.001° to the

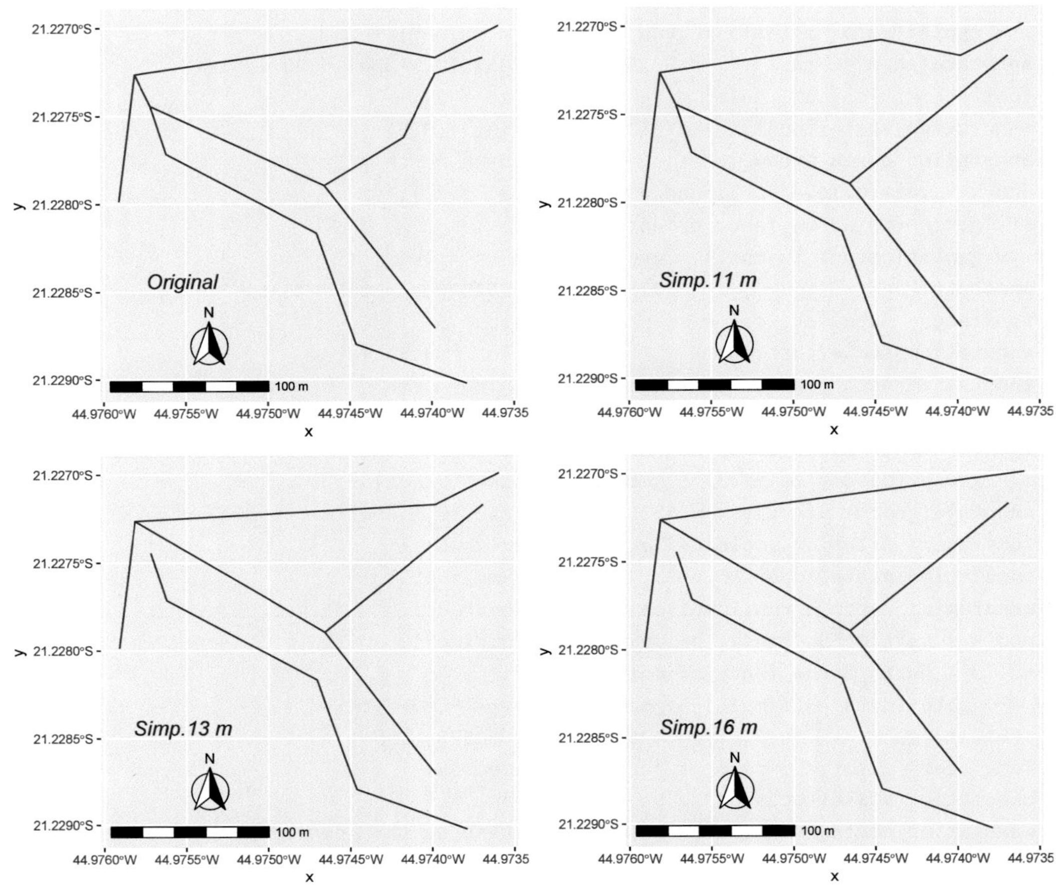

FIGURE 10.10 Simplification of simulated drainage network with simplification tolerance threshold of 11, 13 and 16 m.

originally created polygon, respectively. The `st_sfc` function is used to join the polygon geometries into a 3-line file. A WGS-84 geographic projection system is assigned to the polygons using the `st_crs` function. Since the `st_centroid` function did not provide correct centroids for data with longitude and latitude coordinates, the data are transformed to the UTM map projection, 23S zone using the `st_transform` function. The `st_centroid` function is used to determine the centroid within each of the three polygons.

```
apoly <- st_polygon(list(rbind(c(-44.97537, -21.22805), c(-44.97537, -21.22722),
    c(-44.97417, -21.22722),c(-44.97417, -21.22805),
    c(-44.97537, -21.22805)))) # Create polygon a
bpoly <- apoly + 0.001 # Create polygon b
cpoly <- apoly - 0.001 # Create polygon c
mp <- st_sfc(apoly, bpoly, cpoly) # Combining geometries in the sfc class
st_crs(mp) <- 4326 # Assigning projection
mpp <- st_transform(mp, crs = "+proj=utm +zone=23 +south") # Transform crs
mp_centroid <- st_centroid(mpp) # Centroid calculation
```

The results are mapped using functions from the `ggplot2` and `ggspatial` packages (Figure

10.11).

```
a1 <- ggplot(data = mpp) + geom_sf(fill="white", colour= "black") +
  annotate(geom = "text", x = 502720, y = 7652575, label = "Original",
  fontface = "italic", color = "grey22", size = 5) +
  annotation_scale(location = "br", width_hint = 0.5) +
  annotation_north_arrow(location = "br", which_north = "true",
  pad_x = unit(0.75, "in"), pad_y = unit(0.5, "in"),
  style = north_arrow_fancy_orienteering)
a2 <- ggplot(data = mpp) + geom_sf(fill="white", colour= "black") +
  geom_sf(data = mp_centroid, colour= "red", shape = 1, size = 3) +
  annotate(geom = "text", x = 502720, y = 7652575, label = "Centroid",
  fontface = "italic", color = "grey22", size = 5) +
  annotation_scale(location = "br", width_hint = 0.5) +
  annotation_north_arrow(location = "br", which_north = "true",
  pad_x = unit(0.75, "in"), pad_y = unit(0.5, "in"),
  style = north_arrow_fancy_orienteering)
grid.arrange(a1, a2, ncol = 2)
```

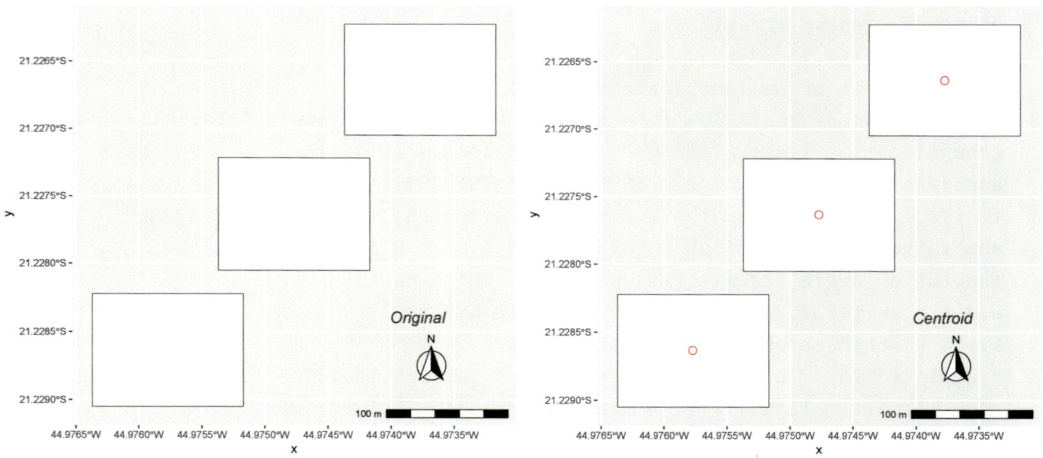

FIGURE 10.11 Polygon features (left) used to illustrate the concept of centroids calculated as a function of the center of mass of each polygon (right).

10.5.4 BUFFER

Buffers are applied to points, lines and polygons to define an area, given a distance from a geometric feature. Buffers of 10 and 20 m are applied to the simulated drainage network, as well as to the polygons and centroids defined in the previous item using the `st_buffer` function.

```
pbuf10 <- st_buffer(mp_centroid, dist = 10) # 10m buffer at the point
pbuf20 <- st_buffer(mp_centroid, dist = 20) # 20m buffer at the point
lbuf10 <- st_buffer(l14sf, dist = 10) # 10m buffer on the line
lbuf20 <- st_buffer(l14sf, dist = 20) # 20m buffer on the line
```

```
polbuf10 <- st_buffer(mpp, dist = 10) # 10m buffer in the polygon
polbuf20 <- st_buffer(mpp, dist = 20) # 20m buffer in the polygon
```

The results are mapped using functions from the `ggplot2`, `ggspatial` and `scales` packages (Figure 10.12).

```
a1 <- ggplot(data = pbuf10) + geom_sf(fill = "white", colour = "black") +
    geom_sf(data = mp_centroid, fill = "black", colour = "red") +
    annotate(geom = "text", x = 502700, y = 7652600, label = "Buffer 10m",
    fontface = "italic", color = "grey22", size = 5) +
    annotation_scale(location = "br", width_hint = 0.5) +
    annotation_north_arrow(location = "br", which_north = "true",
    pad_x = unit(0.75, "in"), pad_y = unit(0.3, "in"),
    style = north_arrow_fancy_orienteering)
a2 <- ggplot(data=pbuf20) + geom_sf(fill = "white", colour = "black") +
    geom_sf(data = mp_centroid, fill = "black", colour = "red") +
    annotate(geom = "text", x = 502700, y = 7652600, label = "Buffer 20 m",
    fontface = "italic", color = "grey22", size = 5) +
    annotation_scale(location = "br", width_hint = 0.5) +
    annotation_north_arrow(location = "br", which_north = "true",
    pad_x = unit(0.75, "in"), pad_y = unit(0.3, "in"),
    style = north_arrow_fancy_orienteering)
a3 <- ggplot(data = lbuf10) + geom_sf(fill = "white", colour = "black") +
    geom_sf(data = l14sf, fill = "black", colour = "blue") +
    annotate(geom = "text", x = 502530, y = 7652580, label = "Buffer 10 m",
    fontface = "italic", color = "grey22", size = 5) +
    annotation_scale(location = "bl", width_hint = 0.5) +
    annotation_north_arrow(location = "bl", which_north = "true",
    pad_x = unit(0.75, "in"), pad_y = unit(0.5, "in"),
    style = north_arrow_fancy_orienteering)
a4 <- ggplot(data = lbuf20) + geom_sf(fill = "white", colour = "black") +
    geom_sf(data = l14sf, fill = "black", colour = "blue") +
    annotate(geom = "text", x = 502530, y = 7652580, label = "Buffer 20 m",
    fontface = "italic", color = "grey22", size = 5) +
    annotation_scale(location = "bl", width_hint = 0.5) +
    annotation_north_arrow(location = "bl", which_north = "true",
    pad_x = unit(0.75, "in"), pad_y = unit(0.5, "in"),
    style = north_arrow_fancy_orienteering)
a5 <- ggplot(data = polbuf10) + geom_sf(fill = "white", colour = "black") +
    geom_sf(data = mpp, fill = "antiquewhite", colour = "red") +
    annotate(geom = "text", x = 502720, y = 7652580, label = "Buffer 10 m",
    fontface = "italic", color = "grey22", size = 5) +
    annotation_scale(location = "br", width_hint = 0.5) +
    annotation_north_arrow(location = "br", which_north = "true",
    pad_x = unit(0.75, "in"), pad_y = unit(0.3, "in"),
    style = north_arrow_fancy_orienteering) +
  scale_x_continuous(labels = label_number(accuracy = 0.01))
a6 <- ggplot(data = polbuf20) + geom_sf(fill = "white", colour = "black") +
    geom_sf(data = mpp, fill = "antiquewhite", colour = "red") +
```

```
    annotate(geom = "text", x = 502720, y = 7652580, label = "Buffer 20 m",
    fontface = "italic", color = "grey22", size = 5) +
    annotation_scale(location = "br", width_hint = 0.4) +
    annotation_north_arrow(location = "br", which_north = "true",
    pad_x = unit(0.75, "in"), pad_y = unit(0.2, "in"),
    style = north_arrow_fancy_orienteering) +
    scale_x_continuous(labels = label_number(accuracy = 0.01))
grid.arrange(a1, a3, a5, a2, a4, a6, ncol = 3)
```

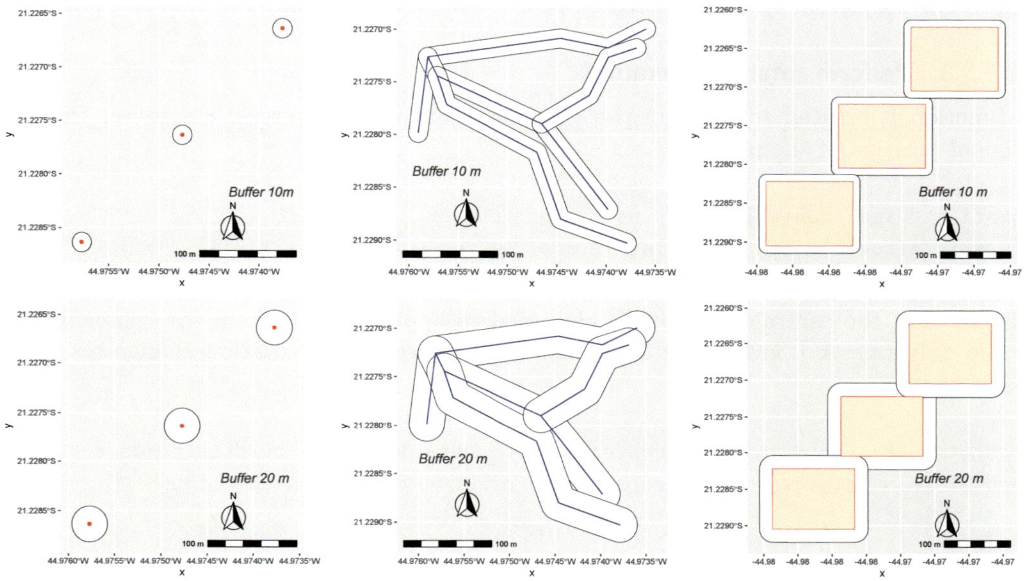

FIGURE 10.12 Different buffer types applied to vector features of points (left), lines (center) and polygons (right) with buffers of 10 (top) and 20 m (bottom), respectively.

10.5.5 AFFINE

The affine functions are used to perform position change, scale, and rotation on vector features. These operations are applied to the three polygons created earlier as well as to the centroid created at the center of mass of the polygons.

10.5.5.1 Perform shifting operation

The shifting operation is applied to the previously created polygon, with a position shift of 10 m to the north and to the east, respectively. The UTM projection system is assigned to the polygon position change results for mapping purposes.

```
bordUtm <- mpp
bordsfc <- st_geometry(bordUtm)
bordshift <- bordsfc + c(10, 10) # 10x10m shift
bordshift <- st_set_crs(bordshift, "+proj=utm +zone=21 +south +datum=WGS84
                        +units=m +no_defs")
```

10.5.5.2 Perform scaling operation

To perform scaling, the position of the centroid inside the polygon is used to halve the border of the original rectangle according to the following function:

```
bordcentroidsfc <- mp_centroid
bordscale <- (mpp - bordcentroidsfc) * 0.5 + bordcentroidsfc
 bordscale <- st_set_crs(bordscale, "+proj=utm +zone=21 +south +datum=WGS84
                         +units=m +no_defs")
```

The UTM projection system is assigned to the scale change results for mapping purposes.

10.5.5.3 Perform rotation operation

A function is created to perform the rotation operation in degrees, clockwise, from the original position of the polygon.

```
rotation = function(a){r = a * pi / 180 # Convert degrees to radians
  matrix(c(cos(r), sin(r), -sin(r), cos(r)), nrow = 2, ncol = 2)}
```

In this case, the centroid position is also used to apply rotation from the center of mass of the polygon. A UTM projection system is assigned to the rotation change results for mapping purposes.

```
bordrotate <- (bordsfc - bordcentroidsfc) * rotation(30) + bordcentroidsfc
bordrotate <- st_set_crs(bordrotate, "+proj=utm +zone=21 +south +datum=WGS84
                         +units=m +no_defs")
```

The results are mapped using functions from the `ggplot2`, `ggspatial` and `scales` packages (Figure 10.13).

```
a1 <- ggplot(data = bordUtm) + geom_sf(fill = "antiquewhite") +
  annotate(geom = "text", x = 502650, y = 7652550, label = "Original",
  fontface = "italic", color = "grey22", size = 5) +
  annotation_scale(location = "br", width_hint = 0.5) +
  annotation_north_arrow(location = "br", which_north = "true",
  pad_x = unit(0.75, "in"), pad_y = unit(0.5, "in"),
  style = north_arrow_fancy_orienteering) +
  scale_x_continuous(labels = label_number(accuracy = 0.01))
a2 <- ggplot(data = bordUtm) + geom_sf(data = bordUtm, fill = "antiquewhite") +
  geom_sf(aes(geometry = bordshift), fill = "NA", linetype ="dashed",
  size=0.7) + annotate(geom = "text", x = 502650, y = 7652550,
  label = "Shifting", fontface = "italic", color = "grey22", size = 5) +
  annotation_scale(location = "br", width_hint = 0.5) +
  annotation_north_arrow(location = "br", which_north = "true",
  pad_x = unit(0.75, "in"), pad_y = unit(0.5, "in"),
  style = north_arrow_fancy_orienteering) +
  scale_x_continuous(labels = label_number(accuracy = 0.01))
a3 <- ggplot(data = bordUtm) + geom_sf(fill = "antiquewhite") +
  annotate(geom = "text", x = 502650, y = 7652550, label = "Scaling",
```

```
  fontface = "italic", color = "grey22", size = 5) +
  geom_sf(aes(geometry = bordscale), fill = "NA", linetype = "dashed",
  size=0.7) + annotation_scale(location = "br", width_hint = 0.5) +
  annotation_north_arrow(location = "br", which_north = "true",
  pad_x = unit(0.75, "in"), pad_y = unit(0.5, "in"),
  style = north_arrow_fancy_orienteering) +
  scale_x_continuous(labels = label_number(accuracy = 0.01))
a4 <- ggplot(data = bordUtm) + geom_sf(fill = "antiquewhite")+
  annotate(geom = "text", x = 502650, y = 7652550, label = "Rotation",
  fontface = "italic", color = "grey22", size = 5) +
  geom_sf(aes(geometry = bordrotate), fill = "NA", linetype = "dashed",
  size = 0.7) + annotation_scale(location = "br", width_hint = 0.5) +
  annotation_north_arrow(location = "br", which_north = "true",
  pad_x = unit(0.75, "in"), pad_y = unit(0.5, "in"),
  style = north_arrow_fancy_orienteering) +
  scale_x_continuous(labels = label_number(accuracy = 0.01))
grid.arrange(a1, a2, a3, a4, ncol = 2)
```

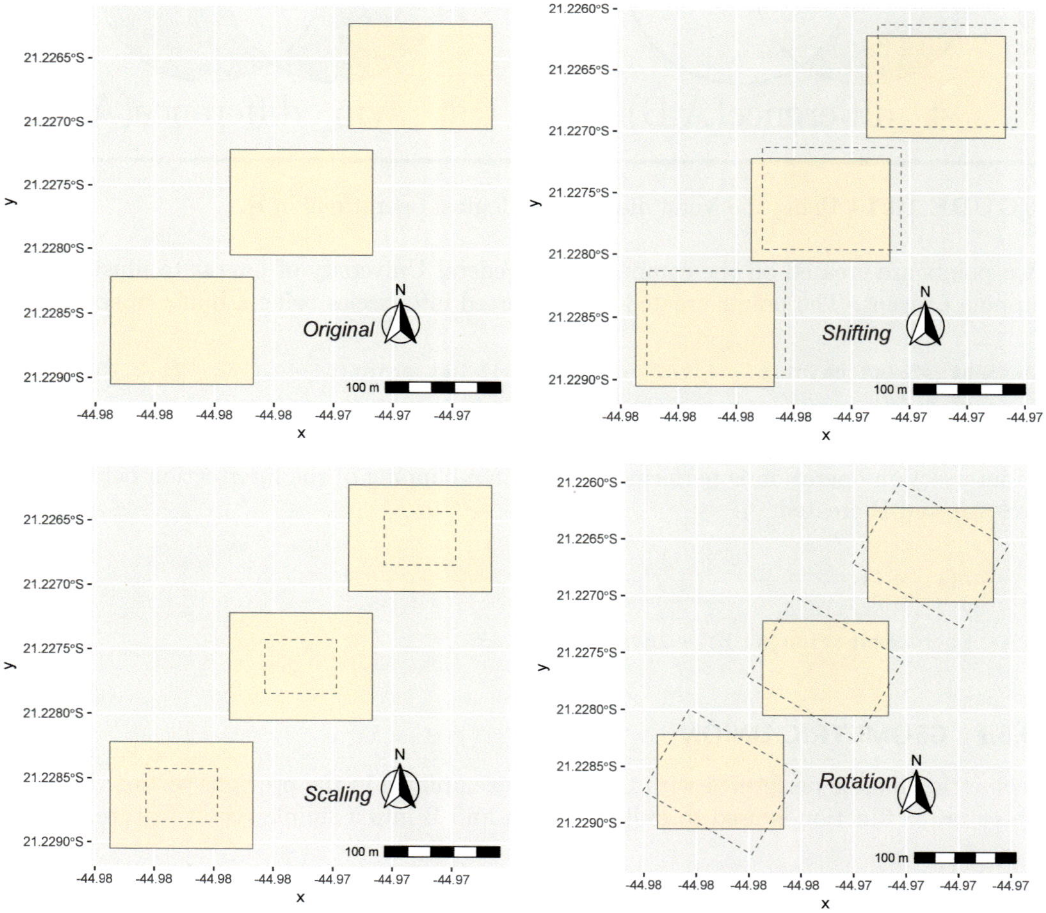

FIGURE 10.13 Shifting, scaling and rotating operations performed on polygon features.

10.5.6 CLIPPING

Different combinations of the Venn diagram can be used to perform clipping, joining, intersection, and difference between geospatial objects (Figure 10.14).

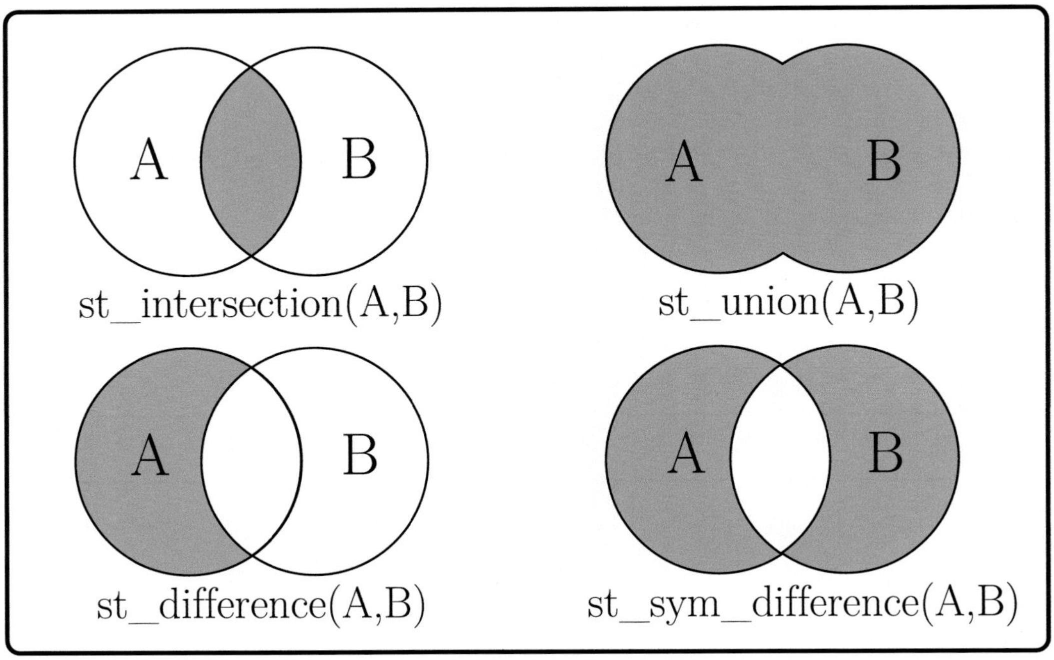

FIGURE 10.14 Using the Venn diagram for logical operations in R.

Two points are created on the campus of the Federal University of Lavras to illustrate the clipping concept. The points created are converted into circles using a buffer operation.

```
p2 <- st_sfc(st_point(c(-44.976, -21.227613)), st_point(c(-44.975, -21.227613)))
c2 <- st_buffer(p2, dist = 0.001) # Converting points to circles
```

An intersection operation is performed to obtain a clipping of the intersection between the circles A and B created.

```
A <- c2[1] # Circle A
B <- c2[2] # Circle B
AB <- st_intersection(A, B) # Intersection AB
```

10.5.7 GEOMETRIC UNION

Geometric union is performed with the circles exemplified in the previous section; however, the `st_union` function is used to unite circles A and B into a simple vector feature.

```
uAB <- st_union(A, B)
```

The WGS-84 coordinate system is assigned to the intersection and union results and both geometries are mapped to illustrate the concepts presented.

```
st_crs(c2) <- 4326 # Points
st_crs(A) <- 4326 # Buffer A
st_crs(B) <- 4326 # Buffer B
st_crs(AB) <- 4326 # Intersection AB
st_crs(uAB) <- 4326 # Union AB
```

The results are mapped using functions from the `ggplot2` and `ggspatial` packages (Figure 10.15).

```
a1 <- ggplot(data = c2) + geom_sf(fill = "NA", size = 1, colour = "black") +
  geom_sf(aes(geometry = p2), colour = "black",  shape = 1, size = 5) +
  annotate(geom = "text", x = -44.9765, y = -21.2275, label = "A",
  fontface = "italic", color = "grey22", size = 5) +
  annotate(geom = "text", x = -44.9745, y = -21.2275, label = "B",
  fontface = "italic", color = "grey22", size = 5)
a2 <- ggplot(data = c2) + geom_sf(fill = "NA", linetype = "dashed",
  size = 1, colour = "black") + geom_sf(aes(geometry = AB), fill = "antiquewhite")
a3 <- ggplot(data = uAB) + geom_sf(fill = "antiquewhite", size = 1,
                                   colour = "black")
a4 <- ggplot(data = c2) + geom_sf(fill = "NA", size = 1, colour = "NA") +
  geom_sf(aes(geometry = p2), colour = "black",  shape = 1, size = 5) +
  annotate(geom = "text", x = -44.976, y = -21.2275, label = "p1",
  fontface = "italic", color = "grey22", size = 5) +
  annotate(geom = "text", x = -44.975, y = -21.2275, label = "p2",
  fontface = "italic", color = "grey22", size = 5)
grid.arrange(a4, a1, a2, a3, ncol=2)
```

10.5.8 TRANSFORMATION BETWEEN GEOMETRIC TYPES

The transformation between geometric types can be performed with the `st_cast` function. The use of this function varies depending on the type of data model used in the file, and there may be differences when performing the operation if the geo-object consists of single simple geometric feature (`sfg`), columnar simple geometric feature (`sfc`), and simple feature objects (`sf`) (Lovelace et al., 2019).

A `multipoint` array is created in a two-column array to illustrate the concept of transformation between point, line, and polygon geometry types.

```
pts <- st_multipoint(matrix(c(-44.97537, -44.97537, -44.97417, -44.97417,
              -21.22805, -21.22722, -21.22722, -21.22805), ncol = 2))
```

The `multipoint` feature is transformed into a feature of line and polygon types using the `st_cast` function.

```
lns <- st_cast(pts, "LINESTRING")
plg <- st_cast(pts, "POLYGON")
```

The transformation process can be reversed from line to points and from polygon to `multipoint` as follows:

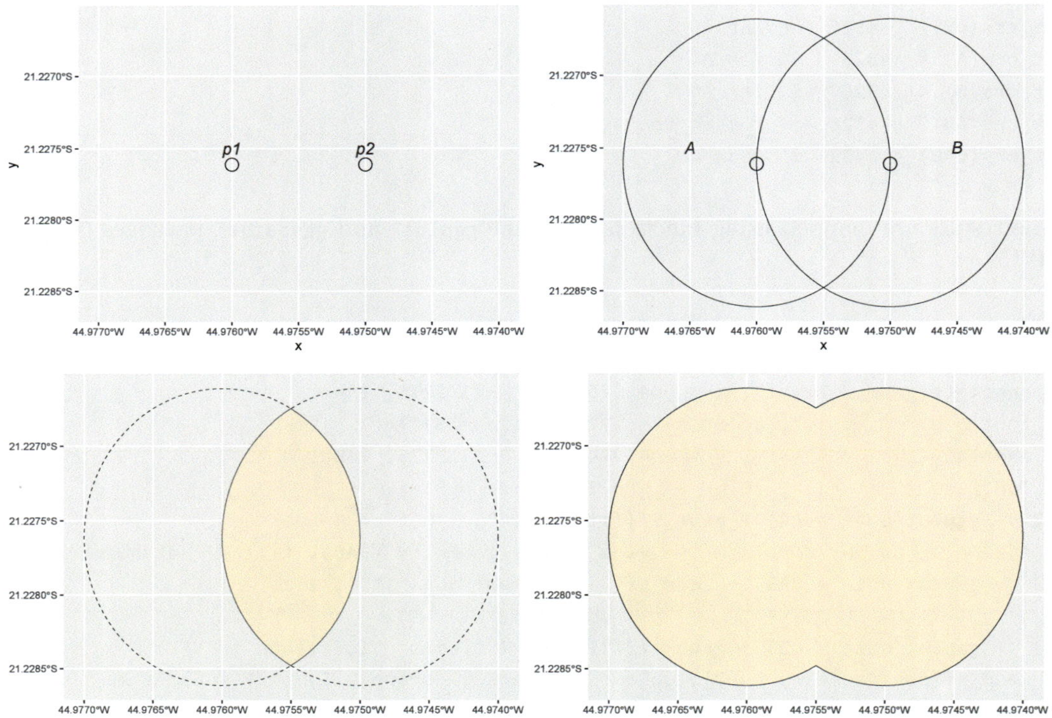

FIGURE 10.15 Creation of the points, p1 and p2 (top-left), buffering of 0.001 degrees around the points (top-right), in order to illustrate the concepts of intersection (bottom-left) and geometric union (bottom-right).

```
pts1 <- st_cast(lns, "MULTIPOINT")
pts2 <- st_cast(plg, "MULTIPOINT")
all.equal(pts1, pts2)
# [1] TRUE
```

The created objects are converted to `sfc` class geometry using the `st_sfc` function, and then assigned a WGS-84 reference coordinate system using the `st_crs` function.

```
ptsfc <- st_sfc(pts) # Convert points to sfc
lnsfc <- st_sfc(lns) # Convert lines to sfc
plgsfc <- st_sfc(plg) # Convert polygon to sfc
st_crs(ptsfc) <- 4326 # Points
st_crs(lnsfc) <- 4326 # Lines
st_crs(plgsfc) <- 4326 # Polygon
```

The results are mapped using functions from the `ggplot2` package (Figure 10.16).

```
a1 <- ggplot(data = ptsfc) + geom_sf(fill = "NA", size = 5, shape = 1,
                                      colour = "black")
a2 <- ggplot(data = lnsfc) + geom_sf(size = 1, colour = "black")
```

```
a3 <- ggplot(data = plgsfc) + geom_sf(fill = "antiquewhite", size = 1,
                                      colour = "black")
grid.arrange(a1, a2, a3, ncol = 2)
```

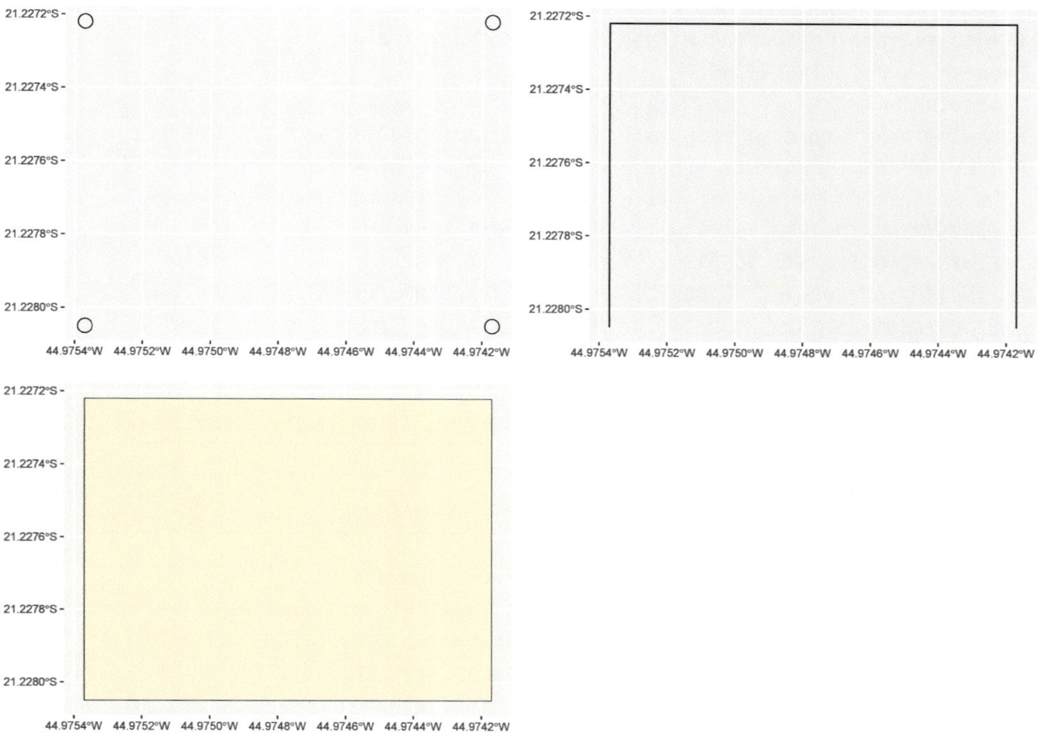

FIGURE 10.16 Conversion between geometry types, from point (top-right) to line (top-left) and from point to polygon (bottom).

The `st_cast` function is also used in the drainage network example where simplification thresholds are applied to the sf class object with `multiline` geometry in the columns of the vector feature. In this case, the initially created sfc object consists of four `linestrings` that defined the entire network of lines created. The `st_cast` function is used in that situation to convert the network of lines into four multilinearities. In this example, the vector network with four `multilinestrings` is converted into `linestrings` to make it possible to create new attributes and record measurements required in the database of the simple vector features of each line.

```
lsfc <- st_cast(l14, "LINESTRING") # Convert multilinestring to linestring
lsf <- st_sf(lsfc) # Convert to sf
```

The UTM map projection zone 23S, datum WGS-84, is assigned to the line vectors.

```
st_crs(lsf) <- "+proj=utm +zone=23 +south +datum=WGS84 +units=m +no_defs"
```

The registration of the name of four rivers is performed for each `linestring`, as well as the calculation of the length of the lines using of the `st_length` function.

```
lsf$name <- c("River 1", "River 2", "River 3", "River 4")
lsf$length <- st_length(lsf)
lsf
# Simple feature collection with 4 features and 2 fields
# Geometry type: LINESTRING
# Dimension:     XY
# Bounding box:  xmin: 502500 ymin: 7652505 xmax: 502740 ymax: 7652730
# Projected CRS: +proj=utm +zone=23 +south +datum=WGS84 +units=m +no_defs
#                              lsfc    name        length
# 1 LINESTRING (502500 7652620,... River 1 317.7555 [m]
# 2 LINESTRING (502510 7652700,... River 2 257.2087 [m]
# 3 LINESTRING (502520 7652680,... River 3 287.6380 [m]
# 4 LINESTRING (502630 7652630,... River 4 134.6537 [m]
```

The results are mapped using functions from the `ggplot2` package (Figure 10.17).

```
a1 <- ggplot(data = l14sf) + geom_sf(size = 1, shape = 1, colour = "blue") +
  ggtitle("Multilinestring") +
  scale_x_continuous(labels = label_number(accuracy = 0.01))
a2 <- ggplot(data = lsf) + geom_sf(size = 1, aes(colour = name)) +
  ggtitle("Linestring") + theme(legend.position = c(0.2, 0.2)) +
  scale_x_continuous(labels = label_number(accuracy = 0.01)) +
  scale_colour_discrete(name = "Rivers")
grid.arrange(a1, a2, ncol = 2)
```

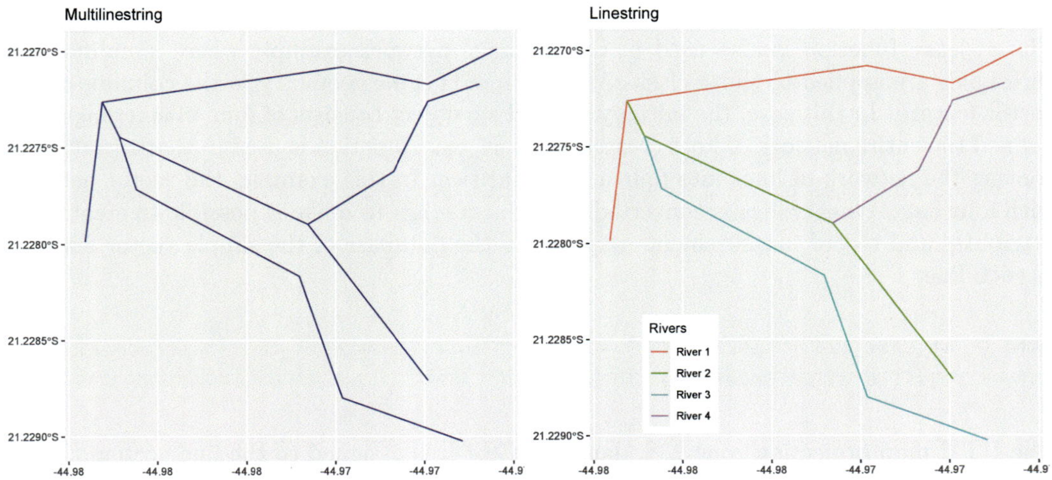

FIGURE 10.17 Example of transformation between geometric types to obtain multilinestring and linestring type features from simulated drainage network.

10.5.9 DENSIFICATION

The drainage network created earlier is densified using the `densify` function, dividing each segment into two (n = 2) and four (n = 4) segments.

```
l <- lsf # Create a new geobject l from lsf
l_dense <- densify(l, n = 2) # Divide each segment into 2
l_dense4 <- densify(l, n = 4) # Divide each segment into 4
```

The size of the file with the geographic simple features after simplification is evaluated with the `object.size` function.

```
object.size(l) # Original
# 10392 bytes
object.size(l_dense) # n = 2
# 11256 bytes
object.size(l_dense4) # n = 4
# 11672 bytes
```

The `st_cast("MULTIPOINT")` function is used to convert the densified line to multipoint in order to view the comparative results of the densification in relation to the original file.

The results are mapped using functions from the `sf` package (Figure 10.18).

```
par(mfrow = c(2, 2))
plot(st_geometry(l), col = "blue", lwd = 5, bgc = 'gray92', axes = TRUE,
     main = "Original linestring")
plot(st_geometry(l), col = "gray", lwd = 5, bgc = 'gray92', axes = TRUE,
     main = "Original densification")
plot(st_geometry(l) |> st_cast("MULTIPOINT"), col = "blue", pch = 19, add = TRUE)
plot(st_geometry(l), col = "gray", lwd = 5, bgc = 'gray92', axes = TRUE,
     main = "Densification (n = 2)")
plot(st_geometry(l_dense) |> st_cast("MULTIPOINT"), col = "blue", pch = 19,
     add = TRUE)
plot(st_geometry(l), col = "gray", lwd = 5, bgc = 'gray92', axes = TRUE,
     main = "Densification (n = 4)")
plot(st_geometry(l_dense4) |> st_cast("MULTIPOINT"), col = "blue", pch = 19,
     add = TRUE)
```

This densification option can be useful for increasing the density of sample points along the drainage network of the watershed or a transect, devised for soil sampling in a systematic way.

10.5.10 SMOOTHING METHODS

The smoothing methods by Chaikin's corner cutting algorithm, by Gaussian kernel regression and by spline interpolation are implemented in the drainage network used in the previous example using the `smooth` function (Figure 10.19).

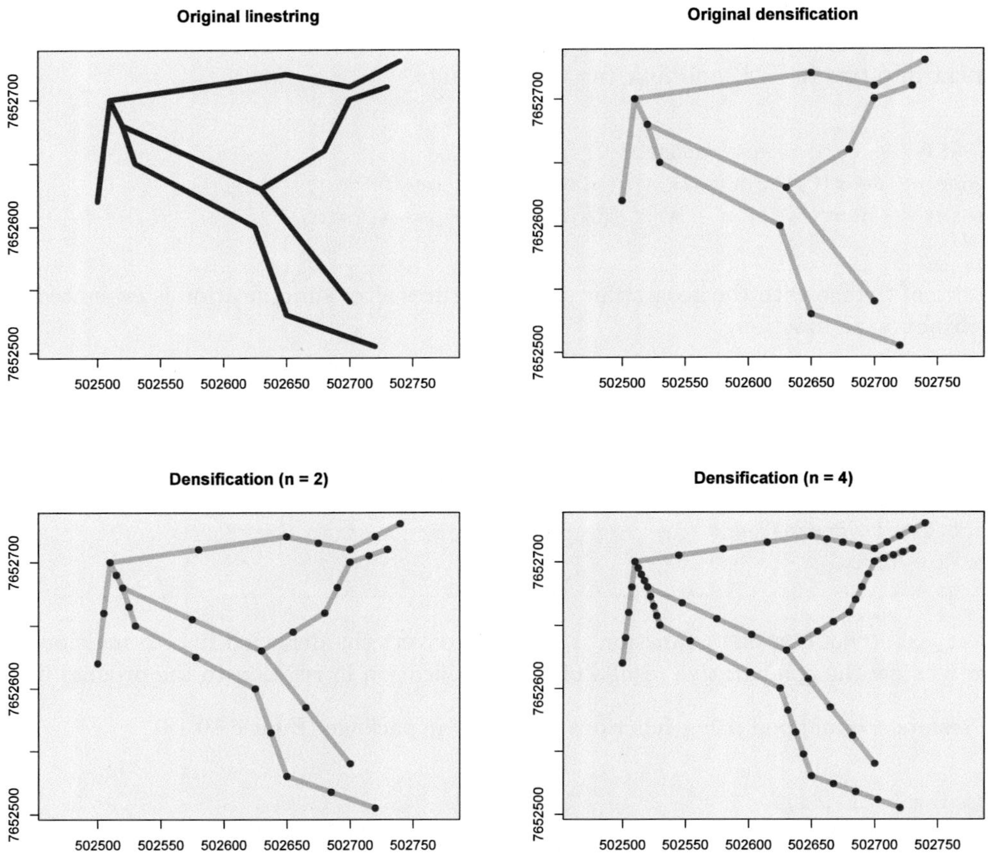

FIGURE 10.18 Densification of original simulated drainage network and conversion of the lines into densified versions with n = 2, n = 4 and converted to multipoint for the geovisualization of the density increase of the lines converted to points.

```
l_chaikin <- smooth(lsf, method = "chaikin")
l_kernel <- smooth(lsf, method = "ksmooth")
l_spline <- smooth(lsf, method = "spline")
```

The results are mapped using functions from the `sf` package (Figure 10.19).

```
par(mfrow = c(2, 2))
plot(st_geometry(lsf), col = "blue", lwd = 5, bgc = 'gray92',
     axes = TRUE, main = "Original linestring")
plot(st_geometry(lsf), col = "gray", lwd = 5, bgc = 'gray92', axes = TRUE,
     main = "Chaikin")
plot(st_geometry(l_chaikin), col = "blue", lwd = 2, lty = 2, add = TRUE)
plot(st_geometry(lsf), col = "gray", lwd = 5, bgc = 'gray92', axes = TRUE,
     main = "ksmooth")
```

```
plot(st_geometry(l_kernel), col = "blue", lwd = 2, lty = 2, add = TRUE)
plot(st_geometry(lsf), col = "gray", lwd = 5, bgc = 'gray92', axes = TRUE,
     main = "spline")
plot(st_geometry(l_spline), col = "blue", lwd = 2, lty = 2, add = TRUE)
```

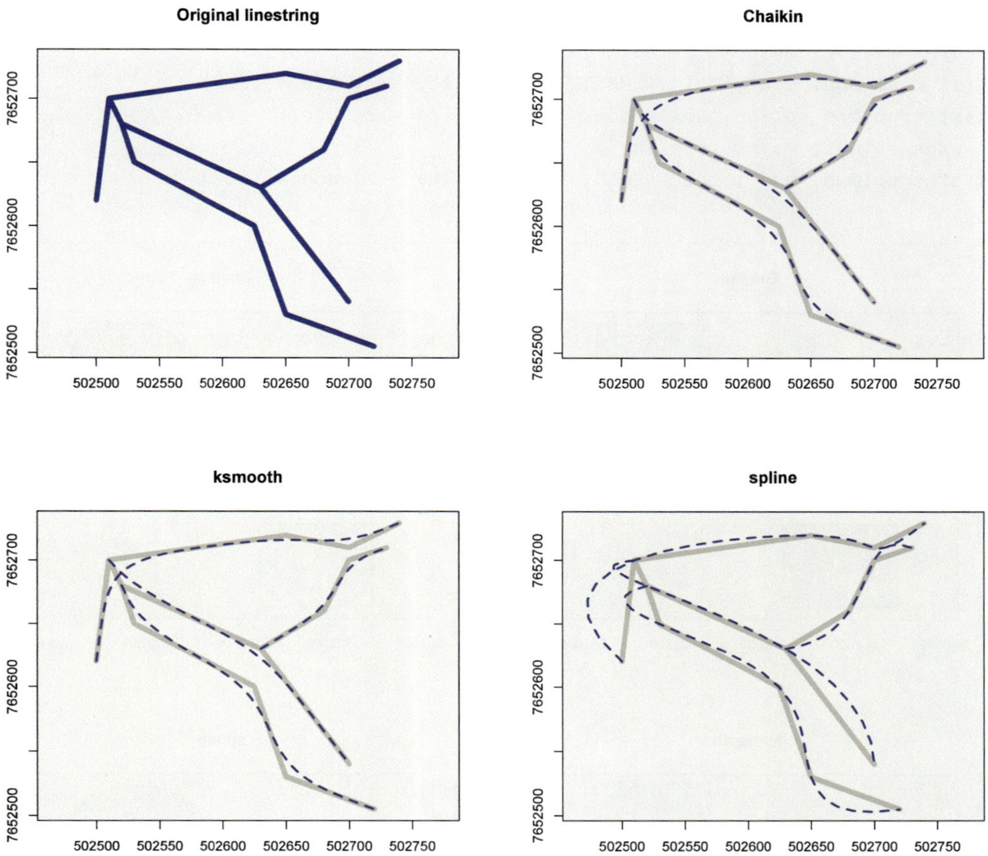

FIGURE 10.19 Smoothing of original simulated drainage network by Chaikin's corner cutting, Gaussian Kernel regression and spline interpolation methods.

We can see that Chaikin's corner cutting method is the smoothing that comes closest to the original line network, followed by Gaussian kernel regression and spline interpolation.

Polygon smoothing can also be performed. The same smoothing methods used previously are also applied to the polygon features created on the UFLA campus for comparative evaluation of the results (Figure 10.20).

```
mpp_chaikin <- smooth(mpp, method = "chaikin")
mpp_kernel <- smooth(mpp, method = "ksmooth")
mpp_spline <- smooth(mpp, method = "spline")
```

The results are mapped using functions from the `sf` package (Figure 10.20).

```
par(mfrow = c(2, 2))
plot(st_crop(mpp, box), col = "gray", lwd = 2, bgc = 'gray92', axes = TRUE,
     main = "Original")
plot(st_crop(mpp_chaikin, box), col = "gray", lwd = 2, bgc = 'gray92',
     axes = TRUE, main = "Chaikin")
plot(st_crop(mpp, box), col = "NA", lwd = 2, lty = 2, add = TRUE)
plot(st_crop(mpp_kernel, box), col = "gray", lwd = 2, bgc = 'gray92',
     axes = TRUE, main = "ksmooth")
plot(st_crop(mpp, box), col = "NA", lwd = 2, lty = 2, add = TRUE)
plot(st_crop(mpp_spline, box), col = "gray", lwd = 2, bgc = 'gray92',
     axes = TRUE, main = "spline")
plot(st_crop(mpp, box), col = "NA", lwd = 2, lty = 2, add = TRUE)
```

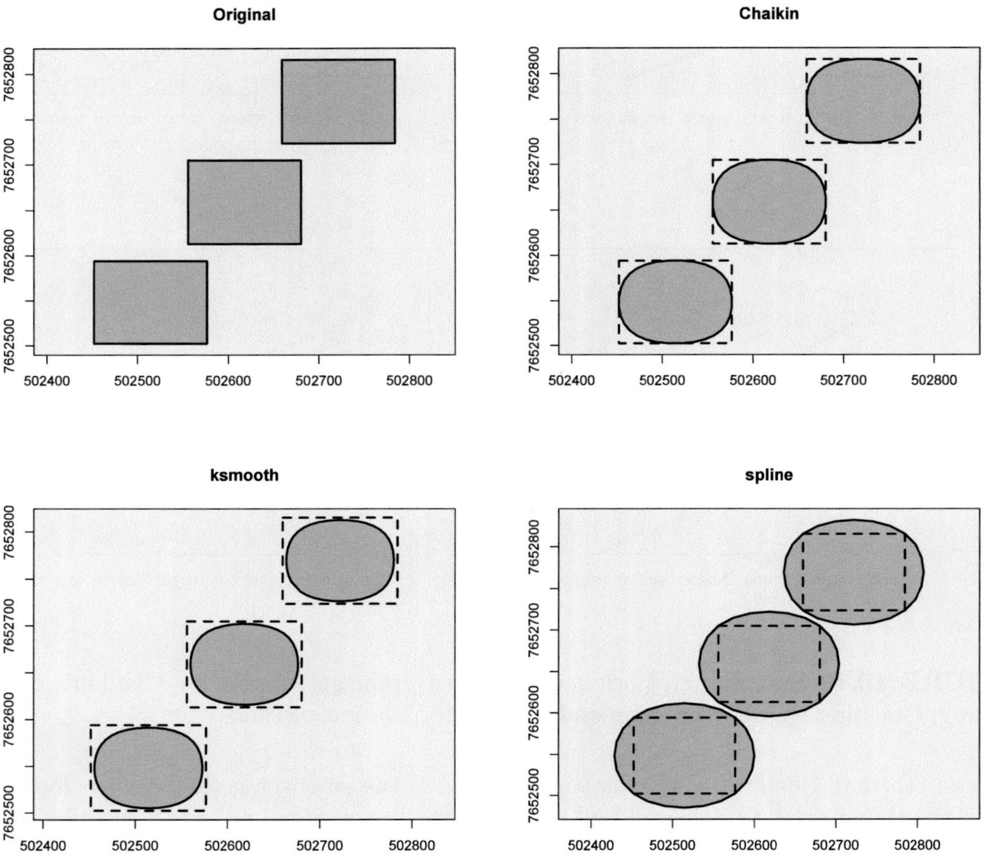

FIGURE 10.20 Smoothing of polygons by Chaikin's corner cutting, Gaussian Kernel regression and spline interpolation methods.

In the case of polygon smoothing, the spline method smoothes the edges of the polygon beyond the inner region defined by the rectangular geometry, while in the other methods evaluated, the smoothing occurs at the geometric limits of the inner region of the polygon.

11 Raster Geometric Analysis with Environmental Geocomputation and R

11.1 LEARNING QUESTIONS

The learning questions answered by reading the chapter are:

- What types of geometric analysis operations on raster data can be used?
- How to explore the interaction between raster data and vector-type features.
- How to use geometric analysis of raster data in scientific investigations.
- How can the results of geometric analysis of raster data be used in decision-making?

11.2 LEARNING OUTCOMES

The learning outcomes answered by reading the chapter are:

- Verify the use of types of geometric raster data analysis operations.
- Understand how the interaction between raster data and vector type features is performed.
- Understand how geometric analysis of raster data is used in scientific research.
- Evaluate the use of geometric analysis results of raster data in decision-making.

11.3 INTRODUCTION

In the domain of data analysis in geocomputation, regardless of whether the data model used is vector feature or raster data, the goal is to explore geospatial relationships, patterns and trends implicit in the data source for use in problem-solving and decision-making (Lo and Yeung, 2007).

Raster data analysis methods are classified by different ways identified by categories of spatial analysis operations such as (Lo and Yeung, 2007):

- Logical operations <- Logical operators to create new data layers;
- Arithmetic operations <- Arithmetic operators to create new data layers;
- Overlay operations <- Joining attribute values from more than one data layer by means of logical and arithmetic operators;
- Geometric properties operations <- Determining indices to describe geometric features and topological relationships of geospatial data;
- Geometric transformations operations <- Processes of modifying geometric features of a data layer;
- Geometric derivation operations <- Creation of new data by generalization, filtering and interpolation techniques.

Considering geometric transformation analysis of raster data, some geometric operations used are (Lovelace et al., 2019):

- Geometric intersection;

DOI: 10.1201/9781003724377-11

- Change of origin and extent;
- Aggregation and disaggregation;
- Resampling.

Geometric operations are also performed considering the interaction between raster type data and vector type features, such as (Lovelace et al., 2019):

- Raster clipping from vector;
- Raster extraction from vector;
- Rasterization of vector feature;
- Geospatial vectorization.

The choice of analysis operations for geometric transformation of raster data and interaction of raster data with vector features are dependent on the functions available for use in a data model and the applications used with raster data. Further details of operations for geometric analysis of raster data and raster-vector interaction are presented below.

11.4 GEOMETRIC OPERATIONS ON RASTER DATA

Geometric operations on raster data are required in different applications, such as georeferencing by co-registering an image based on another image, in terms of coordinate reference system, origin and resolution. Changing the resolution of the raster can be helpful in reducing the time it takes to process and store data on the computer. Geometric intersection can be used to obtain raster subsets. Changing the origin and extent of the raster can make it easier to perform joint analysis on exactly matched raster layers.

11.4.1 GEOMETRIC INTERSECTION

Geometric intersection operations between raster data can be used to obtain a new raster as a function of the overlap between two rasters. With this, raster subsets can be obtained relative to the raster of interest (Lovelace et al., 2019). A geometric intersection operation between raster data is used to obtain a new raster as a function of the overlap of two rasters: one is the Shuttle Radar Topography Mission (SRTM) elevation obtained in Chapter 9, and the other is a raster created in the extension of municipalities to the south of Minas Gerais, Brazil. Increasing numbers to the raster brightness values are assigned starting from 1 to 4599 brightness values. However, the raster subset is created with the brightness and resolution information of the original raster.

The `raster` package (Hijmans et al., 2020) is loaded in R with the `library` function to perform this example.

```
library(raster)
# Loading required packages: sp
```

A raster from the SRTM mission obtained from the `geodata` package (Hijmans et al., 2024a) is imported into R using the `raster` function. An intersection raster is created with extent, spatial resolution, and brightness values using the `raster` and `values` functions.

```
bra <- raster("C:/geo/C9/bra.tif") # Import SRTM data
bra # See header
# class      : RasterLayer
# dimensions : 4716, 5388, 25409808  (nrow, ncol, ncell)
```

```
# resolution : 0.008333333, 0.008333333  (x, y)
# extent     : -74.1, -29.2, -33.9, 5.4  (xmin, xmax, ymin, ymax)
# crs        : +proj=longlat +datum=WGS84 +no_defs
# source     : bra.tif
# names      : bra
# values     : -127, 2665  (min, max)
r <- raster(xmx = -44.479124, xmn = -45.090629, ymn = -21.817136,
            ymx = -21.289025, res = 0.008333333) # Create raster
values(r) <- 1:ncell(r) # Create raster values
r # See header
# class      : RasterLayer
# dimensions : 63, 73, 4599  (nrow, ncol, ncell)
# resolution : 0.008333333, 0.008333333  (x, y)
# extent     : -45.09063, -44.4823, -21.81402, -21.28902  (xmin, xmax, ymin, ymax)
# crs        : +proj=longlat +datum=WGS84 +no_defs
# source     : memory
# names      : layer
# values     : 1, 4599  (min, max)
```

An output raster is created from a subset of the input raster using a bracket operator. The argument `drop = FALSE` is used so that the result is a `RasterLayer` maintaining spatial properties of extent, resolution, and coordinate reference system.

```
rmg <- bra[r, drop = FALSE]
rmg
# class      : RasterLayer
# dimensions : 63, 73, 4599  (nrow, ncol, ncell)
# resolution : 0.008333333, 0.008333333  (x, y)
# extent     : -45.09167, -44.48333, -21.81667, -21.29167  (xmin, xmax, ymin, ymax)
# crs        : +proj=longlat +datum=WGS84 +no_defs
# source     : memory
# names      : bra
# values     : 856, 1534  (min, max)
```

The results are mapped side-by-side with the `plot` function (Figure 11.1).

```
par(mfrow = c(1, 3))
plot(bra, col = grey(1:100/100))
plot(r, col = grey(1:100/100))
plot(rmg, col = grey(1:100/100))
```

11.4.2 ORIGIN AND EXTENT

When performing geospatial analysis operations between rasters, such as algebraic analysis, the origin and extent characteristics between the analyzed data must be the same (Lovelace et al., 2019).

As an example, two rows and four columns are added on each side of the output raster of the previous example, with the altitude value equal to 2. This increased the number of cells

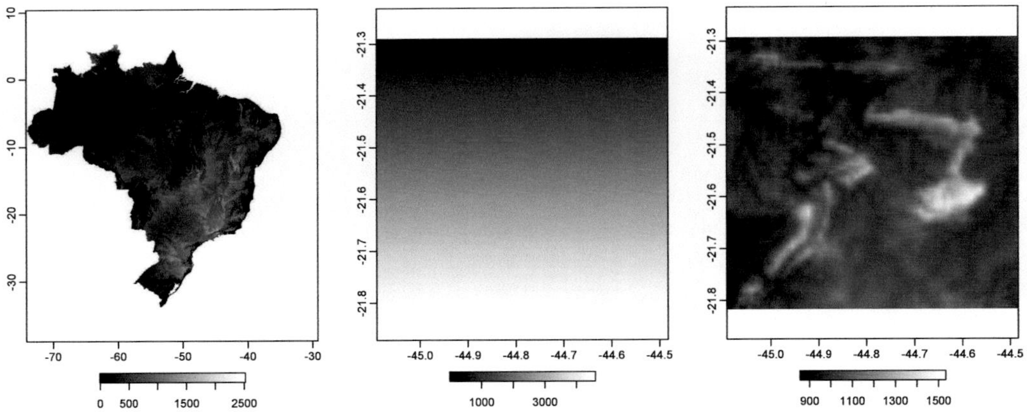

FIGURE 11.1 Geometric intersection operation used to define the subset (right) of a given raster (Brazil SRTM) (left) from an intersection raster created over municipalities in Minas Gerais (center).

in the original raster from 4599 to 5427.

The `extend` function is used to modify the extension of the output raster from the previous example.

```
rmg1 <- extend(rmg, c(2, 4), value = 2)
rmg1
# class       : RasterLayer
# dimensions : 67, 81, 5427  (nrow, ncol, ncell)
# resolution : 0.008333333, 0.008333333  (x, y)
# extent      : -45.125, -44.45, -21.83333, -21.275  (xmin, xmax, ymin, ymax)
# crs         : +proj=longlat +datum=WGS84 +no_defs
# source      : memory
# names       : bra
# values      : 2, 1534  (min, max)
```

The results are mapped using the `plot` function (Figure 11.2).

```
par(mfrow = c(1, 2))
plot(rmg, col = grey(1:100/100), horizontal = TRUE)
plot(rmg1, col = grey(1:100/100), horizontal = TRUE)
```

The origin of a raster starts at a corner of the raster and can vary according to the resolution of the raster. When performing algebraic operations between rasters, the same origin of the rasters is required. The origin (x, y) of the elevation raster subset created, at the spatial resolution of 0.008333333°, over municipalities in Minas Gerais, is 0.000000e+00, 3.552714e-15.

A new raster `rmg2` is created from the raster `rmg`. The `origin` function is used to change the x, y origin of the new raster.

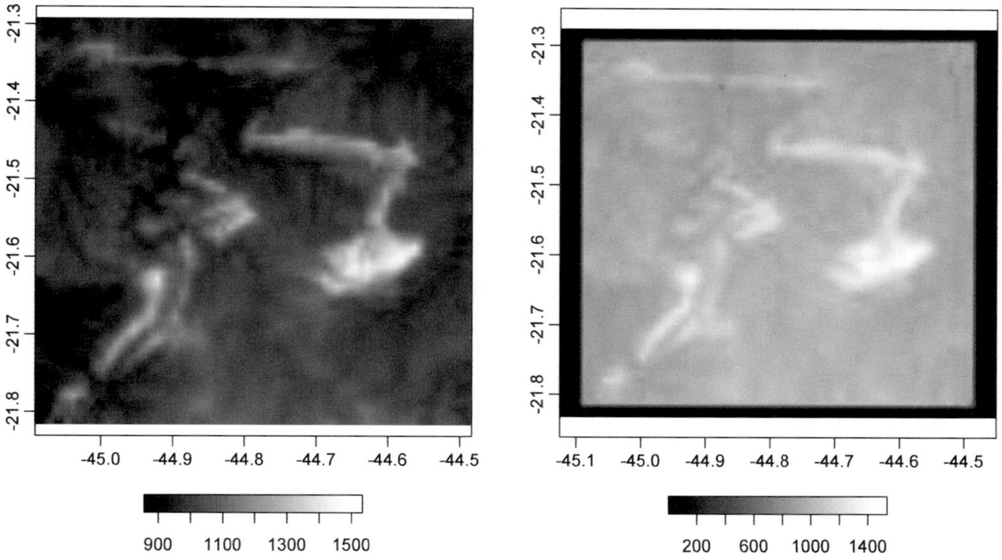

FIGURE 11.2 Operation to extend an elevation raster (left) by two rows and four columns with altitude value equal to 2 (right).

```
rmg2 <- rmg
origin(rmg2) <- c(1, 1)
origin(rmg)
# [1]  7.105427e-14 -6.039613e-14
origin(rmg2)
# [1] 0.000000e+00 3.552714e-15
```

11.4.3 AGGREGATION AND DISAGGREGATION

The spatial resolution of a raster can be reduced (aggregation) or increased (disaggregation) according to an algebraic condition and method of analysis of neighboring cells. The aggregation is used to reduce the spatial resolution of the digital elevation model in municipalities of Minas Gerais, with aggregation factors 2, 4 and 6, using the mean of neighboring cells, to create an output raster at different resolutions.

The `projectRaster` function is used to transform the cartographic projection of the `rmg` raster from the previous example from geodetic to plane with 1000 m spatial resolution.

```
rmgprj <- projectRaster(rmg, crs="+proj=utm +zone=23 +south", res = 1000)
rmgprj
# class       : RasterLayer
# dimensions  : 68, 73, 4964  (nrow, ncol, ncell)
# resolution  : 1000, 1000  (x, y)
# extent      : 485491.8, 558491.8, 7582572, 7650572  (xmin, xmax, ymin, ymax)
# crs         : +proj=utm +zone=23 +south +datum=WGS84 +units=m +no_defs
# source      : memory
```

```
# names       : bra
# values      : 859.6598, 1529.069  (min, max)
```

The aggregate function is used to aggregate the rmgprj and rmg rasters with different factors based on the mean of neighboring cells. Note that this function can be applied to both plane and geodetic coordinate reference systems.

```
ag2prj <- aggregate(rmgprj, fact = 2, fun = mean) # Plane projection
ag4prj <- aggregate(rmgprj, fact = 4, fun = mean) # Plane projection
ag6prj <- aggregate(rmgprj, fact = 6, fun = mean) # Plane projection
ag2 <- aggregate(rmg, fact = 2, fun = mean) # Geodetic projection
ag4 <- aggregate(rmg, fact = 4, fun = mean) # Geodetic projection
ag6 <- aggregate(rmg, fact = 6, fun = mean) # Geodetic projection
```

The results of cell aggregation in the geodesic projection are mapped comparatively with the plot function (Figure 11.3).

```
par(mfrow = c(2, 2))
plot(rmg, col=grey(1:100/100), horizontal = TRUE)
plot(ag2, col=grey(1:100/100), horizontal = TRUE)
plot(ag4, col=grey(1:100/100), horizontal = TRUE)
plot(ag6, col=grey(1:100/100), horizontal = TRUE)
```

Using the res function, we can see the different resolutions of the original and created rasters with aggregation of 2, 4 and 6, are 0.008333333°, 0.01666667°, 0.03333333° and 0.05°, respectively. Reprojecting the image in UTM projection, 23 South zone, at 1 km spatial resolution, the resolutions of the original and created rasters with aggregation of 2, 4 and 6, are 1000 m, 2000 m, 4000 m and 6000 m, respectively.

```
res(rmg)
# [1] 0.008333333 0.008333333
res(ag2)
# [1] 0.01666667 0.01666667
res(ag4)
# [1] 0.03333333 0.03333333
res(ag6)
# [1] 0.05 0.05
res(rmgprj)
# [1] 1000 1000
res(ag2prj)
# [1] 2000 2000
res(ag4prj)
# [1] 4000 4000
res(ag6prj)
# [1] 6000 6000
```

In the case of disaggregation, the spatial resolution of a raster is increased, determining an increase in the number of cells in the image to represent the same geographic objects.

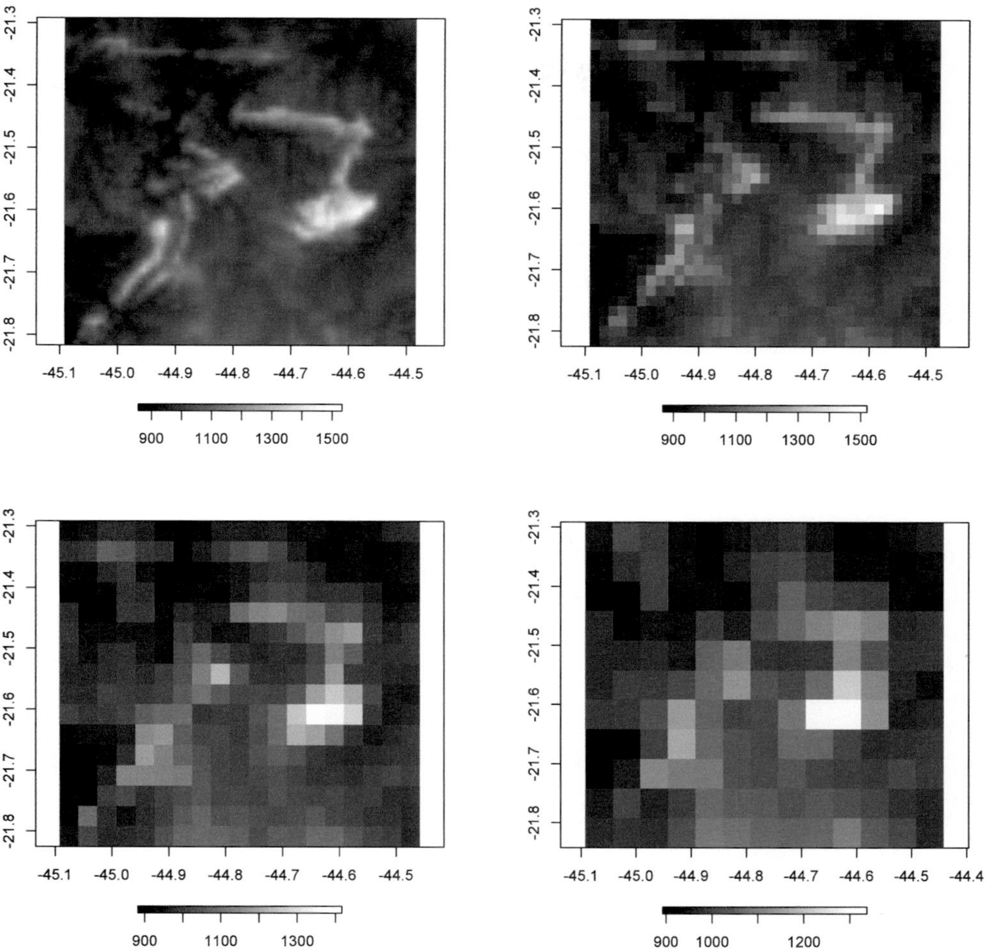

FIGURE 11.3 Aggregation operation of a digital elevation model of municipalities in Minas Gerais (top-left), with aggregation factors 2 (top-right), 4 (bottom-left) and 6 (bottom-right).

To perform this operation, it is necessary to specify the interpolation method used, either nearest neighbor interpolation or the bilinear method.

11.4.4 RESAMPLING

Aggregation and disaggregation methods are suitable for changing the resolution of the raster. Resampling is a process of calculating cell values at new cell locations in a target raster with custom resolution and origin. When you have two or more rasters with different resolutions and origins, resampling can be used to change the geometry of one raster based on the resolution of another raster by applying a resampling method (Lovelace et al., 2019).

There are several methods to estimate values for a raster with different resolution and origin, such as (Lovelace et al., 2019):

- Nearest neighbor <- The value of the closest cell of the original raster is assigned to the target cell, being fast and suitable for categorical raster;

- Bilinear interpolation <- A weighted average of the four closest cells of the original raster is assigned to the target raster cell, and the processing is faster for continuous raster;
- Cubic interpolation <- Values of 16 closest cells of the original raster are used to determine the value of the target cell by applying third-order polynomial functions. It is a method used for continuous raster resulting in smoother surface than bilinear interpolation despite higher computational requirement;
- Cubic spline interpolation <- Values of 16 cells of the original raster are used to determine the target cell value by applying cubic spline methodology which is third-order polynomial functions applied to continuous raster;
- Lanczos windowed sinc resampling <- Values of 36 cells closest to the original raster are used to determine the target cell value, being applied to continuous raster.

An extension in a region near the Carrancas mountain range is defined to evaluate different raster resampling methods. The `terra` (Hijmans et al., 2024b) package is used in this case, as there are more options available to perform nearest neighbor, bilinear interpolation, cubic interpolation, cubic spline interpolation and Lanczos windowed sinc resampling compared to the original raster cropped in the same region. The original raster `rmg` is resampled based on a new raster created in the `SpatRaster` class.

To perform this example, a target raster `rr` is created at a higher spatial resolution of 0.0002777778° using the `raster` and `values` functions. The objective is to resample the source raster `rmg` at lower spatial resolution of 0.008333333° based on the raster with higher spatial resolution.

```
rr <- raster(xmx = -44.50886, xmn = -44.86566, ymn = -21.67172, ymx = -21.41721,
  res = 0.0002777778,
  crs = "+proj=longlat +datum=WGS84 +ellps=WGS84 +towgs84=0,0,0")
values(rr) <- 1:ncell(rr)
```

The `crop` function is used to crop the source raster to the extent of the raster with the highest spatial resolution.

```
original <- crop(rmg, rr)
```

The `terra` package (Hijmans et al., 2024b) is loaded in R with the `library` function to perform the resampling methods.

```
library(terra)
# terra 1.7.60
```

The `rast` function is used to convert the class of the `original`, `rmg`, and `rr` geo-fields from `RasterLayer` to `SpatRaster`.

```
orig <- rast(original)
ext(orig)
# SpatExtent : -44.8666666666667, -44.5083333333333, -21.675,
# -21.4166666666667 (xmin, xmax, ymin, ymax)
res(orig)
```

```
# [1] 0.008333333 0.008333333
rmg_rast <- rast(rmg)
ext(rmg_rast)
# SpatExtent : -45.0916666666667, -44.4833333333333, -21.8166666666667,
# -21.2916666666667 (xmin, xmax, ymin, ymax)
res(rmg_rast)
# [1] 0.008333333 0.008333333
rr_rast <- rast(rr)
ext(rr_rast)
# SpatExtent : -44.86566, -44.5089933048, -21.6716544648,
# -21.41721 (xmin, xmax, ymin, ymax)
res(rr_rast)
# [1] 0.0002777778 0.0002777778
```

The `resample` function is used to resample the raster with lower spatial resolution in relation to the raster with higher spatial resolution using the `near`, `bilinear`, `cubic`, `cubicspline`, and `lanczos` methods.

```
res_near <- terra::resample(rmg_rast, y = rr_rast, method = "near")
res_bil <- terra::resample(rmg_rast, y = rr_rast, method = "bilinear")
res_cub <- terra::resample(rmg_rast, y = rr_rast, method = "cubic")
res_cubsp <- terra::resample(rmg_rast, y = rr_rast, method = "cubicspline")
res_lanczos <- terra::resample(rmg_rast, y = rr_rast, method = "lanczos")
```

The results of the resampled rasters are visually compared with the original raster using the `plot` function (Figure 11.4). Note that the `nearest neighbor` method preserves the brightness characteristics of the original raster when compared to the other methods, where the results are smoothed. The `lanczos` method is closer to `cubicspline`, better highlighting the lowest and highest regions of the relief.

```
par(mfrow = c(2, 3), cex=0.9)
plot(orig, col = grey.colors(100, start = 0, end = 1), legend = TRUE,
     main = "Original")
plot(res_near, col = grey.colors(100, start = 0, end = 1),
     legend = TRUE, main = "Nearest neighbor")
plot(res_bil, col = grey.colors(100, start = 0, end = 1),
     legend = TRUE, main = "Bilinear interpolation")
plot(res_cub, col = grey.colors(100, start = 0, end = 1),
     legend = TRUE, main = "Cubic interpolation")
plot(res_cubsp, col = grey.colors(100, start = 0, end = 1),
     legend = TRUE, main = "Cubic spline")
plot(res_lanczos, col = grey.colors(100, start = 0, end = 1),
     legend = TRUE, main = "Lanczos wind. sinc resampling")
```

`SpatRaster` class rasters can be transformed back to `RasterLayer` class using the `as(object_name, “Raster”)` function.

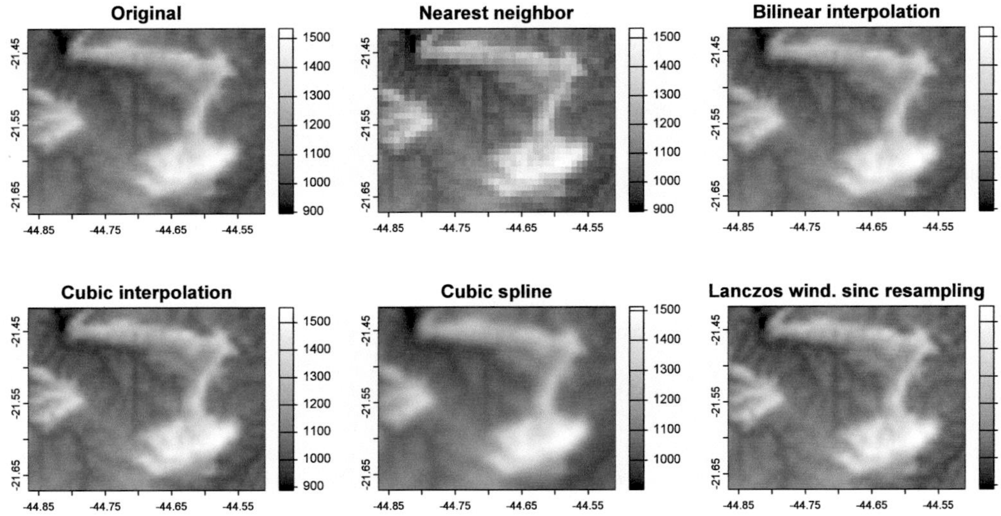

FIGURE 11.4 Resampling operation of a region near the Carrancas mountain range, Minas Gerais, Brazil, used to evaluate different methods of raster resampling by nearest neighbor, bilinear interpolation, cubic interpolation, cubic spline interpolation and Lanczos windowed sinc resampling.

```
origsp <- as(orig, "Raster")
res_nearsp <- as(res_near, "Raster")
res_bilsp <- as(res_bil, "Raster")
res_cubspp <- as(res_cub, "Raster")
res_cubsppp <- as(res_cubsp, "Raster")
res_lanczossp <- as(res_lanczos, "Raster")
```

The spplot function can be used as an alternative for comparative mapping of results (Figure 11.5).

```
a <- spplot(origsp, "bra", main = "Original")
b <- spplot(res_nearsp, "bra", main = "Nearest neighbor")
c <- spplot(res_bilsp, "bra", main = "Bilinear interpolation")
d <- spplot(res_cubspp, "bra", main = "Cubic interpolation")
e <- spplot(res_cubsppp, "bra", main = "Cubic spline")
f <- spplot(res_lanczossp, "bra", main = "Lanczos wind. sinc resampling")
grid.arrange(a, b, c, d, e, f, ncol = 3)
```

11.5 INTERACTION BETWEEN RASTER AND VECTOR SIMPLE FEATURES DATA

The main techniques used with interaction between raster geo-field and vector simple feature geo-object are (Lovelace et al., 2019):

- Performing raster cropping and masking from vector features;

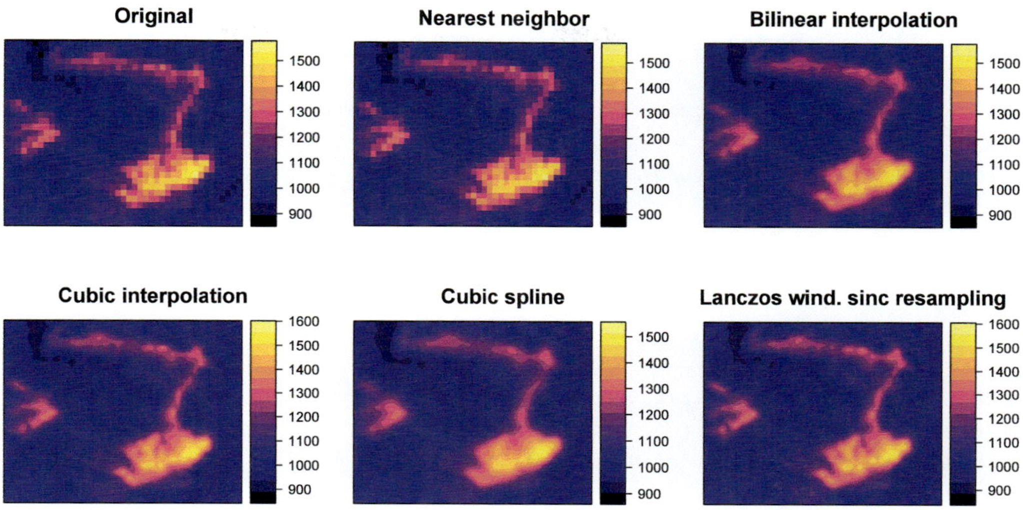

FIGURE 11.5 Resampling operation of a region near the Carrancas mountain range, Minas Gerais, Brazil, used to evaluate different methods of raster resampling by nearest neighbor, bilinear interpolation, cubic interpolation, cubic spline interpolation and Lanczos windowed sinc, mapped using spplot package.

- Extracting raster values from vector features;
- Conversions between raster and vector.

The packages `rgdal` (Bivand et al., 2021), `raster`, `sf` (Pebesma et al., 2021a), `dplyr` (Wickham et al., 2021a), `purr` (Wickham and Henry, 2023), `geosphere` (Hijmans et al., 2019), `gridExtra` (Auguie and Antonov, 2017), and `RStoolbox` (Leutner et al., 2019) required for the demonstration of examples are loaded into R using the `library` function.

```
library(rgdal)
# Loading required packages: sp
library(raster)
library(sf)
# Linking to GEOS 3.11.2, GDAL 3.7.2, PROJ 9.3.0; sf_use_s2() is TRUE
library(dplyr)
library(purrr)
library(geosphere)
library(gridExtra)
library(RStoolbox)
# package 'RStoolbox' was built under R version 3.5.3
```

11.5.1 RASTER CROP

The `crop` function is used to reduce the extent of a raster based on the rectangular extent of a vector feature. The `mask` function is used to assign `NA` values to the edges of a vector feature. The raster masking can be performed on the outer or inner boundary of the vector feature. Cropping and masking are applied to define a digital elevation model of a coffee crop in Santo Antônio do Amparo, Minas Gerais.

The geo-object with the coffee field boundary in geodetic and plane projection considered in the analysis is imported into R using the `readOGR` function.

```
bord <- readOGR(dsn = "C:/geo/artigoML/shape/borda.shp")
bord
# class       : SpatialPolygonsDataFrame
# features    : 1
# extent      : -44.8845, -44.88082, -20.89132, -20.8863  (xmin, xmax, ymin, ymax)
# crs         : +proj=longlat +datum=WGS84 +no_defs
# variables   : 1
# names       :     ID
# value       : buffer
bordPrj <- readOGR(dsn = "C:/geo/artigoML/shape/bordaPrj.shp")
bordPrj
# class       : SpatialPolygonsDataFrame
# features    : 1
# extent      : 512013.2, 512395.5, 7689875, 7690432  (xmin, xmax, ymin, ymax)
# crs         : +proj=utm +zone=23 +south +datum=WGS84 +units=m +no_defs
# variables   : 1
# names       :     ID
# value       : buffer
```

The SRTM digital elevation model of the area is imported into R using the `raster` function.

```
srtm <- raster("C:/geo/SRTM/s21_w045_1arc_v3.tif")
srtm
# class      : RasterLayer
# dimensions : 3601, 3601, 12967201  (nrow, ncol, ncell)
# resolution : 0.0002777778, 0.0002777778  (x, y)
# extent     : -45.00014, -43.99986, -21.00014, -19.99986  (xmin, xmax, ymin, ymax)
# crs        : +proj=longlat +datum=WGS84 +no_defs
# source     : s21_w045_1arc_v3.tif
# names      : s21_w045_1arc_v3
```

The `crop` and `mask` functions are used to perform a subset of the SRTM model within the area of interest, except in the case of the inverted mask, where the results are obtained outside the area.

```
srtmCrop <- crop(srtm, bord) # Crop
srtmWGS <- mask(srtmCrop, bord) # Mask
srtmWGS1 <- mask(srtmCrop, bord, inverse = TRUE) # Inverse mask
```

The results are mapped using the `plot` function (Figure 11.6).

```
par(mfrow = c(2, 2))
e <- extent(c(xmin = -44.94615, xmax = -44.80951, ymin = -20.937,
              ymax = -20.83542))
plot(srtm, col = grey(1:100/100), ext = e, horizontal = TRUE)
plot(bord, add = TRUE, border = "red", lwd = 2)
```

```
plot(srtmCrop, col = grey(1:100/100), horizontal = TRUE)
plot(bord, add = TRUE, border = "red", lwd = 2)
plot(srtmWGS, col = grey(1:100/100), horizontal = TRUE)
plot(bord, add = TRUE, border = "red", lwd = 2)
plot(srtmWGS1, col = grey(1:100/100), horizontal = TRUE)
plot(bord, add = TRUE, border = "red", lwd = 2)
```

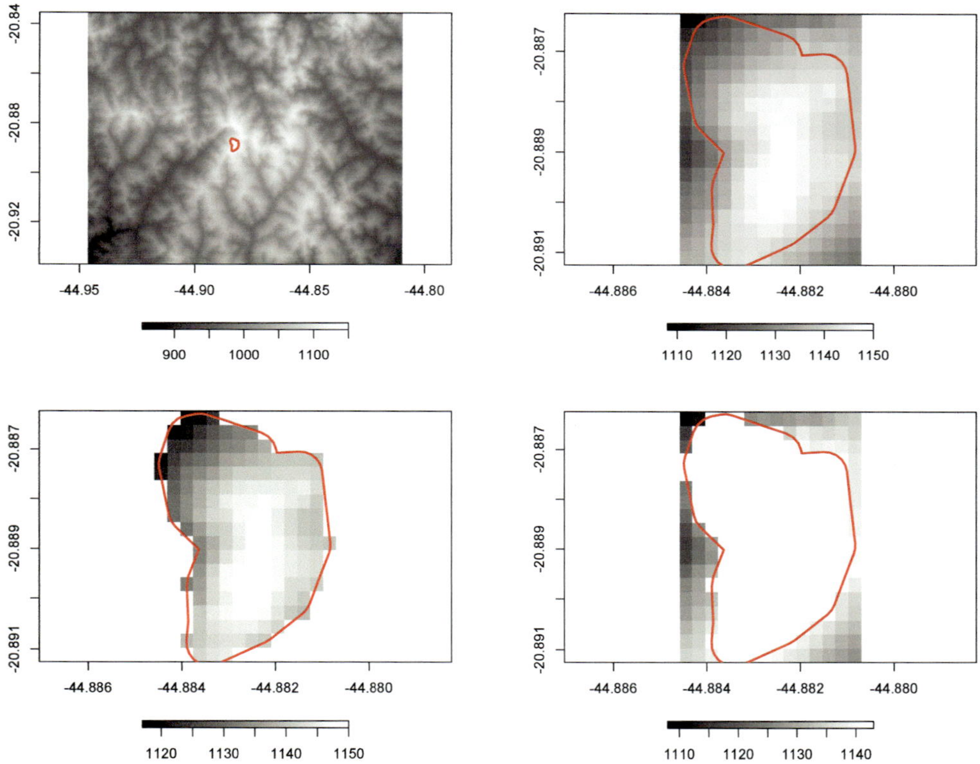

FIGURE 11.6 Original SRTM tile in a combined raster-vector plot (top-left), raster crop (top-right), raster mask (bottom-left), and inverted raster mask (bottom-right) in a coffee field in Santo Antônio do Amparo, Minas Gerais.

11.5.2 RASTER EXTRACTION

Raster data extraction is the process used to identify and return the values associated with a raster, at specific locations, based on a geographic geo-object, usually vector, used to extract the values of interest. The extracting vector geo-object can be of point features, lines, or polygons. One of the most commonly used examples is the extraction of cells from a raster using points. Altitude values from a digital elevation model are extracted from georeferenced sample points in a coffee field.

To perform this example, a sample grid of points and the boundary of the area where the research (de Carvalho Alves et al., 2022) was conducted are imported into R using the `read_sf` function.

```
msf <- read_sf("C:/geo/artigoML/shape/malhaWGS84.shp")
msf
# Simple feature collection with 85 features and 5 fields
# Geometry type: POINT
# Dimension:     XY
# Bounding box:  xmin: -44.88392 ymin: -20.89088 xmax: -44.88137 ymax: -20.88687
# Geodetic CRS:  WGS 84
# A tibble: 85 × 6
#          x        y     z    xo    yo                geometry
#      <dbl>    <dbl> <dbl> <dbl> <dbl>             <POINT [°]>
# 1 512125. 7689925. 1136. -44.9 -20.9 (-44.88342 -20.89088)
# 2 512153. 7689937. 1138. -44.9 -20.9 (-44.88315 -20.89077)
# 3 512142. 7689964. 1137. -44.9 -20.9 (-44.88325 -20.89052)
# 4 512177. 7689949. 1140. -44.9 -20.9 (-44.88292 -20.89066)
# 5 512195. 7689973. 1142. -44.9 -20.9 (-44.88275 -20.89044)
# 6 512163. 7689982. 1140. -44.9 -20.9 (-44.88306 -20.89036)
# 7 512213. 7689981. 1142. -44.9 -20.9 (-44.88257 -20.89037)
# 8 512178. 7690008. 1142. -44.9 -20.9 (-44.88291 -20.89013)
# 9 512140. 7690052. 1138. -44.9 -20.9 (-44.88327 -20.88973)
# 10 512156. 7690037. 1140. -44.9 -20.9 (-44.88312 -20.88986)
bsf <- read_sf("C:/geo/artigoML/shape/borda.shp")
# Simple feature collection with 1 feature and 1 field
# Geometry type: POLYGON
# Dimension:     XY
# Bounding box:  xmin: -44.8845 ymin: -20.89132 xmax: -44.88082 ymax: -20.8863
# Geodetic CRS:  WGS 84
# A tibble: 1 × 2
#   ID                                                      geometry
# <chr>                                                <POLYGON [°]>
# 1 buffer ((-44.88211 -20.89081, -44.88221 -20.89086, -44.88323  ...
```

An identifier field `id` is created in the database with values from 1 to 85. The field `id` is added to the database with the function `st_sf(data.frame)`.

```
id <- 1:85 # Create id field
msfid <- st_sf(data.frame(msf, id)) # Create sf
msfid
# Simple feature collection with 85 features and 6 fields
# Geometry type: POINT
# Dimension:     XY
# Bounding box:  xmin: -44.88392 ymin: -20.89088 xmax: -44.88137 ymax: -20.88687
# Geodetic CRS:  WGS 84
# First 10 features:
#          x       y        z        xo        yo id                   geometry
# 1 512124.8 7689925 1136.276 -44.88342 -20.89088  1 POINT (-44.88342 -20.89088)
# 2 512152.6 7689937 1138.442 -44.88315 -20.89077  2 POINT (-44.88315 -20.89077)
# 3 512142.3 7689964 1137.292 -44.88325 -20.89052  3 POINT (-44.88325 -20.89052)
# 4 512177.1 7689949 1139.901 -44.88292 -20.89066  4 POINT (-44.88292 -20.89066)
# 5 512194.7 7689973 1141.520 -44.88275 -20.89044  5 POINT (-44.88275 -20.89044)
```

```
# 6 512162.6 7689982 1139.760 -44.88306 -20.89036  6 POINT (-44.88306 -20.89036)
# 7 512213.2 7689981 1142.336 -44.88257 -20.89037  7 POINT (-44.88257 -20.89037)
# 8 512177.8 7690008 1141.937 -44.88291 -20.89013  8 POINT (-44.88291 -20.89013)
# 9 512140.5 7690052 1138.201 -44.88327 -20.88973  9 POINT (-44.88327 -20.88973)
#10 512156.2 7690037 1140.032 -44.88312 -20.88986 10 POINT (-44.88312 -20.88986)
```

A field with elevation values extracted from the SRTM raster is created with the `$` operator and the `raster::extract` function. The first ten extracted elevation values are obtained with the `print` function.

```
msfid$elev <- raster::extract(srtmWGS, msfid)
names(msfid)
# [1] "x"     "y"     "z"     "xo"    "yo"    "id"    "geometry" "elev"
print(msfid$elev[1:10])
# [1] 1143 1144 1144 1145 1146 1147 1148 1149 1143 1146
```

The `as_Spatial` function is used to convert the sample grid geo-objects and coffee field boundary into the `sp` class.

```
msfidsp <- as_Spatial(msfid)
class(msfidsp)
# [1] "SpatialPointsDataFrame"
# attr(,"package")
# [1] "sp"
bordsp <- as_Spatial(bsf)
class(bordsp)
# [1] "SpatialPolygonsDataFrame"
# attr(,"package")
# [1] "sp"
```

The results are mapped using the `plot` function (Figure 11.7).

```
par(mfrow = c(1, 2))
plot(srtmWGS, col = grey(1:100/100), axes = TRUE, horizontal = TRUE)
plot(bordsp, col = "transparent", add = TRUE)
plot(msfidsp, add = TRUE, pch = 1, colour = "black")
text(msfidsp, labels = msfid$id, pos = 3, halo = TRUE, offset = 0.5, cex = 1)
plot(msfidsp$id, msfid$elev, xlab = "ID", ylab = "Altitude")
```

The values of a raster can also be extracted by means of a transect defined by a line. In this case, a line is created between the sample points with id, 15 and 67, of the sample grid defined on the coffee field with `cbind`, `st_linestring` and `st_sf` functions, in order to extract altitude values along the transect. It should be noted that sampling by transect does not necessarily have to be done along a straight line, as in this example.

```
tr <- cbind(c(-44.882205707, -44.883592683), c(-20.890239393, -20.886869872)) |>
st_linestring() |> st_sfc(crs = projection(srtmWGS)) |> st_sf()
tr
```

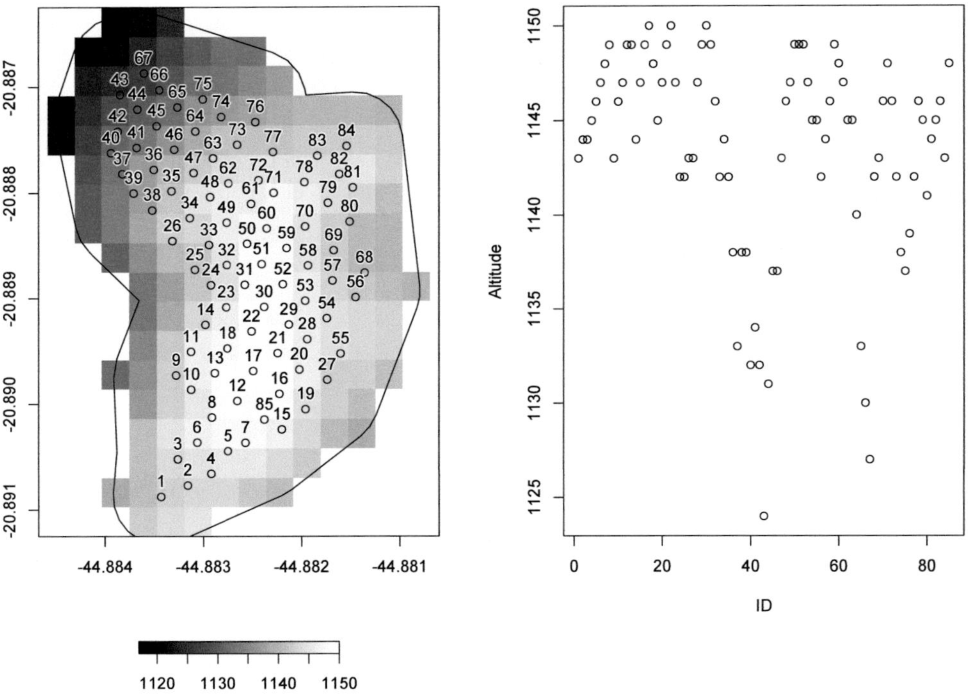

FIGURE 11.7 Altitude value extraction operation from a digital elevation model (left) and the altitude values extracted for each identifier code (ID) of the georeferenced point in a coffee crop (right) in Santo Antônio do Amparo, Minas Gerais.

```
# Simple feature collection with 1 feature and 0 fields
# Geometry type: LINESTRING
# Dimension:     XY
# Bounding box:  xmin: -44.88359 ymin: -20.89024 xmax: -44.88221 ymax: -20.88687
# Geodetic CRS:  +proj=longlat +datum=WGS84 +no_defs
#                          geometry
# 1 LINESTRING (-44.88221 -20.8...
```

The `raster::extract` function is used to extract values from the SRTM raster along the created transect.

```
transect <- raster::extract(srtmWGS, tr, along = TRUE, cellnumbers = TRUE)
transect
# [[1]]
#        cell s21_w045_1arc_v3
#   [1,]  205             1147
#   [2,]  191             1149
#   [3,]  190             1149
#   [4,]  176             1150
```

```
#  [5,]  162            1150
#  [6,]  148            1150
#  [7,]  147            1147
#  [8,]  133            1145
#  [9,]  119            1146
# [10,]  118            1142
# [11,]  104            1144
# [12,]   90            1146
# [13,]   76            1143
# [14,]   75            1141
# [15,]   61            1137
# [16,]   47            1133
# [17,]   46            1131
# [18,]   32            1127
```

With the determination of the coordinates associated with each extracted cell and with the determination of the geodetic distance along the transect, it is possible to determine the altitude profile along the created line using `map_dfr`, `xyFromCell`, `distGeo` and `cumsum` functions in the following chunk code.

```
transect_df <- purrr::map_dfr(transect, as_data_frame, .id = "ID")
transect_coords <- xyFromCell(srtmWGS, transect_df$cell)
pair_dist <- geosphere::distGeo(transect_coords)[-nrow(transect_coords)]
transect_df$dist <- c(0, cumsum(pair_dist))
transect_df
# A tibble: 18 × 4
#    ID     cell s21_w045_1arc_v3  dist
#    <chr> <dbl>            <dbl> <dbl>
# 1 1        205             1147    0
# 2 1        191             1149   30.8
# 3 1        190             1149   59.7
# 4 1        176             1150   90.4
# 5 1        162             1150 121.
# 6 1        148             1150 152.
# 7 1        147             1147 181.
# 8 1        133             1145 212.
# 9 1        119             1146 242.
# 10 1        118             1142 271.
# 11 1        104             1144 302.
# 12 1         90             1146 333.
# 13 1         76             1143 363.
# 14 1         75             1141 392.
# 15 1         61             1137 423.
# 16 1         47             1133 454.
# 17 1         46             1131 483.
# 18 1         32             1127 514.
```

The results are mapped using the `plot` function (Figure 11.8).

```
par(mfrow = c(1, 2), mar = c(2, 2, 0.3, 0.3), mgp = c(1.5, 0.6, 0))
plot(srtmWGS, col = grey(1:100/100), axes = TRUE, horizontal = TRUE)
plot(bordsp, col = "transparent", add = TRUE)
plot(tr, add = TRUE, lwd = 2, colour = "black")
text(msfidsp[67,], labels = c("end"), halo = TRUE, pos = 4, offset = 0.5)
text(msfidsp[15,], labels = "start", halo = TRUE, pos = 4, offset = 0.5)
plot(transect_df$dist, transect_df$s21_w045_1arc_v3, xlab = "Distance (m)",
     ylab = "Altitude (m)")
```

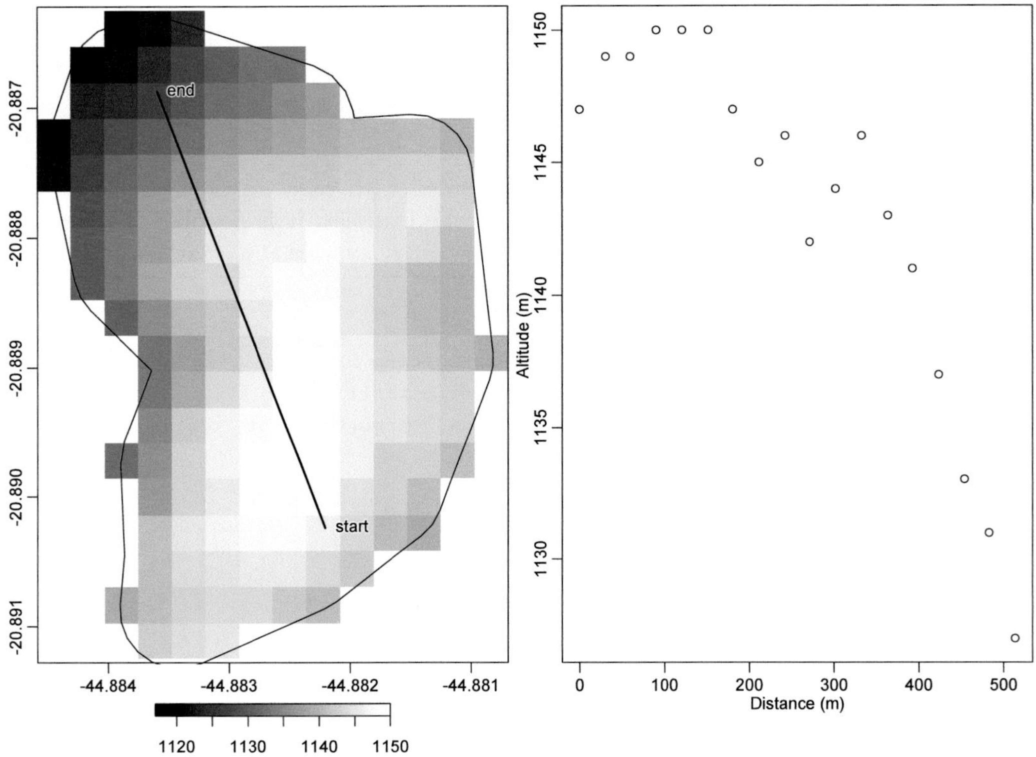

FIGURE 11.8 Extraction of altitude value from a digital elevation model using a transect (right) defined in coffee field to obtain the altitude values at each cell of the raster, in Santo Antônio do Amparo, Minas Gerais.

The extraction of raster values is also evaluated in terms of raster-polygon interaction. The analyzed raster dataset is extracted by a polygon feature delimited around the crop.

```
srtm_values <- raster::extract(x = srtmWGS, y = bord, df = TRUE)
```

The altitude values from the digital elevation model of the coffee crop are extracted, to obtain information about the minimum-maximum and mean elevation of the coffee field using `group_by` and `summarize_at` functions.

```
group_by(srtm_values, ID) |>
  summarize_at(vars(s21_w045_1arc_v3), list(~min(.), ~mean(.), ~max(.)))
# A tibble: 1 × 4
#      ID   min  mean   max
#   <dbl> <dbl> <dbl> <dbl>
# 1     1  1117 1140.  1150
```

We observed minimum, mean, and maximum altitudes of the evaluated coffee field of 1117, 1140, and 1150 m, respectively.

The extraction of a categorical raster can also be performed. To illustrate this operation, a categorical raster obtained by random forest classification of the incidence of coffee bacterial blight on coffee trees under the influence of environmental variables is extracted from a polygon vector feature. The categorical classes of the disease incidence raster are extracted according to the vector feature defined around the coffee field, in Santo Antônio do Amparo, Minas Gerais (de Carvalho Alves et al., 2022) (Figure 11.9).

The `readRSTBX` function is used to import the results of the analysis into R, including the machine learning model fit and the classified raster with attributes. It should be noted that in version 1.0 of the `RStoolbox` package, there are problems importing these results, probably due to program updates, and it is only possible to import them in version 3.5.3 of R, where the package was installed at the time.

```
SC <- readRSTBX("C:/geo/c11/SC")
SC
# superClass results
# ************ Validation **************
# $validation
# [1] "No independent validation was performed!"
# *************** Map *****************
# $map
# class       : RasterLayer
# dimensions : 20, 15, 300  (nrow, ncol, ncell)
# resolution : 30, 30  (x, y)
# extent      : 511973.8, 512423.8, 7689867, 7690467  (xmin, xmax, ymin, ymax)
# crs    : +proj=utm +zone=23 +south +datum=WGS84 +units=m +no_defs +ellps=WGS84
# source      : C:/geo/c11/SC.grd
# names       : id
# values      : 1, 5  (min, max)
# attributes :
#  ID value
#    1     1
#    2     2
#    3     3
#    4     4
#    5     5
```

The `levels` function and the `<-` operator are used to create new attribute categories `Disease` and `Class` in the classified raster.

```
rat <- levels(SC$map)[[1]]
rat$Disease <- c('0%', '1-25%', '26-50%', '51-75%', '76-100%' )
rat$Class <- c('No', 'Low', 'Average', 'High', 'Very high')
levels(SC$map) <- rat
SC$map
# class       : RasterLayer
# dimensions : 20, 15, 300  (nrow, ncol, ncell)
# resolution : 30, 30  (x, y)
# extent      : 511973.8, 512423.8, 7689867, 7690467  (xmin, xmax, ymin, ymax)
# crs      : +proj=utm +zone=23 +south +datum=WGS84 +units=m +no_defs +ellps=WGS84
#             +towgs84=0,0,0
# source      : C:/geo/c11/SC.grd
# names       : id
# values      : 1, 5  (min, max)
# attributes :
#   ID value Disease     Class
#    1     1      0%        No
#    2     2   1-25%       Low
#    3     3  26-50%   Average
#    4     4  51-75%      High
#    5     5 76-100% Very high
```

Each cell class is counted according to the percentage of disease incidence using `extract`, `select`, `gather`, `group_by`, `tally` and `spread` functions in the following chunk code below.

```
coffee_blight <- raster::extract(SC$map, bordPrj, df = TRUE, factors = TRUE)
dplyr::select(coffee_blight, ID, Disease) |> tidyr::gather(key, value, -ID) |>
group_by(ID, key, value) |> tally() |> tidyr::spread(value, n, fill = 0)
# A tibble: 1 x 8
# Groups:   ID, key [1]
#      ID key       `0%` `1-25%` `26-50%` `51-75%` `76-100%` `<NA>`
#   <dbl> <chr>    <dbl>   <dbl>    <dbl>    <dbl>     <dbl>  <dbl>
# 1     1 Disease     69      15        8       11        65      2
```

We observed 65 cells in the class 76 to 100% incidence of the coffee bacterial blight, which can be used to evaluate the epidemic's progress. There was a high incidence of the disease in much of the field, when compared to the lower classes of disease incidence in the field (0% = 69 cells).

```
par(mfrow = c(1, 2), mar = c(2, 2, 0.3, 0.3), mgp = c(1.5, 0.6, 0))
plot(srtmWGS, col = grey(1:100/100), axes = TRUE, horizontal = TRUE)
plot(bord, col = "transparent", add = TRUE)
plot(SC$map, "Disease", col = rev(grey(1:100/100)), axes = TRUE, horizontal = TRUE)
plot(bordPrj, col = "transparent", add = TRUE)
```

11.5.3 RASTERIZATION

Rasterization of points, lines and polygons can be performed differently according to the type of geometry used in rasterization. A raster template must be created to assign values

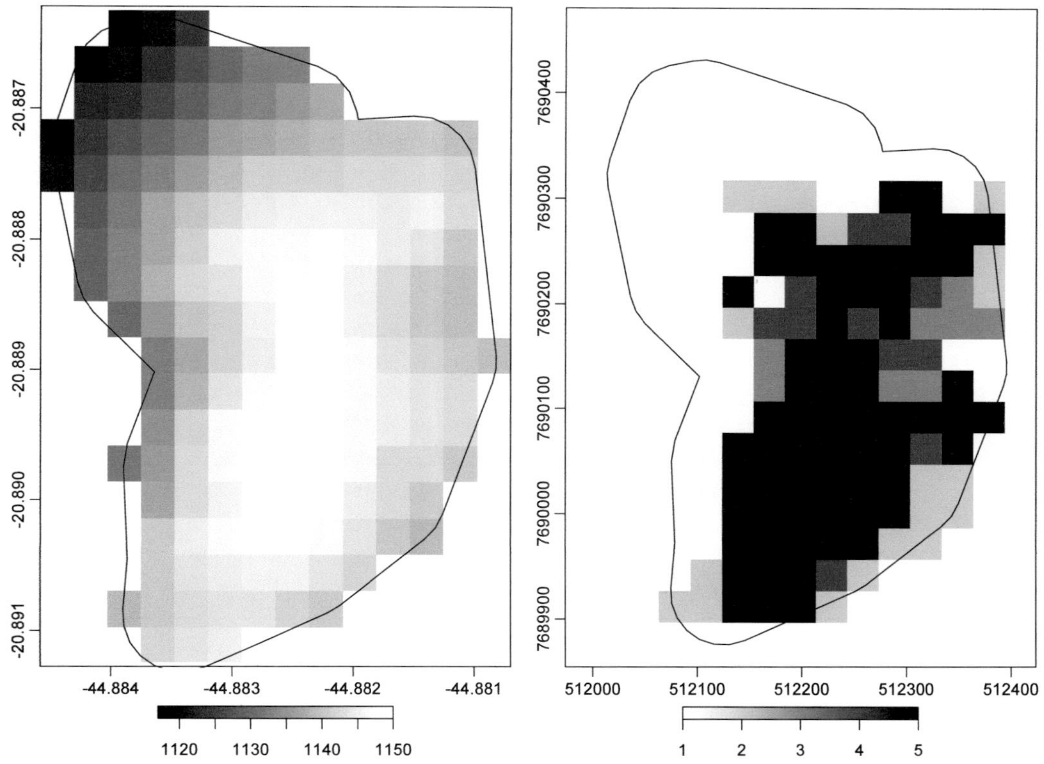

FIGURE 11.9 Extraction of raster with continuous values of altitude (m) (left) and categorical random forest classification of the incidence of bacterial blight on coffee plantations (right), extracted from vector features.

to the cells that are touched by the point, line or polygon in the raster and perform the conversion at these locations. The polygon rasterization criterion is the condition that the centroid of the cell lies within the vector feature that bounded the polygon (Lovelace et al., 2019).

Rasterization is used to rasterize points with altitude values in georeferenced coffee crop sample locations in Santo Antônio do Amparo, Minas Gerais. The spatial resolution of the raster is set to 5 m.

The `msfid` geo-object from the previous example is reprojected to the UTM zone 23 South system based on EPSG code 32723 using the `st_transform` function. The `raster` function is used to create a raster template geo-field with a spatial resolution of 5 m, in the same cartographic projection as the sample grid, using the `extent` and `st_crs` functions, respectively.

```
grid_projected <- st_transform(msfid, 32723) # Plane projection
grid_projected
# Simple feature collection with 85 features and 7 fields
# Geometry type: POINT
# Dimension:     XY
# Bounding box:  xmin: 512072.9 ymin: 7689925 xmax: 512337.7 ymax: 7690368
```

```
# Projected CRS: WGS 84 / UTM zone 23S
# First 10 features:
#      x        y        z        xo       yo id                 geometry elev
# 1  512124.8 7689925 1136.276 -44.8834 -20.8908 1 POINT (512124.8 7689925) 1143
# 2  512152.6 7689937 1138.442 -44.8831 -20.8907 2 POINT (512152.6 7689937) 1144
# 3  512142.3 7689964 1137.292 -44.8832 -20.8905 3 POINT (512142.3 7689964) 1144
# 4  512177.1 7689949 1139.901 -44.8829 -20.8906 4 POINT (512177.1 7689949) 1145
# 5  512194.7 7689973 1141.520 -44.8827 -20.8904 5 POINT (512194.7 7689973) 1146
# 6  512162.6 7689982 1139.760 -44.8830 -20.8903 6 POINT (512162.6 7689982) 1147
# 7  512213.2 7689981 1142.336 -44.8825 -20.8903 7 POINT (512213.2 7689981) 1148
# 8  512177.8 7690008 1141.937 -44.8829 -20.8901 8 POINT (512177.8 7690008) 1149
# 9  512140.5 7690052 1138.201 -44.8832 -20.8897 9 POINT (512140.5 7690052) 1143
# 10 512156.2 7690037 1140.032 -44.8831 -20.889 10 POINT (512156.2 7690037) 1146
raster_template1 <- raster(extent(grid_projected), resolution = 5,
                    crs = st_crs(grid_projected)$proj4string) # Create raster
# raster_template1
# class      : RasterLayer
# dimensions : 89, 53, 4717  (nrow, ncol, ncell)
# resolution : 5, 5  (x, y)
# extent     : 512072.9, 512337.9, 7689923, 7690368  (xmin, xmax, ymin, ymax)
# crs   : +proj=utm +zone=23 +south +datum=WGS84 +units=m +no_defs +ellps=WGS84
```

The `rasterize` function is used to create raster elevation points based on the `z` elevation data column.

```
points_raster <- rasterize(grid_projected, raster_template1, field = "z")
points_raster
# class      : RasterLayer
# dimensions : 89, 53, 4717  (nrow, ncol, ncell)
# resolution : 5, 5  (x, y)
# extent     : 512072.9, 512337.9, 7689923, 7690368  (xmin, xmax, ymin, ymax)
# crs   : +proj=utm +zone=23 +south +datum=WGS84 +units=m +no_defs +ellps=WGS84
# source     : memory
# names      : layer
# values     : 1121.273, 1145.9  (min, max)
```

The `as_Spatial` function is used to convert the geo-object from class `sf` to `sp` in order to map the altitude values in the vector and rasterized sample grid using the `spplot` and `grid.arrange` functions (Figure 11.10).

```
b1 <- as_Spatial(grid_projected) # Convert to sp class
a <- spplot(b1[3])
b <- spplot(points_raster, colorkey = list(space = "bottom"))
grid.arrange(a, b, ncol = 2)
```

Rasterization is applied to the polygon-type vector simple feature around coffee field. To perform the line conversion, the borders of the area are transformed from polygon to lines, followed by rasterization. A raster template is created at a resolution of 0.0002777778°, to the same extent as the vector feature, in order to characterize the results obtained.

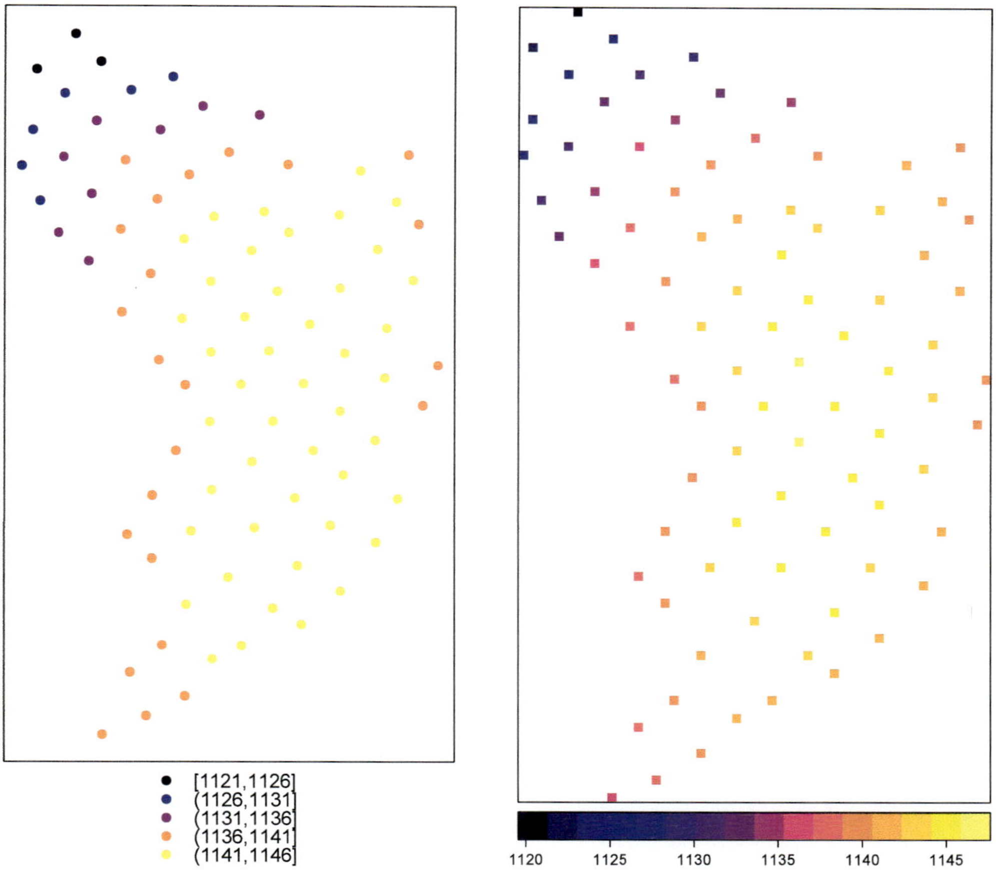

FIGURE 11.10 Rasterization of points (right) with altitude attribute (m) (left) of coffee field in Santo Antônio do Amparo, Minas Gerais.

Changing the spatial resolution may be necessary depending on the level of detail required by the user in the rasterization.

The `st_cast` function is used to create a line geo-object based on the `bsf` polygon of the coffee field. The `raster` function is used to create a raster template in the same geographic extent and cartographic projection as the polygon with a spatial resolution of 0.0002777778°.

```
line_borders <- st_cast(bsf, "MULTILINESTRING")
line_borders
# Simple feature collection with 1 feature and 1 field
# Geometry type: MULTILINESTRING
# Dimension:     XY
# Bounding box:  xmin: -44.8845 ymin: -20.89132 xmax: -44.88082 ymax: -20.8863
# Geodetic CRS:  WGS 84
# A tibble: 1 x 2
#   ID                                                   geometry
#   <chr>                                       <MULTILINESTRING [°]>
```

```
# 1 buffer ((-44.88211 -20.89081, -44.88221 -20.89086, -44.88323 -20.89129, ...
raster_template <- raster(extent(bsf), resolution = 0.0002777778,
          crs = st_crs(bsf)$proj4string)
raster_template
# class      : RasterLayer
# dimensions : 18, 13, 234  (nrow, ncol, ncell)
# resolution : 0.0002777778, 0.0002777778  (x, y)
# extent     : -44.8845, -44.88088, -20.8913, -20.8863  (xmin, xmax, ymin, ymax)
# crs        : +proj=longlat +datum=WGS84 +no_defs +ellps=WGS84 +towgs84=0,0,0
```

The `rasterize` function is used to convert the vector line and polygon into a raster.

```
l_raster <- rasterize(line_borders, raster_template)
l_raster
# class      : RasterLayer
# dimensions : 18, 13, 234  (nrow, ncol, ncell)
# resolution : 0.0002777778, 0.0002777778  (x, y)
# extent     : -44.8845, -44.88088, -20.8913, -20.8863  (xmin, xmax, ymin, ymax)
# crs        : +proj=longlat +datum=WGS84 +no_defs +ellps=WGS84 +towgs84=0,0,0
# source     : memory
# names      : layer
# values     : 1, 1  (min, max)
# attributes :
#  ID   ID.2
#   1 buffer
p_raster <- rasterize(bsf, raster_template)
p_raster
# class      : RasterLayer
# dimensions : 18, 13, 234  (nrow, ncol, ncell)
# resolution : 0.0002777778, 0.0002777778  (x, y)
# extent     : -44.8845, -44.88088, -20.8913, -20.8863  (xmin, xmax, ymin, ymax)
# crs        : +proj=longlat +datum=WGS84 +no_defs +ellps=WGS84 +towgs84=0,0,0
# source     : memory
# names      : layer
# values     : 1, 1  (min, max)
# attributes :
#  ID   ID.2
#   1 buffer
```

The results are mapped using the `plot` function (Figure 11.11).

```
par(mfrow = c(1, 2), mar = c(2, 2, 0.3, 0.3), mgp = c(1.5, 0.6, 0))
plot(l_raster, col = grey(1:100/100), axes = TRUE, legend = FALSE)
plot(bord, col = "transparent", add = TRUE)
plot(p_raster, col = grey(1:100/100), axes = TRUE, legend = FALSE)
plot(bord, col = "transparent", add = TRUE)
```

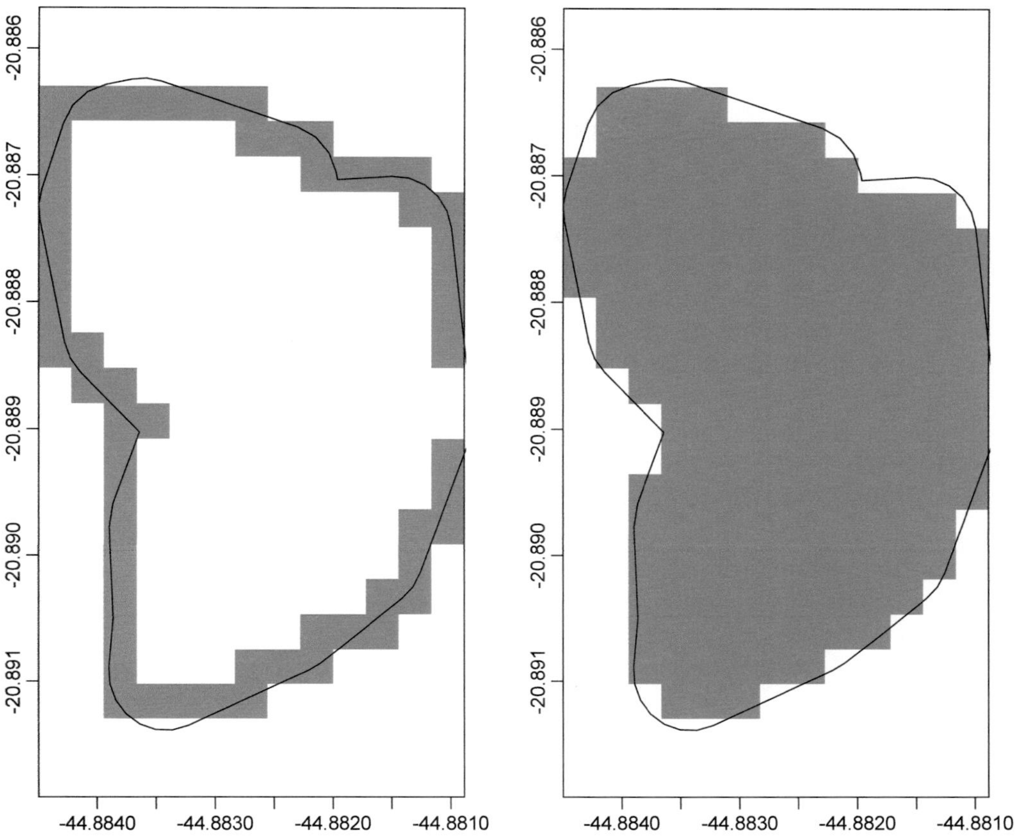

FIGURE 11.11 Line (left) and polygon (right) rasterization in the coffee field.

11.5.4 VECTORIZATION

Spatial vectorization is used as the opposite of rasterization. In vectorization, continuous raster data can be converted into discrete vectors such as points, lines and polygons (Lovelace et al., 2019). The vectorization of an altitude raster of the coffee field is performed to obtain point feature with altitude values in order to illustrate the raster-to-point vectorization operation. The centroid of each cell of the raster is converted into a point feature. The result can be obtained in a point simple feature or through a data matrix, according to the configuration of the algorithm used.

The `rasterToPoints` and `st_as_sf` functions are used to convert the SRTM model of the coffee field into a point geo-object.

```
elev_pt <- rasterToPoints(srtmWGS, spatial = TRUE) |> st_as_sf()
elev_pt
# Simple feature collection with 168 features and 1 field
# Geometry type: POINT
# Dimension:     XY
# Bounding box:  xmin: -44.88444 ymin: -20.89111 xmax: -44.88083 ymax: -20.88639
# CRS:           +proj=longlat +datum=WGS84 +no_defs +ellps=WGS84 +towgs84=0,0,0
# First 10 features:
```

```
#     s21_w045_1arc_v3                     geometry
# 1              1117 POINT (-44.88389 -20.88639)
# 2              1121 POINT (-44.88361 -20.88639)
# 3              1124 POINT (-44.88333 -20.88639)
# 4              1119 POINT (-44.88417 -20.88667)
# 5              1121 POINT (-44.88389 -20.88667)
# 6              1124 POINT (-44.88361 -20.88667)
# 7              1128 POINT (-44.88333 -20.88667)
# 8              1131 POINT (-44.88306 -20.88667)
# 9              1133 POINT (-44.88278 -20.88667)
# 10             1133  POINT (-44.8825 -20.88667)
```

The `as_Spatial` function is used to convert the geo-object from class `sf` to `sp` in order to map the altitude values in the vector and rasterized sample grid using the `spplot` and `grid.arrange` functions (Figure 11.12).

```
elev_pts <- as_Spatial(elev_pt)
a <- spplot(srtmWGS, colorkey = list(space = "bottom"))
b <- spplot(elev_pts)
grid.arrange(a, b, ncol = 2)
```

Another option is to map the results with the `sf` package using the following code.

```
par(mfrow = c(1, 2), mar = c(2, 2, 0.3, 0.3), mgp = c(1.5, 0.6, 0))
plot(srtmWGS, col = grey(1:100/100), axes = TRUE, legend = TRUE, horizontal = TRUE)
plot(bord, col = "transparent", add = TRUE)
plot(elev_pt[1], key.pos = NULL, reset = FALSE, main = "", axes = TRUE)
plot(bord, col = "transparent", add = TRUE)
```

Another type of vectorization used is raster-to-line conversion in order to produce contour lines to represent climate variables, altitude by contour lines, soil, water and vegetation attributes with continuous variation.

The `disaggregate` function is used to create a new raster with higher spatial resolution using the `bilinear` interpolation method. Note that the raster now has a spatial resolution of 100 m.

```
rd <- disaggregate(rmgprj, 10, method = "bilinear")
rd
# class       : RasterLayer
# dimensions : 680, 730, 496400  (nrow, ncol, ncell)
# resolution : 100, 100  (x, y)
# extent      : 485491.8, 558491.8, 7582572, 7650572  (xmin, xmax, ymin, ymax)
# crs         : +proj=utm +zone=23 +south +datum=WGS84 +units=m +no_defs
# source      : memory
# names       : bra
# values      : 861.1861, 1526.113  (min, max)
```

A `hillShade` map is created and overlayed with a colored altitude raster with the contour

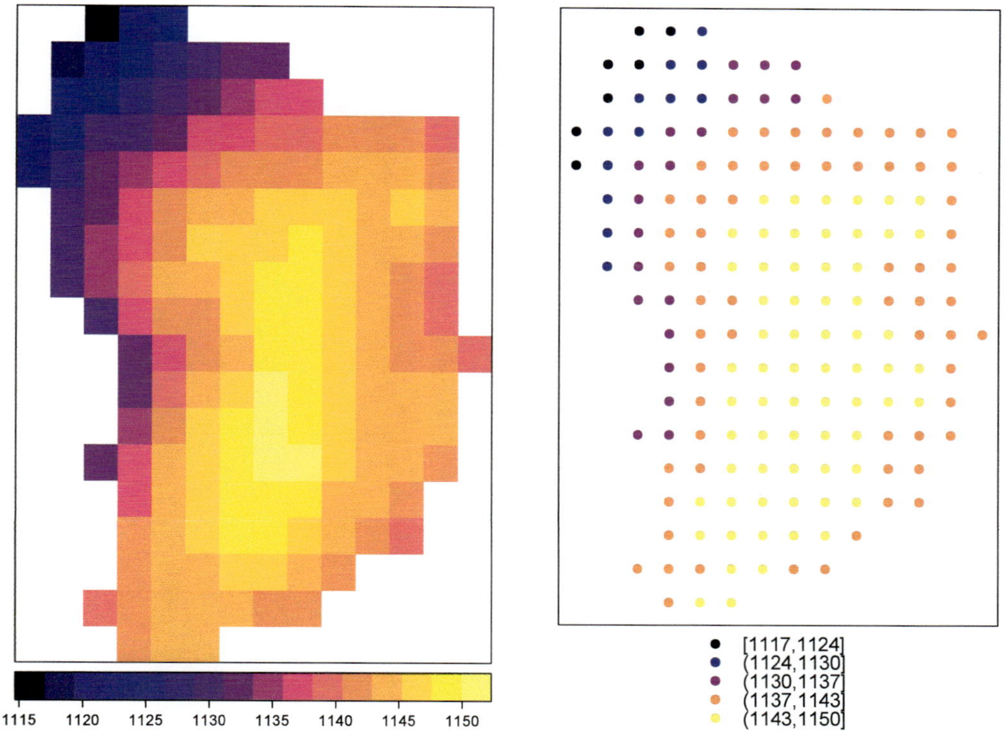

FIGURE 11.12 Vectorization of altitude cells (left) and conversion into spatial points with attributes (right) in the coffee field.

lines vectorized on top of the raster layers, in Carrancas, Minas Gerais, using the `plot` and `contour` functions (Figure 11.13).

```
# create hillshade
hs <- hillShade(slope = terrain(rd, "slope", unit="radians"),
                aspect = terrain(rd, "aspect", unit="radians"))
par(mfrow = c(1, 1), mar = c(2, 2, 0.3, 0.3), mgp = c(1.5, 0.6, 0))
plot(hs, col = gray(0:100 / 100), legend = FALSE) # Plot
# overlay with DEM
plot(rd, col = terrain.colors(25), alpha = 0.5, legend = FALSE, add = TRUE)
contour(rd, col = "white", add = TRUE, labcex = 1.0) # add contour lines
```

Raster vectorization into polygons can also be performed. The coffee field altitude raster is rasterized into square polygons to illustrate this operation. Then an operation is used to group raster polygons with the same altitude values in each polygon.

The `rasterToPolygons` function is used to convert each raster cell to a vector, and the `st_as_sf` function is used to convert the result to a geo-object of class `sf data.frame`. The `group_by` and `summarize` functions are used to group neighboring polygons with the same altitude value into a single polygon geometry.

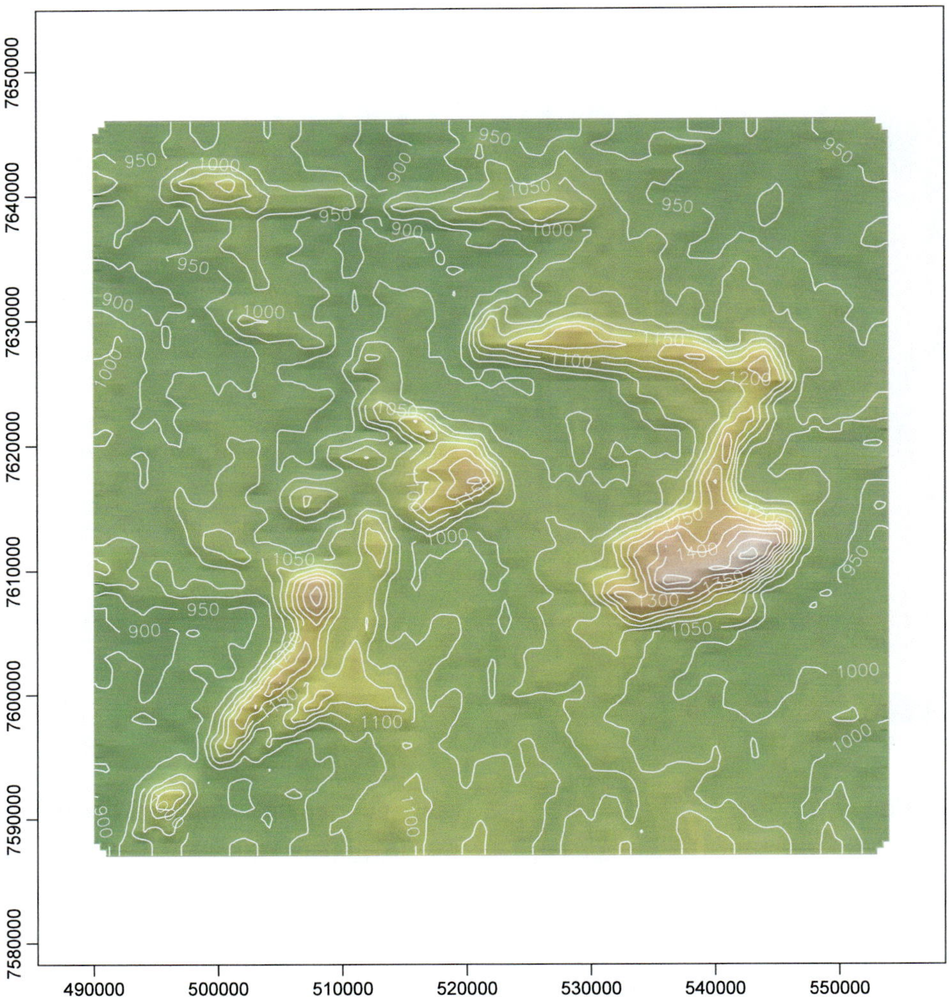

FIGURE 11.13 Vectorization of altitude cells (left) on contour lines overlayed with relief and altitude shadow raster of Carrancas, Minas Gerais, at 1 km spatial resolution.

```
raster_poly <- rasterToPolygons(srtmWGS) |> st_as_sf()
raster_poly2 <- raster_poly |> group_by(s21_w045_1arc_v3) |> summarize()
```

The `as_Spatial` function is used to convert the geo-object from class `sf` to `sp` in order to map the altitude values using the `spplot` and `grid.arrange` functions (Figure 11.14).

```
a1 <- as_Spatial(raster_poly ) # Convert to sp
a2 <- as_Spatial(raster_poly2 ) # Convert to sp
a <- spplot(srtmWGS, colorkey = list(space = "bottom"))
b <- spplot(a1, colorkey = list(space = "bottom"))
c <- spplot(a2, colorkey = list(space = "bottom"))
grid.arrange(a, b, c, ncol = 3)
```

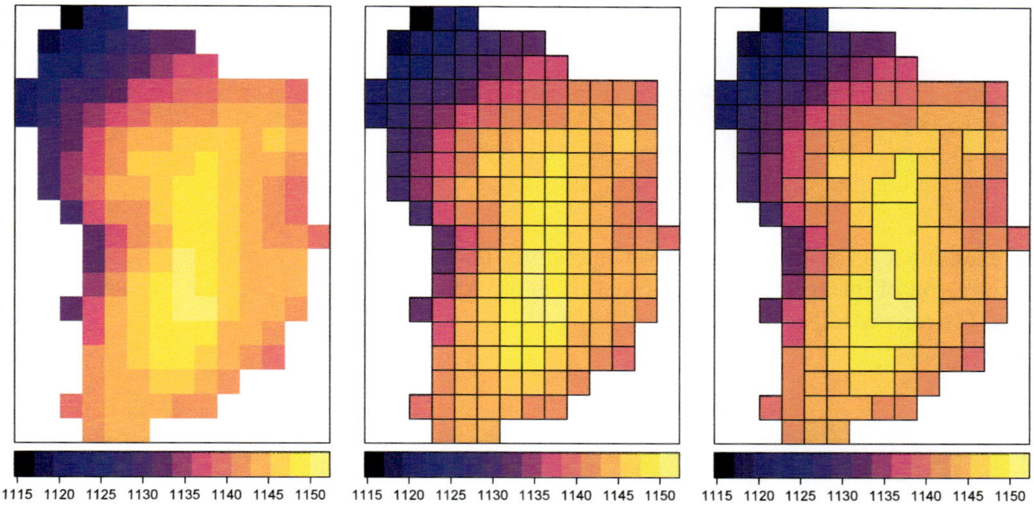

FIGURE 11.14 Vectorization of altitude raster cells (left) of the coffee field into polygons (center) and aggregation of polygons with the same altitude value (right).

11.6 RASTER THREE-DIMENSIONAL GEOMETRY

The bidimensional and three-dimensional geometry and aspects of relief shading, water, and clouds can be explored with environmental geocomputation and R. Using the `rayshader` package (Morgan-Wall, 2024), it is possible to perform detailed mapping for enhanced geovisualization of the relief using digital elevation models. SRTM data with a spatial resolution of 30 m or ALOS PALSAR data with a spatial resolution of 12.5 m can be used for better geovisualization of the results.

Data from the ALOS PALSAR RTC Project[1] at 12.5 m spatial resolution are obtained to perform this example. The `rayshader` and `terra` packages required for analysis are loaded into R with the `library` function. It should be noted that the same analysis can also be performed with the `raster` package.

```
library(terra)
# terra 1.7.60
library(rayshader)
# package 'rayshader' was built under R version 4.3.3
```

A tile of ALOS PALSAR data from the Carrancas region is imported into R using the `rast` function. Note that the data has 12.5 m spatial resolution.

```
alos <- rast(
  "C:/sr/AlosPalsar/zap/AP_27029_FBS_F6750_RT1/AP_27029_FBS_F6750_RT1.dem.tif")
alos
```

[1]https://www.earthdata.nasa.gov/

```
# class       : SpatRaster
# dimensions  : 5981, 6593, 1  (nrow, ncol, nlyr)
# resolution  : 12.5, 12.5  (x, y)
# extent      : 467714.8, 550127.3, 7599073, 7673835  (xmin, xmax, ymin, ymax)
# coord. ref. : WGS 84 / UTM zone 23S (EPSG:32723)
# source      : AP_27029_FBS_F6750_RT1.dem.tif
# name        : AP_27029_FBS_F6750_RT1.dem
```

The `plot` function is used to preview the region. The `draw` function is used to define a subset in the region of interest. The `crop` function is used to crop the area from the subset defined in the Carrancas hill region (Figure 11.15).

```
plot(alos) # Map
e <- draw() # Define extent
e # See extent coordinates
# SpatExtent : 520912.305722892, 549483.065323795, 7605209.12989458,
# 7629698.35240964 (xmin, xmax, ymin, ymax)
alos_crop <- crop(alos, e) # Crop
alos_crop # See header
# class       : SpatRaster
# dimensions  : 1959, 2285, 1  (nrow, ncol, nlyr)
# resolution  : 12.5, 12.5  (x, y)
# extent      : 520914.8, 549477.3, 7605210, 7629698  (xmin, xmax, ymin, ymax)
# coord. ref. : WGS 84 / UTM zone 23S (EPSG:32723)
# source(s)   : memory
# varname     : AP_27029_FBS_F6750_RT1.dem
# name        : AP_27029_FBS_F6750_RT1.dem
# min value   :                        914
# max value   :                       1574
```

The raster defined in the area of interest is converted to matrix using the `raster_to_matrix` function.

```
elevation_matrix <- raster_to_matrix(alos_crop)
# [1] "Dimensions of matrix are: 1959x2285"
```

The functions `sphere_hade`, `add_water`, `add_shadow`, and `plot_map` are used to generate a hillshade map in the analysis region. Note that no water was detected in this subset of data, requiring a smoother surface to generate these results, which occur further north of this region (Figure 11.16).

```
elevation_matrix |> sphere_shade(texture = "desert") |>
    add_water(detect_water(elevation_matrix), color = "desert") |>
    add_shadow(ray_shade(elevation_matrix), 0.5) |>
    add_shadow(ambient_shade(elevation_matrix), 0) |> plot_map()
```

The functions `sphere_shade`, `add_water`, `add_shadow`, `plot_3d`, and `render_snapshot` are used for three-dimensional mapping of the same region (Figure 11.17). Note that `rayshader` also includes the option to add a procedurally generated cloud layer and, optionally, shadows.

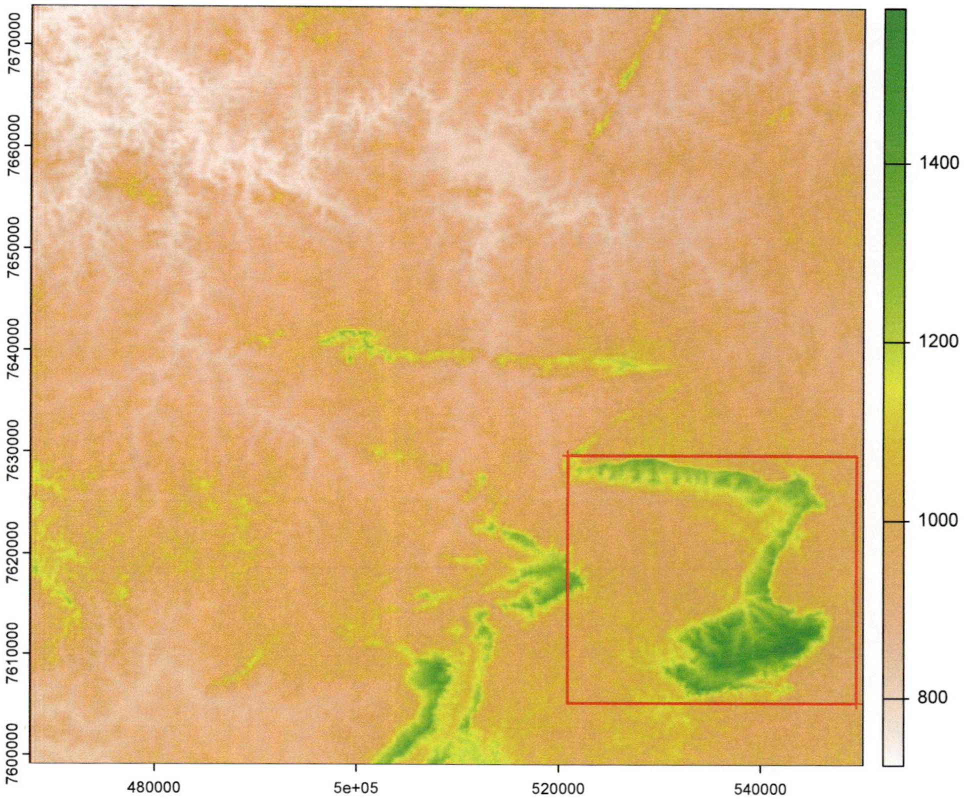

FIGURE 11.15 Definition of raster subset of interest based on raster-vector interaction in the Carrancas hill region, Minas Gerais, Brazil.

```
elevation_matrix |> sphere_shade(texture = "desert") |>
  add_water(detect_water(elevation_matrix), color = "desert") |>
  add_shadow(ray_shade(elevation_matrix, zscale = 3), 0.5) |>
  add_shadow(ambient_shade(elevation_matrix), 0) |>
  plot_3d(elevation_matrix, zscale = 10, fov = 0, theta = 135, zoom = 0.75,
  phi = 45, windowsize = c(1000, 800))
render_snapshot()
```

11.7 COMPUTATIONAL PRACTICE

Geometric operations on raster data and the interaction between raster data and vector features are evaluated from functions available in R packages. Examples are created in this computational practice to exemplify codes and functions applied to environmental geographic solutions.

FIGURE 11.16 Bidimensional mapping of hillshade in the Carrancas hill region, Minas Gerais, Brazil.

11.7.1 PACKAGES USED

The `install.packages` function is used to install the `sf`, `raster`, `dplyr`, `purrr`, `geosphere` and `gridExtra` packages needed to perform the computational practice activities. If the packages are already installed, it is not necessary to install the packages again for each computational practice.

```
install.packages("sf")
install.packages("raster")
install.packages("dplyr")
install.packages("purrr")
install.packages("geosphere")
install.packages("gridExtra")
```

The `library` function is used to load the packages needed to perform the computational practice.

FIGURE 11.17 Three-dimensional mapping of hillshade in the Carrancas hill region, Minas Gerais, Brazil.

```
library(sf)
library(raster)
library(dplyr)
library(purrr)
library(geosphere)
library(gridExtra)
```

11.7.2 GEOMETRIC OPERATIONS ON RASTER DATA

Geometric operations on raster data to obtain geometric intersection, change of extent and origin, aggregation, disaggregation and resampling are evaluated.

11.7.3 GEOMETRIC INTERSECTION

A geometric intersection operation between raster data is used to obtain a new raster as a function of the overlap of two simulated rasters. A raster subset is obtained relative to a raster with a larger extent. For this, a raster between municipalities in Minas Gerais, Brazil is created. Brightness values to the raster are assigned according to the count of the number of cells in the raster at the spatial resolution of 0.008333333° (∼ 1 km).

```
r <- raster(xmx = -44.479124, xmn = -45.090629, ymn = -21.817136,
            ymx = -21.289025, res = 0.008333333)
values(r) <- 1:ncell(r)
```

Then a second raster is created to be used as a geometric region to perform a subset in the Carrancas region, Minas Gerais, Brazil, with a spatial resolution of 0.0002777778° (∼ 30 m). Brightness values are assigned to the raster with random numbers obtained from counting the number of cells in the raster.

```
r1 <- raster(xmx = -44.526692, xmn = -44.718586, ymn = -21.635500,
             ymx = -21.430372, res = 0.0002777778)
values(r1) <- runif(ncell(r1))
```

The geometric intersection is performed using `[` associated with the parameter `drop = FALSE` to obtain a subset, `r2`, of the original raster, `r`, near Carrancas, `r1` (Figure 11.18).

```
r2 <- r[r1, drop = FALSE]
r2
# class       : RasterLayer
# dimensions : 25, 23, 575  (nrow, ncol, ncell)
# resolution : 0.008333333, 0.008333333  (x, y)
# extent     : -44.71563, -44.52396, -21.63902, -21.43069 (xmin, xmax, ymin, ymax)
# crs         : +proj=longlat +datum=WGS84 +no_defs
# source      : memory
# names       : layer
# values      : 1287, 3061  (min, max)
```

It can be seen that the raster subset created is obtained at the resolution of the original raster, with 0.008333333° spatial resolution.

```
par(mfrow = c(1, 3))
plot(r, col = grey(1:100/100), main = "Input raster", legend = FALSE)
plot(r, col = grey(1:100/100), main = "Input and intersection raster",
     legend = FALSE)
plot(r1, col = grey(1:100/100), add = TRUE, legend = FALSE)
plot(r, col = grey(1:100/100), main = "Input and output raster", legend = FALSE)
plot(r2, col = grey(1:100/100), add = TRUE, legend = FALSE)
```

The results are mapped using the `plot` function.

11.7.4 ORIGIN AND EXTENT

When performing raster geospatial analysis operations, the origin and extent must match between the analyzed maps. This kind of problem can occur when working with multiresolution remote sensing data, usually with different extent.

As an example, four rows and four columns are added on each side of the output raster of the previous example, with the brightness value equal to 1, using the `extend` function

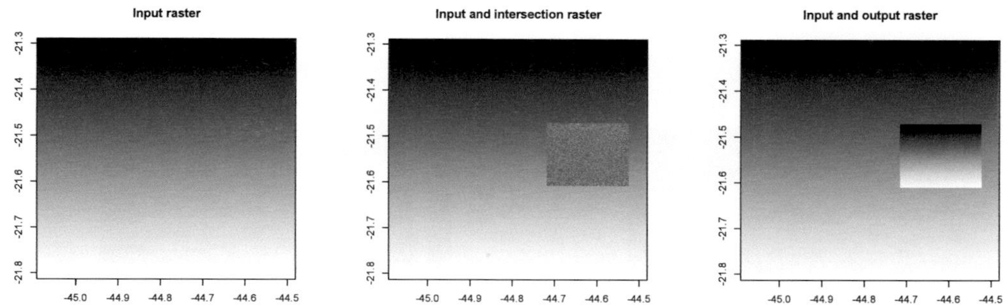

FIGURE 11.18 Geometric intersection operation used between an input raster (left) and an intersection raster (center) to define an output subset (right) with the same characteristics as the input raster.

(Figure 11.19).

```
r3 <- extend(r2, c(4, 4), value = 1)
r3
# class       : RasterLayer
# dimensions : 33, 31, 1023  (nrow, ncol, ncell)
# resolution : 0.008333333, 0.008333333  (x, y)
# extent   : -44.74896, -44.49063, -21.67236, -21.39736  (xmin, xmax, ymin, ymax)
# crs         : +proj=longlat +datum=WGS84 +no_defs
# source      : memory
# names       : layer
# values      : 1, 3061  (min, max)
```

This increases the number of cells in the original raster from 575 to 1023. The results are mapped using the `plot` function.

```
par(mfrow = c(1, 2), mar = c(3, 3, 5, 5), mgp = c(1.5, 0.6, 0))
plot(r2, col = grey(1:100/100), main = "Input raster", horizontal = TRUE)
plot(r3, col = grey(1:100/100), main = "Output raster", horizontal = TRUE)
```

Performing an algebraic operation between rasters under different origins can determine results for the area of intersection between rasters. Using the `extend` function, we can extend the dimension of one raster with respect to another. As an example, a sum algebraic operation is performed on rasters `r2` and `r3` of different sizes, resulting in the raster `r4`, referring to the intersection between the two input rasters.

```
r4 <- r2 + r3
# Warning message:
# In r2 + r3 :
# Raster objects have different extents. Result for their intersection is returned
```

The `extend` function is used to increase the geometric cell extent of the `r2` raster relative to `r3`, creating the `r5` raster, with new rows and columns and no brightness value, `NA`.

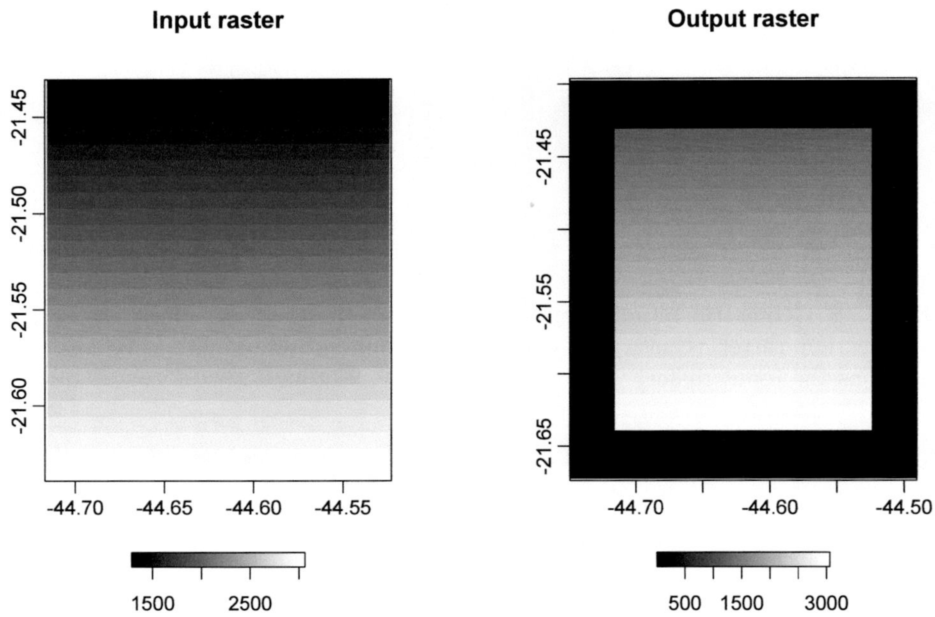

FIGURE 11.19 Operation to extend an input raster (left) into a new output raster with brightness value equal to 1, in four new rows and columns (right), respectively.

```
r5 <- extend(r2, r3)
```

Then the sum between rasters `r3` and `r5` with the same length is performed without any problems of overlapping rasters.

```
r6 <- r3 + r5
```

The origin of a raster is set to a point on the raster obtained by using the `origin` function. In the case of the created raster, for example, `r2`, the origin is (0.001035863, 0.002640815).

```
origin(r2)
# [1] 0.001035863 0.002640815
```

The change of origin can be performed in situations where it is necessary to perform algebraic operations between rasters. The origin of raster `r2` is illustratively changed to (0,0), creating raster `r7`, according to the following code:

```
r7 <- r2
origin(r7) <- c(0.0, 0.0) # Change of origin to (0,0)
origin(r2)
# [1] 0.001035863 0.002640815
origin(r7)
# [1] 0 0
```

11.7.5 AGGREGATION AND DISAGGREGATION

The aggregation and disaggregation operations are used to reduce and increase the spatial resolution of a raster, respectively. The aggregation function is used to reduce the spatial resolution of the previously created raster `r` with aggregation factors 2, 4, and 6, using the average of the neighboring cells to create the output raster (`fun = mean`) (Figure 11.20).

```
agr2 <- aggregate(r, fact = 2, fun = mean)
agr4 <- aggregate(r, fact = 4, fun = mean)
agr6 <- aggregate(r, fact = 6, fun = mean)
```

The results are mapped using the `plot` function.

```
par(mfrow = c(2, 2), mar = c(2, 2, 0.3, 0.3), mgp = c(1.5, 0.6, 0))
plot(r, col=grey(1:100/100), horizontal = TRUE)
plot(agr2, col=grey(1:100/100), horizontal = TRUE)
plot(agr4, col=grey(1:100/100), horizontal = TRUE)
plot(agr6, col=grey(1:100/100), horizontal = TRUE)
```

Using disaggregation, the spatial resolution of a raster is increased, determining an increase in the number of cells in the image. The raster `r` is disaggregated to perform this operation by setting the `bilinear` interpolation method and disaggregation factor 6. With this, the spatial resolution of the new raster increased from 0.008333333° to 0.001388889°.

```
desr <- disaggregate(r, fact = 6, method = "bilinear")
```

A useful way to perform alignment of a raster, based on another raster at once, in terms of source, extent and resolution, can be performed with the `resample` function of the `raster` package. In this case, the bilinear interpolation method is set as default for applying this function. As an example, the raster `r1`, created earlier, is resampled as a function of the raster `r`, by means of the `resample` function.

```
rres <- resample(r1, r)
```

The results are mapped using the `plot` function.

```
par(mfrow = c(1, 2), mar = c(2, 2, 0.3, 0.3), mgp = c(1.5, 0.6, 0))
plot(r, col=grey(1:100/100), legend = FALSE)
plot(rres, col=grey(1:100/100), legend = FALSE)
```

Note that the cell brightness is filled only where there is an analyzed match between the two rasters analyzed; however, the origin, extent and resolution between the two rasters is the same (Figure 11.21).

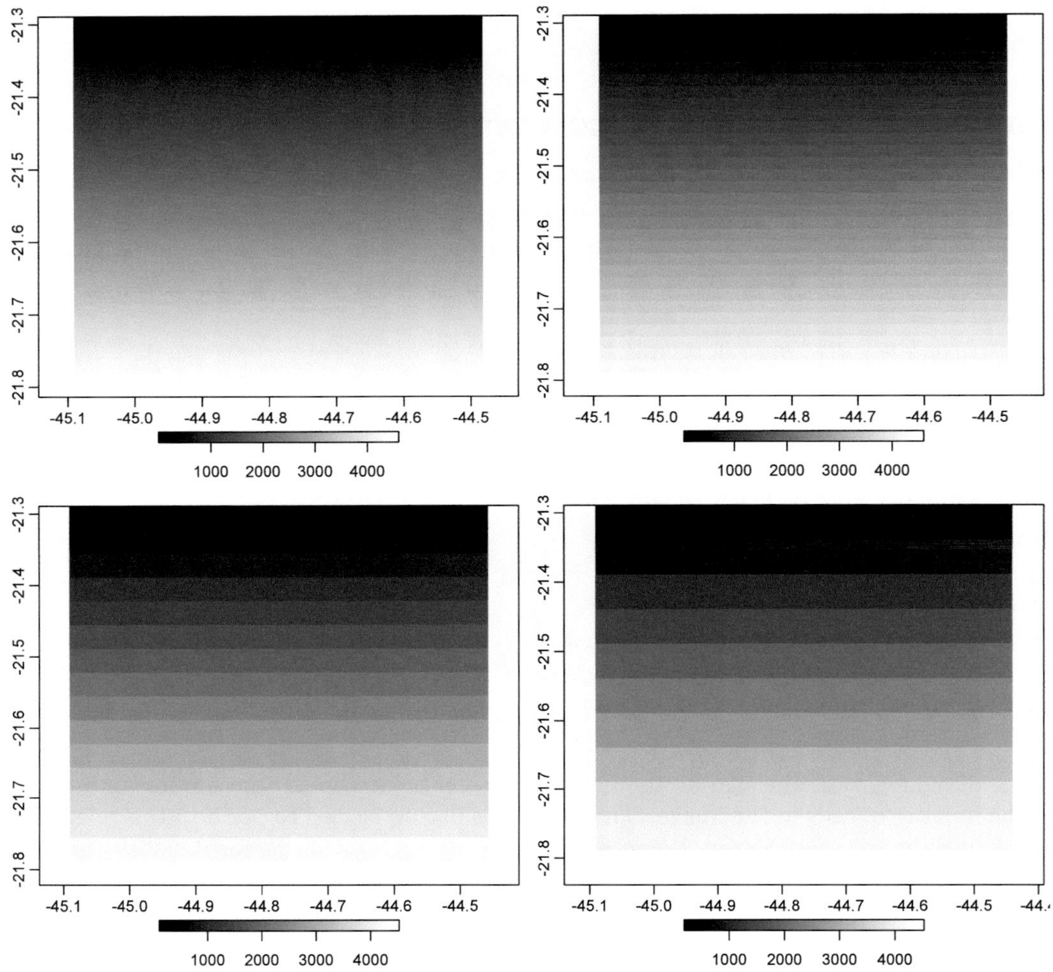

FIGURE 11.20 Operation to aggregate the spatial resolution of a raster (r) (top-left) with factors 2 (top-right), 4 (bottom-left) and 6 (bottom-right), respectively.

11.7.6 INTERACTION BETWEEN RASTER DATA AND VECTOR FEATURES

The main techniques used for interaction between raster objects and vector features are performing cropping, masking, and extraction of raster values from vector features, as well as raster to vector conversions (Lovelace et al., 2019).

11.7.7 RASTER CROPPING AND MASKING

Cropping, masking, and inverted masking are illustrated from the creation of a vector feature within the raster extension `r`. The WGS-84 geographic coordinates of a polygon with four irregular vertices is created to illustrate this concept.

```
pol <- st_polygon(list(rbind(c(-44.90981, -21.55543), c(-44.77387, -21.43829),
                             c(-44.60594, -21.54427),c(-44.70790, -21.62794),
                             c(-44.90981, -21.55543))))
```

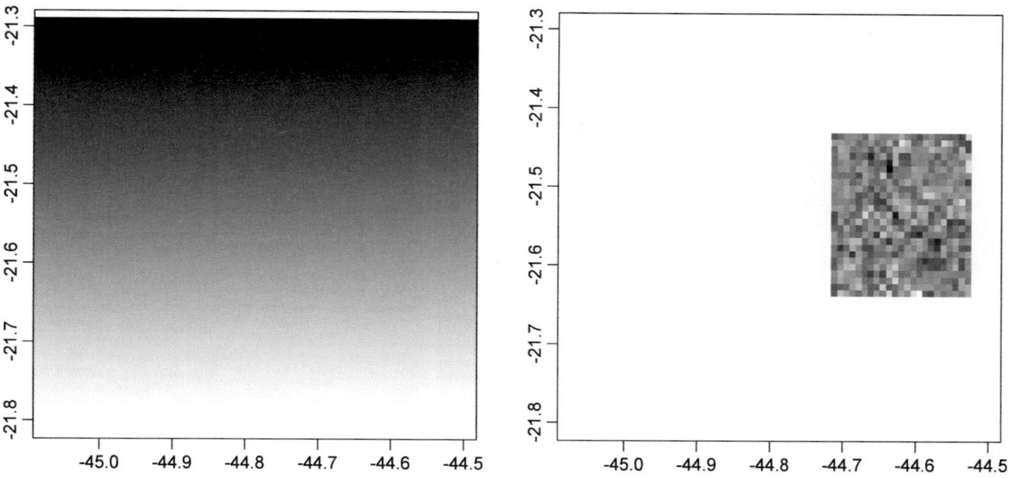

FIGURE 11.21 Use the resample function to align the raster, r1, based on the raster, r.

The polygon `pol` created in class `sfg` is transformed into a polygon of class `sfc`, using the `st_sfc` function.

```
polsfc <- st_sfc(pol)
```

With this, it is possible to assign the WGS-84 geographic projection system to the polygon.

```
st_crs(polsfc) <- 4326
```

Next, the polygon, `polsfc`, is transformed into a polygon, `p`, of class `sf`, using the `st_as_sf` function.

```
p <- st_as_sf(polsfc)
```

Thus, it is possible to perform crop, mask and inverse mask on the previously created raster `r` from the created polygonal vector feature, `p` (Figure 11.22).

```
rCrop <- crop(r, p) # Crop
rWGS <- mask(rCrop, p) # Mask
rWGS1 <- mask(rCrop, p, inverse = TRUE) # Inverse mask
```

The results are mapped using the `plot` function.

```
par(mfrow = c(2, 2), mar = c(2, 2, 0.3, 0.3), mgp = c(1.5, 0.6, 0))
plot(r, col = grey(1:100/100), legend = FALSE)
plot(p, add = TRUE, border = "red", lwd = 2)
plot(rCrop, col=grey(1:100/100), legend = FALSE)
plot(p, add = TRUE, border = "red", lwd = 2)
plot(rWGS, col=grey(1:100/100), legend = FALSE)
plot(p, add = TRUE, border = "red", lwd = 2)
```

```
plot(rWGS1, col = grey(1:100/100), legend = FALSE)
plot(p, add = TRUE, border = "red", lwd = 2)
```

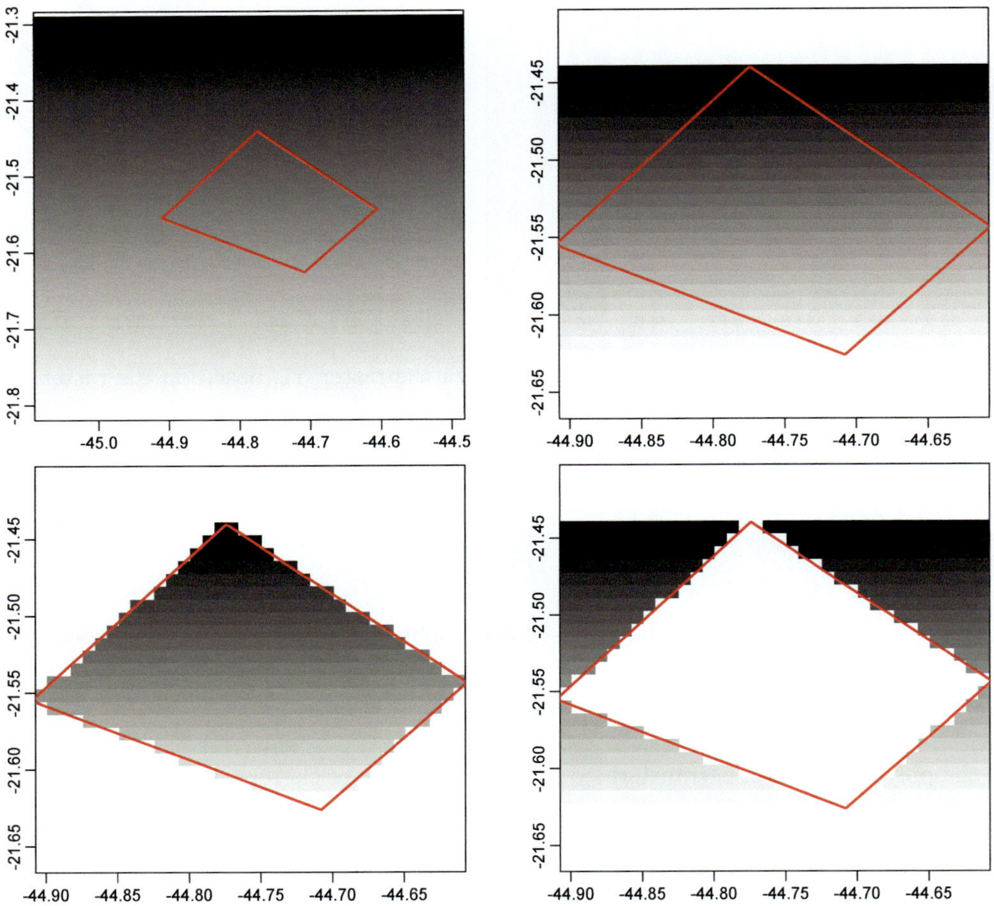

FIGURE 11.22 Operation of cropping (top-right), masking (bottom-left) and inverted masking (bottom-right) obtained from interaction between raster and vector features (top-left).

11.7.8 RASTER EXTRACTION

11.7.8.1 Point raster extraction

Raster data extraction is used to extract values from a raster based on vector features. To exemplify the method, points are created at the vertices of the polygon in the previous example.

```
mpt <- st_multipoint(matrix(c(-44.90981, -44.77387, -44.60594, -44.70790,
          -21.55543, -21.43829, -21.54427, -21.62794), ncol = 2))
```

The `multipoint`, `mpt`, created in the `sfg` class are transformed into `point` of the `sfc` class by the `st_sfc` function and converted to `point` by the `st_cast` function.

```
pt <- st_cast(x = st_sfc(mpt), to = "POINT")
```

The WGS-84 geographic projection system is assigned to the points by the `st_crs` function.

```
st_crs(pt) <- 4326
```

A column to identify the points is created and assigned to the `data.frame` of `pt`.

```
id <- 1:4
pt <- st_sf(data.frame(pt, id))
```

The points are transformed into geo-objects of class `sp` using the `as_Spatial` function.

```
pts <- as_Spatial(pt)
```

The extraction of raster brightness values, `r`, is performed from the points, `pts`, created (Figure 11.23).

```
pts$r <- raster::extract(r, pts)
```

The results are mapped using the `plot` function.

```
par(mfrow = c(1, 2), mar = c(2, 2, 0.3, 0.3), mgp = c(1.5, 0.6, 0))
plot(r, col = grey(1:100/100), axes = TRUE, horizontal = TRUE)
plot(pts, add = TRUE, pch = 1, colour = "black")
text(pts, labels = pts$id, pos = 3, halo = TRUE, offset = 0.5, cex = 1)
plot(pts$r, xlab = "ID", ylab = "r value")
```

11.7.8.2 Raster extraction by line

The values of a raster, `r`, are extracted using a transect defined by a line, `l`. In this case, a line is created between sample points with ID, 4, and 2, from the previous example, in order to extract raster values along a transect. In this example, a transect defined as a straight line was chosen. The projection of the line is defined as the same as in the raster and the line is converted to class `sf`.

```
l <- cbind(c(-44.7079, -44.77387), c(-21.62794, -21.43829)) |>
  st_linestring() |>
  st_sfc(crs = projection(r)) |>
  st_sf()
```

The extraction of brightness data from the raster with the transect is performed by the `extract` function. The parameters `along = TRUE` and `cellnumbers = TRUE` are used to obtain the `ID` of the cells along the path traversed between the start and end of the line.

```
lr <- raster::extract(r, l, along = TRUE, cellnumbers = TRUE)
```

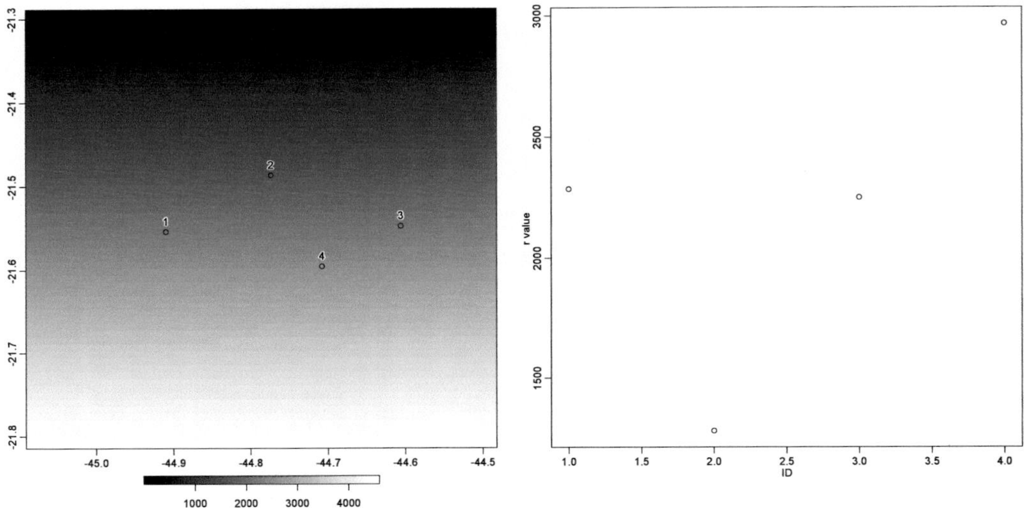

FIGURE 11.23 Raster value extraction operation for four points (left) and the graph of extracted values for each point's identifier (ID) code (right).

The purr and geosphere packages are used to create a database with the determination of the coordinates associated with each extracted cell and with the determination of the geodetic distance along the transect. Thus, it is possible to determine the profile of raster cell brightness values, r, along the line, l, created (Figure 11.24).

```
transect_df1 <- purrr::map_dfr(lr, as_data_frame, .id = "ID")
transect_coords1 <- xyFromCell(r, transect_df1$cell)
pair_dist1 <- geosphere::distGeo(transect_coords1)[-nrow(transect_coords1)]
transect_df1$dist1 = c(0, cumsum(pair_dist1))
```

The results are mapped using the plot function.

```
par(mfrow = c(1, 2), mar = c(2, 2, 0.3, 0.3), mgp = c(1.5, 0.6, 0))
plot(r, col = grey(1:100/100), axes = TRUE, horizontal = TRUE)
plot(l, add = TRUE, lwd = 2, colour = "black")
text(pts[2,], labels = "End", halo = TRUE, pos = 4, offset = 0.5)
text(pts[4,], labels = "Start", halo = TRUE, pos = 4, offset = 0.5)
plot(transect_df1$dist1, transect_df1$layer, xlab = "Distance (m)",
     ylab = "r value")
```

11.7.8.3 Raster-polygon extraction

The extraction of raster values is evaluated in terms of raster-polygon interaction. The raster values, r, are extracted in order to obtain information about the minimum, maximum and mean raster value within a polygon feature, p.

```
r_values <- raster::extract(x = r, y = p, df = TRUE)
```

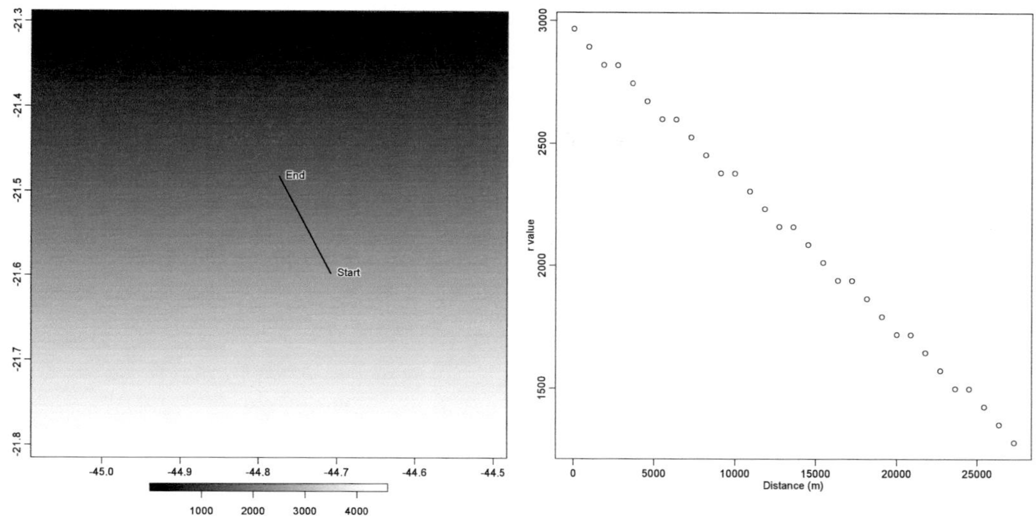

FIGURE 11.24 Extraction of raster value using a transect (left) to obtain cell brightness values (right) according to the variation of distance along a line-type vector feature.

The minimum, mean and maximum brightness values within the polygon are 1352, 2192 and 2966, respectively.

```
group_by(r_values, ID) |>
  summarize_at(vars(layer), list(~min(.), ~mean(.), ~max(.)))
# A tibble: 1 x 4
#      ID   min  mean   max
#   <dbl> <dbl> <dbl> <dbl>
# 1     1  1352 2192.  2966
```

The extraction of a categorical raster can also be performed. A categorical raster is created from the `r` raster. The values of `r` are classified into five brightness classes in a categorical raster, defined as `rc`.

```
rc <- r
rc[] <- 1
rc[1000:2000] <- 2
rc[2001:3000] <- 3
rc[3001:4000] <- 4
rc[4001:4599] <- 5
```

The `ratify` function is used to create an attribute table with the `ID` referring to the raster classes, `rc`.

```
rc <- ratify(rc)
```

Two columns are created in the `rat1` attribute table with the `class` and `code` variables, referring to nominal class and cell brightness value code, respectively. The created items are added to the raster `rc` attribute table with the `levels` function.

```
rat1 <- levels(rc)[[1]]
rat1$class <- c('a', 'b', 'c', 'd', 'e')
rat1$code <- c(1, 2, 3, 4, 5)
levels(rc) <- rat1
rc
# class       : RasterLayer
# dimensions : 63, 73, 4599  (nrow, ncol, ncell)
# resolution : 0.008333333, 0.008333333  (x, y)
# extent      : -45.09063, -44.4823, -21.81402, -21.28902  (xmin, xmax, ymin, ymax)
# crs         : +proj=longlat +datum=WGS84 +no_defs
# source      : memory
# names       : layer
# values      : 1, 5  (min, max)
# attributes :
# ID class code
#  1     a    1
#  2     b    2
#  3     c    3
#  4     d    4
#  5     e    5
```

With this, it is possible to extract the cells from the raster, `rc`, within the polygon, `p`, in order to obtain the count of each cell class within the polygon. We observed 116 and 304 cells with codes 2 and 3, respectively (Figure 11.25).

```
rc_p <- raster::extract(rc, p, df = TRUE, factors = TRUE) # Extract raster values
dplyr::select(rc_p, ID, code) |> tidyr::gather(key, code, -ID) |> # Count
  group_by(ID, key, code) |>  tally() |> tidyr::spread(code, n, fill = 0)
# A tibble: 1 x 4
# Groups:   ID, key [1]
#      ID key     `2`   `3`
#   <dbl> <chr> <dbl> <dbl>
# 1     1 code    116   304
```

The results are mapped using the `plot` function.

```
par(mfrow = c(1, 2), mar = c(2, 2, 0.3, 0.3), mgp = c(1.5, 0.6, 0))
plot(r, col = grey(1:100/100), axes = TRUE, horizontal = TRUE)
plot(p, col = "transparent", add = TRUE)
plot(rc, "code", col = rev(grey(1:100/100)), axes = TRUE, horizontal = TRUE)
plot(p, col = "transparent", add = TRUE)
```

11.7.9 RASTERIZATION

Rasterization of points, lines, and of polygons is performed using a raster template created to determine values to the cells that are touched by the vector feature and performing the conversion at those locations.

Rasterization is used to rasterize raster points, `agr6`, with brightness values at locations

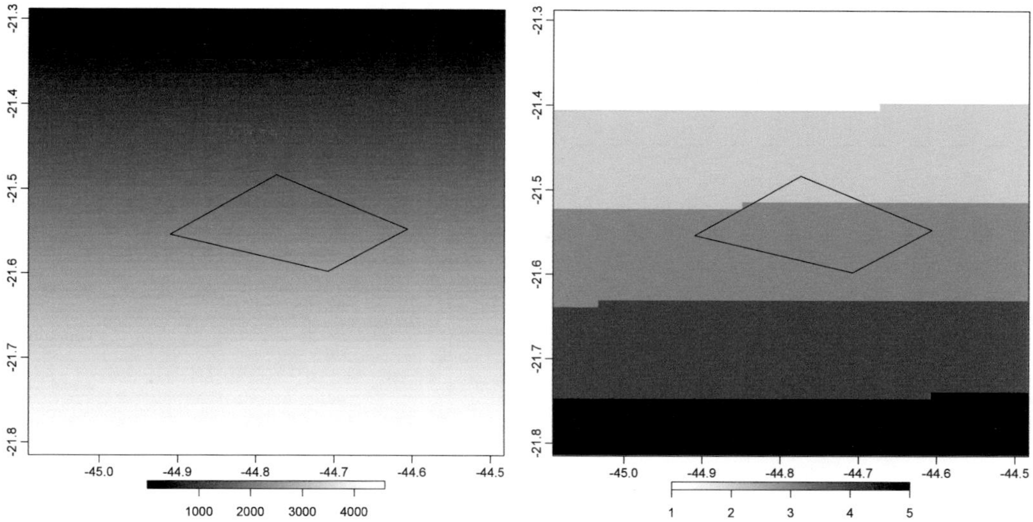

FIGURE 11.25 Extraction of raster with continuous (left) and categorical (right) values, extracted from polygon-type vector feature.

defined by the points, `pt`. A raster is created to represent the output results of the rasterization with a spatial resolution of 0.05° (Figure 11.26).

```
raster_temp <- raster(extent(agr6), resolution = 0.05,
                      crs = st_crs(agr6)$proj4string)
```

The points are rasterized using the `rasterize` function.

```
pts_raster1 <- rasterize(pt, raster_temp, field = "id")
```

The results are mapped using the `spplot` function.

```
a <- spplot(pts[1])
b <- spplot(pts_raster1, colorkey = list(space = "bottom"))
grid.arrange(a, b, ncol = 2)
```

Similarly, rasterization is used on vector features of the line and polygon type, using the `rasterize` function (Figure 11.27).

```
l_raster1 <- rasterize(l, raster_temp) # Line rasterization
p_raster1 <- rasterize(p, raster_temp) # Polygon rasterization
```

The results are mapped using the `plot` function.

```
psp <- as_Spatial(p) # Convert to sp
lsp <- as_Spatial(l) # Convert to sp
par(mfrow = c(1, 2), mar = c(2, 2, 0.3, 0.3), mgp = c(1.5, 0.6, 0))
```

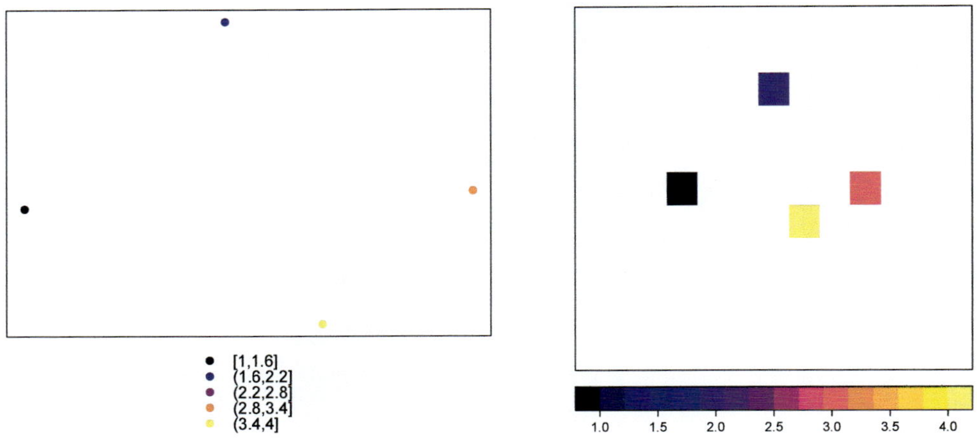

FIGURE 11.26 Rasterization of points at spatial resolution of 0.05 ° (right) with altitude attributes (m) (left) of the coffee crop in Santo Antônio do Amparo, Minas Gerais State, Brazil.

```
plot(l_raster1, col = grey(1:100/100), axes = TRUE, legend = FALSE)
plot(lsp, col = "black", add = TRUE)
plot(p_raster1, col = grey(1:100/100), axes = TRUE, legend = FALSE)
plot(psp, col = "transparent", add = TRUE)
```

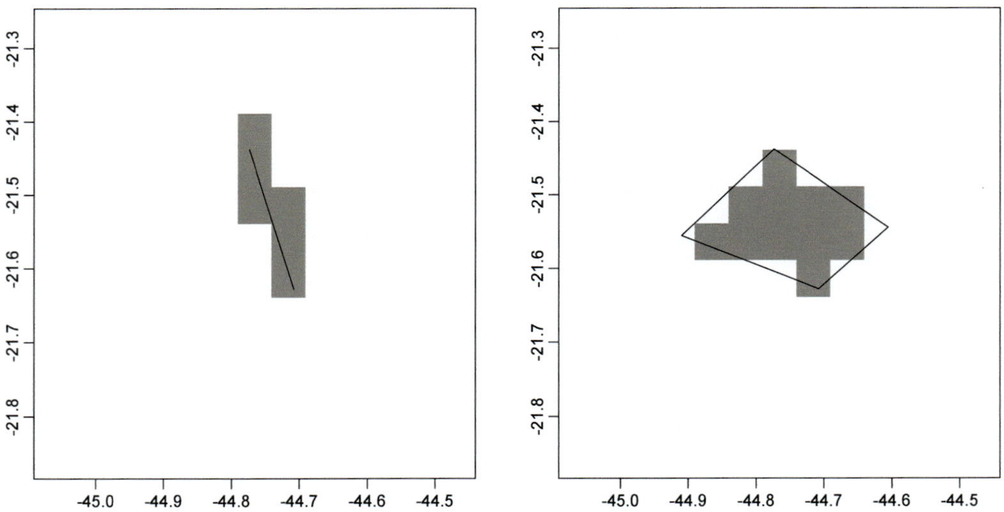

FIGURE 11.27 Line (left) and polygon (right) rasterization at 0.05 ° spatial resolution.

11.7.10 VECTORIZATION

Vectorization of the `agr6` raster is used to obtain point features with raster brightness values using the `rasterToPoints` function. In this case, the centroid of each cell in the raster is converted to a point feature. The `spatial = TRUE` parameter is used to get the result into a point geospatial feature (Figure 11.28).

```
r_pt <- rasterToPoints(agr6, spatial = TRUE) |> st_as_sf()
```

The results are mapped using the `spplot` function.

```
r_pts <- as_Spatial(r_pt) # Convert to sp
a <- spplot(agr6, colorkey = list(space = "bottom"))
b <- spplot(r_pts)
grid.arrange(a, b, ncol = 2)
```

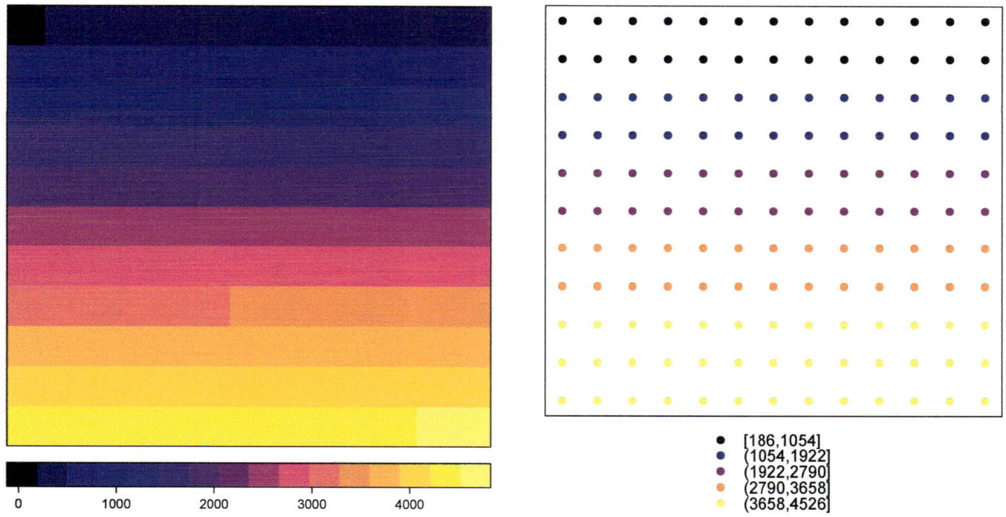

FIGURE 11.28 Vectorization of raster cells (left) and conversion into points with attributes (right).

Another type of vectorization used is raster-to-line conversion. This type of vectorization is performed with the `contour` function (Figure 11.29).

```
plot(agr6, col = terrain.colors(25), alpha = 0.5, legend = FALSE)
contour(agr6, col = "black", add = TRUE, labcex = 1.0)
```

Raster vectorization into polygons is also performed. The raster `agr6` is rasterized into square polygons and an operation to group raster polygons with the same brightness value `grou_by` is used (Figure 11.30).

```
raster_pol <- rasterToPolygons(rc) |> st_as_sf()
raster_pol1 <- raster_pol |> group_by(layer) |> summarize()
```

The results are mapped using the `plot` function.

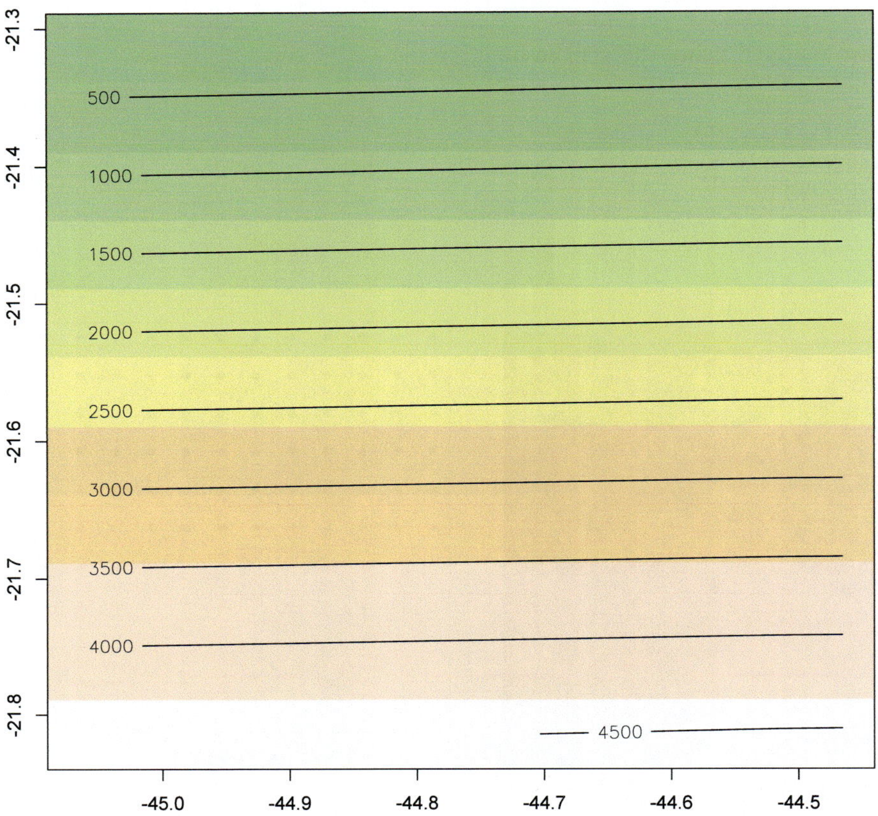

FIGURE 11.29 Vectorization of raster cells overlayed by lines with cell brightness values.

```
a1 <- as_Spatial(raster_pol) # Convert to sp
a2 <- as_Spatial(raster_pol1) # Convert to sp
a <- spplot(rc, "code", colorkey = list(space = "bottom"))
b <- spplot(a1, colorkey = list(space = "bottom"))
c <- spplot(a2, colorkey = list(space = "bottom"))
grid.arrange(a, b, c, ncol = 3)
```

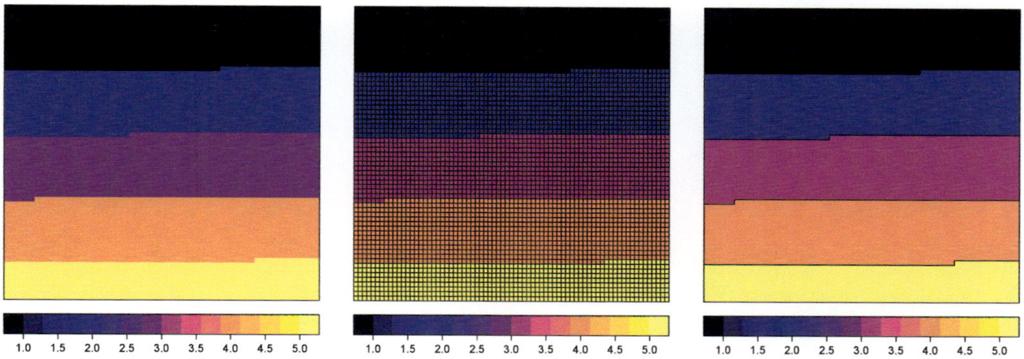

FIGURE 11.30 Vectorization of cells from a categorical raster with five classes of cell values (left) into polygons (center) and aggregation of polygons with the same attribute value (right).

12 Creating Continuous Surfaces with Environmental Geocomputation and R

12.1 LEARNING QUESTIONS

The learning questions answered by reading the chapter are:

- What are the requirements for interpolating attributes of vector features?
- What methods can be used to create continuous surfaces from vector feature attributes?
- How can interpolation of vector attributes be performed?
- What is the difference between classification and interpolation when generating continuous surfaces from vector feature attributes?
- How can point features and area features be used to interpolate attributes from vector features?

12.2 LEARNING OUTCOMES

The learning outcomes answered by reading the chapter are:

- Understand the requirements needed to interpolate attributes of vector features;
- Verify which methods are used to create continuous surfaces from vector feature attributes;
- Learn how to perform interpolation of vector feature attributes to create a digital elevation model;
- Understand the difference between classification and interpolation applied to generate continuous surfaces from the attributes of vector features;
- Check the difference of using features of points or area to interpolate attributes of vector features.

12.3 INTRODUCTION

Geospatial analysis methods are applied to geographic data to add utility value to data in order to reveal patterns and anomalies that are not immediately obvious, contributing to decision-making and scientific findings. Data interpolation is used as a process to estimate a point at an unobserved location from information contained in the neighborhood. Geospatial interpolation applications can be performed in different situations (Longley et al., 2001):

- Estimating rainfall, temperature and other climate variables obtained from weather stations;
- Estimating terrain elevation to create a digital elevation model and making contour maps;
- Creating sample grids for spatial prediction of attributes and resampling of other raster data;

DOI: 10.1201/9781003724377-12

- Estimating water, vegetation, soil, and socioeconomic attributes.

In the absence of sampling information about an attribute at a given location, the principle of Tobler's law can be used, i.e., there is a greater chance that nearby points are more similar than distant points. According to this principle, in the absence of better information, it can be assumed relatively smooth variation between attributes in a field with spatial autocorrelation of a given attribute (Longley et al., 2001).

Methods used to create continuous surfaces from attribute data are used in studies of hydrology, erosion, land degradation, agricultural and forest management, air, soil, and water pollution, as well as other geographic problems of determining exposure of areas to human disease, progress, economic well-being, and marketing studies (Table 12.1) (Burrough and McDonnell, 1998).

TABLE 12.1 Applications used in geocomputation to create continuous surfaces from attribute features.

Applications	Description
Geometric	Conversion of map coordinates from one projection to another
Interactive	Update and edit data attributes
Classification	Classifying attributes of geographic data
Localization	Localization of entities with set of attributes
Summary	Summarize the geometry of geographic attributes by aggregating data
Statistics	Calculate summary statistics for vector feature attributes
Proximity	Perform nearest neighbor analysis; create buffer zones
Interpolation	Interpolate point data on regular grids and contour lines
Diagram	Calculate block diagrams for three-dimensional data
Overlay	Overlay vector and raster maps, perform Boolean logic, arithmetic functions, filtering, attribute data manipulation
Conversion	Conversion from polygon to raster
Detection	Edge detection in a raster image
Network analysis	Determine the shortest distance to travel along a route
DEM analysis	Representation of landscape on 2.5-D and 3-D surfaces. Calculation of slope, aspect, hillshade, flow direction, watershed edges
Models	Simulation models

Classical geospatial analysis is deterministic in nature and used to represent the variation of a given attribute, discretized, at a given time. In modern geospatial analysis, spatial relationships between objects are evaluated in terms of association, neighborhood, proximity, and autocorrelation of the analyzed variable. Modern geospatial analysis is more complex than classical analysis in terms of the concept and technique used. Unlike classical analysis, in modern analysis, geospatial heterogeneity is emphasized in the geo-field. In addition, the use of spatial analysis techniques across different disciplines is incorporated, consisting of three basic components for analysis (Lo and Yeung, 2007):

- A data model of data conception, manipulation and recording;
- A set of statistical and graphical data analysis to evaluate information that can be gleaned from the data;
- A set of visualization tools to assist in interpreting results, making logical decisions, and generating scientific knowledge.

Methodologies for geospatial analysis of vector features to create geo-fields are incorporated in different ways, from applications with specific functions, geocomputation applications in statistical software, to fully integrated geospatial analysis in geocomputation software (Lo and Yeung, 2007).

12.4 DATA INTERPOLATION

Interpolation is the procedure used for predicting attribute values at unobserved locations from point locations within the same geographic region. The geospatial prediction of points outside a given area encompassed by observations is called extrapolation. Interpolation is used to convert attributes of vector features into geo-fields. The spatial pattern obtained in the geo-field can be compared with the attributes of points of the same variable and of other variables correlated with a given attribute throughout space. Interpolation is used in the following situations (Burrough and McDonnell, 1998):

- To increase the spatial resolution of a raster, as in bilinear interpolation;
- To represent a variable with a data model other than point patterns;
- To estimate values at unobserved locations.

Continuous surfaces obtained from interpolation can be used as map layers associated with contour lines and sample points. The data used in interpolation can be obtained by regular or irregular sampling in the field, by sampling remote sensing images from a regular grid, or by scanning photographs in a scanner. The attributes predicted by the interpolator are expressed by the same unit of measurement as the source attribute or by probability estimates indicating a value exceeding a threshold or class of the data attribute used in the interpolation (Burrough and McDonnell, 1998).

12.5 GEOVISUALIZATION OF CONTINUOUS SURFACES

Continuous surfaces are represented in the form of images or lines. Imaging methods include regular grids, irregular grids, and tessellations, where the variation of the value of a mapped attribute is indicated by a colored grid zone or in gray levels. The 2.5-D representation is an example obtained by representing an attribute on a continuous surface (Burrough and McDonnell, 1998).

Line representation can be done by isolines with the same elevation value along the line to characterize hillshade and relief gradients, as well as watercourses. Image and feature vector representation methods can be combined to improve the perception of the mapping performed. True 3-D or 4-D surfaces can be used to visualize continuous volume variation by block diagrams (fence) of attributes such as volume of gold or of a pollution plume (Burrough and McDonnell, 1998).

12.6 PRINCIPLES OF INTERPOLATION

The principle used in interpolation is that points closer in space have more similar attributes than points further away. We can think of the case of measuring altitude points, where points a few meters closer may show very similar values when compared to points kilometers away. Many interpolation methods use the available local information to provide a better description of the available information about the variation of an attribute in an area. If the variation is successfully captured, it can be expected that the spatial prediction of the attribute at an unobserved location will be better than if the prediction was made by classification. In this case, there will be less error in the resulting maps that will be used in subsequent modeling and mapping. In the classification methods used for value prediction at unobserved locations, the estimates are based on class taxonomy concepts, with no spatial association between the measured value within the class. With this, all the short variation information observed in a classified water, vegetation or soil attribute is lost in spatial prediction by classification (Burrough and McDonnell, 1998).

12.7 DATA SOURCES FOR INTERPOLATION

The data sources used to obtain continuous surfaces used in interpolation are (Burrough and McDonnell, 1998):

- Images obtained by remote sensing and photogrammetry;
- Scanned documents;
- Vector features with attributes structured in data models in regular, irregular, random arrays or linear patterns;
- Polygon features used in choropleth mapping.

Data obtained by field sampling to perform interpolation can present complex patterns both as a function of the distance at which the samples are obtained and the physical process represented by the attribute sampled in the field. In many situations, the nature of the spatial variation is not known, and assumptions need to be made about the smoothing of the attribute samples, as well as the statistical assumptions of probability distribution and statistical stationarity of the variable (Burrough and McDonnell, 1998).

12.7.1 SPATIAL SAMPLING

The location of sample points can be critical for performing subsequent geo-field creation analyses from vector feature attribute data. A completely regular sampling network may be preferred to determine unbiased mean and variance, since there will be samples at all locations in a region, separated by a fixed distance from the sampling grid. Sampling can also be performed randomly within a region of interest. A middle way between regular and random sampling is stratified random sampling. In this case, sample points are randomized within blocks, strata, or sample plots. Cluster or nested sampling can be used to assess spatial variation at different scales. Transect sampling is used to assess rivers, beaches, and hillshade, and contour line digitizing can be used to create digital elevation model from printed maps (Figure 12.1).

12.7.2 SAMPLE DATA SUPPORT

The support is the technique used in geostatistics to define an area or volume in which the measurement of a physical sample is performed. When data sampled at a given support are used to map attributes at unobserved locations, the predictions are relative to the original support, unless some methodology has been used to relativize the information to another area or larger volume (upscaling). In the sampling of coffee trees in Santo Antônio do Amaparo, Minas Gerais, Brazil, a support of five coffee plants sampled in a cross shape, and represented by the geographic coordinates of the central plant is used in the interpolation of geodetic altitude (m) and mean incidence (%) of bacterial blight in the five coffee plants sampled at the point.

The packages `raster` (Hijmans et al., 2020), `sf` (Pebesma et al., 2021a), and `tmap` (Tennekes et al., 2020) are loaded in R to illustrate the data support used for coffee crop data sampling.

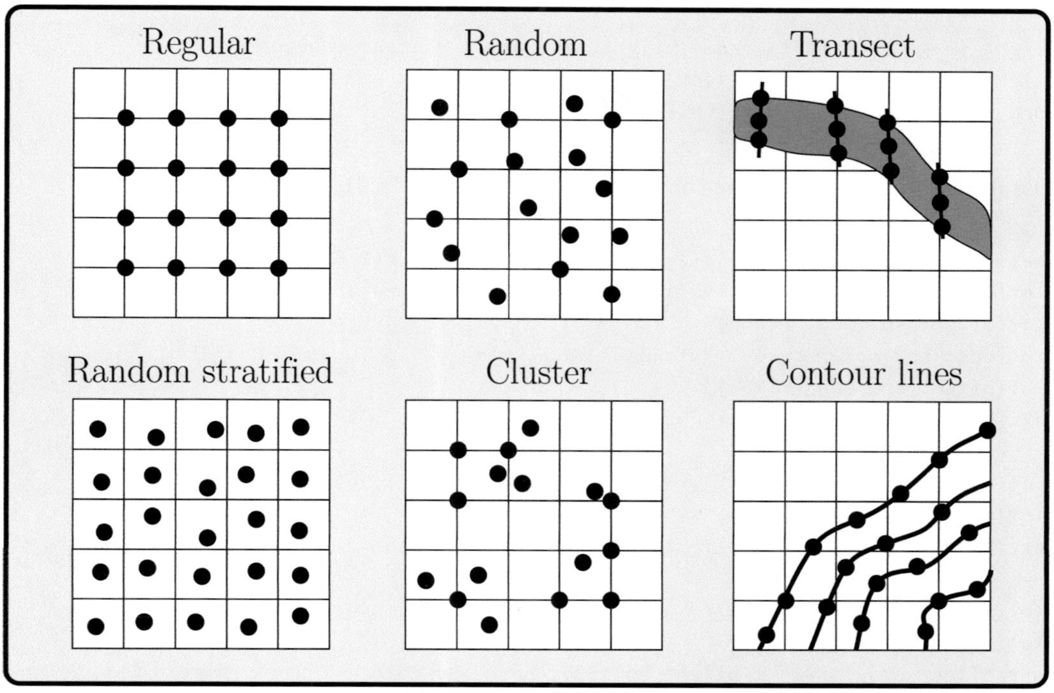

FIGURE 12.1 Different ways of collecting attributes by sampling meshes.

```
library(raster)
# Loading required packages: sp
library(sf)
# Linking to GEOS 3.11.2, GDAL 3.7.2, PROJ 9.3.0; sf_use_s2() is TRUE
library(tmap)
# package 'tmap' was built under R version 4.3.3
```

The bacterial blight incidence data, the area boundaries, and the digital elevation model are imported into R using the `st_read` and `raster` functions, respectively.

```
blight <- st_read("C:/geo/artigoML/shape/mancha.shp",
                  quiet = TRUE) # Bacterial blight
bord <- st_read("C:/geo/artigoML/shape/bordaPrj.shp", quiet = TRUE) # Border
srtm30 <- raster("C:/geo/SRTM/srtm30.tif") # SRTM data
```

A database with attributes is created with point identification data in the field `id` and altitude data obtained with a global positioning system. The altitude data (`alt`) is configured without decimal places using the `round` function. The `st_sf` function is used to join the `id` and `alt` data with the disease incidence data `blight`. The `as_Spatial` function is used to transform the `sf` `data.frame` data class into `sp`.

```
id <- 1:85 # Create id
alt <- round(blight$Z, digits = 0) # Round altitude
```

```
blight_id <- st_sf(data.frame(blight, id, alt)) # Create database
blight_sp <- as_Spatial(blight_id) # Transform sf in sp
bord_sp <- as_Spatial(bord) # Transform sf in sp
```

The results are mapped using the `plot` function (Figure 12.2).

```
par(mfrow = c(2, 2), mar = c(2, 2, 1.5, 1.5), mgp = c(1.5, 0.6, 0))
plot(srtm30, col = grey(1:100/100), main = "SRTM elevation (m)")
plot(blight_sp, add = TRUE)
plot(bord_sp, col = "grey92", axes = TRUE, main = "ID")
plot(blight_sp, add = TRUE)
text(blight_sp, labels = blight_sp$id, pos = 3, halo = TRUE, offset = 0.5,
     cex = 0.7)
plot(bord_sp, col = "grey92", axes = TRUE, main = "Elevation (m)")
plot(blight_sp, add = TRUE)
text(blight_sp, labels = blight_sp$alt, pos = 3, halo = TRUE, offset = 0.5,
     cex = 0.7)
plot(bord_sp, col = "grey92", axes = TRUE, main = "Incidence (%)")
plot(blight_sp, add = TRUE)
text(blight_sp, labels = blight_sp$i280, pos = 3, halo = TRUE, offset = 0.5,
     cex = 0.7)
```

The `tmap` package can be used as an alternative option for mapping results with different configuration alternatives. Note that symbol maps are obtained for the altitude and incidence of the sample points (Figure 12.3).

```
s1 <- tm_shape(srtm30) + tm_raster(n = 10,palette = "Greys",
  title = "SRTM elevation (m)") + tm_shape(bord_sp) + tm_polygons(alpha = 0) +
  tm_legend(legend.outside = FALSE)
s2 <- tm_shape(bord_sp) + tm_polygons() + tm_shape(blight_sp) +
  tm_dots(col = "id", palette = "black",
  title = "ID", size = 0.5, legend.show = TRUE) +
  tm_text("id", just = "left", xmod = .5, size = 0.7, remove.overlap = TRUE) +
    tm_legend(legend.outside = FALSE)
s3 <- tm_shape(bord_sp) + tm_polygons() + tm_shape(blight_sp) +
  tm_dots(col = "Z", palette = "Greys",
  title = "Elevation (m)", size = 0.7) + tm_legend(legend.outside = FALSE)
s4 <- tm_shape(bord_sp) + tm_polygons() + tm_shape(blight_sp) +
  tm_dots(col = "i280", palette = "Greys",
  title = "Incidence (%)", size = 0.7) + tm_legend(legend.outside = FALSE)
 tmap_arrange(s1, s2, s3, s4, ncol = 2)
```

In demographic studies, the support refers to irregular polygons used to represent the boundaries of municipalities or zip codes, giving the data different types of support. In the case of studies of reference flow in watersheds, the support can be defined by the edges of sub-basins or by Voronoi polygons defined as a function of the centroid of points with flow attributes (m^3s^{-1}).

The packages `maptools` (Tennekes, 2021), `rgeos` (Bivand et al., 2018), `gridExtra` (Auguie and

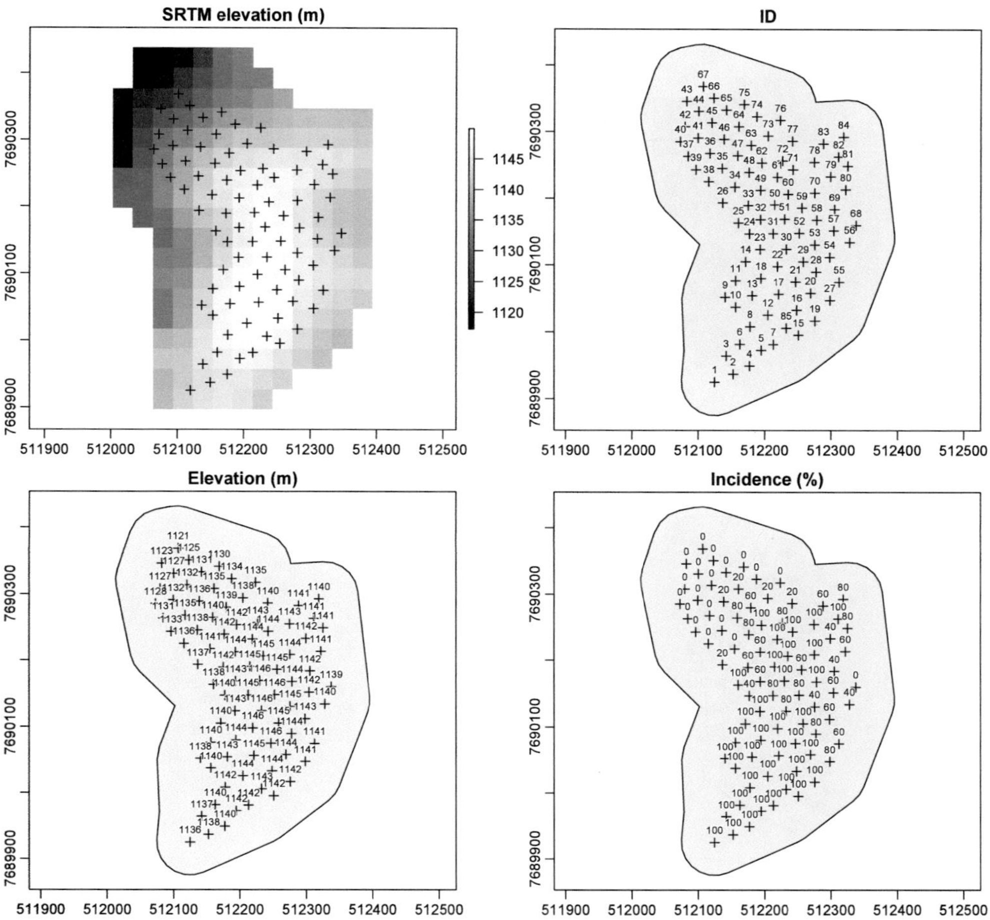

FIGURE 12.2 Digital elevation model SRTM at 30 m spatial resolution with sample points (top-left) representing the central sampling support point of five coffee plants with ID identification (top-right), geodetic altitude with rounded values obtained with GNSS on the central plant (bottom-left) and assessment of the mean bacterial blight incidence in five coffee plants evaluated on April 16, 2014, represented by the georeferenced central point, in coffee crop, in Santo Antônio do Amaparo, Minas Gerais.

Antonov, 2017), `sf` (Pebesma et al., 2021a), `ggplot2` (Wickham et al., 2020), and `ggspatial` (Dunnington and Thorne, 2021) packages are loaded into R using the `library` function.

```
library(maptools)
# Loading required packages: sp
library(rgeos)
# GEOS runtime version: 3.11.2-CAPI-1.17.2
library(ggplot2)
library(sf)
# Linking to GEOS 3.11.2, GDAL 3.7.2, PROJ 9.3.0; sf_use_s2() is TRUE
library(gridExtra)
library(ggspatial)
```

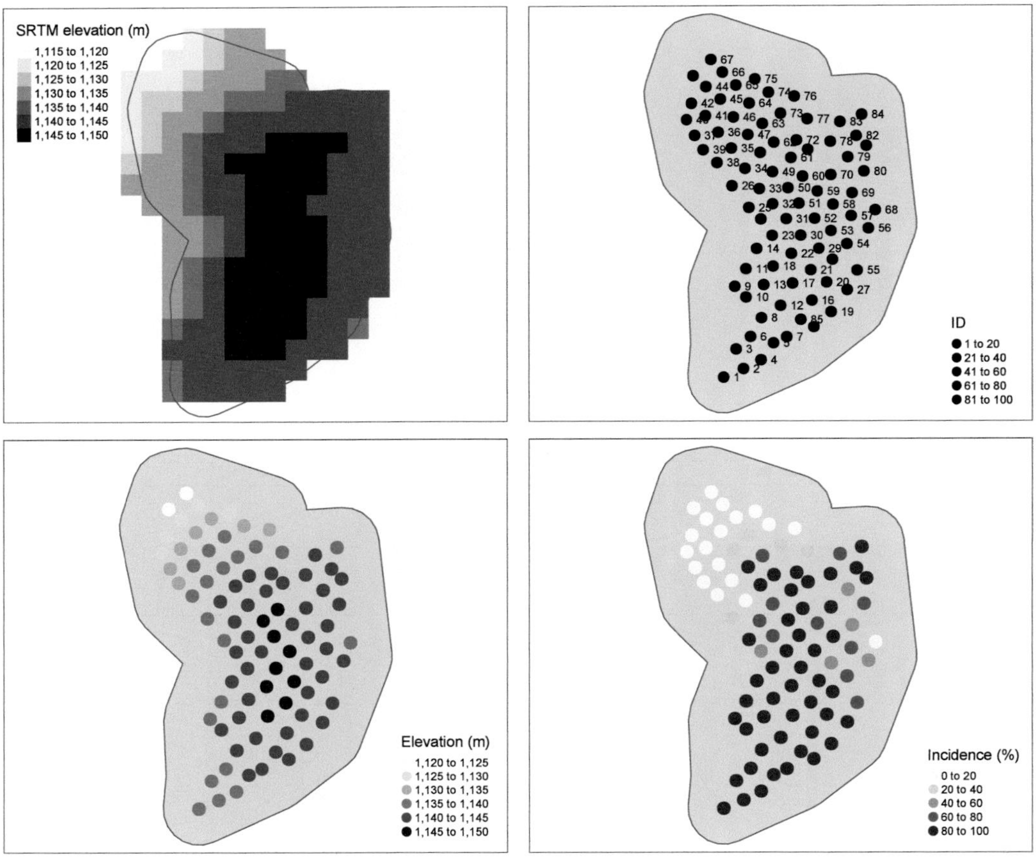

FIGURE 12.3 Digital elevation model SRTM at 30 m spatial resolution (top-left), sample points representing the central sampling support point of five coffee plants with ID identification (top-right), symbol maps with geodetic altitude using rounded values obtained with GNSS on the central plant (bottom-left) and assessment of the mean bacterial blight incidence in five coffee plants evaluated on April 16, 2014, represented by the georeferenced central point, in coffee crop, in Santo Antônio do Amaparo, Minas Gerais.

```
# package 'ggspatial' was built under R version 4.3.3
```

Sub-basin and municipality data are imported into R using the `st_read` function.

```
qm <- st_read("C:/Projetos/Uniselva/shape/vazessema/Otto_SEMA.shp",
      quiet = TRUE) |> st_transform(crs = "EPSG:32721") # Sub-basin data
mun <- st_read("C:/Aulas/2021_1/Geoprocessamento/shape/municipiosMT.shp",
      options = "ENCODING=WINDOWS-1252", quiet = TRUE) |>
      st_transform(crs = "EPSG:32721") # Municipalities data
```

The `st_union` function is used to produce the watershed boundary from the municipality polygons. The `st_centroid` function is used to determine the centroids for each sub-basin polygon. The `as_Spatial` and `st_as_sf` functions are used to convert the data into `sp` and `sf data.frame` class geo-objects, respectively.

```
limsp <- st_union(mun) |> as_Spatial() # Union and sf to sp conversion
class(limsp) # See class
# [1] "SpatialPolygons"
# attr(,"package")
# [1] "sp"
qmcsp <- st_centroid(qm) |> as_Spatial() # Centroid and sf to sp conversion
# st_centroid assumes attributes are constant over geometries
class(qmcsp) # See class
# [1] "SpatialPointsDataFrame"
# attr(,"package")
# [1] "sp"
qmcsf <- st_as_sf(qmcsp)
class(qmcsf)
# [1] "sf"         "data.frame"
```

The watershed boundary data is converted to `SpatialPolygonsDataFrame` and `sf data.frame` after creating a geo-object database.

```
# Extract polygon ID's
(pid <- sapply(slot(limsp, "polygons"), function(x) slot(x, "ID")))
# Create dataframe with correct rownames
(p.df <- data.frame( ID = 1:length(limsp), row.names = pid))
# Convert to SpatialPolygonsDataFrame
bor <- SpatialPolygonsDataFrame(limsp, p.df)
class(bor) # See class
# [1] "SpatialPolygonsDataFrame"
# attr(,"package")
# [1] "sp"
borsf <- st_as_sf(bor) # Convert sp boundaries to sf
class(borsf) # See class
# [1] "sf"         "data.frame"
```

A `voronoipolygons` function is created to determine Voronoi polygons from point geometry. Voronoi diagram and Delaunay triangles can also be performed with the `terra` package[1].

```
voronoipolygons <- function(x, poly) {
  require(deldir)
  if (.hasSlot(x, 'coords')) {crds <- x@coords} else crds <- x
  bb = bbox(poly)
  rw = as.numeric(t(bbox(poly)))
  z <- deldir(crds[,1], crds[,2], rw = rw)
  w <- tile.list(z)
  polys <- vector(mode = 'list', length = length(w))
  require(sp)
  for (i in seq(along=polys)) {
  pcrds <- cbind(w[[i]]$x, w[[i]]$y)
```

[1]https://rdrr.io/cran/terra/man/voronoi.html

```
  pcrds <- rbind(pcrds, pcrds[1,])
  polys[[i]] <- Polygons(list(Polygon(pcrds)), ID = as.character(i))}
  SP <- SpatialPolygons(polys)
  SpatialPolygonsDataFrame(SP, data.frame(x = crds[,1], y = crds[,2],
  row.names = sapply(slot(SP, 'polygons'), function(x) slot(x, 'ID'))))}
```

Voronoi polygons are determined and clipped to the watershed region. Flow data are associated with the created polygons, and the final geo-object is converted to `sf data.frame`.

```
vor <- voronoipolygons(qmcsp, bor) # Voronoi calculation
proj4string(vor) <- proj4string(bor) # Assign crs
vorclip <- gIntersection(bor, vor, byid = TRUE) # Clip
vorqm <- over(vor, qmcsp[,"QM"]) # Get flow data over Voronoi
vorclip$QM <- vorqm$QM # Add flow to the attribute table
vorclipsf <- st_as_sf(vorclip) # Convert to sf
```

The results are mapped using functions from the `ggplot2` package (Figure 12.4).

```
a1 <- ggplot(data = borsf) + geom_sf(fill = "antiquewhite") +
 annotation_scale(location = "br", width_hint = 0.5) +
 annotation_north_arrow(location = "br", which_north = "true",
 pad_x = unit(0.75, "in"), pad_y = unit(0.5, "in"),
 style = north_arrow_fancy_orienteering)
a2 <- ggplot(data = borsf) + geom_sf(fill = "antiquewhite") +
 geom_sf(data = qm, fill = "antiquewhite", size = 0.5) +
 annotation_scale(location = "br", width_hint = 0.5) +
 annotation_north_arrow(location = "br", which_north = "true",
 pad_x = unit(0.75, "in"), pad_y = unit(0.5, "in"),
 style = north_arrow_fancy_orienteering)
a3 <- ggplot(data = borsf) + geom_sf(fill = "antiquewhite") +
 geom_sf(data = qmcsf,  colour = "black", size = 0.5) +
 annotation_scale(location = "br", width_hint = 0.5) +
 annotation_north_arrow(location = "br", which_north = "true",
 pad_x = unit(0.75, "in"), pad_y = unit(0.5, "in"),
 style = north_arrow_fancy_orienteering)
a4 <- ggplot(data = borsf) + geom_sf(fill = "antiquewhite") +
 geom_sf(data = vorclipsf, fill = "antiquewhite", size = 0.5) +
 annotation_scale(location = "br", width_hint = 0.5) +
 annotation_north_arrow(location = "br", which_north = "true",
 pad_x = unit(0.75, "in"), pad_y = unit(0.5, "in"),
 style = north_arrow_fancy_orienteering)
grid.arrange(a1, a2, a3, a4, ncol = 2)
```

12.7.3 TERMINOLOGY

The terminology used to describe point features with attribute values is represented by (Burrough and McDonnell, 1998):

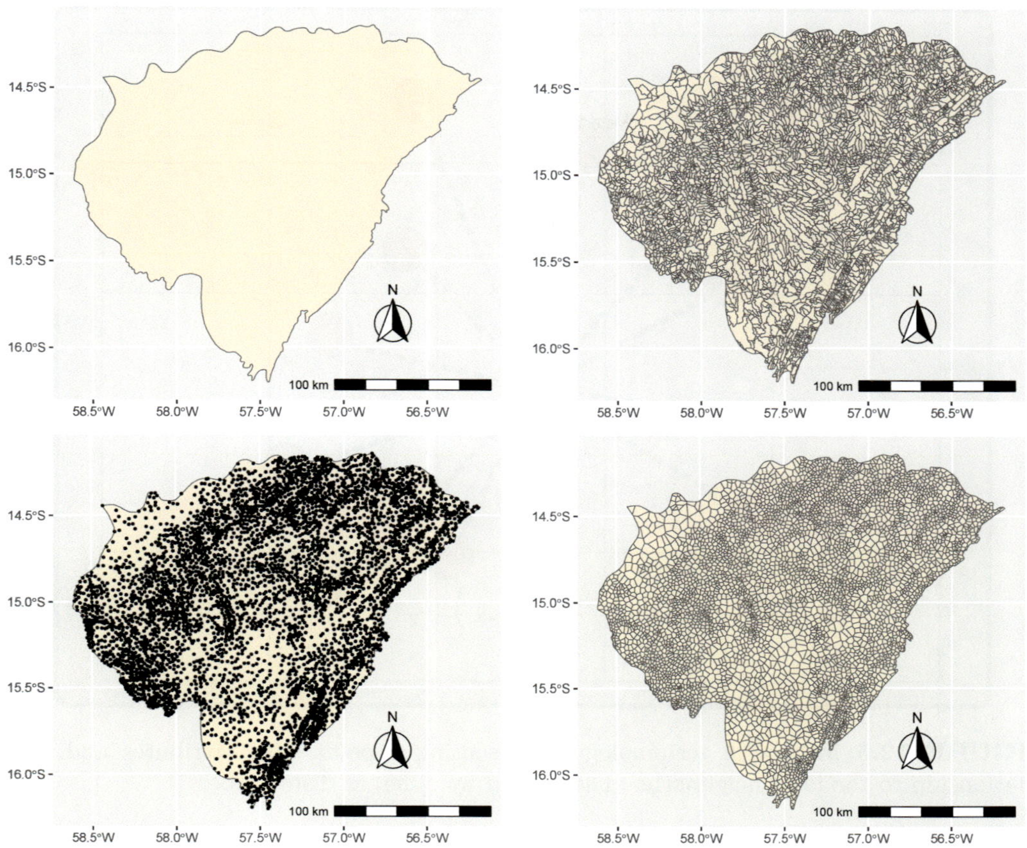

FIGURE 12.4 Watershed boundary (top-left), Agência Nacional das Águas (ANA) ottobasins with sub-basin polygons (top-right) and centroid of polygons obtained as a function of ottobasins (bottom-left) to generate Voronoi polygons with different support (bottom-right).

$$z(x_i) \tag{12.1}$$

where i is denoted as a number of n possible georeferenced measurements, represented on a Cartesian coordinate grid, x.

A predicted value at an unsampled location is defined by (Figure 12.5) (Burrough and McDonnell, 1998):

$$\hat{z}(x_0) \tag{12.2}$$

An interpolation method used to predict values of an attribute at a location with the identical value as measured is called an exact interpolator. Considering that an exact interpolator is an ideal method unlikely to occur in nature, in view of the impossibility of knowing the value of an attribute in all locations, all methods can be considered inexact

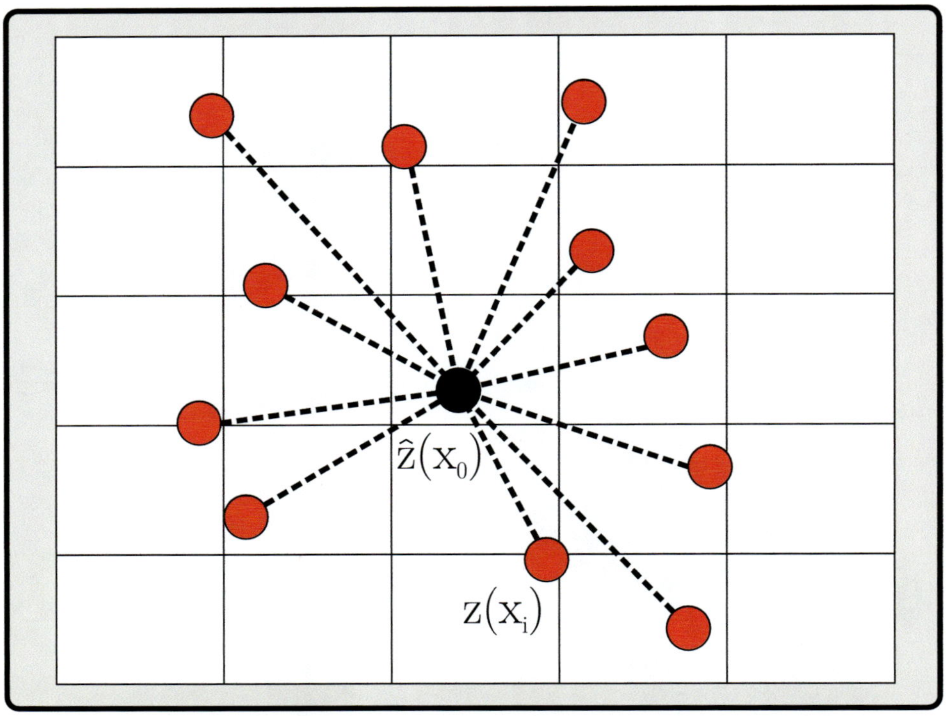

FIGURE 12.5 Map with terminology representing a point with attributes and their relationship to the local neighborhood separated by different distances.

interpolators. The statistic of the absolute difference between measured (observed) and predicted values is used to evaluate the quality of an interpolation method is the residual:

$$residual = z(x_i) - \hat{z}(x_i) \tag{12.3}$$

The mean squared error (MSE) statistic between the observed and predicted values is:

$$MSE = \frac{1}{n}\sum_{i=1}^{n}[z(x_i) - \hat{z}(x_i)]^2 \tag{12.4}$$

12.8 INTERPOLATION METHODS

Interpolation methods can be divided into global and local types. In global interpolation are included classification methods, use of trend surfaces in coordinates, regression models and spectral analysis methods. In local interpolation the Thiessen polygon, pycnophylactic, linear weighting by inverse distance interpolation, and thin plate splines methods are included. The kriging methods refer to geostatistical methods where it is necessary to understand the principle of spatial autocorrelation for its application. In some situations of irregular sampling and low sample density, some simple interpolation methods can lead to poor quality results. In this case, with the use of kriging methods better results can be obtained. However, considering the greater theoretical complexity of the method, the kriging method will be discussed in more detail in the next chapter (Burrough and McDonnell, 1998).

12.9 GLOBAL INTERPOLATION

In global interpolation methods, all data are used to make predictions over the entire area of interest. Global interpolation methods can be used not only for direct interpolation, but also to examine and remove the possible effects of global variations caused by trends due to larger spatial variation, with different mean values in subsets of data at locations throughout the geographic region evaluated. The global interpolation methods are easy to calculate and are based on standard statistical ideas of analysis of variance and regression. In regression methods, a possible functional relationship between easily measured attributes and those that can be predicted is used to make the prediction. These methods can be based on geographic coordinate analysis of sample points by trend analysis or on an empirical regression model determined in relation to the attributes by a transfer function. Fourier transform and wavelet methods are also used in interpolation of remotely sensed image data. Classification methods can also be used to divide a region into areas characterized by statistical moments, such as mean and variance, and to perform attribute prediction within these areas (Burrough and McDonnell, 1998).

12.10 GLOBAL PREDICTION WITH CLASSIFICATION MODELS

Under sparse data situations for modeling, it may be convenient to assume that the observations are obtained from a population with stationarity, i.e., the mean and variance are independent of location and support size. In homogeneous polygon classification, it is assumed that the unit variance within the polygon is smaller than the outer variance defined by the spatial units. This conceptual model is used in mapping soil and landscape units that are considered homogeneous. The simplest classification model used scientifically is the analysis of variance (ANOVA) model (Burrough and McDonnell, 1998):

$$z(x_0) = \mu + \alpha_k + \epsilon \tag{12.5}$$

where z is the value of the attribute in the location, x_0; μ, the general mean of z in the domain of interest, α_k, the deviation between μ and the mean of the unit k; and ϵ, the residual error, also called noise.

In the ANOVA model, the ideal situation is that each class k presents the attribute value with a normal distribution and distinct mode. The average attribute value per class k is defined by $\mu + \alpha_k$, as a function of a set of samples assumed to be spatially independent. The mean within-class variance is obtained from ϵ and is assumed to be the same for all classes.

The quality of the classification can be obtained as a function of relative variance (σ_R), calculated by (Burrough and McDonnell, 1998):

$$\sigma_R = \frac{\sigma_W^2}{\sigma_T^2} \tag{12.6}$$

where σ_W^2 is the variance obtained within the class and σ_T^2, the total variance of the sample. These values can be estimated in map units with more than one sample point. The smaller

σ_R, the better the classification. The significance of the classification test can be performed by applying the F-test, considering the variance ratio and the number of degrees of freedom (Burrough and McDonnell, 1998).

When all attributes are measured exactly, there can be a natural variation of values around the mean. In some situations, this variation can be measured by the bell-shaped curve. Assuming a normal distribution of the attribute, approximately 65% of the population values will occur within the range defined by 1 standard deviation (1σ). Similarly, 95% of the data will occur within the 2σ range. Therefore, the larger the σ interval in the normal distribution, the less precision can be associated with the sample values for a given attribute. The mean (m) and standard deviation (s) of observed attribute samples can be calculated by (Burrough and McDonnell, 1998):

$$m = \frac{1}{n}\sum_{i}^{n} Z_i \tag{12.7}$$

$$s = [\frac{1}{n}\sum_{i}^{n}(Z_i - m)^2]^{0,5} \tag{12.8}$$

The square of the standard deviation is called the variance (σ^2). The variance is useful for evaluating the different sources of variation that can be grouped into combined uncertainty estimates from different sources. With this, it may be possible to determine a polygon map, p_k, containing n_i observations per polygon, used to calculate the total variance of observations (σ_T^2), composed of the variances between classes, σ_B^2, and, within the class σ_W^2 (Burrough and McDonnell, 1998):

$$s_T^2 = s_B^2 + s_W^2 \tag{12.9}$$

The degrees of freedom are obtained by $p_k - 1$ and $n - p_k$ for s_B^2 and s_W^2, respectively. The magnitude of the variance explained by the rank at p_k is denoted R^2, and can be obtained by (Burrough and McDonnell, 1998):

$$R^2 = \frac{s_B^2}{s_T^2} \tag{12.10}$$

where the larger the R^2, the more variation explained by the p_k classes and the mean value of each class, m_k, can be used as a good overall predictor.

The normality of the residuals of geodetic altitude (m) data relative to the total mean value of the data is determined using a frequency distribution histogram to illustrate an example of data distribution relative to the normal distribution. The altitude data are obtained from topographic survey to obtain sample grid ($n = 85$) in coffee field in Santo Antônio do Amparo, Minas Gerais.

Altitude data is extracted from the geographic database of bacterial blight samples taken from a coffee field and imported into R using the `st_read` function. The mean and residuals are calculated.

```
z <- blight$Z # Select the altitude data
m <- mean(z) # Calculate mean
v <- z - m # Determine residuals
```

The frequency distribution histogram of the residuals is calculated, and the normal distribution curve is fitted to the data and plotted (Figure 12.6).

```
# Calculate and plot the histogram
hv <- hist(v, breaks = 10, col = "grey90", xlab = "Altitude residual (m)",
           main = "", xlim = c(-20, 20), ylim = c(0, 20))
# Fit and plot the normal distribution line on the histogram
xvfit <- seq(min(v), 20, length = 100)
yvfit <- dnorm(xvfit, mean = mean(v), sd = sd(v))
yvfit <- yvfit * diff(hv$mids[1:2]) * length(v)
lines(xvfit, yvfit, col = "black", lwd = 2)
```

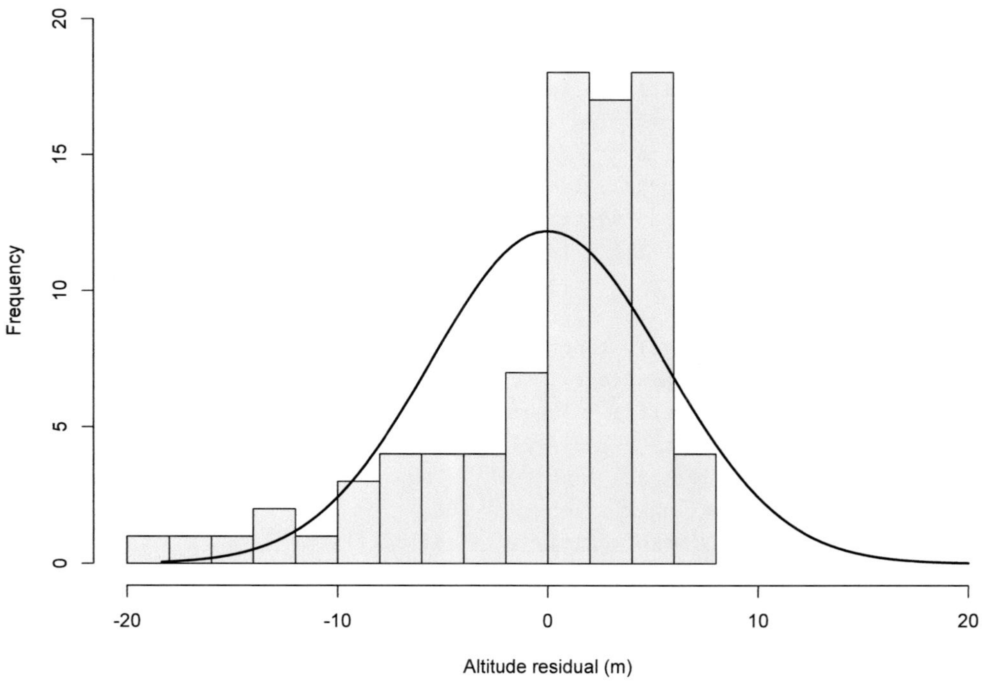

FIGURE 12.6 Frequency distribution histogram of residuals based on georeferenced altitude values in a coffee field in Santo Antônio do Amparo, Minas Gerais.

12.10.1 ASSUMPTIONS ABOUT SPATIAL VARIATION FROM CLASSIFICATION MODELS

The assumptions made about spatial variation using classification, such as ANOVA, are (Burrough and McDonnell, 1998):

- The attribute variations on the map are random, but not continuous in space;
- All units on the map have the same value of variance within the class (noise);
- All the evaluated attributes have a normal distribution;
- All variations occur from class edges, discrete and not gradual.

It is worth pointing out that not all these assumptions are necessarily verified when using the classification methodology, because the data may not present a normal distribution, requiring the use of transformations to better suit the use of the method.

As an example, the logarithm and square root transformations are applied to altitude data measured in coffee crop.

```
sqrtz <- sqrt(z) # sqrt
logz <- log(z) # log
log1z <- log1p(z) # log(1+x)
```

The altitude transformation results are represented graphically with a frequency distribution histogram comparing the transformed data with the original data (Figure 12.7).

```
par(mfrow = c(2, 2), mar = c(3, 3, 1.5, 1.5), mgp = c(1.5, 0.6, 0))
hv <- hist(z, breaks = 10, col = "grey90", xlab = "Altitude (m)", main = "")
xvfit <- seq(min(z), max(z), length = 100)
yvfit <- dnorm(xvfit, mean = mean(z), sd = sd(z))
yvfit <- yvfit * diff(hv$mids[1:2]) * length(z)
lines(xvfit, yvfit, col = "black", lwd = 2)
hv <- hist(sqrtz, breaks = 10, col = "grey90", xlab = "sqrt(Altitude)", main = "")
xvfit <- seq(min(sqrtz), max(sqrtz), length = 100)
yvfit <- dnorm(xvfit, mean = mean(sqrtz), sd = sd(sqrtz))
yvfit <- yvfit * diff(hv$mids[1:2]) * length(sqrtz)
lines(xvfit, yvfit, col = "black", lwd = 2)
hv <- hist(logz, breaks = 10, col = "grey90", xlab = "log(Altitude)", main = "")
xvfit <- seq(min(logz), max(logz), length = 100)
yvfit <- dnorm(xvfit, mean = mean(logz), sd = sd(logz))
yvfit <- yvfit * diff(hv$mids[1:2]) * length(logz)
lines(xvfit, yvfit, col = "black", lwd = 2)
hv <- hist(log1z, breaks = 10, col = "grey90", xlab = "log1p(Altitude)", main = "")
xvfit <- seq(min(log1z), max(log1z), length = 100)
yvfit <- dnorm(xvfit, mean = mean(log1z), sd = sd(log1z))
yvfit <- yvfit * diff(hv$mids[1:2]) * length(log1z)
lines(xvfit, yvfit, col = "black", lwd = 2)
```

In a spatial analysis, the study of the data can begin with the use of summary statistics to evaluate measures of central tendency and dispersion of the data. The values of mean, median and mode can be used to evaluate attribute data to assess central tendency. The median can be calculated by arranging the data values in ascending or descending order. The middle value from this arrangement of data (50% of the data) is the median. The

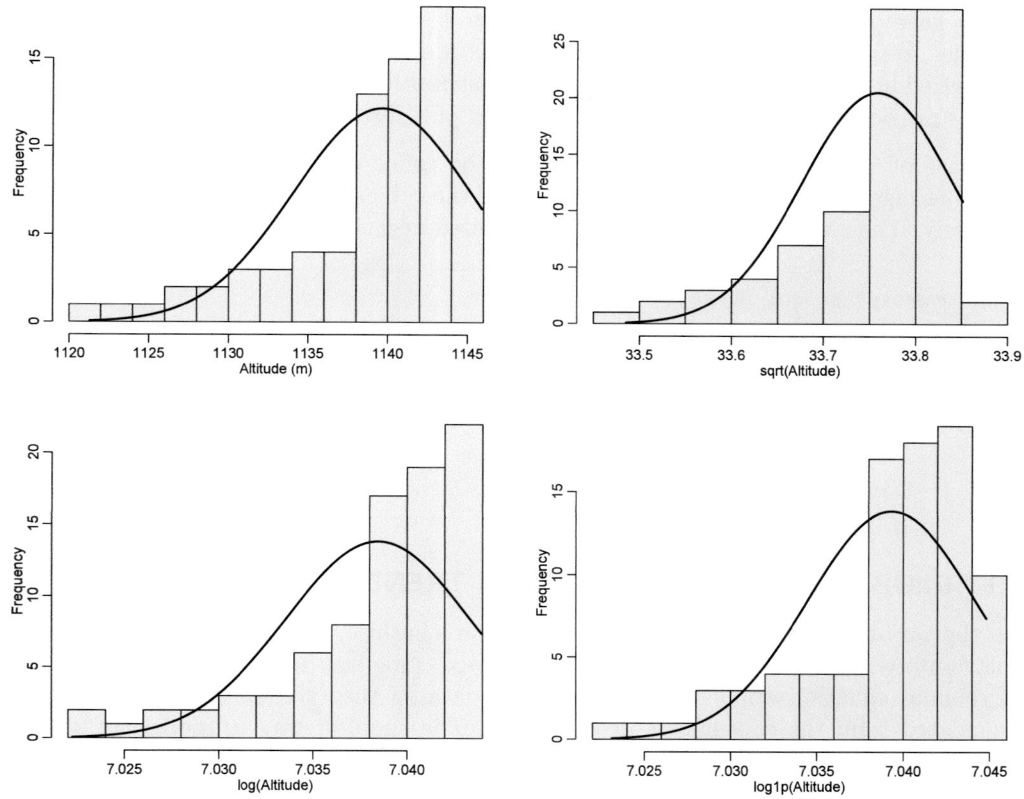

FIGURE 12.7 Example of square root (sqrt), logarithm (log) logarithm log(1+x) (log1p) transformations applied to geodetic altitude (m) data in an attempt to obtain normality of attribute frequency distributions.

mode is the value that occurs most often in the distribution. The skewness value (sk) is a measure of the skewness of the distribution as a function of the difference between the mean and the median (mna) from the standard deviation (Lo and Yeung, 2007):

$$sk = \frac{(m - mna)}{\sigma} \tag{12.11}$$

where skewness equal to zero indicates symmetrical distribution, and there may be negative and positive values of variation when the median is greater than the mean and less than the mean, respectively.

The kurtosis (kt) is a measure used to indicate the peak of the frequency distribution of an attribute and can be calculated by (Lo and Yeung, 2007):

$$kt = \frac{\sum_{i=1}^{n} (x_i)^4 / n}{(\sigma^2)^2} \tag{12.12}$$

where a kurtosis value equal to 3 represents a curve with a standard normal distribution, and may have values greater than 3 and less than 3 for curves with more and less sharp frequencies, respectively. Mesocurtic distributions are defined by having $kt = 0$ and are characterized in data with normal and binomial distributions. Values of kurtosis above and below the mesocurtic normal distribution are called platicurtic and leptocurtic, respectively.

In the case of the altitude data obtained from the coffee crop sampling grid, using the `moments` package (Komsta and Novomestky, 2022), the values of sk and kt are -1.36 and 4.4, respectively. Thus, the altitude data analyzed showed negative and leptokurtic asymmetry.

```
library(moments) # Load package
sk <- skewness(blight$Z) # Skewness
sk
# [1] -1.365904
kt <- kurtosis(blight$Z) # Kurtosis
kt
# [1] 4.414775
```

12.11 GLOBAL INTERPOLATION WITH TREND SURFACES

When the variation of an attribute is continuous on a surface, it may be possible to model the attribute with a smoothed mathematical surface. One way to accomplish this is with a polynomial equation applied to the observed data so that the unsampled values can be calculated from the coordinates of the point. The simplest way to model the spatial variation of an attribute is with the use of multiple regression of the attribute as a function of geographic coordinates. The idea is to fit a polynomial line or least squares surface on the point data. The x, y coordinates are assumed to be independent variables and, z, the attribute of interest as the dependent variable with normal distribution. Independence of regression errors from location is also assumed. Given the value of an attribute of an environment variable, z, measured along points, x_1, x_2, x_n, on a transect, the variation of z along location x can be obtained by the regression model (Burrough and McDonnell, 1998):

$$z(x) = b_0 + b_1 x + \epsilon \tag{12.13}$$

where b_0 and b_1 are the polynomial model parameters known as the intercept and slope of a simple regression. The residual, ϵ, is assumed to have a normal distribution regardless of the values of x.

In some situations, z may not be a linear function of x. In this case, a quadratic or higher order polynomial model can be used as the regression model (Burrough and McDonnell, 1998):

$$z(x) = b_0 + b_1 x + b_2 x^2 + \epsilon \tag{12.14}$$

By increasing the number of parameters in the above equation, it may be possible to fit a complex curve to a set of points in order to reduce ϵ to zero.

Trend surfaces can be used to estimate the value of $z(x)$ for all points on a regular grid. In the trend surface model, the existence of a process with mean $\mu(x)$ is implied, where

$x = (x1, x2)$ define the coordinates at the location, x. In the trend model, the trend argument can be defined as (Diggle and Ribeiro, 2007):

- Constant (cte), in which the mean is considered constant throughout the evaluated region, that is, $\mu(x) = \mu$;
- First-order (1st), where the mean is considered as a first-order polynomial applied to the coordinates, $\mu(x) = b_0 + b_1 x1 + b_2 x2$;
- Second-order (2nd), where the mean is considered a second-order polynomial applied to the coordinates, $\mu(x) = b_0 + b_1 x1 + b_2 x2 + b3(x1)^2 + b4(x2)^2 + b5x1x2$;
- Another higher-order trend modeling or a new covariate entered into the model as an external trend to be modeled.

It should be noted that trend surface analysis is an easy technique to understand, but it can be difficult to describe the physical meaning of modeling a variable with a complex higher-order trend surface. In addition, trend surfaces are susceptible to discrepant data, which can determine undesirable results in spatial prediction (Burrough and McDonnell, 1998).

Trend surface analysis associated with quartile map visualization methodology and variable fitting as a function of x and y coordinates is applied to geodetic altitude (m) data obtained in a coffee field. First, the analysis is applied on the original data (Figure 12.8) and then compared with the trend modeling with constant mean in the region (Figure 12.9) and under first-order polynomial trend surface fit to the coordinates with smoothing line (lowess) adjusted to the data (Figure 12.10) (Diggle and Ribeiro, 2007).

The `geoR` package (Ribeiro Jr et al., 2020) is used as an example of exploratory data analysis in a coffee field.

```
library(geoR) # Load package
# geoR version 1.9-4 (built on 2024-02-14) is now loaded
# package ‘geoR’ was built under R version 4.3.3
```

The `read.geodata` function is used to import altitude data obtained from a sample grid of a coffee field.

```
z <- read.geodata("C:/geo/c12/ima1.txt", header = TRUE, coords.col = 1:2,
                  data.col = 3)
```

The `plot` function is used to characterize the exploratory analysis of the original data and under first-order trend and constant mean modeling.

```
plot(z, ylab = "Altitude (m)")
plot(z, trend = "cte", lowess = TRUE, ylab = "Altitude (trend cte)")
plot(z, trend = "1st", lowess = TRUE, ylab = "Altitude (trend 1st)")
```

12.12 GLOBAL PREDICTION WITH REGRESSION MODELS

Regression models can be used as global prediction methods associated with the use of abundant, easily measured and low-cost variables in order to perform spatial prediction. Empirical first- and second-order regression models are used to estimate the mean air temperature in Minas Gerais, Brazil, as a function of parameters obtained by the

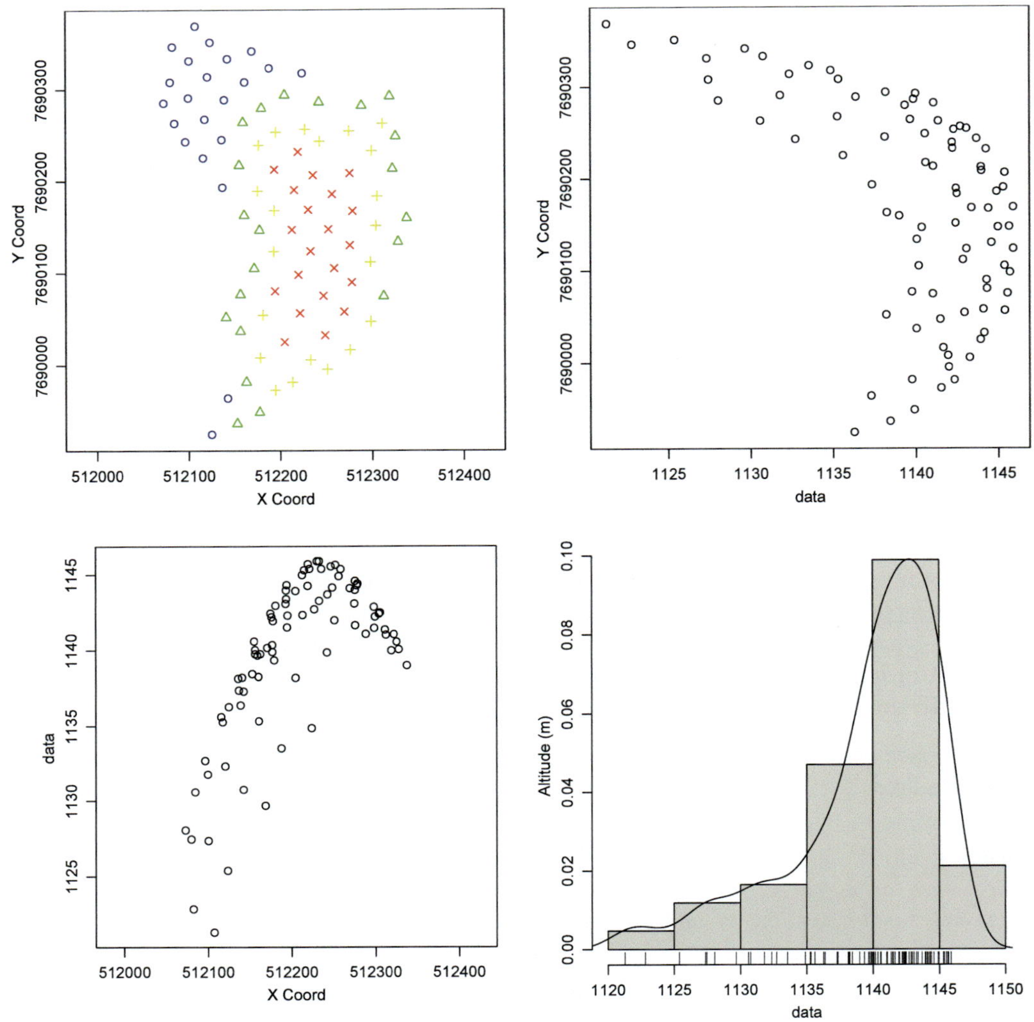

FIGURE 12.8 Exploratory analysis of geodetic altitude (m) obtained at a coffee field using a map of quartile symbols (top-right), graph with the variation of the attribute as a function of the latitude (top-right) and longitude (bottom-left) coordinate axis, and the histogram of the frequency distribution of altitude at 85 points of a stratified irregular sampling grid.

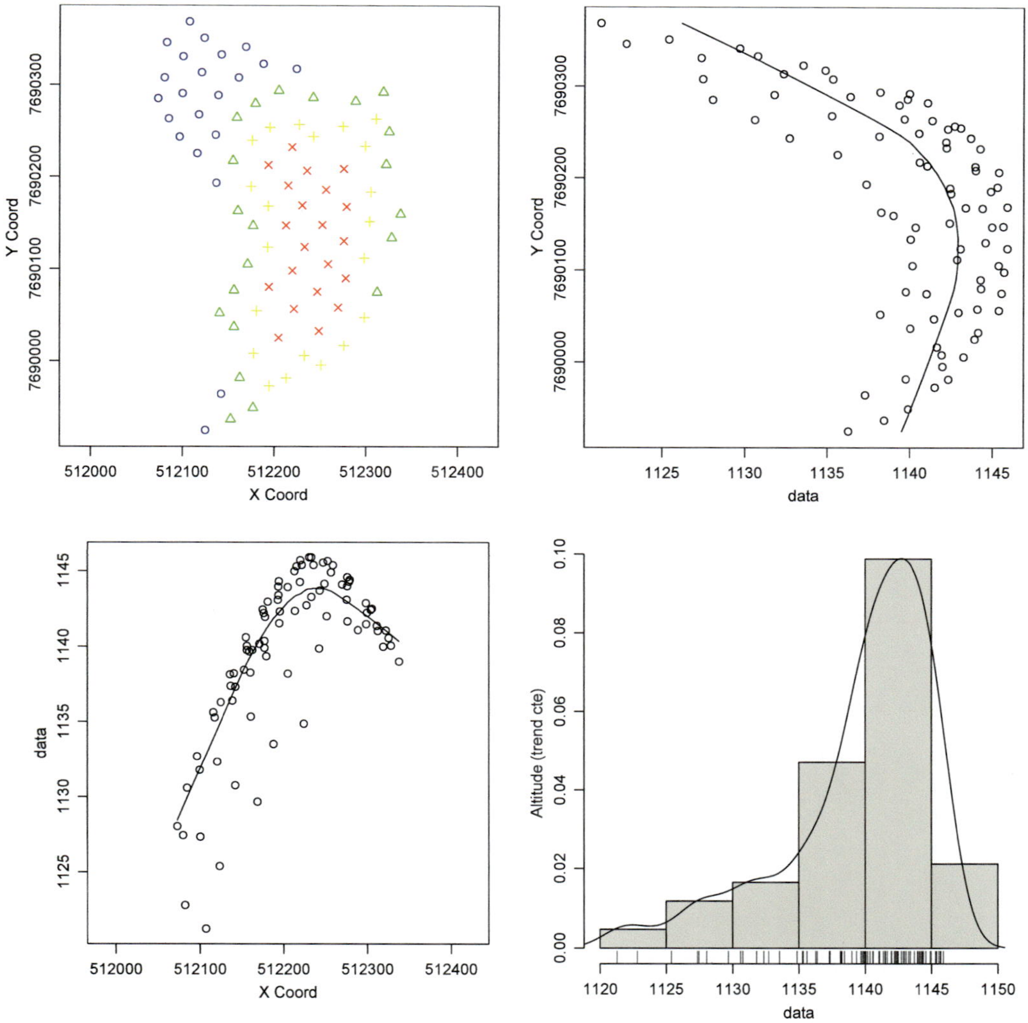

FIGURE 12.9 Exploratory analysis of geodetic altitude (m) obtained at a coffee field using the quartile symbol map (top-right), graph with the variation of the attribute as a function of the latitude (top-right) and longitude (bottom-left) coordinate axis under trend surface fit with constant mean variation in the region, and the frequency distribution histogram with density line fitted to the altitude data from 85 points of a stratified irregular sampling grid.

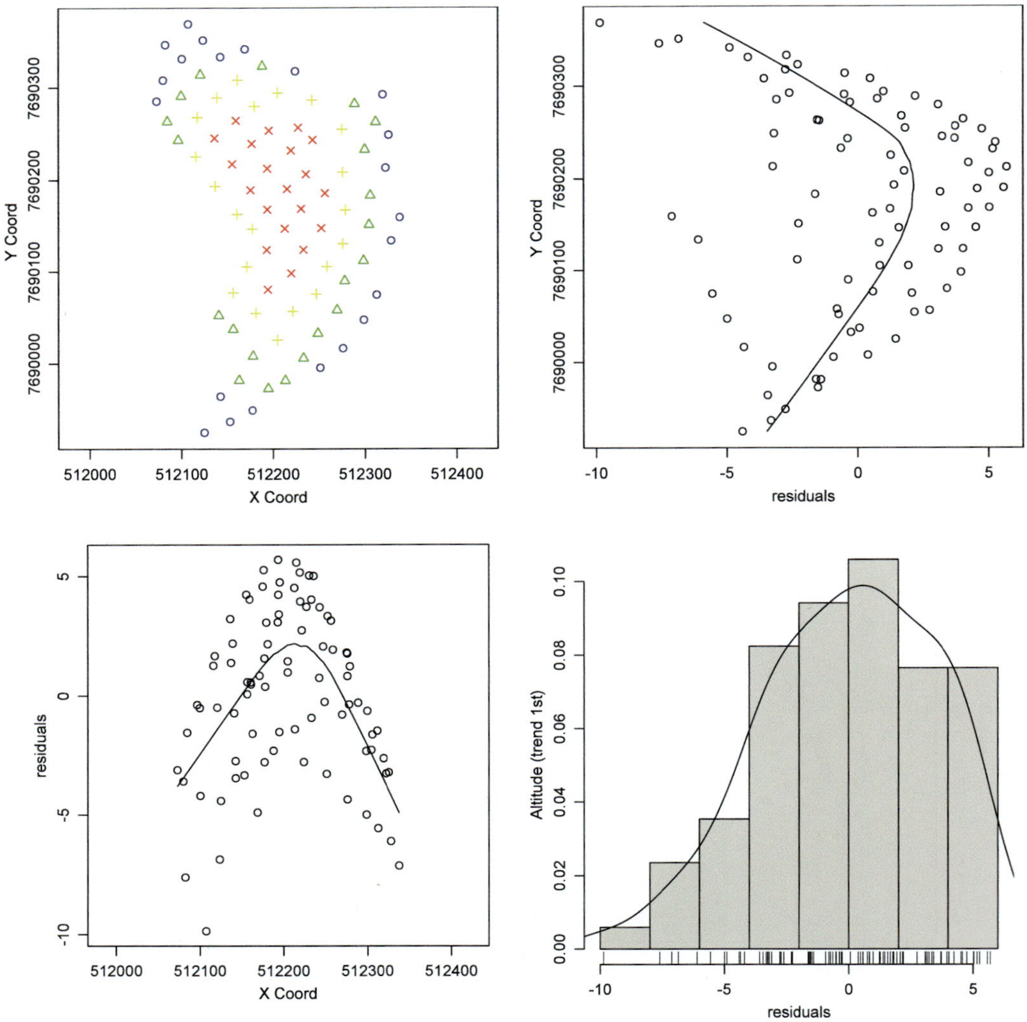

FIGURE 12.10 Exploratory analysis of geodetic altitude (m) obtained at a coffee field using a map of quartile symbols (top-right), graph with the variation of the attribute as a function of the latitude (top-right) and longitude (bottom-left) coordinate axis under first-order polynomial trend fit applied to the x and y coordinates with smoothing line, and the frequency distribution histogram with density line fitted to the altitude residuals of 85 points of a stratified irregular sampling grid.

relationship between altitude (m)(A) (independent variable) and air temperature (T)(°C) (dependent variable) from INMET main climatological stations and climatological normals (1960-1990).

The `readr` (Wickham et al., 2021c), `ggpubr` (Kassambara, 2025), `gridExtra`(Auguie and Antonov, 2017), `raster` (Hijmans et al., 2020), and `sf` (Pebesma et al., 2021a) packages are loaded in R for analysis.

```
library(readr)
# package 'readr' was built under R version 4.3.3
library(ggpubr)
# Loading required packages: ggplot2
library(gridExtra)
library(raster)
# Loading required packages: sp
library(sf)
# Linking to GEOS 3.11.2, GDAL 3.7.2, PROJ 9.3.0; sf_use_s2() is TRUE
```

The database containing the variables, altitude and mean air temperature is imported into R using the `read_csv2` function.

```
mgAT1 <- read_csv2("C:/geo/c9/mgAT1.csv", col_names = TRUE,
                   locale = locale(encoding = 'ISO-8859-1'))
# Rows: 32 Columns: 9
# Column specification
# Delimiter: ";"
# chr (1): Cidade
# dbl (8): Y, X, Z, T, PAnual, iu, Eto, Deficit
```

Functions from the `ggpubr` package are used to create regression graphs of mean air temperature as a function of altitude at the weather station (Figure 12.11).

```
# First-order polynomial
formula <- y ~ poly(x, 1, raw = TRUE)
p1 <- ggscatter(mgAT1, x = "Z", y = "T", add = "reg.line") +
stat_cor(label.x = 1000, label.y = 25) +
geom_text(aes(label = Cidade), hjust = 0, vjust = 0, check_overlap = TRUE) +
stat_regline_equation(label.x = 900, label.y = 24.5,
aes(label =  paste(..eq.label.., ..adj.rr.label.., sep = "~~~~")),
formula = formula)
# Second-order polynomial
formula <- y ~ poly(x, 2, raw = TRUE)
p2 <- ggscatter(mgAT1, x = "Z", y = "T", add = "reg.line") +
stat_cor(label.x = 1000, label.y = 25) +
geom_text(aes(label = Cidade), hjust = 0, vjust = 0, check_overlap = TRUE) +
stat_regline_equation(label.x = 900, label.y = 24.5,
aes(label =  paste(..eq.label.., ..adj.rr.label.., sep = "~~~~")),
formula = formula)
grid.arrange(p1, p2, ncol = 1)
```

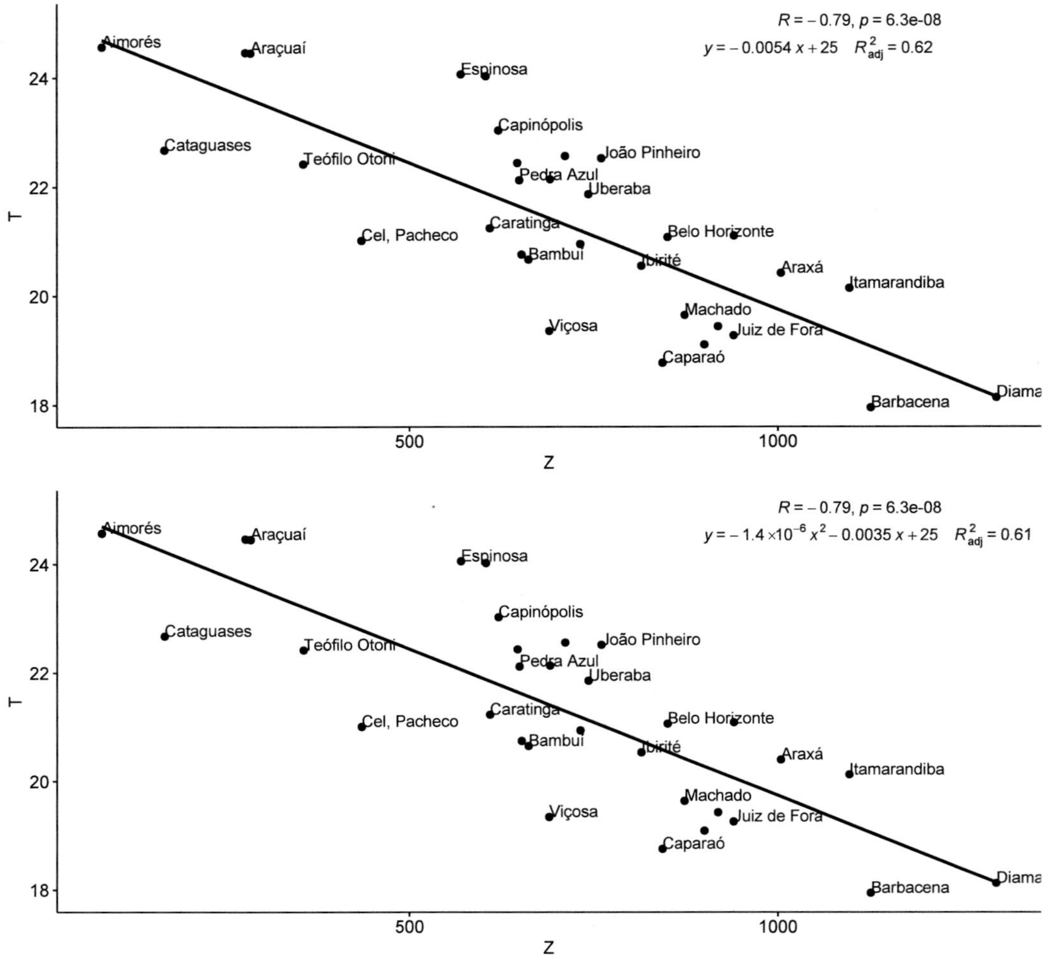

FIGURE 12.11 Using first- and second-order polynomial models to estimate the mean air temperature (°C) in Minas Gerais, Brazil, as a function of altitude (m) from INMET's main weather stations and climatological normals (1960-1990).

The regression equations obtained for first- and second-order polynomials and quality coefficients are described below.

$$T = 25 - 0.0054 * A \quad (R^2adj = 0.62) \tag{12.15}$$

$$T = 25 - 0.0035 * A - 1.4x10^{-6}x^2 \quad (R^2adj = 0.61) \tag{12.16}$$

where T is the mean air temperature (°C) and A, the altitude (m). In this case, the first-order fit gives the best results, given the linear pattern observed in the data used in the modeling, as well as the adjusted R^2 value.

The GADM database with administrative boundary polygons for Brazil is imported into R using the `read_sf` function.

```
br1 <- read_sf("C:/geo/GADM/Brazil/gadm36_BRA_1.shp", quiet = TRUE)
```

A subset of the Minas Gerais polygon is obtained based on the name of the state of interest in the attribute table.

```
brmg1 <- br1[br1$NAME_1 == "Minas Gerais",]
mg <- as_Spatial(brmg1)
```

The SRTM digital elevation model of Minas Gerais is imported into R using the `raster` function. The `names` function is used to rename the layer as `Altitude`.

```
srtmMG <- raster("C:/geo/SRTM/srtmMG.tif")
names(srtmMG) <- c("Altitude")
```

First- and second-order polynomial functions are used to determine the mean air temperature as a function of altitude.

```
mgT <- 25 - 0.0054 * srtmMG$Altitude
mgT2 <- 25 - 0.0035 * srtmMG$Altitude - ((1.4 * 10^-6) * (srtmMG$Altitude)^2)
```

The results are mapped using the `plot` function (Figure 12.12).

```
par(mfrow = c(1, 2), mar = c(2, 2, 1, 0.3), mgp = c(1.5, 0.6, 0))
plot(mgT, col = grey(1:100/100), main = "1st", horizontal = TRUE)
plot(mg, col = "transparent", add = TRUE)
plot(mgT2, col = grey(1:100/100), main = "2nd", horizontal = TRUE)
plot(mg, col = "transparent", add = TRUE)
```

In multiple regression, other independent variables can be used (Burrough and McDonnell, 1998; Sediyama and Melo Júnior, 1998):

$$z(x) = b_0 + b_1P_1 + b_2P_2 + b_3P3 + \epsilon \tag{12.17}$$

where $b_0, ..., b_n$ are the regression parameters and $P_1, ..., P_n$, the independent variables used in the modeling. In this case, the variables P_1, P_2, P_3 are defined by the geodetic longitude, latitude and altitude of INMET weather stations used in the prediction of the annual mean air temperature, $z(x)(°C)$, as a function of the independent variables, longitude $(P_1)(°)$, latitude $(P_2)(°)$ and altitude $(P_3)(m)$ of the weather station where the temperature is measured (observed).

The `lm` function is used to fit multiple regression to the data. The results are checked with the `summary` function.

```
fit <- lm(T ~ X + Y + Z, data = mgAT1) # Fit multiples regression
summary(fit) # show results
# Call: lm(formula = T ~ X + Y + Z, data = mgAT1)
```

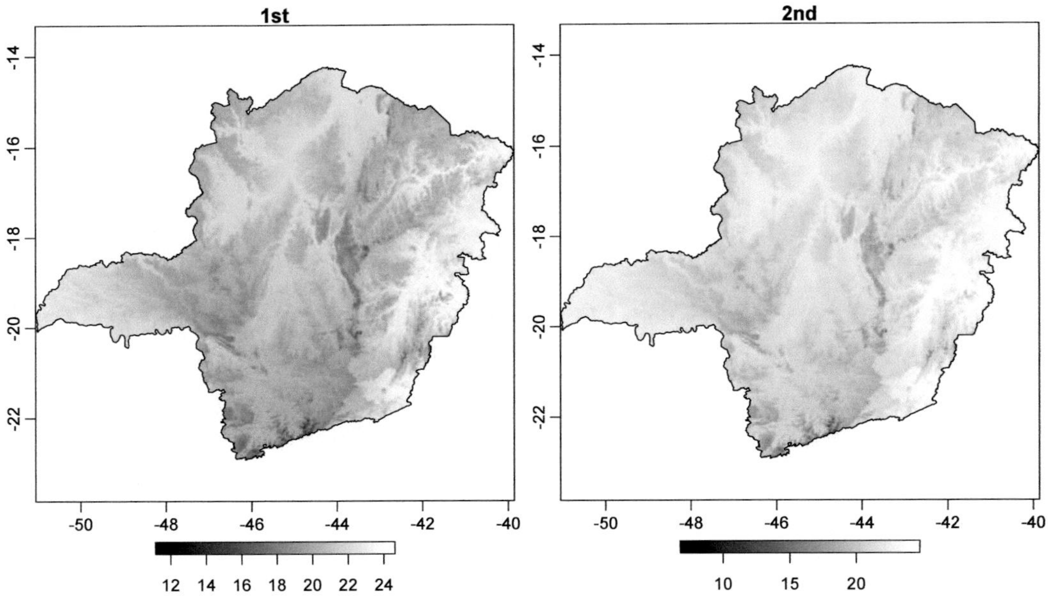

FIGURE 12.12 Example of the use of first- and second-order regression models to estimate the mean air temperature (°C) in Minas Gerais, Brazil, as a function of parameters obtained by the relationship between altitude (m) and air temperature from INMET's main weather stations and climatological normals (1960-1990).

```
# Residuals:
#       Min       1Q   Median       3Q      Max
# -1.00010 -0.37027  0.08312  0.26544  1.13031
# Coefficients:
#                Estimate Std. Error t value Pr(>|t|)
# (Intercept) 24.9821533  2.3571283  10.599 2.63e-11 ***
# X           -0.2208373  0.0526477  -4.195 0.000249 ***
# Y            0.5001462  0.0528832   9.458 3.25e-10 ***
# Z           -0.0053889  0.0003975 -13.556 7.97e-14 ***
# ---
# Signif. codes:  0 '***' 0.001 '**' 0.01 '*' 0.05 '.' 0.1 ' ' 1
# Residual standard error: 0.557 on 28 degrees of freedom
# Multiple R-squared:  0.9195,  Adjusted R-squared:  0.9109
# F-statistic: 106.6 on 3 and 28 DF,  p-value: 2.004e-15
```

After including the coordinates in the multiple regression fit to the annual mean air temperature data, the quality of model fit is significant for all parameters, with better quality of the R^2 value:

$$T_{year} = 24.9821533 - 0.2208373(Long) + 0.5001462(Lat) - 0.0053889(Alt) \quad (R^2adj = 0.91) \tag{12.18}$$

A chunk code is used to create a raster with attributes of independent variables of longitude, latitude, and altitude.

```
r <- srtmMG # Create raster r
rlon <- rlat <- r # Copy r to rlon and rlat rasters
xy <- xyFromCell(r,1:length(r)) # Get matrix of longitudes (x) and latitudes (y)
rlon[]<-xy[,1] # Create raster of longitudes
rlat[]<-xy[,2] # Create raster of latitudes
Long <- mask(rlon, r) # Mask longitudes in MG
Lat <- mask(rlat, r) # Mask latitudes in MG
srtm <- stack(r, Long, Lat) # Stack longitudes, latitudes and altitudes
names(srtm) <- c("Alt", "Long", "Lat") # Rename raster layers
srtm # See header
# class      : RasterStack
# dimensions : 1043, 1343, 1400749, 3  (nrow, ncol, ncell, nlayers)
# resolution : 0.008333333, 0.008333333  (x, y)
# extent     : -51.05, -39.85833, -22.925, -14.23333  (xmin, xmax, ymin, ymax)
# crs        : +proj=longlat +datum=WGS84 +no_defs
# names      :       Alt,       Long,        Lat
# min values :  63.00000, -51.04583, -22.92083
# max values : 2599.0000,  -39.8625,  -14.2375
```

The multiple regression model determined previously is applied to the raster with altitude, longitude, and latitude layers to determine the dependent variable temperature.

```
Tyear <- 24.9821533 - (0.2208373 * srtm$Long) + (0.5001462 * srtm$Lat) -
         (0.0053889 * srtm$Alt)
```

The `list` function is used to create a `SpatialPolygonsDataFrame` object with the boundaries of the region of interest.

```
borders <- list("sp.polygons", mg, fill = "transparent", col = 'black', first = FALSE)
```

The results are mapped using the `spplot` function (Figure 12.13).

```
spplot(Tyear, sp.layout = borders, scales = list(draw = TRUE))
```

This type of regression model, also called a transfer function, can be used for other sets of independent variables, such as rainfall as a function of distance from the sea, zinc content in soil as a function of distance from a river, vigor of vegetation as a function of humidity, and number of customers as a function of income. In all these situations, the importance of the physical sense of the model should be considered as one of the most important features of regression. It should also be noted that regression is considered an inexact interpolator (Burrough and McDonnell, 1998).

12.13 DETERMINISTIC LOCAL INTERPOLATION METHODS

In local interpolation operations, a local, smaller area around the value to be interpolated is used as the source of information for interpolation to ensure that all estimates are made from the local neighborhood with the best possible fit. When using local interpolation

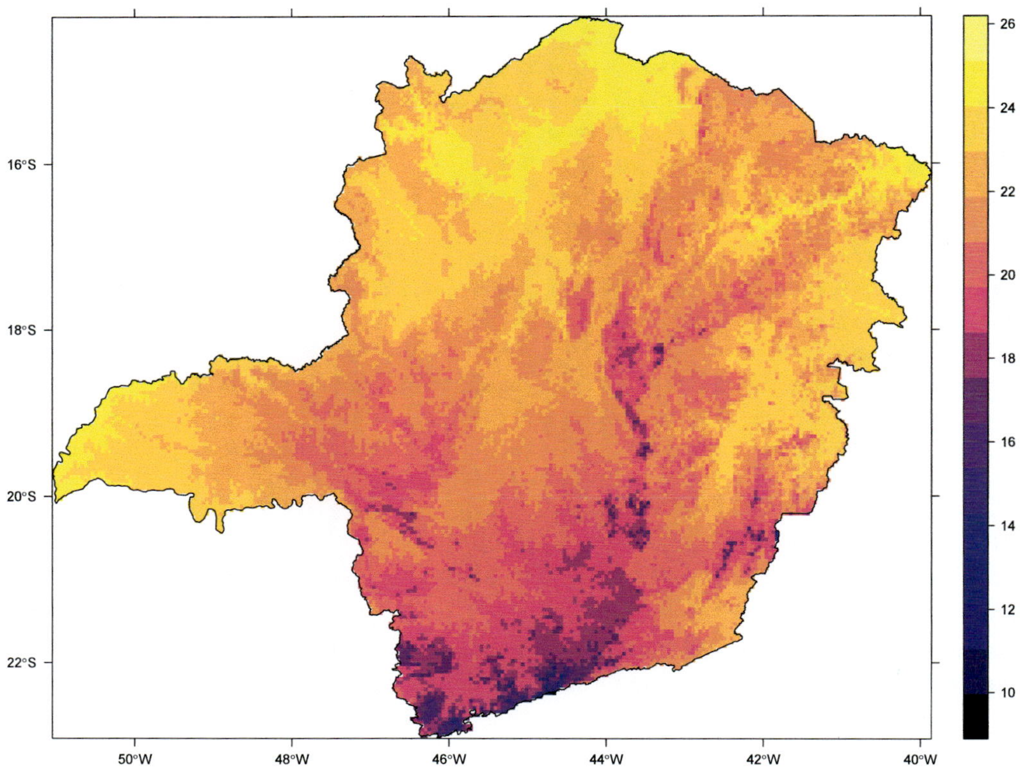

FIGURE 12.13 Example of the use of multiple regression to estimate the mean air temperature in Minas Gerais (°C), as a function of parameters obtained by the relationship between longitude (degrees), latitude (degrees), altitude (m) and air temperature from INMET's main climatological stations and climatological normals (1960-1990).

methods, it is necessary to use a systematic process for spatial prediction that is repeated until all points on a grid are predicted, with the following steps (Burrough and McDonnell, 1998):

- Define a neighborhood area around the point to be predicted;
- Locate the point data within this neighborhood;
- Choose a mathematical function to represent the variation of the attribute considering a limited number of points;
- Evaluate the interpolation method as a function of the spatial prediction on a regular grid.

When performing local interpolation, the following aspects should be considered (Burrough and McDonnell, 1998):

- The type of interpolation function used;
- The size, shape and orientation of the neighborhood;
- The number of point data;
- The distribution of the point data on a mesh with regular or irregular distribution;

- The possibility of using an external source of information as a source for trend modeling.

The methods used in local interpolation are (Burrough and McDonnell, 1998):

- Nearest neighbors;
- Inverse distance weighting;
- Splines and other types of nonlinear functions;
- Optimized functions for using spatial covariance.

All these local functions are used for a given degree of smoothing of the attribute data according to a search window or distance. The first three types of local interpolators mentioned above are covered in this chapter, and the last one in the next chapter (Burrough and McDonnell, 1998).

12.14 THIESSEN INTERPOLATION BY NEAREST NEIGHBOR

The Thiessen polygons, also called Dirichlet or Voronoi, are used in the prediction of attributes in non-sampled locations based on the neighborhood of points closest to the geographical location where the attribute prediction will be made. The geometric arrangement of Thiessen polygons is dependent on the distance from the grid and the irregularity of the mesh in which the attributes are obtained in a given region under analysis. Delaunay triangulation is the principle used to join the lines that define the topological tessellation of Thiessen polygons. This type of triangulation is the same as the one used in digital terrain modeling with the triangulated irregular network (TIN) data model. In the triangulation procedure, the weights assigned to each attribute value are calculated based on the geometry of three neighboring samples and the point to be estimated.

Considering the location of three samples with attributes, $z(x_1), z(x_2), z(x_3)$, the point $\hat{z}(x_0)$ to be estimated as a function of the area (A) of three triangles defined by triangulation between the three neighboring points and the point to be predicted is (Figure 12.14) (Isaaks and Srivastava, 1989):

$$\hat{z}(x_0) = \frac{\sum_{i=1}^{n} z(x_i) A_i}{A_T} \tag{12.19}$$

where A_i is the area defined by the triangle opposite the vertex of attribute $z(x_i)$ in the neighborhood of the point to be predicted, $\hat{z}(x_0)$, and A_T, the total area defined by the vertices of the attributes neighboring $\hat{z}(x_0)$.

The triangulation can be considered a linear combination of weights where each value is weighted according to the area of the opposite triangle. Thus, nearby points receive the highest weights in the interpolation (Alves and Silva, 2016).

The Thiessen polygon methodology is used to define Thiessen polygon network and line network used in Delaunay interpolation from irregular sample mesh used to support point attribute analysis in a coffee crop in Santo Antônio do Amparo.

The `deldir` (Turner, 2024) package is used as an example to create a network of Thiessen polygons and lines from Delaunay interpolation.

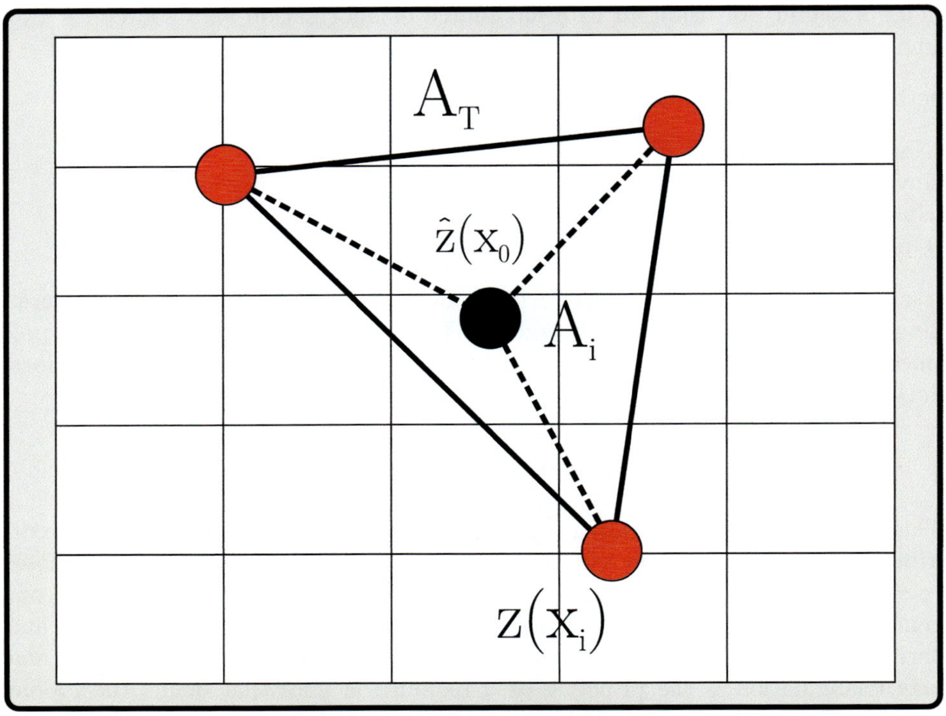

FIGURE 12.14 Principle of interpolation by the Delaunay triangulation method used to define the topology of Thiessen polygons and the TIN topology.

```
library(deldir)
# deldir 2.0-4       Nickname: "Idol Comparison"
# package 'deldir' was built under R version 4.3.3
```

The functions `deldir`, `triang.list`, and `tile.list` are used to create objects of class `triang.list` and `tile.list` in order to generate a network of Thiessen polygons and lines by Delaunay interpolation.

```
x <- blight_sp$X
y <- blight_sp$Y
z <- blight_sp$Z
d <- deldir(x, y, z = z)
ttt <- triang.list(d)
sss <- tile.list(d)
```

The results are mapped using the `plot` function (Figure 12.15).

```
par(mfrow = c(1, 1), mar = c(2, 2, 0.3, 0.3), mgp = c(1.5, 0.6, 0))
plot(sss)
plot(ttt, add = TRUE, border = "blue")
plot(blight_sp, add = TRUE, border = "blue")
```

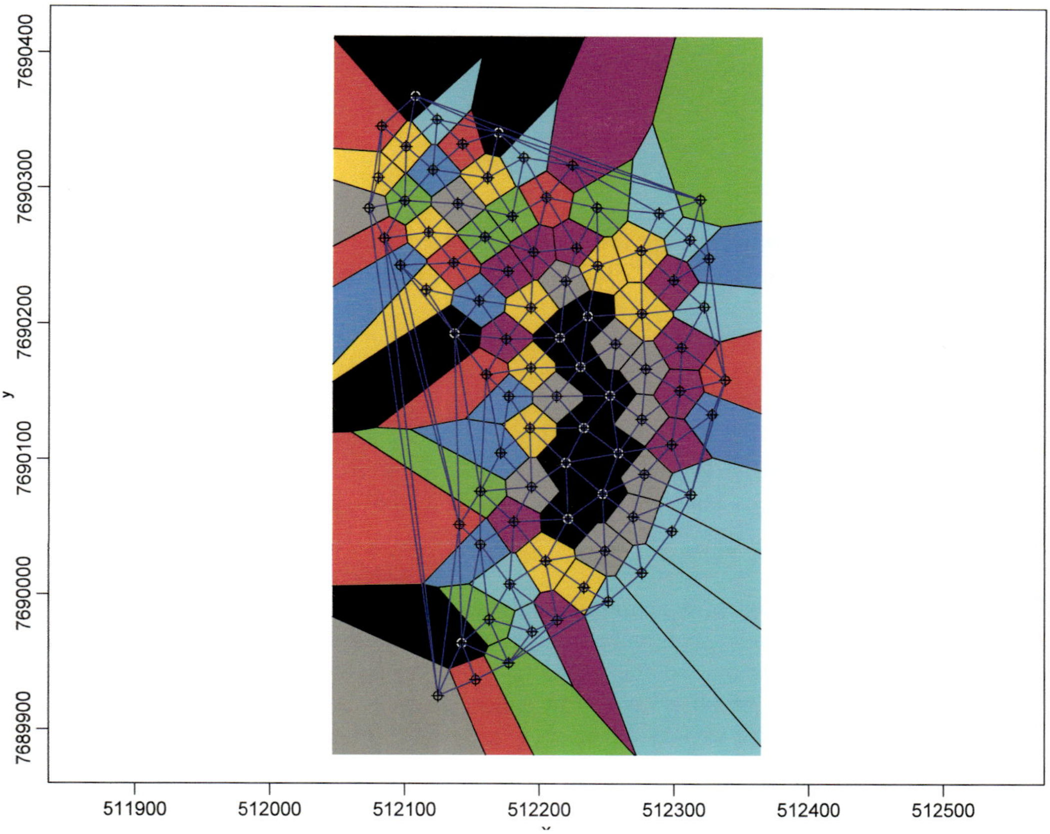

FIGURE 12.15 Thiessen polygon network and line network layers used in Delaunay interpolation from sample mesh used as support for point attribute analysis.

Point geometry with altitude attributes (m) is mapped on the coffee field in order to visualize the pattern of altitude variation in the geographic region. The point attribute values are assigned to the determined Thiessen polygon features. The polygons are then clipped as a function of the edge with the georeferenced crop boundaries, redefining the outer geometry of the Thiessen polygons. The mapping of the geodetic altitude is performed on the polygons with points used in the modeling considering homogeneous areas defined by Thiessen polygons in the coffee field and the points used in the definition of the geometry of the polygons.

The packages `dismo` (Hijmans et al., 2024c), `gridExtra` (Auguie and Antonov, 2017), `deldir` (Turner, 2024), and `rgeos` (Bivand et al., 2018) are loaded into R to perform this example.

```
library(dismo)
library(gridExtra)
library(deldir)
library(rgeos)
```

The `extent`, `voronoi`, `gIntersection`, and `over` functions are used to determine Thiessen polygons with altitude attributes in a coffee field.

```
e <- extent(bord_sp) # Get extent
voronoi <- voronoi(blight_sp, ext = e) # Voronoi calculation
vor <- gIntersection(bord_sp, voronoi, byid = TRUE) # Clip at the intersection
vo <- over(vor, blight_sp[,"Z"]) # Get the Z value of the mesh
vor$z <- vo$Z # Add the result to the vor attribute table
```

The results are mapped with the `spplot` function using the `sp` (Pebesma et al., 2021b) and `latticeExtra` (Sarkar and Andrews, 2022) packages (Figure 12.16).

```
p1 <- spplot(blight_sp["Z"])
p2 <- spplot(voronoi["Z"])
p3 <- spplot(vor["z"])
p4 <- spplot(vor, 'z') +
  latticeExtra::layer(sp.polygons(blight_sp, col = "white", lwd = 1.5)) +
  latticeExtra::layer(sp.points(blight_sp, pch = 1, col = 'black'))
grid.arrange(p1, p2, p3, p4, ncol = 2)
```

Thiessen polygons are used in geocomputation as a rapid analysis methodology to relate point attributes to geo-fields, such as in weather station data applications. However, there can be mapping problems due to the geometric design of the polygon that is influenced by the distribution of point observations in space, as well as the problem of discretizing continuous data with abrupt variation from polygon edges. The advantage of using Thiessen polygons is that in the absence of more precise information about the edges of the studied phenomenon, the polygons defined in relation to the sample points can be used to elaborate choropleth maps of vegetation, land use class, with the same value as the obtained samples. Thus, since the predicted values in the polygons are equal to the attributes measured in the field, Thiessen polygons can be considered as an interpolation method with exact prediction (Burrough and McDonnell, 1998).

Interpolation by Thiessen polygons is considered an extreme case of attribute mapping in geo-field with discretized polygons. Since in this approach the measured attribute is considered homogeneous within polygon units, with values changing only from the edges of the geometric features, it can be considered that the estimation of continuous attributes, such as, population density, meteorological variables, and altitude and flow rate may not be adequately represented by this methodology.

12.15 TOBLER PYCNOPHYLACTIC INTERPOLATION

A continuous attribute interpolation methodology is developed for attribute smoothing from discrete edges of geometric features, as in the case of population counts within administrative regions. The assumption of this method, called pycnophylactic interpolation, is that the population is not spread evenly across administrative areas, but with a tendency to aggregate in certain localities (Tobler, 1979). In pycnophylactic interpolation, the sum of the estimated values associated with all pixels in any polygon is equal to the population given for that polygon. Another property of this type of interpolator is that all pixel values are greater than or equal to zero. This method is generally used to obtain pixel-based population estimates when the total populations for a set of irregular polygons are known (Brunsdon, 2020). Using this methodology, a more realistic representation of the density of people or rainfall can be obtained in a given area with high values of the attribute when compared to another area where the attribute had low values. Pycnophylactic interpolation

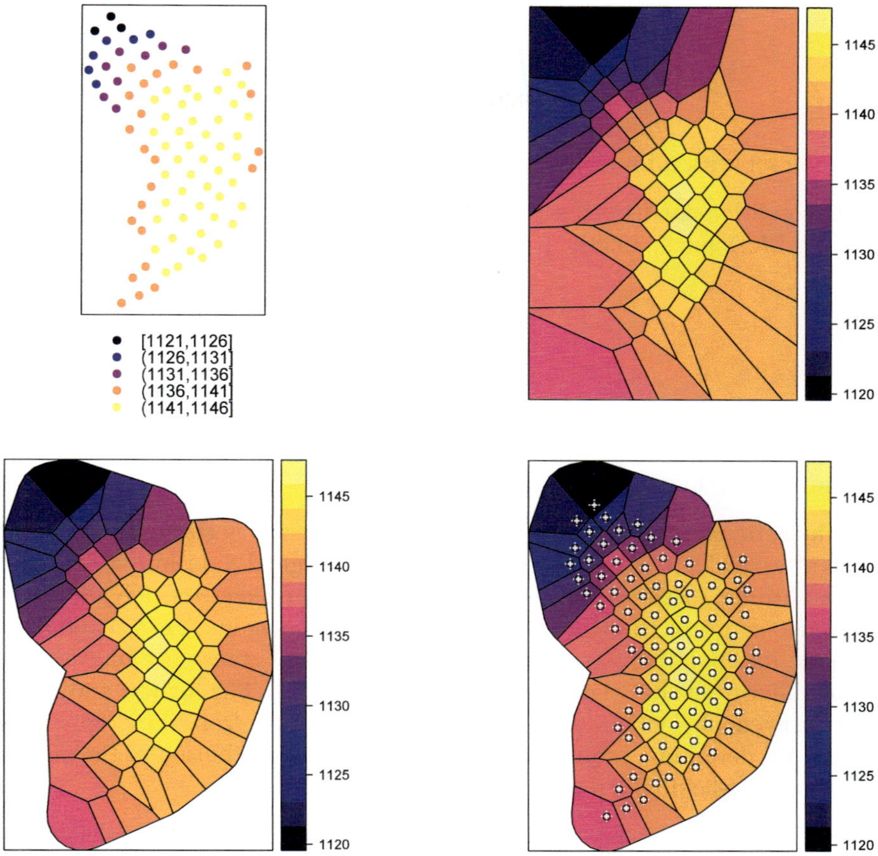

FIGURE 12.16 Points with altitude attributes (m) (top-left), determination of Thiessen polygons with altitude attribute based on the location of the points (top-right), clipping of the outer geometry of the polygons (bottom-left) and mapping of the polygons with points used in the modeling (bottom-right).

is considered a method of mass preservation and reallocation of primary data ensuring that the volume of the attribute in a spatial entity remains the same regardless of whether the attribute is represented in discretized polygons or on a continuous surface. The primary condition for mass preservation is the invertibility condition (Burrough and McDonnell, 1998):

$$V_i = \int_{R_i} \int f(x, y)dxdy \tag{12.20}$$

where V_i represents the value of the attribute in a region, R_i, for all i.

With this, the total volume of an attribute per polygon is invariant to the measurement obtained within discrete edges or by a smoothed surface over neighboring areas. The density of neighboring areas can be grouped by a continuous surface fitted in the region by a Laplacian minimization condition (Burrough and McDonnell, 1998).

$$\int_R \int (\frac{\delta f^2}{\delta x} + \frac{\delta f^2}{\delta y}) dx dy \tag{12.21}$$

where R, is the set with all regions considered in the interpolation.

The general condition used by the interpolator on the edges of the discretized regions is (Burrough and McDonnell, 1998):

$$\frac{\delta f}{\delta \eta} = 0 \tag{12.22}$$

where this condition constrains the gradient of the fitted surface normal to the region border as plane, and *meta* defines the boundary between regions.

The pycnophylactic method can be applied to variables such as soil zinc content; however, in this type of variable, it is necessary to convert the original data into a density function calculated according to the total amount of zinc within Thiessen polygons in order to obtain a pseudo-count of the zinc content within each polygon. Thus, it is obvious that pycnophylactic interpolation is not an exact interpolator, because the interpolated values at each pixel differ from the original value at the same pixel. The highest and lowest values obtained with pycnophylactic interpolation can differ greatly from the originally observed variable, and can determine problems in interpreting the distribution of the variable when represented as a geo-field (Burrough and McDonnell, 1998).

The representation of reference flow rate ($m^3 s^{-1}$) in ottobasins by polygons delimited by sub-basins is compared with the centroid representation of these polygons converted into Voronoi polygons in order to compare different supports for Tobler's pycnophylactic interpolation. The spatial distribution of the attributes in the different polygons used as sample support is compared with the respective pycnophylactic predictions of the reference flow rate in a watershed in southwestern Mato Grosso state. The data are interpolated at a spatial resolution of 2.5 km in order to obtain a magnitude close to the original data when compared to the interpolation result. With the use of both evaluated supports, it is possible to perform the flow prediction by the pycnophylactic method in the evaluated region. The estimation with the use of sub-basins as support gave magnitude of results closer to the original data. However, with the use of Thiessen polygons as support, it is possible to define patterns of higher and lower values of the attribute in the area, facilitating the identification of key points of variation of the attribute in the region. To obtain spatial prediction with a magnitude closer to the original data, it may be necessary to adjust the pixel spatial resolution in order to obtain better prediction interpretation results.

The `pycno` package (Brunsdon, 2025) is used in this example with watershed flow data obtained at the beginning of this chapter. The `pycno` package is loaded into R with the `library` function.

```
library(pycno)
```

The flow data in sub-basin polygons are transformed into the `sp` class using the `as_Spatial` function.

```
qm_sp <- as_Spatial(qm)
```

The pycno function is used to perform pycnophylactic interpolation at a spatial resolution of 2.5 km on two types of geographic support, i.e., for sub-basin polygons and Voronoi polygons determined from the centroid of the sub-basins.

```
picnoA <- pycno(qm_sp, qm_sp$QM, r = 0.2, celldim = 2500, converge = 5)
# Maximum Change:      70.89443 - will stop at       0.00725
# Maximum Change:      15.33235 - will stop at       0.00725
# Maximum Change:       6.38851 - will stop at       0.00725
# Maximum Change:       2.58793 - will stop at       0.00725
# Maximum Change:       1.65174 - will stop at       0.00725
# Maximum Change:       0.85654 - will stop at       0.00725
# Maximum Change:       0.49881 - will stop at       0.00725
# Maximum Change:       0.31312 - will stop at       0.00725
# Maximum Change:       0.18523 - will stop at       0.00725
# Maximum Change:       0.11822 - will stop at       0.00725
# Maximum Change:       0.07196 - will stop at       0.00725
# Maximum Change:       0.04524 - will stop at       0.00725
# Maximum Change:       0.02789 - will stop at       0.00725
# Maximum Change:       0.01743 - will stop at       0.00725
# Maximum Change:       0.01080 - will stop at       0.00725
# Maximum Change:       0.00673 - will stop at       0.00725
picnoB <- pycno(vorclip, vorclip$QM, r = 0.2, celldim = 2500, converge = 5)
# Maximum Change:      80.41431 - will stop at       0.00578
# Maximum Change:      22.09458 - will stop at       0.00578
# Maximum Change:       7.56081 - will stop at       0.00578
# Maximum Change:       2.29276 - will stop at       0.00578
# Maximum Change:       1.03917 - will stop at       0.00578
# Maximum Change:       0.39702 - will stop at       0.00578
# Maximum Change:       0.22991 - will stop at       0.00578
# Maximum Change:       0.08550 - will stop at       0.00578
# Maximum Change:       0.05449 - will stop at       0.00578
# Maximum Change:       0.03111 - will stop at       0.00578
# Maximum Change:       0.02054 - will stop at       0.00578
# Maximum Change:       0.01226 - will stop at       0.00578
# Maximum Change:       0.00787 - will stop at       0.00578
# Maximum Change:       0.00480 - will stop at       0.00578
```

The results are mapped with the spplot function and arranged side-by-side with the grid.arrange function (Figure 12.17).

```
a1 <- spplot(qm_sp, "QM", main = "Otto basins", border = "grey")
a2 <- spplot(vorclip, "QM", main = "Voronoi", border = "grey")
a3 <- spplot(picnoA, main = "Tobler Otto")
a4 <- spplot(picnoB, main = "Tobler Voronoi")
grid.arrange(a1, a2, a3, a4, nrow = 2)
```

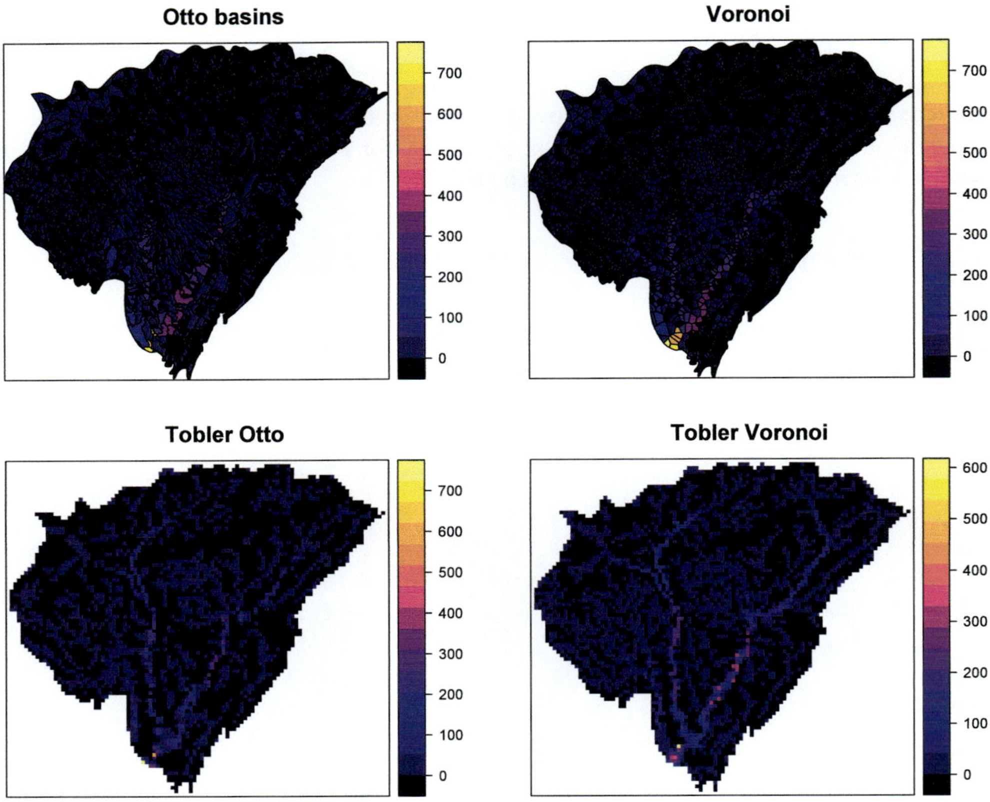

FIGURE 12.17 Representation of reference flow rate in ottobasin polygons (top-left), in Voronoi polygons (top-right), flow rate interpolated at 2.5 km spatial resolution by Tobler's pycnophylactic method, considering the ottobasins as support (bottom-left) and Voronoi polygons as support (bottom-right), in a watershed in southwestern Mato Grosso, Brazil.

12.16 LINEAR INTERPOLATION BY INVERSE DISTANCE WEIGHTING

In the inverse distance weighting (IDW) interpolation method, the ideas of Thiessen polygon proximity are combined with those of trend surface to perform linear interpolation of point attributes in geo-field. The IDW hypothesis is that the value of an attribute, z, at an unsampled location varies according to the inverse of the weighted distance of point features with attributes within a neighborhood or moving window around the point. Typically point data is used on regular or irregular grids in a given area where the interpolation method is used to make predictions on a denser regular grid used for attribute mapping. The IDW is considered a linear interpolator because the weights used in the spatial prediction of the attribute are determined using a linear function of the distance between a dataset of the attribute in the neighborhood and the point to be predicted. Since the prediction is usually performed on a regular grid, contour lines can be defined as a vector on the map to facilitate visualization of the attribute, as in a shaded relief map (Burrough and McDonnell, 1998).

The term inverse of distance is used because attribute points are interpolated using a function weighted in relation to the inverse of the distance between the attribute and the point to be predicted on the geo-field grid (Slocum et al., 2009; Alves and Silva, 2016):

$$\hat{z}(x_0) = \frac{\sum_{i=1}^{n} z(x_i)/d_i^k}{\sum_{i=1}^{n} 1/d_i^k} \tag{12.23}$$

where k is the power defined in the interpolation, d_i, the Euclidean distance between attribute points relative to the grid point to be predicted and, n, the number of points with attributes used to estimate a grid point.

IDW can be considered a weighted moving window method according to the variation of weights, λ, according to the neighborhood of attribute points, $z(x_i)$ (Burrough and McDonnell, 1998):

$$\hat{z}(x_0) = \sum_{i=1}^{n} \lambda_i z(x_i) \tag{12.24}$$

where

$$\sum_{i=1}^{n} \lambda_i = 1 \tag{12.25}$$

The map resulting from IDW interpolation is dependent on the value of k, and can present very different maps by varying, for example, $k = 1$ to $k = 4$. The inverse distance interpolator is forced to be an exact interpolator by producing the same attribute value when $d = 0$. The quality of the final map produced by IDW may vary depending on the grouping of attribute data into clusters and the presence of discrepant data in the set of sample values. In the case of sparse data, there may be formation of "duck egg" shaped spatial patterns around the sparse sample points with large variation from the sample neighborhood (Burrough and McDonnell, 1998).

The weights defined by IDW (λ) can be defined by Shepard's method, by the inverse of the distance between the location to be interpolated (x) and the sample neighbor (x_i), raised to the power, k (Tarroso, 2021):

$$\lambda(x) = \frac{1}{d(x, x_i)^k} \tag{12.26}$$

Another option may be to define the weights using the modified Shepard method, where the weighting is performed as a function of a search radius, r, used to calculate the interpolation weights (Tarroso, 2021):

$$\lambda(x) = [\frac{r - d(x, x_i)}{rd(x, x_i)}]^k \tag{12.27}$$

In Shepard's method with neighborhood, the maximum amount of n neighbors is used in determining weights (Tarroso, 2021).

Interpolation by the inverse distance weighting is applied to altitude sample data on a georeferenced irregular grid in a coffee field. The power values are defined as k = 1, 2, 3, 4, in order to compare the results on the obtained maps. With varying power, different altitude maps are produced in the area, with different degrees of attribute smoothing. The smoothness of the interpolated maps increased with increasing power, making clearer the effect of the interpolator in representing the three-dimensional variation of the attribute in the area.

The packages `raster` (Hijmans et al., 2020), `gstat` (Pebesma and Graeler, 2021), `gridExtra` (Auguie and Antonov, 2017), and `rgl` (Adler and Murdoch, 2025) are loaded in R as examples for performing IDW interpolation.

```
library(raster)
# Loading required packages: sp
library(gstat)
# package 'gstat' was built under R version 4.3.3
library(gridExtra)
library(rgl)
# package 'rgl' was built under R version 4.3.3
```

The `extent` function is used to define the extent of the area in which the interpolation grid will be created from data obtained from a coffee field loaded into R at the beginning of this chapter.

```
e <- extent(bord_sp) # extent
e # See extent
# class       : Extent
# xmin        : 512013.2
# xmax        : 512395.5
# ymin        : 7689875
# ymax        : 7690432
```

The `expand.grid` function is used to create the 5x5 m interpolation grid in the previously defined extension.

```
grid <- expand.grid(x = seq(from = round(e@xmin), to = round(e@xmax), by = 5),
                    y = seq(from = round(e@ymin), to = round(e@ymax), by = 5))
```

The `coordinates` function is used to convert the previously created grid into `SpatialPoints`.

```
coordinates(grid) <- ~x + y
```

The `proj4string` function is used to assign the coordinate system of the geo-object that defines the boundaries of the region in the created grid.

```
proj4string(grid) <- proj4string(bord_sp)
```

A subset of the `SpatialPoints` grid is created within the boundaries of the geographic region considered in the analysis using the bracket operator.

```
g <- grid[bord_sp,]
```

The `gridded` function is used to convert the g object from `SpatialPoints` to `SpatialPixels`.

```
gridded(g) <- TRUE
```

The IDW model is specified with the `gstat` function for application to the altitude data obtained from the sample grid performed on a coffee field loaded into R at the beginning of this chapter, within the geographic region, using interpolation weights 1, 2, 3, 4.

```
idw1 <- gstat(formula = Z ~ 1, data = blight_sp , nmax = length(blight_sp),
              set = list(idp = 1))
idw2 <- gstat(formula = Z ~ 1, data = blight_sp , nmax = length(blight_sp),
              set = list(idp = 2))
idw3 <- gstat(formula = Z ~ 1, data = blight_sp , nmax = length(blight_sp),
              set = list(idp = 3))
idw4 <- gstat(formula = Z ~ 1, data = blight_sp , nmax = length(blight_sp),
              set = list(idp = 4))
```

The `predict` function is used for IDW prediction in pixels for each of the evaluated interpolation weights.

```
pred1 <- predict(object = idw1, newdata = g)
# [inverse distance weighted interpolation]
pred2 <- predict(object = idw2, newdata = g)
# [inverse distance weighted interpolation]
pred3 <- predict(object = idw3, newdata = g)
# [inverse distance weighted interpolation]
pred4 <- predict(object = idw4, newdata = g)
# [inverse distance weighted interpolation]
```

The `spplot` and `grid.arrange` functions are used to map the results (Figure 12.18).

```
i1 <- spplot(pred1[1], main = "k = 1", sp.layout = bord_sp,
             scales = list(draw = TRUE))
i2 <- spplot(pred2[1], main = "k = 2", sp.layout = bord_sp,
             scales = list(draw = TRUE))
i3 <- spplot(pred3[1], main = "k = 3", sp.layout = bord_sp,
             scales = list(draw = TRUE))
i4 <- spplot(pred4[1], main = "k = 4", sp.layout = bord_sp,
             scales = list(draw = TRUE))
grid.arrange(i1, i2, i3, i4, ncol = 2)
```

Three-dimensional altitude mapping is performed using functions from the `rgl` package in order to verify the degree of altitude smoothing with the weights analyzed in IDW interpolation.

The `as.matrix` function is used to convert the spatial prediction results into a matrix.

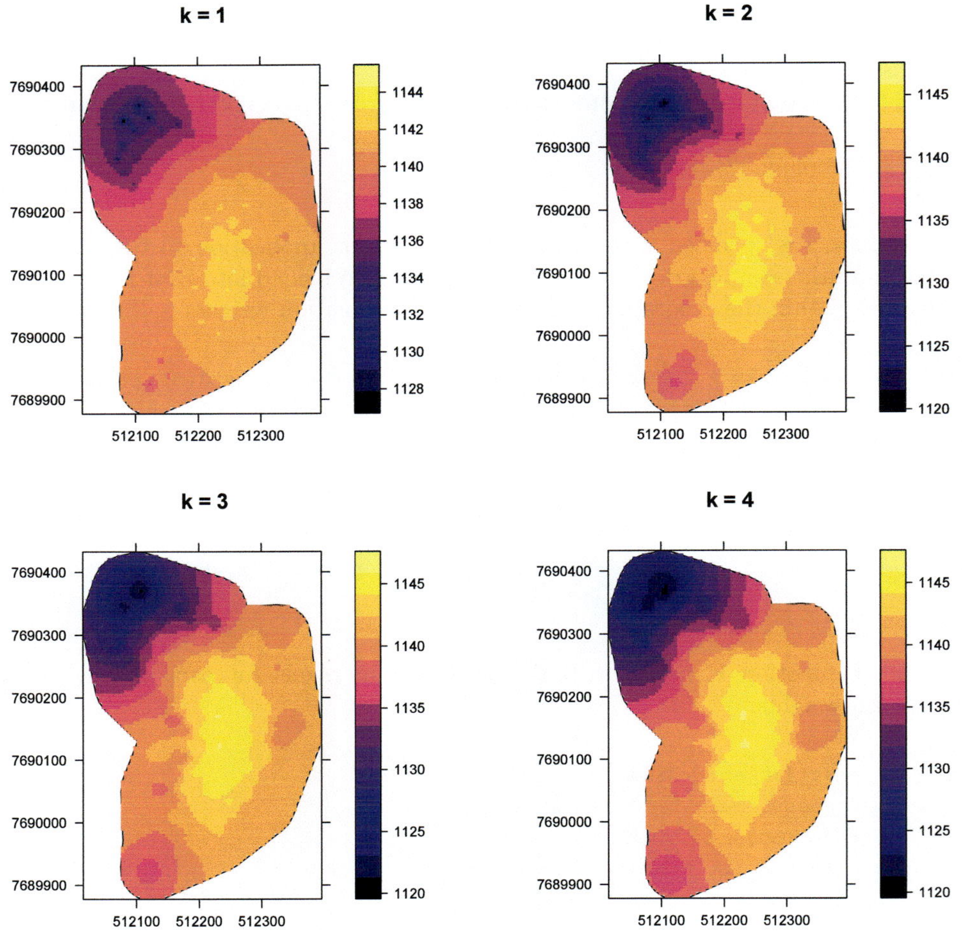

FIGURE 12.18 Altitude interpolation by the inverse distance weighting, obtained by surveying georeferenced points in a coffee field, with power values k = 1, 2, 3 and 4.

```
k1 <- as.matrix(pred1)
k2 <- as.matrix(pred2)
k3 <- as.matrix(pred3)
k4 <- as.matrix(pred4)
```

The results are mapped using the `persp` function (Figure 12.19).

```
par(mfrow = c(2, 2), mar = c(2, 2, 1.5, 1.5), mgp = c(1.5, 0.6, 0))
persp(k1)
persp(k2)
persp(k3)
persp(k4)
```

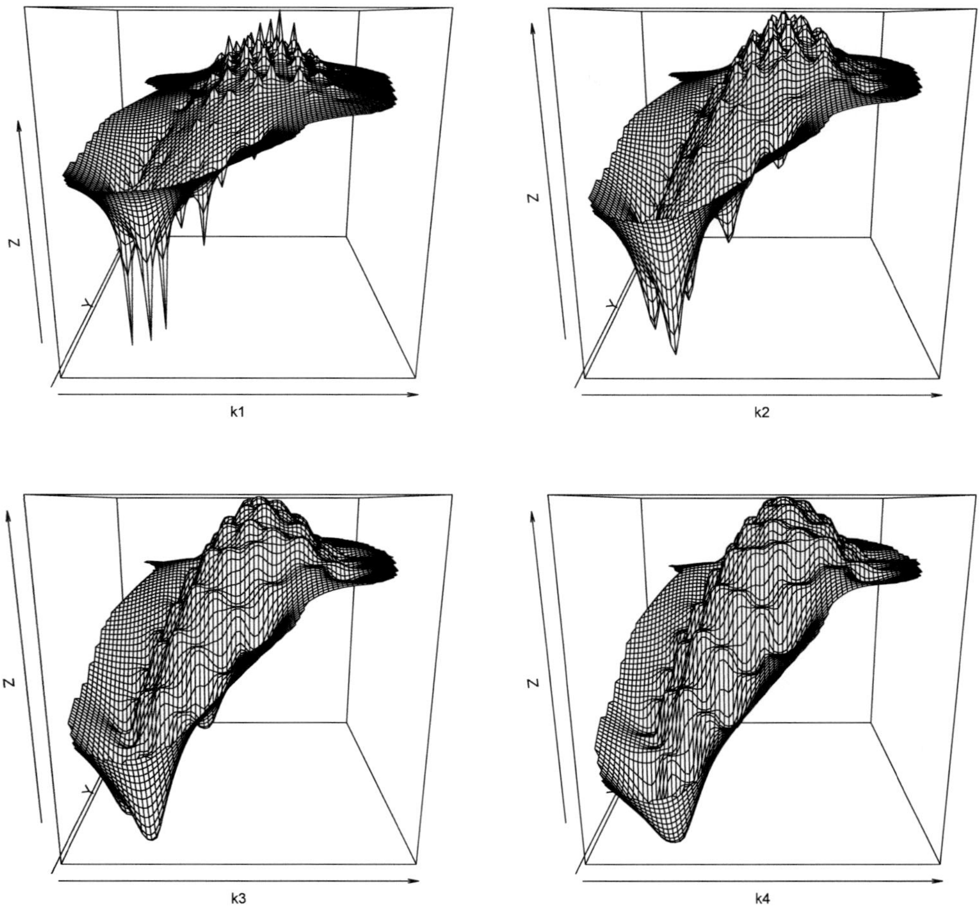

FIGURE 12.19 Three-dimensional representation of the altitude surface interpolated by inverse distance weighting with power values k = 1, 2, 3 and 4, at 5 m spatial resolution.

12.17 SPLINE INTERPOLATION

Splines are considered functions in which accurate prediction of point attributes can be performed by ensuring continuity of predicted values at unobserved locations between attribute points. Under conditions of natural attribute variation, artifacts can be produced in the mapping as a function of low or high attribute values, which can be removed using local average smoothing by thin plate splines. As in line interpolation, in spline methods applied to interpolation surfaces, an approximate function passes close to the observed points and can be used to smooth the attribute values, considering the existence of a random error (Burrough and McDonnell, 1998):

$$y(x_i) = z(x_i) + \epsilon(x_i) \tag{12.28}$$

where z is the attribute value of a point associated to a random error, ϵ.

The spline function $p(x)$, should not exhibit values too different from the original data, and it is necessary to use a smoothing function $A(f)$ with weights, w_i^2 obtained inversely proportional to the error variance (Burrough and McDonnell, 1998):

$$A(f) + \sum_{i=1}^{n} w_i^2 [f(x_i) - y(x_i)]^2 \tag{12.29}$$

where

$$w_i^2 = \frac{p}{Var[\epsilon(x_i)]} = \frac{p}{s_i^2} \tag{12.30}$$

where p determines the relative importance of the attribute characteristics smoothed by the spline.

The advantage of using splines is that univariate and multivariate interpolation of attributes can be performed quickly and with interpolation results very close to the originally measured attributes, with prediction errors associated with an approximate function. The disadvantages of using spline are related to the indirect estimation of errors associated with the interpolator and the possibility of performing a smoothed estimate of the reality of observed attributes from vector features of sample points in a given geographical region (Burrough and McDonnell, 1998).

The thin plate splines methodology is used to predict the altitude determined from sample points in a coffee field, at a spatial resolution of 5 m. The prediction uncertainties with respect to the smoothing function used are also mapped. The spline smoothing representation is compared with the IDW method ($k = 2$), and more smoothing of the altitude can be observed when interpolated by spline. The thin plate splines and IDW methods presented good results for altitude interpolation in a coffee crop; however, we should choose the best interpolation method according to cultural preferences and the surface smoothing limit tolerated in relation to reality.

The `fields` (Nychka et al., 2021) and `tmap` (Tennekes et al., 2020) packages are loaded into R to perform this example with thin plate spline interpolation.

```
library(fields)
# Loading required packages: spam
# Spam version 2.11-1 (2025-01-20) is loaded.
# Loading required packages: viridisLite
# package ‘fields’ was built under R version 4.3.3
# package ‘spam’ was built under R version 4.3.3
library(tmap)
# package ‘tmap’ was built under R version 4.3.3
```

The `matrix` function is used to create a matrix with the coordinates of points from the same sampling grid used in the previous example. The `Tps` function is used to perform thin plate spline fitting to the altitude data sampled in the area from the coordinates of each sampling point. The `summary` function is used to obtain information about the fit and uncertainty of the results. The smoothing parameter is obtained with the `print` function.

```
coords <- matrix(c(blight_sp$X, blight_sp$Y), byrow = FALSE, ncol = 2)
surf <- Tps(coords, blight_sp$Z)
# Grid searches over lambda (nugget and sill variances) with  minima at the
# endpoints:
#   (GCV) Generalized Cross-Validation
#    minimum at  right endpoint  lambda  =  1.517267e-05 (eff. df= 80.75 )
summary(surf)
# CALL: Tps(x = coords, Y = blight_sp$Z)
# Number of Observations:                85
# Number of unique points:               85
# Number of parameters in the null space 3
# Parameters for fixed spatial drift     3
# Effective degrees of freedom:          80.7
# Residual degrees of freedom:           4.3
# MLE tau                                0.1084
# GCV tau                                0.06771
# MLE sigma                              775
# Scale passed for covariance (sigma)    <NA>
# Scale passed for nugget (tau^2)        <NA>
# Smoothing parameter lambda             1.517e-05
#Residual Summary:
#       min       1st Q     median      3rd Q         max
#-0.0503900 -0.0063160  0.0009429  0.0092640  0.0511300
# Covariance Model: Rad.cov
#  Names of non-default covariance arguments:
#       p
# DETAILS ON SMOOTHING PARAMETER:
# Method used:   GCV    Cost:  1
#   lambda        trA        GCV   GCV.one GCV.model    tauHat
# 1.517e-05 8.075e+01 9.169e-02 9.169e-02        NA 6.771e-02
# Summary of all estimates found for lambda
#               lambda   trA     GCV  tauHat -lnLike Prof converge
# GCV        1.517e-05 80.75 0.09169 0.06771        84.09       NA
# GCV.model         NA    NA      NA      NA           NA       NA
# GCV.one    1.517e-05 80.75 0.09169 0.06771           NA       NA
# RMSE              NA    NA      NA      NA           NA       NA
# pure error        NA    NA      NA      NA           NA       NA
# REML       1.517e-05 80.75 0.09169 0.06771        84.09       NA
print(surf$lambda)
# [1] 1.517267e-05
```

The prediction grid used in the previous example is converted to `RasterLayer` class with the `raster` function.

```
gr <- raster(g)
```

The `interpolate` function is used to perform thin plate spline interpolation on the prediction grid. The `mask` function is used to mask the results within the boundaries of the coffee field.

```
ps <- interpolate(gr, surf)
psmask <- mask(ps, bord_sp)
```

The interpolate function is also used to predict the standard deviation of the altitude prediction in the analyzed region.

```
pse <- interpolate(gr, surf, fun = predictSE)
psemask <- mask(pse, bord_sp)
```

The results are mapped using functions from the tmap package (Figure 12.20).

```
s1 <- tm_shape(bord_sp) + tm_polygons() + tm_shape(blight_sp) +
  tm_dots(col="Z", palette = "Greys",
  title="Altitude (m)", size=0.7) + tm_legend(legend.outside=TRUE)
s2 <- tm_shape(ps) + tm_raster(n = 10,palette = "Greys",
  title="Predicted altitude (m)") + tm_shape(bord_sp) + tm_polygons(alpha = 0) +
  tm_legend(legend.outside = TRUE)
s3 <- tm_shape(bord_sp) + tm_polygons() + tm_shape(psmask) +
  tm_raster(n = 10, palette = "Greys",
  title="Altitude mask") + tm_legend(legend.outside = TRUE)
s4 <- tm_shape(bord_sp) + tm_polygons() + tm_shape(psemask) +
  tm_raster(n = 10, palette = "Greys",
  title = "Standard deviation") + tm_legend(legend.outside = TRUE)
tmap_arrange(s1, s2, s3, s4, ncol = 2)
```

The as.matrix function is used to convert the results produced into a matrix in order to compare the thin plate spline prediction with the IDW prediction ($k = 2$).

```
spline <- as.matrix(psmask)
idw_k2 <- as.matrix(pred2)
```

The persp function is used to perform three-dimensional mapping of the results (Figure 12.21).

```
par(mfrow = c(1, 2), mar = c(2, 2, 1.5, 1.5), mgp = c(1.5, 0.6, 0))
persp(spline)
persp(idw_k2)
```

12.18 DIGITAL ELEVATION MODEL INTERPOLATION APPLICATIONS

Unlike other types of quantitative data, terrain elevation is an easy and inexpensive type of data to measure. The types of measurements are not only topographic measurements in the field with simple equipment like tape measure, level and theodolite, but also indirect electronic measurements with GNSS, digital photogrammetry, digital terrestrial and aerial laser scanning, Radar data, in different resolutions, like those of the Shuttle Radar Topography Mission (SRTM).

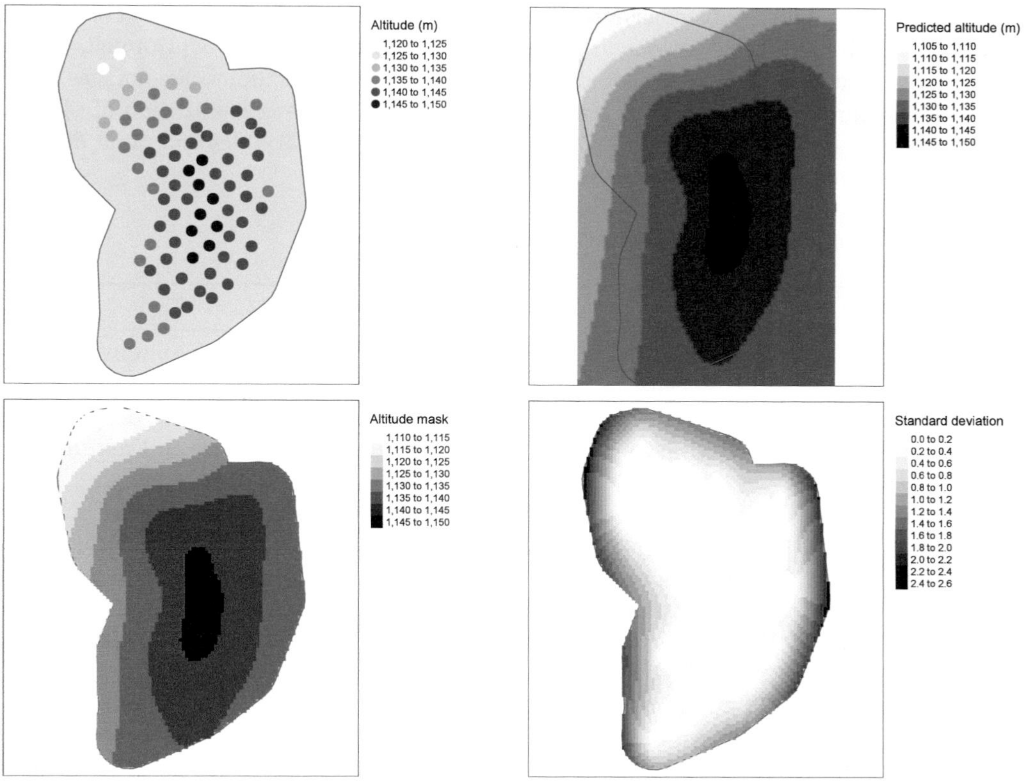

FIGURE 12.20 Altitude prediction from points (top-left) by thin plate spline (top-right), predicted altitude mask (bottom-left) and standard deviation of the prediction (bottom-right), at 5 m spatial resolution.

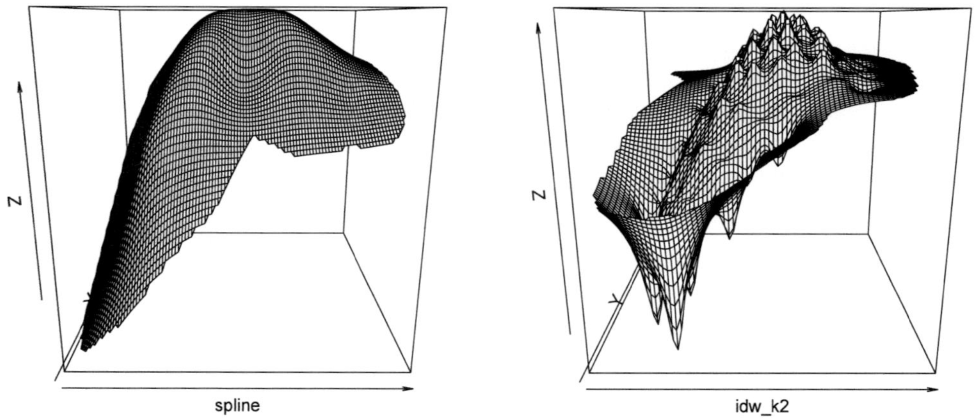

FIGURE 12.21 3D comparison of altitude prediction from points by thin plate spline (left) and by inverse distance weighting (right), in coffee crop in Santo Antônio do Amparo, at 5m spatial resolution.

A digital elevation model can be considered as a special type of continuous surface created by interpolation, with many applications in geocomputation (Burrough and McDonnell, 1998). Digital elevation models are used not only to generate stereo orthomosaic images, but also to perform three-dimensional representations of the landscape, visualization of terrain geomorphology, and modeling as a co-variable for meteorological variables and plant disease epidemiology, among other traditional uses (Table 12.2).

TABLE 12.2 Traditional digital elevation model applications.

Application	Traditional use
Storage	Elevation data for digital topographic maps in national database.
Creation	Data from digital orthomosaic and orthophotos.
Cut and fill	Design and execution of civil and military engineering projects.
3-D visualization	Three-dimensional landscape visualization for civil and military planning purposes.
Field of view	Determination of terrain view with `viewshed` analysis.
Planning	Route planning, utility location, among others.
Statistical analysis	Comparison of different terrain types with digital elevation model.
Data source	Variable determination: aspect, profile curvature, relief shadow, insolation, hydrological and ecological modeling.
Frame	Can be used as a background for presenting thematic information in mapping combining relief data with thematic data on water, soil, vegetation, and land use.
Template	Can be used as a template for geocomputation operations with raster data with the same size and spatial resolution characteristics.
Modeling	Provide data for simulation models and models of processes that occur in the landscape.

The thin plate spline technique is used to perform a smoothed prediction of coffee bacterial blight epidemic by exploring the relationship of the disease with the altitude covariate, used as a linear predictor of the disease incidence, at 5 m spatial resolution. Furthermore, the standard deviation of the disease incidence prediction is determined, in order to assess the uncertainty about the epidemic in the terrain, according to the variation of the relief. In the region located by the delimitation of the sample points, the error in determining the interpolated disease incidence ranged from $\pm 5 - 10\%$. In regions with little neighborhood density, as in the borders delimited by a buffer around the crop, the maximum error was $\pm 45\%$ (Figure 12.22).

The `fields`, `raster`, `sp`, `sf`, and `gridExtra` packages loaded in previous examples and the thin plate spline predicted altitude raster in a coffee field are used as examples for multivariate thin plate spline interpolation.

The `matrix` and `Tps` functions are used to convert the sample grid coordinate data into a matrix and perform the spline fit for multivariate interpolation, respectively. The results of the model fit are verified with the `summary` function.

```
coord <- matrix(c(blight_sp$X, blight_sp$Y), byrow = FALSE, ncol = 2) # Coords
surf1 <- Tps(coord, blight_sp$i280, Z = blight_sp$Z) # Fit spline model
```

```
summary(surf1) # See fit results
# CALL: Tps(x = coord, Y = blight_sp$i280, Z = blight_sp$Z)
# Number of Observations:                85
# Number of unique points:               85
# Number of parameters in the null space: 4
# Parameters for fixed spatial drift:     3
# Effective degrees of freedom:          29.4
# Residual degrees of freedom:           55.6
# MLE tau                                14.9
# GCV tau                                15.42
# MLE sigma                              170800
# Scale passed for covariance (sigma)    <NA>
# Scale passed for nugget (tau^2)        <NA>
# Smoothing parameter lambda             0.0013
# Residual Summary:
#     min    1st Q   median    3rd Q      max
#-35.6000  -3.8230  -0.1967   5.3700  40.7800
# Covariance Model: Rad.cov
#  Names of non-default covariance arguments:
#       p
# DETAILS ON SMOOTHING PARAMETER:
# Method used:   GCV    Cost:  1
#   lambda        trA        GCV    GCV.one GCV.model     tauHat
#   0.0013    29.3870   363.5690   363.5690        NA    15.4231
# Summary of all estimates found for lambda
#              lambda   trA   GCV tauHat -lnLike Prof converge
# GCV        0.001300 29.39 363.6  15.42        358.8        6
# GCV.model        NA    NA    NA     NA           NA       NA
# GCV.one    0.001300 29.39 363.6  15.42           NA        6
# RMSE             NA    NA    NA     NA           NA       NA
# pure error       NA    NA    NA     NA           NA       NA
# REML       0.001642 26.81 364.3  15.79        358.8        5
```

A function is created to perform the prediction of the fitted model on the raster.

```
pfun <- function(model, x, ...) {predict(model, x[,1:2], Z = x[,3], ...)}
```

The `interpolate` function is used to interpolate the altitude raster determined by thin plate spline in the previous example, but now with the aim of interpolating the incidence of coffee bacterial blight in the field.

```
ps1 <- interpolate(psmask$layer, surf1, xyOnly = FALSE, fun = pfun)
```

A function is created to predict the standard deviation of the interpolation of disease incidence in the field.

```
pfunSE <- function(model, x, ...) {predictSE(model, x[,1:2], Z=x[,3], ...)}
```

The `interpolate` function is used for spatial prediction of the standard deviation of disease incidence.

```
psel <- interpolate(psmask$layer, surf1, xyOnly = FALSE, fun = pfunSE)
```

The results are mapped using the `spplot` and `grid.arrange` functions. Note that the multivariate prediction determined a negative trend in disease prediction and a very high error in regions close to the coffee field border (Figure 12.22).

```
p1 <- spplot(blight_sp, "i280", main = "i280", sp.layout = bord_sp,
             scales = list(draw = TRUE),  key.space = "right",
             xlim = c(512013.2, 512395.5), ylim = c(7689875, 7690432))
p2 <- spplot(psmask, "layer", main = "Spline altitude (m)", sp.layout = bord_sp,
             scales = list(draw = TRUE))
p3 <- spplot(ps1, "layer", main = "Predicted disease incidence (%)",
             sp.layout = bord_sp, scales = list(draw = TRUE))
p4 <- spplot(psel, "layer", main = "Standard deviation (%)", sp.layout = bord_sp,
             scales = list(draw = TRUE))
grid.arrange(p1, p2, p3, p4, ncol = 2)
```

The thin plate spline technique is also used to perform a smoothed prediction of annual mean air temperature (°C) of Minas Gerais, based on data of longitude (°), latitude (°) and altitude (m) of main climatological stations of INMET, used in the measurement of data referring to the climatological normals ($1960 - 1990$). The objective of the analysis is to perform the downscaling of air temperature at sparse climatological stations throughout the state of Minas Gerais, from SRTM data at the spatial resolution of 0.008333333°. The multivariate spline function fitting is fast and simple to use, determining a predicted temperature map on the SRTM digital elevation model data. The standard deviation of the prediction is also determined relative to the SRTM elevation model used in the temperature prediction. The smallest temperature prediction errors ($< \pm 0.1^\circ$) occurred mainly in the central region of the state, while larger errors, around $\pm 0.5^\circ$ occurred west of the Triangulo Mineiro and in other sparse regions on the edges of the state and in some central regions of the state.

A similar example to the previous one is used for spatial prediction of mean air temperature using the `matrix` and `Tps` functions.

```
coord <- matrix(c(mgAT1$X, mgAT1$Y), byrow = FALSE, ncol = 2)
surf2 <- Tps(coord, mgAT1$T, Z = mgAT1$Z)
summary(surf2)
# CALL: Tps(x = coord, Y = mgAT1$T, Z = mgAT1$Z)
# Number of Observations:                  32
# Number of unique points:                 32
# Number of parameters in the null space: 4
# Parameters for fixed spatial drift:      3
# Effective degrees of freedom:            4
# Residual degrees of freedom:             28
# MLE tau                                  0.521
# GCV tau                                  0.557
# MLE sigma                                0.002338
```

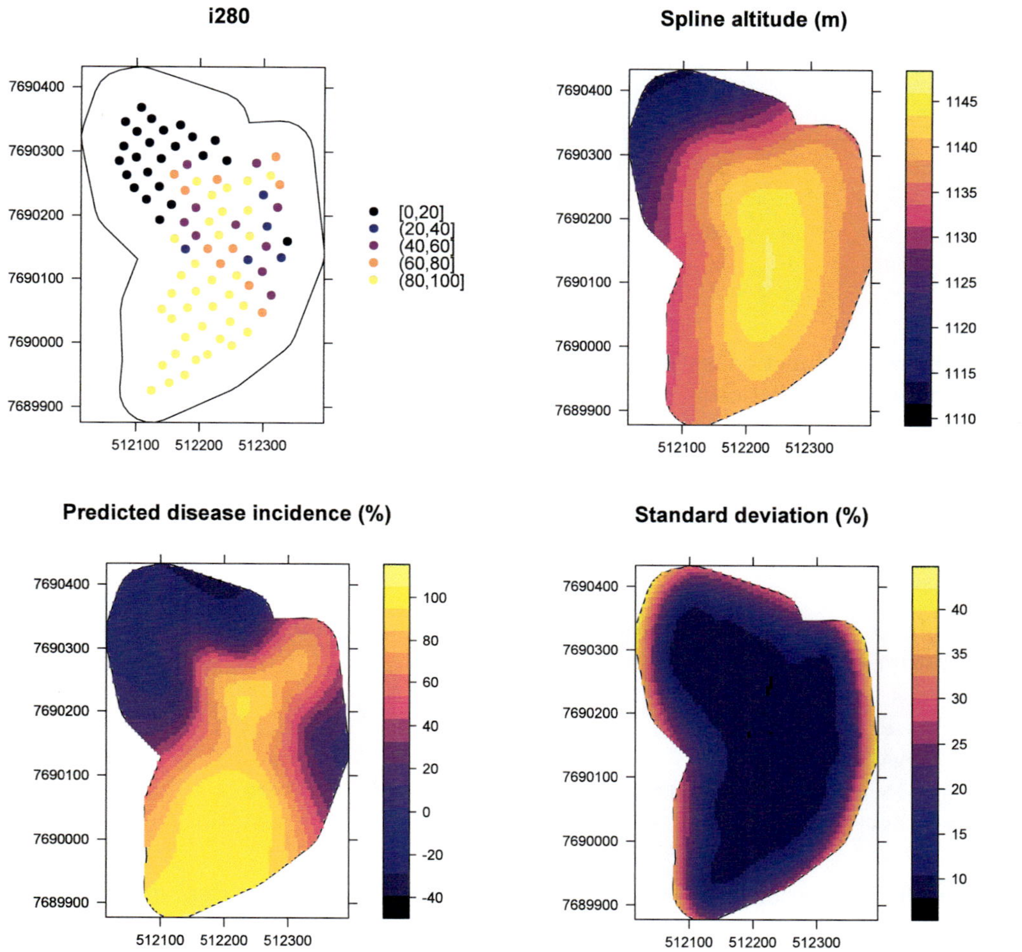

FIGURE 12.22 Epidemic prediction of coffee bacterial blight (top-left) by thin plate spline using an altitude covariate as linear predictor (top-right), predicted disease incidence (bottom-left) and standard deviation of the prediction (bottom-right), at 5 m spatial resolution.

```
# Scale passed for covariance (sigma)     <NA>
# Scale passed for nugget (tau^2)         <NA>
# Smoothing parameter lambda              116.1
# Residual Summary:
#    min    1st Q   median    3rd Q      max
# -1.00000 -0.37030  0.08312  0.26540  1.13000
# Covariance Model: Rad.cov
#  Names of non-default covariance arguments:
#       p
# DETAILS ON SMOOTHING PARAMETER:
# Method used:   GCV    Cost:  1
#   lambda        trA       GCV   GCV.one GCV.model    tauHat
```

```
# 116.1121     4.0010     0.3546     0.3546          NA     0.5570
# Summary of all estimates found for lambda
#             lambda    trA     GCV tauHat -lnLike Prof converge
# GCV           116.1 4.001 0.3546  0.557          23.34        NA
# GCV.model        NA    NA     NA     NA             NA        NA
# GCV.one       116.1 4.001 0.3546  0.557             NA        NA
# RMSE             NA    NA     NA     NA             NA        NA
# pure error       NA    NA     NA     NA             NA        NA
# REML          116.1 4.001 0.3546  0.557          23.34        NA
```

The `interpolate` function is used to perform spatial prediction of temperature and standard deviation of the prediction.

```
ps2 <- interpolate(srtmMG$Altitude, surf2, xyOnly = FALSE,
                   fun = pfun) # Predict temp.
pse2 <- interpolate(srtmMG$Altitude, surf2, xyOnly = FALSE,
                    fun = pfunSE)  # Predict sd
```

A geo-object of type `SpatialPointsDataFrame` is created for mapping points with the temperature attribute.

```
coords <- mgAT1[,c("X", "Y")]
temp <- mgAT1[,c("T")]
t_spdf <- SpatialPointsDataFrame(coords = coords, data = temp)
proj4string(t_spdf) <- CRS("+proj=longlat +datum=WGS84 +no_defs")
```

The results of spatial temperature prediction using multivariate thin plate splines are displayed using the `spplot` and `grid.arrange` functions (Figure 12.23).

```
a1 <- spplot(t_spdf, "T", main = "Temperature (Degrees Celsius)", sp.layout = mg,
             scales = list(draw = TRUE),  key.space = "bottom",
             xlim = c(-51.04588, -39.85676), ylim = c(-22.92275, -14.23343))
a2 <- spplot(srtmMG, "Altitude", main = "SRTM altitude (m)", sp.layout = mg,
             scales = list(draw = TRUE), colorkey = list(space = "bottom"))
a3 <- spplot(ps2, sp.layout = mg, main = "Predicted temperature (Degrees Celsius)",
             scales = list(draw = TRUE))
a4 <- spplot(pse2, sp.layout = mg, main = "Standard deviation (Degrees Celsius)",
             scales = list(draw = TRUE))
grid.arrange(a1, a2, a3, a4, ncol = 2)
```

The interpolation of altitude of a coffee field sample grid, in Santo Antônio do Amparo, Minas Gerais, Brazil, is used to generate a digital elevation model of the field, based on the results of thin plate splines and IDW (k = 2). To compare the results obtained by both methods, a `hillShade` map is determined and overlayed on the altitude raster with contour lines, in order to explore the perception of the relief of the coffee crop, at 5 m spatial resolution. With the use of thin plate splines, besides the low interpolation errors determined by the method in the areas close to the sampling sites (<0.4 m), the relief smoothing determined better perception of the terrain than the IDW interpolation. With the use of interpolation by the inverse distance weighting, there was formation of some

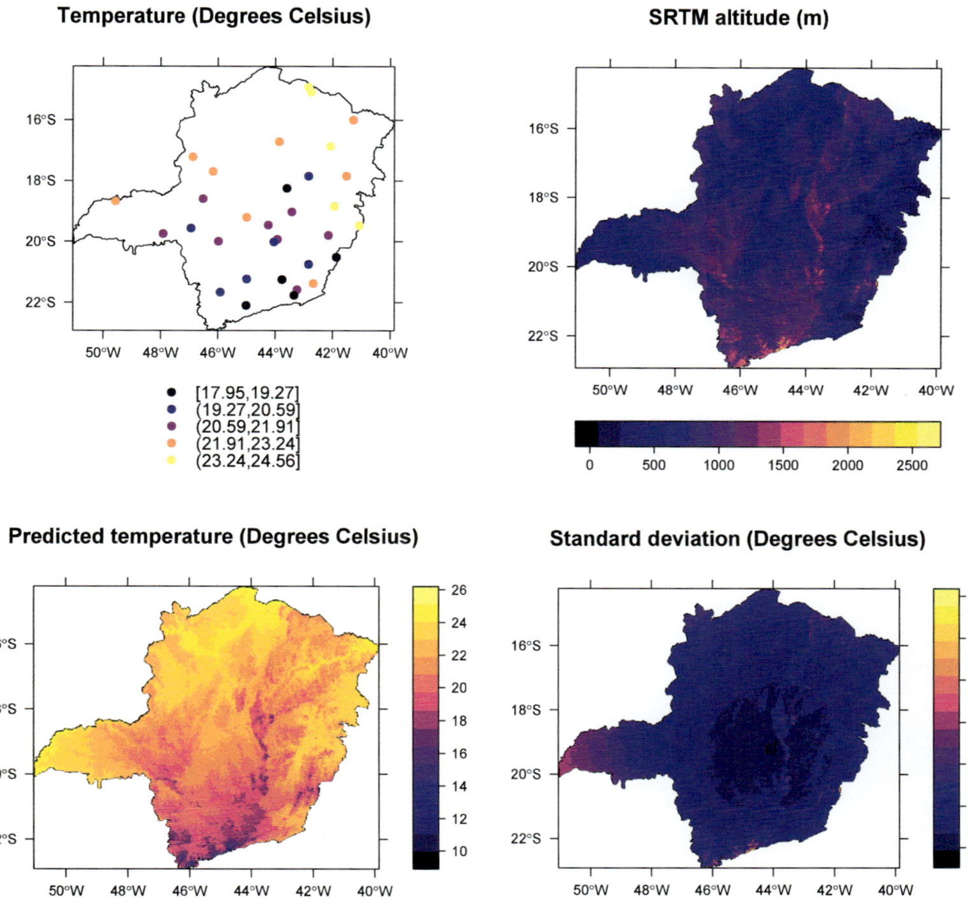

FIGURE 12.23 Prediction of air temperature in Minas Gerais by performing a downscaling of sparse point station data against the SRTM digital elevation model. The uncertainty of the prediction is determined to evaluate the uncertainty of multivariate temperature modeling by thin plate splines.

regions with `duck egg` shape that should be evaluated with more caution in case of adopting this digital elevation model for precision agriculture purposes.

The georeferenced altitude data in the coffee field interpolated by thin plate splines and IDW (k = 2) are mapped using hillshade and contour line geovisualization techniques in order to comparatively evaluate the interpolation methods used with the following chunk code (Figure 12.24).

```
par(mfrow = c(1, 2), mar = c(2, 2, 0.3, 0.3), mgp = c(1.5, 0.6, 0))
# Thin plate splines
## create hillshade
hs <- hillShade(slope = terrain(psmask, "slope"),
                aspect = terrain(psmask, "aspect"))
```

```
plot(hs, col = gray(0:100 / 100), legend = FALSE)
## overlay with DEM
plot(psmask, col = terrain.colors(25), alpha = 0.5, legend = FALSE, add = TRUE)
plot(bord_sp, col = NA, add = TRUE) # Plot limits
## add contour lines
contour(psmask, col = "white", add = TRUE, labcex=0.8)
# IDW
## Convert SpatialPixelsDataFrame in raster
predidw <- raster(pred2)
## create hillshade
hs1 <- hillShade(slope = terrain(predidw, "slope"),
                 aspect = terrain(predidw, "aspect"))
plot(hs1, col = gray(0:100 / 100), legend = FALSE)
## overlay with DEM
plot(predidw, col = terrain.colors(25), alpha = 0.5, legend = FALSE, add = TRUE)
plot(bord_sp, col = NA, add = TRUE) # Plot limits
## add contour lines
contour(predidw, col = "white", add = TRUE, labcex=0.8)
```

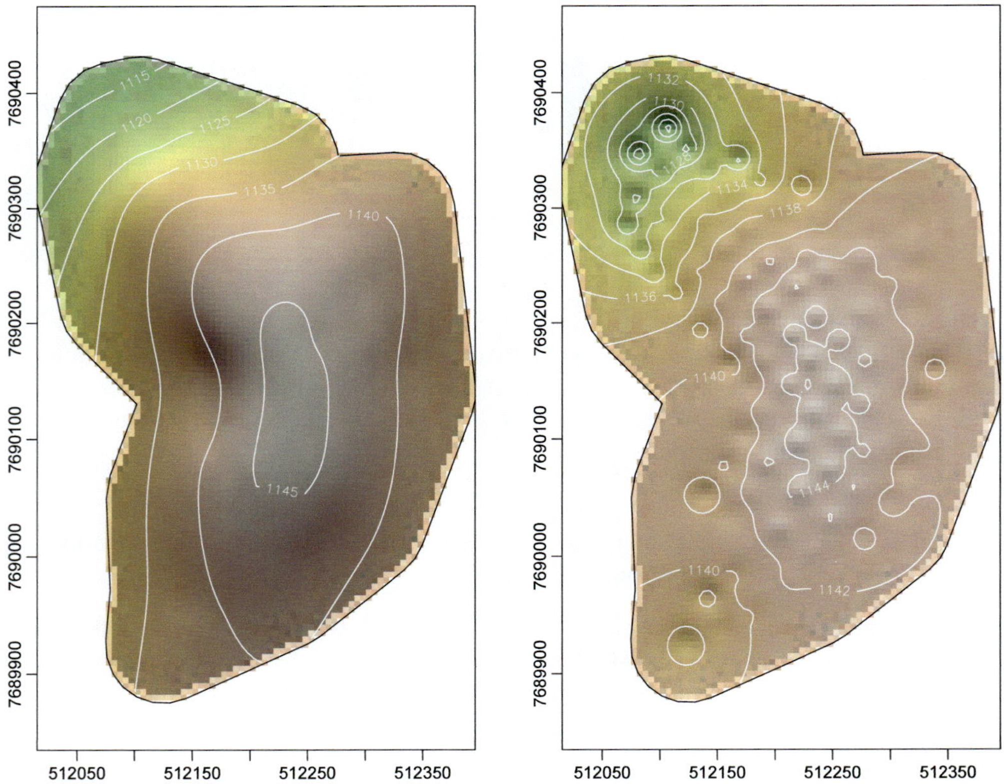

FIGURE 12.24 Comparison of a hillshade map overlaid on an altitude raster with contour lines to explore the perception of relief from coffee farming, in Santo Antônio do Amparo, Minas Gerais. The digital elevation model is created by thin plate spline interpolation (left) and inverse distance weighting (k = 2) (right), at 5 m spatial resolution.

12.18.1 DIGITAL ELEVATION MODEL REPRESENTATION WITH TRIANGULATED IRREGULAR NETWORK

A `Triangulated Irregular Network` (TIN) is a continuous digital terrain model with facets of triangles connected based on Delaunay triangulation from observations of points spaced on an irregular mesh. With the use of TIN, it is possible to store accurate elevation data effectively, as well as perform some types of spatial analysis based on geometric and topological surface variation. With the use of TIN structure, it is possible to determine slope maps, shaded relief and contour maps, block diagrams and line-of-sight maps, among others. Data stored in TIN can be transformed into raster using a reference grid and linear interpolation to determine height values as pixel brightness information (Burrough and McDonnell, 1998).

In the `lawn` package (Chamberlain and Hollister, 2021) there is an approach for generating and mapping by TIN triangulation in a given geographical region in order to return as output a collection of TIN polygons in `featurecollection` class with a given number of features and points.

The `lawn` package v.0.6.0 is loaded into R for analysis using the `library` function.

```
library(lawn)
```

The `lawn_tin` and `lawn_random` functions are used to perform TIN triangulation from geodetic coordinates defined in a region of Brazil. The results are mapped with the `view` function (Figure 12.25).

```
a <- lawn_tin(lawn_random(bbox = c(-55, -26, -36, -12))) # Perform TIN
view(a) # Map
class(a) # See class
# [1] "featurecollection"
a # See header
# <FeatureCollection>
#   Bounding box: -53.7 -24.0 -36.4 -15.4
#   No. features: 13
#   No. points: 104
#   Properties: NULL
```

12.18.2 GEOMETRIC CORRECTION OF RASTER DATA

In image data obtained by aerial or orbital survey, there are distortions due to the variation of the relief at the imaged location. This type of distortion can be corrected or correctly registered in a georeferenced database with geometric accuracy. Convolution methods can be used to correct the geometry of a given surface and modify resolution, pixel size, and orientation characteristics of the original data. For non-coplanar situations, distortion of an image can be performed in several polynomial orders in order to obtain a satisfactory fit of an image to a curved surface of the Earth, in the case of orbital survey, using a conventional projection such as Universal Transverse Mercator. For coplanar situations, the first-order distortion equations for performing translation, rotation and scaling operations are (Burrough and McDonnell, 1998):

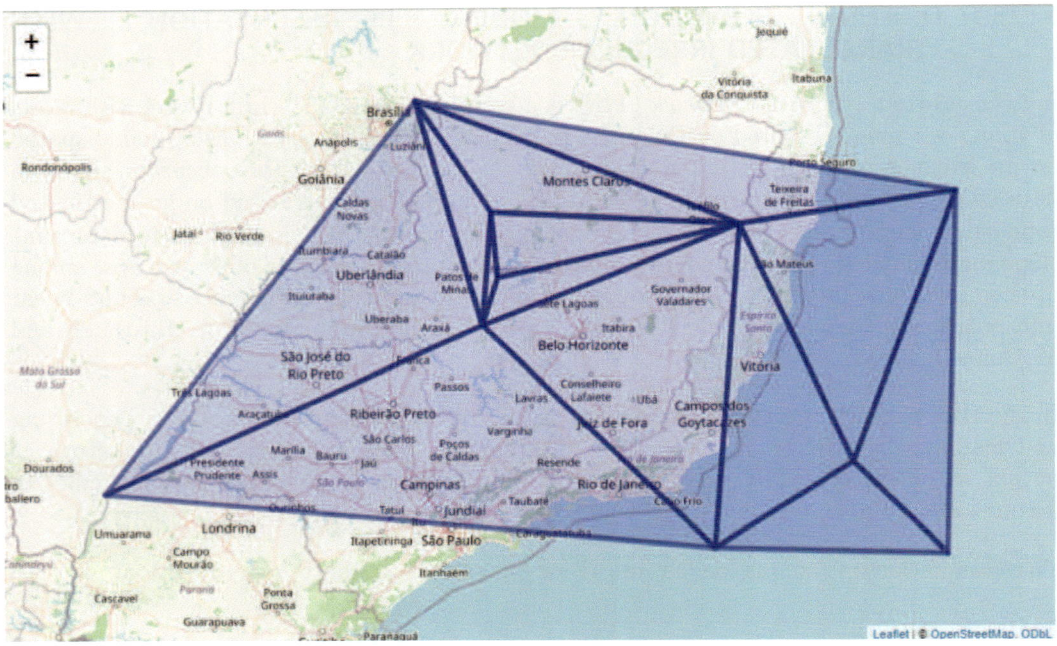

FIGURE 12.25 Random triangulation of points to generate featurecollection class TIN polygons by the lawn package in R.

$$u = a_0 + a_1 a_2 x + a_1 a_3 y \tag{12.31}$$

$$v = b_0 + b_1 b_2 x + b_1 b_3 y \tag{12.32}$$

where x, y are the pixels with original coordinates, u, v, the new coordinates, a_0, b_0, the translation parameters, a_1, b_1, the x, y scale values, respectively. The parameters a_2, b_2, a_3, b_3 are dependent on the angle of rotation, θ:

$$\begin{pmatrix} a_2 = cos\theta & b_2 = -sen\theta \\ a_3 = sen\theta & b_3 = cos\theta \end{pmatrix} \tag{12.33}$$

When interpolating the pixel value on the surface with geometric distortion, nearest neighbor interpolation by the bilinear method can be used to determine a new value as a function of four neighboring input pixels around the output pixel. In some situations it may be necessary to use cubic convolution, with a neighborhood weighting of 16 pixels and two-dimensional function $x(senx/x)$.

12.18.3 IMAGE ORTHORECTIFICATION

Digital photogrammetry has been used to obtain geometrically corrected terrain and land use images, often in high detail, for use in digital management and environmental applications. The geometric correction of the image can be corrected using reliable topographic survey data to ensure scale and map projection preservation, generating orthorectified image products.

The thin plate splines methodology is used to perform the distortion of a mesh into another configuration using landmarks with known coordinate values on the surface (Schlager et al., 2021), being necessary to evaluate which methodology would be of best interest to use in each specific situation.

12.19 COMPUTATIONAL PRACTICE

Several packages in R and other geocomputation software can be used to perform the interpolation methods. Some packages and functions are listed as options for use, with different degrees of difficulty, types of data models, and different ways of applying the same theoretical analysis methodology. The R packages and functions used in various point pattern analysis and geospatial interpolation applications are described below (Table 12.3).

TABLE 12.3 R packages and functions used in different applications of point data pattern analysis and geospatial interpolation.

Package	Function	Description
sf[2]	st_triangulate, st_voronoi, st_centroid	Perform triangulation on a set of points, create Voronoi tessellation, determine the centroid of a geometry
moments[3]	skewness, kurtosis	Determining skewness, kurtosis
geoR[4]	trend.spatial	Trend matrix determination
stats[5]	lm	Determination of analysis of variance, regression, and fitting linear and multiple regression models
pycno[6]	pycno	Perform Tobler's pycnophylactic interpolation
gstat[7]	gstat	Perform IDW and kriging interpolation
phylin[8]	idw	Perform IDW interpolation
raster[9]	interpolate	Perform IDW prediction and kriging with rasters
fields[10]	Tps	Interpolation by thin plate splines
morpho[11]	warp.mesh	Using thin plate splines for control point-based mesh distortion
geometry[12]	delaunayn, distmesh2d	Perform Delaunay interpolation in N dimensions and mesh creation

[2]https://r-spatial.github.io/sf/reference/geos_unary.html

[3]https://cran.r-project.org/web/packages/moments/index.html

[4]https://www.rdocumentation.org/packages/geoR/versions/1.8-1/topics/trend.spatial

[5]https://www.rdocumentation.org/packages/stats/versions/3.6.2/topics/lm

[6]https://cran.r-project.org/web/packages/pycno/index.html

[7]https://cran.r-project.org/web/packages/gstat/gstat.pdf

[8]https://www.rdocumentation.org/packages/phylin/versions/2.0.2/topics/idw

[9]https://www.rdocumentation.org/packages/raster/versions/3.4-5/topics/interpolate

[10]https://rdrr.io/cran/fields/man/Tps.html

[11]https://www.rdocumentation.org/packages/Morpho/versions/2.2/topics/warp.mesh

[12]https://cran.r-project.org/web/packages/geometry/index.html

Package	Function	Description
spatstat[13]	ppp, delaunay, dirichlet, idw	Point pattern analysis, statistical analysis, data interpolation
rgl[14]	persp3d	Mapping surfaces in 3-D
lawn[15]	lawn_tin	Geospatial analysis with GeoJSON data objects

Using the R package `spatstat` (Baddeley and Turner, 2005; Baddeley et al., 2013, 2015), it is possible to perform a number of statistical analyses and applications of interpolation methods, focusing on point pattern analysis, model fitting and simulations. We decided to perform a computational practice using the `spatstat` and `lawn` (Chamberlain and Hollister, 2021) packages.

The `install.packages` function is used to install the packages needed to perform the hands-on computational activity. If the packages are already installed, it is not necessary to install the packages again for each computational practice.

```
install.packages("spatstat")
install.packages("lawn")
```

The `library` function is used to enable the packages needed to perform the tasks.

```
library(spatstat)
# Loading required packages: spatstat.data
# Loading required packages: spatstat.univar
# spatstat.univar 3.1-4
# Loading required packages: spatstat.geom
# spatstat.geom 3.6-0
# Loading required packages: spatstat.random
# spatstat.random 3.4-2
# Loading required packages: spatstat.explore
# Loading required packages: nlme
# spatstat.explore 3.5-3
# Loading required packages: spatstat.model
# Loading required packages: rpart
# spatstat.model 3.4-2
# Loading required packages: spatstat.linnet
# spatstat.linnet 3.3-2
# spatstat 3.4-1
library(lawn)
```

12.19.1 CREATE POINT PATTERN

The `runif` function is used to create a point pattern from a set of random coordinates, x, y.

[13]https://cran.r-project.org/web/packages/spatstat/index.html

[14]https://cran.r-project.org/web/packages/rgl/index.html

[15]https://cran.microsoft.com/snapshot/2020-04-20/web/packages/lawn/index.html

```
x <- runif(20)
y <- runif(20)
```

Next, the ppp function is used to create a pattern of points along the two-dimensional space, x, y.

```
X <- ppp(x, y)
```

The point pattern is mapped with the plot function (Figure 12.26).

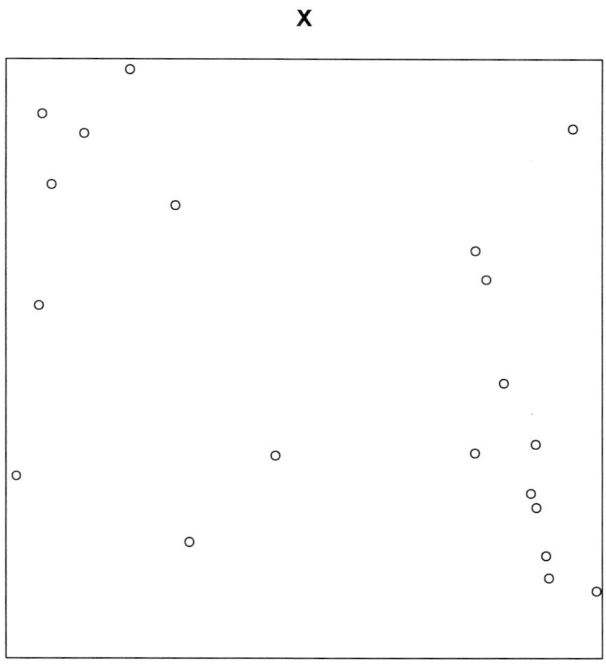

FIGURE 12.26 Spatial pattern of random points created with the ppp function from the spatstat package.

12.19.2 PERFORM DELAUNAY AND DIRICHLET INTERPOLATION

From the defined point pattern, comparison maps are performed with the plot function of the Delaunay triangulation, compared to the Dirichlet or Voronoi polygons, determined by the delaunay and dirichlet functions, respectively (Figure 12.27).

```
par(mfrow = c(1, 2), mar = c(2, 2, 1.5, 1.5), mgp = c(1.5, 0.6, 0))
plot(delaunay(X))
plot(X, add = TRUE)
plot(dirichlet(X))
plot(X, add = TRUE)
```

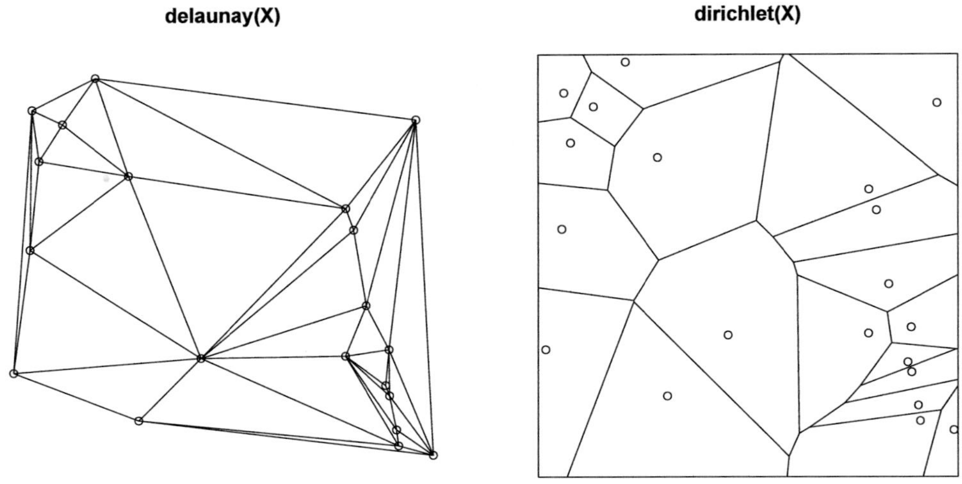

FIGURE 12.27 Delaunay triangulation (left) and Dirichlet polygons (right) with the points used to determine the surfaces.

12.19.3 PERFORMING TIN TRIANGULATION

An approach for generating and mapping by TIN triangulation in a given geographical region is performed in order to return GeoJSON objects as output when creating a collection of TIN random polygons of class `featurecollection` with a given number of features and points around Minas Gerais, Brazil (Figure 12.28).

The `lawn_tin` and `lawn_random` functions are used to create random polygons in Minas Gerais. The `view` function is used to map the polygons and points in the region of interest.

```
tin <- lawn_tin(lawn_random(bbox = c(-55, -26, -36, -12))) # Create
view(tin) # View
tin
# <FeatureCollection>
#   Bounding box: -53.7 -25.1 -39.9 -12.1
#   No. features: 12
#   No. points: 96
#   Properties: NULL
```

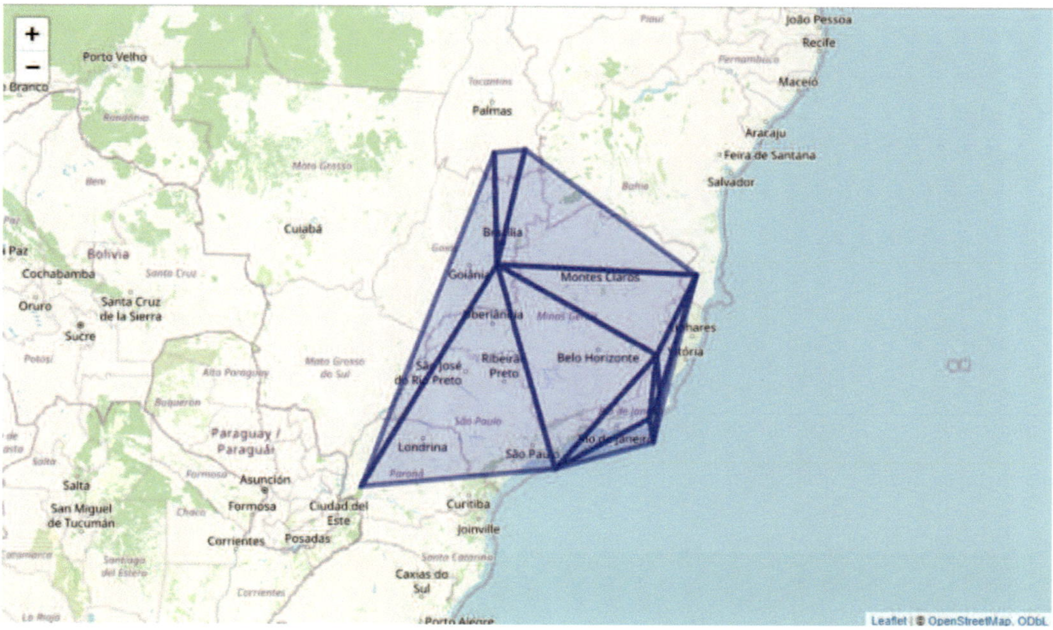

FIGURE 12.28 Random triangulation of points to generate TIN polygons.

13 Geostatistical Analysis with Environmental Geocomputation and R

13.1 LEARNING QUESTIONS

The learning questions answered by reading the chapter are:

- How was the kriging geostatistical interpolation method created and what is its principle?
- How can the prediction and uncertainty of kriging be useful in decision-making?
- How can sampling be optimized using variograms?
- What are the different types of variogram and kriging models created?
- In which geostatistical software has variogram modeling and kriging interpolation been implemented?
- How can an interpolation method be chosen based on the advantages, disadvantages and characteristics of using different methodologies, data and available data models?

13.2 LEARNING OUTCOMES

The learning outcomes answered by reading the chapter are:

- Understand the principle of the geostatistical method of interpolation by kriging;
- Check how to predict and determine the uncertainty of kriging;
- Check how interpolation by kriging can be useful in decision-making;
- Check how to optimize sampling using variograms;
- Evaluate the different types of variogram and kriging models used;
- Learn about some software in which geostatistics has been implemented to model variograms and interpolation by kriging;
- Check how to choose an interpolation method based on the advantages, disadvantages and characteristics of using different methodologies, data and available data models.

13.3 INTRODUCTION

The term geostatistics was etymologically considered to be the statistical study of natural phenomena (Matheron, 1962; Journel and Huijbregts, 1991). Geostatistics has been used as an alternative to describe the spatial continuity of geographic data of natural phenomena, as well as providing options for adapting modeling with classical regression techniques to better exploit the spatial continuity of attributes (Isaaks and Srivastava, 1989). Geostatistics can be used in agriculture to detail the amount of application of a given input to the soil from detailed knowledge of the distribution of a particular chemical element, or to estimate probabilities of a given attribute value exceeding a specified threshold to avoid yield losses (Webster and Oliver, 2007).

DOI: 10.1201/9781003724377-13

Optimal interpolation methods have been used in mining industry applications, water, soil, vegetation, meteorology, the study of relief and demographic indicators, in order to provide detailed information for mapping. With the use of geostatistics, it has been possible to refine the interpolation of attributes, taking into account a series of method characteristics, such as (Burrough and McDonnell, 1998):

- The number of points needed to calculate a local mean;
- The size, orientation and shape of the neighborhood of points;
- Definition of interpolation weights as a function of distance;
- Determining errors associated with the interpolated values (uncertainties).

In geostatistical interpolation methods, the spatial variation of any attribute with continuous variation on the surface was described by means of a stochastic surface, but not just a simple mathematical smoothing function. The attribute used in interpolation was called a regionalized variable and the interpolator was called kriging, after the publications of French geomathematician Georges Matheron and mining engineer D. G. Krige. The use of kriging methods made it possible to improve on the limitations of deterministic interpolation methods by providing direct estimates of the quality of predictions in terms of estimating the variance of predicted values in unsampled locations.

Among geospatial interpolation methods, kriging was the most convincing scientific interpolation method in terms of theoretical complexity and statistical methodology used (Longley et al., 2001). Kriging has been one of the methods used to create continuous surfaces from attribute data in different types of applications, such as meteorological studies.

The first step in performing a geostatistical analysis is to define the spatial sampling support for the data, so that each observation represents a signal of the observed variable in a local area that represents the variable in the field. The sampling support for studies of spatial variability of climate variables is the weather stations installed in the field for continuous monitoring. As an example, the steps for preparing data for climate studies from weather stations are demonstrated step-by-step. The packages `readr` (Wickham et al., 2021c), `rgdal` (Bivand et al., 2021), `sf` (Pebesma et al., 2021a), `raster` (Hijmans et al., 2020), `prettymapr` (Dunnington, 2024), `gridExtra` (Auguie and Antonov, 2017), and `dplyr` (Wickham et al., 2021a) required for analysis are loaded into R with the `library` function.

```
library(readr)
# package ‘readr’ was built under R version 4.3.3
library(rgdal)
# Loading required packages: sp
library(sf)
# Linking to GEOS 3.11.2, GDAL 3.7.2, PROJ 9.3.0; sf_use_s2() is TRUE
library(raster)
library(prettymapr)
# package ‘prettymapr’ was built under R version 4.3.3
library(gridExtra)
library(dplyr)
```

Data from the sample grid with weather stations are imported into R using the `read_csv2` function.

```
mgAT1 <- read_csv2("C:/geo/c9/mgAT1.csv", col_names = TRUE,
                    locale = locale(encoding = 'ISO-8859-1'))
# Rows: 32 Columns: 9
#── Column specification ────────────────────
# Delimiter: ";"
# chr (1): Cidade
# dbl (8): Y, X, Z, T, PAnual, iu, Eto, Deficit
```

Longitude and latitude coordinates are assigned to the data using the `coordinates` function.

```
coordinates(mgAT1) <- c("X", "Y")
```

The cartographic projection of the coordinates is assigned to the data using the `proj4string` function. Using the `class` function, you can verify that the data class is `SpatialPointsDataFrame`.

```
proj4string(mgAT1) <- CRS("+init=epsg:4326")
class(mgAT1)
# [1] "SpatialPointsDataFrame"
# attr(,"package")
# [1] "sp"
```

The `st_as_sf` and `st_transform` functions are used to reproject the mesh to the Albers equal-area conic projection plane coordinate system.

```
mesh_Albers <- st_as_sf(mgAT1) |>
  st_transform("+proj=aea +lat_1=-5 +lat_2=-42 +lat_0=-32
               +lon_0=-60 +x_0=0 +y_0=0 +ellps=aust_SA +units=m +no_defs")
```

An identifying column for the attribute table is created and added to the geo-object using the `st_sf` function and converted to class `sp` using the `as_Spatial` function. The `coordinates` function is used to create columns of longitude `X` and latitude `Y` attribute data in the geographic database.

```
id <- 1:32 # Create id data
mesh_Albers <- st_sf(data.frame(mesh_Albers, id)) |> as_Spatial() # Join data
mesh_Albers$X <-  coordinates(mesh_Albers)[,1] # Create longitude data column
mesh_Albers$Y <-  coordinates(mesh_Albers)[,2] # Create latitude data column
```

In addition to the sample grid, it is also necessary to obtain the boundary of the geographic region from the GADM database. The functions `read_sf`, `filter`, `st_transform`, and `as_Spatial` are used to import the polygons of Brazilian municipalities, perform a subset in the state of Minas Gerais, reproject the geometry to Albers equal-area conic projection, and convert the data model to class `sp`, respectively.

```
bord_Albers <- read_sf("C:/geo/GADM/Brazil/gadm36_BRA_1.shp") |>
  filter(NAME_1 == "Minas Gerais") |>
```

```
  st_transform("+proj=aea +lat_1=-5 +lat_2=-42 +lat_0=-32
  +lon_0=-60 +x_0=0 +y_0=0 +ellps=aust_SA +units=m +no_defs") |> as_Spatial()
```

The elevation raster layer of the area obtained from Shuttle Radar Topography Mission (SRTM) data is imported and renamed using the `raster` and `setNames` functions, respectively.

```
srtmMG <- raster("C:/geo/SRTM/srtmMG.tif") |> setNames("Alt")
```

Longitude and latitude raster layers are obtained from the elevation raster, masked and stacked together with the elevation file in a new `srtm` object using the `xyFromCell`, `xy`, `mask`, and `stack` functions, according to the following chunk code.

```
r <- srtmMG
rlon <- rlat <- r
xy <- xyFromCell(r, 1:length(r))
rlon[] <- xy[,1]
rlat[] <- xy[,2]
Long <- mask(rlon, r)
Lat <- mask(rlat, r)
srtm <- stack(r, Long, Lat)
```

The geo-field `srtm_Albers` is created by reprojecting the elevation raster to Albers equal-area conic projection at a spatial resolution of 1 km, renaming the elevation, longitude, and latitude layers to Z, X, Y, using the `projectRaster` and `setNames` functions, from the following chunk code.

```
srtm_Albers <- srtm |>
  projectRaster(crs=CRS("+proj=aea +lat_1=-5 +lat_2=-42 +lat_0=-32 +
  lon_0=-60 +x_0=0 +y_0=0 +ellps=aust_SA +units=m +no_defs"), res = 1000) |>
    setNames(c("Z", "X", "Y"))
```

The results are mapped using the `plot` function, the labels with station city names are displayed using the `text` function, and the graphic scale and compass rose for north are included in the mapping using the `addnortharrow` and `addscalebar` functions, respectively (Figure 13.1).

```
par(mfrow = c(1, 1), mar = c(2, 2, 1.5, 1.5), mgp = c(1.5, 0.6, 0))
plot(srtm_Albers$Z, col = grey(1:100/100), main = "")
plot(bord_Albers, col = "transparent", add = TRUE)
plot(mesh_Albers, add = TRUE, col = "black", bg = "white", pch = 21)
text(mesh_Albers, labels = mesh_Albers$Cidade, pos = 3, halo = TRUE, offset = 0.5,
     cex = 0.7)
addnortharrow(scale = 0.5, pos = "topleft", padin = c(0.95, 0.15))
addscalebar(pos = "bottomleft", widthhint = 0.1, padin = c(0.95, 0.15))
```

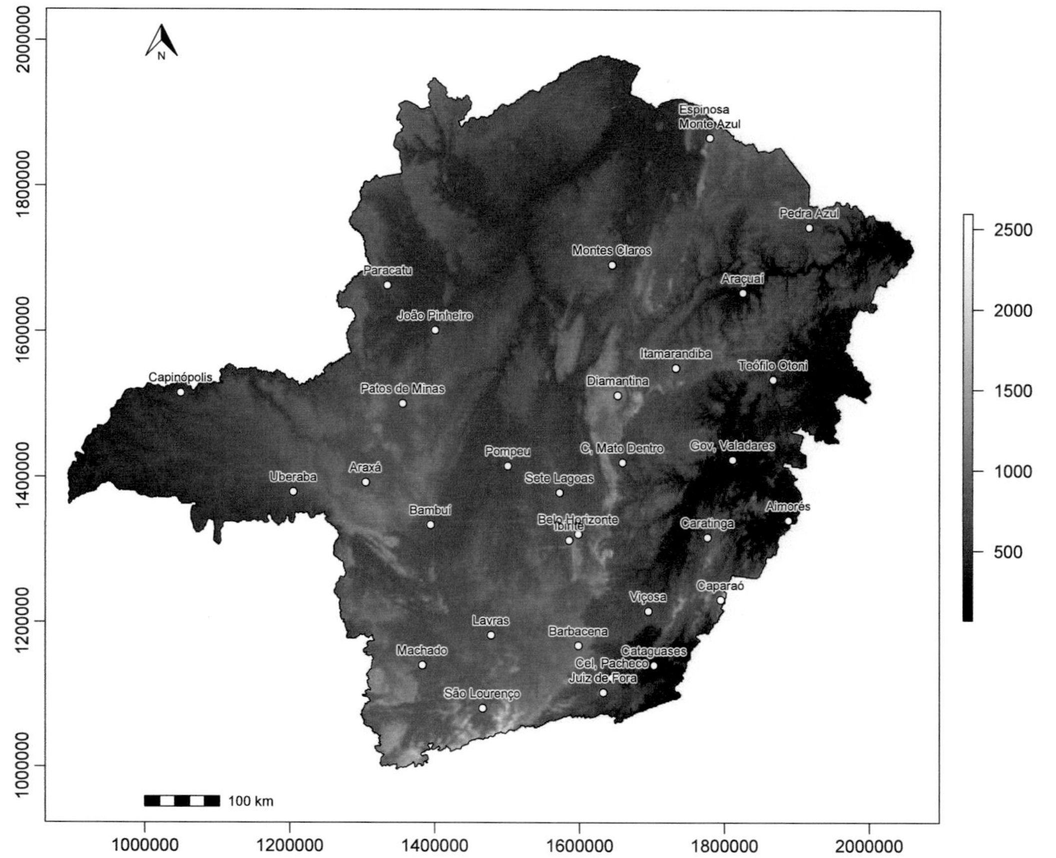

FIGURE 13.1 Location of main INMET meteorological stations used in a study of mean air temperature in the period of climatological normals (1960-1990) overlaid on the SRTM digital elevation model obtained by remote sensing.

13.4 THEORY OF REGIONALIZED VARIABLES

Natural phenomena can be characterized by the distribution of one or more variables in space, using the theory of regionalized variables (Journel and Huijbregts, 1991). In the theory of regionalized variables, spatial variation is expressed as the sum of three main components (Figure 13.2):

- Structural, with constant mean or trend;
- Random, but with spatial correlation, also known as variation of the regionalized variable;
- Random noise without spatial correlation or residual error.

Considering x a position in one, two or three dimensions, the value of the random variable in Z is:

$$Z(x) = m(x) + \epsilon'(x) + \epsilon'' \tag{13.1}$$

where $m(x)$ is a deterministic function used to describe a structural component of Z in

x, ϵ', the local, stochastic variation with spatial dependence in the residuals of $m(x)$, the regionalized variable, and ϵ'', a residual referring to a Gaussian noise, independent of space, with zero mean and variance, σ^2. The capital letter Z is used to indicate a random function and not a measured attribute, z.

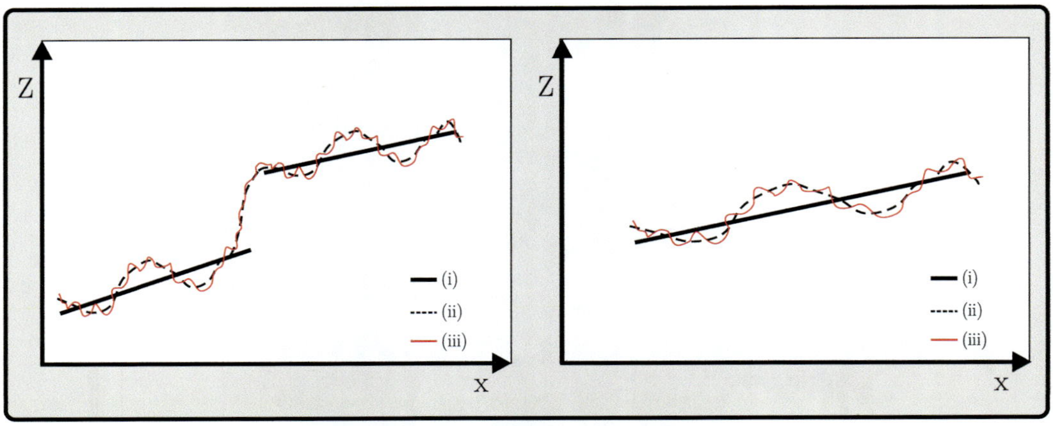

FIGURE 13.2 Division of spatial variation into the components, constant mean (left) or trend (right), (i) spatially correlated random irregular variation (ii) and uncorrelated random local variation (iii) caused by measurement error and spatial variation over short distances.

The variance of a random variable, Z, is the expected value of the square deviation of the mean of Z, $\mu = E[Z]$:

$$Var(Z) = E[(Z - \mu)^2] \tag{13.2}$$

Variance can also be thought of as the covariance of a random variable with itself:

$$Var(Z) = Cov(Z, Z) \tag{13.3}$$

In geostatistical analysis, the first step is to decide whether to use a function to describe $m(x)$. In the simplest case, where there is no trend in the data, $m(x)$ is equal to the mean value sampled in the area and the mean or expected difference between two locations x and $x + h$, separated by the distance vector, h (Burrough and McDonnell, 1998; Webster and Oliver, 2007):

$$E[Z(x) - Z(x + h)] = 0 \tag{13.4}$$

where $Z(x), Z(x + h)$ are the values of the random variable, Z, at the locations, $x, x + h$.

The autocorrelation, $\rho(h)$, of the attribute can be obtained by:

$$\rho(h) = \frac{C(h)}{C(0)} \tag{13.5}$$

where $C(0)$ is the convariance at lag 0, or σ^2 and, $C(h)$ is:

$$C(h) = cov[Z(x), Z(x+h)] = E[\{Z(x) - \mu\}\{Z(x+h) - \mu\}] = E[\{Z(x)\}\{Z(x+h) - \mu^2\}] \quad (13.6)$$

In the intrinsic or first-order stationarity hypothesis, it is assumed that the variance of the differences depends only on the distance between the locations (Burrough and McDonnell, 1998; Webster and Oliver, 2007):

$$var[Z(x) - Z(x+h)] = E[\{z(x) - Z(x+h)\}^2] = 2\gamma(h) \quad (13.7)$$

where $\gamma(h)$ is defined by semivariance.

The conditions of stationarity of the difference and variance of differences define the requirements for the intrinsic hypothesis of the theory of regionalized variables. Thus, once structural effects on the variable are taken into account, the variation is homogeneous so that the differences obtained from the variable are merely a function of distance, h.

The variance of the random component, $Z(x)$, can be described by the semivariance:

$$var[\epsilon'(x) - \epsilon'(x+h)] = 2\gamma(h) \quad (13.8)$$

Considering the intrinsic hypothesis, the semivariance can be estimated from the sample data:

$$\hat{\gamma}(h) = \frac{1}{2n}\sum_{i=1}^{n} z(x_i) - z(x_i + h)^2 \quad (13.9)$$

where n is the number of pairs of observed points of values of the attribute z, separated by the distance h. The plot of $\hat{\gamma}(h)$ against h is known as the experimental variogram.

The experimental variogram is the first step used to provide a quantitative description of the variation in the regionalized variable. With the variogram, it is possible to obtain useful information for interpolation, sampling optimization and determining patterns throughout space. To do this, it is necessary to fit a theoretical model to the experimental variogram (Burrough and McDonnell, 1998).

In the second-order stationarity process, the variogram and covariance are equivalent (Webster and Oliver, 2007):

$$\hat{\gamma}(h) = C(0) - C(h) \quad (13.10)$$

In this case, the semivariance can be obtained from the autocorrelation coefficient (Webster and Oliver, 2007):

$$\hat{\gamma}(h) = \sigma^2\{1 - \rho(h)\} \quad (13.11)$$

Under the intrinsic hypothesis, the covariance function is non-existent, and the variogram can be used in different applications.

13.5 USE OF VARIOGRAMS IN GEOSPATIAL ANALYSIS

In geospatial modeling studies, different attributes of points can be modeled considering a univariate or multivariate phenomenon. In the study of air temperature in Minas Gerais, longitude, latitude and altitude variables have been used as co-variables in temperature modeling, as can be seen in the mapping of these attributes in the region studied (Figure 13.3).

Using data from weather station point grids in Minas Gerais, it is possible to perform exploratory analysis to analyze the spatial pattern of points with attributes. The `spplot` mapping functions from the `sp` package are used to map point patterns of spatial variation in altitude (m), mean air temperature (°), longitude (m), and latitude (m).

```
a <- spplot(mesh_Albers[2], sp.layout = bord_Albers, main = "Altitude (m)",
  scales = list(draw = TRUE), xlim = c(850000, 2100000), ylim = c(950000, 2000000))
b <- spplot(mesh_Albers[3], sp.layout = bord_Albers,
  main = "Temperature (Degrees Celcius)",
  scales = list(draw = TRUE), xlim = c(850000, 2100000), ylim = c(950000, 2000000))
c <- spplot(mesh_Albers[9], sp.layout = bord_Albers, main = "Longitude (m)",
  scales = list(draw = TRUE), xlim = c(850000, 2100000), ylim = c(950000, 2000000))
d <- spplot(mesh_Albers[10], sp.layout = bord_Albers, main = "Latitude (m)",
  scales = list(draw = TRUE), xlim = c(850000, 2100000), ylim = c(950000, 2000000))
grid.arrange(a, b, c, d, ncol = 2)
```

In plant disease epidemiology, the variables of longitude, latitude and altitude are used as co-variables in modeling the incidence of bacterial blight spot in a coffee field, Santo Antônio do Amparo, Minhas Gerais, as can be observed in the mapping of attributes at points where the disease was evaluated in the region studied.

The sample mesh data for bacterial blight spot and the boundary of the coffee field used in Chapter 12 are shown in this example. The spatial points with attributes are mapped using the `spplot` and `grid.arrange` functions (Figure 13.4).

```
a <- spplot(blight_sp[13], sp.layout = bord_sp, main = "Incidence (%)",
            key.space = "right",
  scales = list(draw = TRUE), xlim = c(512013.2, 512395.5),
  ylim = c(7689875, 7690432))
b <- spplot(blight_sp[1], sp.layout = bord_sp, main = "Altitude (m)",
            key.space = "right",
  scales = list(draw = TRUE), xlim = c(512013.2, 512395.5),
  ylim = c(7689875, 7690432))
c <- spplot(blight_sp[2], sp.layout = bord_sp, main = "Longitude (m)",
            key.space = "right",
```

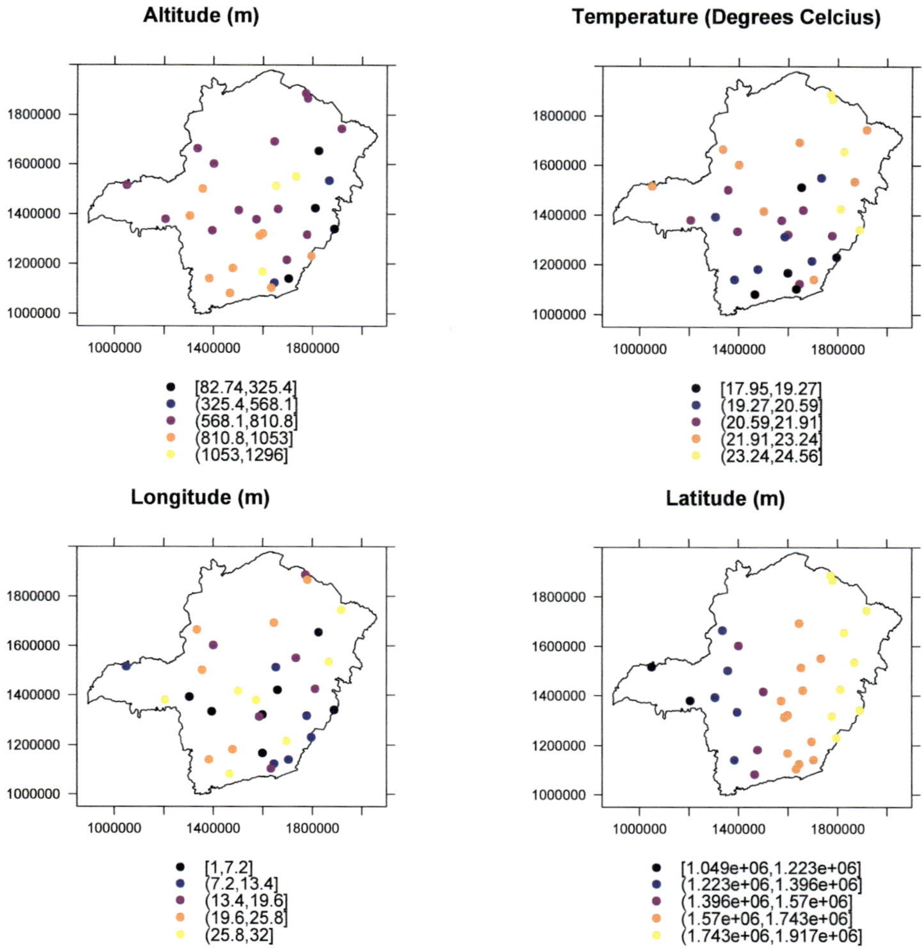

FIGURE 13.3 Variation point pattern of mean air temperature attributes (1960-1990), longitude, latitude and altitude of INMET climate stations in Minas Gerais.

```
  scales = list(draw = TRUE), xlim = c(512013.2, 512395.5),
  ylim = c(7689875, 7690432))
d <- spplot(blight_sp[3], sp.layout = bord_sp, main = "Latitude (m)",
            key.space = "right",
  scales = list(draw = TRUE), xlim = c(512013.2, 512395.5),
  ylim = c(7689875, 7690432))
grid.arrange(a, c, b, d, ncol = 2)
```

The four-year-old Catucaí Amarelo 2SL cultivar was planted at a spacing of 3.7 m between rows and 0.7 m between plants in the same row, for a total of 3862 plants ha^{-1}. Thus, at each sampling point there was a rectangular support area with an approximate width and length of 7.4x1.4 m, respectively. The support area was therefore 10.36 m^2. Considering that the square root of 10.36 is 3.21 m, it was decided to adopt a prediction surface in square blocks of 5x5 m as the kriging interpolation prediction grid. The percentage of disease incidence was assessed on April 16, 2014.

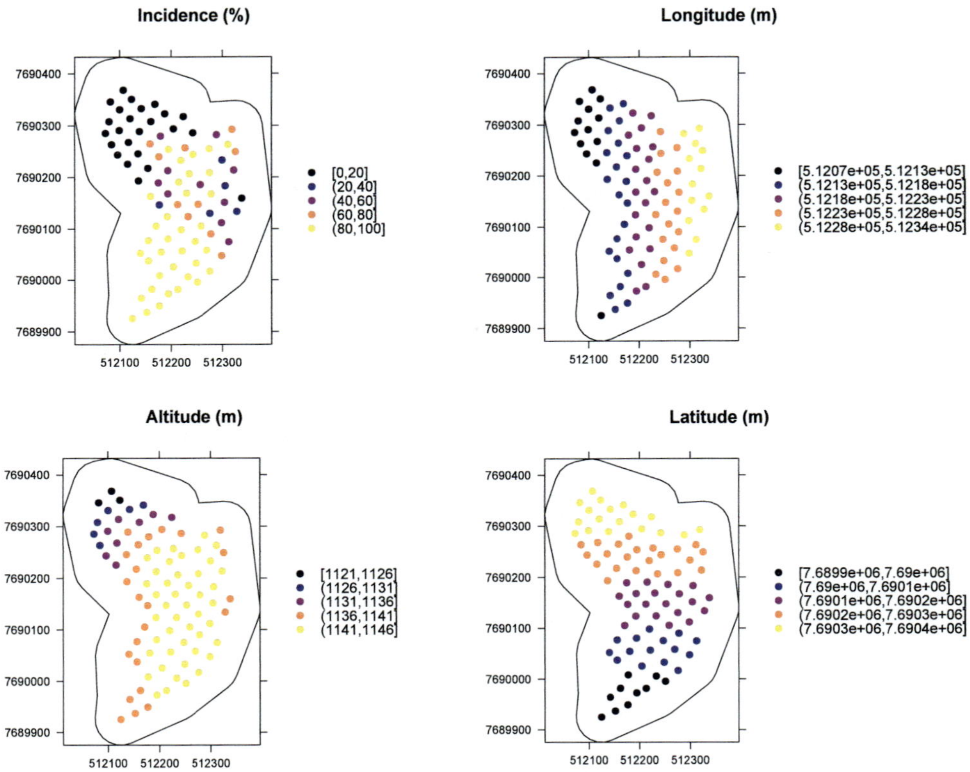

FIGURE 13.4 Variation pattern of longitude, latitude, altitude and incidence of bacterial disease in a coffee field, Santo Antônio do Amparo, Minas Gerais.

When fitting variogram models to attribute data obtained through georeferenced or referenced sampling, a curve can be fitted to the semivariance points, $\gamma(h)$, as a function of a lag distance, h. In the variogram, a plateau value (sill) is defined with the stabilization of the value of $\gamma(h)$ in relation to h. After defining the plateau, the values of h after this threshold are considered to be independent of distance. With the definition of the threshold, the range of an attribute is defined up to the same distance h that characterizes the threshold. The range is important for describing the differences in an attribute according to a distance at which nearby points have similar values for the attribute. Once you have the range value, you can define the size of a moving window used in interpolation. The fitted model may not pass through the origin of the y axis when $h = 0$. The positive value of $\gamma(h)$ when $h = 0$ is used to determine the residual, with noise without spatial correlation, ϵ'', called the nugget effect. The nugget effect is the variance of measurement errors combined with spatial variation at distances closer than the sampling spacing (Figure 13.5) (Burrough and McDonnell, 1998).

Fitting variograms is an essential procedure for determining optimum interpolation weights. A variogram without a good fit can be the result of using too few samples to determine the experimental variogram. One suggestion when analyzing environmental variables may be to use at least 50 to 100 points with attribute data to obtain a good variogram. The number of points required for sampling also depends on the type of attribute analyzed, the variation of the attribute in the area, and the search window for neighboring points used to

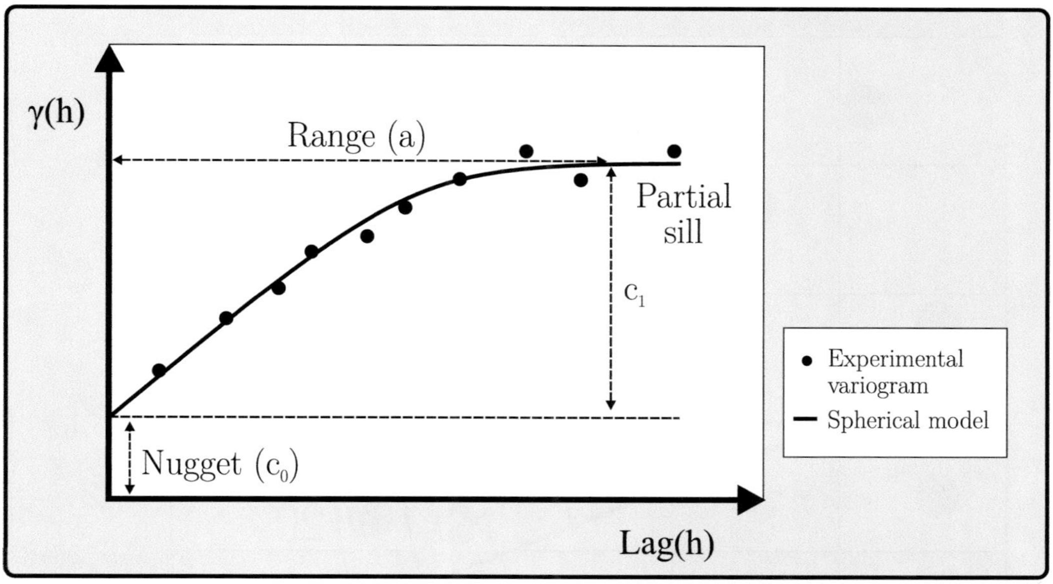

FIGURE 13.5 Determination of a simple transitive variogram.

determine the variogram (Burrough and McDonnell, 1998; Webster and Oliver, 2007).

As the attributes georeferenced in the area can occur at irregular distances, linear and angular tolerance values can be used to obtain points that will be included in the variogram calculation (Figure 13.6) (Isaaks and Srivastava, 1989; Slocum et al., 2009).

With information on the range of the variogram, the size of the search window used in local interpolation can be known. Variograms can show periodicity, in the case of a pattern of variation over long distances of the attribute being assessed, anisotropy, in the case of variation in specific directions, and trend, in the case of an external factor interfering in the spatial distribution of the phenomenon being assessed.

In practice, when there are a large number of control points, some strategies can be used to select an appropriate number of control points to estimate each grid value. In a simpler strategy, the points were placed within a circle of fixed size used to make an estimate. The quadrant and octant strategy can also be used to divide the points into four and eight sectors, respectively (Figure 13.7) (Alves and Silva, 2016; Slocum et al., 2009).

13.5.1 ISOTROPIC AND ANISOTROPIC VARIATION

To calculate the experimental variogram, a circular search radius is used to delimit a zone in which the central point is the value of h from the center. This circular radius is used to estimate the $(z_i - z_j)^2$ contribution of all pairs of data. The variogram surface of mean air temperature for Minas Gerais, referring to the normals (1960 to 1990), is determined considering external trend of the square root of altitude in the area, cutoff of 700,000 and width of 60,000. By ignoring directional effects, the resulting variogram is called isotropic. However, it is possible to determine the variogram for specific directions, called anisotropic variogram. When determining the variogram for specific directions, an interpretation of the phenomenon can arise according to the azimuth variation for different directions.

The `gstat` package (Pebesma and Graeler, 2021) used in this example is loaded into R with

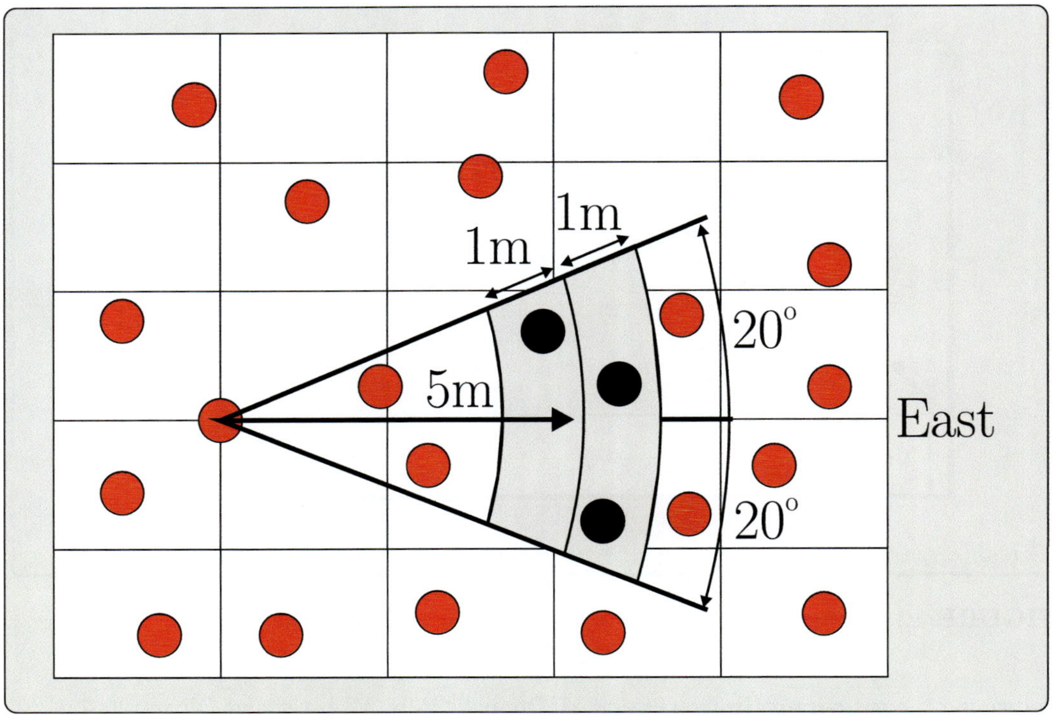

FIGURE 13.6 Selection of control points used to calculate semivariance at a fixed distance of 5 m to the west, with a tolerance of 1 m and 20 degrees of horizontal angle variation.

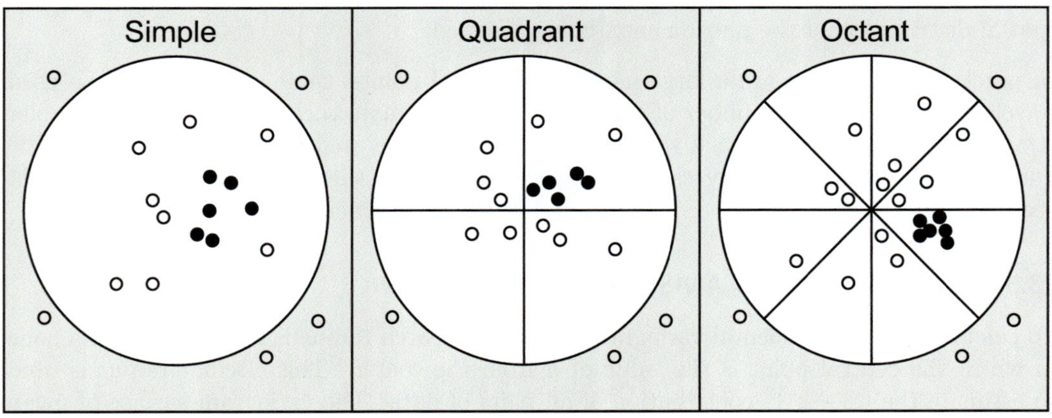

FIGURE 13.7 Strategies used to define the number of neighboring points to estimate a point.

the `library` function.

```
library(gstat)
# package 'gstat' was built under R version 4.3.3
```

The `variogram` function is used to determine the experimental variogram from temperature data considering external trend of the square root of altitude in the area. The argument `map = TRUE` is used to determine the variogram map.

```
bv <- variogram(T ~ sqrt(Z), mesh_Albers, cutoff = 700000, width = 60000,
                map = TRUE)
```

The `plot` function is used to map the isotropic variogram (Figure 13.8).

```
plot(bv)
```

In the case of mean air temperature, to evaluate the anisotropy of temperature in Minas Gerais, anisotropic variograms are evaluated in the directions, 0°, 45°, 90°, 135°, 270°, and 315°, considering external trend of the square root of altitude in the area.

The `variogram` function is used to map anisotropic variograms for different directions with the same configuration of the isotropic variogram determined previously.

```
bva <- variogram(T ~ sqrt(Z), mesh_Albers, cutoff = 700000, width = 60000,
                 alpha = c(0, 45, 90, 135, 270, 315))
```

The results are displayed using the `plot` function (Figure 13.9).

```
plot(bva)
```

In this case, as there is no considerable variation for a specific direction, it was decided to assume isotropic temperature variation.

Anisotropy can be evaluated for specific zones of variation of the horizontal angle, clockwise in relation to north (azimuth), in the case of zonal anisotropy. Anisotropy can also be geometric, or elliptical, and can have elliptical zones of influence, with the distribution of the range following an ellipse (Andriotti, 2013).

13.5.2 NESTED VARIOGRAMS

In some situations, such as sparse data, it may be possible and useful to distinguish more than one variogram component in the analysis, in order to detect two random patterns, with different ranges, by a complex variogram, $\gamma_T(h)$. The complex variogram can be divided into several components according to a linear corregionalization model (Isaaks and Srivastava, 1989; Burrough and McDonnell, 1998):

$$\gamma_T(h) = \gamma_1(h) + \gamma_2(h) + ... + \gamma_n(h) \tag{13.12}$$

The variogram with two levels of nesting can be defined by:

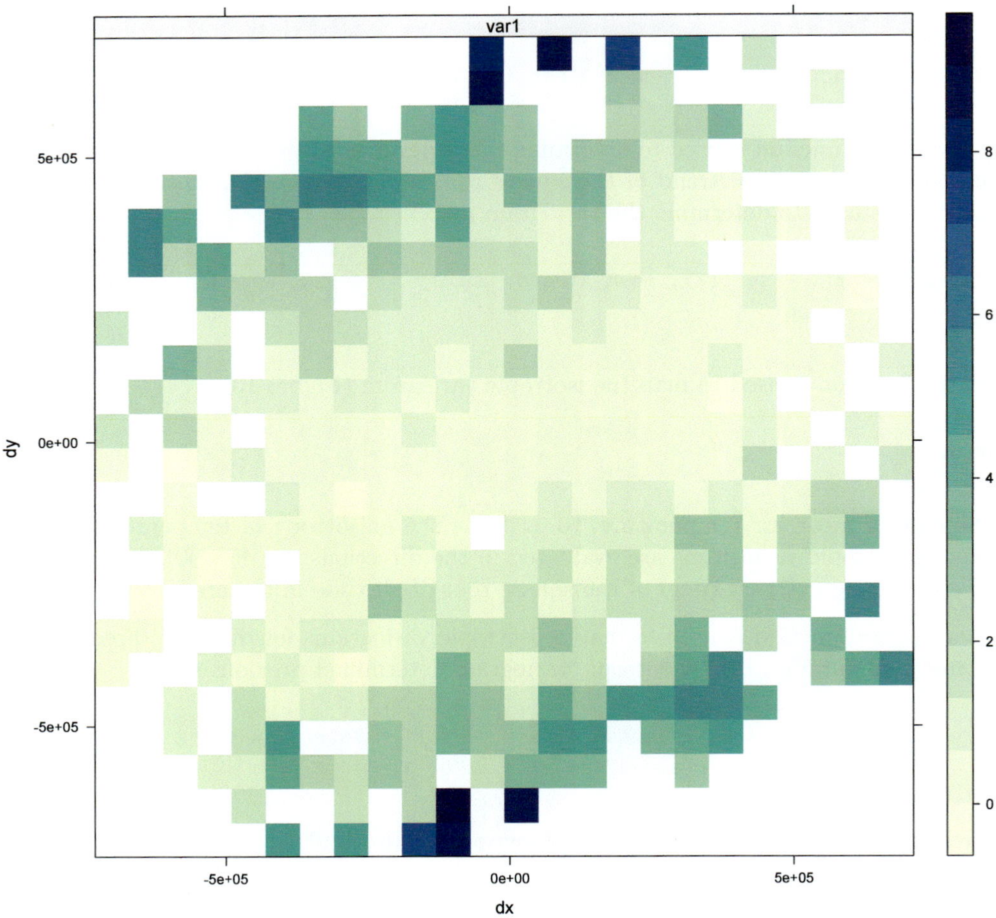

FIGURE 13.8 Circular window used to estimate the semivariance of mean air temperature in Minas Gerais, in any direction, considering external trend of the square root of altitude in the area.

$$var[\epsilon'(x) - \epsilon'(x+h)] = \gamma_1(h) + \gamma_2(h) \tag{13.13}$$

where each subvariogram is defined by its own set of parameters.

13.5.3 GEOSPATIAL VARIATION WITHIN CLASSES

Different parts of a terrain with differences in land cover, types of rocks, soils, frequency of flooding, types of agricultural cultivation, can present a unique spatial correlation structure or pattern in each part of the area. In this case, the area can be modeled using specific variograms for each part of the area, rather than using an overall variogram. One problem with using this approach can be insufficient data determining variograms with poor quality of fit. However, local models can be used for each spatial polygon or map unit to model

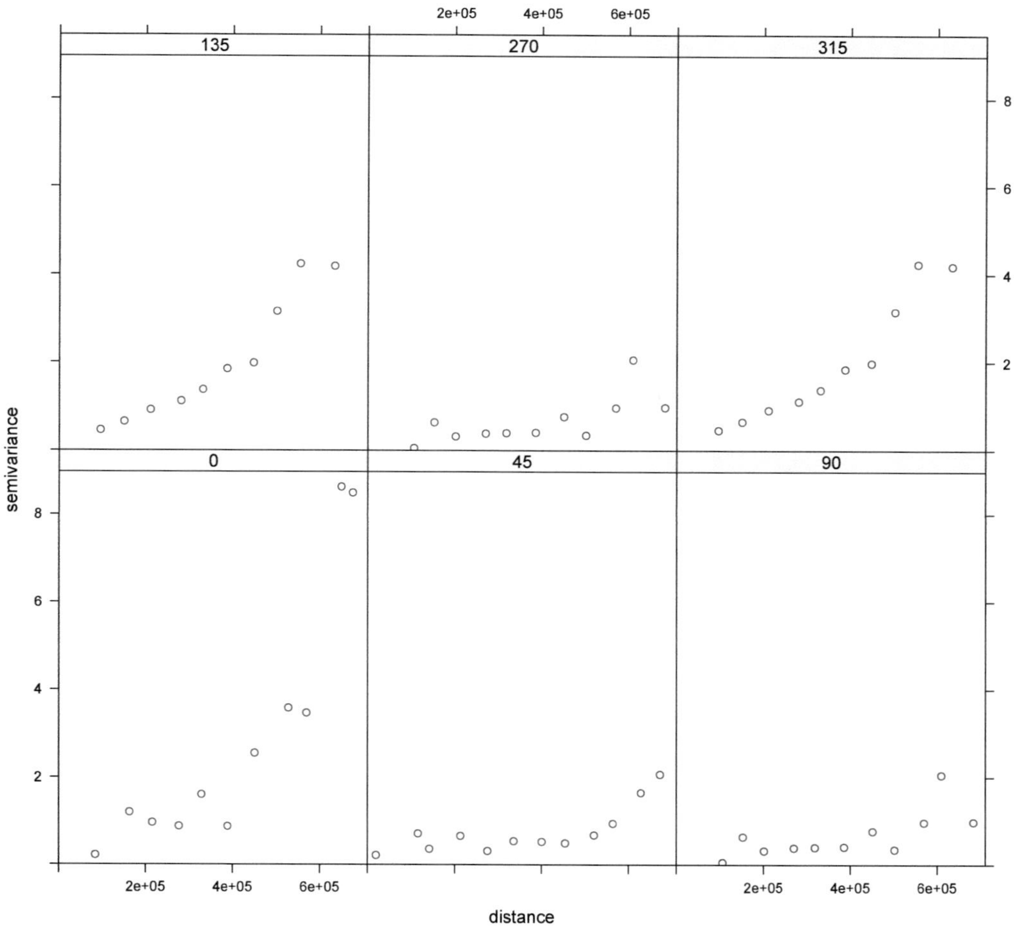

FIGURE 13.9 Determination of anisotropic variograms for azimuth directions 0, 45, 90, 135, 270 and 315 degrees, considering external trend of the square root of altitude in the area.

any attribute in the area and store the parameters in a relational database (Burrough and McDonnell, 1998).

13.6 VARIOGRAM MODEL FITTING

The shape of the variogram can reveal the type of variation of an attribute that occurs in a given area. Various theoretical models have been developed which can be fitted to the experimental variogram points or to the data using different methods: least squares, weighted least squares, maximum likelihood and restricted maximum likelihood (Diggle and Ribeiro, 2007). Among the theoretical models used, the equations of the spherical (sph), exponential (exp), Gaussian (gau) and linear (lin) models are presented:

$$\gamma(h)_{sph} = c_0 + c_1\{\frac{3h}{2a} - \frac{1}{2}(\frac{h}{a})^3\} \qquad for\ 0 < h < a \tag{13.14}$$

$$\gamma(h)_{exp} = c_0 + c_1\{1 - exp(\frac{-h}{a})\} \tag{13.15}$$

$$\gamma(h)_{gau} = c_0 + c_1\{1 - exp(\frac{-h^2}{a^2})\} \tag{13.16}$$

where c_0 and c_1 are the nugget effect and the model contribution (partial sill), respectively.

All these models described are considered transitive variograms, since the spatial correlation structure varies along h. In non-transitive models, with no defined plateau, the variogram can be linear:

$$\gamma(h)_{lin} = c_0 + bh \tag{13.17}$$

where b is the slope of the model's fit line.

However, other variogram models can be used, and can be evaluated in order to obtain less error in the mapping and a better characterization of the phenomenon studied (Table 13.1).

TABLE 13.1 Acronyms and names of variograms used in geostatistics.

Acronym	Variogram model name
Nug	nugget
Exp	exponential
Sph	spherical
Gau	Gaussian
Exc	Exclass (Exponential class/stable)
Mat	Matern
Ste Mat	Matern, M. Stein's parameterization
Cir	circular
Lin	linear
Bes	bessel
Pen	pentaspherical
Per	periodic
Wav	wave
Hol	hole
Log	logarithmic
Pow	power
Spl	spline
Leg	Legendre
Err	Measurement error
Int	Intercept

The `show.vgms` function is used to demonstrate types of variograms used in the `gstat` package (Figure 13.10).

```
show.vgms()
```

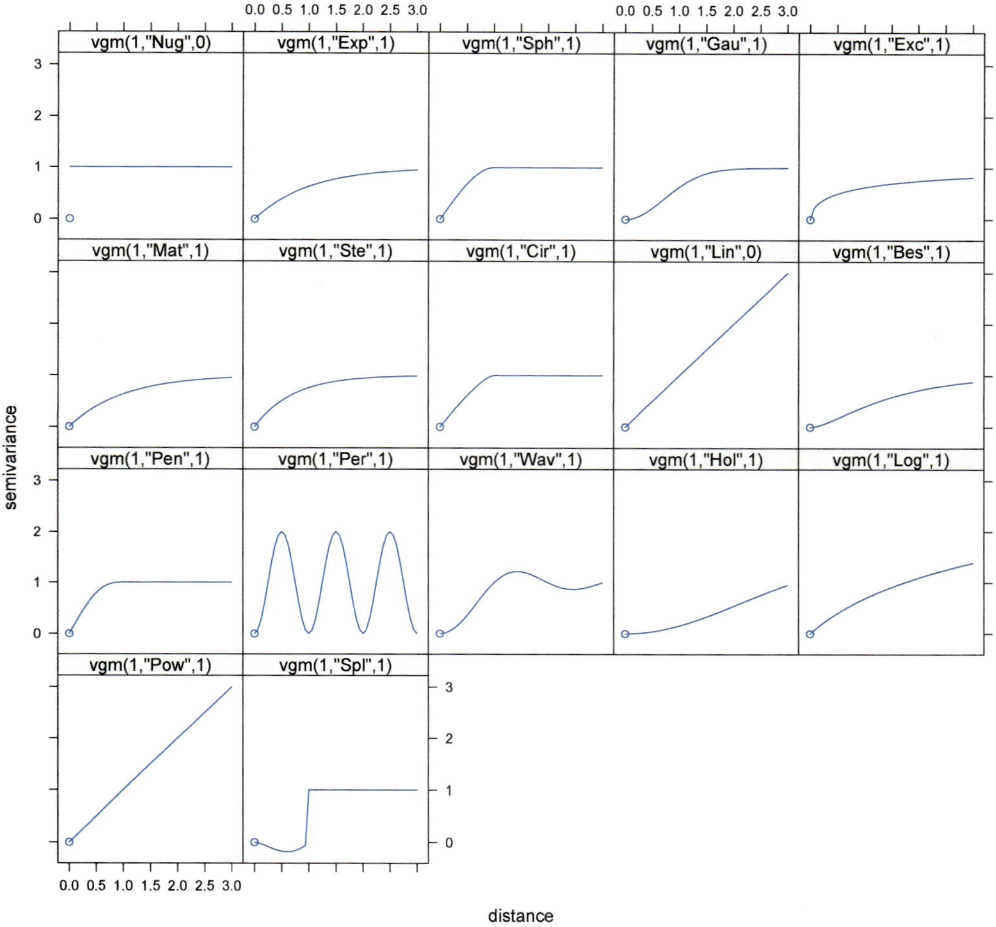

FIGURE 13.10 Examples of variogram models used in geostatistics by the `gstat` package.

The variograms can show different forms of smoothing based on the variation in the kappa parameter, as can be observed in the Matern (Mat) (Figure 13.11).

```
show.vgms(kappa.range = c(.1, .2, .5, 1, 2, 5, 10), max = 10)
```

The variograms can show a set of exponential class models (Exc) with different shape parameter with kappa variation (Figure 13.12).

```
show.vgms(kappa.range = c(.05, .1, .2, .5, 1, 1.5, 1.8, 1.9, 2), models = "Exc",
          max = 10)
```

The variograms can show a set of Matern, M. Stein's (Ste) models with different shape parameter with kappa variation (Figure 13.13).

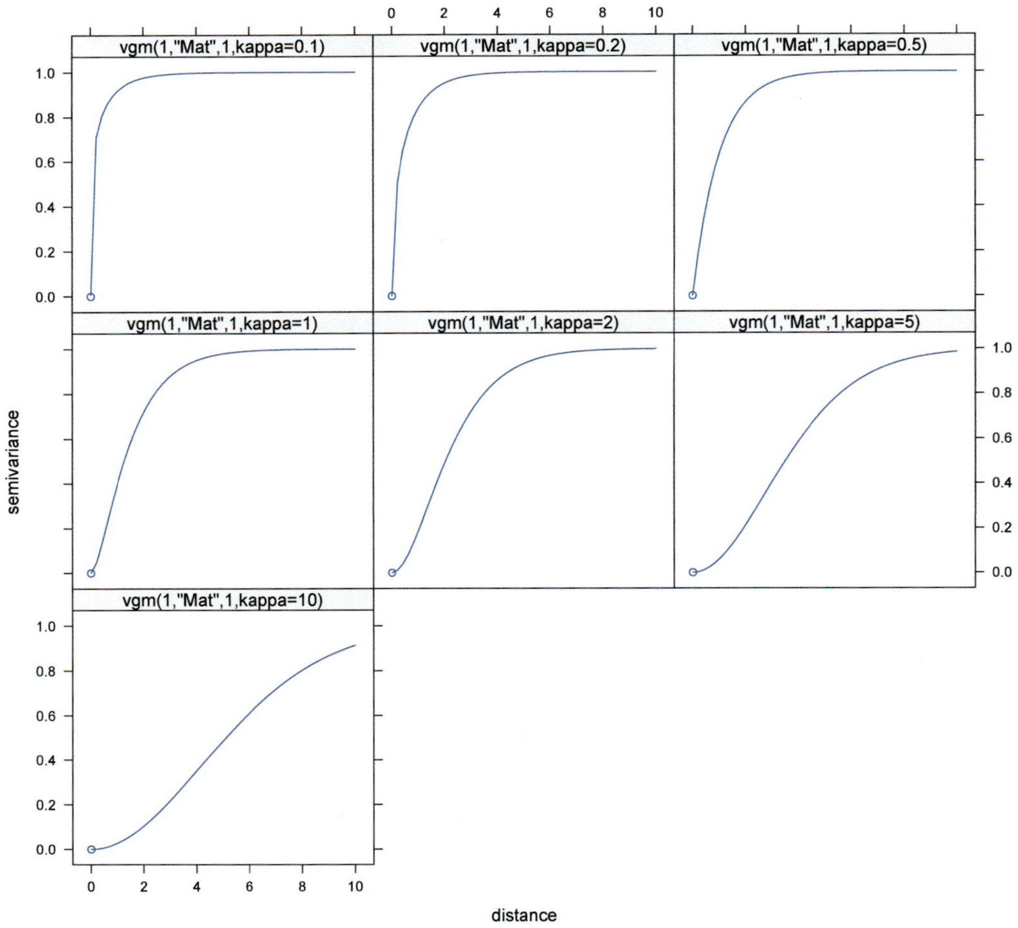

FIGURE 13.11 Examples of Matern variogram models with different smoothness based on kappa parameter variation.

```
show.vgms(kappa.range = c(.01, .02, .05, .1, .2, .5, 1, 2, 5, 1000), models = "Ste",
          max = 2)
```

When developing theoretical variogram models, certain mathematical conditions and prerequisites must be considered for use in variography and interpolation, and models cannot be used indiscriminately (Isaaks and Srivastava, 1989; Burrough and McDonnell, 1998).

13.7 INTERPOLATION BY ORDINARY KRIGING

The fitted variogram model can be used to determine the weights, λ_i, needed in local interpolation. This procedure is similar to that used in weighted moving average interpolation, except that the weights are derived from geostatistical analysis. The value $\hat{z}(x_0)$ is interpolated at an unobserved location by ordinary kriging (Burrough and McDonnell, 1998). The kriging estimator is classified as best linear unbiased estimator

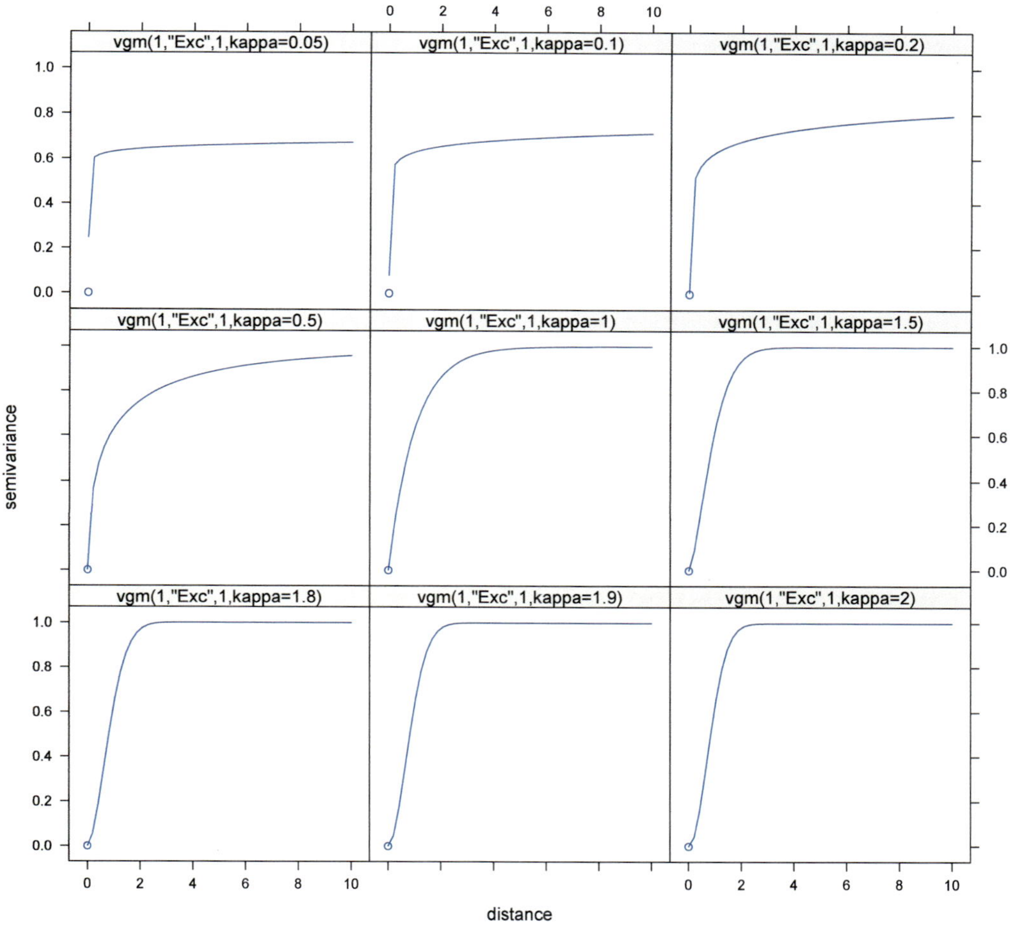

FIGURE 13.12 Examples of exponential variogram models with different shape parameter with kappa variation.

(BLUE) type method, formed by a combination of data (Burrough and McDonnell, 1998; Andriotti, 2013):

$$\hat{z}(x_0) = \sum_{i=1}^{n} \lambda_i z(x_i) \tag{13.18}$$

where the estimator is unbiased, i.e., without bias, according to the universality condition (Burrough and McDonnell, 1998; Andriotti, 2013):

$$\sum_{i=1}^{n} \lambda_i = 1 \tag{13.19}$$

The weights, λ_i, are determined so that there is no bias in the estimate of $\hat{z}(x_0)$ and to

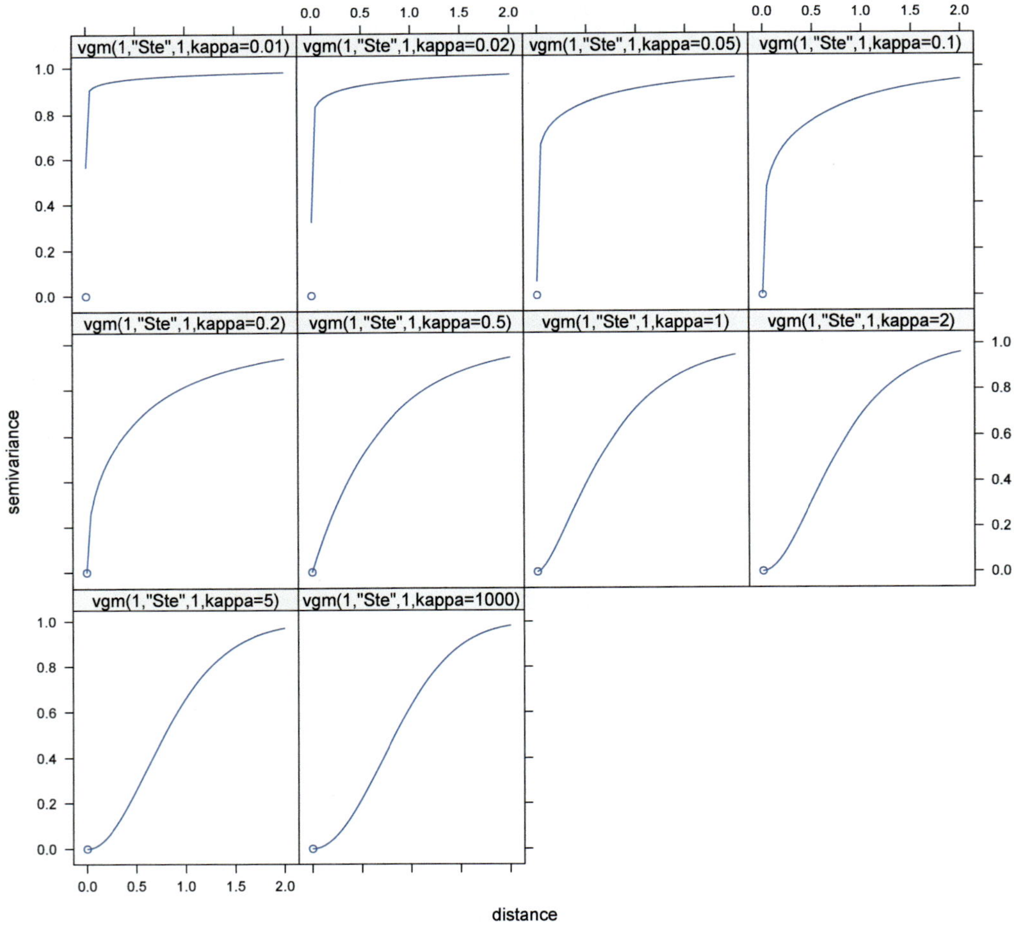

FIGURE 13.13 Examples of M. S. Matern (Ste) variogram models with different shape parameter with kappa variation.

obtain a minimum variance value (best), σ_e^2, $[\hat{z}(x_0) - z(x_0)]$, than any linear combination of observed values, as an optimality condition (Burrough and McDonnell, 1998; Andriotti, 2013). The value of the minimum prediction variance, prediction error or kriging variance is defined by (Burrough and McDonnell, 1998):

$$\hat{\sigma}_e^2 = \sum_{i=1}^{n} \lambda_i \gamma(x_i, x_0) + \phi \tag{13.20}$$

and, obtained by:

$$\sum_{i=1}^{n} \lambda_i \gamma(x_i, x_j) + \phi = \gamma(x_j, x_0) \quad \mathit{for\ all}\ j \tag{13.21}$$

where $\gamma(x_i, x_j)$ is the semivariance of z between the sample points, x_i and x_j, $\gamma(x_i, x_0)$, the semivariance between x_i and the unobserved point, x_0. These values are obtained from the fitted variogram. The value, ϕ, is a Lagrange multiplier used in minimization by ordinary kriging.

In ordinary kriging, the values mapped onto a regular grid that is more refined than the original distance of the attribute data can be used to make contour maps. The kriging variance can also be mapped to obtain information on the quality of the interpolated attribute values in the region being evaluated. However, kriging variance has been mapped in terms of kriging standard deviation or kriging error, with the same units as the attribute, for a direct evaluation of the estimates (Burrough and McDonnell, 1998).

13.8 BLOCK KRIGING

Using kriging, it is possible to obtain satisfactory interpolation weights and information on errors. However, the map obtained through interpolation may not be exactly what you want, as point kriging can be used to predict values in a different area or volume to the one in which the samples were originally obtained. Considering the variation over short distances of a given attribute, point kriging can result in interpolation problems with undesirable shapes and values. Block kriging can be used to solve this problem when estimating the value of the attribute z in a block of land, B (Figure 13.14). Block kriging can be useful for estimating a mean value by modifying the point kriging equations. The value of $z(B)$ is obtained by (Burrough and McDonnell, 1998):

$$z(B) = \int \frac{z(x)dx}{areaB} \tag{13.22}$$

and, estimated by (Burrough and McDonnell, 1998):

$$\hat{z}(B) = \sum_{i=1}^{n} \lambda_i z(x_i) \tag{13.23}$$

with, $\sum_{i=1}^{n} \lambda_i = 1$.

The minimum variance is defined by (Burrough and McDonnell, 1998):

$$\hat{\sigma}(B)^2 = \sum_{i=1}^{n} \lambda_i \bar{\gamma}(x_i, B) + \phi - \bar{\gamma}(B, B) \tag{13.24}$$

and, obtained according to:

$$\sum_{i=1}^{n} \lambda_i \gamma(x_i, x_j) + \phi = \gamma(x_j, B) \quad for\ all\ j \tag{13.25}$$

Block kriging variance estimates are usually smaller and can result in a smoother surface when compared to point kriging (Burrough and McDonnell, 1998).

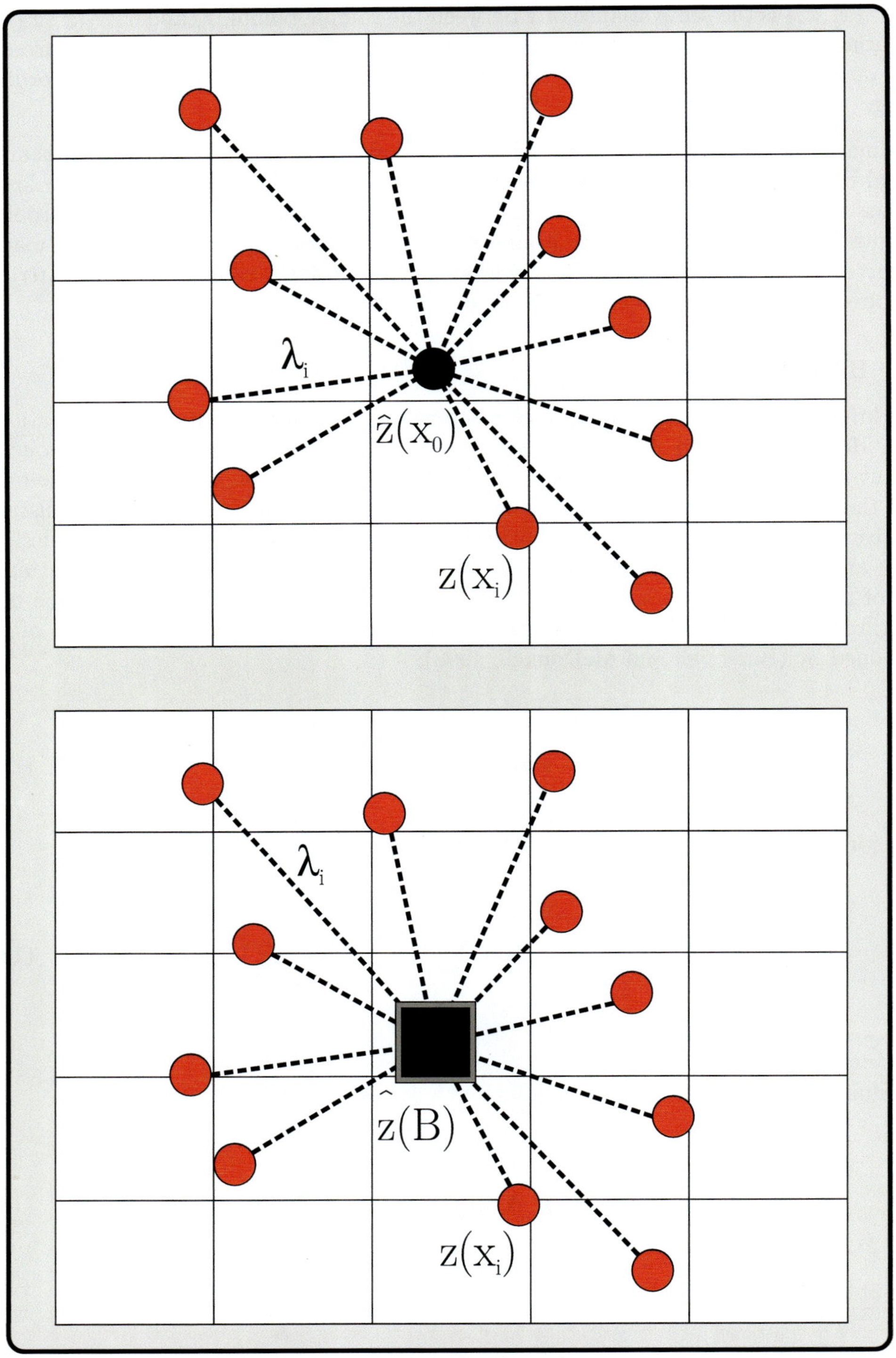

FIGURE 13.14 Ordinary kriging with point prediction (top) and block prediction (bottom) under the influence of points neighboring the value of the predicted attribute.

13.8.1 MATRIX REPRESENTATION OF KRIGING EQUATIONS

Kriging equations can be represented in matrix format for point kriging and block kriging. In point kriging, the matrix is defined by (Webster and Oliver, 2007):

$$A\lambda = b \tag{13.26}$$

where

$$\begin{bmatrix} \gamma(x_1, x_1) & \gamma(x_1, x_2) & \cdots & \gamma(x_1, x_n) & 1 \\ \gamma(x_2, x_1) & \gamma(x_2, x_2) & \cdots & \gamma(x_2, x_n) & 1 \\ \vdots & \vdots & & \vdots & \vdots \\ \gamma(x_n, x_1) & \gamma(x_n, x_2) & \cdots & \gamma(x_n, x_n) & 1 \\ 1 & 1 & \cdots & 1 & 0 \end{bmatrix} \tag{13.27}$$

$$\begin{bmatrix} \lambda_1 \\ \lambda_2 \\ \vdots \\ \lambda_n \\ \phi \end{bmatrix} \tag{13.28}$$

and,

$$\begin{bmatrix} \gamma(x_1, x_0) \\ \gamma(x_2, x_0) \\ \vdots \\ \gamma(x_n, x_0) \\ 1 \end{bmatrix} \tag{13.29}$$

The matrix A is inverted, and the weights are obtained by (Webster and Oliver, 2007):

$$\lambda = A^{-1}b \tag{13.30}$$

The kriging variance is obtained by (Webster and Oliver, 2007):

$$\hat{\sigma}^2(x_0) = b^T\lambda \tag{13.31}$$

The difference from block kriging occurs in, b and, $\hat{\sigma}^2$, by (Webster and Oliver, 2007):

$$\begin{bmatrix} \bar{\gamma}(x_1, x_B) \\ \bar{\gamma}(x_2, x_B) \\ \vdots \\ \bar{\gamma}(x_n, x_B) \\ 1 \end{bmatrix} \tag{13.32}$$

and,

$$\hat{\sigma}^2(B) = b^T\lambda - \bar{\gamma}(B, B) \tag{13.33}$$

13.9 UNIVERSAL KRIGING

In some situations, point attribute data is not the only type of information on the distribution of z, and it is possible to use other data sources in the same area, at different spectral resolutions or a cheap and easy-to-measure co-variable, such as a digital elevation model. In this case, co-kriging can be used to provide estimates of a sparse attribute as a function of another attribute with more data and correlated with the attribute of interest. However, co-kriging is a difficult technique to perform given the need to obtain a co-regionalization model with cross-variogram fit. An alternative used by environmental scientists was to obtain knowledge of transfer models by empirical regression as part of the kriging procedure. This method is called universal kriging or kriging with external drift. In many situations where there is trend, it may be necessary to assess the understanding of the genesis of the phenomenon studied, as well as whether the spatial prediction is a satisfactory representation (Isaaks and Srivastava, 1989; Burrough and McDonnell, 1998).

Considering the possibility of deterministic trend modeling by regression, the variogram performed on mean air temperature data is fitted from the residuals determined with trend modeling of the square root of altitude in the area.

The `automap` package (Hiemstra, 2013) is used in this example for variogram fitting.

```
library(automap)
# package 'automap' was built under R version 4.3.3
```

The `autofitVariogram` function is used to fit the `Ste` model to the mean air temperature data, considering the external trend of the square root of altitude. The `summary` function enables to observe details of the fit performed.

```
t_ste <- autofitVariogram(T ~ sqrt(Z), mesh_Albers, model = c("Ste"))
summary(t_ste)
# Experimental variogram:
#    np       dist     gamma dir.hor dir.ver   id
# 1 21   75561.46 0.1917022       0       0 var1
# 2 25 123844.36 0.5869716       0       0 var1
# 3 47 180037.88 0.6724962       0       0 var1
# 4 62 239453.39 0.7309150       0       0 var1
# 5 63 301257.83 0.7585008       0       0 var1
# 6 73 369602.63 1.2437146       0       0 var1
# Fitted variogram model:
#   model   psill    range kappa
# 1   Nug 0.00000      0.0     0
# 2   Ste 1.21873 268046.1     1
# Sums of squares betw. var. model and sample var.[1] 9.7244e-11
```

The results are displayed using the `plot` function (Figure 13.15).

```
plot(t_ste)
```

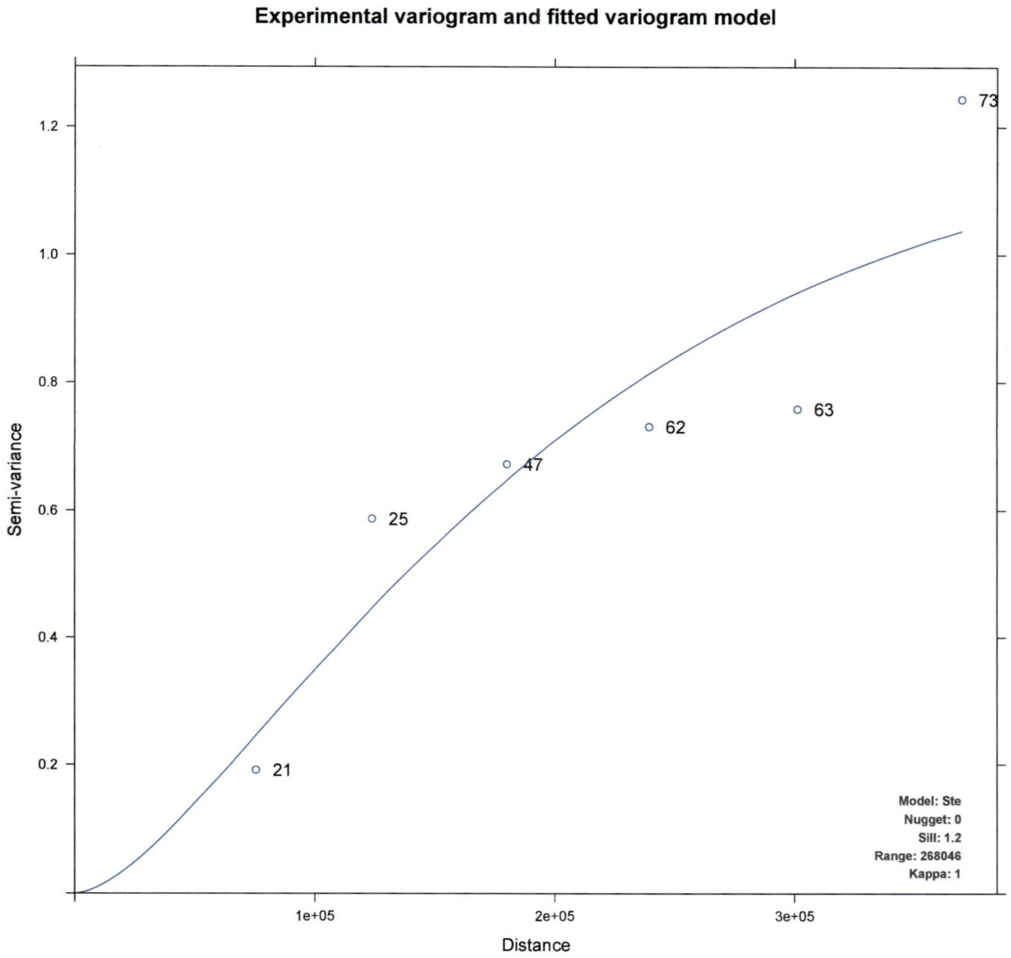

FIGURE 13.15 Experimental and fitted variogram of the mean air temperature in Minas Gerais (1960-1990) using trend residuals of the square root of altitude.

In this case, the $m(x)$ term used to model the regionalized variable in x is obtained by (Burrough and McDonnell, 1998):

$$Z(x) = \sum_{j=1}^{p} f_j(x)\beta_j + \epsilon'(x) + \epsilon'' \qquad (13.34)$$

The variogram, $\gamma'(h)$ is applied to the residuals, based on the modeling of the temperature trend as a function of altitude.

Universal kriging is performed on a grid with SRTM altitude values with a mean aggregated to 1000 m, in order to model the trend of the external variable altitude on the attribute temperature in a more refined neighborhood to obtain the downscaling effect of the variable.

The prediction is performed in point form and in blocks of 100,000x100,000 m, in order to evaluate the variation in mean air temperature (1960-1990) in Minas Gerais with an interpolation method with optimal prediction.

The SRTM elevation raster used at the beginning of this chapter is converted to `SpatialPixelsDataFrame` in order to be used as a reference grid for interpolating the mean air temperature in Minas Gerais. The function `as("SpatialPixelsDataFrame")` is used to create the interpolation grid.

```
srtmpdf <- as(srtm_Albers, "SpatialPixelsDataFrame")
```

The `autoKrige` function is used to perform universal kriging of points and blocks of 100,000x100,000 m using trend residuals of the square root of altitude.

```
tk1 <- autoKrige(T ~ sqrt(Z), mesh_Albers, srtmpdf) # Point kriging
# [using universal kriging]
tk2 <- autoKrige(T ~ sqrt(Z), mesh_Albers, srtmpdf,
                          block = c(100000, 100000)) # Block kriging
# [using universal kriging]
```

The results are mapped using the `spplot` function (Figure 13.16).

```
a <- spplot(tk1$krige_output["var1.pred"], sp.layout = bord_Albers,
            scales = list(draw = TRUE), main = "Point kriging")
b <- spplot(tk1$krige_output["var1.var"], sp.layout = bord_Albers,
            scales = list(draw = TRUE), main = "Point kriging sd")
c <- spplot(tk2$krige_output["var1.pred"], sp.layout = bord_Albers,
            scales = list(draw = TRUE), main = "Block kriging")
d <- spplot(tk2$krige_output["var1.var"], sp.layout = bord_Albers,
            scales = list(draw = TRUE), main = "Block kriging sd")
grid.arrange(a, b, c, d, ncol = 2)
```

The results of block kriging for predicting mean air temperature are similar to those of point kriging; however, there is a greater smoothing of the standard deviation of block kriging when compared to point kriging.

As an example of epidemic modeling in the coffee field, the bacterial blight spot interpolation grid `g` used in Chapter 12 of class `SpatialPoints` is prepared in this example for geostatistical analysis. A `gdf` object of class `SpatialPointsDataFrame` is created with the addition of an `ID` column in the attribute table.

```
gdf <- SpatialPointsDataFrame(g, data.frame(ID = 1:length(g)))
```

Columns with longitude and latitude coordinates are added to the attribute table using the `coordinates` function.

```
gdf$X <- coordinates(gdf)[,1]
gdf$Y <- coordinates(gdf)[,2]
```

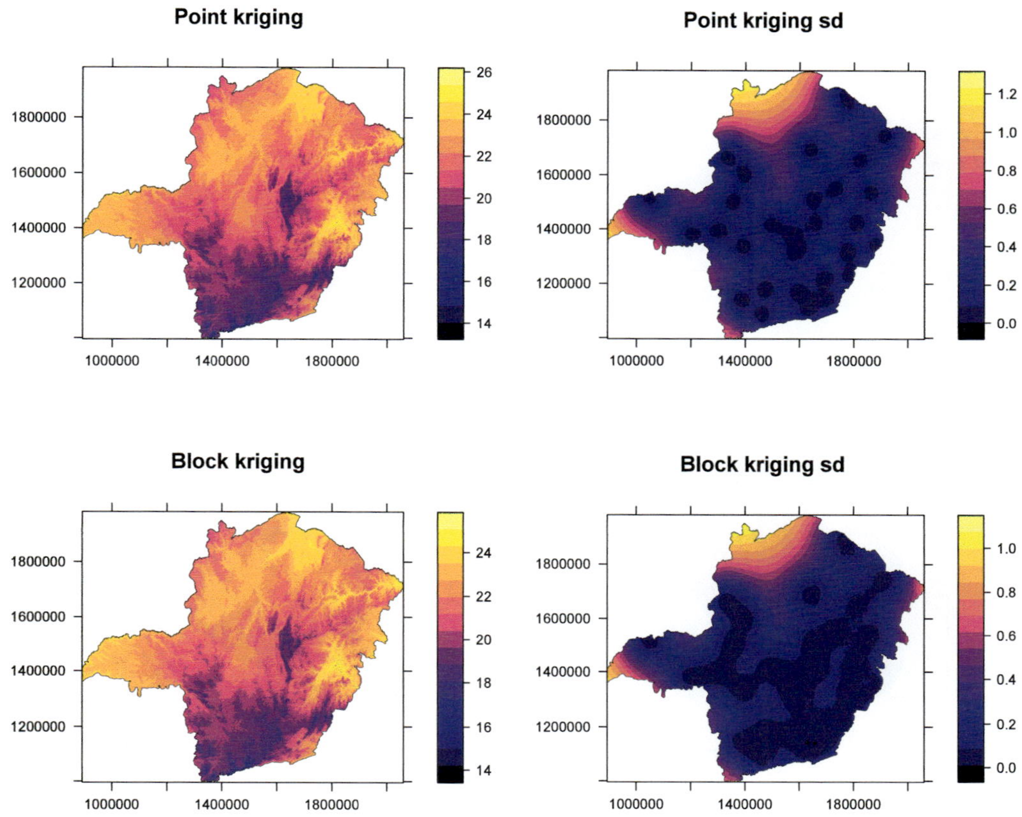

FIGURE 13.16 Universal kriging of mean air temperature in Minas Gerais predicted using point kriging (top-left), standard deviation of point kriging (top-right), temperature predicted using block kriging (bottom-left) and standard deviation of temperature predicted using block kriging (bottom-right).

The `gridded` function is used to convert the `SpatialPointsDataFrame` object into a `SpatialPixelsDataFrame`.

```
gridded(gdf) <- TRUE
```

The altitude of the coffee field is modeled using universal kriging to obtain a 5x5 m elevation model of the field. This elevation model with the refined altitude data will be used for the spatial prediction of bacterial blight spot, considering the altitude trend over short distances.

Universal kriging with the `autokrige` function is used to validate the best variogram for interpolation of altitude in the coffee field by comparing the spherical and Matern models with M. Stein's parameterization.

```
zks <- autoKrige(Z ~ log(X) + log(Y), blight_sp, gdf, model = c("Sph")) # Sph
zkm <- autoKrige(Z ~ log(X) + log(Y), blight_sp, gdf) # Ste
```

The results are mapped using the `plot` function. Note that the Matern variogram model with M. Stein's parameterization determined a lower error in the spatial prediction of the area's altitude when compared to the spherical model (Figures 13.17 and 13.18).

```
plot(zks) # Sph
plot(zkm) # Ste
```

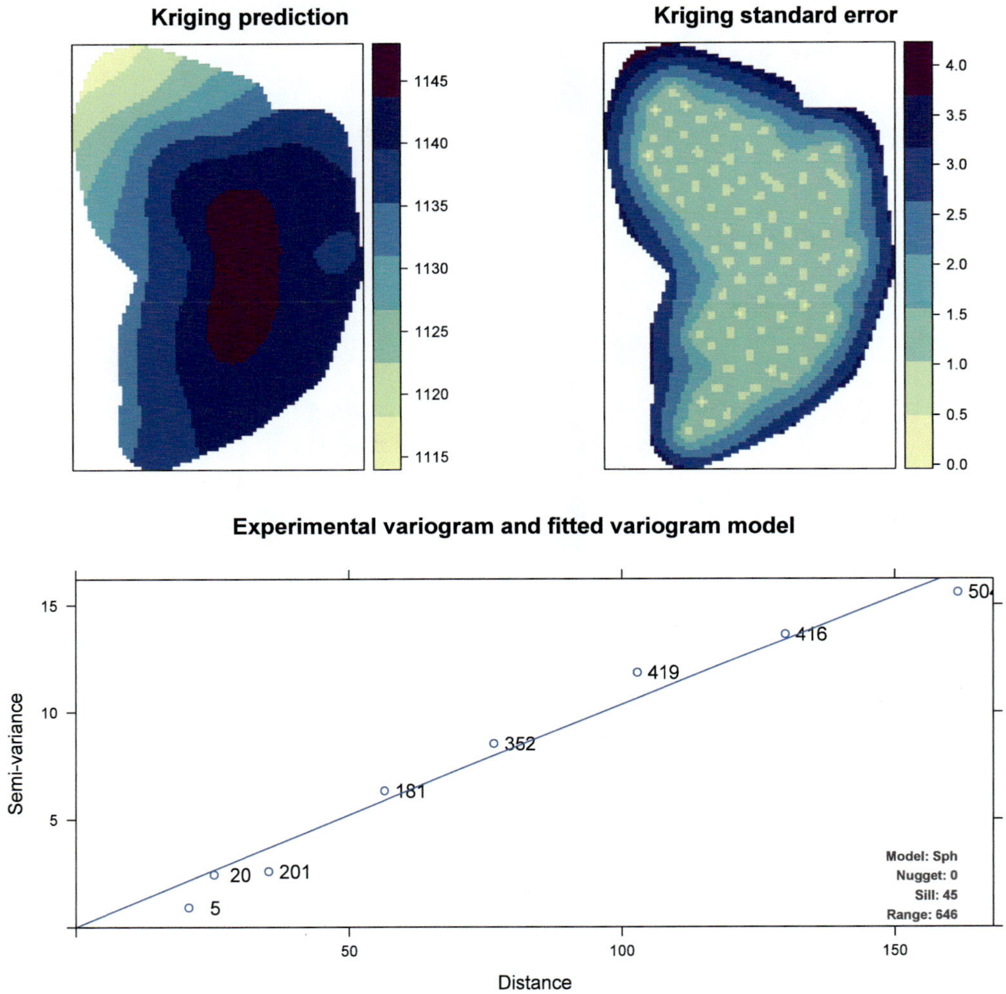

FIGURE 13.17 Spherical variogram model of altitude in a coffee field, with residual modeling of the logarithm of longitude and the logarithm of latitude (bottom) and point prediction of kriging (top-left) with the standard deviation of kriging (right).

The result of the spatial prediction of altitude with the `Ste` variogram is converted to `RasterLayer` with the `raster` function.

```
zkmr <- raster(zkm$krige_output["var1.pred"])
```

A chunk code is used to create the `elev` prediction grid with raster layers of altitude,

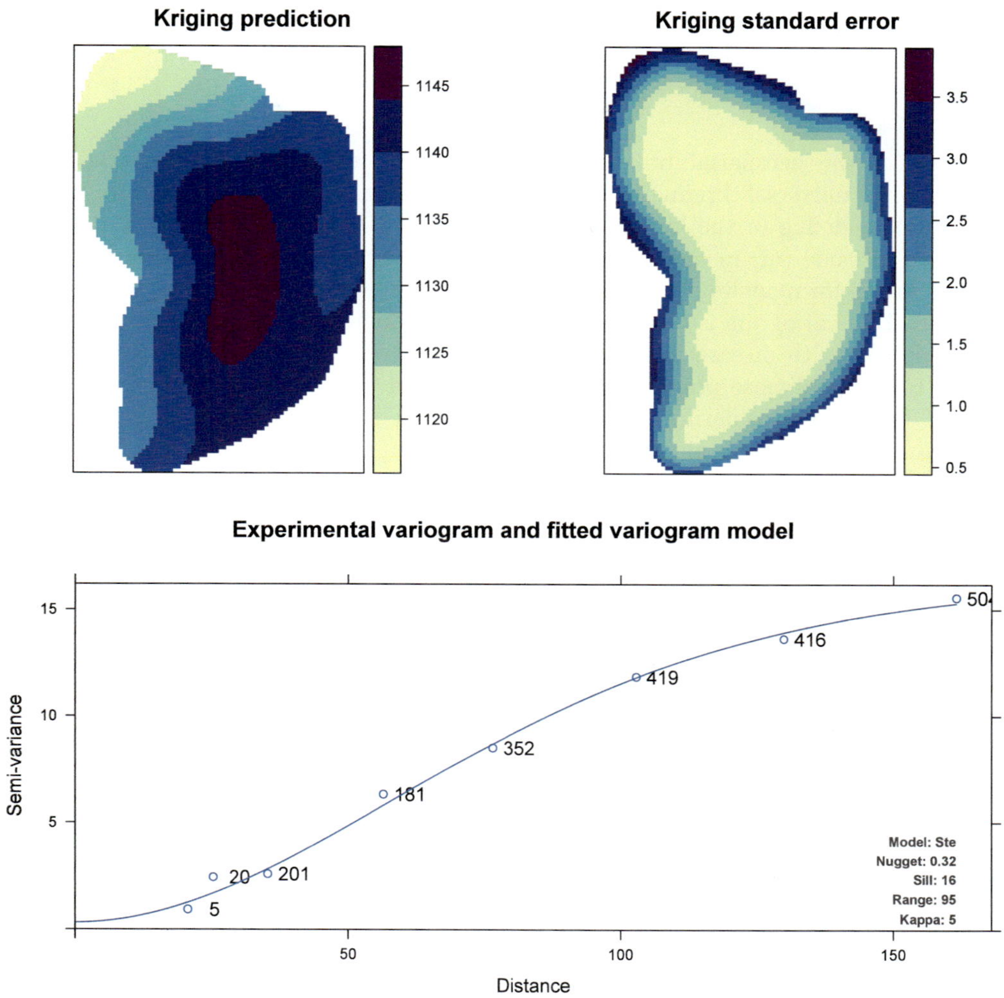

FIGURE 13.18 Matern with M. Stein's parameterization variogram model of altitude in a coffee field, with residual modeling of the logarithm of longitude and the logarithm of latitude (bottom) and point prediction of kriging (top-left) with the standard deviation of kriging (right).

longitude, and latitude.

```
lon <- lat <- zkmr
xy <- xyFromCell(zkmr, 1:length(zkmr))
lon[] <- xy[,1]
lat[] <- xy[,2]
Long <- mask(lon, zkmr)
Lat <- mask(lat, zkmr)
elev <- stack(zkmr, Long, Lat)
names(elev) <- c("Z", "X", "Y")
```

The prediction grid is converted to `SpatialPixelsDataFrame` with the function as("SpatialPixelsDataFrame") in order to generate the geo-field elevdf.

```
elevdf <- as(elev, "SpatialPixelsDataFrame")
```

In the case of the geospatial progress of plant disease epidemics, ordinary kriging is compared with universal kriging. Using universal kriging, it is possible to perform deterministic modeling of the logarithm of longitude (m), logarithm of latitude (m) and logarithm of altitude (m) in order to obtain a Matern variogram with Matern, M. Stein's parameterization for modeling the bacterial blight spot epidemic over short distances, with 51 m range. This variogram is compared with trend modeling of the logarithm of longitude and latitude and with disease modeling without trend, obtaining variograms with ranges of 66 and 676 m of disease progress. Therefore, when modeling the spread of the disease over short distances, considering the effect of the area's altitude, a reference sample mesh distance can be the 51 m obtained in this study.

The autokrige function is used to compare the results of ordinary kriging with universal kriging by modeling trends in logarithm of longitude and latitude and logarithm of longitude, latitude, and altitude.

```
mk <- autoKrige(i280 ~ 1, blight_sp, elevdf)
# [using ordinary kriging]
mk1 <- autoKrige(i280 ~ log(X) + log(Y), blight_sp, elevdf)
# [using universal kriging]
mk2 <- autoKrige(i280 ~ log(X) + log(Y) + log(Z), blight_sp, elevdf)
# [using universal kriging]
```

The results are mapped using the plot function (Figures 13.19, 13.20, and 13.21).

```
plot(mk) # Ordinary kriging
plot(mk1) # Universal kriging logxy
plot(mk2) # Universal kriging logxyz
```

Using universal block kriging, we can predict disease incidence by considering the influence of longitude, latitude and altitude on the epidemic progress in the field. Bacterial blight spot prediction is performed using 30x30 m blocks. The use of 30 m blocks has the effect of smoothing out the kriging errors, which are low in the vicinity of sampling points and higher in the regions around the boundary and sparse vicinity of the irregular sampling mesh proposed for mapping the disease in the field (Figure 13.22).

```
mk3 <- autoKrige(i280 ~ log(X) + log(Y) + log(Z), blight_sp, elevdf,
                 block = c(30, 30))
# [using universal kriging]
plot(mk3) # Map
```

13.10 OTHER KRIGING METHODS

In some situations, it may be of interest to map the value of the attribute that exceeded a certain threshold. In this case, the interest is not in estimating the best values for $z(x_0)$, but in what the probability is of the value of an attribute being below or above

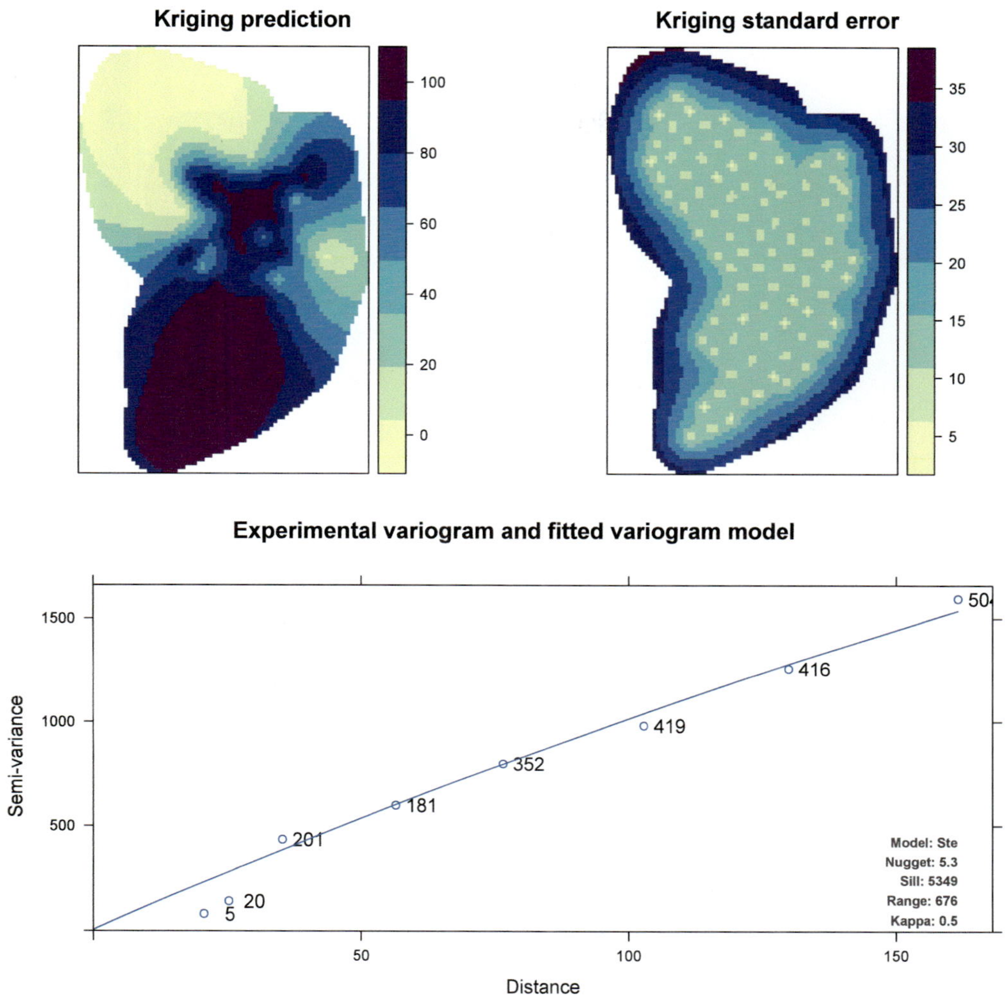

FIGURE 13.19 Modeling bacterial blight spot percentage in a coffee field on April 16, 2014 using ordinary kriging.

a certain threshold. A threshold value can be useful in decision-making in different mining applications, soil pollution, marketing studies, disease management and the application of inputs in agriculture.

13.10.1 INDICATOR KRIGING

Indicator kriging is a non-linear form of ordinary kriging with binary transformation of the original continuous data. The calculation of variograms performed on the binary data can be transformed or undergo trend modeling to indicate the prediction of the variable of interest on a probability scale between 0 and 1, according to the chosen threshold. By calculating the variograms, insights can be gained into the variable's continuous pattern within interval classes selected as the threshold value for indicator kriging.

Indicator kriging is used to map coffee bacterial blight spot considering the disease incidence thresholds on the leaves at intensity threshold values greater than 1%, 26%, 50% and

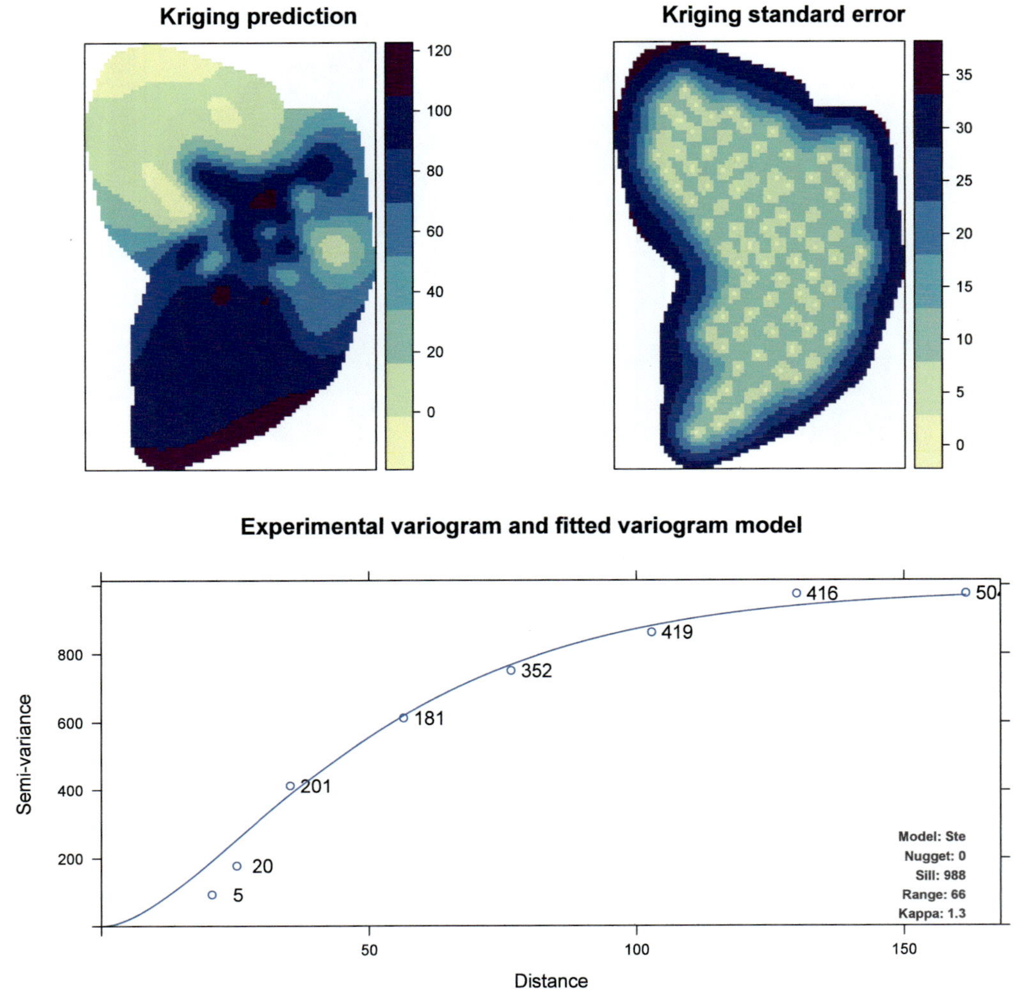

FIGURE 13.20 Modeling bacterial blight spot percentage in a coffee field on April 16, 2014 using the trend of the logarithm of longitude and latitude.

75%. Using ordinary kriging, the variograms are influenced by the coordinates, showing a variation in range according to the threshold chosen. With the use of disease modeling at each threshold, under trend, $log(X) + log(Y) + log(Z)$, it is possible to obtain Matern, M. Stein's parameterization models with ranges close to 30 m for the thresholds used.

```
mik1 <- autoKrige(I(i280 > 1) ~ log(X) + log(Y) + log(Z), blight_sp, elevdf,
                  block = c(30, 30)) # 1%
# [using universal kriging]
mik26 <- autoKrige(I(i280 > 26) ~ log(X) + log(Y) + log(Z), blight_sp, elevdf,
                  block = c(30, 30)) # 26%
# [using universal kriging]
mik51 <- autoKrige(I(i280 > 51) ~ log(X) + log(Y) + log(Z), blight_sp, elevdf,
                  block = c(30, 30)) # 51%
```

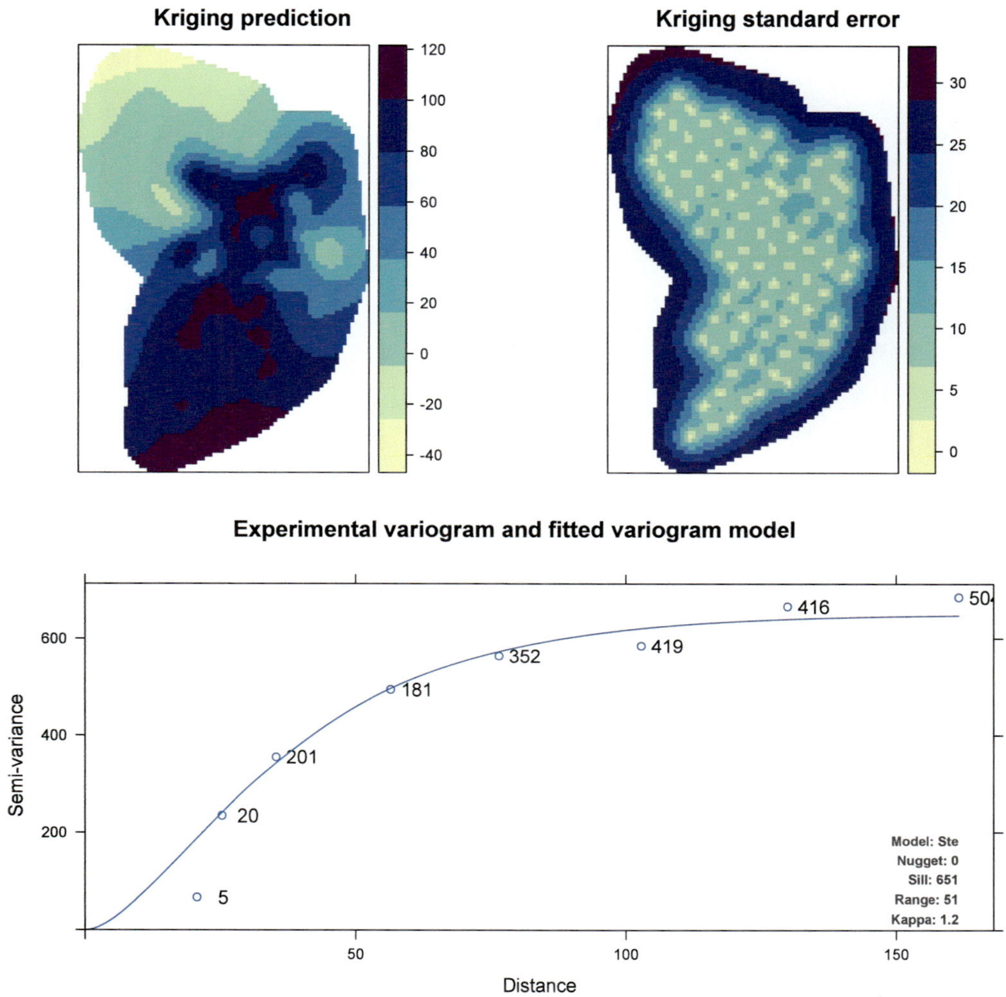

FIGURE 13.21 Modeling bacterial blight spot percentage in a coffee field on April 16, 2014 (top) using the trend of the logarithm of longitude, latitude and altitude.

```
# [using universal kriging]
mik75 <- autoKrige(I(i280 > 75) ~ log(X) + log(Y) + log(Z), blight_sp, elevdf,
                   block = c(30, 30)) # 75%
# [using universal kriging]
```

The results are mapped using the `plot` function (Figures 13.23, 13.24, 13.25, and 13.26).

```
plot(mik1)
plot(mik26)
plot(mik51)
plot(mik75)
```

Thus, it is possible to map the geospatial pattern of disease spatial variation, with continuous

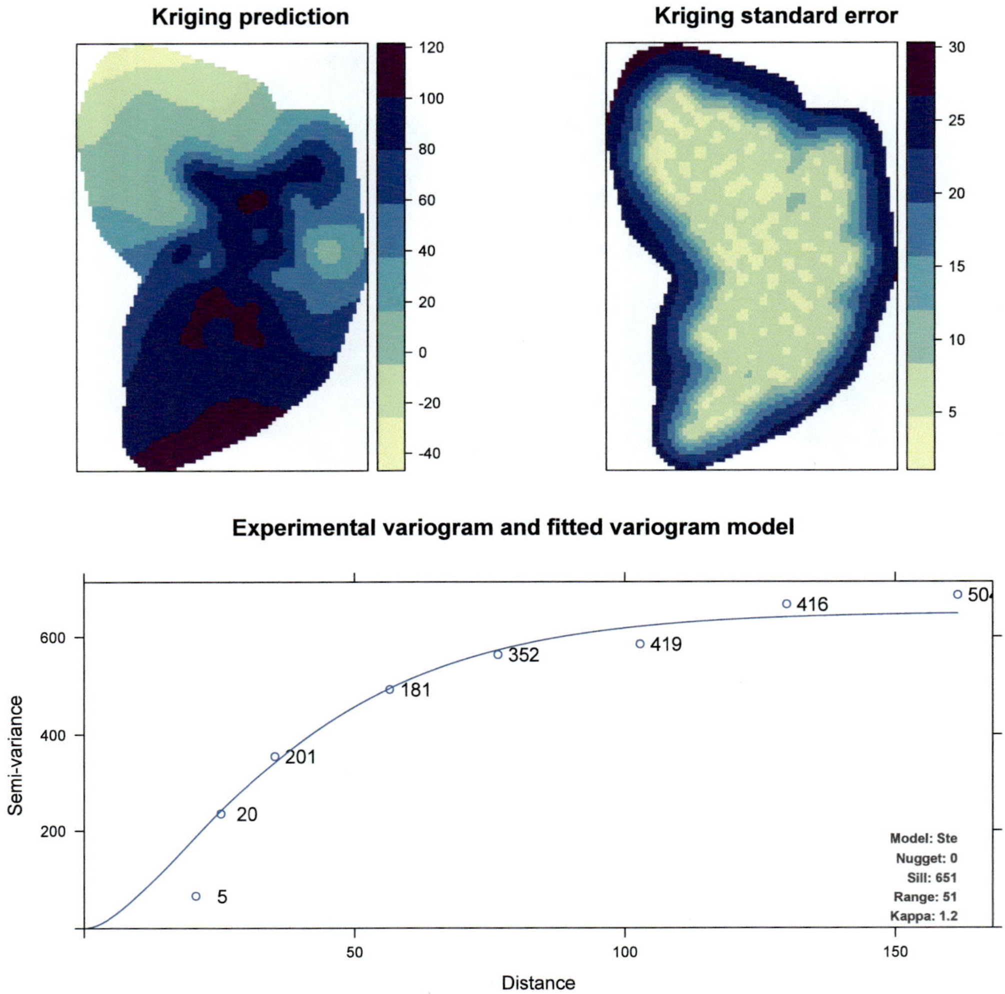

FIGURE 13.22 Universal 30x30m block kriging in a coffee field used for bacterial blight spot prediction in the field using the trend of the logarithm of longitude, latitude and altitude.

borders between classes of disease occurrence, from the perspective of probability theory. Despite the prediction of values above and below the probability scale (0 to 1), due to the universal indicator kriging, with blocks measuring 30x30 m, we can characterize the dissemination pattern with the most intense outbreaks and the aggregate pattern of the disease with contagion capacity and spatial dependence structure in range of approximately 30 m.

The quality of the universal indicator kriging prediction is assessed in the kriging standard deviation maps for each threshold used to map the probability of disease incidence in the crop. The lowest prediction errors are observed with the disease incidence threshold greater than 26%. In general, the highest prediction errors are observed in border regions and with more sparse observation data, in all the thresholds evaluated.

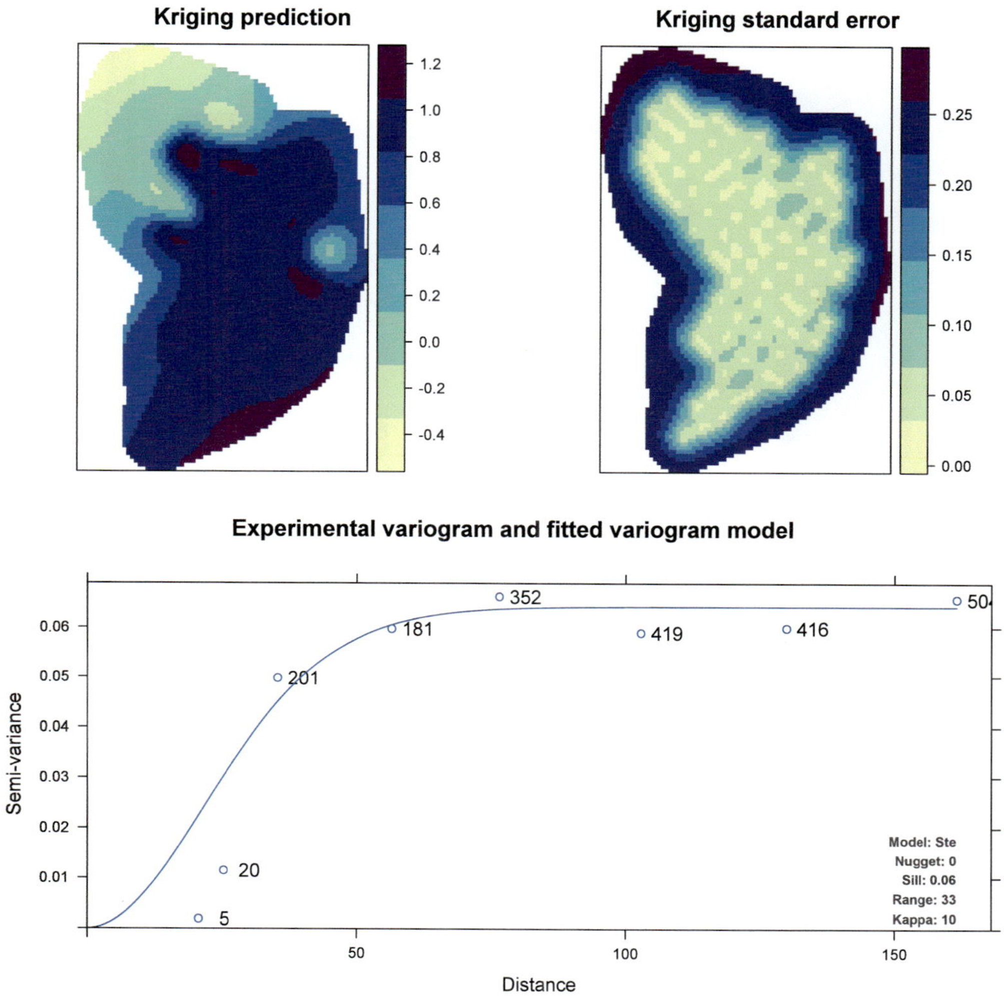

FIGURE 13.23 Indicator universal 30x30m block kriging with trend modeling of the incidence of bacterial blight spot in a coffee field (log(X) + log(Y) + log(Z)), considering 1% threshold of disease intensity.

13.11 VARIOGRAM VALIDATION WITH KRIGING

In cross-validation, kriging equations can be used to assess the quality of the variogram modeling. This method is based on calculating the moments of the residue distribution $(\hat{z}(x_i) - z(x_i))$ for all the point data, so that each point is removed and predicted again from the remaining data. This procedure can be used to test the consistency of the variogram in terms of bias, so that the mean difference of the residuals should be 0 and the variance, 1, under ideal modeling conditions (Burrough and McDonnell, 1998) (Figure 13.27).

Considering that kriging can be used to validate the use of a given variogram model, the Matern model with M. Stein's parameterization (Ste) is compared with the spherical model (Sph) to estimate the altitude of an irregular sampling grid of points defined in the coffee field (See Figures 13.17 and 13.18).

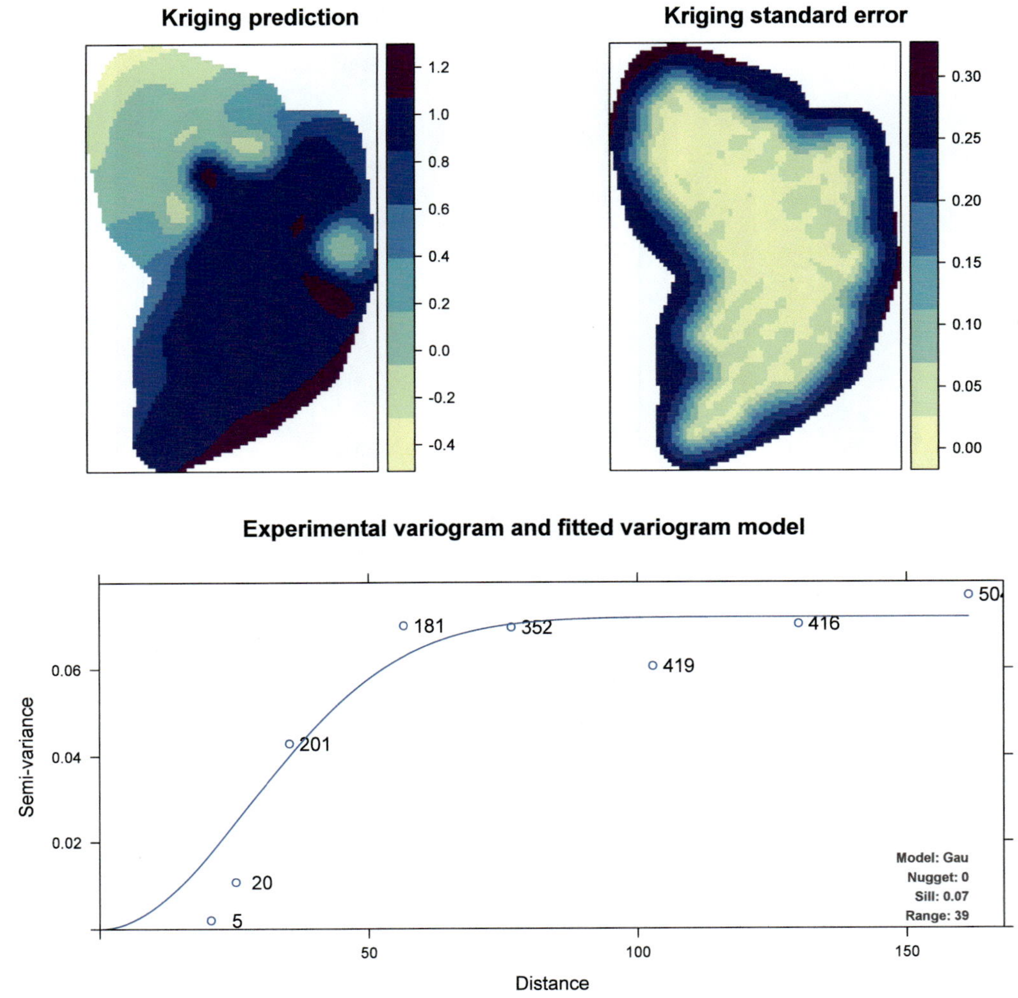

FIGURE 13.24 Indicator universal 30x30m block kriging with trend modeling of the incidence of bacterial blight spot in a coffee field (log(X) + log(Y) + log(Z)), considering 26% threshold of disease intensity.

13.12 KRIGING CROSS-VALIDATION

The types of errors used in cross-validation are defined in Table 13.2 (Hiemstra, 2013).

TABLE 13.2 Types of errors determined in cross-validation.

Error type	Description
mean_error	Mean Error. The mean of the cross-validation residuals. Ideally, it should be close to 0. Positive values indicate underestimation, while negative values indicate overestimation of the observed data.

Error type	Description
me_mean	Mean Error of Kriging Variance. It represents the average theoretical uncertainty (standard error) predicted by the model. It should be close to the RMSE. A large discrepancy between me_mean and RMSE suggests that the semivariogram model parameters (nugget, sill, or range) need adjustment to better reflect the actual prediction errors.
MSE	Mean Squared Error. It measures the average of the squared differences between the observed and predicted values. It should be as small as possible. It evaluates the accuracy of the predictions, where smaller values indicate a better fit to the observed data.
MSNE	Mean Squared Normalized Error. It checks if the kriging variance is consistent with the actual estimation errors. Values greater than 1 indicate underestimation of variability, while values less than 1 indicate overestimation. Ideally, it should be close to 1.
cor_obspred	Correlation between Observed and Predicted values. It measures the strength of the linear relationship between the measured data and the model estimates. Should be close to 1. High correlation indicates that the model follows the spatial trend of the observed data and accurately replicates the measured values.
cor_predres	Correlation between Predicted and Residuals. Measures the linear relationship between the predicted values and the residuals. It should be close to 0. A high correlation here suggests a systematic bias where the error depends on the predicted value (e.g., the model performs worse for higher values), indicating that the model has not captured all the spatial structure.
RMSE	Root Mean Squared Error. Provides a measure of the standard deviation of the residuals. It should be as small as possible. It measures the accuracy of the predictions; smaller values indicate a better fit between the observed and predicted values.
URMSE	Unbiased Root Mean Squared Error. It represents the random component of the error after removing the systematic bias (mean_error). It should be as small as possible, indicating that the remaining prediction errors are purely stochastic and not caused by model distortion.
iqr	Interquartile Range. Interquartile range of the residuals. It is used to assess the spread of the residuals by focusing on the middle 50% of the distribution. Smaller values generally indicate more precise and consistent predictions. Ideally, it should be small.

The Mean Error (mean_error) evaluates the bias of the predictions (Isaaks and Srivastava, 1989):

$$mean\_error = \frac{1}{n}\sum_{i=1}^{n}(\hat{z}(x_i) - z(x_i)) \tag{13.35}$$

Where n is the number of observation points, $\hat{z}(x_i)$ is the predicted value at location x_i, and $z(x_i)$ is the observed value at location x_i.

The Mean Error of Kriging Variance (me_mean) represents the average kriging standard error (Hiemstra et al., 2009):

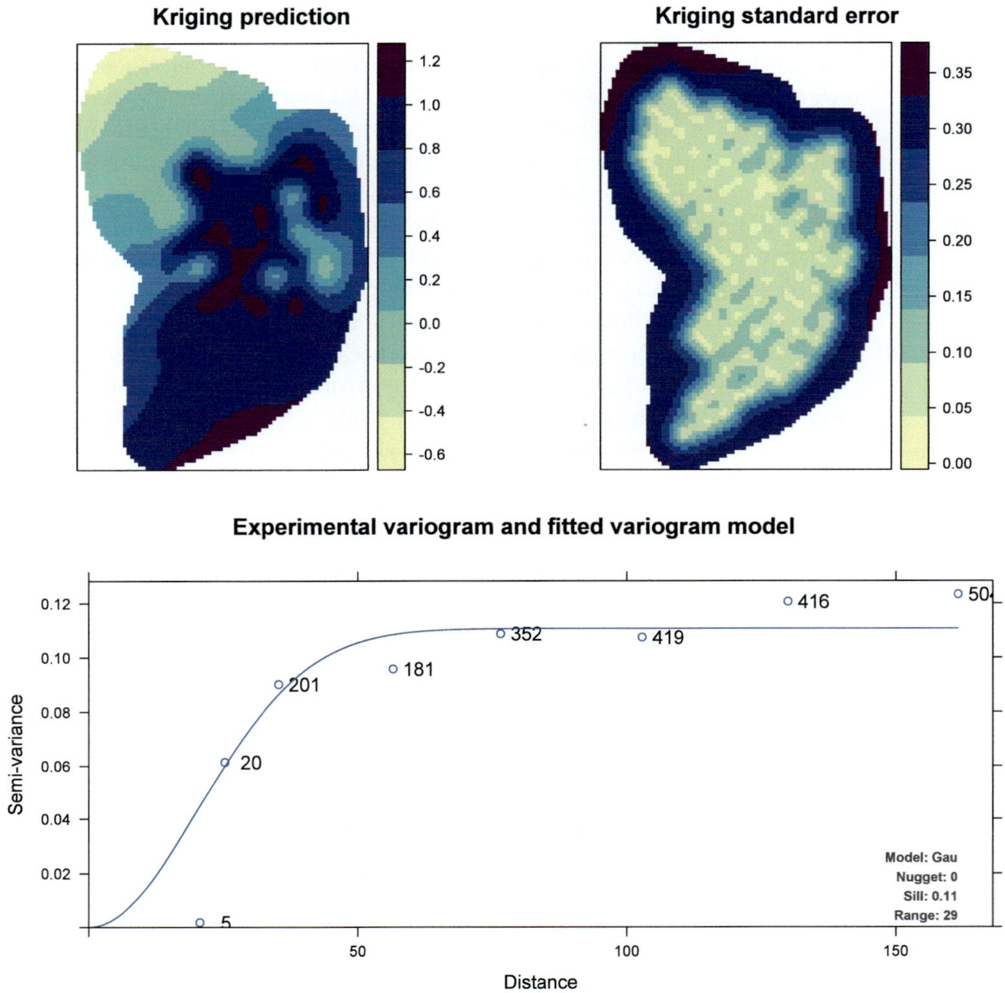

FIGURE 13.25 Indicator universal 30x30m block kriging with trend modeling of the incidence of bacterial blight spot in a coffee field (log(X) + log(Y) + log(Z)), considering 51% threshold of disease intensity.

$$me\_mean = \sqrt{\frac{1}{n}\sum_{i=1}^{n}\sigma^2(x_i)} \tag{13.36}$$

Where n is the number of observation points and $\sigma^2(x_i)$ is the kriging variance at location x_i.

The Mean Squared Error (MSE) measures the average squared difference between the observed and predicted values (Webster and Oliver, 2007):

$$MSE = \frac{1}{n}\sum_{i=1}^{n}(\hat{z}(x_i) - z(x_i))^2 \tag{13.37}$$

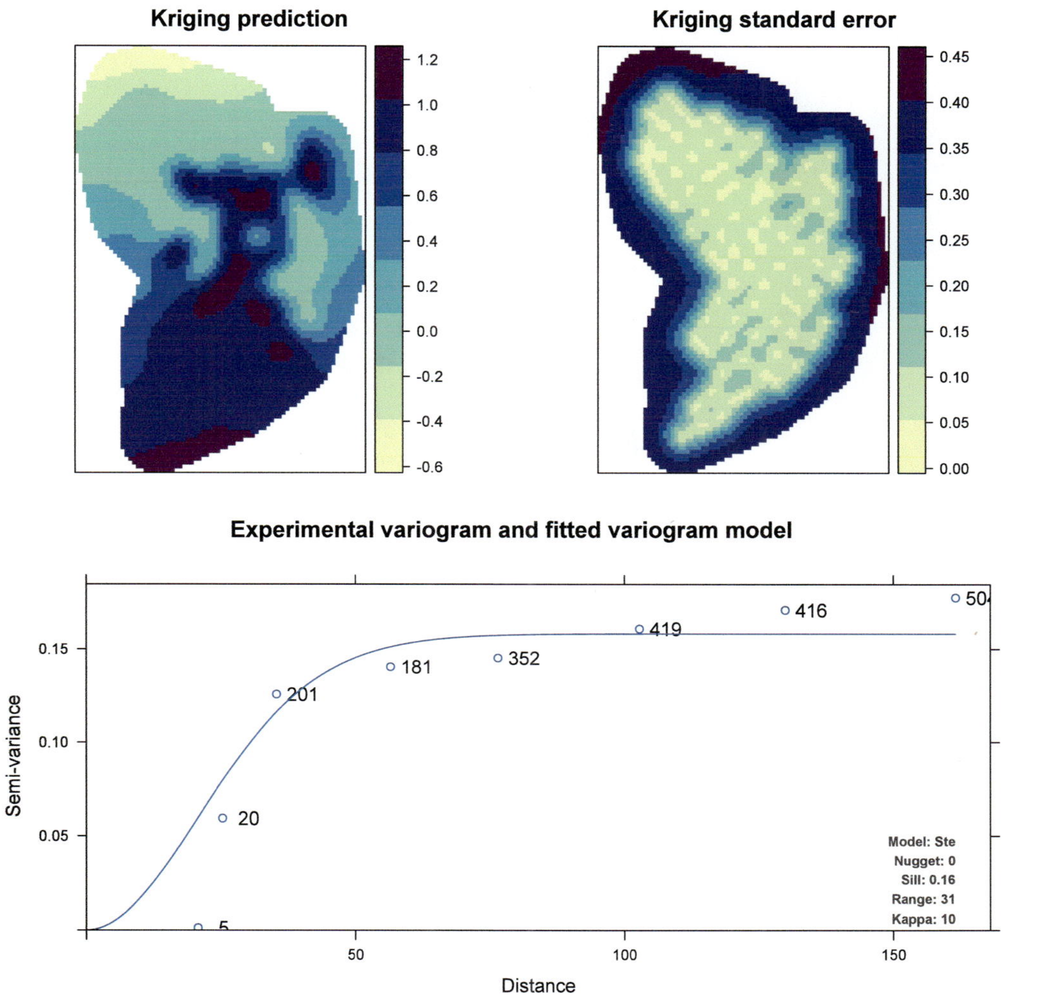

FIGURE 13.26 Indicator universal 30x30m block kriging with trend modeling of the incidence of bacterial blight spot in a coffee field (log(X) + log(Y) + log(Z)), considering 75% threshold of disease intensity.

Where n is the number of observation points, $\hat{z}(x_i)$ is the predicted value at location x_i, and $z(x_i)$ is the observed value at location x_i.

The Mean Squared Normalized Error (MSNE) standardizes the squared errors by the kriging variance (Lark, 2000):

$$MSNE = \frac{1}{n} \sum_{i=1}^{n} \left[\frac{(\hat{z}(x_i) - z(x_i))^2}{\sigma^2(x_i)} \right] \tag{13.38}$$

Where n is the number of observation points, $\hat{z}(x_i)$ is the predicted value at location x_i, $z(x_i)$ is the observed value at location x_i, and $\sigma^2(x_i)$ is the kriging variance at location x_i.

The Root Mean Squared Error (RMSE) provides a measure of the standard deviation of the residuals (Webster and Oliver, 2007):

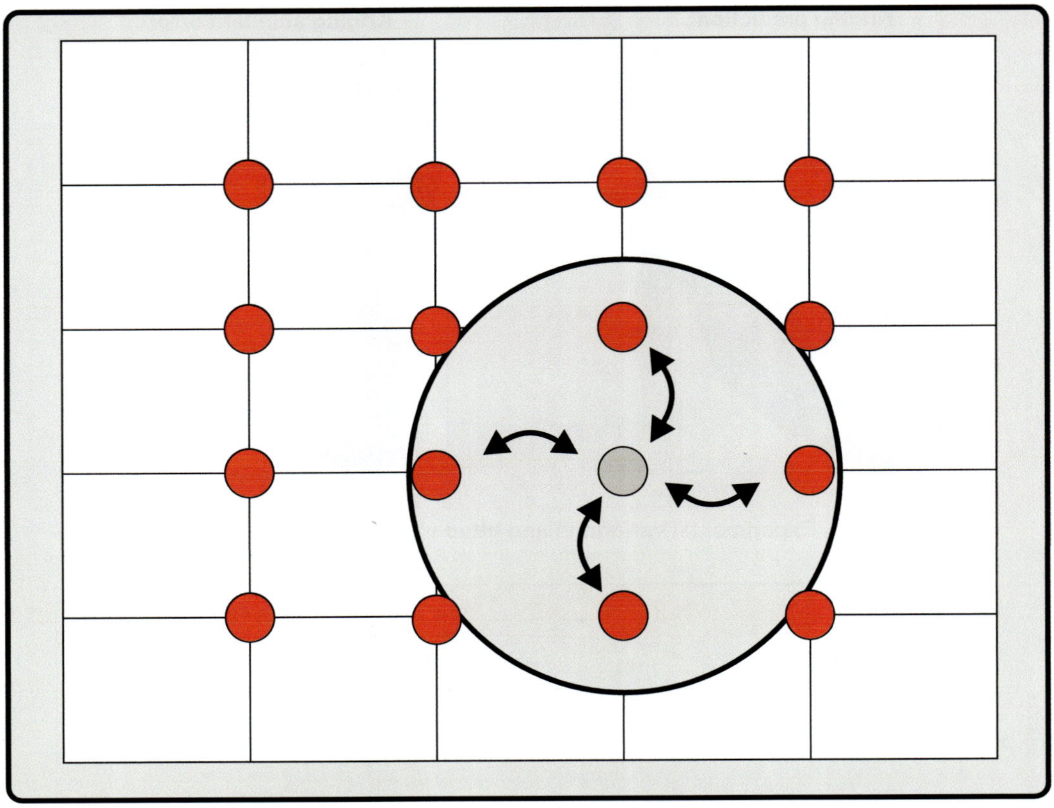

FIGURE 13.27 Using cross-validation to check the variogram against the original data.

$$RMSE = \sqrt{\frac{1}{n}\sum_{i=1}^{n}(\hat{z}(x_i) - z(x_i))^2} \qquad (13.39)$$

Where n is the number of observation points, $\hat{z}(x_i)$ is the predicted value at location x_i, and $z(x_i)$ is the observed value at location x_i.

The Unbiased Root Mean Squared Error (URMSE) represents the random component of the error after removing the systematic bias (mean_error) (Willmott and Matsuura, 2005):

$$URMSE = \sqrt{RMSE^2 - mean\_error^2} \qquad (13.40)$$

Where $RMSE$ is the Root Mean Squared Error and $mean\_error$ is the Mean Error.

The Correlation between Observed and Predicted values (cor_obspred) measures the strength of the linear relationship between the measured data and the model estimates (Isaaks and Srivastava, 1989):

$$cor\_obspred = \frac{\sum_{i=1}^{n}(z_i - \bar{z})(\hat{z}i - \bar{\hat{z}})}{\sqrt{\sum i = 1^n (z_i - \bar{z})^2 \sum_{i=1}^{n}(\hat{z}_i - \bar{\hat{z}})^2}} \qquad (13.41)$$

Where $z(x_i)$ is the observed value, $\bar{z}$ is the mean of observed values, $\hat{z}(x_i)$ is the predicted value, and $\bar{\hat{z}}$ is the mean of predicted values.

The Correlation between Predicted values and Residuals (cor_predres) measures the linear relationship between the predicted values and the residuals (Cressie, 1993):

$$cor\_predres = \frac{\sum_{i=1}^{n}(\hat{z}i - \bar{\hat{z}})(\epsilon_i - \bar{\epsilon})}{\sqrt{\sum i = 1^n(\hat{z}i - \bar{\hat{z}})^2 \sum i = 1^n(\epsilon_i - \bar{\epsilon})^2}} \tag{13.42}$$

Where $\hat{z}(x_i)$ is the predicted value, $\bar{\hat{z}}$ is the mean of predicted values, ϵ_i is the residual, and $\bar{\epsilon}$ is the mean of the residuals.

The Interquartile Range (iqr) is used to assess the spread of the residuals by focusing on the middle 50% of the distribution (Hyndman and Athanasopoulos, 2021):

$$iqr = Q_3 - Q_1 \tag{13.43}$$

Where Q_3 is the third quartile and Q_1 is the first quartile of the data.

13.12.1 EVALUATION OF MODEL ACCURACY AND RELIABILITY

To choose the best interpolation model, we must look beyond a single number and evaluate how the model performs in different ways. In this study, we use a set of traditional statistical measures to check if our predictions are biased, accurate, and if the uncertainty estimates are trustworthy.

13.12.1.1 Checking for systematic bias

First, we examine the Mean Error (ME) and the me_mean. According to Isaaks and Srivastava (1989), these values tell us if the model has a tendency to consistently over- or underestimate the real values. Ideally, the Mean Error should be zero. When we look at the me_mean, as suggested by Hiemstra et al. (2009), we are simply looking at that bias as a percentage of the average value, which helps us understand if the error is large or small compared to the overall magnitude of the data.

13.12.1.2 Measuring prediction accuracy

To see how close our estimates are to the real observations, we use the Root Mean Squared Error (RMSE). As explained by Webster and Oliver (2007), the RMSE is the most common way to measure accuracy because it uses the same units as our data. However, because the RMSE can be heavily affected by a few large errors, we also report the Interquartile Range (IQR) of the residuals. Following the principles of Hyndman and Athanasopoulos (2021), the IQR is a more "robust" measure that tells us how the majority of our predictions are performing, without being distracted by a few extreme outliers.

13.12.1.3 Validating the uncertainty of the model

A crucial part of this evaluation is checking the quality of the variogram model. We use the Mean Squared Normalized Error (MSNE) for this purpose. According to Lark (2000), the MSNE, also known as the Mean Squared Deviation Ratio, is the definitive diagnostic for the variogram. Ideally, the MSNE should be equal to 1. As Lark (2000) points out, if the MSNE is significantly higher than 1, it indicates that the variogram model is underestimating the actual spatial variability. If it is lower than 1, the model is overestimating the uncertainty. Therefore, a value close to 1 confirms that the theoretical variogram chosen by the software correctly represents the true spatial structure of the data, making the resulting kriging variance maps reliable for decision-making.

Finally, we look at the Correlation (cor_obspred) to see how well the predicted and observed values follow a straight line (Cressie, 1993). By balancing all these traditional metrics; bias, accuracy, and especially the variogram validation, we can confidently select the most appropriate model.

13.12.1.4 Uncertainties in geostatistical modeling using environmental variables

Cross-validation 10-fold is used to compare the variograms used to map bacterial blight spot in coffee crop using ordinary and universal kriging. The ordinary kriging estimates are compared with the universal kriging estimates with regressive trend of the model with logarithm of longitude, latitude, and with logarithm of longitude, latitude and altitude (Table 13.3).

```
cv1 <- autoKrige.cv(i280 ~ 1, blight_sp,  nfold = 10) # Ordinary kriging
summary(cv1) # See results
cv2 <- autoKrige.cv(i280 ~ log(X) + log(Y), blight_sp,
                    nfold = 10) # Univ. kriging logxy
summary(cv2) # See results
cv3 <- autoKrige.cv(i280 ~ log(X) + log(Y) + log(Z), blight_sp,
                    nfold = 10) # Universal kriging logxyz
summary(cv3) # See results
```

TABLE 13.3 Cross-validation used to compare kriging results according to error indicators for ordinary kriging (cv1), universal kriging with regressive trend of the model with logarithm of longitude and latitude (cv2), and with logarithm of longitude, latitude and altitude (cv3).

Indicator	cv1	cv2	cv3
mean_error	-0.5674	-1.367	-0.4502
me_mean	-0.0091	-0.02193	-0.007221
MAE	11.33	12.37	12.15
MSE	327.5	395.6	354.3
MSNE	1.126	1.611	1.341
cor_obspred	0.8948	0.874	0.8865
cor_predres	0.0004115	-0.124	-0.08683
RMSE	18.1	19.89	18.82
RMSE_sd	0.444	0.488	0.4618
URMSE	18.09	19.84	18.82
iqr	8.091	8.126	13.07

The prediction errors are similar, so that in all situations the correlation values between the observed data and the predicted values (cor_obspred) and the correlation between the predicted values and the residual values (cor_predres) are close to 1 and 0, respectively. In the case of universal kriging, there was greater error in spatial prediction; however, the effect of altitude is considered in the spatial prediction of the disease, adding external information available for mapping the disease. Other variables, such as surface temperature, should be evaluated in the future in the universal kriging model in order to consider the effect of new external variables on disease progress.

The `compare.cv` function is used to create bubbleplots for cross-validations `cv1`, `cv2`, and `cv3` (Figure 13.28).

```
compare.cv(cv1,cv2,cv3, bubbleplots = TRUE)
#                    cv1       cv2       cv3
# mean_error     -0.5674    -1.367   -0.4502
# me_mean        -0.0091 -0.02193 -0.007221
# MAE              11.33     12.37     12.15
# MSE              327.5     395.6     354.3
# MSNE             1.126     1.611     1.341
# cor_obspred     0.8948     0.874    0.8865
# cor_predres 0.0004115    -0.124  -0.08683
# RMSE              18.1     19.89     18.82
# RMSE_sd          0.444     0.488    0.4618
# URMSE            18.09     19.84     18.82
# iqr              8.091     8.126     13.07
```

The maps with disease prediction by ordinary kriging are defined as reference for evaluating the disease prediction by universal kriging. The result of the squared ordinary kriging prediction residuals minus the universal kriging residuals with trend $log(X)+log(Y)$ squared or with trend $log(X) + log(Y) + log(Z)$ squared, obtained in the cross-validation at each point observed in the field, are compared to assess the locations in the field that are most modified (red bubble) with the different trends added to the prediction model. We can observe that the biggest changes occurred in the mapping by universal kriging considering the trend modeling in X, Y, Z (Figures 13.29 and 13.30).

```
compare.cv(cv1,cv2, bubbleplots = TRUE, plot.diff = TRUE) # cv1 x cv2
compare.cv(cv1,cv3, bubbleplots = TRUE, plot.diff = TRUE) # cv1 x cv3
```

With the use of altitude in trend modeling by universal kriging, there was a greater difference in disease prediction mainly in the north-central part of the area. With trend modeling in X and Y, the greatest differences in relation to ordinary kriging occurred mainly in the central-northeastern part of the area.

13.13 COMPUTATIONAL PRACTICE

In the computational practice, some options of the geostatistical package `gstat` (Pebesma and Graeler, 2021) are evaluated. The `gstat` R package is loaded with the `library` function.

```
library(gstat)
# package 'gstat' was built under R version 4.3.3
```

The variogram models available in `gstat` are accessed with the `vgm` function.

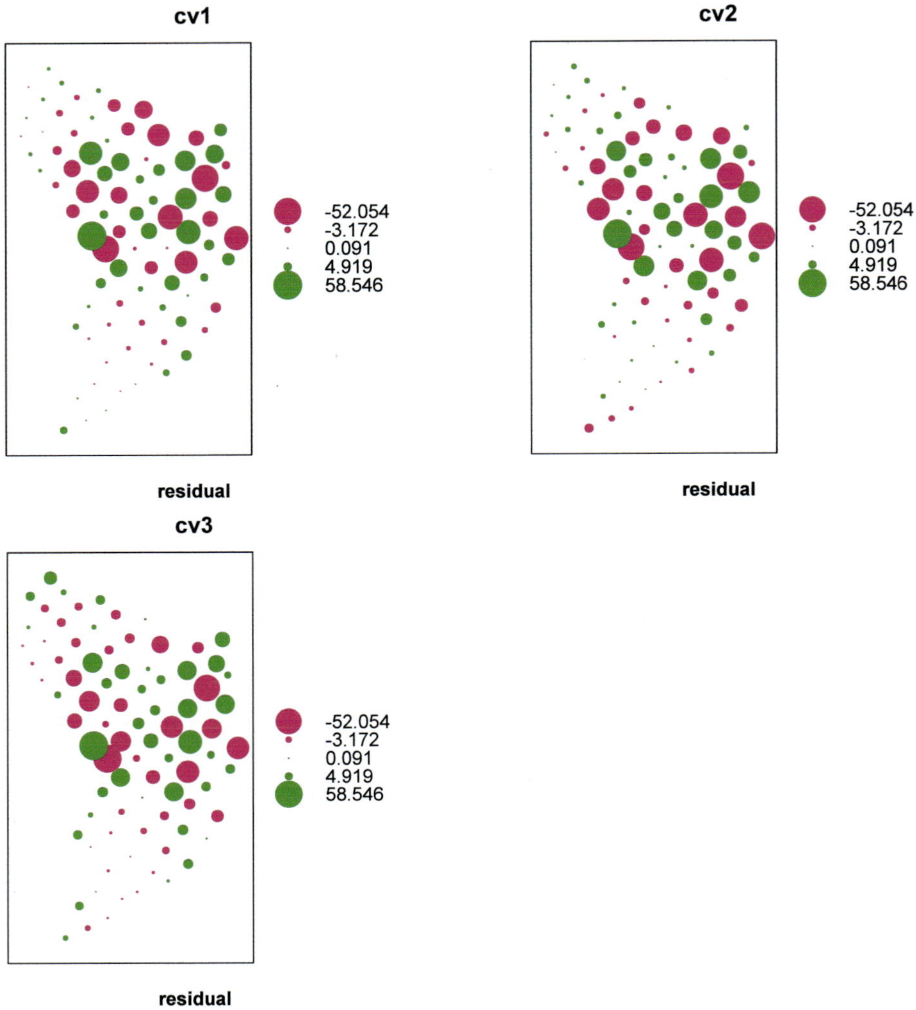

FIGURE 13.28 Using bubble plots to evaluate differences in the prediction of bacterial blight spot in a coffee field from cross-validation when comparing spatial predictions of ordinary kriging, universal kriging with log(X) + log(Y) (cv2) trend modeling and with log(X) + log(Y) + log(Z) (cv3) trend modeling.

```
vgm()
# short                                            long
# 1     Nug                                  Nug (nugget)
# 2     Exp                             Exp (exponential)
# 3     Sph                               Sph (spherical)
# 4     Gau                                Gau (gaussian)
# 5     Exc          Exclass (Exponential class/stable)
# 6     Mat                                  Mat (Matern)
# 7     Ste Mat (Matern, M. Stein's parameterization)
# 8     Cir                                Cir (circular)
# 9     Lin                                  Lin (linear)
```

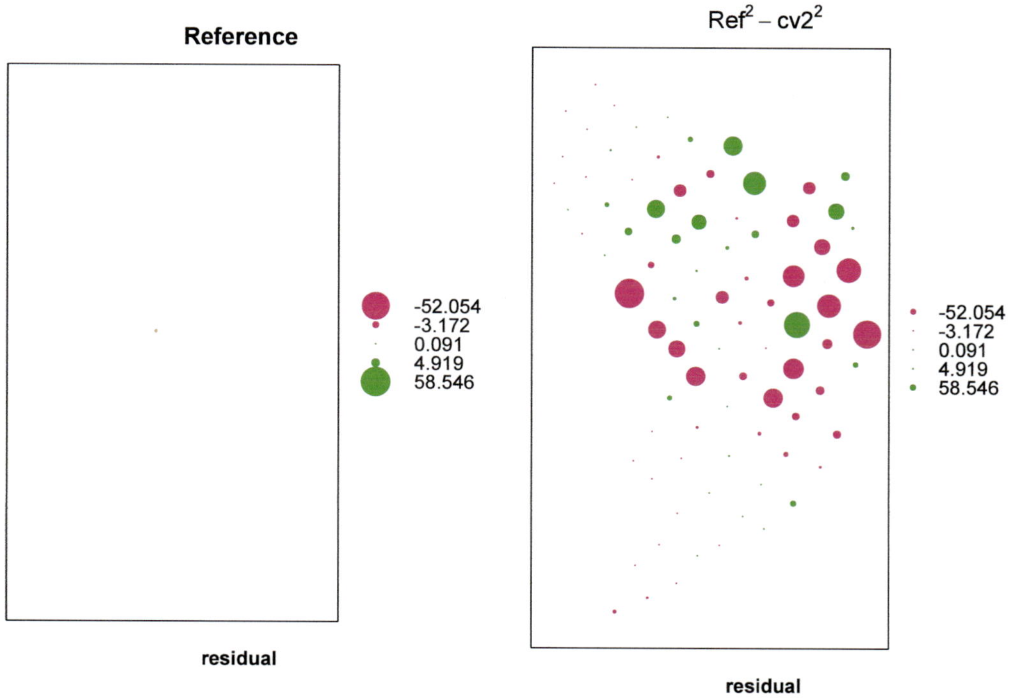

FIGURE 13.29 Using bubble plots to evaluate differences in the prediction of bacterial blight spot in a coffee field from cross-validation residues when comparing the residues of squared ordinary kriging predictions minus the residues of universal kriging predictions with squared log(X) + log(Y) (cv2) trend modeling.

```
# 10    Bes                    Bes (bessel)
# 11    Pen           Pen (pentaspherical)
# 12    Per                 Per (periodic)
# 13    Wav                     Wav (wave)
# 14    Hol                     Hol (hole)
# 15    Log              Log (logarithmic)
# 16    Pow                    Pow (power)
# 17    Spl                   Spl (spline)
# 18    Leg                 Leg (Legendre)
# 19    Err      Err (Measurement error)
# 20    Int                Int (Intercept)
```

Graphs of spherical, exponential, Gaussian and linear variograms are determined with the `show.vgms` function (Figure 13.31).

```
show.vgms(models = c("Sph", "Exp", "Gau", "Lin"), nugget = 0.1)
```

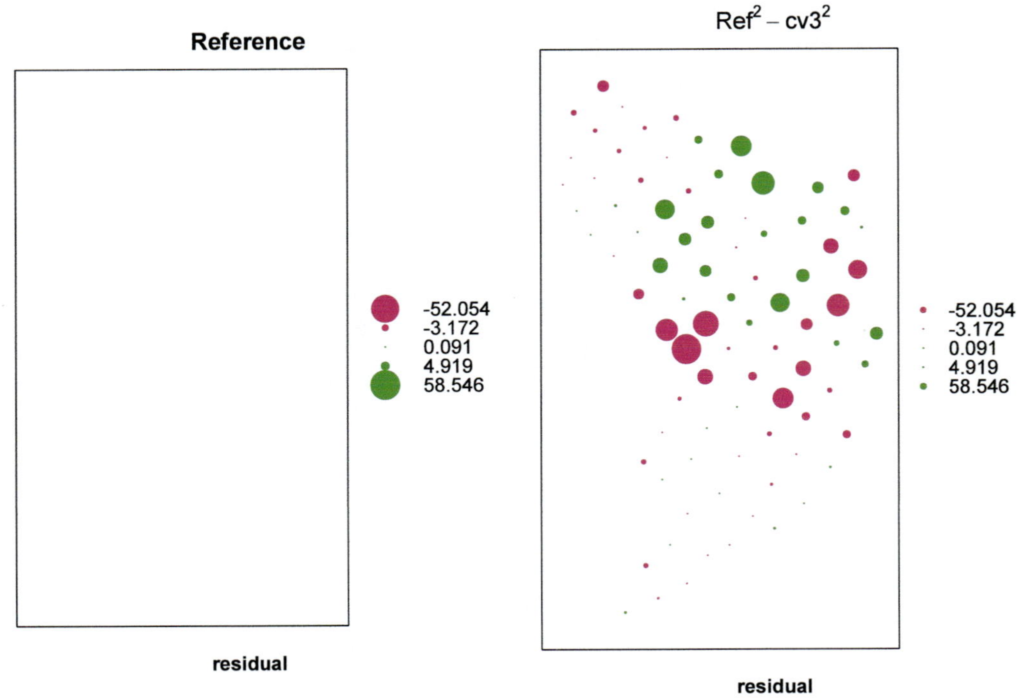

FIGURE 13.30 Using bubble plots to evaluate differences in the prediction of bacterial blight spot in a coffee field from cross-validation residues when comparing the residues of squared ordinary kriging predictions minus the residues of universal kriging predictions with squared log(X) + log(Y) + log(Z) (cv3) trend modeling.

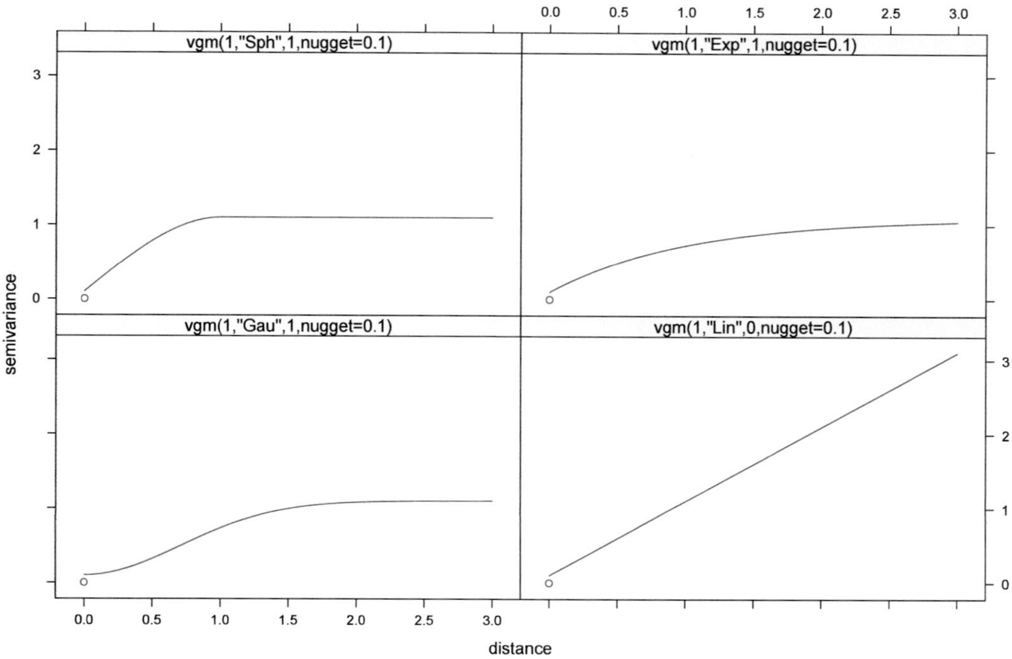

FIGURE 13.31 Variogram simulation with parameters.

14 Article Elaboration with Environmental Geocomputation and R

14.1 LEARNING QUESTIONS

The learning questions answered by reading the chapter are:

- How to perform application of science with geocomputation to prepare articles.
- What is the relationship between geocomputation applications with science practice?
- How to solve geographic problems with geocomputation science in an objective way.

14.2 LEARNING OUTCOMES

The learning outcomes answered by reading the chapter are:

- Understand scientific applications of geocomputation with article writing;
- Identify hypotheses in scientific articles in geocomputation;
- Verify an example of writing a scientific paper in geocomputation with R;
- Check an example of scientific dissemination of geocomputation with R.

14.3 INTRODUCTION

Articles are written communications, published in specialized journals, with the objective of disseminating to the scientific community the results, even if partial, of research in a specific area. According to the Brazilian Association of Technical Standards (ABNT - Associação Brasileira de Normas Técnicas, 2003), a scientific article can be defined as part of a publication with declared authorship, which presents and discusses ideas, methods, techniques, processes and results in various areas of knowledge.

The scientific article can be:

- **Original** <- works resulting from scientific research presenting original data of findings with respect to experimental or observational aspects of medical, biochemical and social characteristics and includes descriptive analysis and/or inferences of own data;

- **Review** <- critical synthesis of available knowledge on a given topic, through analysis and interpretation of pertinent bibliography that discusses the limits and methodological reaches, allowing to indicate perspectives of continuity of studies in that line of research, i.e., works that have the objective of summarizing, analyzing, evaluating or synthesizing already published research works, bibliographic reviews, among others.

DOI: 10.1201/9781003724377-14

14.4 IMPLEMENTING SCIENTIFIC GEOCOMPUTATION PROJECTS

In the implementation cycle of scientific projects in geocomputation, the development occurs as a continuous and recursive process in which modifications and improvements can be made according to adaptations needed to new scientific and technological advances. With the implementation of the projects, results are obtained in order to enable the elaboration of articles for scientific dissemination.

The model is divided into different working phases of the geographic information systems development process (Figure 14.1) (Lo and Yeung, 2007):

- Planning;
- Analysis;
- Design;
- Implementation and data collection;
- Support.

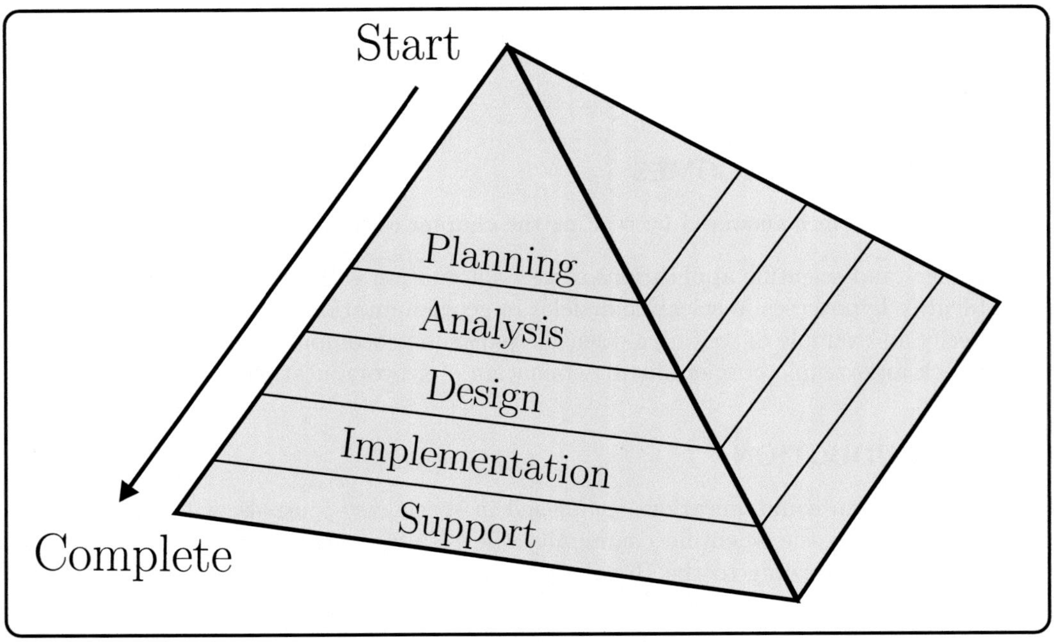

FIGURE 14.1 Pyramidal model of the geographic information systems implementation cycle.

Currently, the geographic information system implementation project is comprehensive with different project management and support activities (Lo and Yeung, 2007):

- Planning, localization and control;
- Formal review of business cases and project proposals;
- Availability of studies by simulation or prototypes;
- Data and software quality control;
- Software evaluation and standard setting;
- Software acquisition, installation and version control;
- Preparation and production of technical documents and user guides;
- Risk analysis with contingency plan to recover data and system failure.

14.5 SCIENTIFIC DISSEMINATION IN GEOCOMPUTATION

In scientific dissemination, considerations are presented about scientific writing in the form of a scientific paper and scientific presentation in the format of a slide show.

14.5.1 SCIENTIFIC WRITING

Scientific writing, although an indispensable step in the scientific process, is often overlooked in undergraduate courses in favor of maximizing class time devoted to scientific concepts. However, the ability to effectively communicate research findings is crucial to success in science. Students and professional scientists are judged by the number of papers published and the number of citations those papers have received. Therefore, a solid foundation in scientific writing can better prepare undergraduate and graduate students for productive academic careers (Turbek et al., 2016).

In writing a scientific paper, the structure of the manuscript should be similar to that of an hourglass, with an open and broad beginning, tapering off as it narrows down to the relevant literature that determined the relevance of the paper as unpublished, definition of the hypothesis, material and methods used. The knowledge gap that will be evaluated should be clear by this point. In the results phase, corresponding to the middle of the hourglass, the objectives are achieved using geocomputation and statistical techniques used to evaluate the acceptance of the hypothesis. In the discussion phase of the results obtained, the hourglass is again enlarged to interpret the results based on the existing literature in order to make clear that the knowledge gap detected previously is filled, culminating in conclusions and implication of the work (Figure 14.2).

Therefore, in the Introduction, one should start from broad ideas to specific questions, using a structure that resembles the image of a funnel or an inverted pyramid. The operation of the Discussion and Introduction sections should be opposite and mirrored. In the Discussion, the pyramid presents the conventional format, starting from specific questions (study findings) to more comprehensive elaborations (Figure 14.3) (Cáceres et al., 2011).

The immutable characteristics of good scientific writing that distinguish it from other literature are (Lindsay, 2011):

- Accuracy;
- Clarity;
- Brevity.

A vague text cannot be considered a scientific text, because it lacks clarity or is ambiguous. A prolix and unnecessarily discursive scientific manuscript is considered poor in terms of scientific writing. Therefore, a clear, precise and brief text can be read and understood by a larger number of readers (Lindsay, 2011).

When writing the scientific article in the field of geocomputation, often the topics should be written with the verb in the past tense, because the text was written in the context of presenting the results of the work with relevant scientific conclusions about a given geographic problem. In summary, topics ranging from the title to the bibliographic references are contemplated in the preparation of a scientific article (Table 14.1).

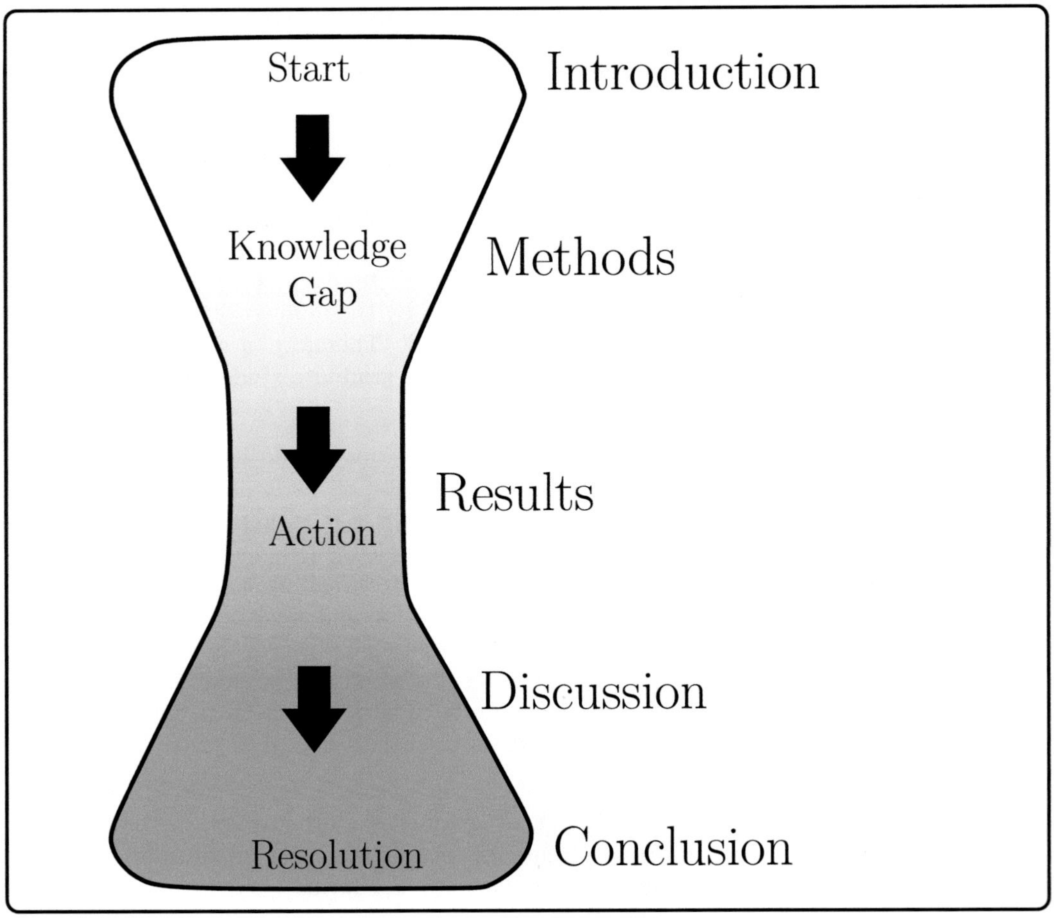

FIGURE 14.2 Structure of a scientific paper compared to the flow of an hourglass.

TABLE 14.1 Aspects considered in the development of scientific paper topics in geocomputation.

Topic	Aspects considered
Title	It should be succinct and contain in a few words what it is intended to accomplish with the work.
Authors	Include sequence of authors, advisors and affiliations according to the norms.
Abstract and keywords	Include index words for the Abstract that have not yet been placed in the title.
Introduction	Point out the problem and the demand, and show ability to diagnose the problem and solve it. Get information about what has already been done in other work on a similar subject. At the end of the Introduction, briefly present the objective and the hypothesis.
Methods	Put as many details as possible, but without exaggerating, e.g., variable studied, date the radiometric data obtained, and analysis performed on the data.

Topic	Aspects considered
Results	Present the results obtained based on the data analysis.
Discussion	Discuss the results obtained based on the available literature.
Conclusion	Address what was obtained with the execution of the work within what was proposed in the objective.
References	Include the bibliographical references used. Prioritize citations of scientific articles in journals with editorial board and renowned in the area.

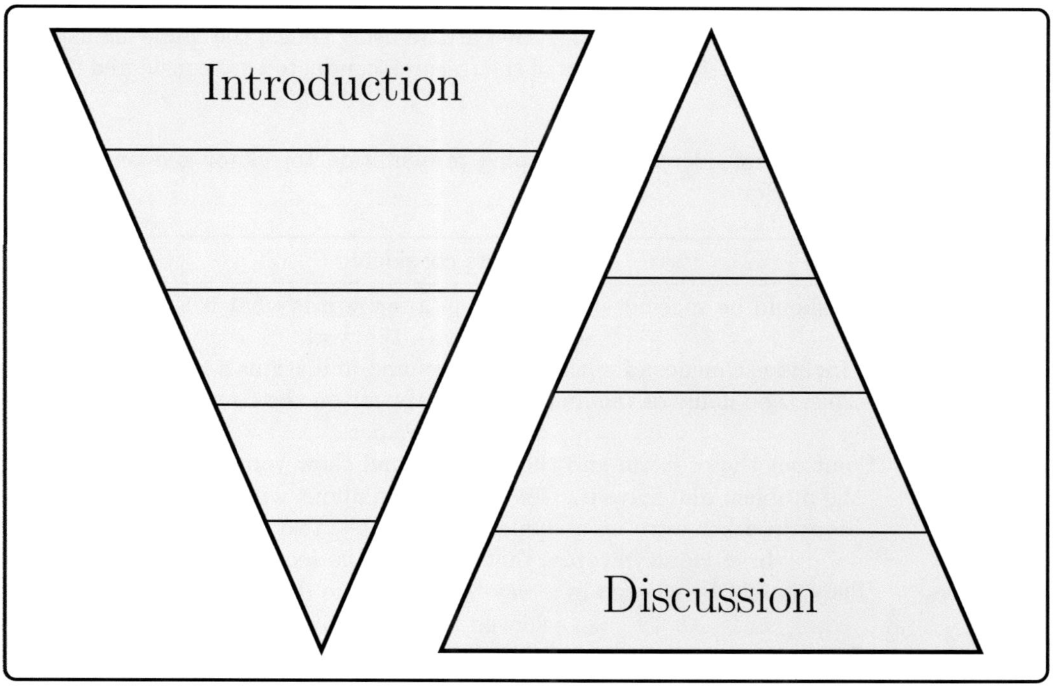

FIGURE 14.3 Mirror structure of introduction and discussion sections.

14.5.2 SCIENTIFIC DISSEMINATION

Formal scientific dissemination is accomplished through presentations in different types of media. The use of short video presentations has been a form of presentation used by researchers and in scientific journals. However, transforming a scientific publication into a short video with accessible language can be a complex mission for researchers with little affinity and availability of these resources.

Dissemination of science means allowing other people to learn about the research conducted. Scientists working in academia (universities, research centers) and in research and development in large corporations and small companies need to disseminate research results to gain respect and credibility in society and careers. This can create a market for products and attract talented employees, as well as build a network of collaborations with other research groups. This can increase the chances of obtaining funding for projects and research with public and private investors (de Carvalho Alves and Sanches, 2023).

Scientists should inform contributors and investors about accountability and if possible the positive impact of research on health and socio-economic progress of society. An effective communication strategy can increase the likelihood that it will attract the attention of decision-makers, and that science will be used to support decisions about strategic evidence-based policy priorities to meet the needs of the population (de Carvalho Alves and Sanches, 2023).

Researchers can disseminate scientific work through articles, review papers, workshops, posters, and talks at conferences, seminars, and in reports. In addition, written and visual materials (video and infographics) can be used for flyers, brochures, press releases, websites, newsletters, blogs, and the entire broad spectrum of social media (de Carvalho Alves and Sanches, 2023).

When preparing the presentation of the scientific article using slides, the following aspects on each topic proposed in the dissemination of the research conducted are considered (Table 14.2).

TABLE 14.2 Aspects considered in developing presentation topics for geocomputation scientific papers with slides.

Topic	Aspects considered
Title	It should be succinct and contain in a few words what it is intended to accomplish with the work.
Authors	Include sequence of authors, advisors and affiliations according to the norms. The name of the authors must appear on the cover page, below the title of the paper.
Introduction	Point out the problem and the demand, and show your ability to diagnose the problem and solve it. Get information about what has already been done in other work on a similar subject. The text should be presented briefly and in topics that facilitate the sequence of ideas.
Objectives	They should be written in a way that leaves no doubt as to what is to be achieved in the work.
Methods	Put as many details as possible, but without exaggerating, e.g., variable studied, date the radiometric data obtained, and analysis performed on the data.
Results and discussion	Present the results obtained based on the data analysis. Images, graphs, and tables should be legible and avoid too much text for participants to read the slides. The discussion can be held immediately after the presentation of each result.
Conclusions	Address what was obtained with the execution of the work within what was proposed in the objective. Open space for discussion and debate with the public.
References	Include the bibliographical references used.

14.5.3 PREDATORY JOURNALS

With today's technological development, one can publish a book or scientific article on a website with all the necessary characteristics to simulate the image and procedures of a major publisher. In predatory journals, classified as open access, the scientific quality of the publications is highly questionable. In these journals, peer review of low-quality academic manuscripts is requested from authors in order to collect payments for publication. The number of articles in these journals that have actually been read, cited, or have any significant research impact in the same area of knowledge may be low. Citation statistics over a five-year period in Google Scholar for 250 random articles published in predatory journals in 2014 averaged 2.6 citations per article, with 56% of the articles having no citations at all. Based on a comparative random sample of articles published in approximately 25,000 peer-reviewed journals included in the Scopus index, an average of 18.1 article citations was observed over the same period, with only 9% of articles having no citations at all. Therefore, it was concluded that in articles published in predatory journals, there was little scientific impact (Björk et al., 2020).

With the subversion of the scientific publication process, not validated by peers, the basic foundation of communication in science can be discredited, ultimately turning it into mere opinion pieces disguised as scientific articles, without any validation of the published content. Therefore, choosing the journal in which to publish a paper can be a challenge full of pitfalls for the less and more experienced authors, and it is important to evaluate whether some knowledge about the credibility of the journal and the publisher has been made available, in addition to exercising the authors' critical sense in choosing a journal with credibility (Penedo and Borges, 2017).

14.6 PRACTICE FOR WRITING AN ARTICLE

In the practical activity, a scientific article abstract was made as an example of geocomputation applications.

14.6.1 PREPARATION OF A SCIENTIFIC ARTICLE ABSTRACT

The article abstract was prepared as follows:

Potential effects of spatio-temporal temperature variation for monitoring coffee leaf rust progress under CMIP6 climate change scenarios

Marcelo de Carvalho Alves - Federal University of Lavras, Agricultural Engineering Department, email: *marcelo.alves@ufla.br*

Abstract

Plant diseases occur in all regions of the globe where there are susceptible hosts, aggressive and virulent pathogens, and a favorable environment. We aimed to evaluate potential effects of temperature variation on the monocyclic and polycyclic processes of coffee leaf rust in susceptible Arabica coffee cultivars cultivated in Minas Gerais state, Brazil. Historical monthly mean air temperature data, from 1970-2000, was downscaled and used as the temperature reference period and maximum and minimum temperature for four future scenarios, 2021-2040, 2041-2060, 2061-2080, and 2081-2100 (SSP126 scenario) from Coupled Model Intercomparison Project phase 6 (CMIP6) were used to characterize the future mean air temperature variation were used in the modeling of areas favorable to the progress of coffee leaf rust in Minas Gerais using spatial data techniques. Digital elevation model was considered for downscaling climate data. A non-linear regression model simulating the monocyclic process of coffee leaf rust was used to simulate the potential progress of the disease in susceptible cultivars under the different scenarios evaluated. In general, coffee leaf rust progress increased in susceptible cultivars located in areas with higher ground elevation, with emphasis in the south of the state, as well as in the main Arabica coffee producing regions. There was a reduction of areas favorable to rust in the north of the state due to temperature increase considering climate change scenario; however, new areas in the south of the state became more favorable to the disease. Temperature increase from 2021 to 2040, 2041 to 2060, 2061 to 2081, and from 2081 to 2100, was 1.2°, 0.6°, 0.2°, and 0.1°C, respectively.

Keywords: Epidemiology, WorldClim, Shared Socioeconomic Pathways, CNRM-ESM2-1 model.

When preparing the Abstract, some journals offer the option of using a graphic abstract to illustrate what was done in the article in summary form. As an example, we have created a graphic abstract for one of the articles published in the laboratory manual in this book (Figure 14.4).

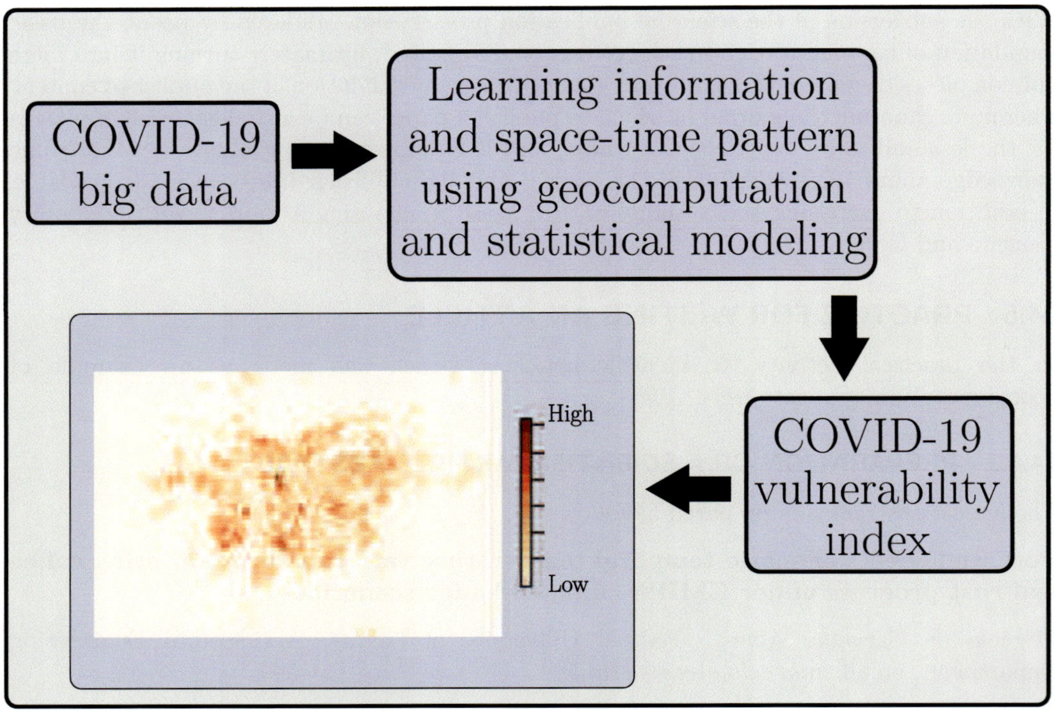

FIGURE 14.4 Example of a graphic abstract used to illustrate what was accomplished in an article in summary form.

References

ABNT - Associação Brasileira de Normas Técnicas (2003). NBR 6022 Informação e documentação - Artigo em publicação periódica científica impressa - Apresentação. Technical report, Associação Brasileira de Normas Técnicas, Rio de Janeiro.

Adler, D. and Murdoch, D. (2025). *rgl: 3D Visualization Using OpenGL*. R package version 1.3.18.

Alphan, H. and Aşur, F. (2021). Geospatial analysis of lake scenery as an indicator for the environment: The city of Van (Turkey) and its surroundings. *Environmental and Sustainability Indicators*, 9:100091.

Alves, M. C., Carvalho, L. G., Barbosa, H. A., Sanches, L., Oliveira, M. S., Ferreira, D. F., and Silva, S. S. (2020). Human progress and drought sensitivity behavior. *Science of The Total Environment*, 702:134966.

Alves, M. C. and Silva, F. M. (2016). *Geomática para Levantamento de Ambientes: Base para Aplicações em Topografia, Georreferenciamento e Agricultura de Precisão*. Editora UFLA, Lavras, 1st edition.

Andriotti, J. L. S. (2013). *Fundamentos de estatística e geoestatística*. Editora Unisinos.

Appelhans, T., Detsch, F., Woellauer, C. R. S., Forteva, S., Nauss, T., Pebesma, E., Russell, K., Sumner, M., Darley, J., Roudier, P., Schratz, P., Marburg, E. I., and Busetto, L. (2020). *mapview: Interactive Viewing of Spatial Data in R*.

Arvanitou, E.-M., Ampatzoglou, A., Chatzigeorgiou, A., and Carver, J. C. (2021). Software engineering practices for scientific software development: A systematic mapping study. *Journal of Systems and Software*, 172:110848.

Auguie, B. and Antonov, A. (2017). *gridExtra: Miscellaneous Functions for Grid Graphics*.

Aybar, C., Qiusheng, W., Bautista, L., Yali, R., Barja, A., Ushey, K., Ooms, J., Appelhans, T., Allaire, J., Tang, Y., Roy, S., Adauto, M., Carrasco, G., Bengtsson, H., Hollister, J., Donchyts, G., and Appel, M. (2022). *rgee: R Bindings for Calling the Earth Engine API*. R package version 1.1.4.

Baddeley, A., Lawrence, T., and Rubak, E. (2017). *globe: Plot 2D and 3D Views of the Earth, Including Major Coastline*.

Baddeley, A., Rubak, E., and Turner, R. (2015). *Spatial Point Patterns: Methodology and Applications with R*. Chapman & Hall/CRC Press, London.

Baddeley, A. and Turner, R. (2005). spatstat: An R package for analyzing spatial point patterns. *Journal of Statistical Software*, 12(6):1–42.

Baddeley, A., Turner, R., Mateu, J., and Bevan, A. (2013). Hybrids of gibbs point process models and their implementation. *Journal of Statistical Software*, 55(11):1–43.

Bailey, J., Hafen, R., Nowosad, J., and Zehnder, L. S. (2018). *geogrid: Turn Geospatial Polygons into Regular or Hexagonal Grids*.

Beven, K. J., Wood, E. F., and Sivapalan, M. (1988). On hydrological heterogeneity—catchment morphology and catchment response. *Journal of hydrology*, 100(1-3):353–375.

Bivand, R., Keitt, T., Rowlingson, B., Pebesma, E., Sumner, M., Hijmans, R., Baston, D., Rouault, E., Warmerdam, F., Ooms, J., and Rundel, C. (2021). *rgdal: Bindings for the Geospatial Data Abstraction Library.*

Bivand, R., Lewin-Koh, N., Pebesma, E., Archer, E., Baddeley, A., Bearman, N., Bibiko, H.-J., Brey, S., Callahan, J., Carrillo, G., Dray, S., Forrest, D., Friendly, M., Giraudoux, P., Golicher, D., Rubio, V. G., Hausmann, P., Hufthammer, Ove, K., Jagger, T., Johnson, K., Lewis, M., Luque, S., MacQueen, D., Niccolai, A., Pebesma, E., Lam, O. P., and Turner, R. (2020a). *maptools: Tools for Handling Spatial Objects.*

Bivand, R., Nowosad, J., Lovelace, R., Monmonier, M., and Snow, G. (2020b). *spData: Datasets for Spatial Analysis.*

Bivand, R., Rundel, C., Pebesma, E., Stuetz, R., Hufthammer, K. O., Giraudoux, P., Davis, M., and Santilli, S. (2018). *rgeos: Interface to Geometry Engine - Open Source ('GEOS').*

Bivand, R. S., Pebesma, E., and Gómez-Rubio, V. (2013). *Applied Spatial Data Analysis with R.* Springer New York, New York, NY, 2nd edition.

Björk, B.-C., Kanto-Karvonen, S., and Harviainen, J. T. (2020). How frequently are articles in predatory open access journals cited. *Publications*, 8(2):17.

Brenning, A., Bangs, D., Becker, M., Schratz, P., and Polakowski, F. (2018a). *RSAGA: SAGA Geoprocessing and Terrain Analysis.*

Brenning, A., Polakowski, F., Becker, M., and Muenchow, J. (2018b). *RPyGeo: ArcGIS Geoprocessing via Python.*

Brunsdon, C. (2020). Estimation of Population from Pycnophylatic Interpolation (pycno).

Brunsdon, C. (2025). *pycno: Pycnophylactic Interpolation.* R package version 1.4.1.

Burrough, P. A. and McDonnell, R. (1998). *Principles of Geographical Information Systems.* Oxford University Press, Oxford.

Cáceres, A. M., Gândara, J. P., and Puglisi, M. L. (2011). Scientific writing and the quality of papers: towards a higher impact. *Jornal da Sociedade Brasileira de Fonoaudiologia*, 23(4):401–406.

Câmara, G. and Medeiros, J. S. (1998). Princípios básicos em geoprocessamento. In Assad, E. D.; Sano, E. E., editor, *Sistema de Informações Geográficas*, chapter 1, pages 3–11. Embrapa, Brasília, 2nd edition.

Casaca, J., Matos, J., and Baio, M. (2007). *Topografia Geral.* LTC Editora, Rio de Janeiro, 4th edition.

Chamberlain, S. (2018). *getlandsat: Get Landsat 8 Data from Amazon Public Data Sets.*

Chamberlain, S., Anderson, B., Salmon, M., Erickson, A., Potter, N., Stachelek, J., Simmons, A., Ram, K., and Edmund, H. (2021). *rnoaa: 'NOAA' Weather Data from R.*

Chamberlain, S. and Hollister, J. (2021). *lawn: Client for 'Turfjs' for 'Geospatial' Analysis.* R package version 0.6.0.

Chamberlain, S., Zhu, H., Jahn, N., Boettiger, C., and Ram, K. (2022). *rcrossref: Client for Various CrossRef 'APIs'*. R package version 1.2.0, https://github.com/ropensci/rcrossref.

Chang, K.-T., Merghadi, A., Yunus, A. P., Pham, B. T., and Dou, J. (2019). Evaluating scale effects of topographic variables in landslide susceptibility models using gis-based machine learning techniques. *Scientific reports*, 9(1):12296.

Chang, W., Cheng, J., Allaire, J., Sievert, C., Schloerke, B., Xie, Y., Allen, J., McPherson, J., Dipert, A., Borges, B., RStudio, Foundation, Otto, M., Thornton, J., Khan, P. N., Tsaran, V., Lembree, D., Srinivasu Chakravarthula, O'Conno, C., Petre, S., Rowls, A., Gandy, D., Reavis, B., Bejaoui, S., Ineshin, D., Samhuri, S., Limited, S., Gruber, J., Sagalaev, I., and Team, R. C. (2021). *shiny: Web Application Framework for R.*

Chaves, M., de Carvalho Alves, M., de Oliveira, M., and Sáfadi, T. (2018). A geostatistical approach for modeling soybean crop area and yield based on census and remote sensing data. *Remote Sensing*, 10(5):680.

Cheng, J., Karambelkar, B., Xie, Y., Wickham, H., Russell, K., Johnson, K., Schloerke, B., Foundation, J., Contributors, Agafonkin, V., and Leafle, R. (2021). *leaflet: Create Interactive Web Maps with the JavaScript 'Leaflet' Library.*

Conrad, O., Bechtel, B., Bock, M., Dietrich, H., Fischer, E., Gerlitz, L., Wehberg, J., Wichmann, V., and Böhner, J. (2015). System for automated geoscientific analyses (SAGA) v. 2.1. 4. *Geoscientific Model Development*, 8(7):1991–2007.

Cooley, D., Barcelos, P., and Rstudio (2020). *googleway: Accesses Google Maps APIs to Retrieve Data and Plot Maps.*

Cressie, N. (1993). *Statistics for Spatial Data.* John Wiley & Sons.

Daniel Baston (2023). *exactextractr: Fast Extraction from Raster Datasets using Polygons.* R package version 0.10.0, https://github.com/isciences/exactextractr.

de Carvalho Alves, M., Pozza, E. A., Sanches, L., Belan, L. L., and de Oliveira Freitas, M. L. (2022). Insights for improving bacterial blight management in coffee field using spatial big data and machine learning. *Tropical Plant Pathology*, 47(1):118–139.

de Carvalho Alves, M. and Sanches, L. (2022). *Surveying with Geomatics and R.* CRC Press.

de Carvalho Alves, M. and Sanches, L. (2023). *Remote Sensing and Digital Image Processing with R.* CRC Press.

de Montaigne, M. (1987). *Ensaios.* Os Pensadores. Abril Cultural, São Paulo, 4 edition. Tradução de: Essais.

Decorps, J. P. (2021). *GADMTools: Easy Use of 'GADM' Maps.*

Diggle, P. J. and Ribeiro, P. J. (2007). *Model-based Geostatistics.* Springer Series in Statistics. Springer New York, New York, NY.

Dorman, M. (2024). *nngeo: k-Nearest Neighbor Join for Spatial Data.* R package version 0.4.8, https://github.com/michaeldorman/nngeo/.

Douglas, D. H. and Peucker, T. K. (1973). Algorithms for the reduction of the number of points required to represent a digitized line or its caricature. *Cartographica: The International Journal for Geographic Information and Geovisualization*, 10(2):112–122.

Dunnington, D. (2024). *prettymapr: Scale Bar, North Arrow, and Pretty Margins in R.* R package version 0.2.5.

Dunnington, D., Pebesma, E., Rubak, E., Ooms, J., and Google, Inc. (2021). *s2: Spherical Geometry Operators Using the S2 Geometry Library.* R package version 1.0.7.

Dunnington, D. and Thorne, B. (2021). *ggspatial: Spatial Data Framework for ggplot2.*

European Space Agency (2021). New GOCE geoid.

Fick, S. E. and Hijmans, R. J. (2017). WorldClim 2: new 1-km spatial resolution climate surfaces for global land areas. *International Journal of Climatology*, 37(12):4302–4315.

Fitz, P. R. (2008). *Cartografia Básica.* Editora Oficina de Textos, São Paulo, 1st edition.

GADM database of Global Administrative Areas (2021). GADM Maps and Data, Version 4.0. Accessed: 2023-10-15.

Garnier, S., Ross, N., Rudis, B., Sciaini, M., Camargo, A., and Scherer, C. (2024). *viridis: Colorblind-Friendly Color Maps for R.*

Gesmann, M., de Castillo, D., Cheng, J., Baldry, A., and Ingeniería, D. (2021). *googleVis: R Interface to Google Charts.*

Giraud, T. (2023). *linemap: Line Maps.* R package version 0.3.0.

Giraud, T. (2025). *mapsf: Thematic cartography.* R Package Version 1.0.0.

Giraud, T., Lambert, N., Hernangómez, D., and Fellows, I. (2021). *cartography: Thematic Cartography.*

Goodchild, M. F. (1992). Geographical information science. *International Journal of Geographical Information Systems*, 6(1):31–45.

Gorelick, N., Hancher, M., Dixon, M., Ilyushchenko, S., Thau, D., and Moore, R. (2017). Google Earth Engine: Planetary-scale geospatial analysis for everyone. *Remote Sensing of Environment*, 202:18–27.

Goslee, S. C. (2011). Analyzing remote sensing data in R: the landsat package. *Journal of Statistical Software*, 43(4):1–25.

Graves, S., Dorai-Raj, S., and Francois, R. (2020). *sos: Search Contributed R Packages, Sort by Package.*

Hafen, R. and Schloerke, B. (2020). *geofacet: 'ggplot2' Faceting Utilities for Geographical Data.*

Harris, I., Jones, P. D., Osborn, T. J., and Lister, D. H. (2014). Updated high-resolution grids of monthly climatic observations - the CRU TS3.10 Dataset. *International Journal of Climatology*, 34(3):623–642.

Hart, E. (2014). *rWBclimate: A package for accessing World Bank climate data.*

Hiemstra, P. (2013). *automap: Automatic interpolation package.*

Hiemstra, P. H., Pebesma, E. J., Twenhöfel, C. J., and Heuvelink, G. B. (2009). Real-time automatic interpolation of ambient gamma dose rates. *Computers & Geosciences*, 35(8):1630–1641.

Hijmans, R. J., Barbosa, M., Ghosh, A., and Mandel, A. (2024a). *geodata: Download Geographic Data.*

Hijmans, R. J., Bivand, R., Dyba, K., Pebesma, E., and Sumner, M. (2024b). *terra: Spatial Data Analysis.*

Hijmans, R. J., Phillips, S., Leathwick, J., and Elith, J. (2024c). *dismo: Species Distribution Modeling.* R package version 1.3-16.

Hijmans, R. J., van Etten, J., Sumner, M., Cheng, J., Baston, D., Bevan, A., Bivand, R., Busetto, L., Canty, M., Fasoli, B., Forrest, D., Ghosh, A., Goliche, D., Gray, J., Greenberg, J. A., Hiemstra, P., Hingee, K., Geosciences, I. f. M. A., Karney, C., Mattiuzzi, M., Mosher, S., Nowosad, J., Pebesma, E., Shortridge, O. P. L., Ashton, E. B. R., Rowlingson, B., Venables, B., and Wueest, R. (2020). *raster: Geographic Data Analysis and Modeling.*

Hijmans, R. J., Williams, E., and Vennes, C. (2019). *geosphere: Spherical Trigonometry.*

Horn, B. K. (1981). Hill shading and the reflectance map. *Proceedings of the IEEE*, 69(1):14–47.

Hyndman, R. J. and Athanasopoulos, G. (2021). *Forecasting: Principles and Practice.* OTexts, Melbourne, Australia, 3 edition.

Instituto Brasileiro de Geografia e Estatística - IBGE (1995). *Tabelas para Cálculos no Sistema de Projeção Universal Transverso de Mercator (UTM).* Rio de Janeiro, 2nd edition.

Isaaks, E. H. and Srivastava, R. M. (1989). *Applied Geostatistics*, volume 561. Oxford University Press, Oxford.

ISO (2000). ISO 8601 – An International Standard for Date and Time Formats. Technical report.

Jensen, J. R. and Jensen, R. R. (2012). *Introductory Geographic Information Systems.* Prentice Hall, 1st edition.

Jeworutzki, S., Giraud, T., Lambert, N., Bivand, R., Pebesma, E., and Nowosad, J. (2020). *cartogram: Create Cartograms with R.*

Journel, A. G. and Huijbregts, C. J. (1991). *Mining geostatistics.* Academic Press.

Karney, C. F. (2013). Algorithms for geodesics. *Journal of Geodesy*, 87(1):43–55.

Kassambara, A. (2025). *ggpubr: ggplot2 Based Publication Ready Plots.* R package version 0.6.1.

Komsta, L. and Novomestky, F. (2022). *moments: Moments, Cumulants, Skewness, Kurtosis and Related Tests.* R package version 0.14.1.

Krehbiel, C., Friesz, A. M., Jami, M., and Bolch, E. (2024). Making geospatial data easier to access: Nasa earthdata's application for extracting and exploring analysis ready samples (appeears). In *AGU Fall Meeting Abstracts*, volume 2024, pages IN23E–05.

Kwok, R. (2018). Ecology's remote sensing revolution. *Nature*, pages 137–138.

Lamigueiro, O. P. (2018). *Displaying Time Series, Spatial, and Space-Time Data with R.* Chapman & Hall/CRC, New York, 2nd edition.

Lamigueiro, O. P. and Hijmans, R. (2020). *rasterVis: Visualization Methods for Raster Data.*

Lark, M. R. (2000). Estimating variogram parameters by maximum likelihood. *Mathematical Geology*, 32(3):317–339.

Leutner, B., Horning, N., Schwalb-Willmann, J., and Hijmans, R. J. (2019). *RStoolbox: Tools for Remote Sensing Data Analysis.*

Lindsay, D. (2011). *Scientific Writing = Thinking in Words.* CSIRO Publishing, Collingwood, Australia.

Lo, C. P. and Yeung, A. K. W. (2007). *Concepts and Techniques of Geographic Information Systems.* Pearson Prentice Hall, Upper Saddle River, NJ, 2nd edition.

Longley, P. A., Goodchild, M. F., Maguire, D. J., and Rhind, D. W. (2001). *Geographic Information Systems and Science.* John Wiley & Sons, Inc, Chichester.

Lorenzo, P. D. (2025). *usmap: US Maps Including Alaska and Hawaii.* R Package Version 0.8.0.

Lovelace, R., Nowosad, J., and Muenchow, J. (2019). *Geocomputation with R.* CRC Press, Boca Raton, 1st edition.

Massicotte, P. and South, A. (2023). *rnaturalearth: World Map Data from Natural Earth.* R package version 1.0.1, https://github.com/ropensci/rnaturalearth.

Matheron, G. (1962). *Traité de géostatistique appliquée.* Number 14. Editions Technip.

McCormac, J. C., Sarasua, W. A., and Davis, W. J. (2012). *Surveying.* John Wiley & Sons.

Medeiros, C. B. and Pires, F. (1998). Banco de dados e sistemas de informações geográficas. In Eduardo Delgado Assad, E. E. S., editor, *Sistemas de Informações Geográficas: Aplicações na Agricultura*, chapter 3, pages 31–45. Embrapa-SPI/Embrapa-CPAC, Brasília, 2nd edition.

Mondello, C., Hepner, G., and Medina, R. M. (2008). ASPRS 10 Year Remote Sensing Industry Forecast: Phase V. *Photogrammetric Engineering & Remote Sensing*, 74(11):1297–1305.

Morgan-Wall, T. (2024). *rayshader: Create Maps and Visualize Data in 2D and 3D.* R package version 0.37.3, https://github.com/tylermorganwall/rayshader.

Mullen, L. and Bratt, J. (2025). *USAboundaries: Historical and Contemporary Boundaries of the United States of America.* R package version 0.4.0, https://github.com/ropensci/USAboundaries.

Neves, M. C., Spadotto, C. A., Luiz, A. J. B., and Quirino, T. R. (1998). Caracterização Espaço-temporal do Uso de Agrotóxicos para o Estado de São Paulo. In Assad, E. D. and Sano, E. E., editors, *Sistema de informações geográficas. Aplicações na agricultura*, chapter 12, pages 233–240. Embrapa-SPI/Embrapa-CPAC, Brasília, 2nd edition.

Nychka, D., Furrer, R., Paige, J., and Sain, S. (2021). fields: Tools for spatial data. R package version 16.3.1.

Ooms, J., Kornel Lesiński, and Authors of the dependency Rust crates (2025). *gifski: Highest Quality GIF Encoder.* R package version 1.32.0-2.

Open Geospatial Consortium (2021). OGC Standards and Technical Documents. Accessed: 2023-10-15.

Openshaw, S. and Abrahart, R. J. (2001). *GeoComputation*. Taylor & Francis, London, 1st edition.

Padgham, M., Rudis, B., Lovelace, R., Salmon, M., Smith, A., Smith, J., Gilardi, A., Spinielli, E., Kalicinski, M., Noam, F., and Lukasz, B. (2021). *osmdata: Import 'OpenStreetMap' Data as Simple Features or Spatial Objects.*

Pawley, S. (2024). *Rsagacmd: Linking R with the Open-Source SAGA-GIS Software.* R package version 0.4.3.

Pebesma, E. and Bivand, R. (2021a). Spatial Data Science with applications in R.

Pebesma, E. and Bivand, R. (2021b). Spatial Econometrics Models. In *Spatial Data Science: With Applications in R*, chapter 16. Chapman & Hall/CRC, Boca Raton. Online pre-print version.

Pebesma, E. and Bivand, R. (2023). *Spatial Data Science: With Applications in R.* The R Series. Chapman & Hall/CRC, Boca Raton.

Pebesma, E., Bivand, R., Racine, E., Sumner, M., Cook, I., Keitt, T., Lovelac, R., Wickham, H., Ooms, J., Müller, K., Pedersen, T. L., and Baston, D. (2021a). *sf: Simple Features for R.*

Pebesma, E., Bivand, R., Rowlingson, B., Gomez-Rubio, V., Hijmans, R., Sumner, M., MacQueen, D., Lemon, J., Lindgren, F., O'Brien, J., and O'Rourke, J. (2021b). *sp: Classes and Methods for Spatial Data.*

Pebesma, E. and Bivand, R. S. (2005). S classes and methods for spatial data: the sp package. *R news*, 5(2):9–13.

Pebesma, E. and Graeler, B. (2021). *gstat: Spatial and Spatio-Temporal Geostatistical Modelling, Prediction and Simulation.*

Pebesma, E., Mailund, T., and Hiebert, J. (2016). Measurement Units in R. *The R Journal*, 8(2):486–494.

Pebesma, E., Nüst, D., and Bivand, R. (2012). The R software environment in reproducible geoscientific research. *Eos, Transactions American Geophysical Union*, 93(16):163–163.

Pebesma, E., Rundel, C., Teucher, A., and Developers, L. (2020). *lwgeom: Bindings to Selected 'liblwgeom' Functions for Simple Features.*

Pebesma, E., Sumner, M., Racine, E., Fantini, A., and Blodgett, D. (2021c). *stars: Spatiotemporal Arrays, Raster and Vector Data Cubes.*

Penedo, J. and Borges, N. (2017). The challenge of choosing where to publish: The Predatory Journals! *Revista Portuguesa de Cirurgia*, II(42):5–6.

Pereira, R. H. M., Goncalves, C. N., de Araujo, P. H. F., Carvalho, G. D., de Arruda, R. A., Nascimento, I., da Costa, B. S. P., Cavedo, W. S., Andrade, P. R., da Silva, A., Braga, C. K. V., Schmertmann, C., Samuel-Rosa, A., Ferreira, D., Research, and Economic, I. I. f. A. (2021). *geobr: Download Official Spatial Data Sets of Brazil.*

Planchon, O. and Darboux, F. (2002). A fast, simple and versatile algorithm to fill the depressions of digital elevation models. *CATENA*, 46(2-3):159–176.

R Core Team (2023). *R: A Language and Environment for Statistical Computing.* R Foundation for Statistical Computing, Vienna, Austria.

Ramirez, M. R. (2007). Sistema gerenciador de banco de dados em sistemas de informações geográficas. In *Geomática: Modelos e Aplicações Ambientais*, chapter 2nd, pages 57–106. Embrapa Informação Tecnológica, Brasília, 1st edition.

Ribeiro Jr, P. J., Diggle, P. J., Christensen, O., Schlather, M., Bivand, R., and Ripley, B. (2020). *geoR: Analysis of Geostatistical Data.*

Rogers, E. M. (1995). *Diffusion of Innovations.* Free Press, 4th edition.

Sagan, C. (2013). *Cosmos.* Ballantine Books, New York.

Santos, H. G., Jacomine, P. K. T., Anjos, L. H. C., Oliveira, V. A., Lumbreras, J. F., Coelho, M. R., Almeida, J. A., Araujo Filho, J. C., Oliveira, J. B., and Cunha, T. J. F. (2018). *Sistema Brasileiro de Classificação de Solos.* Embrapa-Solos, Brasília, 5th edition.

Sarkar, D. and Andrews, F. (2022). *latticeExtra: Extra Graphical Utilities Based on Lattice.* R package version 0.6-30.

Schlager, S., Jefferis, G., and Ian, D. (2021). *Morpho: Calculations and Visualisations Related to Geometric Morphometrics.*

Sediyama, G. C. and Melo Júnior, J. C. F. (1998). Modelos para estimativa das temperaturas normais mensais médias, máximas, mínimas e anual no estado de Minas Gerais. *Engenharia na Agricultura*, 6(1):57–61.

Séférian, R., Nabat, P., Michou, M., Saint-Martin, D., Voldoire, A., Colin, J., Decharme, B., Delire, C., Berthet, S., Chevallier, M., Sénési, S., Franchisteguy, L., Vial, J., Mallet, M., Joetzjer, E., Geoffroy, O., Guérémy, J. F., Moine, M. P., Msadek, R., Ribes, A., Rocher, M., Roehrig, R., Salas-y Mélia, D., Sanchez, E., Terray, L., Valcke, S., Waldman, R., Aumont, O., Bopp, L., Deshayes, J., Éthé, C., and Madec, G. (2019). Evaluation of CNRM Earth System Model, CNRM-ESM2-1: Role of Earth system processes in present-day and future climate. *Journal of Advances in Modeling Earth Systems*, 11(12):4182–4227.

Sievert, C., Parmer, C., Hocking, T., Chamberlain, S., Ram, K., Corvellec, M., Despouy, P., and Brüggemann, S. (2021). *plotly: Create Interactive Web Graphics via 'plotly.js'.*

Silva, F. A. M. and Assad, E. D. (1998). Análise espaço-temporal do potencial hídrico climático do estado de Goiás. In Assad, E. D. and Sano, E. E., editors, *Sistema de informações geográficas. Aplicações na agricultura*, chapter 15, pages 273–309. Embrapa-SPI/Embrapa-CPAC, Brasília.

Silva, I. and Segantine, P. (2015). *Topografia para Engenharia: Teoria e Prática de Geomática.* Elsevier, Rio de Janeiro, 1st edition.

Siqueira, R. P. (2017). *brazilmaps: Brazilian Maps from Different Geographic Levels.*

Siqueira, R. P. (2021). *sidrar: An Interface to IBGE's SIDRA API.*

Slocum, T. A., McMaster, R. B., Kessler, F. C., and Howard, H. H. (2009). *Thematic Cartography and Geovisualization.* Pearson Prentice Hall, Essex, 3rd edition.

Slowikowski, K. (2024). *ggrepel: Automatically Position Non-Overlapping Text Labels with ggplot2.* R package version 0.9.6.

Smith, D. D. and Wischmeier, W. H. (1957). Factors affecting sheet and rill erosion. *Transactions, American Geophysical Union*, 38(6):889.

South, A., Michael, S., and Massicotte, P. (2024). *rnaturalearthdata: World Vector Map Data from Natural Earth Used in rnaturalearth.* R package version 1.0.0, https://github.com/ropensci/rnaturalearthdata.

Sparks, A. H. (2018). nasapower: A nasa power global meteorology, surface solar energy and climatology data client for r. *The Journal of Open Source Software*, 3(30):1035.

Sparks, A. H., Hengl, T., and Nelson, A. (2025). *GSODR: Global Surface Summary of the Day (GSOD) Weather Data Client.* R package version 4.1.4, https://github.com/ropensci/GSODR.

Strimas-Mackey, M. (2021). *smoothr: Smooth and Tidy Spatial Features.*

Tarroso, P. (2021). idw: Inverse Distance Weighting interpolation.

Tennekes, M. (2018). tmap: Thematic Maps in R. *Journal of Statistical Software*, 84(6).

Tennekes, M. (2021). *tmaptools: Thematic Map Tools.* R package version 3.1-1.

Tennekes, M. and Ellis, P. (2023). *treemap: Treemap Visualization.* R Package Version 2.4-4.

Tennekes, M., Nowosad, J., Gombin, J., Jeworutzki, S., Russell, K., Zijdeman, R., Clouse, J., Lovelace, R., and Muenchow, J. (2020). *tmap: Thematic Maps.*

Teucher, A. and Russell, K. (2023). *rmapshaper: Client for mapshaper for Geospatial Operations.* R package version 0.5.0.

Tobler, W. R. (1979). Smooth pycnophylactic interpolation for geographical regions. *Journal of the American Statistical Association*, 74(367):519–530.

Turbek, S. P., Chock, T. M., Donahue, K., Havrilla, C. A., Oliverio, A. M., Polutchko, S. K., Shoemaker, L. G., and Vimercati, L. (2016). Scientific Writing Made Easy: A Step-by-Step Guide to Undergraduate Writing in the Biological Sciences. *The Bulletin of the Ecological Society of America*, 97(4):417–426.

Turner, R. (2024). *deldir: Delaunay Triangulation and Dirichlet (Voronoi) Tessellation.* R package version 2.0-4.

Vasques, G. M., Dart, R. O., Baca, J. F. M., Ceddia, M. B., and Mendonça Santos, M. L. (2017). Mapa de estoque de carbono orgânico do solo (COS) a 0-30 cm do Brasil. Technical report, Embrapa Solos, Rio de Janeiro.

Visvalingam, M. and Whyatt, J. D. (1993). Line generalisation by repeated elimination of points. *The Cartographic Journal*, 30(1):46–51.

Volpato, G. L. (2007). Como escrever um artigo científico. In *Anais da Academia Pernambucana de Ciência Agronômica*, volume 4, pages 97–115.

Walker, K. (2025). *tigris: Load Census TIGER/Line Shapefiles.* R package version 2.2.1.

Wang, L. and Liu, H. (2006). An efficient method for identifying and filling surface depressions in digital elevation models for hydrologic analysis and modelling. *International Journal of Geographical Information Science*, 20(2):193–213.

Webster, R. and Oliver, M. A. (2007). *Geostatistics for Environmental Scientists.* John Wiley & Sons, 2nd edition.

Wickham, H. (2019). *Advanced R.* Chapman & Hall/CRC, Boca Raton.

Wickham, H., Chang, W., Henry, L., Pedersen, T. L., Takahashi, K., Wilke, C., Woo, K., Yutani, H., Dunnington, D., and RStudio (2020). *ggplot2: Create Elegant Data Visualisations Using the Grammar of Graphics.*

Wickham, H., François, R., Henry, L., Müller, K., and RStudio (2021a). *dplyr: A Grammar of Data Manipulation.*

Wickham, H. and Henry, L. (2023). *purrr: Functional Programming Tools.* R package version 1.0.2.

Wickham, H., Hester, J., Chang, W., and RStudio (2021b). *devtools: Tools to Make Developing R Packages Easier.*

Wickham, H., Hester, J., Francois, R., RStudio, Jylänki, J., and Jørgensen, M. (2021c). *readr: Read Rectangular Text Data.*

Wickham, H., Pedersen, T. L., and Seidel, D. (2023). *scales: Scale Functions for Visualization.* R package version 1.3.0.

Wickham, H. and RStudio (2019). *stringr: Simple, Consistent Wrappers for Common String Operations.*

Wickham, H. and RStudio (2021). *tidyverse: Easily Install and Load the 'Tidyverse'.*

Wilke, C. O. (2024). *cowplot: Streamlined Plot Theme and Plot Annotations for ggplot2.* R package version 1.1.3.

Willmott, C. J. and Matsuura, K. (2005). Advantages of the mean absolute error (mae) over the root mean square error (rmse) in assessing average model performance. *Climate Research*, 30:79–82.

Wischmeier, W. H. and Smith, D. D. (1978). *Predicting Rainfall Erosion Losses: A Guide to Conservation Planning.* Department of Agriculture, Science and Education Administration, Washington, D.C.

Wise, M. (1986). Literary Sources for the History of Palestine and Syria: The Dead Sea Scrolls: Part 1. Archaeology and Biblical Manuscripts. *The Biblical Archaeologist*, 49(3):140–154.

Xie, Y. (2015). *Dynamic Documents with R and knitr.* Chapman & Hall/CRC, Boca Raton, Florida, 2nd edition. ISBN 978-1498716963.

Xie, Y. (2018). knitr: A Comprehensive Tool for Reproducible Research in R. In *Implementing Reproducible Research*, pages 3–31. Chapman & Hall/CRC, Boca Raton.

Xie, Y. (2023). *knitr: A General-Purpose Package for Dynamic Report Generation in R.* R package version 1.45.

Zevenbergen, L. W. and Thorne, C. R. (1987). Quantitative analysis of land surface topography. *Earth surface processes and landforms*, 12(1):47–56.

Index

For Product Safety Concerns and Information please contact our
EU representative GPSR@taylorandfrancis.com Taylor & Francis
Verlag GmbH, Kaufingerstraße 24, 80331 München, Germany